Control and Management of Pests in Stored Products

Stored commodities are human-made ecosystems, and interactions of biological agents with their surrounding physical environment could result in significant economic losses if the physical environment is not manipulated to make it lethal or at least difficult for the survival of biological agents. ***Control and Management of Pests in Stored Products*** is based on 18 invited presentations by world-renowned experts on topics of relevance to the control and management of pests in stored products. Each chapter synthesizes the state-of-the-art knowledge on the selected topics dealing with fumigation, fumigants, and other methods of controlling insects such as low temperature, diatomaceous earth, integrated pest management, and provides recommendations for future research. It also includes two chapters on practical aspects of fumigation: engineering considerations and safety. The contents of the chapters were presented as the keynote addresses at the International Conference on Controlled Atmosphere and Fumigation in Stored Products.

This book serves as a reference book for graduate students, researchers, and facility managers and can also be useful as a textbook for courses dealing with aspects of grain storage for students in agricultural engineering, agricultural entomology, and food science.

Control and Management of Pests in Stored Products

Edited by
Digvir S. Jayas

CRC Press
Taylor & Francis Group
Boca Raton London New York

CRC Press is an imprint of the
Taylor & Francis Group, an **informa** business

First edition published 2024
by CRC Press
2385 NW Executive Center Drive, Suite 320, Boca Raton, FL 33431

and by CRC Press
4 Park Square, Milton Park, Abingdon, Oxon, OX14 4RN

CRC Press is an imprint of Taylor & Francis Group, LLC

Library of Congress Cataloging-in-Publication Data

Names: Jayas, Digvir S., 1958- editor.
Title: Control and management of pests in stored products / edited by Digvir S. Jayas
Description: Boca Raton, FL : CRC Press, 2024 | Includes bibliographical references and index
Identifiers: LCCN 2023057314 | ISBN 9781032314662 (hardback) | ISBN 9781032314679 (paperback) | ISBN 9781003309888 (ebook)
Subjects: LCSH: Insect pests--Control. | Food storage pests--Control. | Pests--Control.
Classification: LCC SB937 .C668 2024 | DDC 632/.7--dc23/eng/20240129
LC record available at https://lccn.loc.gov/2023057314]

ISBN: 978-1-032-31466-2 (hbk)
ISBN: 978-1-032-31467-9 (pbk)
ISBN: 978-1-003-30988-8 (ebk)

DOI: 10.1201/9781003309888

Typeset in Century Old Style
by KnowledgeWorks Global Ltd.

Dedication

Dedicated to my wife, Manju Jayas, for inspiring, encouraging, and supporting me throughout my career

Contents

Preface

The International Conferences on Controlled Atmosphere and Fumigation in Stored Products (CAF) are the major international forums for reporting advances in research and development of gaseous treatments for the preservation of stored commodities. The CAF2024 is the 12th in a series of conferences held over the past 44 years in Rome, Italy (1980); Perth, Australia (1983); Singapore (1989); Winnipeg, Canada (1992); Nicosia, Cyprus (1996); Fresno, United States (2000); Gold Coast, Australia (2004); Chengdu, China (2008); Antalya, Turkey (2012); New Delhi, India (2016); and Winnipeg, Canada (2020). History and details of past CAF conferences can be found at this website: http://ftic.co.il/caf-en.php.

Stored commodities are human-made ecosystems, and interactions of biological agents with their surrounding physical environment could result in significant economic losses if the physical environment is not manipulated to make it lethal or at least difficult for the survival of biological agents. Surroundings of stored commodities are manipulated using physical and chemical means based on research and development by stored-products researchers around the globe. The CAFs, which are held every four years, synthesize the research and development dealing with the treatment of a broad range of durable and other commodities held in storage, utilizing hermetic storage technology, controlled atmospheres (CA), modified atmosphere (MA), extreme temperatures, aeration and chilled aeration, or chemical fumigation for the control of pests. Some of the techniques, such as aeration or chilling, are used to keep pest populations under an economic threshold or at low levels until other techniques can be used for complete control. Over the years, the conference objectives have been refined to report on advances in research and development; the current status of CA, MA, and fumigation for insect pests and microflora; and quality control in stored products. Treatment of products such as cereal grains, oilseeds, legumes, root crops, dried fruits, nuts, vegetables, and other agricultural products including bulbs and flowers are covered by these conferences. The professional background for the CAF conference is characterized by (a) the need to ensure the continued use of phosphine against the pressures of increasing resistance to this valuable fumigant, (b) the sustainable availability of novel CA/MA technologies, and (c) reports on new fumigants including ethyl formate, ethanedinitrile, and nitric oxide.

Therefore, to make CAF2024 memorable, the Winnipeg CAF Conference Organizing Committee decided to publish an edited book based on 18 invited papers from the experts in the field on topics of significance to CAF. Each group of co-authors has attempted to synthesize knowledge from the past four or five decades and propose needs for future research and development related to their assigned topics. Thus, this book is a series of chapters on the state-of-the-art research and development of technologies for managing pests of stored products so as to attain high quality for the food supplies of the world. The Taylor & Francis Group/CRC Press agreed to publish this commemorative book, and I have had the privilege to work with amazing authors of each chapter in editing this book.

The book begins by introducing the biological pests of stored commodities and other products of interest to humans. It next highlights the resistance of pests to common chemicals, and then following chapters present a synthesis of state-of-the-art knowledge about different techniques for pest control and management: hermetic storage, silo bag storage, CA and MA, extreme temperatures, chlorine dioxide, ethanedinitrile, ethyl formate, nitric oxide, sulfuryl fluoride, propylene oxide, natural products, contact insecticides, diatomaceous earth, and amorphous silica. The book concludes with engineering, mathematical models, and safety aspects. This book will be globally useful for undergraduate and graduate students in stored-product entomology and biosystems engineering programs, researchers and professors in these areas, and consultants and managers of stored commodities and facilities used for storage. In addition, CAFs are typically attended by scientists from government agencies or academic institutions who represent the disciplines of agronomy, agricultural economics, biology, chemistry, food science, and plant science, as well as personnel from the private sectors of pest control and agricultural chemical manufacturing. They will appreciate using this book as a reference on a regular basis.

In addition to this book, the CAF Conference Local Organizing Committee will produce a high-quality refereed proceedings based on submitted papers and presentations covering the latest research and development in the following sessions: (1) Biological Response to Fumigants,

(2) Engineering and Modeling of Fumigants, (3) Insect Resistance to Fumigants, (4) Hermetic Storage, (5) Controlled and Modified Atmospheres, (6) Quarantine and Alternatives to Methyl Bromide, (7) Alternate Fumigants, (8) Bio-fumigants, (9) Alternative Approaches, (10) Fumigation Overview and Processes, (11) Fumigation Monitoring, and (12) Role of Internet of Things, Artificial Intelligence, Machine Learning, Large Data Analytics for Management of Stored Commodities.

It is my hope that this book and research and development papers contributed during the 12th CAF will play a significant role now and well into the future in managing stored commodities and in preserving the quality of the stored commodities to provide nutritious and wholesome food to the growing world population.

Digvir S. Jayas, OC, PhD, DSc, PEng, PAg, FRSC

Acknowledgments

I sincerely thank the authors of each chapter. It has been an immense pleasure to work with the authors of each chapter. It was an honor to edit their work, and their timely response in making the required changes during revisions was greatly appreciated.

I cannot thank my wife, Manju Jayas, enough for her inspiration, support, and encouragement during the editing of this book. She, however, mentioned more than a few times that I spend too much time sitting in front of the computer and do not take care of myself or complete many items that need to be done around the house. There were many occasions when I had a great desire to spend more time with my grandkids (Priya Jayas, Isabella Jayas, Rohan Jayas, Gabriel Jayas, and Leon Jayas) but pressing timelines to complete this book prevented me from doing so. I ask them for their forgiveness and hope when they see the finished book, they will appreciate my absences. I love them so much and thank them for their hugs and smiles.

I thank my current employer, the University of Lethbridge, Lethbridge, Alberta, Canada, and my former employer, the University of Manitoba, Winnipeg, Canada, for allowing me to focus on editing this book and for providing the required computer, secure data storage, and Internet access to work from home.

Finally, I thank CRC Press/Taylor & Francis Group for printing this book and for their due diligence in the production process.

Digvir S. Jayas, OC, PhD, DSc, PEng, PAg, FRSC

About the Editor

Dr. Digvir S. Jayas is the President and Vice-Chancellor of the University of Lethbridge, Lethbridge, Alberta, Canada. He is also Vice-President (Research and International) Emeritus and Distinguished Professor Emeritus at the University of Manitoba. He was educated at the G.B. Pant University of Agriculture and Technology in Pantnagar, India; the University of Manitoba; and the University of Saskatchewan. He completed a 11.75-year term as Vice-President (Research and International) at the University of Manitoba. Before assuming that position, he was Vice-President (Research) for two years and Associate Vice-President (Research) for eight years. Prior to this, he was Associate Dean (Research) in the Faculty of Agricultural and Food Sciences, Head of the Department of Biosystems Engineering, and Interim Director of the Richardson Centre for Food Technology and Research. He also served as Interim President of the Natural Sciences and Engineering Research Council of Canada (NSERC) for one year and served as Interim Director (CEO) of TRIUMF—Canada's particle accelerator center—for 4.5 months. He is a Registered Professional Engineer and a Registered Professional Agrologist.

Dr. Jayas is a former Tier I (Senior) Canada Research Chair in Stored-Grain Ecosystems. He conducts research related to drying, handling, and storing of cereal grains, pulses, and oilseeds and digital image processing for grading and processing operations in the agri-food industry. He has collaborated with researchers in several countries and has had significant impact on the development of efficient grain storage, handling, and drying systems in Canada, China, India, Ukraine, and the United States. He has authored or co-authored over 1,000 technical articles in scientific journals, conference proceedings, and books dealing with issues of storing, drying, handling, and quality monitoring of grains and foods.

Dr. Jayas has received awards in recognition of his research and professional contributions from the Agriculture Institute of Canada, Applied Zoologists Research Association (India), American Society of Agricultural and Biological Engineers (ASABE) (formerly American Society of Agricultural Engineers), Canadian Institute of Food Science and Technology, Canadian Academy of Engineering, Canadian Society for Bioengineering (formerly Candian Society for Agriciltural Engineering), Engineers Canada (formerly Canadian Council of Professional Engineers), Engineers Geoscientists Manitoba (formerly Association of Professional Engineers and Geoscientists of Manitoba), Engineering Institute of Canada, Indian Society of Agricultural Engineers, Manitoba Institute of Agrologists, National Academy of Agricultural Sciences (India), National Academy of Sciences (India), Sigma Xi, and The Society of Tropical Agriculture. He was the recipient of the 2017 Sukup Global Food Security Award from ASABE and the 2008 Brockhouse Canada Prize from NSERC. In 2009, he was inducted as a Fellow of the Royal Society of Canada (RSC), and in 2018, he was appointed as an Officer of the Order of Canada for "his advancements to agricultural practices worldwide, and for his promotion of academic and scientific research in Canada." In 2019, he received the RSC Sir John William Dawson Medal for "important contributions of knowledge in multiple domains." In 2022, Dr. Jayas was inducted into both the Manitoba Agricultural Hall of Fame and the Canadian Agricultural Hall of Fame in recognition of his engineering contributions to grain preservation. In 2023, he received the John Deere Gold Medal from the ASABE, the Lieutenant Governor's Award for Excellence in Public Administration from the Institute for Public Administration Canada (IPAC) Manitoba, and the Fellowship from Sigma Xi.

Dr. Jayas serves or has served on the boards or committees of many organizations such as ArcticNet, Churchill Marine Observatory (CMO), Centre for Innovative Sensing of Structures (SIMTReC), Genome Prairie, GlycoNet, Manitoba Centre for Health Policy, North Forge Technology Exchange, NRC Council, NSERC Council, Research Manitoba, and TRIUMF. He has served as the President of the Agriculture Institute of Canada, the Canadian Institute of Food Science and Technology, the Canadian Society for Bioengineering, Engineers Canada, Engineers Geoscientists Manitoba, and the Manitoba Institute of Agrologists. He chaired the NSERC Council; the board of directors of TRIUMF; the board of RESOLVE, a prairie research network on family violence; and the Smartpark (University of Manitoba's Research and Technology Park) Advisory Committee.

Contributors

Manjree Agarwal
ChemCentre
Bentley, Western Australia

Paraskevi Agrafioti
Laboratory of Entomology and
    Agricultural Zoology
Department of Agriculture, Crop
    Production and Rural Environment
University of Thessaly
Nea Ionia, Greece

Adeyemi Oluseye Akinyemi
Department of Agronomy
Osun State University
Ejigbo Campus
Nigeria

Christos G. Athanassiou
Laboratory of Entomology and
    Agricultural Zoology
Department of Agriculture, Crop
    Production and Rural Environment
University of Thessaly
Nea Ionia, Greece

Samuel Adelani Babarinde
Department of Crop and
    Environmental Protection
Ladoke Akintola University of
    Technology
Ogbomoso, Nigeria

Ricardo Bartosik
Instituto de Innovación para la
    Producción Agropecuaria y el
    Desarrollo Sostenible (IPADS)
    Balcarce (INTA-CONICET)
Balcarce, Buenos Aires, Argentina

Vimala S. K. Bharathi
Department of Biosystems
    Engineering
University of Manitoba
Winnipeg, Manitoba, Canada

Edmond L. Bonjour
Department of Entomology and
    Plant Pathology
Oklahoma State University
Stillwater, Oklahoma

CONTRIBUTORS

Rania Marie Buenavista
Department of Grain Science
   and Industry
Kansas State University
Manhattan, Kansas

Leandro Cardoso
Instituto de Innovación para la
   Producción Agropecuaria y el
   Desarrollo Sostenible (IPADS)
   Balcarce (INTA-CONICET)
Balcarce, Buenos Aires, Argentina

Watcharapol Chayaprasert
Department of Agricultural
   Engineering
Kasetsart University – Kamphaeng
   Saen Campus
Nakhon Pathom, Thailand

Tom de Bruin
Postharvest Industries West Africa
   Ltd (PIWA)
Tema Industrial Area, Ghana

Patrick Ducom
Captsystemes
Lormont, France

Thierry Ducom
Captsystemes
Paris, France

Valerie Ducom
Ceres Innovation
Molleville, France

Xinyi E
Department of Biological and
   Agricultural Engineering
Kansas State University
Manhattan, Kansas

Marina Gourgouta
Laboratory of Entomology and
   Agricultural Zoology
Department of Agriculture,
   Crop Production and Rural
   Environment
University of Thessaly
Nea Ionia, Magnesia, Greece

David W. Hagstrum
Department of Entomology
Kansas State University
Manhattan, Kansas

Nadav Inbari
Green Storage Ltd.
Rishon Letsion, Israel

Rajeswaran Jagadeesan
Postharvest Commodity Protection
   Unit
Queensland Department of
   Agriculture and Fisheries
Dutton Park, Queensland,
   Australia

Digvir S. Jayas
Department of Biosystems
   Engineering
University of Manitoba, Winnipeg,
Manitoba, Canada

Fuji Jian
Department of Biosystems
   Engineering
University of Manitoba
Winnipeg, Manitoba, Canada

Carol L. Jones
Department of Biosystems and
   Agricultural Engineering
Oklahoma State University
Stillwater, Oklahoma

Efstathios Kaloudis
Laboratory of Computer
   Simulation, Genomics and
   Data Analysis
Department of Food Science and
   Nutrition
School of the Environment
University of the Aegean
Lemnos, Greece

Dr. Dimitrios Kateris
Institute for Bio-Economy
   and Agri-Technology
   (IBO),
Centre for Research and
   Technology – Hellas
   (CERTH),
Volos, Greece

Yong-Biao Liu
USDA-ARS
Crop Improvement and
   Protection Unit
Salinas, California

Dirk E. Maier
Department of Agricultural and
   Biosystems Engineering
Iowa State University
Ames, Iowa

Selladurai Manivannan
Department of Grain Science
   and Industry
Kansas State University
Manhattan, Kansas

Hagit Navarro
Green Storage Ltd.
Rishon Letsion, Israel

Shlomo Navarro
Green Storage Ltd.
Rishon Letsion, Israel

Manoj Nayak
Postharvest Commodity Protection
   Unit
Queensland Department of
   Agriculture and Fisheries
Dutton Park, Queensland, Australia

Chris Newman
Stored Grain Services
Perth, Western Australia

Ronald T. Noyes
Grain Storage Engineering, LLC
Hoschton, Georgia

Thomas W. Phillips
Department of Entomology
Kansas State University
Manhattan, Kansas

YongLin Ren
College of Environmental and Life
   Sciences
Murdoch, Western Australia

Maria K. Sakka
Laboratory of Entomology and
   Agricultural Zoology
Department of Agriculture,
   Crop Production and Rural
   Environment
University of Thessaly
Nea Ionia, Magnesia, Greece

Kaliramesh Siliveru
Department of Grain Science and
   Industry
Kansas State University
Manhattan, Kansas

Katie Stevens
Draslovka Agricultural Solutions
Sunshine, Victoria, Australia

CONTRIBUTORS

Bhadriraju Subramanyam
Department of Grain Science
  and Industry
Kansas State University
Manhattan, Kansas

Swaminathan Thalavaisundaram
Draslovka Agricultural Solutions
Sunshine, Victoria, Australia

Hernán Urcola
Instituto de Innovación para la
  Producción Agropecuaria
  y el Desarrollo Sostenible
  (IPADS) Balcarce
  (INTA-CONICET)
Balcarce, Buenos Aires,
  Argentina

Stored-Product Insects

Biology and Management Relevant to Controlled Atmospheres and Fumigation

Thomas W. Phillips and David W. Hagstrum

1.1 Introduction

Stored products are defined as any postharvest agricultural products that are durable and therefore do not require refrigeration or other special preservation during storage over many months to years. The common stored food products for which the Controlled Atmosphere and Fumigation (CAF) conferences report the relevant research are stored grains, nuts, dried fruits, and any value-added products from these that do not require refrigeration. Other stored food products beyond raw and processed grains include dried meats, dried dairy products, and aged cheeses. Nonfood animal products at risk of infestation by stored-product pests include silk, wool, and some animal furs, skins, and feathers. Ecosystems have a simple structure and an abundant but nonregenerating food supply often with several trophic levels, including granivores, fungivores, omnivores, and natural enemies.

Postharvest durable agricultural products of all kinds are at risk worldwide of severe damage and losses from infestation by insects, mites, microbes, and vertebrates that are adapted to storage habitats (Table 1.1).

DOI: 10.1201/9781003309888-1

Table 1.1 Common Stored-Product Insect Species by Taxonomic Grouping, Number of Geographical Regions Where Present, Habits, and Habitats

Order[a], Family	Species	No. Geog. Regions[b]	Habits[c]	Habitats[d]
Class Insecta				
Coleoptera				
Anobiidae	*Lasioderma serricorne* (Fabricius)	7	S	EMC
Anobiidae	*Stegobium paniceum* (Linnaeus)	7	S	MC
Anthribidae	*Anthicus floralis* (Linnaeus)	7	S	MC
Anthribidae	*Araecerus fasciculatus* (DeGeer)	7	F	C
Bostrichidae	*Dinoderus minutus* (Fabricius)	7	S	C
Bostrichidae	*Prostephanus truncatus* (Horn)	6	F	C
Bostrichidae	*Rhyzopertha dominica* (Fabricius)	7	F	EMC
Chrysomelidae[e]	*Acanthoscelides obtectus* (Say)	7	F	
Chrysomelidae[e]	*Callosobruchus analis* (Fabricius)	6	F	
Chrysomelidae[e]	*Callosobruchus chinensis* Linnaeus	7	F	
Chrysomelidae[e]	*Callosobruchus maculatus* (Fabricius)	7	F	C
Chrysomelidae[e]	*Callosobruchus phaseoli* (Gyllenhal)	7	F	
Chrysomelidae[e]	*Caryedon serratus* (Olivier)	6	F	
Chrysomelidae[e]	*Zabrotes subfasciatus* (Boheman)	6	S	C
Cleridae	*Necrobia rufipes* (Fabricius)	7	S	MC
Cryptophagidae	*Cryptophagus cellaris* (Scopoli)	6	S	MC
Cryptophagidae	*Cryptophagus pilosus* Gyllenhal	4	S	MC
Curculionidae	*Sitophilus granarius* (Linnaeus)	6	S	EMC

(Continued)

Table 1.1 (Continued) Common Stored-Product Insect Species by Taxonomic Grouping, Number of Geographical Regions Where Present, Habits, and Habitats

Order[a], Family	Species	No. Geog. Regions[b]	Habits[c]	Habitats[d]
Curculionidae	*Sitophilus oryzae* (Linnaeus)	7	F	EMC
Curculionidae	*Sitophilus zeamais* (Motschulsky)	7	F	EMC
Dermestidae	*Anthrenus flavipes* (LeConte)	6	S	C
Dermestidae	*Anthrenus verbasci* (Linnaeus)	6	S	MC
Dermestidae	*Attagenus unicolor* (Brahm)	6	S	EMC
Dermestidae	*Dermestes frischii* Kugelann	6	S	C
Dermestidae	*Dermestes lardarius* Linnaeus	7	S	EMC
Dermestidae	*Dermestes maculatus* Degeer	7	S	C
Dermestidae	*Trogoderma glabrum* (Herbst)	5	S	MC
Dermestidae	*Trogoderma granarium* Everts	7	S	EMC
Dermestidae	*Trogoderma inclusum* LeConte	6	S	EMC
Dermestidae	*Trogoderma simplex* Jayne	2	S	EC
Dermestidae	*Trogoderma variabile* Ballion	6	S	MC
Histeridae	*Carcinops pumilio* (Erichson)	6	P	EMC
Laemophloeidae	*Cryptolestes ferrugineus* (Stephens)	7	S	EMC
Laemophloeidae	*Cryptolestes pusillus* (Schönherr)	7	S	EMC
Laemophloeidae	*Cryptolestes turcicus* (Grouvelle)	5	S	EMC
Languriidae	*Cryptophilus integer* (Heer)	6	S	MC
Lathridiidae	*Cartodere constricta* (Gyllenhal)	7	S	MC

(Continued)

Table 1.1 (Continued) Common Stored-Product Insect Species by Taxonomic Grouping, Number of Geographical Regions Where Present, Habits, and Habitats

Order[a], Family	Species	No. Geog. Regions[b]	Habits[c]	Habitats[d]
Lathridiidae	*Dienerella filiformis* (Gyllenhal)	3	S	E
Lathridiidae	*Enicmus minutus* (Linnaeus)	6	S	EM
Mycetophagidae	*Typhaea stercorea* (Linnaeus)	7	F	EMC
Nitidulidae	*Carpophilus dimidiatus* (Fabricius)	7	F	EMC
Nitidulidae	*Carpophilus hemipterus* (Linnaeus)	7	S	EMC
Nitidulidae	*Carpophilus humeralis* (Fabricius)	7	S	C
Ptinidae	*Gibbium psylloides* Czenpinski	6	S	MC
Ptinidae	*Mezium Americanum* Laporte	6	S	MC
Ptinidae	*Ptinus fur* (Linnaeus)	5	S	EM
Ptinidae	*Ptinus latro* Fabricius	7	S	MC
Ptinidae	*Ptinus tectus* Boieldieu	7	S	
Ptinidae	*Trigonogenius globulum* Solier	7	S	
Silvanidae	*Ahasverus advena* (Waltl)	7	F	EMC
Silvanidae	*Cathartus quadricollis* (Guérin-Méneville)	6	F	MC
Silvanidae	*Oryzaephilus mercator* (Fauvel)	7	S	C
Silvanidae	*Oryzaephilus surinamensis* (Linnaeus)	7	S	EMC
Silvanidae	*Silvanus bidentatus* (Fabricius)	4	S	MC
Tenebrionidae	*Alphitobius diaperinus* Panzer	7	S	EMC
Tenebrionidae	*Alphitobius laevigatus* (Fabricius)	7	S	MC

(Continued)

Table 1.1 (Continued) Common Stored-Product Insect Species by Taxonomic Grouping, Number of Geographical Regions Where Present, Habits, and Habitats

Order[a], Family	Species	No. Geog. Regions[b]	Habits[c]	Habitats[d]
Tenebrionidae	*Alphitophagus bifasciatus* (Say)	6	S	EMC
Tenebrionidae	*Cynaeus angustus* LeConte	5	S	EM
Tenebrionidae	*Gnatocerus cornutus* Fabricius	7	S	EMC
Tenebrionidae	*Latheticus oryzae* Waterhouse	7	S	EMC
Tenebrionidae	*Palorus ratzeburgii* (Wissmann)	7	S	EMC
Tenebrionidae	*Palorus subdepressus* (Wollaston)	6	S	EMC
Tenebrionidae	*Tenebrio molitor* Linnaeus	7	S	EMC
Tenebrionidae	*Tribolium audax* (Halstead)	1	S	EMC
Tenebrionidae	*Tribolium castaneum* (Herbst)	7	F	EMC
Tenebrionidae	*Tribolium confusum* Jacquelin du Val	6	S	EMC
Tenebrionidae	*Tribolium destructor* (Uyttenboogaart)	4	S	MC
Trogositidae	*Lophocateres pusillus* (Kugelann)	7	S	MC
Trogositidae	*Tenebroides mauritanicus* (Linnaeus)	7	S	EMC
Hemiptera				
Anthocoridae	*Lyctocoris campestris* (Fabricius)	6	P	EMC
Anthocoridae	*Xylocoris flavipes* (Reuter)	7	P	EC
Hymenoptera				
Bethylidae	*Cephalonomia tarsalis* (Ashmead)	6	P	
Bethylidae	*Cephalonomia waterstoni* Gahan	5	P	E

(Continued)

Table 1.1 (Continued) Common Stored-Product Insect Species by Taxonomic Grouping, Number of Geographical Regions Where Present, Habits, and Habitats

Order[a], Family	Species	No. Geog. Regions[b]	Habits[c]	Habitats[d]
Bethylidae	*Holepyris sylvanidis* (Brethes)	7	P	E
Braconidae	*Habrobracon hebetor* (Say)	6	P	E
Ichneumonidae	*Venturia canescens* Gravenhorst	7	P	E
Pteromalidae	*Anisopteromalus calandrae* (Howard)	7	P	E
Pteromalidae	*Dinarmus basalis* Rondani	6	P	
Pteromalidae	*Theocolax elegans* (Westwood)	7	P	E
Lepidoptera				
Cosmopterigidae	*Anatrachyntis rileyi* (Walsingham)	6	F	C
Gelechiidae	*Sitotroga cerealella* (Olivier)	7	F	EMC
Oecophoridae	*Endrosis sarcitrella* (Linnaeus)	4	S	EMC
Oecophoridae	*Hofmannophila pseudospretella* (Stainton)	4	S	EMC
Pyralidae	*Apomyelois ceratoniae* (Zeller)	6	F	
Pyralidae	*Cadra calidella* (Guenee)	3	F	
Pyralidae	*Cadra cautella* (Walker)	7	S	EMC
Pyralidae	*Cadra figulilella* (Gregson)	7	S	MC
Pyralidae	*Corcyra cephalonica* (Stainton)	7	S	EC
Pyralidae	*Ephestia elutella* (Hübner)	7	S	EMC
Pyralidae	*Ephestia kuehniella* Zeller	7	S	EMC
Pyralidae	*Paralipsa gularis* (Zeller)	4	S	MC
Pyralidae	*Plodia interpunctella* (Hübner)	7	F	EMC
Pyralidae	*Pyralis farinalis* (Linnaeus)	6	S	EMC
Tineidae	*Nemapogon granella* (Linnaeus)	6	S	EMC

(Continued)

Table 1.1 (Continued) Common Stored-Product Insect Species by Taxonomic Grouping, Number of Geographical Regions Where Present, Habits, and Habitats

Order[a], Family	Species	No. Geog. Regions[b]	Habits[c]	Habitats[d]
Tineidae	*Niditinea fuscella* (Linnaeus)	4	S	M
Tineidae	*Tineola bisselliella* (Hummel)	5	S	MC
Psocodea				
Liposcelidae	*Liposcelis bostrychophila* Badonnel	7	S	EM
Liposcelidae	*Liposcelis decolor* (Pearman)	6	S	E
Liposcelidae	*Liposcelis entomophilus* (Enderlein)	7	S	EM
Trogiidae	*Trogium pulsatorium* (Linnaeus)	6	S	EM
Class Arachnida				
Sarcoptiformes				
Acaridae	*Acarus siro* Linnaeus	4	S	
Acaridae	*Tyrophagus putrescentiae* (Schrank)	5	S	
Cheyletidae	*Cheyletus eruditus* (Schrank)	4	P	
Parasitiformes				
Ascidae	*Blattisocius tarsalis* (Berlese)	6	P	

[a] Insect orders and common names: Coleoptera—beetles; Hemiptera—true bugs; Hymenoptera—ants, bees, and wasps; Lepidoptera—moths; and Psocodea—psocids. One arachnid order: Sarcoptiformes—mites.

[b] Reported in from one to seven geographical regions: Africa, Asia, Central America, Europe, North America, Oceania, and South America. Derived from Hagstrum and Subramanyam (2009).

[c] Can infest in field prior to harvest (F), storage (S), or predator or parasitoid (P). From Hagstrum and Subramanyam (2009).

[c] In 53 studies, insect species found at grain elevators and flat storages (E), flour and feed mills (M), commercial marketing channels (C), or two or all of these. From http://storedproductinsects.com/new-species/comparison-of-species-list/.

[e] Previously family Bruchidae and now subfamily Bruchinae within the family Chrysomelidae.

Durable stored products are those that do not require refrigeration and that can be stored for months to years in bulk storage structures such as concrete or metal bins, silos, godowns, thatched buildings or simply as outdoor piles with or without coverings. Beyond bulk storage, many of these same insects and other pests can infest buildings used for food processing and value-added finished-product storage such as flour or feed mills, bakeries, extrusion food processing. The continuum extends to warehouse storage of finished and packaged high-value food products, retail markets, as well as people's homes, and restaurants. Human civilization, following the time of hunter-gatherers when people began to domesticate crops and animals, developed simple storage structures to protect their stored products and methods for preventing and/or stopping pest infestation. Nevertheless, risk from pests remains and can be overwhelming if not managed properly. Modern science, engineering, and economics have developed, and continue to develop, effective methods and technology to protect stored products.

This chapter reviews some basics of insect biology and describes the diversity of insects and mites found infesting stored products across the globe. Integrated pest management (IPM) is a decision-making process that includes preventive measures to keep pest populations low, methods for pest detection, and control before marketing that mitigates unacceptable pest or damage levels. We discuss potential of controlled atmospheres and fumigation for the world's food supplies in the future.

1.2 Biology, Diversity, and Ecology of Storage Pests

As a review for the nonentomologist, the following description is derived from common entomology textbooks (Chapman, 1998; Gullan and Cranston, 2014).

Insects are small invertebrate animals that belong to the large taxonomic group known as Arthropods and of which over one million species have been described. Insects have hard chitinous exoskeletons, jointed appendages, three major body parts (the head, the thorax, and the abdomen), six legs, and, in most species, two pairs of wings (Figure 1.1). The head has mouth parts made up of several structures that may act to either bite, chew, or pierce and then push food down into and through the digestive system. The head also has two compound eyes for vision, sometimes three light-sensitive omatidea, and antennae for olfaction and gustation. The thorax, which is the middle section of the insect body between the head and the abdomen, is divided into three segments. Each thoracic segment has a pair of walking legs, and there is a pair of wings on the

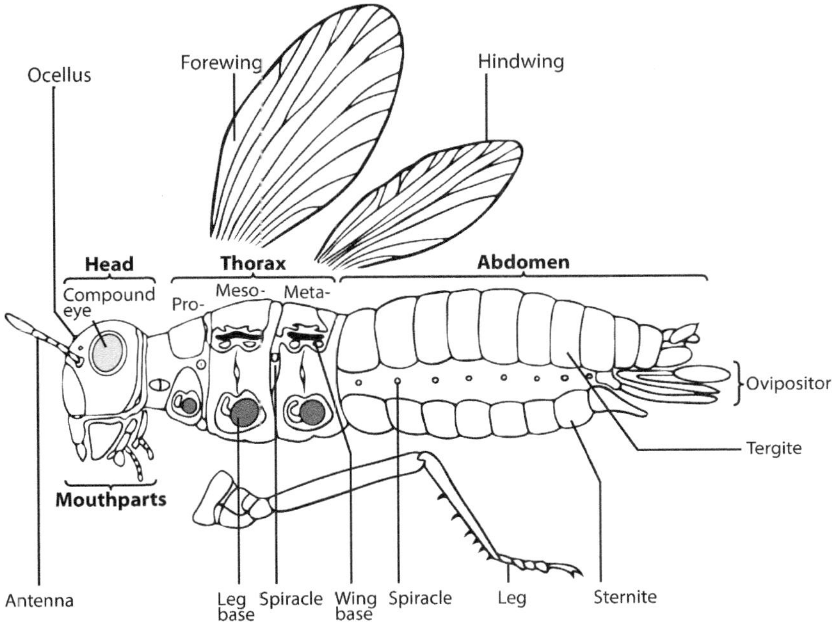

Figure 1.1 Lateral view of the external morphology for a typical insect, for which the model here is a female grasshopper. (From Romoser and Stoffolano, 1998.)

middle and last thoracic segments, for a total of four wings in most insect groups. The forewings in beetles are adapted as hard wing covers for the flying wings called elytra and serve as protection for the entire beetle. The internal anatomy of the insect thorax is predominantly many muscles and ligaments that operate the legs and wings in adults. The abdomen is typically the longest section of the insect body and contains organs for digestion of food, respiration to take in oxygen and remove carbon dioxide, and the reproductive system that generates eggs in the females and sperm in the male. The physiology of insects and other arthropods is very different from that of vertebrates. Insects do not breathe air into the lungs or carry oxygen in blood cells as do vertebrates. Rather, insects have a closed respiratory system that allows air into the body via openings along the sides of the body called spiracles, which allow air into a series of tubes referred to as tracheae (Figure 1.2). Tracheae have repeated branching into the insect's body cavity from the opened spiracles, branching through the body and multiplying into numerous smaller tracheae, resulting in a

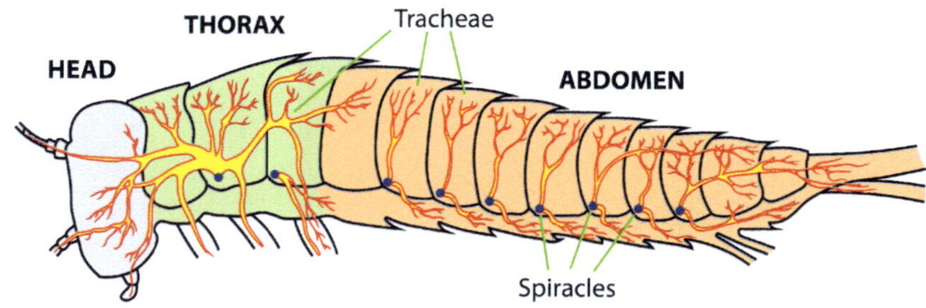

Figure 1.2 A lateral view of the respiratory system of a typical insect. Air passively enters the system through the spiracles, which can be opened or closed and are located on each side of the body, one pair per segment. Air then moves inward via a system of branching tubular trachea. Tracheae deliver oxygen in the air via microscopic tracheales to every tissue and organ in the body, where aerobic metabolism occurs, and the resulting carbon dioxide is delivered in reverse via the same system of trachea to the outside via the spiracles. (From BugWood Wiki, https://wiki.bugwood.org/Insect_Biology.)

body with microscopic tracheoles delivering air with oxygen directly to all living tissues and carrying away carbon dioxide back to the external atmosphere via the same spiracles. Food is taken in and digested by insects via a gastric system functionally similar to that of many other animals (Figure 1.3).

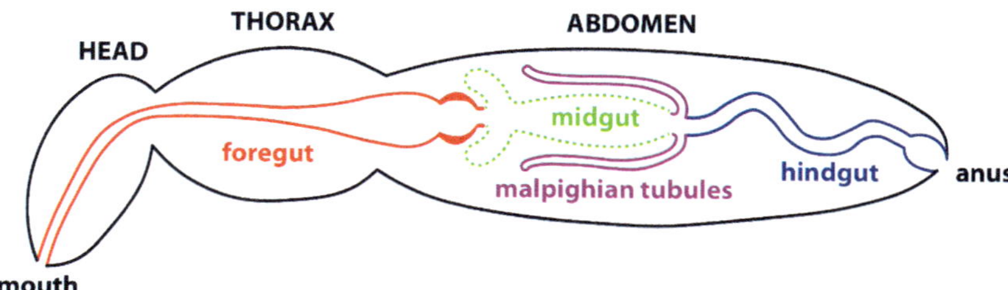

Figure 1.3 Lateral view of a typical insect digestive system showing the basic sections that process food, nutrients, and waste material. The mouth and foregut take in food and physically break it into smaller pieces. Saliva and enzymes begin a predigestive process. The midgut breaks down the food further into fundamental nutritional materials. Malpighian tubules deposit nutrients into the free-moving hemolymph (e.g., blood) in the body cavity and carry waste material from the hemolymph back into the digestive tract. The hindgut removes water and concentrates waste onto small dry pieces of fecal material for excretion via the anus. (From BugWood Wiki, https://wiki.bugwood.org/File:Digestive_system.jpg.)

The mouth and foregut chew food into smaller pieces that are then subjected to predigestive enzymes. Critical digestion and waste removal are in the midgut, ending with concentration and dehydration of waste material in the hindgut before excretion of "frass" via the anus. Insects have an open circulatory system, rather than closed blood vessels, such that the hemolymph, insect blood that does not carry oxygen, fills most of the free interior space of the insect body. Despite the lack of a vertebrate-type circulatory system, insects have a tubular canal at the top of the hemoceol just below the dorsal cuticle, a so-called heart (Figure 1.4). The heart has a peristaltic action that draws blood into paired lateral ostia and pushes it from back to front, resulting in an outflow of blood from lateral vessels that bathe the insect's organs and tissues. Hemolymph carries nutrients and specialized metabolic cells and key molecules as precursors to lipids, carbohydrates, and proteins, as well as waste materials from the tissues that float back to the digestive system via the Malpighian tubules.

The insect life cycle includes development of an embryo in a newly laid tiny egg, followed by the feeding and development of larvae or nymphs that hatch from the egg and molt through a series of small to large larval or nymph stages. Nearly all stored-product insects have what is known as a

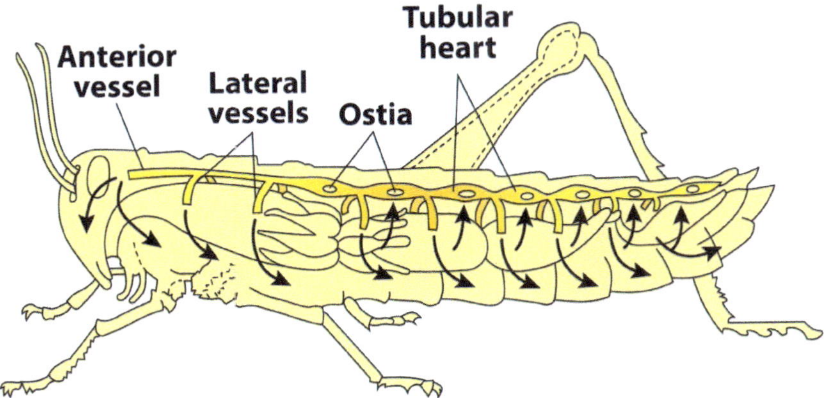

Figure 1.4 The open circulatory system for moving hemolymph throughout the insect's body. The dorsal heart takes in hemolymph from the body cavity via the ostia. Hemolymph is then pushed forward to the anterior aorta to spill over the brain and bathe all the tissues in the main body with nutrients, defensive cells, hormones, and other vital materials. (Image from Britanica.com, https://www.britannica.com/animal/insect/Circulatory-system.)

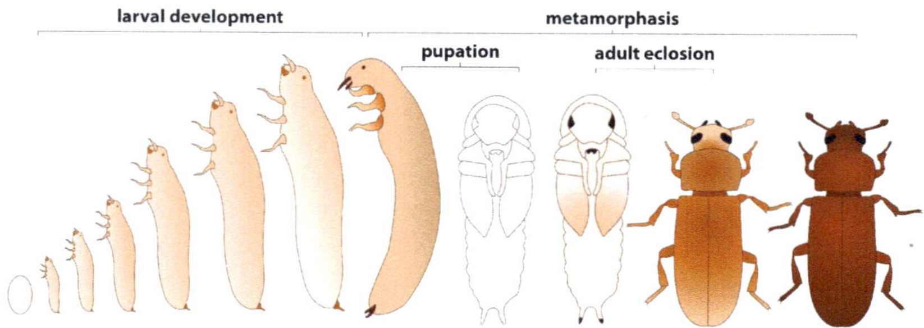

Figure 1.5 Developmental stages of the red flour beetle, *Tribolium castaneum*, from left to right. After eggs hatch, there are six larval instars, each getting larger with feeding before each molt to the next instar. The prepupa is a nonmotile transition stage between the last-stage larva and the early pupa. The pupa is immobile and goes through a metamorphosis to the adult stage. The new adult, which is inactive and careful to avoid damage, passes several days during which the cuticle hardens and darkens to be a fully developed adult ready for dispersal, feeding, and reproduction. The larval feeding stage is the most damaging to stored products. The adult also feeds and is very long lived, lasting up to months or years and able to reproduce for most of its adult life. (Modified from Walski et al., 2016.)

complete life cycle. This so-called holometabolous life cycle has four morphologically and functionally distinct life stages: egg, larva, pupa, and adult. Figure 1.5 shows the life cycle of the red flour beetle, *Tribolium castaneum* (Herbst), one of the most common stored-product insect species worldwide. Life stages of the Indian meal moth, another very common holometabolous insect pest of stored products, are given in Figure 1.6. Unlike the red flour beetle, the Indian meal moth adults will live only 5–10 d, during which time they do not feed, but mate and lay eggs daily that can total 100–200 in their brief lifetime.

Like the red flour beetle and the Indian meal moth, many stored-product insects can develop from egg to adult in a matter of about 4 wk under optimal environmental conditions. Insects and all arthropods are cold-blooded poikilotherms, whose body temperature is dictated by the external temperature and any movement the insect may do to increase or decrease that temperature. The physical environment affecting insect development also includes relative humidity (RH), light intensity, and seasonal changes in day length. Temperature and RH affect the time needed for the genetically controlled sequence in development of an insect from the egg to the reproductive adult representing the "life history" of a given species.

Figure 1.6 Life stages of the Indian meal moth, *Plodia interpunctella*. (a) Recently laid eggs on a U.S. 1-cent coin. (b) Feeding larvae in rearing media, together with pupae in recently spun cocoons near the top of the media. In field conditions, the "wandering" last instar larvae leave the food and move upwards in search of pupation sites while they clear their digestive systems via vomiting and defecating. (c) A single last instar larva (left) pupa (right). (d) A mating pair of two adult moths. The female is the larger moth on the left, copulating with the smaller male moth on the right. (Photos from TWP and USDA.)

Table 1.2 lists the environmental variables that affect the life histories of several species of stored-product insects and mites. Insects survive at temperatures just below their minimum temperature for development, but they will not move much or lay eggs or develop much at these cool temperatures.

Mites are not insects but are in the taxonomic class called Arachnida, and mite pests of stored products come from several taxonomic orders of arachnids. Arachnids also include spiders, the large arachnids that most people are familiar with. Like spiders, mites have two major body sections, eight walking legs, and no wings and therefore are quite different from insects, with three body sections, six legs, and two pair of wings. Stored-product mites are very small but can live several weeks to months, lay many eggs, and

Table 1.2 Optimum Temperature and Relative Humidity (RH) for Development Time from Egg to Adult, Longevity of Adults, and Population Increase of Selected Species of Stored-Product Insects and Mites After 28 d[a]

Species	Minimum Temperature (°C)	Optimal Temperature (°C)	Minimum RH (%)	Optimal RH (%)	Development Time (d)[b]	Adult Longevity[c] (d)	Population Increase after 28 d[d]
Acanthoscelides obtectus	15–17	27–31	30	70–80	28–30	4–37	25
Ahasverus advena	15	30	70	90	16	78–121	
Acarus siro	3	25	63	80	9	30	2,500
Alphitobius diaperinus	15	25–30	30	80–95	35		
Araecerus fasciculatus	22	28–32	50, 65	80	49	40–60	40
Cadra calidella	14	25–30	20–30	70	24		20
Cadra cautella	15, 17	28–32	25	70	25	2–26	50
Cadra figulilella	17	30	70		24	11–16	
Callosobruchus analis	22	30–33	1	70	26	5–59	25
Callosobruchus chinensis	19	28–32	30	70, 90	23		30
Callosobruchus maculatus	22	30–35	10	70	21	5–46	50
Carpophilus dimidiatus	6	33	70		15		
Carpophilus hemipterus	19	31–34	50		12	365	50
Cheyletus eruditus	8	20	56	80	17	35	
Corcyra cephalonica	17, 18	28–32	30	70	28	3–22	10
Cryptolestes ferrugineus	20, 23	32–35	10, 40	70–80	21, 23	180–270	60
Cryptolestes pusillus	18, 22	28–33	48, 60	90	22	80–600	10
Cryptolestes turcicus	18, 21	28–35	45, 50	90	38	180–270	50
Dermestes frischii	22	31–34	50	70	30	90	30

(Continued)

Table 1.2 (Continued) Optimum Temperature and Relative Humidity (RH) for Development Time from Egg to Adult, Longevity of Adults, and Population Increase of Selected Species of Stored-Product Insects and Mites After 28 d[a]

Species	Minimum Temperature (°C)	Optimal Temperature (°C)	Minimum RH (%)	Optimal RH (%)	Development Time (d)[b]	Adult Longevity[c] (d)	Population Increase after 28 d[d]
Dermestes maculatus	20	30–35	30	50–70	44	90	30
Endrosis sarcitrella	10	24–36		80	56		30
Ephestia elutella	10	25	30	70	29	4–16	15
Ephestia kuehniella	10, 12	24–27	1	73, 75	30	7–21	50
Gibbium psylloides	20	31–34	30		45	120–730	4
Gnatocerus cornutus	16	24–30	40		38	365	15
Lasioderma serricorne	20, 22	30	22	70	26	6–42	20
Latheticus oryzae	26	33–37	30	70	29		10
Liposcelis bostrychophilus		27		70	21–23	72–144	22.5
Necrobia rufipes	22	30–34	50		30	420	25
Nemapogon granella	7	25	65	90	70	14–20	
Oryzaephilus mercator	18, 20	30, 34	10	70	25	180–1080	20
Oryzaephilus surinamensis	18, 21	31–35	10	90	20	180–1080	50
Palorus ratzeburgii	17	27–35	20	50–80	25	60–90	
Palorus subdepressus	18	30–32	50	80	30		
Plodia interpunctella	18	26–29	40	70	26	5–13	30
Prostephanus truncatus	18	25–32	40	80	24	45–168	25
Ptinus ocellus	10	20–25	50	80–90	67	180–730	
Pyralis farinalis		27–30		70–80	42	7–21	

(Continued)

Table 1.2 (Continued) Optimum Temperature and Relative Humidity (RH) for Development Time from Egg to Adult, Longevity of Adults, and Population Increase of Selected Species of Stored-Product Insects and Mites After 28 d[a]

Species	Minimum Temperature (°C)	Optimal Temperature (°C)	Minimum RH (%)	Optimal RH (%)	Development Time (d)[b]	Adult Longevity[c] (d)	Population Increase after 28 d[d]
Rhyzopertha dominica	23	32–34	25	50–60	25	120–240	20
Sitophilus granarius	15	26–30	50	70	28	100–432	15
Sitophilus oryzae	17	26–31	45–60	70	24	150–730	25
Sitophilus zeamais	17	27–31	45	70		80–150	
Sitotroga cerealella	16	26–32		75	24	10–30	50
Stegobium paniceum	17	25–28	30	95	40	13–85	8
Tenebrio molitor	18	24–27	60	70	120	15–90	
Tenebroides mauritanicus	15	28–30		70–80	70	180–660	
Tribolium castaneum	20, 22	32–35	1–10	70–75	20	335–540	70
Tribolium confusum	20, 21	30–33	1–10	70	20	730–1080	60
Tribolium destructor	13	30–32	10	70–75	28	290–1460	
Trogoderma granarium	24	33–37	1	25	25	3–19	13
Trogoderma variabile	18	30		70	45	8–58	
Tyrophagus putrescentias	9	25	69	85	10	62	
Zabrotes subfasciatus	22	29–33	10–30	70	24	4–65	20

[a] Based on Boczek (1991), Davey (1965), Howe (1965), Jacob (1996), Sinha and Watters (1985).
[b] Number of days required to develop from egg to adult under ideal conditions.
[c] Adult lifespan in days under optimal conditions.
[d] Population size after 28 d in terms of the number-fold increase from the starting number of adults under optimal conditions. For example, a 20-fold increase from 100 mixed-sex adults can result in a population of 2,000 adults after 28 d.

develop much faster than most insects under optimal conditions, such as 10 d. Thus, the rate of population increase for stored-product mites is substantially more than that for the insects. Table 1.2 reports that the storage mite *Acarus siro* can develop from egg to adult in 9 d at 25°C and 80% RH, resulting in a 2,500-fold population increase after 28 d. That is an extremely large increase in population size compared to the average rate of increase for insects in Table 1.2, which is about 30-fold in 28 d.

In addition to populations increase under ideal conditions, biological factors such as physiology and behavior contribute to the adaptations of insects and mites in their natural habitats and also in their adaptations to be pests of agriculture as with stored products. As discussed above, the life cycle of an insect requires passing through dramatic changes of feeding in one larval instar and then forming a new exoskeleton and molting from the older cuticle to allow the next larval instar to be larger. Thus, the metamorphoses from larva to pupa and pupa to adult are dramatic changes. Molting, development, and metamorphosis are regulated by changes in insect-specific hormones in the insect's body. A terpenoid molecule called juvenile hormone (JH) remains at a high internal concentration during the period of a given larval stage. As a given feeding instar grows and needs to molt, the JH levels decrease in the body and levels of a second hormone called ecdysone increase in concentration, and the molting progresses, followed by hardening of the newly formed cuticle. A lowering or cessation of ecdysone and the return of JH to concentrations of the prior instar will maintain the insect for feeding or dispersal in the new instar.

A change of seasons in temperate climates from summer to autumn and winter comes with seasonal changes in temperature and precipitation and sometimes changes in food quality, food quantity, competitor numbers, or other environmental variables. Many insects adapt to such environmental changes by going into a kind of physiological hibernation called diapause. Diapause is a hormonally directed delay in development and/or changes in behavior in response to regular and recurring periods of adverse environmental conditions (Chapman, 1998). The onset of diapause may be triggered by a shortening of day length preceding the onset of cold temperature, accumulated changes in daily temperature, or any number of environmental variables. Diapause may have a stoppage of development from one stage to the next in some species; others may have a reproductive diapause in adults in which female ovaries do not develop to make eggs or male behavior changes in orientation to females. A good example is the onset of diapause in wandering-stage last instar larvae of the Indian meal moth, *Plodia interpunctella* (Hasan et al., 2020). During spring and summer seasons, when temperatures are warm and day lengths long, last-stage wandering *P. interpunctella* larvae will typically stop feeding, leave their food patch while defecating and vomiting, and spin copious amounts of silk while dispersing away from the food. Wandering larvae will eventually spin a cocoon in a protected hiding place (e.g., behind a warehouse ledge or within corrugation of cardboard food packaging) and change to be a pupa, followed by adult emergence in a few days. Under shortening day lengths with progressively cooler temperature in autumn, the last instar *P. interpunctella* larvae will perform the same behaviors of leaving food and spinning a cocoon in a hidden site. However, these insects will remain in a prepupal stage for many weeks before completing development to adults in the early spring of the next year, when day lengths and air temperature are increasing. Researchers like us, who need active life stages of *P. interpunctella* with reproductively mature adults throughout the year, maintain laboratory colonies of *P. interpunctella* under the long day lengths, such as 16 h light and 8 h dark, typical of summer solstice in temperate regions, with warm temperatures of 25–30°C. Knowledge of diapause in stored-product insects under changing seasons is very important when applying pest control actions such as fumigation and controlled atmospheres because respiration and other physiological processes may be slowed, resulting in one or more life stages being more tolerant of mitigation than when the insects are not in diapause and are more active.

Behaviors for finding mates and food evolved in insects, as with all animals, to be adaptive for maximally effective reproductive success. As an example, short-lived adult female Indian meal moths produce a blend of specific sex pheromones to attract males of their species for mating. The adults are short lived so females need to mate soon after emergence and then lay up to 200 fertile eggs, usually in clutches of 10–20 eggs, laid on or near the appropriate food for their larvae. Orientation to larval food sources by ovipositing adult females is many times in response to chemical and physical cues (Phillips and Strand, 1994; Sambaraju and Phillips, 2008a, 2008b) such as conspecific odors from insects previously on site or odors from the food being sought, as well as to dark vs. light and warm vs. cool locations.

Stored-product insects and mites are adapted to feeding on stored products as described herein, and as such they are associated with and evolutionarily adapted to human habitations in much the way that urban-structural pests like cockroaches, houseflies, and bedbugs have become. Some species are found only infesting stored products in human-made structures or temporary storages and are encountered rarely outside storages, such as during a dispersal flight or infesting dried products still in the fields prior to harvest. Other species may feed on human-made stored products but also exist in wild habitats, feeding and reproducing on wild host materials.

Table 1.1 reported the geographical distributions, habits, and habitats for over 100 species of stored-product insects and mites. The six beetle species from the families Cryptophagidae, Lathridiidae, and Mycetophagidae are primarily fungus feeders. Some common fungi can be found growing in stored products when the moisture content of the commodity exceeds 15% by dry weight. The species *Ahasverus advena*, *Cryptolestes ferrugineus*, *Oryzaephilus mercator*, *Pyralis farinalis*, *Tribolium castaneum*, and *Tribolium confusum* and four species from the order Psocodea can feed and develop on both the commodity and fungi. Many species may vector mold spores. Stored-product insects have clearly been associated with humans for millennia, as 83 stored-product species have been found at archaeological sites and many at more than one site (Hagstrum and Subramanyam, 2009). Over the long coevolution of insects and storage practices, stored-product insects were widely distributed by migration, international commerce and other travel by humans. Therefore, the geographical origin of most stored-product insects was difficult to determine, as precise identification of insects to species and the gathering of taxonomic information on distribution over history did not begin until the early 1880s. Several stored-product insect species (i.e., *Sitophilus granarius*, *Tribolium confusum*, and *Trogoderma granarium*) do not fly, suggesting that moving with commodities is less risky than flying away from them to find others.

Storage insects have adapted to this diversity of habitats by moving to find essential resources, eating many types of food, tolerating a broad range of environmental conditions, and having high population growth rates (Table 1.2). A majority of the species have long-lived adults, while some with short-lived adults have larval or reproductive diapause, which can delay reproduction (see above). In bulk-storage grain and elsewhere at elevators, 61 species were reported in 26 studies (see Hagstrum et al., 2010, for references). At flour or feed mills, 119 species were reported in 15 studies (see Mahroof and Hagstrum, 2012, for references). In commercial marketing channels, 155 species were reported in 12 studies (see Table 1.3 in Hagstrum and Subramanyam, 2006, for references). High population growth rates and large numbers of offspring can ensure a new generation even when insect mortality is high. With an average density of one

Table 1.3 Mortality Responses of Stored-Product Insect and Mite Pests Exposed to Extremes in Temperatures

Condition	Temp. Range (°C)	Effects
Lethal	>62	Death in < 1 min
	50–62	Death in < 1 h
	45–49	Death in < 1 d
	35–44	Populations decrease; mobiles seek cool places
Suboptimal	35	Maximum temperature for reproduction
	32–34	Slow population increase
Optimal	25–31	Maximum rate of population increase
Suboptimal	13–24	Development slows; slow population growth
Lethal	5–12	Slowly lethal or mobile insects migrate to warmth
	1–4	Movement ceases; death in days if not acclimated
	0 to −10	Death in weeks if acclimated, in days if not
	−11 to −25	Death in minutes to hours; insects freeze

Adapted from Fields (1992).

insect per kilogram, wheat is not heavily infested, but there are still many insects in this wheat. Hagstrum and Subramanyam (2006) calculated that a truck carrying 22 t (800 bu) would have 21,600 insects; a railroad hopper car carrying or a farm bin storing 82 t (3,000 bu) would have 81,000 insects. An elevator bin storing 544 t would have 540,000 insects; a barge carrying 1,360 t would have 1,350,000 insects; and a ship carrying 2,720 t would have 2,700,000 insects. Heated warehouses and large bulk storage structures have allowed insects to survive in geographical regions and during seasons in which cold outside temperatures would otherwise kill them by freezing.

Internal feeders result in fragments when infested grain is milled into flour, germ feeders can reduce seed germination, and external feeders may only feed on damaged kernels, cereal dust, dockage, and fungi associated with kernels but may also be important as contaminants or by creating hot spots that can reduce quality and result in economic losses. Storage insects infesting crops in the field before harvest often have been shown to come from nearby storages. In many cases in which stored-product insects have been found in natural habitats (such as bird, rodent, and insect nests or under tree bark), there is little evidence that they are breeding in these habitats; they may simply be migrants from the food-marketing system. The marginal quality of seeds and fruits from grasses, trees, and shrubs as food for stored-product insects, in addition to their small quantities, limits the importance of these foods in breeding source populations. Insects in these families generally develop inside seeds, although only members of the family Curculionidae lay their eggs inside the seeds. For 54 studies on storage insects infesting crops in the field, Hagstrum and Subramanyam (2006) reported that 19 insect species were found to infest 16 crop species in one or more of 18 countries. Seven stored-product insect species are known to infest cowpeas and sorghum in the field, six infest maize, and one to three species infest dried crops before harvest. *Callosobruchus chinensis* infests six crops, and *C. maculatus* and *Sitotroga cerealella* each infest five crops. Other stored-product insect species each infest one to four crops.

Exclusion, inspection of incoming commodities for insects, stock rotation, and sanitation are important methods of minimizing the number and size of source populations. Storage, processing, and marketing facilities and transportation vehicles can be designed to make excluding, finding, and removing source populations easier. Flowering plants have been reported to attract stored-product insects and should not be planted near storage and food processing facilities. Containerized shipping, which began in the late 1960s, can reduce the movement of insects from the commodities in one shipping container to the commodities in another, but if not cleaned shipping containers can be a source of insect infestation of commodities during shipping.

Worldwide invasive species cost billions of dollars annually in loss of commodities, reduce agricultural productivity and increase use of control (Myers and Hagstrum, 2012). Countries have a vested interest in preventing the introduction and spread of invasive insect species. Legislative control can be effective in regulating the introduction and establishment of alien insect pests and limiting their spread. Most countries have legislation restricting the importation of infested commodities. Quarantine insects may or may not be present in an importing country but are not widely distributed in that country.

Legislation and regulation are designed to minimize the introduction and spread of harmful quarantined organisms using inspection, survey, risk assessment, treatment, postentry quarantine, containment campaigns, and eradication. The first step for a new import is to develop a pest risk analysis. Lists can be divided into quarantine pests and regulated nonquarantine pests. Regulated nonquarantine pests are of economic importance but are already widespread in a country. To reduce the burden on port inspectors and the overall number of interceptions at U.S. ports, the United States Department of Agriculture-Animal and Plant Health Inspection Service (USDA-APHIS) has established preclearance programs in a number of exporting countries. Preclearance programs can be beneficial to both trading partners as they allow commodities to move without interruption; however, they are expensive and can be cost prohibitive because most or all of the cost is the responsibility of the exporting country. The khapra beetle, *Trogoderma granarium* (Coleoptera Dermestidae), is listed by the United Nations Food and Agriculture Organization as being among the top 100 most serious quarantined insects among all agricultural pests. The khapra beetle originated in the Indian subcontinent and has been established in much of Asia and Europe. Most dermestid beetles are short lived and act as scavengers, consuming damaged grain and similar stored products. But the khapra beetle can infest high-quality raw cereal grain like wheat and rice and can live months to years in a type of diapause the occurs during shortages of food or times of other limitations. *T. granarium* has been able to establish populations in the United States and other foreign locations (Myers and Hagstrum, 2012). Eradication is often difficult and expensive but must be done to maintain international shipments of major cereal grains around the world.

Hagstrum and Subramanyam (2009) listed 1,663 species of insects and Hagstrum et al. (2013) listed 280 species of mites known to inhabit a diverse array of stored products worldwide. A list of key insect and mite species of stored-product pests, along with several species of natural enemies, all derived from Hagstrum and Subramanyam (2009) and Hagstrum et al. (2013), was given in Table 1.1. Photographs of 16 economically and ecologically important species from Table 1.1 are shown in Figure 1.7. As mentioned earlier and quantified

Figure 1.7 Insect and mite species found commonly in stored products. More details are given in Table 1.2. (a) *Rhyzopertha dominica*, the lesser grain borer, an internal feeder. (b) *Prostephanus truncatus*, the larger grain borer, an internal feeder. (c) *Stophilus oryzae*, the rice weevil, an internal feeder. (d) *Sitotroga cerealella*, the Angoumois grain moth, an internal grain feeder. (e) *Plodia interpuntella*, the Indian meal moth, an external feeder. (f) *Ephestia kuehniella*, the Mediterranean flour moth, an external feeder. (g) *Tribolium castaneum*, the red flour beetle, an external feeder. (h) *Oryzaephilus surinamensis*, the saw-toothed grain beetle, an external feeder. (i) *Lasioderma surinamensis*, the cigarette beetle, an external feeder. (j) *Trogoderma variabile*, the warehouse beetle, an external feeder. (k) *Habrobracon hebetor*, an adult parasitoid wasp, stinging a last instar larva of *P. interpunctella*. (l) A mixed age group of *H. hebetor* larvae feeding on a last instar larva of *P. interpunctella*. (m) *Anisopteromalus calandrae*, an adult parasitoid wasp, stinging or laying eggs on an internal-feeding host beetle larva within a maize kernel. (n) *Tyrophagus putrescentiae*, the mold mite, infesting semi-moist pet food. (o) A closeup of *T. putrescentiae* mites showing eggs, immature nymphs, and adults. (p) *Liposcelis bostrychophila*, a book louse, an external feeder on grain dust and other small food particles. (Photos (a)–(j), (m), and (p) from USDA; (k) and (l) from Jena Johnson, University of Georgia; and (n) and (o) from TWP.)

a bit in Table 1.2, stored-product insects and mites vary in their life histories and ecological roles in stored-product ecosystems. Variations among species that affect their ecological placement in storage systems include short-lived species with complete ovipositon in days vs. long-lived adults that lay small clutches of eggs daily for months; short egg to adult development times with many 100-fold rates of population increase vs. lower levels of reproduction with some parental care and population increase less than 100-fold; broad ranges of host feeding on many species and qualities of foods vs. more restricted host ranges limited to certain crops (e.g., bruchines only on legumes); and internal feeders with protected feeding larvae resulting in severe damage to grains vs. external feeders that infest pre-damaged grains or milled and value-added products. A collection of very specific biological details for 14 species is shown in Figure 1.7.

Ecosystems have biotic components such as animals, plants, and microbes that can interact with their abiotic components such as the air temperature, sunlight, water, geology, and more (Hernandez Nopsa et al., 2015). We can think of stored products in a defined location as comprising an ecosystem that has plants (e.g., the stored grain), microbes such as fungi and bacteria, animals that feed on grain and microbes such as insects, mites and small vertebrates (e.g., mice and birds), and also include natural enemies such as parasitoids and predators feeding insects. All these are affected by air and grain temperature, water from grain moisture, light or lack of light, and other physical variables. The stored product-ecosystem, or food web, depicted by Sinha and Watters (1985) in their book on storage pests in different venues copied in this chapter as Figure 1.8. Sinha and Watters' food web displays a potential mature ecosystem that might occur only in a structure having unprotected grain stored for several years and vulnerable to infestation and the ecological diversity accumulated over time. Nevertheless, properly stored commodities with no detected pest insects still represent an active biological system affected over time by its environment to the point that economic losses result from the organisms and abiotic variables at work.

1.3 Integrated Pest Management for Stored Product Insects

Integrated pest management for stored products is a decision-making process that has as its main components prevention, monitoring, and making decisions for the near future of storage. The practice of IPM acts to avoid the risk of an economically impacting infestation during storage and processing. An attentive manager should decide to either continue monitoring and prevention or to take some action such as selling the product for

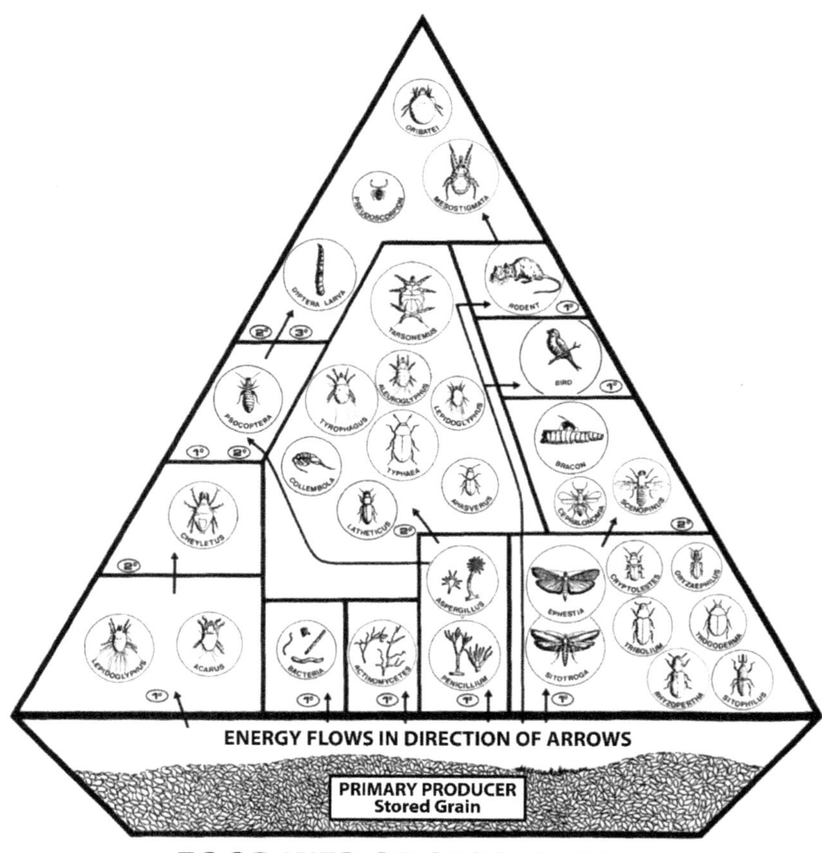

Figure 1.8 Depiction of a food web of stored-product insects, mites, vertebrates, grains, and microbes within a stored-product ecosystem in a model bulk-stored grain environment, as depicted in Sinha and Watters (1985). The stored grain is the primary producer for this system, as it has nutrients derived from the soil and air during photosynthesis by the field crop. Primary consumers are the insects, mites, and microbes supported directly by feeding on the grains. Secondary consumers are the insects and mites feeding on microbes and decayed grain. Another group of secondary consumers is parasitoids and predators that feed on primary consumer insects.

a profit or conducting mitigation via methods like fumigation if increased pest numbers could result in severe economic loss (Figure 1.9). Commodity owners should establish routine practices such as cleaning and cooling to prevent infestations of pests. The manager should have one or more effective methods to detect the presence of pests directly in the commodity or in and around the storage structure. Pest monitoring data can allow the

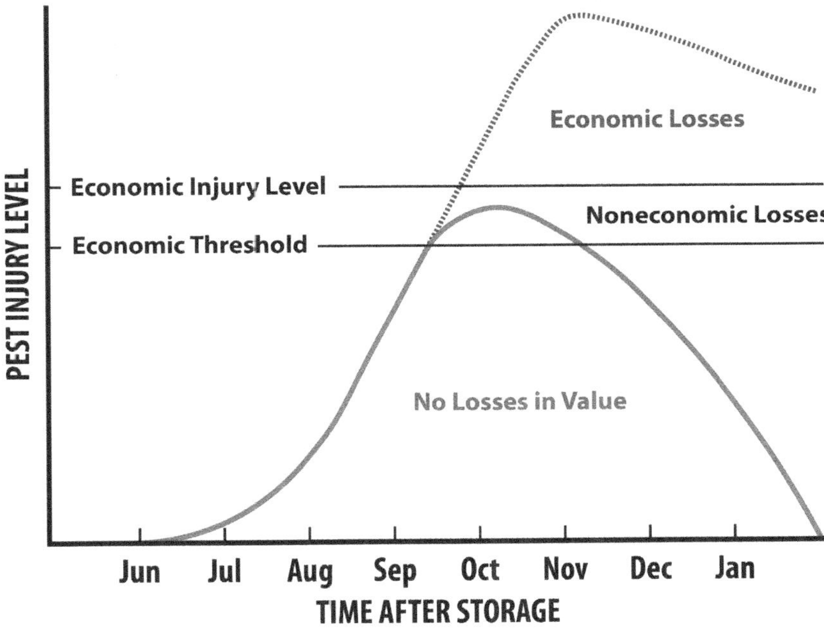

Figure 1.9 A representation of how consistent monitoring can yield information about changes in pest numbers or pest-caused damage to stored products. These variables may increase over time and meet a predetermined lower threshold for doing decisive mitigation (e.g., fumigation) or initiating sale of the product; storage continues if numbers remain below that threshold. The two curved lines report the changes in pest numbers or damage starting in June in temperate North America, at the time when newly harvested cereal crops go into storage. The lower horizontal line represents the number of pests or amount of damage at which profits start to decline, but pest damage and profit losses can be stopped with a decision to sell, move, or fumigate the product. The top horizontal line is the level past which there is no profit and the value and quality of the commodity continue to decrease. The shaded curved line shows a storage for which there was no mitigation; grain damage is extremely high and economic losses are large. (Image from Hagstrum and Flinn, 1995, and modified by DWH.)

manager to determine if the pest population size or the related damage has increased to or past a level considered an economic threshold for pests and/or their damage. Pest increases above the economic threshold occur when the net value, or profit from selling the product, decreases. At an economic threshold the manager could sell or use the product if some profit is guaranteed or do a cost-effective mitigation of the pest populations, perhaps with a controlled atmosphere or fumigant application that abruptly brings pest levels down for no continued loss in profit before a future sale or processing. Figure 1.9 depicts a model for IPM based on the size of the pest populations or amount of product damage, both derived from regular sampling. If pest densities are low and higher critical numbers are not reached, pest monitoring and prevention can continue. Sections 1.4, 1.5, and 1.6 give an overview of the components of a good IPM system for stored-product protection.

1.4 Prevention, Detection, and Monitoring for Pest Infestations

Assuming that the overwhelming majority of dried grain and beans go into storage free of stored-product pests, the most obvious preventive measure is to assure that the storage structure has no live storage insects present inside. To satisfy this need, the manager must conduct a thorough cleaning of the structure by first removing all grains from the most recent storage time. Do not load newly harvested grain on top of old grain. Once the structure is free of remaining grain or commodity from the past season and buildings are well swept, do a thorough cleaning of any remaining grain or grain particles via effective sweeping/removal or vacuum cleaning of remote and ceiling ledges. This will be difficult in a large metal bin with a ventilated floor and very difficult for concrete silos or godowns. Once cleaning has been done as well as possible, seal all obvious leaks or nonclosable openings in the structure. Once cleaning and sealing have been completed, for additional insurance to protect against any pests remaining in the building or against those that might enter the structure subsequently, apply pesticides to surfaces. The manager should spray legally approved and effective residually active insecticides or dusts to indoor cracks and crevices, edges by doors and windows, and outside surfaces of milling machinery and other equipment and around the base and bottom side wall on the outside of the structure (see Chapter 14). The moisture contents of cereal grains and beans should be determined before storage and should not be higher than about 13%. Wheat, rice, and other cereal grains can achieve safe moisture levels of 12% in the field prior to harvest. Maize in temperate climates when harvested mid-autumn could be greater than 15% moisture and should be dried before storage. Of course, since the price of grain will depend on quality and weight per volume, a reduction in weight is not preferred by the seller.

Careful decisions must be made regarding profit and grain quality following storage. Lastly, based on commodity value and risk for infestation soon after storage, a manager might consider applying residual grain protectant insecticide (again, see Chapter 14) directly to the product as it is loaded into the structure. Since storage insects tend to be more prevalent at the top and bottom regions of a grain mass, a cost-effective practice could be application to grain at the very beginning of the bin loading and then only near the end of the loading. Once grain is in storage, a very good nonchemical method to maintain low insect levels in temperate climates for grain harvested in summer or early autumn is to cool the grain using aeration and circulation of cool air throughout the grain mass (see Chapter 10).

The presence of pests must be monitored in a bulk storage structure or any other building used for processing or storing value-added products throughout the entire time stored products are present. If and when pests are detected, the detection method should be continued regularly to monitor changes in pest numbers or damage accumulation that may occur over time and across different locations in the structure (Flinn et al., 2003). Figure 1.10

Figure 1.10 A team collecting samples of stored wheat at different depths from the tops of concrete silos at a regional grain elevator in the state of Oklahoma. (a) A series of connecting metal tubes were pushed to certain depths in the grain via suction through the flexible tube from a vacuum pump secured on the street level below. (b) A sample of grain was sucked through the tube and into a collection vessel from the grain and was then emptied into a plastic bag and labeled. (c) Grain samples were returned to the office, where insects were removed from the sample, sorted by species, and counted. This is a direct sampling method by which one can estimate the actual density of the insect (in insects per kg, for example) throughout that section of the grain bin. (Images by TWP.)

shows a management team systematically taking grain samples from concrete silos at a grain elevator to look for and count stored-product insects in the samples. This is considered a direct sampling of the habitat, the stored grain, to obtain numbers that estimate the number of a given species per known volume of grain. If numbers are low, proper management can continue. However, sampling should be done at regular interval throughout the storage period. Figure 1.11 shows the use of traps to

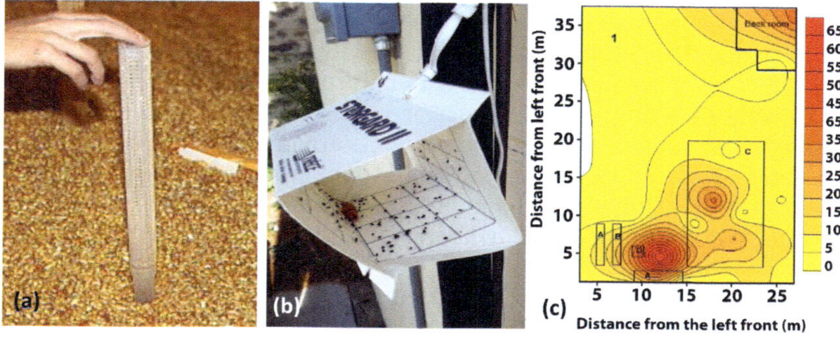

Figure 1.11   These images show methods to detect and monitor stored-product insects with traps. (a) A plastic probe trap being inserted into the top of a grain mass, where it will remain in place over time, perhaps 7 d (photo by TWP). This is a pitfall design such that insects walking through the grain randomly contact the probe and fall through the holes and down into a funnel to a collection tip. This is indirect sampling because it only provides information on the numbers of insects trapped over a certain time period in a particular place in the storage. One can only report "insects per trap," not insects per kg or per cubic meter of grain. (b) A sticky trap baited with the synthetic sex pheromone for the warehouse beetle, *Trogoderma variabile*. The trap has captured males of that species over the preceding week (photo by TWP). This again provides a relative count of the numbers caught in a trap over a week at that location. A relative count from a trap gives the manager information on the presence of that species on site, a "yes" or "no" for its presence. If traps are set out in different locations in a facility and for several weeks, the manager will have information on population variation by location and over time in a building, both of which are very useful for IPM decision-making. (c) A contour map showing relative numbers of a given insect trapped in different locations in a warehouse (courtesy of B. Subramanyam, Kansas State University). The darkest red areas would have the most numbers present, and the lightest yellow areas would have the least numbers of that pest. This information is useful for the manager designing a sanitation or control effort in those high-pest areas of the warehouse.

conduct relative sampling in stored grain using a probe trap and relative sampling for a targeted pest species using pheromone traps. These methods are considered relatively simple because they can provide the number of that insect found at a given time in a given area but do not give information on the number of insects per square meter or per volume of grain at a specific location in the structure. Nevertheless, indirect sampling is valuable because traps or other sampling can be done at more than one location in a structure and over time on a weekly or biweekly routine. Thus, one might learn how the relative numbers of a pest in a space are increasing or decreasing over time or how one location in the building is consistently catching more of that insect than other locations. Therefore, trapping data can be useful for the site manager, who may consider high trap catches a good cause to investigate directly for pest infestation and then use a pest control measure to decrease the pest numbers. If detection and monitoring over time eventually reveal an unacceptably high level of pests or their infestation that could lead to economic loss, mitigation may be needed. The last section of this chapter will touch on such methods.

1.5 Nonpesticide Controls: Extreme Temperatures, Modified and Controlled Atmospheres

The regulatory loss of methyl bromide (MB) fumigation in much of the world due to its role in climate change via damage to the earth's atmospheric ozone layer left many users in dire need of an alternative. MB fumigations were done routinely in flourmills throughout the world, and there were few alternatives in place for the need (Fields and White, 2002). A nonchemical method for structural treatments for pest control is the application of extreme temperatures (Chapter 7). Fields (1992) wrote a comprehensive and timely review of the literature on work to control storage insects and mites with extreme temperatures. Table 1.3 is derived from a similar table in the Fields (1992) paper. As discussed earlier, insects are "cold-blooded" and rely on warm temperatures to feed, develop, reproduce, and disperse. Table 1.3 shows that optimal temperatures for the lives of most insects are about 32–35°C, but all life stages of most species will be killed in one day if exposed to hot temperature above 45°C or very cold temperature below –15°C. The phaseout of MB for allowable uses for stored-product pest control began in 2005 throughout the industrial world, and techniques for heat-treating mills and warehouses were developed soon after that. We were told by a contact in the flour milling industry that 33% of the flourmills in the United States and Canada adopted

Figure 1.12 (a) An electrically powered air heater pushing very hot air, upwards of 60°C, into a flourmill from the lowest floor. (b) A circulating air heater inside a flourmill that uses hot steam from the mill's steam system to heat and blow air into the space at temperatures above 55°C. (Courtesy of B. Subramanyam, Kansas State University.)

heat treatments to replace MB within the first decade of this century. Figure 1.12 shows examples of air heaters used for treating buildings like flour mills to control insect pests. Heat treatments are not regulated by the government as are the treatments with pesticides, and they allow workers to enter treated buildings briefly if need be. Still, killing temperatures need to reach all spaces in a building and remain high for a minimum required time to assure a good kill of the pests. Figure 1.13 shows a diagram of the hours a target temperature of 50°C was maintained on a trial at our university's flourmill. As everyone knows, hot air rises and cold air tends to go downward. The results followed the prediction, as reported in Figure 1.13, as most of the higher floors in the building had broad floor spaces where a 50°C target was held for at least 40 h, while the first floor had only a small floor space held at 50°C for that long.

Freezing temperature are not likely to be applied commercially to buildings as big as a flourmill, but there is definitely a niche for pest control by freezing. Walk-in freezers are widely available for the food industry for use in large restaurants, food-service facilities, and long-term food storage. High-value durable stored products, such as bagged flour or pelleted breakfast cereal or pet foods ready for retail sale, can be infested with storage insects in certain cases. Packages with noticeable infestation would typically be discarded or removed for the retail market, but other packages in the same load or shipment could be frozen as a security to stop small infestations that may be in place. Flinn et al. (2015) reported a commercial-scale

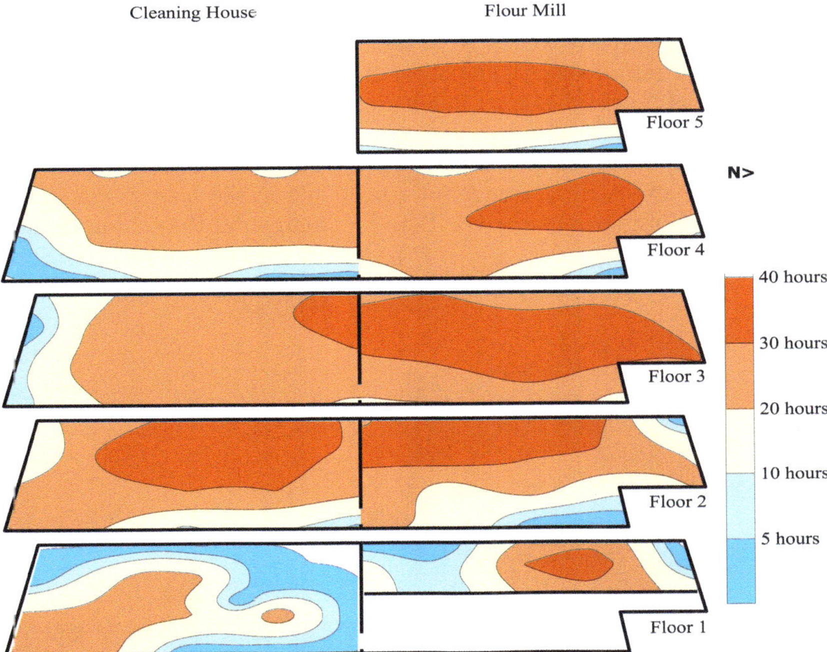

Figure 1.13 The length of time the air temperature was above 50°C in a 48-h heat treatment performed on the cleaning house and flour mill of the Kansas State University Department of Grain Science and Industry in June of 1999. (Image courtesy of Alan Dowdy, USDA ARS, and B. Subramanyam, Kansas State University.)

freeze treatment of bagged flour that had excellent success. Higher-value materials such as museum collections with natural fabrics and glues are subject to infestation by specialized storage pests like clothes moths (Lepidoptera: Tineidae), and cold treatments have been used on them for years (Anonymous, 2018).

Modified atmospheres with low oxygen and high carbon dioxide are naturally generated when a commodity is held in an airtight container or covering, referred to as hermetic sealing (Chapters 3 and 5). Aerobic organisms associated with the commodity at the time of sealing, such as insects, the living dried grain itself, and microbes, will naturally consume the available oxygen and give off carbon dioxide over the course of storage such that the atmosphere becomes toxic, stopping any infestation and preventing future infestations (Figure 1.14). Reaching an insect-killing modified atmosphere may take many weeks at an appropriate warm temperature if the amount of the commodity and its hermetic covering are large. A controlled atmosphere is one displacing as much air as possible with injection of large volumes of nitrogen, or by aerating and essentially fumigating with high concentrations of carbon dioxide. Since carbon dioxide is toxic and is added at very high levels, it is considered a pesticide in the United States and requires legal registration by the U.S. Environmental Protection Agency (Anonymous, 2020).

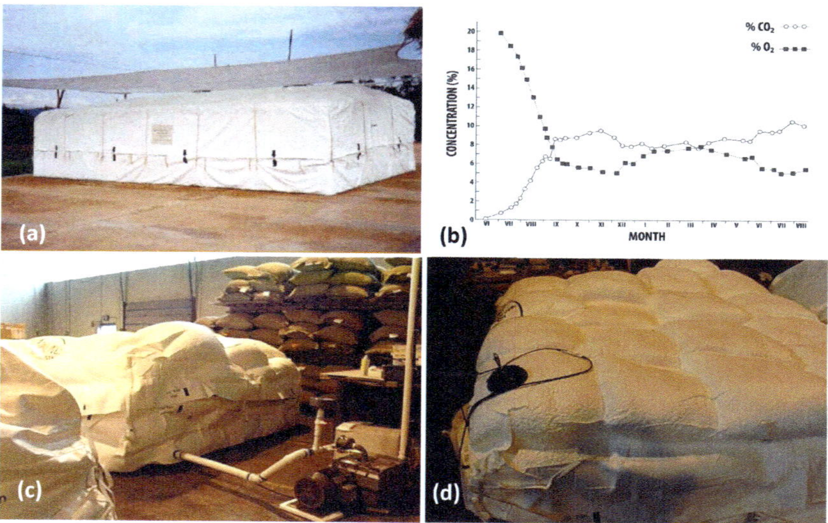

Figure 1.14 Images related to modified and controlled atmospheres. (a) A stack of sacks with 150 mt of corn in an airtight cover in Rwanda. The gastight covering has very limited air, which loses oxygen over time due to aerobic metabolism of the various organisms present, and those same organisms generate carbon dioxide. (b) A hypothetical data plot that shows a decrease of oxygen and increase of carbon dioxide inside a large gastight cover holding a large quantity of grain, similar to that in (a). (c) A gastight fully sealed bag, a "Cocoon" from GrainPro Inc. (https://www.grainpro.com/), holding several sacks of raw organic coffee beans and connected to a vacuum pump. (d) The same load of coffee beans for which the pump has brought the oxygen concentration from 20.09% at 760 mmHg down to close to 1.0% at about 50 mmHg in less than 6 h. All test insects were dead at 24 h. [(a) and (b) from Navarro et al., 2012; (c) and (d) from TWP.]

A low-oxygen atmosphere created with a vacuum, rather than purging with nitrogen, has been found to be very effective at killing pests in small quantities of sealed commodities (Figure 1.14). Chapters 3 and 5 elaborate on the methods and efficacy of controlled and modified atmospheres.

1.6 Fumigation

By far the most commonly used fumigant for stored products worldwide is hydrogen phosphide, or phosphine (Figure 1.15). Phosphine generated from phosphide salt pellets is easy to apply to bins, elevator silos, godowns, warehouses, and covered ground piles. A simple but sometimes significant drawback of the pellets and tablets is that after they are applied and the structure is resealed, the reaction to phosphine is sometimes completed in a matter of hours. The generated gas is lost due to sorption with the commodity as well as slowly leaking out of the storage structure, perhaps resulting in less than optimal kill due to the short exposure time. A very effective alternative to avoid gas losses after application of pellets and tablets is applying phosphine from pressurized cylinders outside the bin (Figure 1.15). High concentrations of phosphine can be delivered into a structure to reach the target dose quickly, and with proper use of remote phosphine detectors the applicator can put more phosphine into the structure when it is needed for an effective application. Unlike MB, phosphine cannot be used in buildings with valuable electronics or machinery due to its corrosive action on metals (Figure 1.15). Thus, phosphine is not a substitute for MB in most cases.

By far the biggest concern about effective use of phosphine to mitigate infestations of stored-product insects is the widespread occurrence of genetically controlled resistance to phosphine in many species of stored-product insects in many places around the world (Nayak et al. 2020, Chapter 2). Phosphine resistance is not new, as it was reported as early as the 1960s, but it was not as widespread as it is at this time. Our experience in the United States grew over the past 30 years from the first U.S. occurrence of phosphine resistance in 1976. Recent work found that 12 out of 25 populations of red flour beetles collected across North America were resistant to phosphine (Cato et al., 2017). Similar work on the lesser grain borer in North America found 32 out of 34 populations were resistant; the only two populations without any resistant beetles were in Canada, where phosphine is used less than in the United States (Afful et al., 2018). In addition to the widespread occurrence of phosphine resistance worldwide, it was recently discovered that there are two genetically distinct phenotypes of resistance across different species. The so-called weak resistance strains can be killed with high doses of phosphine held for long time periods, but strong resistance strains are very difficult to kill (Schlipalius et al., 2012; Afful et al., 2018).

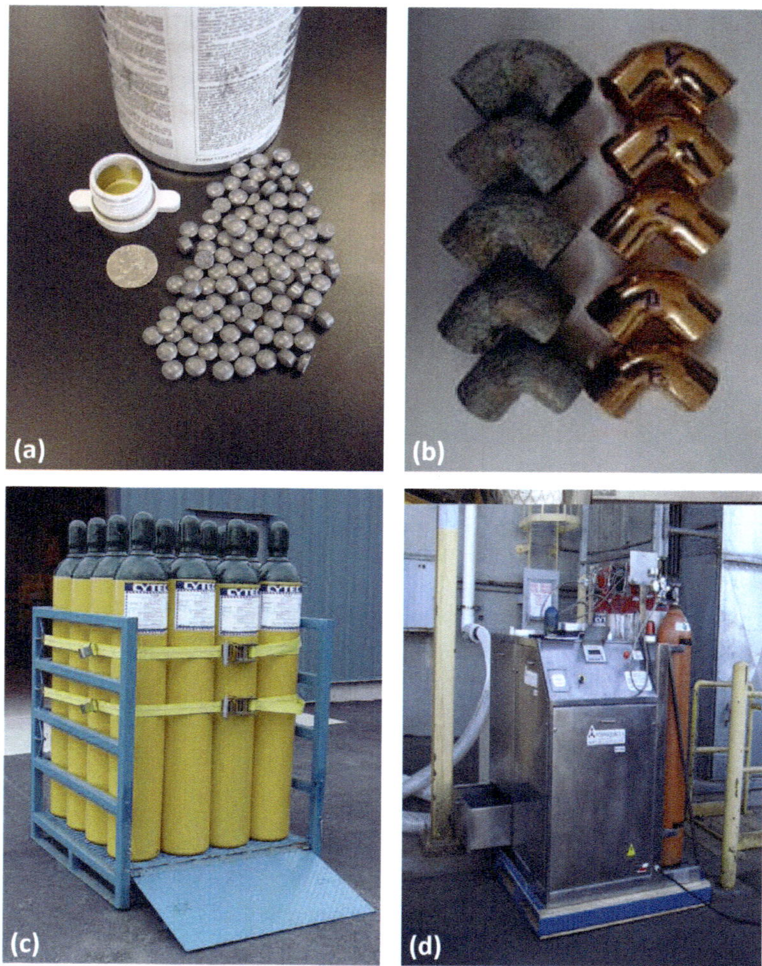

Figure 1.15 Options for and limitations of phosphine fumigation (a) Pellets of aluminum phosphide that react with moisture in warm air to generate hydrogen phosphide gas, PH_3. (b) Copper pipes on the left exposed to the corrosive effect of PH_3 compared to pipes on the right exposed to air. (c) A pressurized cylinder of the ECO_2Fume formulation of 2% PH_3 gas in 98% CO_2. (d) VaporPhos phosphine generator in which 99% PH_3 is quickly added with CO_2 for application to grain bins. [(a) from Degesch America, https://www.degeschamerica.com/; (b) from TWP; (c) from Cytec Inc., now Solvay, https://www.solvay.com/en/chemical-categories/phosphorus-specialties/phosphine-fumigation; and (d) also from Solvay.]

1.7 Conclusions and the Future

Research and commercial implementation of integrated pest management by those attending and contributing to the conferences on Controlled Atmospheres and Fumigation is important now and will be well into the future. Basic and applied research on the biology of stored-product insects and mites has so much more to offer, and the potential for new discoveries is high. There are still many gaps in our knowledge of the behavioral and metabolic response of insects to environmental conditions. Work to date on the population ecology of insects in commodity residues is very fragmentary, and more research in this area has the potential to improve sanitation programs. We have hopes that many commercial pest managers throughout the industry will adopt effective uses of new IPM practices. Economic thresholds are reported for only a few insect and mite pest species that infest limited commodities in very few commercial facilities. Developing sequential sampling plans (e.g., Meikle et al., 2000) for pest infestation in all stored products will greatly increase the cost-effectiveness of monitoring. Prevention, monitoring, and appropriate action can result in more effective use of controlled atmospheres and fumigants into the future.

Acknowledgments

We are very grateful to Digvir Jayas, the editor of this book and chair of the CAF, for inviting us to write this chapter. The Agriculture Research Service and the National Institute for Food and Agriculture, both agencies of the US Department of Agriculture, funded much of our research on stored product insects over the years. We appreciate help with graphics from Crystal Ly. Photographs were graciously provided by Brenda Oppert (USDA) and Jena Johnson (University of Georgia and https://jenajohnson.zenfolio.com/) This article is publication No. 24-208-B of the Kansas Agricultural Experiment Station.

References

Afful E, Elliott B, Nayak MK, Phillips TW (2018) Phosphine resistance in North America field populations of the lesser grain borer, *Rhyzopertha dominica* (Coleoptera: Bostrichidae). J Econ Entomol 111: 463–469.

Anonymous (2018) Integrated Pest Management for Cultural Heritage; Solutions – Low Temperature Treatment (accessed on 2023-11-23 at https://museumpests.net/solutions-low-temperature-treatment/)

Anonymous (2020) Legal registration requirements of chemical in the USA (accessed on 2023-11-21 at https://www3.epa.gov/pesticides/chem_search/ppls/091274-00001-20200331.pdf)

Boczek J (1991) Mite pests in stored food. In: Gorham JR (ed) Ecology and Management of Food Industry Pests, Official Analytical Chem. Assoc., Arlington, pp 57–79.

Cato AJ, Elliott B, Nayak MK, Phillips TW (2017) Geographic variation in phosphine resistance among North American populations of the red flour beetle, *Tribolium castaneum* (Herbst) (Coleoptera: Tenebrionidae). J Econ Entomol 110: 1359–1365.

Chapman RF (1998) The Insects-Structure and Function, 4th ed. Cambridge Univ Press.

Davey PM (1965) Insect pests of stored products in the tropics and the commodities and conditions in which they occur. Trop Stored Prod Info 10: 377–386.

Fields PG (1992) The control of stored-product insects and mites with extreme temperatures. J Stored Prod Res 28: 89–118.

Fields P, White MDG (2002) Alternatives to methyl bromide treatments for stored-product and quarantine insects. Ann Rev Entomol 47: 331–359.

Flinn PW, Arthur FH, Throne JE, Friesen KS, Hartzer KL (2015) Cold temperature disinfestation of bagged flour. J Stored Prod Res 63: 42–46.

Flinn PW, Hagstrum DW, Reed C, Phillips TW (2003) United States Department of Agriculture-Agriculture Research Service stored-grain areawide integrated pest management program. Pest Management Sci 59: 614–618.

Gullan PJ, Cranston PS (2014) The Insects: An Outline of Entomology, 4th ed. Wiley-Blackwell, Oxford, 584 p.

Hagstrum, D. W. and Flinn P. W. (1995). IPM in grain storage and bulk commodities. Stored Product Management. Oklahoma Cooperative Extension Service Circular E-**912**, Stillwater, OK, pp. 201–205.

Hagstrum DW, Flinn PW, Reed CR, Phillips TW (2010) Ecology and IPM of insects at grain elevators and flat storages. Biopestic Intern 6: 1–20.

Hagstrum DW, Klejdysz TZ, Subramanyam B, Nawrot J (2013) Atlas of Stored-Product Insects and Mites, American Association of Cereal Chemists International, St. Paul.

Hagstrum DW, Subramanyam B (2006) Fundamentals of Stored Product Entomology, American Association of Cereal Chemists International, St. Paul.

Hagstrum DW, Subramanyam B (2009) Stored-product insect resource, American Association of Cereal Chemists International, St. Paul.

Hasan MM, Chowdhory SA, Shafiqur Rahman ASM, Athanassiou CG (2020) Development and diapause induction of the Indian meal moth, *Plodia interpunctella* (Hübner) (Lepidoptera: Pyralidae) at different photoperiods. Sci Rep 10: 14707. doi: 10.1038/s41598-020-71659-7

Hernandez Nopsa J, Daglish G, Hagstrum DW, Leslie JF, Phillips TW, Scoglio C, Thomas-Sharma S, Walter GH, Garrett K (2015) Ecological networks in stored grain: Identifying key nodes for emerging pests and mycotoxins in postharvest networks. BioScience 65: 985–1102.

Howe RW (1965) A summary of estimates of optimal and minimal conditions for population increase of some stored products insects. J Stored Prod Res 1: 177–184.

Jacob TA (1996) The effect of constant temperature and humidity on the development, longevity and productivity of *Ahasverus advena* (Waltl,) (Coleoptera: Silvanidae) J Stored Prod Res 32: 115–121.

Mahroof RM, Hagstrum DW (2012) Biology, behavior and ecology of insects in processed commodities. In: Hagstrum DW, Phillips TW, Cuperus G (eds), Stored Product Protection S156, Kansas State University, Manhattan, KS, pp. 33–43.

Meikle WG, Holst N, Degbey N, Oussou R (2000) Evaluation of sequential sampling plans for the larger grain borer (Coleoptera: Bostrichidae) and the maize weevil (Coleoptera: Curculionidae) and of visual grain assessment in west Africa. J Econ Entomol 93: 1822–1831.

Myers SW, Hagstrum DW (2012) Quarantine. In: Hagstrum DW, Phillips TW, Cuperus G (eds), Stored Product Protection S156, Kansas State University, Manhattan, pp. 297–304.

Navarro S, Timlick B, Demianyk CJ, White NDG (2012) Controlled or modified atmospheres. In: Hagstrum DW, Phillips TW, Cuperus G (eds), Stored Product Protection S156, Kansas State University, Manhattan, KS, pp. 191–201.

Nayak MK, Daglish GJ, Phillips TW, Ebert PR (2020) Resistance to the fumigant phosphine and its management in insect pests of stored products: A global perspective. Ann. Rev. Entomol 65: 333–350.

Phillips TW, Strand MR (1994) Larval secretions and food odors affect orientation in female *Plodia interpunctella*. Ent Expt Appl 71: 185–192.

Romoser WS, Stoffolano JG (1998) The Science of Entomology. 4th ed. WCP McGraw-Hill, Boston

Sambaraju KR, Phillips TW (2008a) Effects of physical and chemical factors on oviposition by *Plodia interpunctella* (Lepidoptera: Pyralidae). Ann Entomol Soc Am 101: 955–963.

Sambaraju KR, Phillips TW (2008b) Ovipositional preferences and larval performances of two populations of Indian meal moth, *Plodia interpunctella*. Entomologia Experiemntalis et Applicata 128: 283–293.

Schlipalius DI, Valmas N, Tuck AG, Jagadeesan R, Ma L, et al. (2012) A core metabolic enzyme mediates resistance to phosphine gas. Science 338: 807–810. DOI: 10.1126/science.1224951

Sinha RN, Watters FL (1985) Insect Pests of Flour Mills, Grain Elevators, and Feed Mills and Their Control. Agriculture Canada Publication, Ottawa, 1776 p.

Walski TE, Van Damme JM, Smargiasso N, Christiaens O, DePauw E, Smagg G (2016) Protein N-glycosylation and N-glycan trimming are required for postembryonic development of the pest beetle *Tribolium castaneum* Sci Rep 6: 35151. DOI: 10.1038/srep35151.

Chapter 2

Insect Resistance to Fumigants in Postharvest Commodity Protection

Monitoring and Management

Manoj Nayak and Rajeswaran Jagadeesan

2.1 Introduction

Long before human civilization started cultivation of crops and learned the art of storing grain and food materials, a range of insect species used to live on bark and nests of birds and social insects, as well as wild seeds accumulated by other animals. With the building up of storage, these insects gradually moved to the high-energy food source available in the commodities stored. Over a considerable period, these opportunistic insects adapted well to the storage environment and became what we now know as stored-commodity pests. Archaeological evidence indicates the initiation of the storage of grain as a major source of food for human consumption in mudbrick and straw structures, dating back to several thousand years in ancient civilizations of the Indus Valley and Mesopotamia (Roberts, 1976).

DOI: 10.1201/9781003309888-2

Although mudbrick and straw granaries are still operational in several underdeveloped countries, the rapid colonization of human societies across the globe, industrialization, and technological advancements resulted in modern storage structures that are built with steel, concrete, and durable fabrics (Reed, 1992). These structures enable the industry to store a wide range of durable food and other materials for human consumption and other uses; these include grain (cereals, pulses, oilseeds, and nuts); dried tubers, fruits, herbs, and spices; dried fish and meat products; museum artifacts; leather; and wool. With the diversification in the commodities stored, more and more insect species have shifted their habitat to the storage ecosystem from their natural habitats (Reed, 1992). Several species belonging to three insect orders, Coleoptera (beetles), Lepidoptera (moths), and Psocoptera (psocids), generally infest stored commodities (Rees, 2004). For durable commodities such as grain and its products, insect pests are categorized into primary pests [e.g., the lesser grain borer, *Rhyzopertha dominica* (Fabricius)], which feed on whole grain and are capable of penetrating through the seed coat and pod, and secondary pests [e.g., the red flour beetle, *Tribolium castaneum* (Herbst)], which prefer to feed on grain that has already been damaged or grain products such as flour (Nayak and Daglish, 2018).

There is a wide range of implications from insect infestations in stored commodities, including economic loss due to physical damage, quality degradation, rejection by consumers leading to loss of markets, impact on workplace health and safety, and costs associated with their management (Nayak and Daglish, 2018). Among these, the physical damage and loss of quality and markets due to regulatory controls form a major threat to the biosecurity of food, specifically cereal grains and its products, which continues to be a key international issue. This has been highlighted in an estimated yearly global loss of 1.3 billion tonnes of grain from insect infestations (FAO-World Bank, 2010). The importance of reducing this loss through appropriate management of insects in stored commodities is paramount, in view of a recent projection estimating that food production would need to increase by 60% to feed a global population of 10.5 billion in 2050 (Alexandratos and Bruinsma, 2012).

Currently, an integrated approach involving judicious utilization of available control methods is used to achieve the best outcome from pest management in stored commodities. A typical integrated pest management program for stored commodities involves multiple components, including sample screening and inspections for pests, trapping, hygiene, drying, cooling, heat treatments, controlled atmosphere, and chemical treatments (Daglish et al., 2018). Among the available control methods, insecticide treatments including contact insecticides and fumigants are at the frontline of all major stored-product protection systems. Contact insecticides are either applied directly onto the commodities (protectants) at the time of storage or applied to the storage structures (structural treatments) prior to storing the commodities or dispensed into the storage space (aerosols) during storage. While contact insecticides are used for the provision of protection of commodities from insect attack, fumigants are used to disinfest the commodities when infestations are detected. Although the industry uses both regularly, it heavily depends on the use of specific chemicals that are compatible with the logistical and market requirements, specifically in ensuring food security and market access. Over the last two decades, however, there has been a gradual decline in the use of contact insecticides due to increasing regulatory restrictions, consumer sensitivity toward insecticide residues, development of resistance in major pest species, and high costs associated with the development and registration of new chemicals. Over this period, the use of fumigants, on the other hand, has emerged globally as the most preferred treatment across multiple storage types (e.g., silos, bunkers, sheds, bag stalks, silo bags, shipping containers) (Nayak et al., 2020). Among several fumigants developed over the years, methyl bromide and phosphine have the longest history of use in commodity storage systems. While methyl bromide has undergone a significant phaseout and is restricted to selective treatments only due to its ozone-depleting properties, phosphine is currently the most sought-after fumigant (Nayak et al., 2020). The factors behind the successful use of phosphine over such a long period of time include its global acceptance by markets and regulatory authorities as a residue-free treatment, relative inexpensiveness, versatility in application (e.g., tablets, blankets, cylinderized gas) suiting a range of storage structures, excellent ability to penetrate through commodities, and effectiveness against all life stages of major storage pests (Nayak et al., 2020). The overreliance on phosphine, however, has resulted in the development of resistance in almost all major insect pest species, and the prevalence of phosphine-resistant insects across the globe has been well documented (Lorini et al., 2007; Nayak et al., 2013; Kaur et al., 2015; Kocak et al., 2015; Cato et al., 2017; Afful et al., 2018; Agrafioti et al., 2019; Nayak et al., 2020). This situation has stimulated the search for alternative fumigants as well as the development of strategies to manage ongoing resistance problems so as to extend the usefulness of phosphine into the foreseeable future.

The aim of this chapter is to review recent advances in monitoring and managing resistance to fumigant phosphine in stored commodities. We critically synthesize research published in recent years on phosphine and discuss key elements including resistance development in key stored-product pest species, methods for their diagnosis, resistance monitoring, and their management utilizing available alternative treatments. We also suggest some potential future research in this area.

2.2 Development of Resistance to Fumigants and Its Monitoring

2.2.1 Genetic Basis of Resistance

It is well established that the development of resistance to any insecticide is an evolutionary process, and thus it is best explained in terms of genetics. For fumigants, resistance as a heritable trait evolves in storage insects through selection pressure from treatments and develops through inheritance and fitness costs associated with individuals carrying the resistance genes in a population (Opit et al., 2012a). The spread of resistance generally occurs through the movement of pest populations with resistant individuals across geographic scales through flight activities, transportation of infested commodities, machinery, etc. Therefore, understanding the processes involved in the selection and mechanisms of resistance through which insects evade toxicity is critical for the development of effective resistance management strategies (Collins and Schlipalius, 2018).

Advanced molecular technologies have enabled whole genomic and transcriptome sequencing of major storage pests *T. castaneum* (Richards et al., 2008), *R. dominica* (Schlipalius et al., 2012), and *Liposcelis bostrychophila* (Badonnel) (Dou et al., 2013). It has now been established that resistance to phosphine can occur through multiple variants of resistance genes, and comprehensive genetic and molecular research on key grain insect pests has revealed the presence of two genes, *rph1,* and *rph2*, conferring resistance to phosphine (Schlipalius et al., 2012, 2018). Both genes inherit in incompletely recessive form from parents to progeny and in homozygous isolation confer a low level of (weak) resistance (20 times for *rph1* and 12.5 times for *rph2*), whereas together in an individual insect they interact synergistically and confer a very high level of (strong) resistance (>250 times). The genes *rph1* and *rph2* encode for two key metabolic enzymes, fatty acid desaturase (FADS) and dihydrolipoamide dehydrogenase (DLD), that are involved in aerobic energy metabolism (Tahara et al., 2007). While these two major genes account for most of the resistance phenotype, studies also highlight the involvement of minor genes in parallel, independently, or in combination (Kim et al., 2019).

While resistant individuals enjoy a distinct advantage over their susceptible counterparts in the presence of a toxicant, it is a common assumption that they suffer a fitness cost (e.g., lowered physiological and reproductive success) in the absence of selection pressure (McKenzie, 1996). Although there has been significant research undertaken on the fitness of phosphine-resistant insect pests, so far there has been no consistent conclusion on whether resistant insects incur substantial fitness disadvantages, especially with reproductive traits. For example, fitness investigations carried out on freshly collected field populations indicated that resistant insects incur significant biological costs in terms of reproduction (Pimentel et al., 2007) and development rate (Sousa et al., 2009). However, the conclusions drawn from these studies are very weak, as the observed results did not exclude background genetic effects. In contrast, more recent studies by Daglish et al. (2020) and Singarayan et al. (2021) utilized isogenic-resistant strains and concluded that strongly phosphine-resistant *T. castaneum* and *Cryptolestes ferrugineus* (Stephens) are unlikely to incur substantial biological and reproductive fitness costs. These studies were in alignment with previous studies that examined fitness at phenotypic and genotypic levels (Schlipalius et al., 2008; Jagadeesan et al., 2012), confirming that phosphine resistance does not incur significant biological cost in the event of withdrawal of selection pressure.

Other key areas that need to be better understood include population structure and gene flow in stored-commodity pests that are directly linked with the rate of spread of resistance. Neutral mitochondrial genetic markers have been successfully used for such studies, which assisted in improving understanding of the level of insect movement within the storage environment and across broad landscapes through genetic differentiation in major pest species *T. castaneum*, *C. ferrugineus*, and *R. dominica* (Ridley et al., 2011, 2016; Toon et al., 2018).

2.2.2 Resistance Detection Methods and Monitoring

Early detection of resistance and its characterization form the basis for the diagnosis of control failures, assessment of their likely impact, evaluation of the success of intervention strategies, and development of integrated management strategies. Currently, there are several methods available to detect resistance to phosphine in storage pests collected from the field; these are briefly described below.

2.2.2.1 FAO Test

The earliest method for detecting resistance to phosphine and methyl bromide in several stored-product pests was introduced through a global survey under the auspices of the United Nations Food and Agriculture Organisation (FAO, 1975). This bioassay method, popularly called the FAO test, recommends a diagnostic dose to assess phosphine resistance with a concentration of gas equal to the dose estimated to achieve $LD_{99.9}$ for adult beetles from a susceptible population when exposed for 20 h at 25°C, followed by confirmatory end-point mortality 14 d after fumigation (FAO, 1975). Examples of diagnostic doses recommended for phosphine resistance in common storage pests are as little as 20 ppm (0.03 mg/L) for *R. dominica* and up to 50 ppm (0.07 mg/L) for *Sitophilus granarius* (Linnaeus). This method was modified by Australian researchers to accommodate two different levels of resistance to phosphine, weak and strong. For example, the discriminating doses to determine weak and strong resistance in *R. dominica* are 20 ppm (0.03 mg/L) for 20 h and 180 ppm (0.25 mg/L) for 48 h, respectively (Collins et al., 2017); 30 ppm (0.04 mg/L) and 180 ppm for 20 h, respectively, for *S. oryzae* (Holloway et al., 2016); and 20 ppm and 180 ppm for 20 h, respectively, for *T. castaneum* (Nayak et al., 2017).

2.2.2.2 Rapid Knockdown Tests

Although the FAO test is valuable and discrete in determining resistance phenotypes in field-collected populations, it is time consuming and requires significant numbers to perform the test. There has been substantial interest in developing rapid indicative tests that can reveal results on the same day. The criterion these assays rely on is the proportion of insects knocked down (unable to move in a coordinated manner) with a known phosphine concentration and exposure time, instead of the percentage of dead insects (% mortality) used in FAO tests. Rapid resistance diagnostic tests were developed for several stored-product pests: *T. castaneum* (Reichmuth, 1991; Steuerwald et al., 2006; Cato et al., 2019); *R. dominica* (Afful et al., 2021); *S. oryzae* (Bell et al., 1994; Nayak et al., 2019); *C. ferrugineus* (Nayak et al., 2013); the cigarette beetle, *Lasioderma serricorne* (Fabricius) (Hori and Kasaishi, 2005); the saw-toothed grain beetle, *Oryzaephilus surinamensis* (Linneaus); and the granary weevil, *Sitophilus granarius* (Linneaus) (Steuerwald et al., 2006). These are widely used by researchers and industry as a decision-making tool. Two methodologies are followed: (1) performing discriminating bioassays over time in small gastight glass vials (Nayak et al., 2013, 2019) and (2) utilizing a commercial field test kit (Degesch) with a standard phosphine concentration of 3,000 ppm (4.1 mg/L) (Steuerwald et al., 2006; Cato et al., 2019; Afful et al., 2021). The Degesch test diagnosed resistant insects after 8, 11, 12, and 13 min in *T. castaneum, O. surinamensis, S. granarius,* and *C. ferrugineus,* respectively, whereas the glass vials assays relied on two concentrations, 1,440 (2 mg/L) and 3,600 ppm (5 mg/L), and discriminated the weak and strong resistance phenotypes over 3–5 h in *C. ferrugineus* (Nayak et al., 2013) and *S. oryzae* (Nayak et al., 2019). While the same-day advice to industry using quick tests in the laboratory facilitates timely implementation of necessary management tactics, several improvements to the Degesch field test kit are needed, including means for safe generation and disposal of test gas, with protocols modified to determine both weak and strong levels of resistance, before its broader adoption by industry.

2.2.2.3 Molecular Diagnostics

The identification of the strong resistance gene *rph2* has contributed significantly to the development of a diagnostic molecular assay to accurately determine the strong level of phosphine resistance in major storage pests

(Schlipalius et al., 2012, 2018, 2019; Kaur et al., 2013, 2015; Chen et al., 2015; Kocak et al., 2015; Nayak et al., 2021). Apart from its accuracy, a distinct advantage of molecular diagnostics over phenotypic assays (FAO assays and rapid tests) is its ability to identify heterozygotes (carriers of resistance) in both live and dead insects. Researchers used simple low-throughput DNA marker assays as well as high-throughput strategies that involved tagged primers, multiplexing exon sequencing, and demultiplexing (Schlipalius et al., 2019). The former is preferred in regional-specific studies in which a single variant is prevalent, and the latter can be adopted to gather more specific insight over larger landscapes.

2.3 Monitoring of Resistance to Phosphine and Its Importance

2.3.1 Types of Monitoring

A critical step in the management of resistance is monitoring pest susceptibility to chemical treatments to permit early determination of failure of control measures and resistance developments as well as trends in the frequency of existing resistances. A carefully designed sampling strategy addressing the focal point and monitoring resistance is the key factor that enables assessment of the susceptibility status of a population to a specific treatment. This approach also helps in understanding the potential risks, factors involved in resistance developments, and the geographic spread of existing resistances. For example, Australia has the longest history of monitoring (over three decades) resistance to phosphine and key grain protectants, with a nationally coordinated program that involves collaborative efforts of researchers, farmers, bulk-storage operators, and other stakeholders of the grain industry handling and storing grain (Emery et al., 2011; Holloway et al., 2016; Nayak et al., 2020). A statistically robust nationally agreed on protocol is followed for this program, which currently encompasses the three principal activities of systematic, targeted, and tactical monitoring. Systematic monitoring involves the collection of insect samples from all sectors of the industry in a predetermined survey that provides data on overall trends in the frequency of existing resistances in key pest species both regionally and nationally and also helps in the identification of new resistances. Targeted monitoring, on the other hand, is undertaken where certain storage sites have suspected resistance problems as well as sites where a follow-up visit is necessary to evaluate the success of an intervention strategy that has been implemented to manage a known resistance problem. The third approach is tactical monitoring, in which insect samples are diagnosed on the same day they arrive in the laboratory and results are provided to the industry to allow timely resistance management decisions to be made in the field (Emery et al., 2011; Holloway et al., 2016; Nayak et al., 2020).

2.3.2 Reported Resistance Frequency in Australia and Across the World

The national resistance monitoring program in Australia has helped the grain industry by providing early warning of resistance developments in key pest species and analysis of temporal trends and geographic spread over more than three decades. Resistance data stored in a central database called the Australian Grain Insect Resistance Database (AGIRD) revealed several key factors influencing the development of phosphine resistance (Emery et al., 2011; Nayak et al., 2013, 2017; Daglish et al., 2014, 2015; Falk et al., 2015; Holloway et al., 2016; Collins et al., 2017). Recent results of monitoring in Australia showed that weakly resistant populations are quite common in on-farm as well as bulk storages and contribute to the relative increase in the frequency of strongly resistant populations across the value chain. Such findings highlight the importance of monitoring resistance and how it assists in combatting resistance issues on time. Data from AGIRD recently suggested that, although the frequency of the common weak resistance has increased significantly over the last two decades and is currently recorded at between 60% and 80% (depending on species), the frequency of strong resistance has remained below 10% for *R. dominica* (Collins et al., 2017), *Sitophilus oryzae* (Holloway et al., 2016), and *T. castaneum* (Nayak et al., 2017).

Data on periodic monitoring is also available from several countries across the globe. For example, in the United States, recent resistance surveys have established that strongly resistant populations of major storage pest species are present in grain-growing states (Opit et al., 2012b; Saglam et al., 2015; Gautam et al., 2016, 2020; Cato et al., 2017; Afful et al., 2018). Recent reports from Asia include strongly resistant populations of *R. dominica* in India (Kaur et al., 2015); *S. oryzae* in Vietnam (Nguyen et al., 2015), and *C. ferrugineus* in China (Chen et al., 2021). Lorini et al. (2007) reported a strong resistance frequency of 74% in *R. dominica* populations in bulk storage in Brazil, South America. Other studies from Brazil reported 100% weak resistance frequency in field populations of *T. castaneum*, *R. dominica*, and *O. surinamensis* (Pimentel et al., 2010). Very limited resistance data are available from Europe, with a recent report on strong resistance in *T. castaneum* from Turkey (Kocak et al., 2015) and reported resistance in *S. granarius* and *T. castaneum* from the Czech Republic (Aulicky et al., 2019). More recently, an extensive survey undertaken in grain storage across Greece for the first time recorded resistance in populations of several species (Agrafioti et al., 2019). Only two reports of phosphine resistance are available from Africa; one study was conducted in Morocco, where 50 out of 51 *S. oryzae* populations were detected with resistance (Benhalima et al., 2004), and the other was a case of strong resistance in *T. granarium* in Burkina Faso (Bell and Wilson, 1995). While the majority of the published resistance monitoring data were based on phenotypic tests (bioassays), recent developments in molecular diagnostics have enabled some researchers to be able to screen for the presence of resistance variants in pest populations across the grain value chain (Chen et al., 2015; Schlipalius et al., 2019; Nayak et al., 2021). These studies complemented the phenotypic results and estimated the allelic frequency of heterozygous resistant genotypes for a better understanding of the problem in a region.

2.4 Steps in Monitoring and Its Implications for Managing Pests and Resistance

Monitoring serves as one of the prime components of resistance management, with three interdependent steps: (1) insect sampling, (2) identification of regionally important pest species, and (3) ultimately establishing region-specific pest susceptibility data. The success of monitoring starts with executing an intensive sampling survey at a local scale, targeted to reveal the density and composition of economically important insect pests in a given space and time. Several studies across the world adopted this region-based approach and established pest profiles for major insect species that infest stored products and processed commodities (Pimentel et al., 2010; Afful et al., 2018; Agrafioti et al., 2019; Nayak et al., 2021). These studies emphasize the importance of performing region-specific continual surveillance as part of managing resistance. Once the foundational step revealing species abundance and density has been carried out for a region, susceptibility of the target insect species to different chemical treatments can easily be generated following standard bioassay approaches, as described in Section 2.2.2. In general, randomly chosen test insect cohorts of field populations of each insect species are exposed to a set of known concentrations of fumigants, decided based on two criteria: results of range-finding tests (Robertson et al., 2007) and the relevance of the selected concentrations to the field application rates (Nayak et al., 2013; Gautam et al., 2016). Results of these dose-mortality assays are used to compare the relative susceptibility of species over the study region (spatial) and time (temporal) (Manivannan, 2015; Gautam et al., 2016; Holloway et al., 2016; Collins et al., 2017; Konemann et al., 2017; Nayak et al., 2017) and help researchers to establish comprehensive baseline toxicity information for a specific treatment. Such species-wise baseline toxicity data (White and Lambkin, 1990; Hori and Kasaishi, 2005) are the first defense tool in an integrated pest management program, as they enable tracking of pest susceptibility levels to chemicals (resistance) with a discriminating dose (DD) (Robertson et al., 2007). This multi-step approach to monitoring pest susceptibility is mandatory and was widely followed to diagnose the presence or absence of insects genetically resistant to fumigants (phosphine and sulfuryl fluoride) (Jagadeesan and Nayak, 2023). The direct implications of such programs include identifying the current susceptibility status of field populations, isolating populations that are genetically pure (strains/populations) and characterizing the phenotypic trait (strength of resistance), launching appropriate genetic investigations aimed at rapid resistance diagnosis, and developing and deploying appropriate protocols in an integrated pattern to manage resistant insects.

2.4.1 Establishment of Susceptibility Status of Key Pest Species Within a Region

In a global survey on phosphine resistance carried out in the 1970s (FAO, 1975) in an effort to establish a discriminating dose (DD) for various chemicals used in postharvest grain storage, multiple representative insect populations were collected from 82 countries for diagnostics (Dyte and Halliday, 1985). The study concluded that there was a worldwide increase in the frequency of resistance to phosphine and proposed species-specific DD for monitoring insect resistance globally. This ground-breaking research set the stage for and triggered several subsequent regional surveillance studies, primarily to micro-examine the reported resistance levels and frequency of resistant individuals at higher resolution (Benhalima et al., 2004; Opit et al., 2012b; Daglish et al., 2014, 2015; Kocak et al., 2015; Cato et al., 2017; Agrafioti et al., 2019). Although each of these studies followed a slightly different approach (refer to Section 2.2.2), the results in general indicated that the frequency of resistance to phosphine was high and on the rise in major grain insect pests including *R. dominica, T. castaneum, C. ferrugineus, S. oryzae,* and the psocids, *Liposcelis bostrychophila* (Badonnel) (Chaudhry, 1997; Nayak et al., 2020). The problem is more acute in developing countries, notably in the tropics, with relatively high (>50%) frequency of resistance for selected storage regions (Duong et al., 2016; Schlipalius et al., 2019; Deeksha et al., 2023). These findings reaffirmed the conclusion of the FAO report and emphasized the importance of adopting information (resistance phenotype)–based resistance management programs for fumigants (Collins and Schlipalius, 2018; Daglish et al., 2018). These phenotype-based resistance surveillance programs detected the early onset of resistance or resistance types (weak/strong) and thus played a substantial role in delaying the rate of development of resistance by directing appropriate regions to adopt suitable management approaches (Holloway et al., 2016; Collins et al., 2017; Nayak et al., 2017).

2.4.2 Characterization of the Strength of Resistance

Monitoring also identifies populations that carry novel phenotypic traits such as weak or strong resistance to the fumigant phosphine. Mixed life stages (eggs, larvae, pupae, and adults) of these representative insect populations can be exposed to real-time field treatments (fumigations), reflecting practical field application rates. Results of such studies (Lorini et al., 2007; Nayak et al., 2013; Kaur and Nayak, 2015; Aulicky et al., 2019) revealed effective phosphine concentration regimes (concentration and time) required to control resistant survivors in the field. Subsequently, the data generated on efficacy against key target pests at multiple temperature (°C) regimes contribute to revising or renewing application rates for phosphine (Nayak and Collins, 2008; Kaur and Nayak, 2015).

2.4.3 Provision of a Source of Variation for Genetic Analysis Aimed at Developing DNA-Based Diagnostic Tools

Monitoring provides a variety of insect genetic resources (strains and populations) with novel phenotypic traits. This includes populations carrying unique alleles (variants) conferring phenotypic resistance to phosphine (Schlipalius et al., 2012). These genetic resources were well utilized and led to the identification of the genes responsible for resistance to phosphine in major grain insect pests (Schlipalius et al., 2012, 2018). The functional characterization of these genes also provided new insights into the toxic action of phosphine (Schlipalius et al., 2012) and how resistant insects switch over to biochemical pathways that can avoid phosphine toxicity (Nath et al., 2011). Furthermore, gene complementation studies involving populations and DNA sequences collected from multiple countries confirmed that the phosphine resistance genes, *rph1* and *rph2,* are highly conserved, and a single *rph2* variant (P49/45S) is found prevalently across the world (Chen et al., 2015; Kaur et al., 2015; Nguyen et al., 2016; Jagadeesan et al., 2021a; Yilmaz and Kocak, 2022). Although several other variants of *rph2* also confer resistance to phosphine, the variant P49/45S is unique to grain insect pests and does not exist in other eukaryotes, indicating that pest populations have gone through stringent selection and evolved with specific responses to resist phosphine toxicity. As anticipated, such specific resistance responses remained conserved across the globe (Kaur et al., 2015; Nguyen et al., 2016), irrespective of the fact that pest control practices vary across countries. Utilizing this knowledge base and understanding, researchers developed practical application tools such as gene-specific DNA markers for rapid diagnosis of phosphine resistance in field populations (genotypic testing) (Schlipalius et al., 2019).

2.4.4 Implications of Genetic Tools

Molecular tools to detect phosphine resistance have been well utilized in recent years (Schlipalius et al., 2019), as they provide value-adding insights rather than just detecting resistant insects. Studies relied on either traditional low-throughput approaches (Chen et al., 2015; Yilmaz and Kocak, 2022; Deeksha et al., 2023) or more advanced techniques, utilizing the power of high-throughput sequencing (Schlipalius et al., 2019; Nayak et al., 2021), to estimate accurately the frequency of phosphine resistance in a region by accommodating resistant alleles that are "heterozygous" (*rs*)— insect genotypes having a single copy of the *r* and *s* alleles, inherited from the homozygous resistant (*rr*) and susceptible (*ss*) parents. Phenotypic testing generally lacks the power to detect these heterozygotes (Roush and Tabashnik, 1990), the major carriers of resistance in the field populations (Roush and McKenzie, 1987), and this is quite evident when resistance is inherited in a recessive mode, as with phosphine (Schlipalius et al., 2008; Jagadeesan et al., 2012; Nguyen et al., 2016; Jagadeesan et al., 2021a). In such cases, heterozygote-resistant insects cannot be distinguished from their susceptible counterparts phenotypically using DD. Thus, molecular resistance diagnostics plays a critical role in accurately estimating the resistance allele frequency (including the proportion of heterozygote-resistant insects) in field populations. Studies also proposed frequency thresholds for heterozygotes (Ebert et al., 2003; Lu et al., 2018), which can effectively be used to predict the outbreak of highly resistant homozygote populations. Overall, the generated information allows researchers to map regions of interest for continual monitoring and performing of confirmatory testing. Additionally, the high-throughput monitoring approach allows researchers to screen for novel resistance variants (i.e., the incursion of new resistances) in field populations, and thus the spread of incursions to other regions can be thwarted on time (Schlipalius et al., 2019; Nayak et al., 2021). Neutral mitochondrial DNA markers were also deployed in monitoring programs to better understand the gene flow and population structures of major grain insect pests over a large geographical area, addressing a key question in managing resistance: insect movement (Ridley et al., 2011; Ridley et al., 2016; Thangaraj et al., 2019; McCulloch et al., 2020). These population genetic studies provided vital information such as the active dispersal (flight) of the grain insect pests in the presence and absence of commodities in storage (Holloway et al., 2018; Thangaraj et al., 2018) and the spread of resistant insects over the broader geographical landscapes, both potentially contributing to the development of resistance. This information, along with key ecological investigations into areas such as paternity analysis and the mating behaviour (polyandry) of resistant/susceptible insects (Malekpour et al., 2018; Rafter et al., 2018; Toon et al., 2018), complements molecular resistance diagnostics and could provide a new perspective for modeling the rate of development of resistance and its spread and targeting direct management tactics accordingly.

2.4.5 Development and Deployment of Key Intervention Strategies

The information generated using live insect bioassays (phenotype) and DNA analysis based on gene-specific resistant mutations (genotypes) (Schlipalius et al., 2018; Schlipalius et al., 2019; Nayak et al., 2020) allows researchers to develop and deploy key management strategies that can be implemented stepwise in sequence or following a systematic pattern. In general, the management module includes the following four systematic approaches:

1. Rank or prioritize problematic or economically important pest species for the region (Nayak and Daglish, 2018).
2. Categorize geographical regions into zones (e.g., grain growing, pest, and resistance categories) (Song et al., 2011; Holloway et al., 2016; Nayak et al., 2017).
3. Identify or develop appropriate pest intervention strategies that are suitable for specific storage types within the grain value chain (e.g., silos, bunker storages, open sheds) (Collins et al., 2017; Schlipalius et al., 2019; Nayak et al., 2021).
4. Implement intervention strategies selectively with a special emphasis on rotating treatments spatially and temporally (Collins, 2009).

This four-fold smart grid management approach, in combination with a decision-supporting tree (Figure 2.1), is currently the most preferred and reliable choice for the industry, rather than relying on routine calendar-based treatments. Such an information-based interactive management approach will also avoid indiscriminate dosing; therefore, the proposed steps eliminate all causative factors for the genetic development of resistance in stored-grain insect pests.

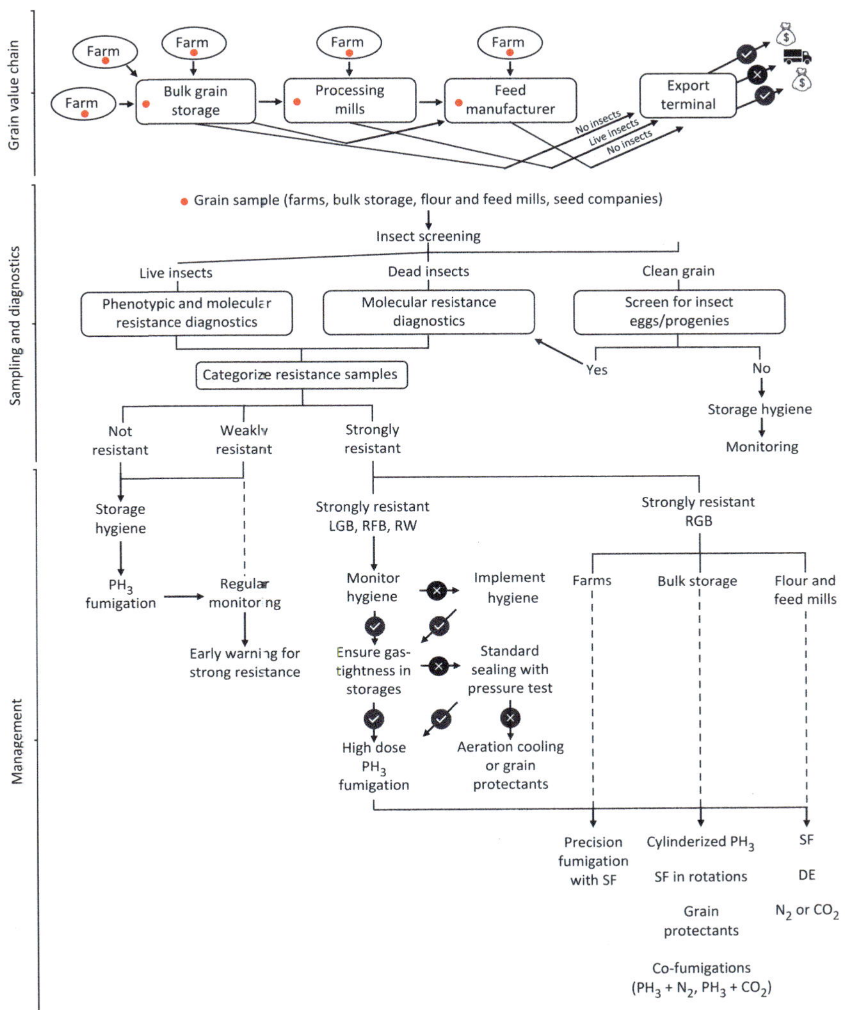

Figure 2.1 A decision-making system for pest and resistance management along the postharvest grain value chain. *Notes:* PH$_3$ = phosphine, SF = sulfuryl fluoride, DE = diatomaceous earth, N$_2$ = nitrogen, CO$_2$ = carbon dioxide, LGB = lesser grain borer (*Rhyzopertha dominica*), RFB = red flour beetle (*Tribolium castaneum*), RW = rice weevil (*Sitophilus oryzae*), RGB = rusty grain beetle (*Cryptolestes ferrugineus*).

2.5 Integrated Strategies for Management of Phosphine Resistance

2.5.1 Development of a Phosphine Resistance Management Strategy

With a successful history of resistance monitoring over three decades, Australia has adopted a well-developed resistance management strategy for both fumigants and contact insecticides used by the industry with stored commodities (Collins, 2009). This strategy has been developed in wider consultation with major storage operators including farmers and bulk handling authorities, chemical companies, regulators, and researchers. It takes the form of a live document adopted by the industry and updated regularly to accommodate changes in protocols as they develop, with the main objective of maintaining the biological efficacy, cost-effectiveness, and useful life of currently registered chemical treatments. For phosphine, the core resistance management strategies adopted in Australia that have global relevance are outlined below.

2.5.2 Early Detection and Characterization of Resistance

While monitoring detects resistant insects early, characterization facilitates establishing homozygous resistant strains and using them as resources to modify fumigation protocols so as to manage them (Daglish et al., 2002; Collins et al., 2005; Nayak et al., 2013). The process usually involves exposing a group of resistant adults to a range of low to high concentrations of phosphine; the dose-mortality data from respective concentrations are then subjected to probit regression analyses (Finney, 1971). Any individuals surviving higher concentrations are reared over 6–8 generations with a series of discriminatory selections promoting the homozygosity of the resistance trait (Collins et al., 2005). This homozygous strain represents the typical field selection and the "worst case" for the resistance development scenario for that species. Key toxicity characteristics such as resistance ratio (RR) and time to population extinction curves (TPE) are established for this strain in each of the insect species and are then used as a reference for the development of practical fumigation strategies in the laboratory (Figure 2.2). Once established in the laboratory, these protocols are validated through industry-scale field trials and finally adopted by industry through modification of the registered label (Collins et al., 2005; Lorini et al., 2007; Kaur and Nayak, 2015).

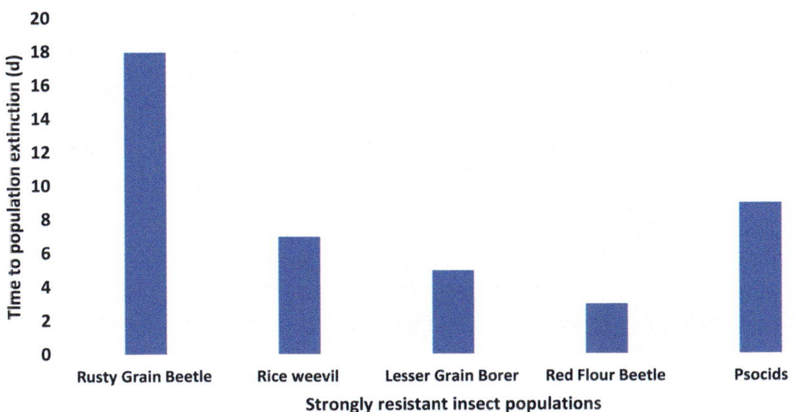

Figure 2.2 The differences in exposure time for the phosphine concentration, 1 mg/L (720 ppm), required to achieve complete extinction of strongly resistant populations of five major stored-grain insect pests at 25°C.

2.5.3 Reducing Selection by Optimizing the Fumigation Regimes and Minimizing Applications

The effect of several biological, nonbiological, and environmental factors on the efficacy of phosphine has been well documented and has helped researchers develop practical approaches to optimizing fumigation protocols so as to manage resistant pest populations. A key biological factor influencing phosphine efficacy is the developmental stage of the pest; generally eggs and pupae were found to be more tolerant of phosphine than were adults and larvae across most of the storage pest species (Daglish et al., 2018; Nayak et al., 2020). Several studies have also reported delayed egg hatching in many pests after exposure to phosphine, requiring amendments to traditional fumigation protocols aimed at achieving complete population extinction (Nayak et al., 2003; Rajendran et al., 2004). It is important, therefore, to develop practical fumigation protocols against resistant pest populations by taking these biological factors into consideration. To address this, recent research has focused on developing fumigation protocols against all life stages of resistant populations that mimic the fumigation in the field situation (Collins et al., 2005; Nayak and Collins, 2008; Kaur and Nayak, 2015). Two key nonbiological factors that influence the efficacy of phosphine significantly are concentration (C) and exposure period (t), both of which have been successfully manipulated over the years to extend the effective life span of phosphine, specifically in controlling strongly phosphine-resistant pests (Daglish et al., 2002; Collins et al., 2005; Nayak and Collins, 2008;

Kaur and Nayak, 2015). These studies have demonstrated that increasing either C or t will enhance phosphine efficacy against strongly resistant pests, though sometimes t plays a more critical role than C. Temperature (°C) also plays a major role in the effectiveness of phosphine, with several studies showing higher efficacy at higher temperatures (30–35°C) (Nayak and Collins, 2008; Kaur and Nayak, 2015).

It is well understood now that underdosing a parcel of grain or commodities or leakage during fumigation not only leads to control failures but also plays a critical role in the selection for resistance (Nayak et al., 2020). Therefore, ensuring the airtightness of the storage structure, following the registered label for the application of the correct dose at the right temperature, and monitoring the gas concentration during the fumigation period to achieve the target dose are critical factors for making fumigation successful (Collins, 2009). For example, in Australia, silos are pressure tested for their sealability/airtightness prior to fumigation following Standards for Sealed Grain Storage Silos (Standard AS2628) (GRDC, 2014). Moreover, the phosphine label has been modified over the last two decades to accommodate new fumigation protocols ($C \times t$) that have been developed against strongly phosphine-resistant *R. dominica* (Collins et al., 2005), *L. bostrychophila* (Nayak and Collins, 2008), and *C. ferrugineus* (Kaur and Nayak, 2015). Active recirculation is also recommended for rapid and uniform distribution of gas throughout the commodities, particularly when phosphine tablets are used in large silos (>150 t capacity) (Collins, 2009).

Optimization of fumigation also needs to address another key factor, sorption by commodities that leads to the loss of the gas and can significantly reduce the effectiveness of the fumigant. Sorption of phosphine and sulfuryl fluoride (SF) by commodities has been well documented through several laboratories (Daglish and Pavic, 2008, 2009; Hwaidi et al., 2015) and field research (Rajendran and Muralidharan, 2001). Daglish and Pavic (2008) established that in wheat, the daily percentage decline in gaseous phosphine was negatively correlated with dose rates and the sorption rate was higher at 25°C than at 15°C (Daglish and Pavic, 2009). Hwaidi et al. (2015) evaluated the sorption of SF in multiple commodities and revealed SF to be four-fold more sorptive than phosphine. These studies have practical implications for the use of phosphine and SF in developing fumigation strategies, taking into consideration the characteristics of commodities, environmental conditions, and storage period.

Repeated fumigation of the same parcel of grain or commodities to control surviving populations in leaky storage is a typical example of selection for resistance (Nayak et al., 2020). To avoid selection for resistance, up to a maximum of three consecutive phosphine fumigations are recommended for the same parcel of grain (Collins, 2009). If insects are not controlled with these fumigations, the use of a break fumigant such as SF or a contact insecticide (if appropriate) is recommended, and storage hygiene is emphasized along with structural treatments (diatomaceous earth) to reduce residual pest populations thriving on spilled grain, grain dust, and dockage (Collins, 2009). Moreover, cooling grain (<15°C) using the principles of aeration (Collins, 2009) at the time of storage is recommended to slow down the growth and multiplication of pest populations, which helps in reducing the number of fumigations, but longer fumigation periods may be needed due to lower phosphine efficacy at lower temperatures (Nayak and Collins, 2008; Kaur and Nayak, 2015).

2.5.4 Isolation and Eradication of Resistant Populations to Restrict Their Spread

In storages where resistant populations have already developed and are well established, it is important that they be isolated and destroyed or eradicated completely so as to restrict their spread to other storage sites/commodities through transportation and other means (Collins, 2009; Nayak et al., 2020). Rotation of chemicals and/or judicious use of alternative treatments such as fumigants and contact insecticides can be employed in such situations. For example, in Australia, *C. ferrugineus* has developed the highest level of resistance recorded for any storage pests, and although phosphine protocols have been established to control this resistance, they are not practically feasible to implement in many of the storage structures due to their lengthy exposure periods (e.g., 21 d) at very high concentration (e.g., 720 ppm) (Nayak et al., 2013; Kaur and Nayak, 2015). In this scenario, SF has proven successful in managing strong resistance, with the advantage that there is no cross-resistance between the two fumigants (Jagadeesan et al., 2015; Nayak et al., 2016; Jagadeesan and Nayak, 2017). Through large-scale field trials involving grain stored in a bunker storages (large, sheeted grain piles) and a concrete silo, Nayak et al. (2016) achieved complete mortality of both natural infestations and caged experimental mixed-age populations including *R. dominica*, *T. castaneum*, *S. oryzae*, and *C. ferrugineus*, using SF at its current registered rate of 1,500 g h/m^3.

Earlier research has shown that the application of high doses to control resistant heterozygotes helps in delaying the evolution of resistance mainly due to the lack of survivors to reproduce further (Roush and Tabashnik, 1990). Care should be taken to implement this tactic with optimization of fumigation in well-sealed storages, as underdosing and leaky structures may lead to the selection of stronger levels of resistance (Daglish et al., 2002; Collins et al., 2005). For example, in a recent work, Afful et al. (2019) recommended a high dose of 400 ppm at 25°C over a 10-d fumigation period to achieve complete control of a strongly phosphine-resistant North American population of *R. dominica*.

2.5.5 Deployment of Suitable Alternatives in Consideration of Operational Logistics

During the 1980s, sixteen potential fumigants were considered for usage in postharvest commodity protection (Fields and White, 2002); of these, only a few are currently used in the field. These include phosphine, methyl bromide (MB), sulfuryl fluoride, ethyl formate (EF), and hydrogen cyanide (HCN). However, the usage of MB has been totally restricted to preshipment and quarantine treatments, due to its ozone-depleting properties (MBTOC, 2006), and EF and HCN both have practical impediments when it comes to field applications (Bell, 2000). Ethyl formate is highly sorptive and holds flammability risks and thus requires a suitable carrier gas such as CO_2 for efficient gas penetration and distribution (Dojchinov et al., 2010), whereas no detailed information on the safety and efficacy of HCN is available, especially for disinfesting stored or processed commodities (Stejskal et al., 2018). This leaves SF as the only available alternative fumigant for disinfesting harvested commodities.

While the usage of SF in postharvest commodity protection is steadily increasing, recent studies on the efficacy of fumigants in the laboratory identified several other promising candidates, primarily to alleviate phosphine-resistant populations and virtually replace MB in quarantine and phytosanitary treatments. These are carbonyl sulphide (CS_2) (Bartholomaeus and Haritos, 2005), chlorine dioxide (Cl_2O_2) (Xinyi et al., 2018; also, see Chapter 8), ethanedinitrile (EDN) (Ramadan et al., 2020; also, see Chapter 9), methyl benzoate (MZ) (Morrison et al., 2019), ozone (O_3) (Velasquez et al., 2017), nitric oxide (NO) (Liu, 2020; also see Chapter 13), allyl isothiocyanate (AITC) (Freitas et al., 2016; de Souza et al., 2018), and propylene oxide (PPO) (Navarro et al., 2004; also see Chapter 11). These candidates are likely to be used selectively in integrated resistance management (IRM) programs in the upcoming years, with consideration of the operational streams and logistics of the industry. However, comprehensive data on biological efficacy and safety need to be established for each, along with the mandatory requirement of clearance from the environmental regulatory authority for each country (Navarro, 2006).

Maintaining hermetic conditions (hypercapnic and/or hypoxic) in selected storage structures, particularly for industries relying on nonchemical control measures, has also been proposed as a viable alternative to phosphine, MB, and SF. Three different CO_2 regimes, 40% over 17 d; 60% over 11 d, and 80% over 8.5 d, were proposed to control insect pests of stored products (Annis, 1987). Recent research by Riudavets et al. (2009) is broadly in agreement with these control regimes and recommended concentrations higher than 50% over 14 d to control 10 pest species. Hypoxia (low oxygen), achieved by purging storage structures with pure nitrogen (N_2), was commonly evaluated as one of the nonchemical alternatives for controlling stored-product pests (Kashi, 1981; Annis, 1987). The effective low O_2 concentration range required was between 0.5% and 1% (Annis, 1987; Navarro and Donahaye, 1990) over a lengthy exposure period of 20–25 d (Sanchez-Molinero et al., 2010). While no cross-resistance has been reported to hypercapnia or hypoxia (Constantin et al., 2020; Sakka and Athanassiou, 2023), these two hermetic treatments can also effectively be used to break resistance to phosphine; however, achieving and maintaining higher CO_2 or lower O_2 concentrations over a lengthy treatment period in practical situations could be challenging (Aulicky et al., 2017). Furthermore, the efficacy of both treatments drops substantially when the ambient temperature drops below 20°C (Kashi, 1981; Annis, 1987).

2.5.6 Role of Sulfuryl Fluoride in Managing Phosphine Resistance

Unlike other treatments, complete withdrawal of phosphine as a result of resistance has never been a viable option for the industry globally, considering the versatility of this noteworthy fumigant and the range of benefits it offers (Nayak et al., 2020). Instead, developing and deploying an effective IRM strategy with viable alternative tools such as SF will prolong the usefulness of the phosphine and ensure the sustainability of chemical treatments in the IRM program.

Several studies on the efficacy of SF have confirmed that phosphine-resistant insects fail to confer resistance to SF (Jagadeesan et al., 2015; Opit et al., 2016). The lack of cross-resistance to SF in phosphine-resistant insects remained consistent, irrespective of inherent differences reported among the species, strains, and life stages (Jagadeesan and Nayak, 2017). This phenomenon was attributed to the differences in the site of action of phosphine and SF in insect cells. Phosphine interferes with two key respiratory enzymes, dihydrolipoamide dehydrogenase (DLD) (Schlipalius et al., 2012) and fatty acid desaturase (FADS) (Schlipalius et al., 2018), and causes mortality via reactive oxygen species (ROS) (Hsu et al., 2000), whereas the fluorine anion (F) in SF inhibits the utilization of cations in various respiratory enzymes and causes mortality via impeding energy production (ATP synthesis) (Mendrala et al., 2003). Thus, SF is widely recognized as a "phosphine resistance breaker" and used by various stakeholders globally. Different SF dose regimes were established for use by the industry, suiting the local operational requirements. For instance, high concentrations of SF (1,196–1,467 g-h/m^3) over short exposures (1–2 d) were commonly adopted to treat flour mills and empty storage structures (Small, 2007; Campbell et al., 2010; Tsai et al., 2011), whereas low-moderate concentrations (400–800 g-h/m^3) over lengthy exposures (6–10 d) were ideal for treating bulk grain storages and warehouses (Nayak et al., 2016; Ling et al., 2008). In both streams, SF was found to be very effective in eliminating phosphine-resistant insect pests; however, fumigations over short exposures (<72 h) hold significant risks for not controlling eggs (Athanassiou et al., 2012; Jagadeesan, et al., 2015) and eventually lead to selecting tolerant insect populations. Thus industry-scale SF field fumigations should be undertaken for longer than 72 h. For instance, SF fumigations performed following the maximal registered label rates over exposure periods of 6–10 d not only achieved complete control of all life stages of strongly phosphine-resistant rusty grain beetles, *C. ferrugineus*, in the field but also prevented quick rebounds in storages from the local environments, favoring long-term economic and environmental benefits (Nayak et al., 2016). Overall, judicious usage of SF in storage will reduce the number of selection events in insects and thereby avoid the subsequent genetic (resistance response) or environmental (residues) consequences. Utilizing SF within the IRM umbrella, as a break fumigant, would augment other chemical treatments, including phosphine, and ensure the long-term viability of both.

2.5.7 Co-fumigation Strategies

Recent studies have explored several approaches to enhancing the toxicity of phosphine and SF, primarily with a focus on overcoming genetic resistance and increased tolerance in eggs, respectively (Misumi et al., 2010; Constantin et al., 2020; Jagadeesan et al., 2021b; Myers et al., 2021). Such studies have clearly demonstrated that the co-fumigation approach holds immense potential for achieving complete control of resistant insect pests with minimal environmental impacts. For instance, co-fumigation of phosphine with SF reduced the required concentration of both the fumigants at least by half in controlling strongly phosphine-resistant *C. ferrugineus*. The half dose rate of phosphine (84 g-h/m^3) and one-fourth label rate of SF (375 g-h/m^3) was sufficient over 168 h (7 d) to eliminate all insect life stages of strongly phosphine-resistant *C. ferrugineus* at 25°C (Jagadeesan et al., 2021b). The efficacy relationship was "additive" and remained consistent irrespective of whether the combo was applied simultaneously or sequentially (Jagadeesan et al., 2018). The efficacy relationship of the mixture PH$_3$ + SF was synergistic (Misumi et al., 2010, 2011) against eggs and adults of the maize [*Sitophilus zeamais* (M.)] and granary (*S. granarius*) weevils; however, relatively higher concentrations of SF were used over short exposures of 3–24 h, targeting quarantine and preshipping treatments.

Some research has also examined the potential of co-fumigating phosphine and SF, each with atmospheric gases such as CO_2 and N_2. These studies have had multiple benefits in addition to enhancing the toxic effect of fumigants; these included reducing environmental hazards, safety, and ease of application in storage. The combination of a moderate concentration of CO_2 (25–30%) with phosphine (Constantin et al., 2020) or SF (Scheffrahn et al., 1995) caused a synergistic increase in toxicity of both the fumigants. The mixtures phosphine + CO_2 and SF + CO_2 have overcome specific weaknesses in phosphine and SF and caused complete mortality of target insects. The key reasons attributed to the enhanced toxicity of these two fumigants with CO_2 include inducing an increase in metabolic rate, opening insect spiracles, facilitating uniform distribution of the fumigants, and targeting hidden insect life stages in the commodities and storages (Rajendran and Muthu, 1989; Leesch, 1992). Co-fumigation of SF with propylene oxide (PPO) was also explored to compensate for the poor ovicidal effect of SF in controlling infestations of a quarantine insect pest, *T. granarium*. Results indicated that both the fumigants, SF and PPO, enhanced the efficacy of each other and achieved a great degree of complementation in killing the most tolerant diapausing larvae and eggs (Myers et al., 2021). Similarly, the hypoxic condition was believed to enhance the toxicity of phosphine against several stored-grain insect pests, particularly in resistant populations; however, the degree of complementation in efficacy was significantly influenced by the species and ambient temperature (Ren et al., 2017). For instance, some studies indicated that internal feeders such as *S. oryzae* and *R. dominica* are more tolerant of hypoxic conditions than external feeders such as *T. castaneum* (Banks and Annis, 1990; Navarro, 2012), whereas others reported the exact opposite of this trend (Hoback and Stanley, 2001; Sakka and Athanassiou, 2023). Thus, more research resolving this phenomenon is required, with the aim of adopting it as a suitable alternative to control phosphine-resistant insect pests. Overall, it is clear that combined treatments for fumigants have practical relevance and their usage in stored-product pest control by the industry is imminent.

2.5.8 Evaluating the Effectiveness of Intervention Strategies Through Monitoring or Trapping

One of the important strategies that insect resistance management relies on for its successful continuum is post-treatment monitoring. As in any other pest management program, evaluating the effect of control measures is vital for postharvest commodity protection, as it provides effective ongoing feedback on key pest intervention strategies. This will allow the industry to plan early, schedule, and select (chemical or nonchemical) specific types of (supporting or subsequent) treatments and apply them if required. In general, three standard approaches are widely followed: (1) repetitive sampling of treated commodities (Nayak et al., 2016), (2) trapping using pheromones or baits over predetermined intervals (Thangaraj et al., 2018), and (3) diagnosing a set of subsamples (Kaur et al., 2013) for validation.

Studies that evaluated the effectiveness of post–phosphine and sulfuryl fluoride fumigations in the field over months provided critical information on population pest management. These include protection period (residual efficacy) (Small, 2007; Campbell et al., 2010), pest rebound (Buckman et al., 2013), residues (Klementz et al., 2002), resistance (change in resistance frequency) (Kaur et al., 2013), insect movements with the storage and environment (Rafter et al., 2021), and population structures (Campbell and Arbogast, 2004). Thus, posttreatment monitoring is a targeted approach, and the synthesized information by and large will be used to reduce chemical footprints in insect resistance management and facilitate continual implementation of environment-friendly sustainable approaches.

2.6 Conclusion and Future Directions

Due to a lack of suitable alternatives, storage sectors across the globe will undoubtedly depend heavily on phosphine to disinfest stored commodities. Sulfuryl fluoride has made significant progress over the last decade as an alternative; its use in bulk storage systems has substantially broadened, specifically in Australia, which encounters a range of resistance problems and is in dire need of alternative fumigants. Although significantly phased out, methyl bromide continues to provide effective pest control; however, its use is restricted to preshipment quarantine and container fumigation. As discussed in preceding sections, other fumigants, including EF, HCN, EDN, and ozone, all have some limitations in terms of their application, narrow efficacy range, residues, and/or workplace health, safety, and environmental concerns. These available alternatives will have some use by the industry but have so far failed to match the combined benefits offered by phosphine and are unlikely to have a broader impact in this field. In this scenario, it is predicted that, with proper management of resistance issues, phosphine will continue to be the fumigant of choice of industry for the foreseeable future. While the cost of SF is limiting its usage on farms and other storages, co-treating phosphine with alternative fumigants (SF or EF) or atmospheric gases (N_2/CO_2) holds promising potential, as it comes with multiple benefits for the industry, consumers, and the environment. However, more research addressing regulatory approval is required in some cases.

We propose the following key areas for future research and development that would help extend the useful life of phosphine:

a. A global survey for strong levels of phosphine resistance. This is critical to our understanding of the global scenario on the development of strong resistance to phosphine in major grain insect species. The last such survey, undertaken in 1976, only aimed to assess the presence of resistance in field populations.

b. Establishing discriminating doses for diagnosing resistance in key grain insect species to alternative fumigants. For instance, the usage of SF has increased significantly in commodity protection in recent years, particularly to control phosphine-resistant insects.

c. Establishing toxicity data and fumigation protocols (concentration × exposure period) for fumigants that are in the registration pipeline (e.g., AITC, EF, EDN, HCN) toward future adoption by the industry.

d. Studies on resistance mechanisms, the effectiveness of phosphine and other alternative fumigants, and establishing management strategies for new insect species including quarantined pests of importance, such as *T. granarium,* through international collaboration.

Acknowledgments

The authors acknowledge the investments from Grains Research and Development Corporation (GRDC) through multiple research projects, including DAQ2209-C01RTX, Australia-India Strategic Research Fund (AISRF 48516), and the Cooperative Research Centre for the Development of Northern Australia (A.1.2021091), toward research and development in resistance monitoring of major storage pests to chemical treatments and their management across Australia.

References

Afful E, Cato A, Nayak MK, Phillips TW (2021) A rapid assay for the detection of resistance to phosphine in the lesser grain borer, *Rhyzopertha dominica* (F.) (Coleoptera: Bostrichidae). J Stored Prod Res 91: 101776.

Afful E, Elliot B, Nayak MK, Phillips TW (2018) Phosphine resistance in North American field populations of the lesser grain borer, *Rhyzopertha dominica* (Coleoptera: Bostrichidae. J Econ Entomol 111: 463–469.

Afful E, Tadesse T, Nayak MK, Phillips TW (2019) High-dose strategies for managing phosphine-resistant populations of *Rhyzopertha dominica* (F.) (Coleoptera: Bostrichidae). Pest Manag Sci 76: 1683–1690.

Agrafioti P, Athanassiou CG, Nayak MK (2019) Detection of phosphine resistance in major stored-product insects in Greece and evaluation of a field resistance test kit. J Stored Prod Res 82: 40–47.

Alexandratos N, Bruinsma J (2012) World Agriculture towards 2030/2050. ESA Working Paper No12-03. Food and Agriculture Organisation of the United Nations.

Annis PC (1987) Towards rational controlled atmosphere dosage schedules - a review of the current knowledge. In: Donahaye E, Navarro S (eds.), Proceedings of 4th international working conference on stored products protection. Maor-Wallach Press, Jerusalem, Israel.

Athanassiou CG, Phillips TW, Aikins MJ, Hasan MM, Throne JE (2012) Effectiveness of sulfuryl fluoride for control of different life stages of stored-product psocids (Psocoptera). J Econ Entomol 105: 282–287.

Aulicky R, Kolar V, Plachy J, Stejskal V (2017) Field efficacy of brief exposure of adults of six storage pests to nitrogen-controlled atmospheres. Plant Prot Sci 53: 169–176.

Aulicky R, Stejskal V, Frydova B (2019) Field validation of phosphine efficacy on the first recorded resistant strains of *Sitophilus granarius* and *Tribolium castaneum* from Czech Republic. J Stored Prod Res 81: 107–13.

Banks HJ, Annis PC (1990) Comparative advantages of high CO_2 and low O_2 types of controlled atmospheres for grain storage. In: Calderon M, Barkai Golan R (eds.), Food Preservation by Modified Atmospheres. CRC Press, Boca Raton, pp. 93–122.

Bartholomaeus AR, Haritos VS (2005) Review of the toxicology of carbonyl sulfide, a new grain fumigant. Food and Chem Toxicol 43: 1687–1701.

Bell CH (2000) Fumigation in the 21st century. Crop Prot 19: 563–569.

Bell CH, Savvidou N, Mills KA, Bradberry S, Barlow ML (1994) A same day test for detecting resistance to phosphine. pp 41–44 In: Highley E, Wright EJ, Banks HJ, Champ BR (eds), Pro 6th Intern Working Conf Stored-Prod Prot, Canberra, Australia.

Bell CH, Wilson SM (1995) Phosphine tolerance and resistance in *Trogoderma granarium* Everts (Coleoptera: Dermestidae). J Stored Prod Res 31: 199–205.

Benhalima H, Chaudhry MQ, Mills KA, Price NR (2004) Phosphine resistance in stored-product insects collected from various grain storage facilities in Morocco. J Stored Prod Res 40: 241–49.

Buckman KA, Campbell JF, Subramanyam B (2013) *Tribolium castaneum* (Coleoptera: Tenebrionidae) associated with rice mills: Fumigation efficacy and population rebound. J Econ Entomol 106: 499–512.

Campbell JF, Arbogast RT (2004) Stored-product insects in a flour mill: Population dynamics and response to fumigation treatments. Entomol Exp et Appl 112: 217–225.

Campbell JF, Toews MD, Arthur FH, Arbogast RT (2010) Long term monitoring of *Tribolium castaneum* populations in two flour mills: Rebound after fumigation. J Econ Entomol 103: 1002–1011.

Cato A, Afful E, Nayak MK, Phillip TW (2019) Evaluation of knockdown bioassay methods to assess phosphine resistance in the red flour beetle, *Tribolium castaneum* (Herbst) (Coleoptera: Tenebrionidae). Insects 10: 140.

Cato AJ, Elliott B, Nayak MK, Phillips TW (2017) Geographic variation in phosphine resistance among North American populations of the red flour beetle (Coleoptera: Tenebrionidae). J Econ Entomol 110: 1359–1365.

Chaudhry MQ (1997) A review of the mechanisms involved in the action of phosphine as an insecticide and phosphine resistance in stored-product insects. Pestic Sci 49: 213–228.

Chen EH, Duan JY, Song W, Wang DX, Tang PA (2021) RNA-seq analysis reveals mitochondrial and cuticular protein genes are associated with phosphine resistance in the rusty grain beetle (Coleoptera: Laemophloeidae). J Econ Entomol 114: 440–453.

Chen Z, Schlipalius D, Opit G, Subramanyam B, Phillips TW (2015) Diagnostic molecular markers for phosphine resistance in US populations of *Tribolium castaneum* and *Rhyzopertha dominica*. PLOS ONE 10(3): e0121343.

Collins PJ (2009) Strategy to manage resistance to phosphine in the Australian grain industry. Rep. CRC70096, Natl. Work. Party Grain Protect., Coop. Res. Cent. Natl. Plant Biosecur. Proj., Bruce, Aust. Phosphine Resistance Strategy.pdf (graintrade.org.au).

Collins PJ, Daglish GJ, Pavic H, Kopittke RA (2005) Response of mixed-age cultures of phosphine-resistant and susceptible strains of lesser grain borer, *Rhyzopertha dominica*, to phosphine at a range of concentrations and exposure periods. J Stored Prod Res 41: 373–385.

Collins PJ, Falk MG, Nayak MK, Emery RN, Holloway JC (2017) Monitoring resistance to phosphine in the lesser grain borer, *Rhyzopertha dominica*, in Australia: A national analysis of trends, storage types, and geography in relation to resistance detections. J Stored Prod Res 70: 25–36.

Collins PJ, Schlipalius DI (2018) Insecticides resistance. In: Arthur FH, Athanassiou CG (eds), Recent Advances in Stored Product Protection, Springer, Berlin, Germany, pp. 169–182.

Constantin M, Jagadeesan R, Chandra K, Ebert PR, Nayak MK (2020) Synergism between phosphine (PH$_3$) and carbon dioxide (CO$_2$): Implications for managing PH$_3$ resistance in rusty grain beetle (Laemophloeidae: Coleoptera). J Econ Entomol 113: 1999–2006.

Daglish GJ, Collins PJ, Pavic H, Kopittke KA (2002) Effects of time and concentration on mortality of phosphine-resistant *Sitophilus oryzae* (L.) fumigated with phosphine. Pest Manag Sci 58: 1015–1021.

Daglish GJ, Jagadeesan R, Nayak MK, McCulloch GA, Singarayan VT, Walter GH (2020) The gene introgression approach and the potential cost of genes that confer strong phosphine resistance in red flour beetle (Coleoptera: Tenebrionidae). J Econ Entomol 113: 1547–1554.

Daglish GJ, Nayak MK, Arthur FH, Athanassiou CG (2018) Insect pest management in stored grain. In: Arthur FH, Athanassiou CG (eds), Recent Advances in Stored Product Protection, Springer, Berlin, Germany, pp. 45–63.

Daglish GJ, Nayak MK, Pavic H (2014) Phosphine resistance in *Sitophilus oryzae* (L.) from eastern Australia: Inheritance, fitness and prevalence. J Stored Prod Res 59: 237–244.

Daglish GJ, Nayak MK, Pavic H, Smith LW (2015) Prevalence and potential fitness cost of weak phosphine resistance in *Tribolium castaneum* (Herbst) in eastern Australia. J Stored Prod Res 61: 54–58.

Daglish GJ, Pavic H (2008) Effect of phosphine dose on sorption in wheat. Pest Manag Sci 64: 513–518.

Daglish GJ, Pavic H (2009) Changes in phosphine sorption in wheat after storage at two temperatures. Pest Manag Sci 65: 1228–1232.

de Souza LP, Faroni LRD, Lopes LM, de Sousa AH, Prates LHF (2018) Toxicity and sublethal effects of allyl isothiocyanate to *Sitophilus zeamais* on population development and walking behavior. J Pest Sci 91: 761–770.

Deeksha MG, Nebapure SM, Kalia VK, Sagar D, Bhattacharya R, Dahuja A, Subramanian S (2023) Comparison of phenotypic and genotypic frequency of phosphine resistance in select field populations of *Tribolium castaneum* from India. Mol Biol Reports 50: 6569–6578.

Dojchinov G, Damcevski KA, Woodman JD, Haritos VS (2010) Field evaluation of vaporised ethyl formate and carbon dioxide for fumigation of stored wheat. Pest Manag Sci 66: 417–424.

Dou W, Shen GM, Niu JZ, Ding TB, Wei DD, Wang JJ, Doucet D (2013) Mining genes involved in insecticide resistance of *Liposcelis bostrychophila* Badonnel by transcriptome and expression profile analysis. PLoS ONE 8(11): e79878.

Duong TM, But NTT, Collins PJ (2016) Status of resistance to phosphine in insect pests of stored products in Vietnam. Indian J Entomol 78: 45–52.

Dyte CE, Halliday D (1985) Problems of development of resistance to phosphine by insect pests of stored grains. EPPO Bulletin 15: 51–57.

Ebert PR, Reilly PEB, Mau Y, Schlipalius D, Collins PJ (2003) The Molecular genetics of phosphine resistance in the lesser grain borer and implications for management. pp. 152–155. In: Wright EJ, Webb MC, Highley E (eds), Stored grain in Australia 2003, Proceedings of the Australian Postharvest Technical Conference, CSIRO Stored Grain Research Laboratory, Canberra.

Emery RN, Nayak MK, Holloway JC (2011) Lessons learned from phosphine resistance monitoring in Australia. Stewart Postharvest Rev 7(3): spr.2011.3.6.

Falk MG, O'Leary R, Nayak MK, Collins PJ (2015) A Bayesian hurdle model for analysis of an insect resistance monitoring database. Environ Ecol Stat 22: 207–226.

FAO (1975) Recommended methods for the detection and measurement of resistance of agricultural pests to pesticides: Tentative method for adults of some major species of stored cereals with methyl bromide and phosphine - FAO Method No 16. FAO Plant Prot Bull 23: 12–25.

FAO-World Bank (2010) Reducing post-harvest losses in grain supply chains in Africa. Report of FAO-World Bank workshop, 18-19 March, 2010, Rome, Italy. 120 p

Fields PG, White NDG (2002) Alternatives to methyl bromide treatments for stored-product and quarantine insects. Annu Rev Entomol 47: 331–359.

Finney DJ (1971) Probit Analysis, 3rd ed. Cambridge University Press, London.

Freitas RCP, Faroni LRD, Haddi K, Jumbo LOV, Oliveira EE (2016) Allyl isothiocyanate actions on populations of *Sitophilus zeamais* resistant to phosphine: Toxicity, emergence inhibition and repellency. J Stored Prod Res 69: 257–264.

Gautam SG, Opit GP, Hosoda E (2016) Phosphine resistance in adult and immature life stages of *Tribolium castaneum* (Coleoptera: Tenebrionidae) and *Plodia interpunctella* (Lepidoptera: Pyralidae) populations in California. J Econ Entomol 109: 2525–2533.

Gautam SG, Opit GP, Konemann C, Shakya K, Hosoda E (2020) Phosphine resistance in saw-toothed grain beetle, *Oryzaephilus surinamensis* in the United States. J Stored Prod Res 89: 101690.

GRDC (2014). Pressure testing sealable silos. Grain Storage Fact Sheet, Grain Res. Dev. Corp., Kingston, Aust. http://storedgrain.com.au/pressure-testing/.

Hoback WW, Stanley DW (2001) Insects in hypoxia. J Insect Physiol 47: 533–542.

Holloway JC, Falk MG, Emery RN, Collins PJ, Nayak MK (2016) Resistance to phosphine in *Sitophilus oryzae* in Australia: A national analysis of trends and frequencies over time and geographic spread. J Stored Prod Res 69: 129–137.

Holloway JC, Mayer DG, Daglish GJ (2018) Flight activity of *Cryptolestes ferrugineus* in southern New South Wales, Australia. J Pest Sci 91: 1353–1362.

Hori M, Kasaishi Y (2005) Development of the new assay method for quickly evaluating phosphine resistance of the cigarette beetle, *Lasioderma serricorne* (Fabricius) (Coleoptera: Anobiidae), based on the knock-down of the adult beetles. Appl Entomol Zool 40: 99–104.

Hsu CH, Han BC, Liu MY, Yeh CY, Casida JE (2000) Phosphine-induced oxidative damage in rats: Attenuation by melatonin. Free Radical Biol Med 28: 636–642.

Hwaidi M, Collins PJ, Sissons M, Pavic H, Nayak MK (2015) Sorption and desorption of sulfuryl fluoride by wheat, flour, and semolina. J Stored Prod Res 62: 65–73.

Jagadeesan R, Collins PJ, Daglish GJ, Ebert P, Schlipalius DI (2012) Phosphine resistance in the rust-red flour beetle, *Tribolium castaneum* (Coleoptera: Tenebrionidae): Inheritance, Gene interactions and fitness costs. PLOS ONE 7(2): e31582.

Jagadeesan R, Nayak MK (2017) Phosphine resistance does not confer cross-resistance to sulfuryl fluoride in four major stored grain insect pests. Pest Manag Sci 73: 1391–1401.

Jagadeesan R, Nayak MK (2023). A two-step approach to monitoring changes in susceptibility to sulfuryl fluoride in red flour beetle, *Tribolium castaneum* (Herbst) (Coleoptera: Tenebrionidae) and its implication for the industry. J Stored Prod Res 104: 102181.

Jagadeesan R, Nayak MK, Pavic H, Chandra K, Collins PJ (2015) Susceptibility to sulfuryl fluoride and lack of cross-resistance to phosphine in developmental stages of the red flour beetle, *Tribolium castaneum* (Coleoptera: Tenebrionidae). Pest Manag Sci 71: 1379–1386.

Jagadeesan R, Schlipalius DI, Singarayan VT, Nath NS, Nayak MK, Ebert PR (2021a) Unique genetic variant in dihydrolipoamide dehydrase (*dld*) gene confers strong resistance to phosphine in the rusty grain beetle, *Cryptolestes ferrugineus* (Stephens). Pestic Biochem Physiol 171: 104717.

Jagadeesan R, Singarayan V, Chandra K, Ebert PR, Nayak MK (2018) Potential of co-fumigation with phosphine (PH$_3$) and sulfuryl fluoride (SO$_2$F$_2$) for the management of strongly phosphine-resistant insect pests of stored grain. J Econ Entomol 111: 2956–2965.

Jagadeesan R, Singarayan VT, Nayak MK (2021b) A co-fumigation strategy utilising reduced rates of phosphine (PH$_3$) and sulfuryl fluoride (SF) to control strongly resistant rusty grain beetle, *Cryptolestes ferrugineus* (Stephens) (Coleoptera: Laemophloeidae). Pest Manag Sci 77: 4009–4015.

Kashi KP (1981) Toxicity of phosphine to five species of stored-product insects in atmospheres of air and nitrogen. Pestic Sci 12: 116–122.

Kaur R, Nayak MK (2015) Developing effective fumigation protocols to manage strongly phosphine resistant Cryptolestes ferrugineus (Stephens) (Coleoptera: Laemophloeidae). *Pest Manag* Sci 71: 1297–1302.

Kaur R, Daniels EV, Nayak MK, Ebert PR, Schlipalius DI (2013) Determining changes in the distribution and abundance of a *Rhyzopertha dominica* phosphine resistance allele in farm grain storages using a DNA marker. Pest Manag Sci 69: 685–688.

Kaur R, Subbarayalu M, Jagadeesan R, Daglish GJ, Nayak MK, Naik HR, Ramasamy S, Subramanian C, Ebert PR, Schlipalius DI (2015) Phosphine resistance in India is characterised by a dihydrolipoamide dehydrogenase variant that is otherwise unobserved in eukaryotes. Heredity 115: 188–194.

Kim K, Yang JO, Sung J, Lee J, Park JS, Lee H, et al., (2019) Minimization of energy transduction confers resistance to phosphine in the rice weevil, *Sitophilus oryzae*. Sci. Rep. 9:14601–14614.

Klementz D, Rassman W, Reichmuth C (2008) Sufuryl fluoride - Efficacy against *Tribolium castaneum* and *Ephestia kuehniella* and residues of the gas in flour after fumigations of mills, In: Daolin G, Navarro S, Jian Y, Cheng T, Zuxun J, Yue L, Yanf L, Haipeng W (eds.), Proceedings of the 8th International Conference on Controlled Atmosphere and Fumigation in Stored Products.

Kocak E, Schlipalius D, Kaur R, Tuck A, Ebert PR, Collins PJ, Yilmaz A (2015) Determining phosphine resistance in rust red flour beetle, *Tribolium castaneum* (Herbst.) (Coleoptera: Tenebrionidae) populations from Turkey. Türk Entomol Derg 39: 129–136.

Konemann CE, Hubhachen Z, Opit GP, Gautam SG, Bajracharya NS (2017) Phosphine resistance in *Cryptolestes ferrugineus* (Coleoptera: Laemophloeidae) collected from grain storage facilities in Oklahoma, USA. J Econ Entomol 110: 1377–1383.

Leesch JG (1992) Carbon dioxide on the penetration and distribution of phosphine through wheat. J Econ Entomol 85: 157–161.

Ling Z, Xinfu Z, Qing X, Jiadong C (2008) Efficacy of sulfuryl fluoride on stored grain pests in a warehouse trial in China. Pp. 579–582. In: Daolin G, Navarro S, Jian Y, Cheng T, Zuxun J, Yue L, Yanf L, Haipeng W (eds), Proceedings of the 8th International Conference on Controlled Atmosphere and Fumigation in Stored Products, Chengdu, China.

Liu Y (2020) Comparison of efficacy of nitric oxide fumigation under nitrogen and carbon dioxide atmospheres in controlling granary weevil (*Sitophilus granrius*) and confused flour beetle (*Tribolium confusum*). J Stored Prod Res 88: 1–7.

Lorini I, Collins PJ, Daglish GJ, Nayak MK, Pavic H (2007) Detection and characterisation of strong resistance to phosphine in Brazilian *Rhyzopertha dominica* (F.) (Coleoptera: Bostrychidae). Pest Manag Sci 63: 358–364.

Lu Y, Zhang C, Wang Z, Yan X, Emery RN (2018) Rapid detection of phosphine resistance in the lesser grain borer, *Rhyzopertha dominica* (Coleoptera: Bostrychidae) from China using ARMS-PCR. pp. 1043–1045. In: Adler CS, Opit G. Fürstenau B, Müller-Blenkle C, Kern P, Arthur FH, Athanassiou CG, Bartosik R, Campbell J, Carvalho MO, Chayaprasert W, Fields P, Li Z, Maier D, Nayak M, Nukenine E, Obeng-Ofori O, Phillips T, Riudavets J, Throne J, Schöller M, Stejskal V, Talwana H, Timlick B, Trematerra P (eds), Proceedings of the 12th International Working Conference on Stored-product Protection, Germany.

Malekpour R, Rafter MA, Daglish GJ, Walter GH (2018) The movement abilities and resource location behavior of *Tribolium castaneum*: Phosphine resistance and its genetic influences. J Pest Sci 91: 739–749.

Manivannan S (2015) Toxicity of phosphine on the developmental stages of rust-red flour beetle, *Tribolium castaneum* Herbst over a range of concentrations and exposures. J Food Sci Tech 52: 6810–6815.

MBTOC (2006) Report of the methyl bromide technical options committee: 2006 assessment United Nations Environment Programme, Nairobi., Kenya.

McCulloch GA, Gurdasani K, Kocak E, Daglish GJ, Walter GH (2020) Significant population genetic structuring in *Rhyzopertha dominica* across Turkey: Biogeographic and practical implications. J Stored Prod Res 85: 1–6.

McKenzie JA (1996) Factors Influencing Selection for Insecticide Resistance. Ecological and Evolutionary Aspects of Insecticide Resistance. R.G. Landes. Co, Academic Press, George Town, Texas, 185 p.

Mendrala AL, Eisenbrandt DL, Markham DA, Clark AJ, Krieger SM, Houtman CE, Rick DL (2003) Sulfuryl fluoride: Pharmacokinetics and metabolism in F344 rats. Toxic Sci 72: 147–147.

Misumi T, Aoki M, Kitamura H (2011) Synergistic and suffocative effects of fumigation with a lower concentration of phosphine and sulfuryl fluoride gas mixture on the mortality of *Sitophilus* species (Coleoperata: Dryophthoridae), a stored-product pest. Part 2: Susceptibility test on *Sitophilus zeamais* for fumigation with a gas mixture, and verification test of a sequential fumigation with two fumigants. Res Bull Plant Prot Japan 47: 1–10.

Misumi T, Aoki M, Tanigawa N, Kitamura H, Suzuki N (2010) Synergistic and suffocative effects of fumigation with a lower concentration phosphine and sulfuryl fluoride gas mixture on mortality of *Sitophilus* species (Coleoptera: Dryophthoridae), a stored product pest. Part 1: Dose-response test on *Sitophilus granarius* and *Sitophilus zeamais* for fumigation with sulfuryl fluoride. Res Bull Plant Prot Japan 46: 1–6.

Morrison WR, Larson N, Brabec D, Zhang A (2019) Methyl benzoate as a putative alternative, environmentally friendly fumigant for the control of stored product insects. J Econ Entomol 112: 2458–2468.

Myers SW, Ghimire MN, Arthur FH, Philips TW (2021) A combination of sulfuryl flouride and propylene oxide treatment for *Trogoderma granarium* (Coleoptera: Dermestidae). J Econ Entomol 114: 1489–1495.

Nath NS, Bhattacharya I, Tuck AG, Schlipalius DI, Ebert PR (2011) Mechanisms of phosphine toxicity. J Toxicol 2011: 494168–494168.

Navarro S (2012) The use of modified and controlled atmospheres for the disinfestation of stored products. J Pest Sci 85: 301–322.

Navarro S, Donahaye E (1990) Generation and application of modified atmospheres and fumigants for the control of storage insects. In: Champ BR, Highley E, Banks HJ (eds) Proceedings of the International Conference on Fumigation and Control Atmosphere Storage of Grain. Singapore, Brown Prior Anderson Pty. Ltd., Burwood.

Navarro S, Isikber AA, Finkelman S, Rindner M, Azrieli A, Dias R (2004) Effectiveness of short exposures of propylene oxide alone and in combination with low pressure or carbon dioxide against *Tribolium castaneum* (Herbst) (Coleoptera: Tenebrionidae). J Stored Prod Res 40: 197–205.

Navarro S (2006) New global challenges to the use of gaseous treatments in stored products. pp. 495–509. In: Lorini I, Bacaltchuk B, Beckel H, Deckers D, Sundfeld E, dos Santos JP, Biagi JD, Celaro JC, D'A Faroni LR, Bortolini LOF, Sartori MR, Elias MC, Guedes RNC, da Fonseca RG, Scussel VM, (eds), Proceedings of the 9th International Working Conference on Stored-product Protection, Campinas, São Paulo, Brazil.

Nayak MK, Collins PJ (2008) Influence of temperature and humidity on toxicity of phosphine against strongly resistant *Liposcelis bostrychophila* Badonnel (Psocoptera: Liposcelididae), a cosmopolitan pest of stored commodities. Pest Manag Sci 64: 971–976.

Nayak MK, Collins PJ, Pavic H, Kopittke RA (2003) Inhibition of egg development by phosphine in the cosmopolitan pest of stored products *Liposcelis bostrychophila* (Psocoptera: Liposcelididae). Pest Manag Sci 59: 1191–1196.

Nayak MK, Daglish GJ (2018) Importance of stored product insects. In: Athanassiou CG, Arthur FH (eds), Recent Advances in Stored Product Protection, Springer-Verlag GmbH Germany, pp. 1–17.

Nayak MK, Daglish GJ, Phillips TW, Ebert PR (2020) Resistance to the fumigant phosphine and its management in insect pests of stored products: A global perspective. Annu Rev Entomol 65: 333–350.

Nayak MK, Falk MG, Emery RN, Collins PJ, Holloway JC (2017) An analysis of trends, frequencies and factors influencing the development of resistance to phosphine in the red flour beetle *Tribolium castaneum* (Herbst) in Australia. J Stored Prod Res 72: 35–48.

Nayak MK, Holloway JC, Emery RN, Pavic H, Bartlet J, Collins PJ (2013) Strong resistance to phosphine in the rusty grain beetle, *Cryptolestes ferrugineus* (Stephens) (Coleoptera: Laemophloeidae): Its characterisation, a rapid assay for diagnosis and its distribution in Australia. Pest Manag Sci 69: 48–53.

Nayak MK, Jagadeesan R, Kaur R, Daglish GJ, Reid R, Pavic H, Smith LW, Collins PJ (2016) Use of sulfuryl fluoride in the management of strongly phosphine resistant insect pest populations in bulk grain storages in Australia. Indian J Entomol 78: 100–107.

Nayak MK, Jagadeesan R, Singarayan VT, Nath NS, Pavic H, Dembowski B, Daglish GJ, Schlipalius DI, Ebert PE (2021) First report of strong phosphine resistance in stored grain insects in a far northern tropical region of Australia, combining conventional and genetic diagnostics. J Stored Prod Res 92: 101813.

Nayak MK, Kaur R, Jagadeesan R, Pavic H, Phillips TW, Daglish GJ (2019) Development of a quick knockdown test for diagnosing resistance to phosphine in *Sitophilus oryzae* (Coleoptera: Curculionidae), a major pest of stored products. J Econ Entomol 112: 1975–1982.

Nguyen T, Collins PJ, Duong TM, Schlipalius D, Ebert P (2016) Genetic conservation of phosphine resistance in the rice weevil, *Sitophilus oryzae* (L). J Heredity 107: 228–237.

Nguyen TT, Collins PJ, Ebert PR (2015) Inheritance and characterization of strong resistance to phosphine in *Sitophilus oryzae* (L.). PLOS ONE 10: e0124335.

Opit GP, Collins PJ, Daglish GJ (2012a) Resistance management. In: Hagstrum WD, Philips TW, Cuperus GW (eds), Stored Product Protection, Kansas State University, Manhattan, Kansas, pp. 143–156.

Opit GP, Phillips TW, Aikins MJ, Hasan MM (2012b) Phosphine resistance in *Tribolium castaneum* and *Rhyzopertha dominica* from stored sheat in Oklahoma. J Econ Entomol 105: 1107–1114.

Opit GP, Thoms E, Philips TW, Payton ME (2016) Effectiveness of sulfuryl fluoride fumigation for the control of phosphine-resistant grain insects infesting stored wheat. J Econ Entomol 109: 930–941.

Pimentel MAG, D'A Faroni LR, Tótola MR, Guedes RNC (2007) Phosphine resistance, respiration rate and fitness consequences in stored-product insects. Pest Manag Sci 63: 876–881.

Pimentel MAG, Faroni LRDA, da Silva FH, Batista MD, Guedes RNC (2010) Spread of phosphine resistance among Brazilian populations of three species of stored product insects. Neotropical Entomol 39: 101–107.

Rafter MA, McCulloch GA, Daglish GJ, Gurdasani K, Walter GH (2018) Polyandry, genetic diversity and fecundity of emigrating beetles: Understanding new foci of infestation and selection. J Pest Sci 91: 287–298.

Rafter MA, Muralitharan V, Chandrasekaran S, Mohankumar S, Daglish GJ, Loganthan M, Walter GH (2021) Behaviour in the presence of resource excess - flight of *Tribolium castaneum* around heavily-infested grain storage facilities. J Pest Sci 92: 1227–1238.

Rajendran S, Muralidharan N (2001) Performance of phosphine in fumigation of bagged paddy rice in indoor and outdoor stores. J Stored Prod Res 37: 351–358.

Rajendran S, Muthu M (1989) The toxic action of phosphine in combination with some alkyl halide fumigants and carbon dioxide against the eggs of *Tribolium castaneum* Herbst (Coleoptera: Tenebrionidae). J Stored Prod Res 25: 225–230.

Rajendran S, Parveen H, Begum K, Chethana R (2004) Influence of phosphine on hatching of *Cryptolestes ferrugineus* (Coleoptera: Cucujidae), *Lasioderma serricorne* (Coleoptera: Anobiidae) and *Oryzaephilus surinamensis* (Coleoptera: Silvanidae). Pest Manag Sci 60: 1114–1118.

Ramadan GRM, Abdelgaleil SAM, Shawir MS, El-bakary AS, Edde PA, Philips TW (2020) Sorption of ethanedinitrile in fumigated commodities and its impact on efficacy for *Rhyzopertha dominica* (Coleoptera: Bostrychidae) and *Lasioderma serricorne* (Coleoptera: Anobidae) control. J Stored Prod Res 86: 1–8.

Reed C (1992) Development of storage techniques: A historical perspective. In: Sauer DB (ed), Storage of Cereal Grains and Their Products, American Association of Cereal Chemists Inc, St Paul, MN, pp. 143–156.

Rees D (2004) Insects of Stored Products. CSIRO Publishing, Collingwood, Victoria, Australia, 192 p.

Reichmuth C (1991) A quick test to determine phosphine resistance in stored products research. GASGA Newsletters 15: 14–15.

Ren Y, Agarwal M, Newman J, Eagling D (2017) Delivery and adoption of nitrogen/low oxygen and nitrogen +phosphine technology for the management of grain storage pests and grain quality. Final project report submitted to the Plant Biosecurity Cooperative Research Centre, Canberra, Australia, pp. 1–36.

Richards S, Gibbs RA, Weinstock GM et al (2008) The genome of the model beetle and pest *Tribolium castaneum*. Nature 452:949–955.

Ridley A, Hereward J, Daglish GJ, Raghu S, Collins PJ, Walter GH (2011) The spatiotemporal dynamics of *Tribolium castaneum* (Herbst): Adult flight and gene flow. Mol Ecol 20: 1635–1646.

Ridley AW, Hereward JP, Daglish GJ, Raghu S, McCulloch GA, Walter GH (2016) Flight of *Rhyzopertha dominica* (Coleoptera: Bostrichidae)-a spatio-temporal analysis with pheromone trapping and population genetics. J Econ Entomol 109: 2561–2571.

Riudavets J, Castane C, Alomar O, Pons MJ, Gabarra R (2009) Modified atmosphere packaging (MAP) as an alternative measure for controlling ten pests that attack processed food products. J Stored Prod Res 45: 91–96.

Roberts JM (1976) The Hutchinson History of the World. Hutchinson & Co., London, 1127 p.

Robertson JL, Russell RM, Preisler HK, Savin NE (2007). Bioassays with Arthropods, 2nd ed. CRC Press, Boca Raton, FL.

Roush RT, McKenzie JA (1987) Ecological genetics of insecticide and acaricide resistance. Annu Rev Entomol 32: 361–380.

Roush RT, Tabashnik BE (Eds) (1990) Pesticide Resistance in Arthropods. Chapman and Hall, New York and London, London, 297 p.

Saglam O, Edde PA, Phillips TW (2015) Resistance of *Lasioderma serricorne* (Coleoptera: Anobiidae) to fumigation with phosphine. J Econ Entomol 108: 2489–2495.

Sakka M, Athanassiou CG (2023) Efficacy of nitrogen against stored product insects with different susceptibility levels to phosphine in industrial applications. Agriculture 13: 1–8.

Sanchez-Molinero F, Garcia-Regueiro JA, Arnau J (2010) Processing of dry-cured ham in a reduced-oxygen atmosphere: Effects on physico-chemical and microbiological parameters and mite growth. Meat Sci 84: 400–408.

Scheffrahn RH, Wheeler GS, Su NY (1995) Synergism of methyl bromide and sulfuryl fluoride toxicity against termites (Isoptera: Kalotermitidae, Rhinotermitidae) by admixture with carbon dioxide. J Econ Entomol 88: 649–653.

Schlipalius DI, Chen W, Collins PJ, Nguyen T, Reilly P, Ebert PR (2008) Gene interactions constrain the course of evolution of phosphine resistance in the lesser grain borer, *Rhyzopertha dominica*. Heredity 100: 506–516.

Schlipalius DI, Tuck A, Jagadeesan R, Nguyen T, Kaur R, Subramanian S, Barrero R, Nayak MK, Ebert PR (2018) Variant linkage analysis using *de novo* transcriptome sequencing identifies a conserved phosphine resistance gene in insects. Genetics 209: 281–290.

Schlipalius DI, Tuck AG, Pavic H, Daglish GJ, Nayak MK, Ebert PR (2019) A high-throughput system used to determine frequency and distribution of phosphine resistance across large geographic regions. Pest Manag Sci 75: 1091–1098.

Schlipalius DI, Valmas N, Tuck AG, Jagadeesan R, Ma L, Kaur R, Goldinger A, Anderson C, Kuang J, Zuryn S, Mau YS, Cheng Q, Collins PJ, Nayak MK, Schirra HJ, Hilliard MA, Ebert PR (2012) A core metabolic enzyme mediates resistance to phosphine gas. Science 338: 807–810.

Singarayan VT, Jagadeesan R, Nayak MK, Ebert PR, Daglish GJ (2021) Gene introgression in assessing fitness costs associated with phosphine resistance in the rusty grain beetle. J Pest Sci 94: 1415–1426.

Small GJ (2007) A comparison between the impact of sulfuryl fluoride and methyl bromide fumigations on stored-product insect populations in UK flour mills. J Stored Prod Res 43: 410–416.

Song XH, Wang PP, Zhang HY (2011) Phosphine resistance in *Rhyzopertha dominica* (Fabricius) (Coleoptera: Bostrichidae) from different geographical populations in China. African J Biotech 10: 16367–16373.

Sousa AH, D'A Faroni LR, Pimentel MAG, Guedes RNC (2009) Developmental and population growth rates of phosphine-resistant and -susceptible populations of stored-product insect pests. J Stored Prod Res 45: 241–246. Springer Nature, Berlin, Germany

Stejskal V, Aulicky R, Jonas A, Hnatek J, Malkova J (2018) Bluefume (HCN) and EDN* as fumigation alternatives to methyl bromide for control of primarily stored product pests. pp. 604–608. In: Adler CS, Opit G, Fürstenau B, Müller-Blenkle C, Kern P, Arthur FH, Athanassiou CG, Bartosik R, Campbell J, Carvalho MO, Chayaprasert W, Fields P, Li Z, Maier D, Nayak M, Nukenine E, Obeng-Ofori O, Phillips T, Riudavets J, Throne J, Schöller M, Stejskal V, Talwana H, Timlick B, Trematerra P (eds), Proceedings of the 12th International Working Conference on Stored-product Protection, Germany.

Steuerwald R, Dierks-Lange H, Schmitt S (2006) Rapid bioassay for determining the phosphine tolerance. pp. 306–311. In: Lorini I, Bacaltchuk B, Beckel H, Deckers D, Sundfeld E, dos Santos JP, Biagi JD, Celaro JC, D'A Faroni LR, Bortolini LOF, Sartori MR, Elias MC, Guedes RNC, da Fonseca RG, Scussel VM, (eds), Proceedings of the 9th International Working Conference on Stored-product Protection, Campinas, São Paulo, Brazil.

Tahara EB, Barros MH, Oliveira GA, Netto LE, Kowaltowski AJ (2007) Dihydrolipoyl dehydrogenase as a source of reactive oxygen species inhibited by caloric restriction and involved in *Saccharomyces cerevisiae* aging. FASEB J 21: 274–283.

Thangaraj S, McCulloc GA, Subramanian S, Chandel RK, Debnath S, Subramanian C, Walter GH, Subbrayalu M (2019) Genetic diversity and its geographic structure in *Sitophilus oryzae* (Coleoptera; Curculionidae) across India - implications for managing phosphine resistance. J Stored Prod Res 84: 1–6.

Thangaraj S, Muralitharan V, Daglish GJ, Mohankumar S, Rafter MA (2018) Flight of three major insect pests of stored grain in the monsoonal tropics of India, by latitude, season and habitat. J Stored Prod Res 76: 43–50.

Toon A, Daglish GJ, Ridley AW, Emery RN, Holloway JC, Walter GH (2018) Significant population structure in Australian *Cryptolestes ferrugineus* and interpreting the potential spread of phosphine resistance. J Stored Prod Res 77: 219–224.

Tsai WT, Mason LJ, Chayaprasert W, Maier DE, Ileleji KE (2011) Investigation of fumigant efficacy in flour mills under real-world fumigation conditions. J Stored Prod Res 47: 179–184.

Velasquez LPG, Faroni LRD, Pimentel MAG, Heleno FF, Prates LHF (2017) Behavioral and physiological responses induced by ozone in five Brazilian populations of *Rhyzopertha dominica*. J Stored Prod Res 72: 111–116.

White GG, Lambkin TA (1990) Base-line responses to phosphine and resistance status of stored-grain beetle pests in Queensland, Australia. J Econ Entomol 83: 1738–1744.

Xinyi E, Li BB, Subramanyam B (2018) Toxicity of chlorine dioxide gas to phosphine-susceptible and -resistant adults of five stored-product insect species: Influence of temperature and food during gas exposure. J Econ Entomol 111: 1947–1957.

Yilmaz A, Kocak E (2022) Comparing bioassay and diagnostic molecular marker for phosphine resistance in Turkish populations of *Rhyzopertha dominica* (F.) (Coleoptera: Bostrichidae). Turkish J Entomol 46: 431–440.

Chapter 3

Insect Biology re Controlled Atmospheres, Modified Atmospheres, and Hermetic Storage

Shlomo Navarro, Hagit Navarro, Tom de Bruin, and Nadav Inbari

3.1 Introduction

The method of altering the surrounding environment of organisms causing damage to stored products has been considered as an alternative to control using residue-leaving toxic chemicals. The idea that such methods could be sustainable and environmentally friendly has increased the interest of many researchers and storage technologists. One such technology is the application of controlled and modified atmospheres (CA/MA). MA includes all cases in which the atmospheric gas composition in the treatment enclosure has been modified to create conditions favorable for insect control. In an MA treatment, the atmospheric composition within the treated enclosure may change during the treatment period. In a CA treatment, the atmospheric

DOI: 10.1201/9781003309888-3

composition within the treated enclosure is controlled or maintained at a level and duration lethal to insects. In practice, converting these ideas to applied technologies has given rise to several challenges that the industry is still struggling with.

The first challenge is to render the enclosure destined for the treatment of CA/MA gastight. A fundamental requirement for the application of MA is a well-sealed storage environment. In such a structure, MA created by the application of carbon dioxide (CO_2) can be maintained at an efficient level by restricting air ingress, which otherwise would eventually cause a decrease in the CO_2 concentration and an increase in the oxygen (O_2) level. In practice, it is rare to find fully gastight storage structures, unless they were constructed purposely. The requirement for gastight storage appears to be more critical for the application of MA than for that of fumigants. Fumigants have been used for many years with minimal requirements for structure gastightness; covering the grain bulk or the storage with plastic sheets was usually considered satisfactory (Bond, 1984). The consequences of poorly sealed storages under fumigation were discussed by Banks and Desmarchelier (1979) and Banks (1981), and the necessity for adequate gastight structures has been recognized as the primary factor responsible for the successful application of MA to control insects (Banks and Annis, 1977). Making a structure gastight requires an additional expense that most companies are reluctant to assume due to the cost implications. It appears there is no simple and cost-effective way to make existing storage structures gastight.

The second challenge is the cost of the inert gas to be applied: nitrogen (N_2) or CO_2 or the on-site generation of N_2. In some countries, naturally generated CO_2 is available, reducing the cost. However, in any case, a comparison to the widely applied phosphine gas reveals the extremely low cost of the fumigant, particularly when applied in very low dosages in well-sealed structures. Unless there is a special requirement that the product be fumigated using CA/MA, the economic incentive always favors the low-cost fumigant.

The beneficial effects of MA treatment, as a safe and environmentally benign alternative to the use of conventional residue-producing chemical fumigants for controlling insect pests attacking stored grain, oilseeds, pulses, processed commodities, and packaged foods, have been well documented (Navarro, 2006). A serious interest in using the technique in a practical, routine manner was not pursued until the 1970s and 1980s, probably due to the success of conventional fumigants and grain protectants in controlling stored-product pests. During this period, awareness began to develop that chemicals, if used improperly, left objectionable residues and were hazardous to apply and that there was a potential for the development of insect resistance to them. Research was initiated during this time in

Australia, the United States, and several other countries on the use of modified atmospheres (Ripp et al., 1984). During the last forty years, MA and CA treatments for the disinfestation of dry stored products have received increasing scientific attention.

Although CA has become well established for the control of storage pests, its commercial use is still limited to a few countries. The widespread scientific activities on this subject resulted in several international conferences, such as the International Conferences on Controlled Atmospheres and Fumigation in Stored Products, with the report of the last meeting by Jayas and Jian (2021), and the International Working Conferences on Stored-Product Protection, with the report of Adler et al. (2018) on MA/CA and hermetic storage (HS). These reports reveal that there is an increasing interest in MA, CA, and HS using flexible containers (Murdock and Lowenberg-DeBoer, 2014). It should be noted that research on HS in rigid structures is missing from the literature. It appears that all the available information on the HS of flexible structures is based on hermetic liners. Only in the last ten years has HS emerged as a significant alternative method of postharvest storage, particularly in tropical climate countries, where several HS methods are used (Villers et al., 2010; Baributsa and Concepcion Ignacio, 2020). In South America, silo bags have been used (Bartosik, 2010), and hermetic SuperGrainbags™ have been employed by small farmers for rice seed since 2004, as reported by the International Rice and Research Institute (IRRI) (Rickman and Aquino, 2011). In Africa, the Purdue Improved Cowpea Storage (PICS) hermetic bags have been used (Murdock et al., 1997, 2003; Baributsa et al., 2010). The objective of this chapter is to review the available knowledge on CA, MA, and HS, to evaluate the gaps and missed concepts, and to point out the needed future research and practical information.

3.2 Insect Response to CA/MA and HS

3.2.1 Insect Response to CA/MA

3.2.1.1 Insect Response to Low O_2

Insect response to low O_2 is not necessarily proportional to the depletion of its concentration, particularly at low concentrations for adult curculionid beetles. Although rapid death at below 2% O_2 is normal for most insects, for *Sitophilus oryzae* (L.) this effect is reversed below 1% O_2 in N_2, where adult rice weevils (Navarro, 1978) showed tolerance, increasing the lethal exposure time by apparently closing their spiracles. This resulted in *S. oryzae* adults being killed more quickly at 1.0% O_2 than at 0.1% or 2% O_2 under the same conditions (Figure 3.1).

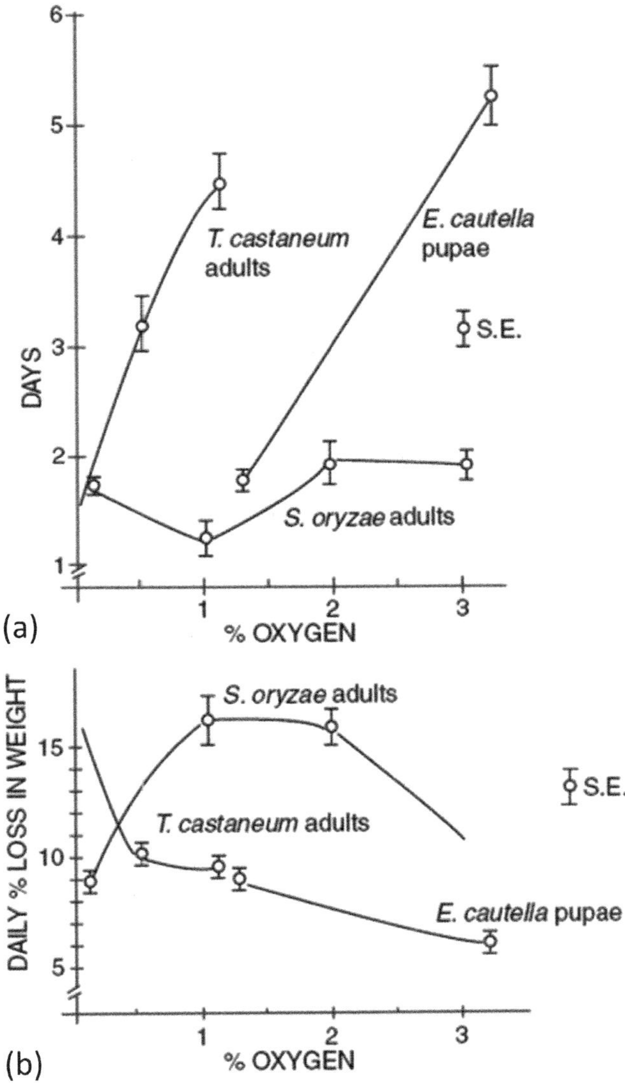

Figure 3.1 (a) Relationship between oxygen concentration and the time required for 95% mortality. (b) The effect on daily percentage loss in weight of three stored-product insects at 54% RH and 26°C. (Redrawn from Navarro, 1978.)

Tribolium castaneum (Herbst) in N_2 showed significant differences in adult mortality between 0.1% and 1.0% O_2 (Navarro, 1978). Adults are generally most susceptible to treatment, and *S. oryzae* and *Rhyzopertha dominica* (F.) were found to be more tolerant than *Tribolium* spp. The lowest level of tolerance to the lack of O_2 was attained around the 1% concentration level. Therefore, Annis (1987) concluded that O_2 levels of 1% are needed to kill insects in 20 d (Table 3.1).

Radek et al. (2017) tested how quickly the mobile adults of six species of storage beetles [*Oryzaephilus surinamensis* (L.), *Cryptolestes ferrugineus* (Stephens), *Tribolium Confusum* (J. du Val), *T. castaneum, Sitophilus granarius* (L.), and *S. oryzae*] are killed after the introduction of the infested commodity to prevent their further spread to the surrounding storage silos. The trials were conducted in a metal silo containing 25 t of seeds using the system of continuous top-down nitrogen filling to replace the O_2. A target concentration ($\leq$ 1% O_2 and 99% N_2) was reached at the bottom of the silo after 12 h of purging nitrogen. Seven days of exposure led to 100% mortality of all tested species except for *S. granarius* (96.7% mortality at the bottom), while 10 d of exposure led to100% mortality of all adults located at both the bottom and the top of the silo. The commodity temperatures ranged from 15.9° to 23.9°C through the grain profile and trials. This experiment showed that 10 d of exposure completely prevented the adult mobile pest stages of all tested species from spreading from the treated silo.

Table 3.1 Suggested Provisional Dosage Regimes for Control of All Stages of the 12 Most Common Insect Species of Stored Grain, Using Modified Atmospheres at Temperatures Between 20 and 29°C

Atmospheric Gas Concentration	Controls Most Common Grain Insects Including *Trogoderma granarium* (Yes/No)	Exposure Period (d)
<1% O_2 (in nitrogen)	yes	20
Constant % CO_2 in air		
40	no	17
60	no	11
80	no	8.5
80	yes	16
CO_2 decay in air from >70% to 35%	no	15
Pressurized CO_2 at >20 bar	*	<0.08

Source: Data, except those on pressurized CO_2, compiled from Annis (1987).
Note: *Available data are based only on *Plodia interpunctella* and *Lasioderma serricorne*.

Sakka et al. (2020) tested the effect of nitrogen on phosphine-susceptible and -resistant populations of stored-product insects in commercial nitrogen chambers with the O_2 level set at 1.0%. Two different temperatures, 28° and 40°C, and three exposure intervals, 2.5, 3, and 9 d, were used in their tests. The trials used adults of the saw-toothed grain beetle, *O. surinamensis*; the red flour beetle, *T. castaneum*; and the rice weevil, *S. oryzae*. Complete parental mortality was observed in all cases for *O. surinamensis* and *S. oryzae*, but there was some survival for *T. castaneum* at 28°C and 3 d of exposure. In general, progeny production was completely (100%) suppressed, with some exceptions for all species and populations.

Navarro and Navarro (2021) tested the efficacy of controlling all life stages of storage insects *T. castaneum* and *O. surinamensis* at 43°C in 1% O_2 and 99% N_2 for 48 h of exposure time. Higher mortality was observed with exposure to 43°C air (control) than with exposure to 28°C air. The adult stage of *O. surinamensis* was the most tolerant, with 99.5% mortality after a 48 h treatment. Larval and pupal stages could not survive the treatment; 100% mortality was recorded. Although in the egg stage in both species, some larvae hatch was observed, the hatched larvae could not survive more than 24 h after the treatment. Average egg mortalities were counted as 97.5% for *T. castaneum* and 98% for *O. Surinamensis*; the young hatched larvae were found dead immediately after treatment. The pupal and larval stages of both species were more susceptible to the treatment than the adult and egg stages.

Sakka et al. (2022) evaluated the efficacy of low O_2 against *Trogoderma granarium* (Everts), *T. castaneum*, and *Callosobruchus maculatus* (F.) in commercial applications. Low O_2 was achieved through the increase of N_2 in different commodities (figs, plums, and sultanas raisins) in commercial N_2 chambers. Two different temperatures were evaluated, 28° and 40°C, with three different exposure periods, 2.5, 3, and 9 d. Adults, diapausing larvae, nondiapausing larvae, and eggs of *T. granarium*, adults of *T. castaneum*, and adults of *C. maculatus* were used in the trials. Larvae of *T. granarium* were found more tolerant than adults and eggs. Adults of *T. castaneum* and *C. maculatus* were found susceptible, as 100% mortality was recorded for all trials. In general, low or no progeny production was recorded for most of the species tested in these trials. Nevertheless, some survival was noted in the case of treatments that were carried out at the lowest temperature level and at the shortest exposure period.

Song et al. (2021) investigated the mortality of different stages of cigarette beetles at different O_2 concentrations ($\leq$2% or 2% to 4%) by applying high concentrations of N_2 to a tobacco warehouse at 18° to 26°C and 55 to 65% RH. The results revealed that the corrected mortality of eggs, larvae, pupae, and adults was 100% under the low O_2 concentration after 27 d exposure. The corrected mortality of larvae exceeded 85% and the hatching rate was less than 10% after 27 d when O_2 concentration was 2% to 4%.

3.2.1.2 Insect Response to High CO_2

Elevated CO_2 levels cause the spiracles to open, resulting in insect death due to water loss. Above 10% CO_2, spiracles remain permanently open. Toxic effects are entirely through the tracheae, not the hemolymph; CO_2 directly affects the nervous system. In some cases, CO_2 can acidify the hemolymph, leading to membrane failure in some tissues (Nicolas and Sillans, 1989). Elevated but sublethal CO_2 levels for prolonged periods can have deleterious effects on insect development, growth, and reproduction (Nicolas and Sillans, 1989; White et al., 1995). Atmospheres containing about 60% CO_2 rapidly kill stored-product insects. At 26°C, about 4 d of exposure would be sufficient to kill all stages (including eggs) of most stored-product insects.

Navarro et al. (2002b), applying various CO_2 concentrations at different temperatures, investigated the mortality of four developmental stages of *Ephestia cautella* (Walker). Results in Table 3.2 summarize the effectiveness of the combination of CO_2 at temperatures in the range of 35° to 45°C. Tests with *E. cautella* showed that the pupa was the most tolerant stage when exposed to 90% CO_2 with an LT_{99} value of 17 h at 35°C and only 3 h when exposed at 45°C. The adult was the most sensitive stage of *E. cautella*, requiring only 4 h of exposure to 90% CO_2 at 35°C.

Riudavets et al. (2021) tested the mortality of the internal feeders *R. dominica* and *Callosobruchus chinensis* (L.) in packages filled with two extreme filling ratios (4% and 96% of chickpeas) and with 50% or 90% CO_2. For both pest species, the exposure time to reach 50% mortality ranged from 7 h (larvae with 90% CO_2) to 2 d (pupae with 50% CO_2) at the lower filling ratio tested (4%). When the filling ratio was increased to 96% of chickpeas, mortality of *R. dominica* eggs and adults decreased significantly but mortality did not vary for the internal developmental stages. A similar effect was observed (a decrease in mortality of external developmental stages) in *C. chinensis* at a 96% filling ratio with 50% CO_2. However, mortality remained the same for the eggs and pupae at 90% CO_2.

Table 3.2 Influence on Various Development Stages of *Ephestia cautella* Exposed to CO_2 Concentrations in Air at Three Different Temperatures

Temp. (°C)	35				40				45			
CO_2 (%)	60	70	80	90	60	70	80	90	60	70	80	90
Eggs	23	23	17	9	16	12	8	5	9	5	3	2
Larvae	60	27	20	12	17	9	6	6	5	4	2	2
Pupae	56	37	17	17	36	10	8	4	7	4	4	3
Adults	20	14	6	4	6	5	3	2	3	2	2	2

Source: Navarro et al., 2003.
Note: Values are expressed in LT99 (hours to obtain 99% mortality).

3.2.1.3 Insect Response to Relative Humidity

Insect mortality increases more rapidly as temperatures rise, and their metabolism speeds up. Cool temperatures slow down the rates of mortality, while lower RH hastens toxic effects, notably in high-CO_2 atmospheres because of the desiccation of insects (Banks and Fields, 1995).

Laboratory studies have shown that lowering the RH increases the effectiveness of MAs. Jay et al. (1971), working with adults of *T. confusum*, *T. castaneum*, and *O. surinamensis*, found that decreasing the RH in atmospheres containing 99% N_2 (balance O_2) from 68% RH to 9% RH increased the mortality of the red flour beetle from 3% to 98.5% in 24 h exposure. The two other insects showed a similar response to reduced RH. These three species also exhibited a similar response to mixtures of CO_2 in air at lowered RH.

Desiccation plays a large role in the mortality of stored-product insects when exposed to some MAs. Jay and Cuff (1981) showed that when larvae, pupae, and adults of the red flour beetle were exposed to varying concentrations of CO_2 or O_2, weight loss was much higher in some of the atmospheres than in others or in the air. Navarro and Calderon (1974) reported a linear relationship of the combined effect of CO_2 and RH in producing a lethal environment for *E. cautella* pupae (Figure 3.2). Navarro and Calderon (1980) also demonstrated the strong dependence of low concentrations of CO_2 and

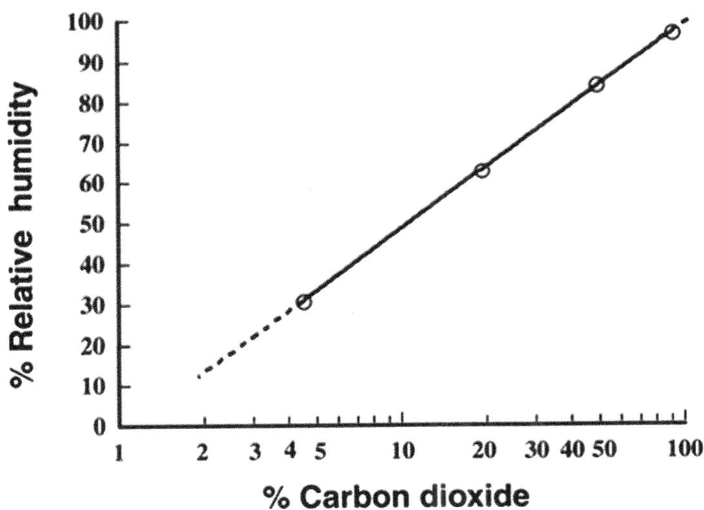

Figure 3.2 The combined effect of carbon dioxide and relative humidity on the time needed to produce 95% mortality for *Ephestia cautella* pupae after four days of exposure. (Redrawn from Navarro and Calderon, 1974.)

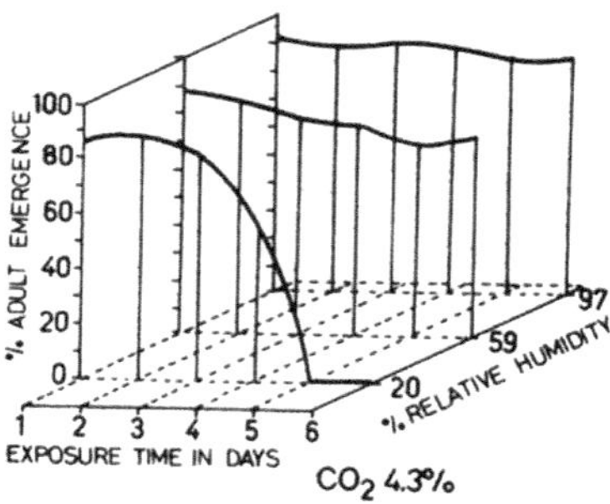

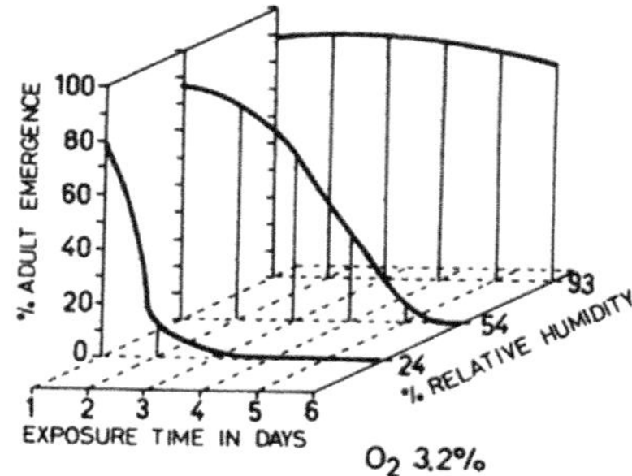

Figure 3.3 Effect of 4.3% CO_2 and 3.2% O_2 on adult emergence from *Ephestia cautella* pupae exposed to different relative humidities at 26°C. (Redrawn from Navarro and Calderon, 1980.)

O_2 on mortality in *E. cautella* pupae (Figure 3.3). Other laboratory studies have shown that the susceptibility of different species or strains of the same species varies considerably when insects are exposed to the same concentrations of MAs (Jay and Pearman, 1971).

The tested RH in a study by Chiappini et al. (2009) was sufficiently low (18%) to result in complete mortality after 120 h at 3% O_2 and 29°C. However, Calderon and Navarro (1979) demonstrated that, after 120 h exposure to 2% O_2 and 30°C at 57% RH, the mortality of *T. castaneum* adults was only 50%.

Navarro (1975) tested the effect of O_2 concentrations on *T. castaneum* adults exposed to different RHs. Adults 0–24 h old were exposed to O_2 concentrations of 0.06%, 0.52%, and 1.11% in combination with RHs in the ranges of 19.6–22.3%, 54.0–54.5%, and 96.2–99.9%, all at 26°C. The combined effect of O_2 and RH producing 95% mortality for *T. castaneum* adults after 4 d of exposure at 26°C is shown in Figure 3.4. Adult mortality and the loss in weight of insects exposed to the above treatments were recorded. To achieve 95% mortality, exposure for 48 h was required at 0.06% O_2 in the highest range of RH tested, but only 30 h of exposure was required for the same effect in the lowest range of RH. Mortality after 10 d was higher than that after 24 h from the end of exposure to the treatments. At 54.5% and lower RH, in combination with low O_2 concentrations, there was a marked loss in weight of the adults.

3.2.1.4 Effect of Air Relative Humidity on Desiccation and Insect Mortality

Murdock et al. (2012) have reported that the cause of death of cowpea bruchids under HS is desiccation resulting from an inadequate supply of water. According to them, when water supply is blocked by a lack of O_2, the

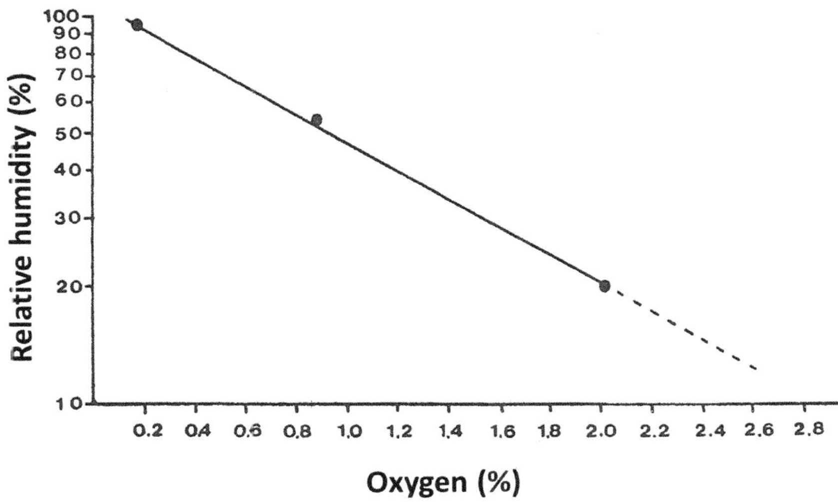

Figure 3.4 The combined effect of O_2 and RH producing 95% mortality for *Tribolium castaneum* (Herbst) adults after 4 days of exposure at 26°C; data were based on the mortality counts carried out 10 days after the end of exposure. (Redrawn from Navarro, 1975.)

Insects gradually desiccate and die as they continue to respire, albeit slowly. Although the statement that the reason for death under HS is desiccation seems to be corroborated by other investigators, the relation of lack of O_2 under the low RH that prevails in grain storage and desiccation was not considered by Murdock et al. (2012).

Conversely, previous works have elaborated on the cause of death as a complex metabolic response to the RH of the storage environment. Although there is evidence that low-humidity environments are more effective in causing mortality (Navarro and Calderon, 1973, 1974; Navarro, 1975), the lethal effects of the low humidity are attributed to desiccation rather than the toxic action of the MAs (Navarro, 1978); the inability of insects to recover after treatment was also attributed to the lack of sufficient triglycerides as substrates for energy metabolism (Navarro and Friedlander, 1974; Friedlander and Navarro, 1978, 1979; Donahaye, 1990).

To prevent commodity deterioration due to high humidity, which favors microflora activity, the ambient RH of storage should be below 65% and preferably in the range of 50% to 60% RH, values at which storage can be considered "safe" (Navarro and Donahaye, 2005). Storage insects keep their water reserves by opening their spiracles for a short duration to allow the expelling of CO_2 and intake of O_2. In this way, storage insects survive the low humidity ambient by maintaining a balance between the metabolic water they can produce and the loss of water. However, when the insect storage environment is depleted (at about 2–3%) of O_2 (Navarro, 1975, 1978; Navarro and Calderon, 1980; see also Figure 3.3) or at a slightly elevated CO_2 of 4.3% (Navarro and Calderon, 1980) and at a low RH, the insect's water deficit is significantly increased. The role of RH in the storage environment in causing water loss was detailed by Navarro (1975, 1978) and Navarro and Calderon (1980). This water loss is accompanied by the inability of storage insects to compensate for the water loss, and therefore they die due to desiccation. In the presence of sufficient humidity in the environment, insects survive until anoxia or hypercarbia becomes the dominant factor in causing insect mortality (Navarro, 1975, 2006) (Figure 3.4).

3.2.1.5 Effects of Combinations of Low O_2 and High CO_2

Researchers have been interested in increasing the efficacy of MA on insects by attempting to combine very low O_2 with very high CO_2 concentrations. Increasing the CO_2 concentration in the normal atmosphere reduces proportionally the partial pressure of the O_2 available to insects. Gas burners or fossil fuel burners also can generate a combination of low O_2 and high CO_2. For example, a typical propane burner would produce an atmosphere of 0.5% O_2, 13.5% CO_2, 1% nitrogen, and 1% argon. Unless a mixture of nitrogen and CO_2 or a gas burner atmosphere is used, the simplest way to achieve a low O_2 and high CO_2 atmosphere is by using CO_2 in the air (Storey, 1975).

In the case of hypoxia (2% to 5% O_2), when a small proportion of CO_2 (5% to 40% CO_2) is added to the initial mixture of N_2/O_2, the mortality rate increases considerably (Calderon and Navarro, 1979). When CO_2 is added to a low O_2 atmospheres, there is a synergistic effect, which is obvious from the significant interaction between the concentrations of these two gases (Calderon and Navarro, 1980).

Krishnamurthy et al. (1986) tested the response of adults of *S. granarius*, *T. castaneum*, *O. surinamensis* (L.), *C. ferrugineus*, and *R. dominica* exposed to simulated combusted atmospheres containing low O_2 (0.5–2.6%) and increased CO_2 (10–30%) with the balance N_2 at 20°C and 70% RH. The mixtures containing 1–1.6% O_2 killed all the insects within 7 d if they also contained 10% or more CO_2. At 2.0–2.6% O_2 *C. ferrugineus* was more tolerant of CO_2 than the other insects tested, whereas at 0.5% O_2 *S. granarius* was the most tolerant species, needing 8–10 d for a complete kill.

Conyers and Bell (2007) carried out laboratory tests on five species of stored-product beetles, *C. ferrugineus*, *O. surinamensis*, *S. granarius*, *S. oryzae*, and *T. castaneum*, to MAs based on simulated burner gas, four different O_2 levels, 3, 4, 5, and 6%, with CO_2 levels of 9.5, 8.5, 7.5, and 6.5%, respectively, at 20° and 25°C, 75% and 85% RH. After exposure to the MAs for 28 d, an assessment was made of the mortality of adults, the number of adults from progeny produced under the MAs, and, for the simulated burner gas, the number of adults from progeny produced in a 28-d period after exposure to the MA. The O_2 content preventing population growth varied with species and temperature. For the simulated burner gas or N_2, it was about 4% for *O. surinamensis, S. granaries,* and *S. oryzae* and about 3% for *C. ferrugineus* and *T. castaneum* at 25°C. At 20°C, it was about 3% for all species tested. When CO_2 was increased to 10% or 20%, reducing O_2 to 5% was sufficient to eliminate the emergence of *S. granaries* at 20°C, but a few individuals emerged at 25°C. For *C. ferrugineus,* there was a 95% reduction with 5% O_2 plus 20% CO_2 at 20°C, but not at 25°C.

Tütüncü (2022) studied 0–2-day-old eggs, 12-day-old mature larvae, and 1-week-old adults of *Carpophilus hemipterus* (L.) exposed to a modified atmosphere with low O_2 and high CO_2 content (2.1% O_2 + 90% CO_2 + 7.9% N_2) for 48–120 h at 20°C and 75 ± 5% RH. LT_{95} values were recorded as 221.83 h, 82.79 h, and 65.31 h for egg, larval, and adult stages, respectively. The statistical difference between LT_{95} values was significant, and the egg stage was found to be more tolerant than the other two life stages.

3.2.1.6 Insect Response to Temperature

At temperatures from 20° to 30°C, most species and developmental stages show 95% mortality in 10 d at both 0% and 1.0% O_2 (Annis, 1987). The only exceptions were *T. granarium* larvae (12 d at 0.1% O_2), *S. oryzae* pupae (20 d at 0.1% O_2 and 14 d at 1% O_2), and *S. granarius* adults (16 d at 1% O_2) (Annis, 1987). The effect of temperature on the length of time necessary to obtain good control with MAs is as important as with conventional fumigants. Jay (1971) states that to obtain good control; "the temperature of the grain should be above 21°C during the application of CO_2." Navarro and Calderon (1980) compared the effect of temperature on exposure time to MAs required to produce mortality of adults of three storage insects. Chiappini et al. (2009) aimed to evaluate the time necessary to obtain total mortality of insects treated in CAs with O_2 percentages higher than those normally used in practice. They estimated the possible positive influence of a temperature increase to compensate for the effects of the reduced anoxia by exposing *T. confusum* adults to various ranges of O_2 percentages (1–10%) and temperatures (23–40°C). Their results showed that total mortality could be achieved within a week, even in quite moderate conditions of temperature (29–37°C) and lowered O_2 percentage (5–8%). They demonstrated that the total mortality achieved in the various exposure periods was due to the effect of the two parameters (temperature and O_2 percentage) together.

Donahaye et al. (1994) reported on responses of larval, pupal, and adult stages of the Nitidulid beetles *C. hemipterus* and *Urophorus humeralis*

(F.) exposed to simulated burner gas concentrations at three temperatures of 26, 30, and 35°C. Comparison of exposure times showed that the effect of temperature on treatment efficacy was most pronounced at the 1% O_2 level, where for the three stages of both species tested, values of LT_{50} at 26°C were about half those at 35°C. However, at 3% O_2 and 35°C, LT_{50} levels were only marginally reduced (Donahaye et al., 1994).

Soderstrom et al. (1992) examined the influence of temperature over the range of 38–42°C on the influence of hypoxia and hypercarbia on *T. castaneum* adults for 60 h exposures. Although the different experimental conditions make comparison difficult, their results indicate that increased temperatures could be used to reduce treatment duration. Donahaye et al. (1996) exposed egg, larvae, pupae, and adults of *T. castaneum* to three low O_2 concentrations (1, 2, and 3%) at 26, 30, and 35°C. At all levels of O_2 in typical respiration atmospheres in hermetic conditions (similar to burner gas atmospheres), the LT_{99} values at 35°C were significantly lower than those at 26°C. Work on all four development stages of *E. cautella* showed the strong influence of temperature on mortality values when the insects were exposed to CO_2 concentrations varying from 60% to 90% in the air (Navarro et al., 2002a) (Table 3.2).

Downes et al. (2008) studied the effect of changing environmental conditions on the metabolic heat rate (MHR) of *S. oryzae* and *T. castaneum* in an attempt to demonstrate that heat treatments for disinfestation performed in a hypoxic or hypercarbic CA may be effective at lower temperatures than are required in air. The MHR of adult *S. oryzae* was changed little over a wide range of O_2 concentrations; the value in the air was close to that in pure O_2 and was only reduced by about 20% and 50% at O_2 concentrations of 5% and 2%, respectively. Heat treatment at 45°C was more effective when the CA was 5% O_2 + 60% CO_2 + N_2 or 60% CO_2 + N_2 than in air, but larval *T. confusum* still survived in the former CA, indicating there were differences in response between species and stages. Tang et al. (2007) reviewed the response of insects to high temperatures during exposure to CAs. They noted that, in general, the higher the temperature, the faster mortality is achieved under a given atmosphere.

3.2.1.7 Insect Movement in Changing Atmospheric Composition

Navarro et al. (1981) tested the effect of O_2 and CO_2 gradients on the vertical dispersion of grain insects in wheat. The study was made on the dispersion of 0–14-d-old adult *O. surinamensis*, *S. oryzae*, and *R. dominica* at 25°C in vertical columns 100 cm tall filled with wheat (MC 12.1%). When the columns contained ordinary air, *O. surinamensis* dispersed from top to bottom within 24 h, whereas *S. oryzae* and *R. dominica* penetrated to a depth of only 50 cm in 72 h. When controlled gas gradients were imposed on the columns from the bottom, the concentration of O_2 in one series of experiments ranged from 0.9% at the bottom to 18.5% at the top, while in another series that of CO_2 ranged from 70.5% at the bottom to 3.3% at the top. In both gas gradients, the downward dispersion of *O. surinamensis* was restricted but the distribution of the other two insect species was unaffected because they did not penetrate deep enough to encounter unfavorable concentrations.

3.2.2 Insect Response to Hermetic Storage

3.2.2.1 Response to Restricted O_2 Supply

Murdock et al. (2012) stated that "depriving insects of O_2 not only limits energy production, but it also deprives them of water—water vital for life. When this water supply is blocked by lack of O_2, the insects gradually desiccate and die as they continue to respire, albeit slowly." Such a statement relates the death of insects to the blocking of the water supply and their inability to get water. Although insects die due to desiccation, this statement is not supported by the rate at which insects lose water. It should be pointed out that the description by Murdock et al. (2012) of insect death is different from those of studies carried out on water deficiency in insects exposed to low-moisture environments. The storage environment RH is obliged to be low, usually at or below 60% equilibrium RH (Navarro and Donahaye, 2005). When insects are exposed to low O_2 or high CO_2 concentrations, their water conservation system (that is, the rate of closing and opening of the spiracular system) is disrupted. Such disruption has been corroborated in several storage insects, including *T. castaneum* adults, *E. cautella* larvae and pupae, and *S. oryzae* adults (Navarro and Calderon, 1973, 1974, 1980; Navarro, 1975; Calderon and Navarro, 1979). Also, previous works have elaborated on the cause of death as a complex metabolic response to the RH of the storage environment. Although there are pieces of evidence that low-humidity environments are more effective in causing mortality (Navarro and Calderon, 1973, 1974; Navarro, 1975) and at low humidity the lethal effects are attributed to desiccation rather than the toxic action of the MAs (Navarro, 1978), the inability of insects to recover after treatment was also attributed to the lack of sufficient triglycerides as substrates for energy metabolism (Navarro and Friedlander, 1974; Friedlander and Navarro, 1978, 1979; Donahaye, 1990).

Sanon et al. (2011) reported that bruchids' numbers and seed damage were significantly reduced when cowpeas were stored within two layers of high-density polyethylene (HDPE) bags of at least 80 µm wall thicknesses. This thickness considerably reduced O_2 concentration in the bag after 5 d of storage and inhibited insect development. However, late instar larvae and pupae were less affected by low O_2 concentration.

Mutambuki et al. (2014) simulated farmer storage practices to evaluate the triple-layer Purdue Improved Cowpea Storage (PICS) airtight bags against two major storage insects. Gas analysis in the PICS bags followed the expected trend, with O_2 levels falling sharply below 10% and CO_2 increasing to almost 10% up to 12 weeks hence, resulting in the death of infesting insects. After 16 weeks, an increase in O_2 levels may be attributed to the perforation of the bags by the larger grain borer, *Prostephanus truncates* (Horn). Values for mean live insects, grain damage, and weight of dust over the six-month trial period showed significant differences between PICS and the other treatments, especially at the 16th and 24th week; differences were even more pronounced at the end of the trial period.

Chigoverah and Mvumi (2018) investigated the performance of four brands of HS bags against adult *P. truncatus* in maize stored for 90 d under simulated resident and incoming infestation. Five treatments were used: four hermetic bag brands—SuperGrainbag (SGB) IV-R, SGB Farm, Purdue Improved Crop Storage (PICS) bag, and Kuraray bag—and an ordinary plastic bag, each containing 50 kg of shelled maize. Externally infested bags had no insect activity; all the introduced insects died of starvation. Internally infested bags were all perforated. The ordinary plastic liner was much more severely perforated, with 151 insect-induced holes, than were hermetic plastic liners (<40 holes). There was no live adult insect infestation in grain samples collected using double-tube multi-slotted sampling probes inserted vertically several times. However, live adult insects were present in the bottom grain layer (5 cm) of all plastic liners. There were no significant differences between hermetic bags regardless of the mode of infestation for all parameters assessed. The results show that the tested hermetic bags are equally susceptible to perforation by resident *P. truncatus*.

Mutambuki et al. (2018) studied AgroZ airtight bags against two major storage insects. Gas analysis in AgroZ bags showed the O_2 level dropping rapidly to 7% within 4 weeks and later increasing gradually to 10% at 12 weeks. Conversely, the CO_2 level increased sharply to 10% and declined gradually to 9% over the same period. The number of insects and percentage of damaged grains with the AgroZ bag and the polypropylene bag significantly differed from the 12th week to the 24th week. The AgroZ bag outperformed the polypropylene bag commonly used by farmers and conveniently protected maize from insect infestation within the six-month storage period.

Kandel et al. (2021) conducted experiments to assess the time required to attain mortality of adult rice weevils when the O_2 levels reached below 5% in airtight containers at a room temperature of $20 \pm 1°C$. Results revealed that it required 69.7, 187.8, and 386.6 h to kill 50% of adult rice weevils exposed to 1%, 3%, and 5% O_2 levels, respectively. No adult emerged from infested grains following exposure to 1% and 3% O_2 levels, but some did at 5% O_2 levels.

Baributsa et al. (2020) evaluated the performance of five postharvest storage methods for maize preservation in Northern Benin. Maize that had been naturally infested by insects was stored in four HS technologies (SuperGrainbag™, AgroZ® bag, EVAL™, and PICS bags), an insecticide-impregnated bag (ZeroFly®), and a regular polypropylene (PP) woven bag as control. Oxygen levels in hermetic bags fluctuated between 0.5% v/v and 1.0% v/v during the seven months of storage. No weight loss or insect damage was observed in grain stored in any of the HS bags after seven months.

However, grain stored in ZeroFly® and PP woven bags had weight losses of 6.3% and 10.3%, respectively.

Goulão et al. (2021) reported that HS was a successful strategy for storing paddy rice under different conditions in Mozambique and Portugal. They showed that under the tropical conditions of Mozambique, even at low moisture (ca. 12%) insect infestation was common, and temperature was high (24 to 36°C), thus increasing insect activity, which changed the atmosphere conditions due to insect respiration. Whereas at the low insect infestation conditions in Portugal, the atmosphere was modified only under high moisture content. In Portugal, when the flux of the grain surpassed the drying capacity, delayed, incomplete, or ineffective drying occurred (moistures >14%); therefore, HS could be successfully used. Under both conditions, at between 4 and 12 months of storage, insect populations were better controlled with HS than with the control or traditional storage (in Portugal, totally suppressed at 24°C; in Mozambique, 96% reduction) but driven by specific underlying mechanisms.

3.2.2.2 Insect Respiration at Low O_2 Concentrations

Emekci et al. (2002) showed that the respiration of *T. castaneum* was suppressed at O_2 levels lower than 5% in eggs and young larvae (Figure 3.5). Pupal respiration of *T. castaneum* decreased as O_2 concentrations fell.

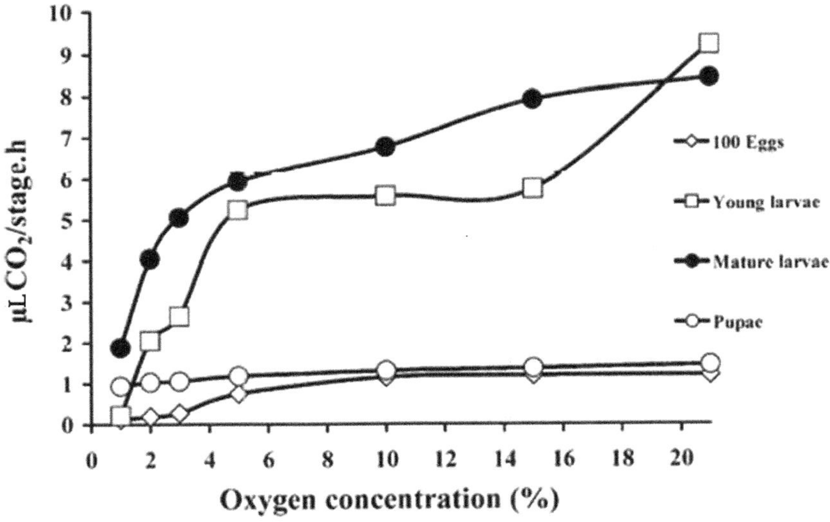

Figure 3.5 Respiration rates of the development stages of *Tribolium castaneum* at different O_2 concentrations in nitrogen at 30°C and 70% RH. (Redrawn and used with permission from Emekci et al, 2002.)

Similar results concerning the respiration of *E. cautella* pupae were reported at reduced O_2 atmospheres at 26°C and 93–99% RH (Navarro and Calderon, 1979). Accordingly, CO_2 production by *E. cautella* was approximately 0.1, 0.2, and 0.35 µL/100 mg·h^{-1} at O_2 concentrations of 1, 3, and 21%, respectively.

The increased respiration rate of *T. castaneum* adults at 3% and 5% O_2 was considered by Emekci et al. (2002) as a compensatory response to the O_2 deficiency. Donahaye (1992) provided data on studies of the physiological differences between strains of adult *T. castaneum* unselected and selected for tolerance to hypoxia and to hypercarbia at 26°C and 95% RH. Carlson (1968), comparing different ratios of percent CO_2 and percent N_2, measured less CO_2 production than in the control in *T. confusum* adults at 30°C in atmospheres of 4.91, 10.15, and 14.90% O_2. His CO_2 measurements for these O_2 levels were between 1.74 and 1.88 µL/mg·h^{-1}, similar to the results of Emekci et al. (2002).

Respiration rates of *R. dominica* pupae increased proportionally as the O_2 concentration increased from 1% to 21% (Figure 3.6) (Emekci et al., 2004). Respiration of eggs was particularly suppressed at low O_2 levels of 3% or less. Carbon dioxide production per hour in eggs in 1% O_2 was found to be 9.7 times less than that in 21% O_2. Emekci et al. (1998) obtained a similar respiratory response in *T. castaneum*, where CO_2 production of eggs in 1% O_2 was nine times less than that in 21% O_2. In young larvae, CO_2 production was adversely affected and respiration decreased by 8.2 times when the O_2

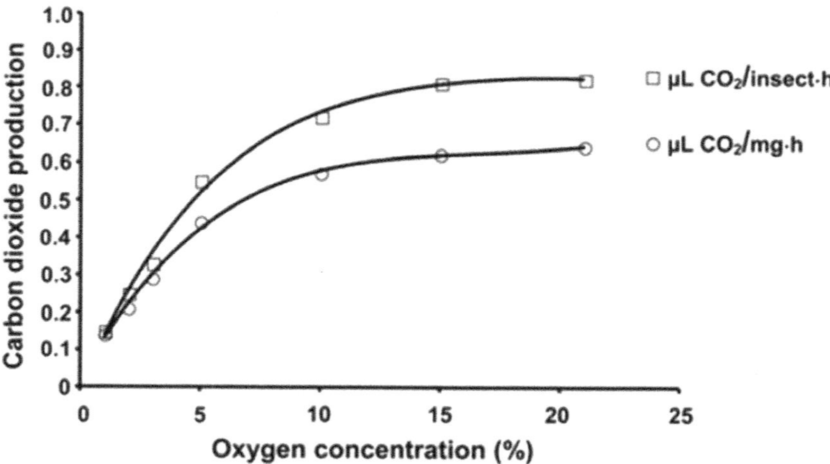

Figure 3.6 Respiration rates of *Rhyzopertha dominica* pupae in response to low oxygen concentrations in nitrogen at 30°C and 70% RH. (Redrawn and used with permission from Emekci et al., 2004.)

concentration decreased from 21% to 1%. Results obtained for pupae showed that *T. castaneum* respiration under the same conditions was more stable than that of *R. dominica*, increasing by only 1.5 times, when O_2 concentrations increased from 1% to 21% (Emekci et al., 1998, 2004).

3.2.2.3 Effects of Hermetic Storage on Mortality of Grain Insects

One of the biological agents that may be present in the grain bulk is insects. In addition to insects, the stored-product fauna contains mite pests. Insects and mites are aerobic organisms that require O_2 for their development. Reduction of the O_2 tension in the grain bulk might be possible either through the natural consumption of O_2 by the biological agents or by artificially producing a low O_2 atmosphere to control insects and mites.

Njoroge et al. (2014) reported on the use of PICS in comparison to polypropylene bags. They were able to monitor O_2 and CO_2 conditions prevailing in PICS bags over the six months of storage of maize. Oxygen concentration dropped steadily from 19.23% to 7.18% and 7.82% within three months in PICS bags with maize infested artificially and PICS bags infested naturally, respectively. Carbon dioxide concentration, on the other hand, increased gradually to 13.17% and 13.93%, respectively, within the same period. In subsequent months, O_2 and CO_2 concentrations, on average, stabilized at 6.66% and 13.84%, respectively. Njoroge et al. (2014) implied that there was no difference in the progression of O_2 and CO_2 profiles in PICS, although the latter had been infested with *P. truncatus*. Evidently, O_2 depletion and CO_2 buildup did not reach extreme levels, probably because of the monthly opening of bags during sampling. Elsewhere, Ognakossan et al. (2013) reported that O_2 levels could be modified by the respiration and metabolism of insects, fungi, and the grain itself to a level of 1% to 2% or below. Other researchers, however, reported failure of the extreme drop in O_2 concentration. Moreno-Martinez et al. (2000), for instance, observed that O_2 was depleted to 0% after 6–9 d in those treatments infested with insects, whereas the same level was reached after 24 d in grain with the storage fungus alone. In the control, a gradual decrease in O_2 eventually stabilized at about 8.4% within 30 d in clean grain stored without insect infestation and fungal infection under hermetic conditions, probably due to an MC that was high enough (15%) to enable moderate respiration (Weinberg et al., 2008).

Similar trials were carried out by Mutungi et al. (2014) on the storage of mung bean and pigeon peas in hermetic triple-layer bags to determine losses caused by *C. maculatus* (F.). Oxygen and CO_2 concentrations in the PICS bags over six months were measured in both pulse types, and O_2 levels dropped rapidly within the first two months.

Baoua et al. (2013) compared side by side GrainPro and PICS bags for postharvest preservation of cowpea grain in Niger. They reported that when bruchid-infested cowpea was stored for more than four months in PICS bags or in commercially available GrainPro SuperGrainbags, preservation of the grain was equally good in both types of bags. In both bag types, O_2 levels dropped rapidly during the first 24 h after closure, eventually reaching levels of 1–3%. Over the four-plus months of the experiment, damage levels did not significantly increase in either type of bag while control grain kept in a conventional woven plastic bag suffered severe damage. Most of the insects found in both GrainPro and PICS bags were dead at the end of the experiment. Surprisingly, in spite of the report that the single-layer SuperGrainbags showed more bruchid holes than did the triple-layer PICS bags, the O_2 level was lower and the CO_2 level was markedly higher in the SuperGrainbags than in the PICS bags.

Murdock and Baoua (2014) described the various features of PICS technology by detailing the background of the development, mode of action, and need for future investigations. They speculated that there are at least sixteen different factors that contribute to the effectiveness of PICS bags. The degree of contribution of each may be substantial or very small. PICS bags stop bruchid populations from expanding and thereby prevent destructive losses. These triple bags seem to be more effective than might be expected based solely on the O_2 barriers presented by the two HDPE liners. Murdock and Baoua (2014) attributed the effectiveness of PICS to the triple layer and to other factors such as the effects on reproduction rate of insects that remain to be investigated in depth. Although Murdock and Baoua (2014) agree that some of their hypotheses are speculative, they point out that the hypotheses should lend themselves to experimental tests.

Baoua et al. (2014) tested the triple-bag hermetic technology for postharvest preservation of Bambara groundnut. They conducted experiments to evaluate the performance of hermetic triple bagging using PICS bags for storage of the Bambara groundnut [*Vigna subterranea* (L.) Verdc.]. Two to five days after the beginning of the experiments, the O_2 level inside the bags averaged about 21% in the controls but had decreased significantly in the PICS bags, reaching 10% with the heavily infested grain but falling only slightly in the lightly infested grain.

A number of studies that were recently published lacked basic parameters to evaluate the successful use of HS at the farm level. The objective of these studies was to test the technology under real field conditions. However, despite the fact that these publications had data on such parameters as gas composition, some failed to incorporate important parameters such as level of gastightness and liner permeability.

Quezada et al. (2006) tested the HS system on preventing the proliferation of *P. truncatus* and storage fungi in maize with different moisture contents. Grain of maize, infested as well as uninfested with *P. truncatus*, was stored for 15 d at 27°C under hermetic and nonhermetic conditions. Grain moisture content (MC) was adjusted to 14, 15, 16, and 17%. Under hermetic conditions at 3 d of storage, O_2 levels were 0.8, 0.6, 0.4, and 0.0% in grain with 14, 15, 16, and 17% MC, respectively. However, at 6, 9, 12, and 15 d, the O_2 level dropped to 0% at all tested MCs. Insect mortality at 3 d of storage under HS conditions ranged from 5% to 30%. At 6 d, insect mortality was 100% in maize with 14, 15, and 17% MC and 95% in grain with 16% MC; however, at 9 d of this MC, mortality was 100%. In this work, the level of gastightness was not reported.

Chigoverah et al. (2014) tested the effect of hermetic facilities on stored-maize insect infestation and grain quality. Artificial infestation involved adding unsexed adult insect species of *S. zeamais* and *P. truncatus*, while in natural infestation no insects were added. Nondestructive sampling was conducted monthly. Collected samples were assessed for insect numbers, species spectrum, insect damage, and weight loss. The researchers concluded that, overall, hermetic facilities can effectively suppress insect development and consequently reduce losses and maintain seed viability during storage without the use of pesticides. In this work, the permeability and the level of gastightness of the tested facilities were not reported.

Baributsa et al. (2017) reported how safely PICS bags store unshelled and shelled groundnuts in Niger. They identified pests present in the stored groundnuts as *T. castaneum*, *Corcyra cephalonica* (Stainton), and *C. ferrugineus*. After 6.7 months of storage, in the woven bag, there was a large increase in the pest population, accompanied by a weight loss of 8.2% for unshelled groundnuts and 28.7% for shelled groundnuts. In PICS bags, for both shelled and unshelled groundnuts, by contrast, the density of insect pests did not increase, there was no weight loss, and the germination rate was the same as that recorded at the beginning of the experiment. In this work, the permeability and the level of gastightness of the PICS bags were not reported.

Singano et al. (2019) evaluated the effectiveness of grain storage facilities and protectants in controlling stored-maize insect pests in a climate risk–prone area of Shire Valley, southern Malawi. Seven grain storage treatments were evaluated for 32 weeks during two storage seasons: Neem leaf powder (NM), Actellic Super dust (ASD), the ZeroFly® bag (ZFB), the Purdue Improved Crop Storage bag (PICS), the SuperGrainbag (SGB), a hermetic metal silo (MS), and untreated grain in a polypropylene bag (PP). Insect pest populations and grain damage increased with storage duration and differed significantly between treatments ($p < 0.05$). Grain stored in hermetic bags (PICS, SGB) sustained significantly less ($p < 0.05$) insect damage and weight loss than the other treatments across sites and seasons. The hermetic bags also outperformed the other treatments in suppressing insect numbers. The hermetic MS, ZFB, ASD, and NM treatments did not effectively protect grain from insect damage. In this work, the permeability and the level of gastightness of the PICS bags and SGB were not reported.

Opoku et al. (2023) evaluated HS bags for the preservation of yellow maize in poultry farms in Dormaa Ahenkro, Ghana. The experiment was set up in a randomized complete block design with ZeroFly®Hermetic (ZFH), Purdue Improved Crop Storage (PICS), and polypropylene (PP) bags as treatments. The number of insects was significantly higher in the PP bag (161) than in the PICS and ZFH bags: 7 and 4.50, respectively. The PICS and ZFH bags had less insect damage and lower weight loss than the PP bags. In this work, the permeability and the level of gastightness of the HS bags were not reported.

3.3 Successful Applications of Hermetic Storage

With recent improvements in materials and the construction of flexible, non-porous bags and liners, a variety of size options offer protection for products from 25–1,000 kg up to 10,000–15,000 t (Navarro, 2010). Commodities including cereals, oilseed grains, pulses, cocoa, and coffee can be stored safely for many months, maintaining high quality and limiting molds and mycotoxins. Plastic structures suitable for long-term storage systems, as well as intermediate storage of grain in bags or bulk, have been developed and applied. Storage systems based on the hermetic principle include the following:

1. Bunker storage in gastight liners for conservation of large bulks of 10,000 to 15,000 t capacity (Navarro et al., 1984, 1994) (Figure 3.7).

Figure 3.7 Bunker storage of 10,000 to 15,000 capacity for the conservation of large bulks of wheat.

2. Flexible gastight silos supported by a weld-mesh frame of 50 to 1,000 t capacity for storage of grain in bulk or bags (Calderon et al., 1989; Navarro et al., 1990, 1998a) (Figure 3.8).

Figure 3.8 Flexible gastight silos of 50 to 1,000 capacity supported by a weld-mesh frame for storage of grain in bulk or bags. (Navarro et al., 1990.)

3. Gastight liners for enclosing stacks of 5 to 1,000 t capacity, called "storage cubes" or Cocoons™ and designed for storage at the farmer-cooperative and small-trader level or larger commercial and strategic storage facilities (Donahaye et al., 1991). These structures are currently in use for bagged storage of cereals (Figure 3.9).

Figure 3.9 Gastight liners for enclosing stacks of 5 to 1,000 t capacity, called storage cubes or Cocoons™. (Donahaye et al., 1991.)

4. Silo bags of 200 t capacity for on-farm grain storage directly in the field. This technique was originally used for grain silage and involves storing dry grain in sealed plastic bags. The silo bag is 60 m long and 2.74 m in diameter (Bartosik, 2012); it has a cover made of three layers (white outside and black inside) with 235 μm of thickness (Cardoso et al., 2008) (Figure 3.10).

Figure 3.10 Silo bags of 200 t capacity for on-farm grain storage directly in the field. (Bartosik, 2012.)

5. A thin liner for jute or woven polypropylene bags, providing a highly efficient method of using airtight bags for masses of 25 to 1,000 kg of product. Originally promoted by IRRI for paddy-seed storage, these liners found their way into the storage and transport of coffee and other commodities and are now commercially available under a range of brand names (Figure 3.11).

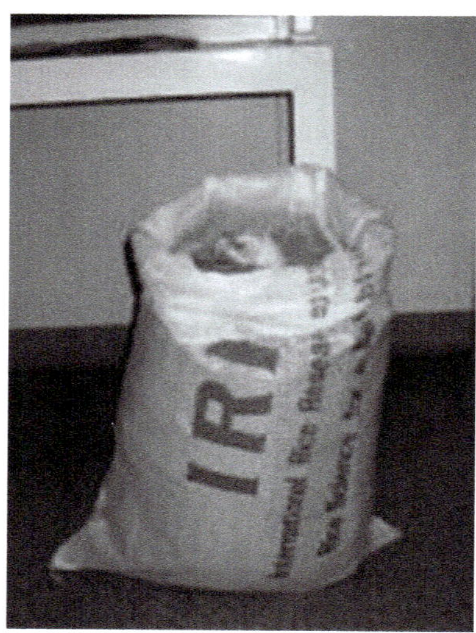

Figure 3.11 SuperGrainbags™, for hermetic storage of paddy seed at IRRI, Los Baños, Philippines. (Courtesy of deBruin.)

6. A granary for use by small-scale farmers, suitable for storing up to 1,000 kg grain, termed GrainSafe. The problem of applying present-day technology of HS for subsistence farmers lies in the need to provide an easily sealable low-cost container. The most recent attempt to address this problem has been through the construction of this granary (Navarro et al., 1998b). The granary was equipped with an upper collapsible sleeve for loading and a lower collapsible sleeve for unloading. The hermetic flexible bag was inserted into a rigid sheath surrounding the vertical sides of the hermetic bag (Figure 3.12).

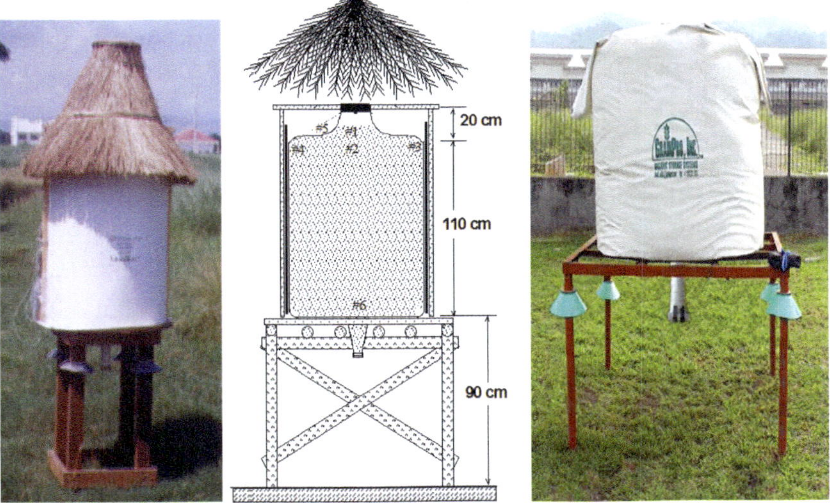

Figure 3.12 Hermetic granary for use by small-scale farmers, suitable for storing up to 1,000 kg, termed GrainSafe™.

7. Purdue Improved Cowpea Storage bags, designed for storage of 50 kg or 100 kg. These bags actually consist of three plastic bags: two 80-μm high-density polyethylene (HDPE) bags, one inserted into the second and then both supported by a third outer bag made of woven polypropylene (Murdock and Baoua, 2014)) Figure 3.13).

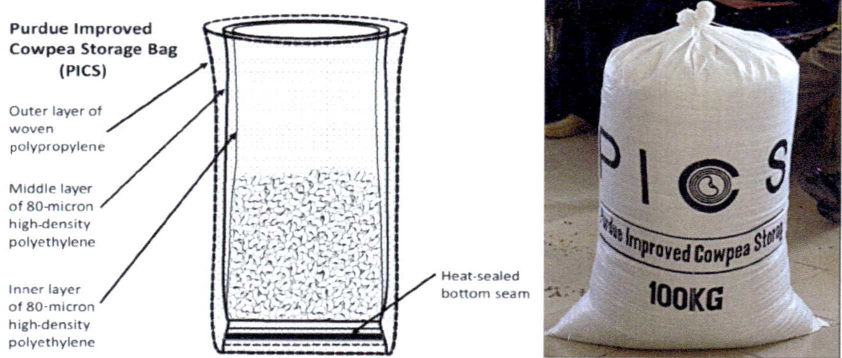

Figure 3.13 Purdue Improved Cowpea Storage bags, designed for storage of 50 kg or 100 kg of cowpea. They consist of three plastic bags: two 80-μm high-density polyethylene bags, one surrounded by the second and both enclosed by a third bag made of woven polypropylene. (Murdock and Baoua, 2014.)

8. Metal silos, also hermetically sealed but physically stronger than flexible liners. These silos have been heavily promoted in Central America (Hellin and Kanampiu, 2008), and their feasibility has been explored in Sub-Saharan Africa (Tefera et al., 2011)) Figure 3.14).

Figure 3.14 Metal silos, hermetically sealed but physically stronger than flexible liners. (Hellin and Kanampiu, 2008.)

9. Container liners. These liners have become a remarkable success in the cocoa shipping industry, which for years was struggling with problems of condensation and infestation, particularly when shipping distances became longer and more problematic due to changes in ambient temperatures. Commercially introduced under the name TranSafeliner (TSL) or Gastight Container Liner (GCL) (Figure 3.15) (Navarro and Navarro, 2020), the liners not only solved the problems of the cocoa trade and industry with condensation and infestation but also helped to stabilize the quality of the cocoa beans by reducing the increase in free fatty acids (FFA) and mold development.

Figure 3.15 TranSafeliner™, suitable for containers. This liner has become a remarkable success in the cocoa shipping industry, preventing condensation and infestation. (Navarro and Navarro, 2020.)

A major challenge that South America is facing is how to minimize quality and quantity losses and improve food safety because of a shortage of permanent storage capacity. As a result, the silo bag system for the temporary storage of dry grain and oilseeds has been adopted in the field. Since 2015, between 40 and 55 million tonnes of grain was stored in these plastic sleeves in Argentina, and the technology subsequently spread worldwide under various commercial names. Commodities stored in silo bags include corn, soybean, wheat, sunflower, malting barley, canola, cotton seed, rice, lentils, sorghum, beans, wet distilled grains with solutes (WDGS), and even fertilizers. The silo bag technology is being adopted in more than 50 countries under very different climatic conditions, including the United States, Australia, Bolivia, Brazil, Canada, Chile, Italy, Kazakhstan, Mexico, Paraguay, Russia, South Africa, Sudan, Ukraine, and Uruguay (Bartosik et al., 2023). Dry grain can be stored in silo bags for more than six months without losing quality (Bartosik, 2012). The use of silo bags is discussed in Chapter 4 of this book.

3.4 Challenges in the Application of HS

In a critical review on HS, Navarro and Navarro (2014) revealed gaps in our knowledge of this modest-looking technology. Their review discusses the contribution of each of the biological components that cause the change in the atmospheric composition and the importance of the physical aspects of permeation to O_2 and the gastightness of the tested enclosures. In that review, it was evident that researchers gave more attention to the specifics of the liner and its thickness than to factors that cause infiltration of air into the tested hermetic facilities. In studies carried out after 2014, researchers continued to test the benefits of hermetic facilities without taking into consideration the level of gastightness of the tested structures (Adler, 2020; Behr et al., 2021; Chigoverah and Mvumi, 2021; Goulão et al., 2021).

Baributsa and Concepcion Ignacio (2020), in a review of developments in the use of hermetic bags for grain storage, emphasized the importance of developing a common international standard for testing and rating hermeticity. Such a standard is needed to protect farmers from buying low-quality bags.

In a recent study by Ridla et al. (2023), the impact of using hermetic packaging and preservatives on the physical properties of rice bran during storage was reported. Not surprisingly, the researchers did not report on the gas concentrations or the level of gastightness of the tested hermetic bags. In future studies, for quality evaluation purposes researchers are strongly encouraged to include the required gastightness-level tests, liner permeability in terms of the oxygen transmission rate and water vapor transmission rate, and liner characteristics and thickness.

The key to the successful implementation of HS is airtightness and control of condensation. Over time, storage size has increased from small family storage to larger bulks representing bigger producers. Several efforts are on the way to use HS for the preservation of national buffer stocks.

Since it is difficult to maintain complete gastightness without any O_2 ingress into the large commercial structures, some tolerances that would permit quality preservation of the grain during HS have been established. We need to consider that the main cause of the deterioration of dry grain is insects, while the main cause of the deterioration of moist grain is microflora. Grain responds differently in the ecosystem of storage when it is at intermediate moisture but close to the critical level where fungi are the dominant microflora. Therefore, HS may be applied both for dry grain and moist grain storage.

For the application of HS to dry grain, an ingress rate of 0.05% O_2/d is sufficient to arrest the theoretical weight loss, caused by insects or microflora, at a level of 0.018% over a one-year storage period (Navarro et al., 1994) (Figure 3.16). For dry grain storage, this level is critical, since even for short storage periods of three to six months, at this ingress rate the possibility of a residual surviving insect population is eliminated at an economical threshold. However, since in most hermetic storages there is some O_2 ingress, the limited surviving population continues to consume the newly infiltrated O_2, causing the O_2 to drop again. This is demonstrated by the concentrations of

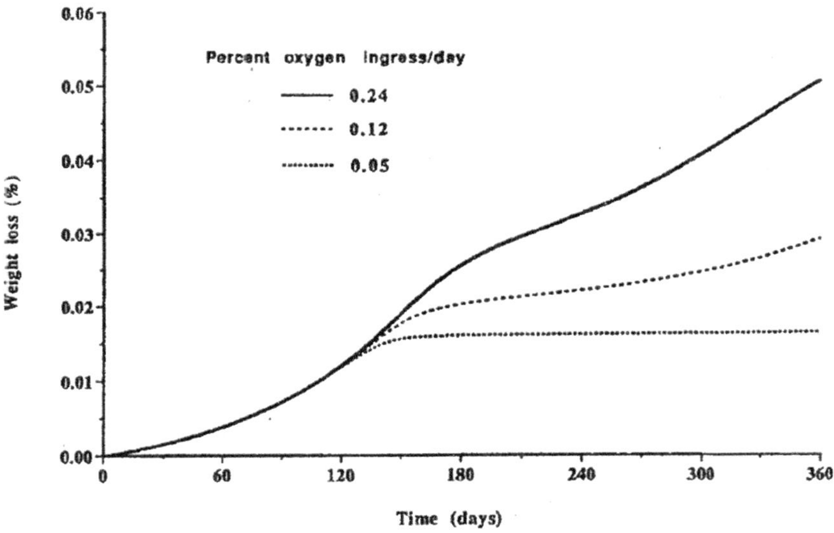

Figure 3.16 Calculated weight loss from a 10 m³ grain mass containing a fixed level of initial infestation of 2 insects/kg, having an O_2 intake of 157 µL/insect/day using a sealed liner at different O_2 ingress rates. (Redrawn from Navarro et al, 1994.)

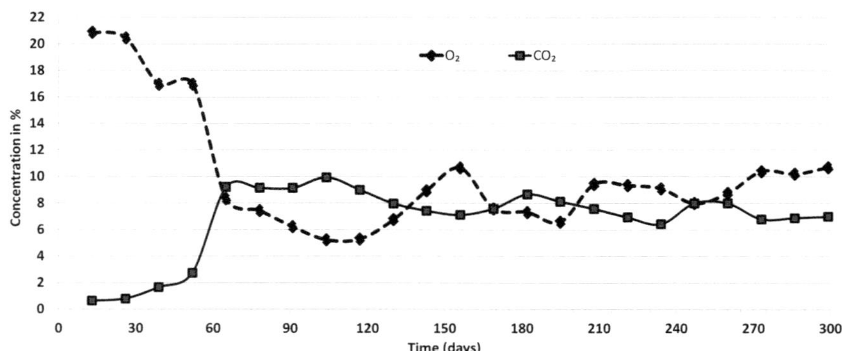

Figure 3.17 Distribution of live insects and change in gas concentrations in the butyl-rubber/EPDM silo of 1,000 t, in which the airtight storage was tested for 10 months. (Redrawn from Navarro, 1977.)

O_2 in Figure 3.17 taken from samples from a large hermetic bulk (Navarro, 1977) and the theoretical calculated lines in Figure 3.18 (Navarro et al., 1994). For higher O_2 ingress rates at a temperature that permits the activity of biological agents, the weight loss continues to rise in proportion to the O_2

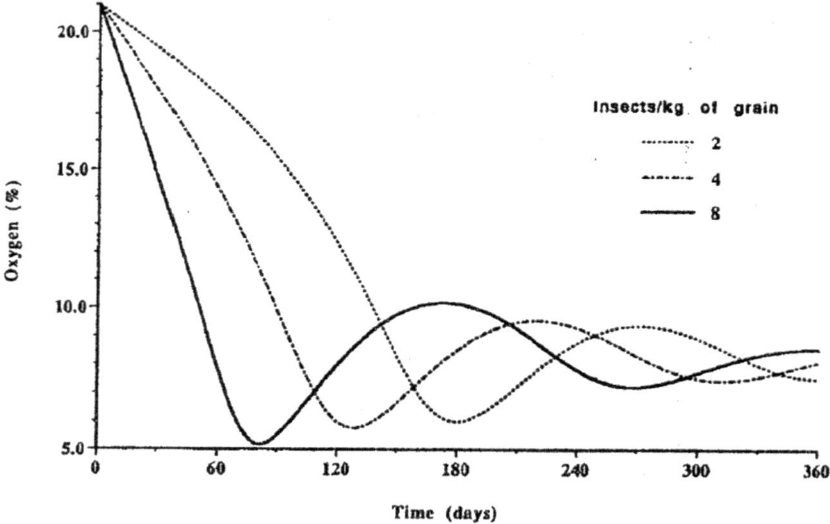

Figure 3.18 Calculated O_2 concentrations in a 10 m³ grain mass containing different infestation levels of insects and having an oxygen intake of 157 μL/insect/day, using a sealed liner with an O_2 ingress rate of 0.24%/day. (Redrawn from Navarro et al., 1994.)

ingress rate, and insect damage might be very significant and unable to be arrested. Thus, ingress rates of up to 0.15% O_2/d can be tolerated. However, for moist grain, higher O_2 ingress rates than 0.15%/d permit grain deterioration, which might lead to the development of mycotoxins (Weinberg et al., 2008). This low O_2 ingress level is difficult to obtain in rigid structures but is achievable in practice using flexible liners. It could serve as a guideline for the sealing specifications of structures appropriate to the HS method.

Experience shows that HS works best for large structures. This is obvious from the lower surface area/volume ratio in large bulks compared with small bulks. A low O_2 ingress rate (should be <0.15%/d to prevent grain deterioration), in practice, is a goal difficult to achieve. Therefore, depending on the commercially available membrane permeability, engineers should aim at designing hermetic structures of sufficiently large dimensions. To emphasize the importance of the size of the structure in HS, calculations were made assuming a permeability level of 200 mL O_2/(m² d⁻¹) for structures of different dimensions ranging from 1 to 1,000 m³ (Navarro et al., 1994) (Figure 3.19). The calculations demonstrate the importance of using low-permeability liners for HS of small-scale storage like individual bags. Indeed, modern multi-layer gas barrier plastics have permeability rates of <1 mL O_2/(m² d⁻¹), thus enabling HS in small volumes, provided the sealing method is adequate.

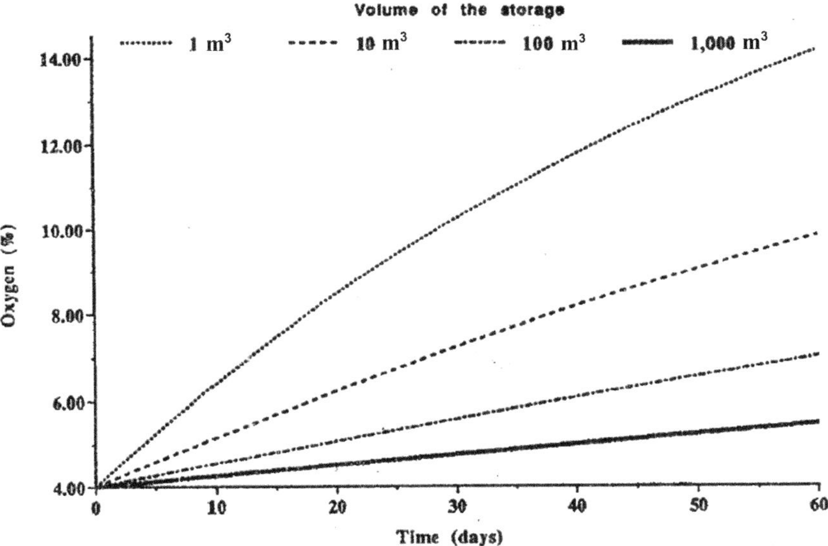

Figure 3.19 Calculated rates of O_2 ingress into grain bulks (without infestation) of different volumes confined within a sealed liner with an oxygen permeability of 200 mL O_2/m² d⁻¹. (Redrawn from Navarro et al., 1994.)

The airtightness of small bags can be negatively affected by improper sealing of the end of the bag and perforations in the plastic cover. A pressure test (or pressure drop test) is one of the methods for evaluating the level of airtightness of a storage structure (Navarro, 1998). Membranes of plastic permit gas permeation and gas exchange. Pressure tests are not capable of measuring the degree of permeability losses.

According to Navarro (2012), although insect respiration causes depletion of the O_2 level of the hermetic storage, to arrest insect development, a sufficiently low O_2 ingress rate is critical to control the insect population or to eliminate the possibility of a residual surviving insect population. The critical residual O_2 level remaining in the hermetic storage structure is exemplified in Figure 3.20, where insect respiration (4 insects/kg grain, based on a respiration rate of 157 µL/insect/d), the O_2 ingress rate, and its difference as the volume of residual O_2 remaining in the HS were plotted on the same graph. From Figure 3.20 it is clear that the residual O_2 concentration would reach about 5% in about 13.5 weeks.

Navarro (2012) analyzed O_2 ingress levels that are achievable in practice using flexible liners. The O_2 ingress levels shown in Figure 3.21 could serve as a guideline for the O_2 permeability specifications of flexible liners appropriate to the HS method. For small volumes, such as bag-size HS structures, low permeability to O_2 is essential; for large volumes, higher permeability levels can be tolerated. To exemplify such tolerances, Figure 3.21 was prepared, which clearly shows the importance of selecting extremely low O_2 permeability

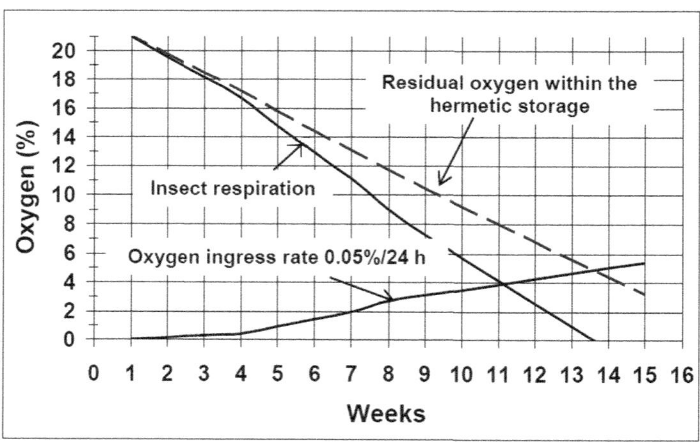

Figure 3.20 Insect respiration (4 insects/kg grain, based on a respiration rate of 157µL/insect/day), the O_2 ingress rate (0.05%/24 h), and its difference as the percentage of residual O_2 remaining in the hermetic storage, to demonstrate the process of obtaining an O_2-depleted atmosphere in hermetic storage of dry grains. (Redrawn from Navarro, 2012.)

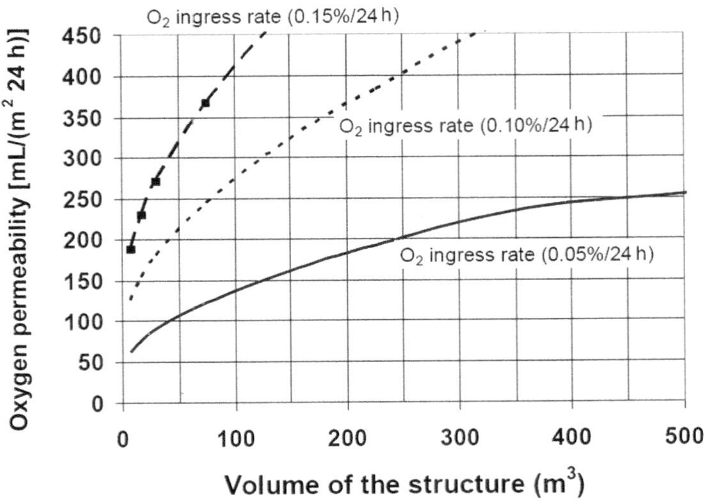

Figure 3.21 Oxygen permeability requirements [mL/(m² 24 h)] of liners in relation to various storage capacities (m³) and O_2 ingress rates (%/24 h) for successful application of hermetic storage of dry grain. (Redrawn from Navarro, 2012.)

liners when using small HS units. According to Figure 3.21, HS structures with capacities similar to bag size (about 100 L) would require liners of a permeability level of <50 mL O_2/(m² d⁻¹) for an ingress rate of 0.05% O_2/d.

The O_2 ingress levels shown in Figure 3.21 could serve as a guideline for the sealing specifications of structures appropriate to the HS method. Flexible structures with O_2 ingress rates higher than 0.15% O_2/d may be used to protect the grain from rain or an increase in moisture, provided the grain is dry and without any infestation. The question is whether these structures should be considered hermetic storage or just simply sealed storage, without the expectation that they will develop a bio-generated atmosphere to protect the grain and with the assumption that fumigation will be used to control the insects.

3.5 Future Research in the Application of CA/MA and HS

3.5.1 Gastightness

Chayaprasert et al. (2015) showed that the pressurizing test and superposition method have potential application in optimizing the structural fumigation process. Similarly, for the successful application of HS, the leakage rate needs to be determined before the start of the storage. For this purpose, a gastightness test needs to be performed. However, the literature on HS trials omits reporting on such tests. Therefore, we have almost no information on the degree of gastightness of the structures tested. Very seldom have researchers reported on the degree of gastightness of the tested structures for either CA/MA or HS.

In an N_2-based CA storage system, a study was carried out on the application of N_2 for controlling stored-grain insects in three concrete silos, each containing 2,400 t of grain, with structural sealing (Navarro, 2012). Airtight valves were installed to improve the airtightness of the bins. Pressure decay tests (250–125 Pa) carried out in full bins showed decay times of from 120 s up to 290 s. Using a pressure swing absorber N_2 generator, the O_2 concentration was reduced in 44–56 h below 0.9%, and after that it was continuously maintained between 0.2% and 0.9% for up to 23.8 d. Treatment for 18.7 and 23.8 d at a grain temperature of 26° and 22°C, respectively, was effective for the control of the adults of several important stored-grain insects.

On the other hand, in studies that evaluate various methods for insect control, tests of the gastightness of HS, such as the pressure decay test, the O_2 depletion test, and the CO_2 accumulation test, are missing. In addition, studies to evaluate the gastightness of HS bags are also missing. Therefore, researchers are encouraged to include the various methods for testing the gastightness of HS systems, along with their advantages and limitations, in their reports.

3.5.2 Condensation

Because HS is a sealed system, it is a technique used to store grains and other food items by creating an airtight environment to prevent the entry of O_2 and other external factors that may cause spoilage. One common method of HS is to use plastic structures commercially sold as "cocoons" or "mobile grain storage." When these structures are used, it is mandatory to watch for the formation of condensation, which can lead to grain spoilage.

Condensation occurs when the air inside the structure comes into contact with the cooler surface of the top section of the liner. This can happen when there is a temperature difference between the grain and the environment, and it is enhanced when the humidity level inside the structure is high. The presence of condensation can lead to the growth of mold and other microorganisms that can cause spoilage and reduce the quality of the grain.

To reduce the effect of condensation in HS, it is important to ensure that the grain is properly dried before it is placed in the structure. The

RH inside the storage structure should also be monitored for moisture buildup. Proper periodic inspections of the grain can help to warn users of the formation of condensation and ensure that the stored grain remains in good condition. A study carried out by Donahaye et al. (1999) tested the possibility of preventing moisture migration in sealed grain stacks stored in the open in the tropics by using reflective covers. The trials in Israel showed that the reflective covers had a strong attenuating influence on the development of temperature gradients and condensation at the top of the Volcani cubes placed in the open, as long as a space for free movement of air was provided between the cover and the plastic liner. This was confirmed in the Philippines, but field trials with 18% MC paddy showed that this insulating effect was not sufficient to prevent a gradual buildup of moisture at the surface layer. However, for dry paddy, after five months of storage under a reflective cover, no perceptible increase in moisture content was found at the top of the stack and the grain remained in good condition. As a result of these findings, a decision was taken to discontinue recommendations for the inclusion of a protective layer of agricultural waste at the top of the stack, to be replaced every three months (Navarro et al., 1996). Instead, a suitably sized reflective cover would be included in the carrying bag of the storage kit, and instructions on setting up the cover would be added to the manual. An additional configuration for positioning the reflective cover is illustrated in Figure 3.22.

However, additional studies on eliminating condensation in flexible structures have yet to be performed, to find an economically and technically feasible solution, taking into account that condensation is not a particular problem of HS only but occurs frequently in metal and concrete silos, especially in tropical climates.

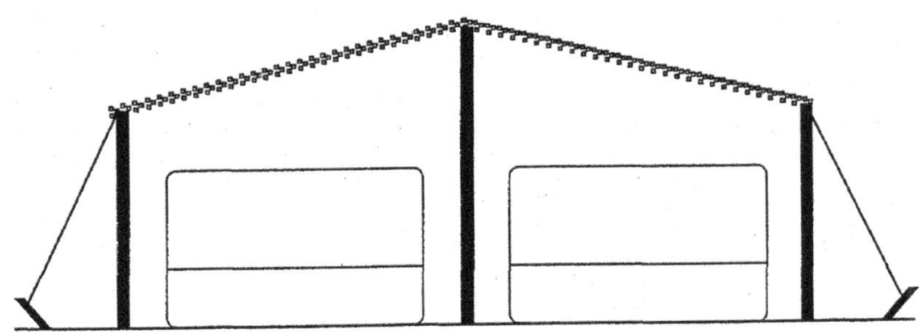

Figure 3.22 Experimental setup of the reflective liner at Muñoz, Philippines. (Redrawn from Donahaye et al., 1999.)

3.5.3 Liner Permeability

Liner permeability refers to the ability of the material used to make the liner of an HS structure impermeable to gases. Liner permeability is an important factor to consider because it affects the gas exchange rate inside the structure.

The liner of HS structures was historically made of PVC and more recently of multi-layer polyethylene (PE), both of which are materials that are to various degrees impermeable to gases such as O_2 and CO_2. The permeability of the liners can be measured by testing the rate of gas transmission through the material.

In HS, the permeability of the liner affects the rates of O_2 depletion and CO_2 buildup inside the gastight structure. A liner that is too permeable may allow for too much gas exchange, which can prevent the depletion of O_2 and the buildup of CO_2, permitting insects to develop and causing the grain to spoil.

In the choice among impermeable liners available on the market, not only does gastightness play a role, but also weight, UV resistance, weldability, allowance for contact with food, and, of course, recyclability.

3.5.4 Recyclable Liners

There are recyclable liners that can be used in HS of grains. The most common type of liner used for large HS structures is made of PVC, to be complemented later by a formulation of multi-layer polyethylene material with gas barriers or laminates. Polyethylene is a food-grade material and very recyclable, provided it is not mixed with materials such as polyamide (PA, nylon) or ethylene vinyl alcohol (EVOH) in excessive amounts.

Used PVC liners, if slightly damaged, can be repaired, giving the liners a life expectancy of ten years or more. If damage is unrepairable, they can be sent to a recycling facility, where they can be processed into new products. Polyethylene is also a commonly recyclable plastic and can be used to make a variety of products such as plastic lumber, plastic bricks, plastic bottles, and other plastic containers.

In addition to being recyclable, liners used in HS can also help to eliminate waste and increase the efficiency of grain storage. By creating an airtight environment, the liners can reduce the need for chemical fumigation and other pest control measures, which can help to reduce the damaging environmental impact of grain storage practices.

It is important to note that not all liners used in HS are recyclable. Some liners may be made of materials that are not recyclable or may have additives that make them difficult or too costly to recycle. Therefore, it is important to choose liners that are made of recyclable materials and to properly dispose of them at the end of their use so as to minimize waste and environmental impact. In addition, the storage facility should consider the expenses involved in handling the recycling process. In many cases, small enterprises that use very limited HS may end up disposing of the liners as garbage.

3.5.5 Remote Sensing

Remote sensing of temperature, humidity, and gas concentration is increasingly used in HS of grains to monitor and control the storage environment. Recent research utilizes the integration of sensors of O_2, CO_2, or both into a storage environment. Remote sensing involves the use of sensors or monitoring equipment that can be located inside the grain storage structure to measure the temperature, humidity, and gas concentration inside the storage space. This information is transmitted wirelessly to a central control unit or computer system, which can be used to monitor storage conditions. Villers et al. (2018) describe progress in the field use of hermetic postharvest storage systems and recent innovations in this technology, which include the introduction of remote monitoring of temperature, humidity, and O_2 or CO_2 levels in large, sealed hermetic containers. Also, introduced in 2018 is the Cocoon Lite™, a second-generation multi-t container featuring major improvements in the permeability, weight, and cost of high-performance, large hermetic storage systems. Early examples of the uses of these innovations and data obtained from their study are expected by year end. The GrainPro EcoWiSe™ is a remote sensing system that enables monitoring of temperature, moisture, and oxygen/carbon dioxide levels, thus providing real-time data on the conditions of the stored commodity involved without manual intervention. One or more low-cost, remote, wireless sensors/transmitters placed inside sealed postharvest hermetic storage units can be read remotely on laptops or cellphones. Data collected and accumulated over time enable development of an algorithm for a stored commodity to define an alarm, whereby the user can be notified immediately of any unsafe humidity or oxygen storage conditions.

In the HS of grains, maintaining the correct gas concentration and humidity levels is essential for preventing spoilage and maintaining grain quality. Remote sensing allows for continuous monitoring of these conditions, which can help to detect any changes or deviations from the desired conditions in real time. This allows for rapid response by discontinuing the storage when any issues arise, such as changes in gas concentration or humidity levels that could lead to mold growth or insect infestations. Overall, remote sensing of gas concentration and humidity is an important tool for ensuring the effectiveness of HS of grains. By providing continuous monitoring and control of storage conditions, remote sensing can help to reduce the risk of grain spoilage and improve the efficiency of grain storage.

3.5.6 Modeling Storability

Mathematical models have been developed that can predict the growth of insects and microorganisms and the rate of deterioration of grains under various storage conditions. These models typically take into account factors such as temperature, humidity, O_2 and CO_2 levels, and the presence of insects and microorganisms, and they use mathematical equations to predict the rate of spoilage or deterioration over time according to the storage structure and location. With input data from sensors or other monitoring systems, these models can serve as a predictive tool for monitoring and managing the storage environment. Overall, the use of mathematical models can provide valuable insights into the effectiveness of HS methods and can help to optimize storage conditions for improved grain quality and reduced spoilage. However, it is important to note that these models are only as accurate as the data that are inputted and may require validation through experimental testing in real-world storage environments.

Agrafioti et al. (2020) used Centaur sensors to test the distribution of phosphine gas in six metal silos with wheat and modeled and compared the results with available distribution data from phosphine sensors. The Centaur model is a mathematical model that is used to predict the rate of gas distribution in stored grain based on the environmental conditions in the storage facility. The model was developed by researchers at the University of Illinois and is based on principles of heat and mass transfer.

3.5.7 Future of Hermetic Storage

As an ancient storage technology revived by the use of low-permeability plastic liners, HS has a promising future in situations where sustainability and lack of storage infrastructure play a role. Especially in tropical climates, storage and transport of dry agricultural commodities are challenging, particularly when taste, aroma, and fat content play an important role. A challenge for the research and industry involved with HS is to develop solutions in which materials have low permeability and are recyclable. No less important is the application of remote control and monitoring equipment with predictive models. Needless to say, all this presupposes a feasible price in order to guarantee gradual adoption of this environmentally friendly and green technology.

References

Adler CS, Opit G, Fürstenau B, et al. (2018) Session 6: Fumigants, Controlled Atmospheres, and Hermetic Storage. Pp. 549–738. Proc 12th Intern Working Stored Prod Prot Berlin, Germany.

Adler C (2020) Hermetic storage: under vacuum, under ambient pressure in bags, or in metal containers - what is most suitable? Pp. 108–115. In: Jayas DS, Jian F (eds) Proceedings of the 11th International Conference on Controlled Atmosphere and Fumigation in Stored Products (CAF2020), CAF Permanent Committee Secretariat, Winnipeg, Canada.

Agrafioti P, Kaloudis E, Bantas S, Sotiroudas V, Athanassiou CG (2020) Modeling the distribution of phosphine and insect mortality in cylindrical grain silos with computational fluid dynamics: Validation with field trials. Comp Elec Agric 173, 105383.

Annis PC (1987) Towards rational controlled atmosphere dosage schedules: a review of current knowledge. pp. 128–148. In: Donahaye J, Navarro S (eds) Proc 4th Intern Working Conf Stored-Prod Prot, Tel Aviv, Israel, 1986.

Banks HJ (1981) Effects of controlled atmosphere storage on grain quality: A review. Food Tech Aust 33, 335–340.

Banks HJ, Annis PC (1977) Suggested procedures for controlled atmosphere storage of dry grain. Tech. Paper 13. Commonwealth Scientific and Industrial Research Organization, Aust. Div. of Entomology, Melbourne. 23 pp.

Banks HJ, Desmarchelier JM (1979) Chemical requirements for a fumigant to be used in partially sealed enclosures. In: Evans DE (ed) Australian Contributions to the Symposium on the Protection of Grain Against Insect Damage During Storage, CSIRO Div. Entomol, Canberra, pp. 92–98.

Banks J, Fields P (1995) Ch. 11: Physical methods for insect control in stored-grain ecosystems. In: Jayas DS, White NDG, Muir WE (eds) Stored Grain Ecosystems, Marcel Dekker, New York, pp. 353–409.

Baoua IB, Amadou L, Lowenberg-DeBoer JD, Murdock LL (2013) Side by side comparison of GrainPro and PICS bags for postharvest preservation of cowpea grain in Niger. J Stored Prod Res 54, 13–16. doi: 10.1016/j.jspr.2013.03.003.

Baoua IB, Amadou L, Ousmane B, Baributsa D, Murdock LL (2014) PICS bags for post-harvest storage of maize grain in West Africa. J Stored Prod Res 58, 20–28. doi: 10.1016/j.jspr.2014.03.001.

Baributsa D, Bakoye ON, Ibrahim B, Murdock LL (2020) Performance of five postharvest storage methods for maize preservation in Northern Benin. Insects 11(8), 541.

Baributsa D, Baoua IB, Bakoye ON, Amadou L, Murdock LL (2017) PICS bags safely store unshelled and shelled groundnuts in Niger. J Stored Prod Res 72, 54–58.

Baributsa D, Concepcion Ignacio MC (2020) Developments in the use of hermetic bags for grain storage. In: Maier D (ed) Advances in Postharvest Management of Cereals and Grains, Burleigh Dodds Science Publishing, Cambridge, UK. pp. 1–28.

Baributsa D, Lowenberg-DeBoer J, Murdock L, Moussa B (2010) Profitable chemical-free cowpea storage technology for smallholder farmers in Africa: opportunities and challenges. pp. 1046–1052. In: Carvalho MO, Fields PG, Adler CS, Arthur FH, Athanassiou CG, Campbell JF, Fleurat-Lessard F, Flinn PW, Hodges RJ, Isikber AA, Navarro S, Noyes RT, Riudavets J, Sinha KK, Thorpe GR, Timlick BH, Trematerra P, White NDG (eds) Proc 10th Intern Working Conf Stored-Prod Prot, Estoril, Portugal.

Bartosik R (2010) Challenges and characteristics of the South American grain and oilseed postharvest system, pp. 57–62, In: Carvalho MO, Fields PG, Adler CS, Arthur FH, Athanassiou CG, Campbell JF, Fleurat-Lessard F, Flinn PW, Hodges RJ, Isikber AA, Navarro S, Noyes RT, Riudavets J, Sinha KK, Thorpe GR, Timlick BH, Trematerra P, White NDG (eds) Proc 10th Intern Working Conf Stored-Prod Prot, Estoril, Portugal.

Bartosik R (2012) An inside look at the silo-bag system, pp. 117–128. In: Navarro S, Banks HJ, Jayas DS, Bell CG, Noyes RT, Ferizli AD, Emekci M, Isikber AA, Alagusundaram K (eds) Proc 9th Intern Conf Controlled Atm Fumi Stored Prod, Antalya, Turkey.

Bartosik R, Urcola H, Cardoso L, Maciel G, Busato P (2023) Silo-bag system for storage of grains, seeds and by-products: A review and research agenda. J Stored Prod Res 100, 1–11. https://doi.org/10.1016/j.jspr.2022.102061

Behr E, Cardoso L, Bartosik R, Marcos Valle F, de la Torre D, Taher H, Maciel G (2021) Evaluation of storage of sunflower pellets in silo bags. pp. 124–131. In: Jayas DS, Jian F (eds) Proceedings of the 11th International Conference on Controlled Atmosphere and Fumigation in Stored Products (CAF2020), CAF Permanent Committee Secretariat, Winnipeg, Canada.

Bond EJ (1984) Manual of Fumigation for Insect Control (Vol. 54). Rome, Italy: FAO.

Calderon M, Donahaye E, Navarro S, Davis R (1989) Wheat storage in a semi-desert region. Trop Sci 29, 91–110.

Calderon M, Navarro S (1979) Increased toxicity of low O_2 atmospheres supplemented with CO_2 on *T. castaneum* adults. Entomologia Experimentalis et Applicata 25, 39–44.

Calderon M, Navarro S (1980) Synergistic effect of CO_2 and O_2 mixture on stored grain insects. Pp 79–84. In: Shejbal J (ed) Proc 1st Intern Conf Controlled Atm Fumi Stored Prod, Castelgandolfo, Italy.,

Cardoso ML, Bartosik RE, Rodriguez JC, Ochandio D (2008) Factors affecting CO_2 concentration in interstitial air of soybean stored in hermetic plastic bags (silobag). pp. 565–568. In: Daolin G, Navarro S, Jian Y, Cheng T, Zuxun J, Yue L, Yang L, Haipeng W (eds) Proc 8th Intern Conf Controlled Atm Fumi Stored Prod, Chengdu, China.

Chayaprasert W, Nukham K, Sukcharoen A (2015) Evaluation of the superposition method for predicting gas leakage rates during fumigations in empty model silos. J Stored Prod Res 64, 13–20.

Carlson S D (1968) Respiration of the confused flour beetle in five atmospheres of varying CO_2:O_2 ratios. J Econ Entomol 61,94–96.

Chiappini E, Molinari P, Cravedi P (2009) Mortality of *Tribolium confusum* J. du Val (Coleoptera: Tenebrionidae) in controlled atmospheres at different O_2 percentages. J Stored Prod Res 45, 10–13.

Chigoverah AA, Mvumi BM (2018) Comparative efficacy of four hermetic bag brands against *Prostephanus truncatus* (Coleoptera: Bostrichidae) in stored maize grain. J Econ Entomol 111(5), 2467–2475.

Chigoverah AA, Mvumi BM (2021) Comparative survival of *Prostephanus truncatus* (Coleoptera: Bostrichidae) and *Sitophilus zeamais* (Coleoptera: Curculionidae) under large-scale hermetic grain storage conditions. pp. 139–145. In: Jayas DS, Jian F (eds) Proceedings of the 11th International Conference on Controlled Atmosphere and Fumigation in Stored Products (CAF2020), CAF Permanent Committee Secretariat, Winnipeg, Canada.

Chigoverah AA, Mvumi BM, Kebede AT, Tefera T (2014) Effect of hermetic facilities on stored maize insect infestation and grain quality. pp. 24–28. In: Arthur FH, Kengkanpanich R, Chayaprasert W, Suthisut D (eds) Proc 11th Intern Working Conf Stored-Prod Prot, Chiang Mai, Thailand.

Conyers ST, Bell CH (2007) A novel use of modified atmospheres: Storage insect population control. J Stored Prod Res 43, 367–374.

Donahaye E (1990). The potential for stored-product insects to develop resistance to modified atmospheres, pp. 989–997. In: Fleurat-Lessard F, Ducom P (eds) Proc 5th Intern Working Conf Stored-Prod Prot, Bordeaux, France.

Donahaye E (1992). Physiological differences between strains of *Tribolium castaneum* selected for resistance to hypoxia and hypercarbia, and the unselected strain. Physiol Entomol, 17(3), 219–229.

Donahaye J, Navarro S, Caliboso F, Sabio G, Mallo G, Dator J (1999) Prevention of Moisture Migration in Sealed Grain Stacks Stored in the Open in the Tropics Using Reflective Covers. pp. 316–323. In: Johnson GI, To LV, Duc ND, Webb MC (eds) Quality Assurance in Agricultural Produce. 19th ASEAN Seminar on Postharvest Technology, Ho Chi Min City, Vietnam 9–12 Nov 1999, ACIAR Proceedings No. 100.

Donahaye E, Navarro S, Rindner M (1994) The influence of temperature on the sensitivity of two nitidulid beetles to low O_2 concentrations. Pp 88–90. In: Highley E, Wright EJ, Banks HJ, Champ BR (eds) Pro 6th Intern Working Conf Stored-Prod Prot, Canberra, Australia.

Donahaye EJ, Navarro S, Rindner M, Azrieli A (1996) The combined influence of temperature and modified atmospheres on *Tribolium castaneum* (Herbst) (Coleoptera: Tenebrionidae). J. Stored Prod Res 32, 225–232.

Donahaye E, Navarro S, Ziv A, Blauschild Y, Weerasinghe D (1991) Storage of paddy in hermetically sealed plastic liners in Sri Lanka. Trop Sci 31, 109–121.

Downes CJ, van Epenhuijsen CW, Lill RE, Downes JE, Carpenter A, Brash D (2008) Calorimetric evaluation of responses of *Sitophilus oryzae* and *Tribolium confusum* to elevated temperatures and controlled atmospheres. J. Stored Prod Res 44, 295–303.

Emekci M, Navarro S, Donahaye J, Rinder M, Azrieli A (1998) Respiration rates of storage insects in airtight conditions. pp. 165–168. In: Adler C, Schoeller M. (eds) International Organization for Biological and Integrated Control of Noxious Animals and Plants: Integrated Protection in Stored Products. University of Gent, Belgium. (www.iobc-wprs.org/)

Emekci M, Navarro S, Donahaye E, Rindner M, Azrieli A (2002) Respiration of *Tribolium castaneum* (Herbst) at reduced O_2 concentrations. J Stored Prod Res 38, 413–425.

Emekci M, Navarro S, Donahaye E, Rindner M, Azrieli A (2004) Respiration of *Rhyzopertha dominica* (F.) at reduced O_2 concentrations. J Stored Prod Res 40, 27–38.

Friedlander A, Navarro S (1978) Influence of Controlled Atmospheres and Relative Humidities on Lipid Content and Water Loss of *Ephestia cautella* (Wlk.) pupae. (Spec. Pub. 117.) Israel Agricultural Research Organization, Division of Scientific Publishing, Bet Dagan, Israel. pp. 7–10.

Friedlander A, Navarro S (1979) The effect of controlled atmospheres on carbohydrate metabolism in the tissue of *Ephestia cautella* (Walker) pupae. Insect Biochem 9, 79–83.

Goulão L, Nguenha R, Carvalho MO (2021) Modified atmospheres in paddy rice HS are biogenerated according to different mechanisms in Portugal and Mozambique. pp. 116–123. In: Jayas DS, Jian F (eds) Proceedings of the 11th International Conference on Controlled Atmosphere and Fumigation in Stored Products (CAF2020), CAF Permanent Committee Secretariat, Winnipeg, Canada.

Hellin J, Kanampiu F (2008) Metal silos and food security in El Salvador. Appropriate Technology 35, 69–70.

Jay EG (1971) Suggested procedures and conditions for using carbon dioxide to control stored-product insects in grain and peanut facilities. Pub. 51-46. U.S. Dep. Agric., Agric. Res. Serv., Washington, DC. 6 pp.

Jay EG, Arbogast RT, Pearman GC Jr (1971) Relative humidity: Its importance in the control of stored-product insects with modified atmospheric gas concentrations. J Stored Prod Res 6, 325.

Jayas DS, Jian F (2021) (eds) Proceedings of the 11th International Conference on Controlled Atmosphere and Fumigation in Stored Products (CAF2020), CAF Permanent Committee Secretariat, Winnipeg, Canada.

Jay EG, Cuff W (1981) Weight loss and mortality of three stages of *Tribolium castaneum* (Herbst) when exposed to four modified atmospheres. J Stored Prod Res 17, 117–124.

Jay EG, Pearman GC (1971) Susceptibility of two species of *Tribolium* (Coleoptera: Tenebrionidae) to alterations of atmospheric gas concentrations. J Stored Prod Res 7, 181–186.

Kandel P, Scharf ME, Mason LJ, Baributsa D (2021) Effect of hypoxia on lethal mortality time of adults *Sitophilus oryzae* L. Insects. 952, 12. https://doi.org/10.3390/insects12100952

Krishnamurthy TS, Spratt E, Bell CH (1986) The toxicity of carbon dioxide to adult beetles in low oxygen atmospheres. J Stored Prod Res 22, 145.

Moreno-Martinez E, Jimenez AS, Vazquez ME (2000) Effect of *Sitophilus zeamais* and *Aspergillus chevalieri* on the O_2 level in maize stored hermetically. J Stored Prod Res 36, 25–36.

Murdock LL, Baoua I (2014) On purdue improved cowpea storage (PICS) technology: Background, mode of action, future prospects. J Stored Prod Res 58, 3–11.

Murdock LL, Kitch L, Moar WW, Chambliss OL, Endodno C, Wolfson JL (1997) Post harvest stored of cowpea in sub-Saharan Africa. pp. 302–312. In:Singh BB, Moham RDR, Dashiell KE, Jackai LEN (eds.) Advances in Cowpea Research. International Institute of Tropical Agriculture and Japan International Research Centre of Agriculture Research.

Murdock LL, Lowenberg-DeBoer JD (2014) Special issue on hermetic storage. J. Stored Prod. Res 58, 1–2.

Murdock LL, Margam V, Baoua I, Balfe S, Shade RE (2012) Death by desiccation: Effects of hermetic storage on cowpea bruchids. J Stored Prod Res 49, 166–170.

Murdock LL, Seck D, Ntoukam G, Kitch L, Shade RE (2003) Preservation of cowpea grain in sub-Saharan Africa - Bean/Cowpea CRSP contributions. Field Crops Res 82, 169–178.

Mutambuki K, Affognon H, Baributsa D (2014) Evaluation of triple layer HS bag (PICS) against *Prostephanus truncatus* and *Sitophilus zeamais*. pp. 355–362. In: Arthur FH, Kengkanpanich R, Chayaprasert W, Suthisut D (eds), Proc 11th Intern Working Conf Stored-Prod Prot, Chiang Mai, Thailand

Mutambuki K, Likhayo P, Mbugua J, Warigia T (2018) Evaluation of AgroZ Hermetic Storage Bag against insect pests on stored maize. pp. 49–55. In: Adler CS, Opit G, Fürstenau B, Müller-Blenkle C, Kern P, Arthur FH, Athanassiou CG, Bartosik R, Campbell J, Carvalho MO, Chayaprasert W, Fields P, Li Z, Maier D, Nayak M, Nukenine E, Obeng-Ofori O, Phillips T, Riudavets J, Throne J, Schöller M, Stejskal V, Talwana H, Timlick B, Trematerra P (eds) Proc 12th Intern Working Conf Stored-Prod Prot, Berlin, Germany.

Mutungi CM, Affognon H, Njoroge AW, Baributsa D, Murdock LL (2014) Storage of mung bean (*Vigna radiata* [L.] Wilczek) and pigeonpea grains (*Cajanus cajan* [L.] Millsp) in hermetic triple-layer bags stops losses caused by *Callosobruchus maculatus* (F.) (Coleoptera: Bruchidae). J Stored Prod Res 58, 39–47.

Navarro S (1975) Effect of oxygen concentrations on *Tribolium castaneum* (Herbst) adults exposed to different relative humidities. Report of the Stored Prod. Res. Lab. 13–21 (in Hebrew with English summary).

Navarro S (1977) Distribution and Abundance of Insects in Butyl-Rubber/EPDM Silos Containing Wheat. pp. 680–687. In Proceedings of the 15th International Congress on Entomology, Washington, DC 1976.

Navarro S (1978) The effects of low oxygen tensions on three stored product insect pests. Phytoparasitica 6, 51–58.

Navarro S (1998) Pressure tests for gaseous applications in sealed storages: Theory and practice. pp. 385–390 (Vol. I) In: Zuxun J, Quan L, Yongsheng L, Xianchang T, Lianghua G (eds) Proc 7th Intern Working Conf Stored-Prod Prot, Beijing, China.

Navarro S (2006) Modified atmospheres for the control of stored-product insects and mites. In: Heaps JW (ed) Insect Management for Food Storage and Processing, Second Edition. AACC International, St. Paul, MN, pp 105–146.

Navarro S (2010) Commercial applications of oxygen depleted atmospheres for the preservation of food commodities. pp. 321–350. In: Doona CJ, Kustin K, Feeherry FE (eds) Case Studies in Novel Food Processing Technologies. Woodhead Publishing Series in Food Science, Technology and Nutrition: Number 197, Woodhead Publishing Limited, Cambridge, UK. 529 p.

Navarro S (2012). Global challenges for the successful application of MA and hermetic storage. Pp: 429–439. In: Navarro S, Banks HJ, Jayas DS, Bell CG, Noyes RT, Ferizli AD, Emekci M, Isikber AA, Alagusundaram K (eds) Proc 9th Intern Conf Controlled Atm Fumi Stored Prod, Antalya, Turkey.

Navarro S, Amos TG, Williams P (1981) The effect of oxygen and carbon dioxide gradients on the vertical dispersion of grain insects in wheat. J Stored Prod Res 17, 101–107.

Navarro S, Calderon M (1973) CO_2 and relative humidity: Interrelated factors affecting the loss of water and mortality of *Ephestia cautella* (Wlk.) (Lepidoptera, Phycitidae). Israel J Entomol 8, 143–152.

Navarro S, Calderon M (1974) Exposure of *Ephestia cautella* (Wlk.) pupae to carbon dioxide concentrations at different relative humidities: The effect on adult emergence and loss in weight. J Stored Prod Res 10, 237–241.

Navarro S, Calderon M (1979) Mode of action of low atmospheric pressures on *Ephestia cautella* (Wlk.) pupae. Experientia 35, 620–621.

Navarro S, Calderon M (1980) Integrated approach to the use of controlled atmospheres for insect control in grain storage. pp 73–78. In: Shejbal, J., (ed) Proc 1st Intern Conf Controlled Atm Fumi Stored Prod, Castelgandolfo, Italy.

Navarro S, Donahaye E (2005) Innovative environmentally Friendly technologies to maintain quality of durable agricultural produce. pp. 205–262. In: Ben-Yehoshua S (ed) Environmentally Friendly Technologies for Agricultural Produce Quality, CRC Press, Taylor & Francis Group, Boca Raton, FL.

Navarro S, Donahaye E, Caliboso FM, Sabio GC (1996) Application of modified atmospheres under plastic covers for prevention of losses in stored grain. Final Report submitted to United States Agency for International Development (USAID), CDR Project NO. C7-053, August 1990-November 1995, 32p.

Navarro S, Donahaye EJ, Caliboso FM, Sabio GC (1998a) Outdoor Storage of Corn and Paddy Using Sealed-Stacks in the Philippines. pp 225–236. In: Villapando I, Ramos CL, Salcedo BGA (eds) Proceeding 18th ASEAN Seminar on Grains Postharvest Technology, 11–13 March 1997, Manila, Philippines,

Navarro S, Donahaye EJ, Ferizli GA, Rindner M, Azrieli A (1998b) A sealed granary for use by small-scale farmers. pp. 434–443. In: Zuxun J, Quan L, Yongsheng L, Xianchang T, Lianghua G (eds) Proc 7th Intern Working Conf Stored-Prod Prot, Beijing, China.

Navarro S, Donahaye E, Fishman S (1994) The future of hermetic storage of dry grains in tropical and subtropical climates. pp. 130–138. In: Highley E, Wright EJ, Banks HJ, Champ BR (eds) Pro 6th Intern Working Conf Stored-Prod Prot, Canberra, Australia.

Navarro S, Donahaye E, Kashanchi Y, Pisarev V, Bulbul O (1984) Airtight storage of wheat in a P.V.C. covered bunker. Pp 601–614. In: Ripp BE, Banks HJ, Calverley DJ, Jay EG, Navarro S (eds) Proc. Int Symp Practical Aspects of Controlled Atm and Fumi in Grain Storages, Amsterdam, Holland.

Navarro S, Donahaye E, Rindner M, Azrieli A (1990) Airtight storage of grain in plastic structures. Hassadeh Q 1(2), 85–88.

Navarro S, Finkelman S, Donahaye E, Dias R, Rindner M, Azrieli A (2002a) Integrated storage pest control methods using vacuum or CO_2 in transportable systems. pp. 207–214. In: Adler C, Navarro S, Scholler M, Stengard-Hansen L (eds) Proceedings of International Organization for Biological Control and Integrated Control of Noxious Animals and Plants, West Palearctic Regional Section (IOBC/WPRS), Working Group "Integrated Protection in Stored Products" meeting, Lisbon, Portugal, September 2001.University of Gent Belgium, 25.

Navarro S, Finkelman S, Sabio G, Isikber A, Dias R, Rindner M, Azrieli A (2002b) Enhanced effectiveness of vacuum or CO_2 in combination with increased temperatures for control of storage insects. pp 818–822. In: Credland PF, Armitage DM, Bell CH, Cogan PM, Highley E (eds) Proc 8th Intern Working Conf Stored-Prod Prot, York, UK.

Navarro S, Friedlander A (1974) Effect of O_2 and carbon dioxide concentrations at different relative humidities on the triglyceride levels in the body of *Ephestia cautella* (Wlk.) pupae. Report of the Stored Prod. Res. Lab. 63–70 (in Hebrew with English summary).

Navarro H, Navarro S (2021) Efficacy of controlling two stored cashew nut insects in 1% O_2 and 99% N_2 at 43°C. pp. 153–159. In: Jayas DS, Jian F (eds) Proceedings of the 11th International Conference on Controlled Atmosphere and Fumigation in Stored Products (CAF2020), CAF Permanent Committee Secretariat, Winnipeg, Canada.

Navarro S, Navarro H (2014) The biological and physical aspects of hermetic storage: a critical review. pp. 24–28. In: Arthur FH, Kengkanpanich R, Chayaprasert W, Suthisut D (eds) Proc 11th Intern Working Conf Stored-Prod Prot, Chiang Mai, Thailand.

Navarro S, Navarro H (2020) Prevention of condensation in shipping containers containing bagged stored products. Integrated Protection of Stored Products, IOBC-WPRS Bulletin, Vol. 148. Pizza, Italy, 135–142.

Nicolas G, Sillans D (1989) Immediate and latent effects of carbon dioxide on insects. Ann Rev Entomol 34, 97–116.

Njoroge AW, Affognon HD, Mutungi CM, Manono J, Lamuka PO, Murdock LL (2014) Triple bag hermetic storage delivers a lethal punch to *P. truncatus* (Horn) (Coleoptera: Bostrichidae) in stored maize. J. Stored Prod Res 58, 12–19. doi: 10.1016/j.jspr.2014.02.005

Ognakossan KE, Tounou AK, Lamboni Y, Hell K (2013) Post-harvest insect infestation in maize grain stored in woven polypropylene and in hermetic bags. Int J Tropic Insect Sci 33, 71–81.

Opoku B, Osekre EA, Opit G, Bosomtwe A, Bingham GV (2023) Evaluation of hermetic storage bags for the preservation of yellow maize in poultry farms in Dormaa Ahenkro, Ghana. Insects 14(2), 141.

Quezada MY, Moreno J, Vázquez ME, Mendoza M, Méndez-Albores A, Moreno-Martínez E (2006) Hermetic storage system preventing the proliferation of *P. truncatus* Horn and storage fungi in maize with different moisture contents. Postharvest Biol Technol 39(3), 321–326.

Radek A, Vlastimil K, Jan P, Vaclav S (2017) Field efficacy of brief exposure of adults of six storage pests to nitrogen-controlled atmospheres. Plant Protect Sci 53(3), 169–76. doi: 10.17221/136/2016-PPS.

Rickman JF, Aquino E (2011) Appropriate technology for maintaining grain quality in small scale storage. pp. 216–224. In: Donahaye EJ, Navarro S, Bell C, Jayas D, Noyes R, Phillips TW (eds) Proc. Int. Conf. Controlled Atmosphere and Fumigation in Stored Products, Gold-Coast Australia 8–13th August 2004. Sichuan Publishing Group, Sichuan, China.

Ridla M, Sabaleku MRB, Nahwori N (2023) Impact of using hermetic packaging and preservative on physical properties of rice bran during storage. Buletin Peternakan 47(1), 22–29.

Ripp BE, Banks HJ, Calverley DJ, Jay EG, Navarro S (1984) (eds.) Proc. Int Symp Practical Aspects of Controlled Atm and Fumi in Grain Storages, Amsterdam, Holland. 789 pp.

Riudavets J, Iturralde-García RD, Castañé C, Bourne R, Wong-Corral FJ (2021) Effect of carbon dioxide sorption in packaged chickpeas on the susceptibility to modified atmospheres of *Rhyzopertha. dominica* and *Callosobruchus chinensis*. Page 175. In: Jayas DS, Jian F (eds)

Proceedings of the 11th International Conference on Controlled Atmosphere and Fumigation in Stored Products (CAF2020), CAF Permanent Committee Secretariat, Winnipeg, Canada.

Sakka MK, Gatzali F, Karathanos VT, Athanassiou CG (2020) Effect of nitrogen on phosphine-susceptible and-resistant populations of stored product insects. Insects 11(12), 885.

Sakka MK, Gatzali F, Karathanos V, Athanassiou CG (2022) Efficacy of low oxygen against *Trogoderma granarium* Everts, *Tribolium castaneum* (Herbst) and *Callosobruchus maculatus* (F.) in commercial applications. J Stored Prod Res 97, 101968.

Sanon A, Dabiré-Binso LC, Ba NM (2011) Triple-bagging of cowpeas within high density polyethylene bags to control the cowpea beetle *Callosobruchus maculatus* F.(Coleoptera: Bruchidae). J Stored Prod Res 47(3), 210–215.

Singano CD, Mvumi BM, Stathers TE (2019) Effectiveness of grain storage facilities and protectants in controlling stored-maize insect pests in a climate-risk prone area of Shire Valley, Southern Malawi. J Stored Prod Res 83, 130–147.

Soderstrom EL, Brandl DG, Mackey B (1992) High temperature combined with carbon dioxide enriched or reduced oxygen atmospheres for control of *Tribolium castaneum* (Herbst) (Coleoptera: Curculionidae). J Stored Prod Res 28, 235–238.

Song S, Du J, Hu B, Yang FY, Song JZ, Wang GY (2021) Low oxygen concentration hinders the occurrence of the cigarette beetle, *Lasioderma serricorne* under nitrogen-controlled atmosphere storage system. pp. 161–166. In: Jayas DS, Jian F (eds) Proceedings of the 11th International Conference on Controlled Atmosphere and Fumigation in Stored Products (CAF2020), CAF Permanent Committee Secretariat, Winnipeg, Canada.

Storey CL (1975) Mortality of adult stored product insects produced by an exotherrnic inert atmosphere generator. J Econ Entomol 68, 316–318.

Tang J, Mitcham EME, Wang SWS, Lurie S (eds) (2007) Heat Treatments for Postharvest Pest Control: Theory and Practice, CABI.

Tefera T, Kanampiu F, De Groote H, Hellin J, Mugo S, Kimenju S, Beyene Y, Boddupalli PM, Shiferaw B, Banziger M (2011) The metal silo: An effective grain storage technology for reducing post-harvest insect and pathogen losses in maize while improving smallholder farmers' food security in developing countries. Crop Protection 30, 240–245.

Tütüncü, Ş. (2022). Effect of a low oxygen and high carbon dioxide modified atmosphere on the mortality of different life stages of Carpophilus hemipterus. J Asia Pac Entomol., 25(2), 101920.

Villers P, de Bruin T, Plijter P (2018). Remote sensing, predictable storage of agricultural commodities and advances in hermetic storage. Julius-Kühn-Archiv, (463).

Villers P, Navarro S, De Bruin T (2010) New Applications of Hermetic Storage for Grain Storage and Transport, pp. 446–452. In: Carvalho MO, Fields PG, Adler CS, Arthur FH, Athanassiou CG, Campbell JF, Fleurat-Lessard F, Flinn PW, Hodges RJ, Isikber AA, Navarro S, Noyes RT, Riudavets J, Sinha KK, Thorpe GR, Timlick BH, Trematerra P, White NDG (eds), Proc 10th Intern Working Conf Stored-Prod Prot, Estoril, Portugal.

Weinberg ZG, Yan Y, Chen Y, Finkelman S, Ashbell G, Navarro S (2008) The effect of moisture level on high-moisture maize (*Zea mays* L.) under hermetic storage conditions—in vitro studies. J Stored Prod Res 44, 136–144.

White NDG, Jayas DS, Muir WE (1995) Toxicity of carbon dioxide at biologically producible levels to stored-product beetles. Environ Entomol 24(3), 640–7. doi: 10.1093/ee/24.3.640.

Silo Bag Storage

*Ricardo Bartosik, Leandro Cardoso,
and Hernán Urcola*

4.1 Introduction

Silo bags offer a flexible and hermetically sealed storage solution (Figure 4.1). Constructed from polyethylene, these bags come in various sizes and are suitable for storing grains and their byproducts for durations that can vary from a few weeks to several months (Bartosik, 2012).

Silo bags were imported to Argentina as a preservation system for forage through the anaerobic silage process in the early 1990s, before being adapted for grain storage in this country (Casini, 2014). Today, the silo bag system comprises a range of specifically developed technological components, including the bagging machine, grain cart with unloading auger, plastic bag with specific features, and equipment for unloading, hermeticity assessment, and monitoring (Santa Juliana and Cardoso, 2014). Private companies collaborated with the National Institute of Agricultural Technology (INTA) of Argentina since the late 1990s to develop these components.

These innovations, together with a strong extension program financed by the public and private sectors and implemented by INTA, led to the massive adoption of the silo bags by the entire agricultural sector of the country, including farmers, grain elevators, grain processing industries, and ports. In 1997 less than 1 Mt per year of different grains were stored in silo bags in Argentina; a few years later, in 2003, this figure had increased to more than 10 Mt; and since 2015 between 40 and 50 Mt are being stored in silo

DOI: 10.1201/9781003309888-4

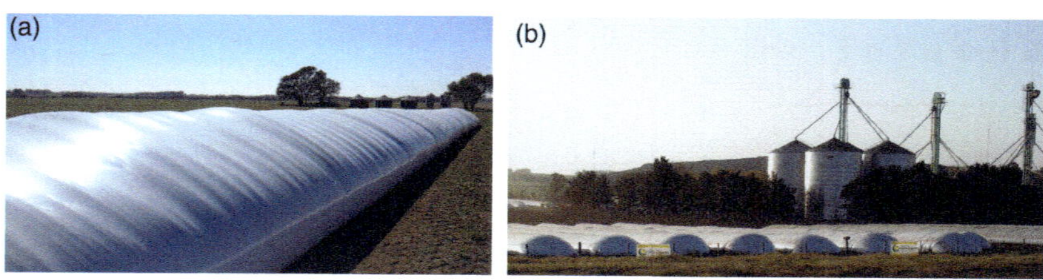

Figure 4.1 Silo bag in a farm (a) and in a grain storage facility (b).

bags (Figure 4.2). The main driving forces for the bag's rapid adoption were the shortage of permanent storage capacity, which created critical logistics problems; the lack of financing for investing in permanent storage structures; the economic crisis, which caused the bankruptcy of many commercial grain elevators; and the structure of the productive system based on land leasing, which discouraged investments in fixed storage facilities (Cardoso et al., 2014). In addition to having been extensively adopted in Argentina, silo bags are used in more than 50 other countries under very different climatic conditions, from the tropics (e.g., Brazil, Colombia, Ecuador, Central

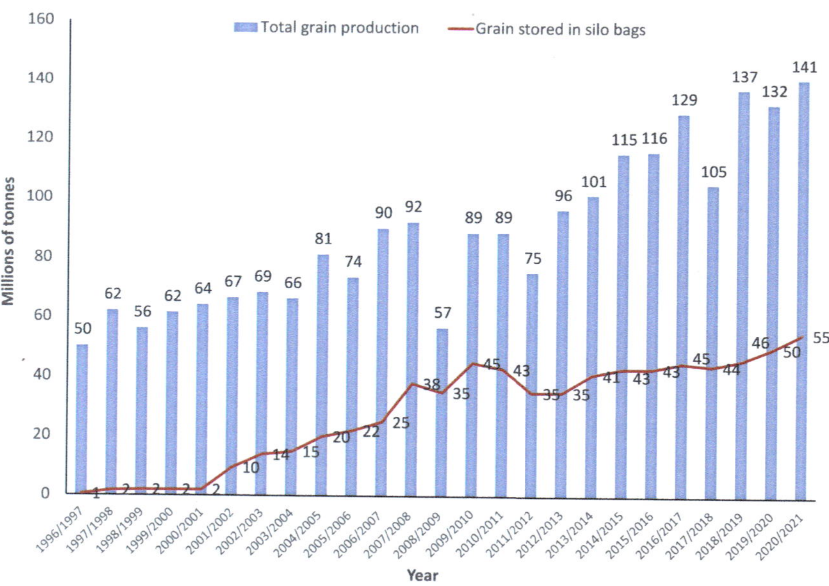

Figure 4.2 Evolution of tonnes of grains produced and stored in silo bags in Argentina. (From INTA, based on compiled information from the Argentine silo bags manufacturing companies.)

America, Sub-Saharan countries) to regions with a temperate climate (e.g., Australia, South Africa, Chile, United States, Turkey) and even regions with cold climates (e.g., Canada, Russia, Ukraine) (CIDETER, 2022). Today, among other products silo bags are used for storing feed grains such as corn and sorghum; oilseeds such as soybean, canola, and sunflower; industrial grains such as rice, wheat, and malting barley; specialty grains such as popcorn, confectionary sunflower, and different kind of beans; organic grains; and byproducts such as wet distiller grains with solubles (WDGS) and pellets of different kinds (Bartosik et al., 2008b, 2012; Faroni et al., 2009; Cardoso et al., 2014; Alvarez et al., 2016; Chelladurai et al., 2016b).

4.2 The Silo Bag System Components

The silo bag system for grain storage is made of several components: the plastic bag, the bagging machine, the unloading machine, and the grain cart. All the components have been meticulously engineered to meet the demanding work capacity requirements, ensuring seamless integration with the combine harvester to avoid work delays. The objective is to optimize labor efficiency, minimize operational expenses, and uphold the grain's quality standards.

4.2.1 The Bag

4.2.1.1 Dimensions

The diameter of the silo bags can vary from 1.22 m (4 ft) to 3.66 m (12 ft) and the length from 30 m (98 ft) to 150 m (492 ft); the most common configuration is 2.74 m (9 ft) in diameter and 75 m (246 ft) in length. This standard silo bag can hold about 220–260 t of wheat, corn, soybean, barley, and sorghum and about 150 t of sunflower seeds.

4.2.1.2 Liner Composition

Silo bags are designed to possess a combination of mechanical properties, including puncture resistance, tear resistance, elasticity, and ductility. These properties are achieved by using a mix of nonlinear low-density polyethylene (NLDPE) and linear low-density polyethylene (LLDPE) in their construction. The plastic liner typically has a thickness of 230 microns and is made up of three to five layers with varying characteristics, depending on the extrusion technology used and the chemical additives incorporated. The outer layer has titanium oxide to provide a white color so as to reflect most of the solar radiation, while the inner layer has a black pigment to avoid translucency. Additives are also used to provide the liner with the necessary ultraviolet light protection and sliding properties during the grain bagging operation, among other characteristics (Cardoso et al., 2014). However, there are some silo bags with a colored outer layer to indicate some particular characteristic (e.g., green indicates that the bag was partially made with recycled plastic, and pink is to support a breast cancer research campaign).

During the filling process, the silo bag stretches, which weakens the mechanical strength of the liner. To prevent excessive stretching, silo bag manufacturers recommend a maximum tangential deformation of 10%. Printed marks on the bags allow for monitoring of the bag's stretch during the filling process.

4.2.1.3 Mechanical Properties and Liner Permeance

There is no published information on the mechanical parameters of a typical silo bag liner. However, a company report (unpublished) indicated the following mechanical properties for a standard silo bag of 235-micron thickness: tensile load at yield of 75 N, displacement at yield from 9.4 to 29.9 mm, elongation at yield from 18.8% to 59.8%, ultimate tensile load from 104.9 to 115.7 N, displacement at break from 461 to 481 mm, and modulus of elasticity from 0.224 to 0.252 GPa. Additionally, the resistance to puncture (dart drop) was 988 g at 20°C. Having reliable data on the mechanical properties of the silo bag plastic liner is important to refine predictive models of the stresses to which the bag is subjected and identify potential improvements in the characteristics of the materials. A numerical analysis conducted by Garg et al. (2019) studied mechanical failure in the polymer of the bags during installation and usage at different temperatures, while Scarabino (2019) developed a model to predict the geometric shape of a filled silo bag and the tensile forces involved. The following mechanical properties were assumed for the model: density of 0.922 kg dm^{-3}, yield tensile strength of 11.2 MPa, elongation at break of 533%, and modulus of elasticity of 0.315 GPa.

The gas permeance of the liner is critical for the successful implementation of modified and controlled atmosphere treatments. Permeance is defined as the gas transmission rate in relation to the difference in the gas partial pressure at both sides of the liner. Abalone et al. (2011) estimated a permeance to O_2 of 4.06×10^{-4} and to CO_2 of 1.34×10^{-3} m^3 d^{-1} m^{-2} atm^{-1} based on the constituent materials of the silo bag (they assumed a film of 240-micron thickness made of 50% high-density polyethylene and 50% low-density polyethylene). Later, Chelladurai et al. (2016a) measured the gas permeability of a new silo bag liner and obtained the following permeance values assuming a thickness of 235 microns: 8.29×10^{-3} and 91.95×10^{-3} m^3 d^{-1} m^{-2} atm^{-1} for O_2 and CO_2, respectively. After 10 months in the field, the permeance of the stretched silo bag liner was reevaluated, revealing an increase of approximately 50% (16.13×10^{-3} and 130.63×10^{-3} m^{3}d^{-1}m^{-2}atm^{-1} for O_2 and CO_2, respectively). To date, this is the only published report of the measured permeability of the silo bag liner to gases. However, a company report (unpublished) indicated significantly lower permeance values of 10.34×10^{-4} m^3 d^{-1} m^{-2} atm^{-1} and 22.48×10^{-4} m^3 d^{-1} m^{-2} atm^{-1} for O$_2$ and CO$_2$, respectively. The methodology used in this report is unknown, and therefore its validity cannot be confirmed. Obtaining reliable data on the permeance of the silo bag liner to gases is crucial for refining current prediction models for the internal atmospheres of silo bags used for storing various grains and byproducts. Additionally, this information is necessary for designing effective pest control strategies based on the use of controlled atmospheres.

There are no reported water vapor permeability coefficients for silo bag films. However, for reference, a permeance of 16.17 g m^{-2} d^{-1} atm^{-1} was derived for a 230-micron-thick LDPE film (Gaikwad and Rubino, 2017). From a practical perspective, this level of water vapor permeance is low enough to prevent any measurable grain moisture variation during storage.

4.2.2 Bagging Machine

The bagging machine (bagger) is used for loading grain into silo bags. It consists of a receiving hopper, aligned with the center of the machine, for grain reception. The filling auger, driven by the tractor's power takeoff (for a standard bagger machine, the tractor's power requirement is of 80 HP), transports the grain from the hopper to the silo bag [Figure 4.3(a)]. It should have a large diameter, ample flight section, and minimal inclination to avoid grain damage. Augers with diameters of up to 450 mm have been incorporated, increasing capacity to over 400 t/h. There is a bagging machine model that operates without auger and without the assistance of a tractor, relying on the specific weight of the grain for filling (named "zero energy") [Figure 4.3(b)]. The tunnel, where the folded silo

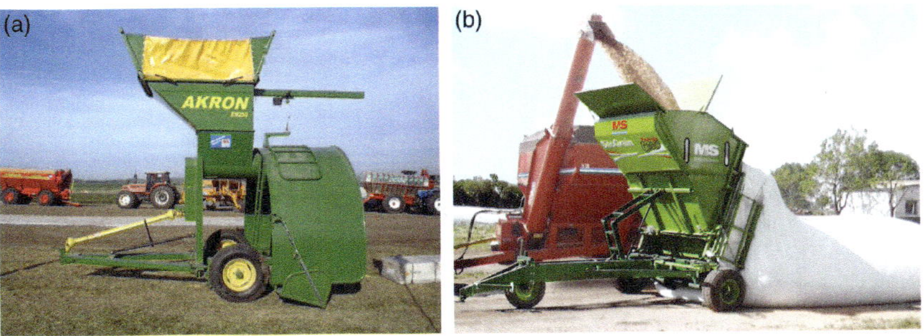

Figure 4.3 (a) Typical bagging machine with a filling auger operated with the PTO of the tractor. (b) A model known as the "zero energy" bagging machine, which operates independently without the need for a tractor's assistance.

bag is mounted, ensures even filling without overload zones. The braking system, either friction-based or disc brakes, ensures proper stretching of the bag. The bagging machine should have complete protection for moving parts, warning labels, handrails, and nonslip platforms for safety. Other technical elements include a winch for lifting the silo bag, lateral feet for stability, and a brake shield to improve compactness and reduce braking effort.

4.2.3 Unloading Machine

The unloading machine is used for unloading (extracting) grain from the silo bag (Figure 4.4). Like the bagging machine, it is characterized by its simplicity in construction and operation. Its operating principle involves a hydraulic roller that winds up the bag, pulling the extractor and the tractor toward the bag at the same time. Meanwhile, two horizontal augers (uncovered), driven by the tractor's power takeoff, collect the grain from the sides toward the center line of the bag (sweep augers), where a third auger collects and elevates the grain to a truck or grain cart (discharge auger). The current average working capacity is about 300–360 t/h. To achieve this capacity, the extractors are equipped with discharge augers of approximately 400 mm diameter and horizontal sweep augers of about

Figure 4.4 Unloading machine with hydraulic roller.

260 mm diameter. Improvements have been made to increase grain flow, reduce turbulence, and facilitate material entry into the discharge auger. Gear transmission modifications have also been implemented to ensure that the increase in working capacity does not require a significant increase in power (estimated at between 60 and 120 HP for maximum working capacity). Efforts have been made to achieve a more uniform grain flow, thereby reducing mechanical damage to the grain. This type of extractor can be adjusted to handle silo bags with diameters of 2.0 m, 2.7 m, and 3.0 m by adding or removing sections of the sweep augers. The working speed is adjusted by controlling the roller speed through centralized hydraulic commands, which also regulate the folding of the discharge tube and the machine's ground clearance. The latest version of this machine incorporates a feature that efficiently compacts the used liner into rolls, significantly simplifying the subsequent collection process for plastic liner recycling.

4.2.4 Grain Cart

The self-unloading grain cart is a trailer with a large hopper used to directly transport grain from the combine to the bagger (Figure 4.5). The cart is equipped with a large auger that transports the grain to the bagger's hopper. The loading capacity has evolved from 8–10 t to 18 t in single-axle self-unloading carts, from 16 t to 26 t in dual-axle carts, and to

Figure 4.5 Grain cart unloading grain into a bagging machine.

over 30 t for single-axle carts with a rear rocker or double rocker. The discharge auger should have a large diameter (300–500 mm) to enable high working capacity (up to 10 t/min) without increasing mechanical grain damage. A large-diameter auger also optimizes tractor power, requiring 120/150 HP in normal soils or 150/180 HP in soft soils for a 25-t capacity cart. For the cart loading operation, it is essential to have a filling indicator in the hopper to prevent grain spillage. There are currently sensor systems located inside the hopper, 25 cm from the edge, that work through contact and indicate grain distribution via a monitor in the tractor cabin. Meanwhile, another monitor in the combine wirelessly receives the data and automatically ends the loading process. It is also useful for the cart to have the option of incorporating an electronic scale to record weight information. High-flotation tires with wide treads facilitate transit on soft terrain.

4.3 Airtightness

Preserving airtightness is crucial for the effective storage of grain in silo bags. Perforations in the plastic cover and failures in the sealing of the beginning or end of the bag cause gas leaks. The primary factors contributing to the loss of airtightness in silo bags are inadequate soil preparation, improper sealing techniques, and the presence of fauna. Previous studies conducted by Cardoso et al. (2012) and Taher et al. (2014) have highlighted these causes. Abadía and Bartosik (2013) have provided a comprehensive summary of the best management practices necessary to achieve and maintain appropriate airtightness. These practices include thorough soil preparation to remove potential puncturing elements like stones, sturdy stems, or weeds, as well as employing heat sealing methods at the ends of the bags. Additionally, preventive measures should be implemented to deter wildlife and domestic animals from damaging the bags. In research conducted by Zufiaurre et al. (2020, 2018), armadillos (*Chaetophractus villosus*), rodents, and dogs were identified as the most damaging species. To minimize vertebrate-related damage, it is recommended to maintain short grass around the bags, install fences as a barrier, and utilize natural repellents [Figure 4.6(a)], as suggested by Abadía and Bartosik (2013).

Cardoso et al. (2012) documented that it is possible to achieve high levels of tightness in standard-size silo bags [pressure decay test (PDT) higher than 5 min] and that airtightness decreases over time due to the appearance of perforations in the liner. The effect of perforations on the internal atmosphere was modeled by Abalone et al. (2011), who concluded that even a small puncture can significantly change the evolution of the

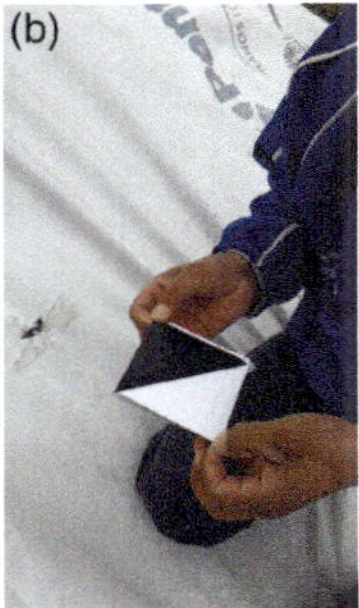

Figure 4.6 Installation of a protective electric fence for the silo bag (a) and silo bag repair (b).

internal gas composition. Consequently, restoring the airtightness is a critical labor for a successful silo bag storage. To address this challenge, specific patching kits are commercially available for both small and large perforations [Figure 4.6(b)].

4.4 Silo Bag Ecosystem

The biotic agents of stored grain ecosystem typically comprises insects, microorganisms, and the grain or seed itself. In the presence of oxygen, these agents primarily rely on aerobic respiration to obtain energy, whereby glucose is oxidized to CO_2, releasing moisture and heat in the process. Generally, respiration is modeled as the breakdown of simple sugars:

$$C_6H_{12}O_6 + 6O_2 \rightarrow 6CO_2 + 6H_2O + 2,835 \text{ kJ/mole} \tag{4.1}$$

The respiration rate of grain is influenced by several factors, including the type of grain (Jian et al., 2014; Ochandio et al., 2014), its condition (such as the presence of foreign matter, mechanical damage, and microbial load) (Al-Yahaya et al., 1993; Bern et al., 2002), and its interaction with abiotic factors such as moisture content (MC), temperature, and gas composition within the storage system (Ochandio et al., 2017; Marcos Valle et al., 2021b). Insects also contribute to the overall respiration rate of the bulk (Cofie-Agblor et al., 1995; Sone, 1999; Chotikasatian et al., 2017). Changes in temperature, MC, microbial load, and oxygen concentration during storage affect the rate of respiration, which is typically quantified as dry matter

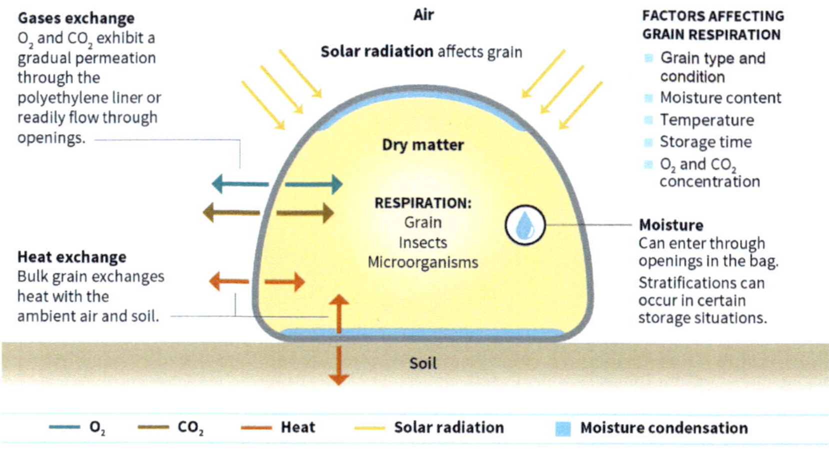

Figure 4.7 Schematic representation of the biotic and abiotic components of the silo bag ecosystem and their interactions.

loss (DML) over time (Steele et al., 1969; Thompson, 1972). Figure 4.7 outlines the biotic and abiotic factors involved during grain storage in silo bags and their interactions.

4.4.1 Heat and Temperature

The temperature of grain stored in a bag mainly depends on the initial temperature of the grain at the time of bagging, the radiation from the sun, and the heat transfer with the surrounding air and soil (Gastón et al., 2009). According to Bartosik et al. (2008b), the temperature of the top layer of grain (0.1 m depth) follows the ambient air temperature, with maximum temperatures at noon and minimum temperatures during the early morning. The daily temperature fluctuations decrease with increasing grain depth, becoming unnoticeable at 0.4 m from the surface (Darby and Caddick, 2007). Several field experiments have characterized the relationship between grain temperature and ambient temperature under different climates and with different grains (Bartosik et al., 2008b; Darby and Caddick, 2007; Ochandio et al., 2010; Ward and Davis, 2013; Chelladurai et al., 2016b). These studies have shown that during storage, the temperature of the grain decreases from summer to winter (Figure 4.8, *top*) and increases from winter to summer (Figure 4.8, *bottom*), gradually approaching and following the ambient temperature of the season.

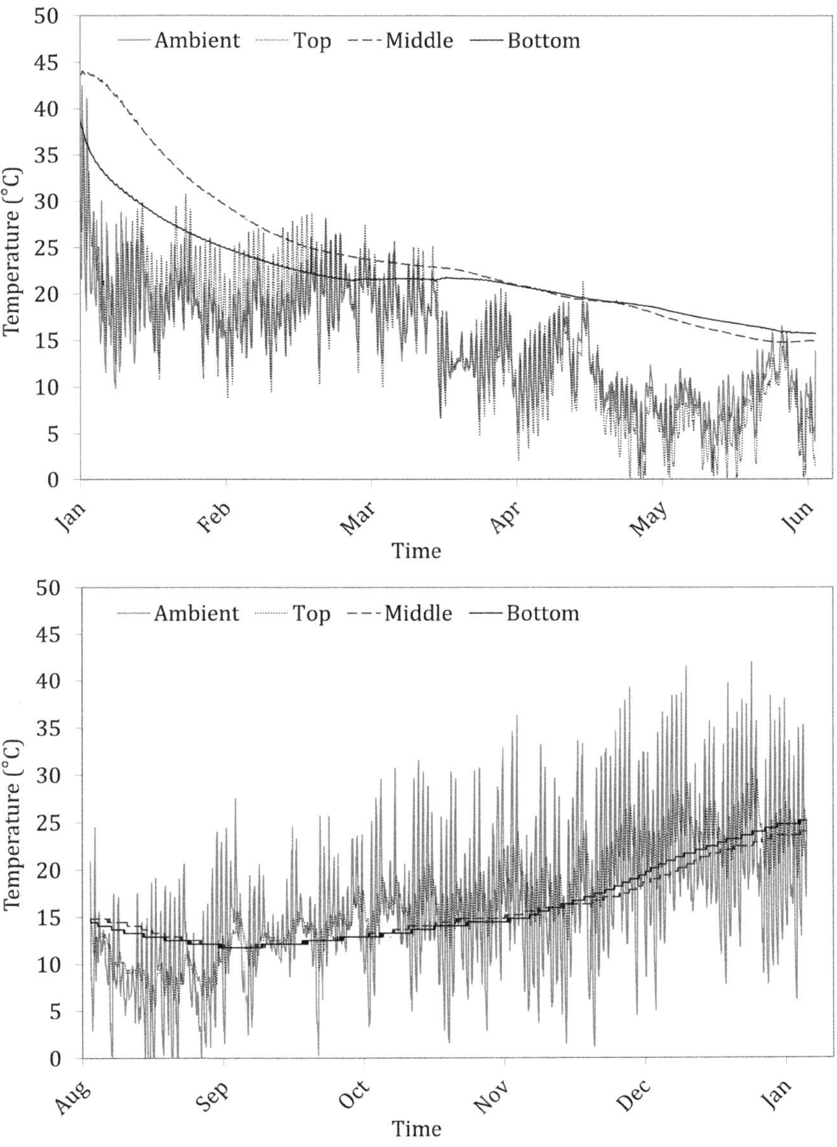

Figure 4.8 Evolution of ambient and grain temperatures (wheat on the top and corn on the bottom) at different layers of a silo bag (top: 0.05 m from the surface, middle: 0.70 m, and bottom: 1.50 m) from January (summer) to June (winter) on the top and from August (winter) to January (summer) on the bottom. (Adapted from Bartosik et al., 2008b.)

In comparison with standard bins equipped with aeration systems, silo bags provide only limited ability to externally control the temperature of grain stored within. However, Clemente et al. (2009) found that covering the silo bag with shade cloth can reduce the temperature of the grain by 4°C. Despite this finding, silo bags are usually placed in the field without any cover. However, the temperature of the grain during storage in a silo bag does not typically reach unsafe levels due to self-heating by respiration, even with slightly moist grain, which can occur if the grain is stored in a metal bin. This is because the surface/volume ratio of a standard silo bag (200 t capacity, 60 m length, and 2.7 m diameter) is almost double that of a regular bin with similar capacity (0.78 m^2/m^3 for a 7-m-diameter and 9-m-height bin of 200 t capacity). Additionally, the distance between the center of the grain mass and the surface is shorter for a silo bag (0.8 m for the central-longitudinal axis of the bag to the soil or the air) than for a standard storage bin (over 3 m). This facilitates a natural heat exchange of the grain mass with the surrounding air and soil, which dissipates most of the heat produced by respiration. Furthermore, if the oxygen concentration substantially decreases during storage, the respiration rate (and heat released) also decreases (Ochandio et al., 2017; Marcos Valle et al., 2021b). The described mechanisms involved in the evolution of the stored grain temperature have been confirmed through modeling by several authors (Gastón et al., 2008, 2009; Arias Barreto et al., 2013, 2017; Jian et al., 2015).

4.4.2 Grain Moisture

In the discussion of permeance properties in Section 4.2.1.3, it was established that the low water vapor permeance prevents significant variations in the average bulk grain MC. However, under certain storage conditions, moisture stratification can occur.

Observations by Ward and Davis (2013) revealed temperature and relative humidity (RH) variations at different locations within the silo bag, displaying a similar daily oscillation pattern. The RH oscillation was particularly noticeable near the surface of the bag, whereas RH remained constant toward the center. Modeling studies conducted by Chelladurai et al. (2016b) and Gastón et al. (2009) on heat and moisture transfer during grain storage in silo bags confirmed that temperature differentials between the top layer and the rest of the bag resulted in moisture migration from the core of the grain mass to the top layer and, to a lesser extent, to the bottom layer. Several field trials supported this observation. Gastón et al. (2009) highlighted that factors such as grain type, grain MC, grain temperature, temperature fluctuation magnitude, and storage time determine the extent of MC stratification. Empirical evidence (Bartosik et al., 2008b; Cardoso et al., 2010; Ochandio et al., 2010) and simulation studies (Gastón et al., 2009) indicate that storing dry grain (with equilibrium RH below 65%) for up to one year does not result in noticeable moisture migration. However, Darby and Caddick (2007) reported moisture stratification during storage of dry barley ($\leq$ 11% MC) under Australian conditions. In this case, the peripheral layer's MC increased to 13% during winter but the layer remained dry during summer with temperatures exceeding 30°C, suggesting that the grain could be stored in optimal condition for up to six months. Conversely, Butts and Ward (2016) reported severe moisture migration in the upper layer (15 cm deep) of a peanut silo bag. Similarly, Bartosik et al. (2008b) observed a substantial moisture increase in the top layer of a sunflower silo bag (16% average MC). These findings suggest that slightly wet and bulky grains may be more susceptible to moisture migration phenomena. Additionally, the formation of air chambers (resulting from improper bagging practices) contributes to moisture condensation in the upper layer of grains. On the other hand, Behr et al. (2021) did not find moisture stratification in bulky materials such as dry sunflower pellets. This indicates that the biotic (grain) and abiotic (temperature, MC) conditions that favor the moisture stratification during silo bag storage have not been fully elucidated.

4.4.3 Internal Atmosphere

Under completely airtight conditions, the biotic components of a bulk tend to consume all the oxygen, resulting in an anoxic atmosphere enriched with CO_2 (Weinberg et al., 2008). The speed at which this process occurs depends on the respiration rate of the grain bulk (Ochandio et al., 2017). However, silo bags in field conditions are not completely airtight (Cardoso et al., 2012), so the internal gas concentration depends on the balance between the respiration rate [oxygen (O_2) consumption and carbon dioxide (CO_2) generation] and the gas exchange rate with the ambient air (entrance of O_2 to the silo bag and exit of CO_2 to the ambient air). The movement of gases in and out of the silo bags depends on the partial pressure differential of the gases and the effective permeability of the silo bag, taking into account the permeance of the plastic film and the contribution of perforations (Bartosik, 2012). As a result, the internal gas concentration at any given time during storage reflects the dynamic equilibrium between respiration and gas exchange rates, which can both fluctuate. The respiration rate can vary due to changes in bulk temperature throughout the seasons, rainwater ingress through perforations in the plastic cover, or condensation buildup in the peripheral grain layer, among other factors. The gas exchange rate, on the other hand, generally increases over time due to the accumulation of perforations in the plastic cover of the silo bag. Therefore, the concentrations of O_2 and CO_2 inside a silo bag change throughout the storage period.

The concentration of gases inside the silo bag is primarily influenced by the MC of the bulk (Cardoso et al., 2008; Rodríguez et al., 2008). As depicted in Figure 4.9, O_2 concentration ranges from 16% to 18% and CO_2 concentration from 1% to 3% for dry grain conditions (MC below 14%). However, as the MC surpasses the limit of biological activity, the O_2 concentration can drop to 0–5%, while CO_2 can spike up to 30%. Bartosik et al. (2008b) report that in some instances, with exceptionally wet grain, CO_2 concentration can even reach levels as high as 70%. Nevertheless, even in dry grain storage, significant fluctuations in the internal atmosphere can occur due to the high biological activity (respiration) of a small amount of spoiling grain, which can be caused by rainwater seeping through a perforation (Bartosik et al., 2008a).

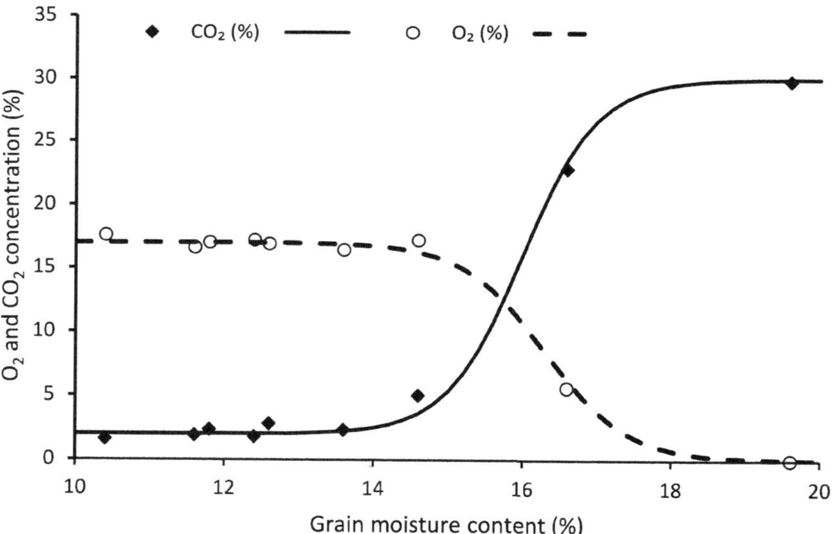

Figure 4.9 Oxygen and carbon dioxide concentration during storage of wheat with different moisture contents in different silo bags. (Adapted from Rodriguez et al., 2008.)

4.5 Relation among Water Activity, Airtightness, and Microbial Activity

The microbiota that develop during the storage of grains in silo bags have been extensively studied for corn (Pacin et al., 2009; Castellari et al., 2010, 2015) and wheat (Darby and Caddick, 2007; Gregori et al., 2013). In these studies, the predominant fungal genera were *Aspergillus*, *Penicillium*, *Fusarium*, and *Eurotium*.

It is widely accepted that mold activity can be prevented when grain is stored at an MC equivalent to $a_w \leq 0.7$ (Table 4.1). Above this threshold, molds become active and temperature plays a crucial role in their regulation and dominance (Marín et al., 1998). Although molds are generally considered obligate aerobes, some species can grow with O_2 concentrations below 3% (Magan and Lacey, 1984). Modified atmospheres are characterized by the formation of hypoxic or anoxic atmospheres that cause a self-regulation effect on biological activity (Navarro et al., 2012). During airtight storage, microbial activity is affected by the triple interaction of a_w, O_2, and temperature. Marcos Valle et al. (2021a) studied these interactions for corn and proposed a general model to explain microbial evolution during hermetic storage in silo bags for grain stored at different MC (Figure 4.10).

For a_w below 0.70 (dry grain), moisture is not available for most microorganisms, biological activity is scarce (respiration is negligible), and the internal atmosphere in the silo bag remains about the same as in the ambient atmosphere. As a_w increases above 0.70, moisture becomes available for

Table 4.1 Moisture Content Equivalent to a_w of 0.7 (Onset of Microbial Activity) and 0.84 (Onset of Mycotoxin Production) at 20°C for Different Grains

Grain	MC (%) Equivalent to a_w = 0.7	MC (%) Equivalent to a_w = 0.84
Wheat	14.9	18.1
Barley	15.3	18.6
Corn	14.0	16.6
Sunflower	8.3	12.5
Sorghum	15.9	18.6
Soybean	12.8	16.5
Canola	9.2	13.8
Rice	14.6	17.4

Note: Values obtained with the INTA-ISU grain aeration and storage app (Maier and Bartosik, 2020), using the equilibrium relative humidity values of 70% and 84% and the moisture relationship equations and parameters from ASAE (2001).

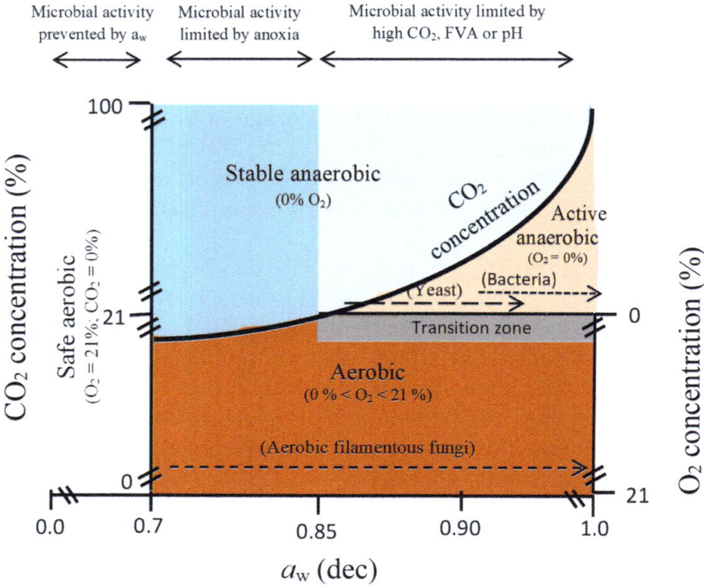

Figure 4.10 Conceptual diagram of the phases of hermetic storage according to the a_w of the environment. (Adapted from Marcos Valle et al., 2021a.)

microbial activity, respiration rate increases, and the modification of the internal atmosphere of the silo bag intensifies. The aerobic phase is characterized by the activity of filamentous fungi, which tends to decrease as O_2 concentration approaches zero. If a_w is lower than 0.85, a stable anaerobic phase follows after O_2 is depleted, which severely limits microbial activity due to the effect of hermetic storage. When a_w is greater than 0.85, a transition to an active anaerobic phase is observed as O_2 is depleted. This transition is characterized by the activity of facultative anaerobic microorganisms such as yeast, and, if a_w is greater than 0.87–0.90, by bacteria. In the active anaerobic phase, microbial activity continues through fermentation processes, which intensify as a_w approaches 1, resulting in an exponential increase of CO_2 concentration. Finally, there is a transition from the active to the stable anaerobic phase. The mechanism that allows this transition could be related to the buildup of CO_2 concentration or the production of volatile fatty acids (Magan and Lacey, 1984; Diawara et al., 1986; Weinberg et al., 2008).

The literature on mycotoxin presence in grain stored in silo bags is limited (Pacin et al., 2009; Subramanyam et al., 2012; Gregori et al., 2013; Castellari et al., 2015; Behr et al., 2021). While these studies report on the concentration of various toxins at the beginning and end of storage, there is a lack of research on the underlying microbiological processes that contribute to their formation in the unique silo bag environment.

A review by Fleurat-Lessard (2017) suggests that above an a_w of 0.84, the risk of mycotoxin formation increases (Table 4.1 shows the equivalent MC for different grains). Storing grain in silo bags below this threshold would disfavor the development of toxins; values above the threshold would favor such development. Modified atmospheres can be effective in preventing mycotoxin development if O_2 concentration drops below 5% (Paster and Bullerman, 1988) or CO_2 concentration increases above 25% (Giorni et al., 2008). Consequently, maintaining a high level of airtightness in the silo bag during storage is crucial to achieve and sustain these critical modifications in the internal atmosphere.

As discussed in previous sections, the factors governing the biological activity of bulk grains in silo bags are dynamic in nature. The average temperature of the stored grain exhibits seasonal variations, moisture migration has the potential to increase the equilibrium a_w in the outer layer, rainwater may infiltrate through openings, and the airtightness of the bags can be compromised by perforations in the plastic cover, thereby affecting the gas composition. Consequently, mold activity is not consistent throughout the storage period. Therefore, it is imperative to comprehend the influence of these factors on mold activity and its progression, as well as to monitor critical variables for effective quality management and assessment of mycotoxin risks associated with grains stored in silo bags. Doing so permits the implementation of comprehensive measures to mitigate potential issues and ensure the preservation of grain quality during storage, with special emphasis on prevention through the storage of dry grain and the implementation of good storage practices in silo bags.

4.6 Quality of Grain During Silo Bag Storage

Bartosik et al. (2023) conducted a comprehensive review of studies examining the quality changes occurring during silo bag storage of various grains. The investigated grains encompassed a wide range of crops, including corn, popcorn, wheat, barley, soybean, canola, beans, rice, sunflower, peanut, cotton seeds, sorghum, and sesame seeds, as well as byproducts such as wet distilled grains with solutes (WDGS) and sunflower pellets.

The quality of grains stored in silo bags is influenced by the interaction between MC and temperature. When the MC is sufficiently low to inhibit microbiological activity, the temperature itself has minimal impact, allowing for storage even during the summer without significant deterioration in quality (Bartosik et al., 2008b; Cardoso et al., 2010), with certain exceptions discussed below. Conversely, when the MC is high enough to allow microbial activity, temperature during storage becomes a major concern. It should be noted that the grain temperature inside silo bags closely reflects the average ambient temperature throughout the year (Figure 4.8 *top* and *bottom*). Consequently, during the winter months, quality parameters are preserved or at least their deterioration is slowed down, due to the combination of low temperature and the modified atmosphere. However, as storage extends into the warmer seasons (spring and summer), microbial activity and other detrimental processes (e.g., biochemical reactions) intensify, leading to a rapid decline in quality parameters that cannot be compensated for by the modified atmosphere (Bartosik et al., 2012).

The available literature primarily comprises studies conducted in mild climatic regions, while a smaller proportion of studies focus on colder climates. However, there is a notable scarcity of research conducted in tropical or subtropical regions. Consequently, caution must be exercised when extrapolating storage recommendations from cold-temperate regions to tropical or subtropical regions. It is important to recognize that temperature alone can have adverse effects on certain quality parameters (e.g., seed viability) (Abadía et al., 2022), emphasizing the need for region-specific investigations and tailored storage practices.

4.7 Insects

4.7.1 Presence of Insects in Silo Bags

According to Massigoge et al. (2010), insects have been documented in silo bags, with live insects found in 6 out of 56 bags during their study on barley storage. While this is the only formal study on the topic, there are anecdotal reports of insects in silo bags. The wheat milling industry in Argentina, which stores dry wheat in silo bags, states that insect presence is rare but more likely in bags with grain previously stored in metal bins and during the summer. Similar anecdotal reports have been mentioned by Darby and Caddick (2007) for dry barley stored in Australia.

The low insect activity observed in silo bags can be attributed to three main factors. First, most of the silo bags are filled with grain coming directly from the harvest. The presence of storage insects in the field depends on environmental conditions at the harvest time (temperature, RH), crop conditions (MC, neighboring crops), and proximity to storage structures (Howlader and Matin, 1988; Trematerra, 2015; Adler et al., 2022). Consequently, grain is usually harvested free of insects (or with an unnoticeable infestation level). Second, silo bags are new and not reused, unlike standard metal or concrete bins, which can have preexisting infestations

(Hagstrum et al., 2012; Abadía and Bartosik, 2013). Additionally, the minimal handling involved in the harvesting and storage process with silo bags reduces the chances of insect colonization below that of storage in traditional grain bins. Finally, the plastic cover of the silo bag acts as a physical barrier, preventing the entry of insects, whereas in conventional storage migrating insects often infest stored grain (Hagstrum, 2001).

4.7.2 Effect of Internal Environment and Self-Modified Atmosphere on Insects

If stored grain becomes infested by insects, their population development will be influenced by the specific environment within the silo bag. The temperature inside the silo bag corresponds to the average ambient temperature throughout the year. In temperate and cold climates during fall and winter, the grain temperature falls below the limit for insect activity (15°C) (Figure 4.8) (Fields, 1992). Achieving complete insect control generally requires reducing the O_2 concentration below 2% or increasing the CO_2 concentration above 35%, although lethal effects were reported with concentrations as low as 20% with long exposure times (Banks et al., 1990; Banks and Annis, 1990; Navarro et al., 2012). Consequently, when the grain retains sufficient moisture to support mold growth (Table 4.1), modification of the internal atmosphere can prove lethal for stored insects (Figure 4.9). However, the internal atmosphere often stabilizes at intermediate concentrations that may not be lethal but can still impact population growth. In a field trial conducted by Subramanyam et al. (2012) in Kansas, silo bags filled with hard red winter wheat were infested with 30 adults of *Rhyzopertha dominica* (F.). They observed that the insect population inside the bags increased 14 to 19 times less than that in the control (growth in a chamber at 28°C and 65% RH). This limitation in population growth was attributed to the particular combination of temperature, RH, and gases inside the silo bag (–3.1° to 32.3°C, 45% to 52% RH, and 2% CO_2).

Considering these factors, conditions that favor insect development (and damage) in silo bags occur when the bag is filled with previously infested grain, the grain is stored during the summer (or when the grain temperature ranges between 25° and 30°C), and the grain has an MC that prevents the formation of a lethal atmosphere for insects. However, even under these conditions, the outbreak of the insect population will be prevented by the hermetic storage provided by the silo bag. Navarro et al. (1994) studied the interdependence between changes in gas concentrations and the dynamics of insect populations under hermetic storage. They determined that the rate of O_2 ingress should be limited to 0.05% per day to arrest insect population growth. Considering the dimensions of a standard silo bag (75 m long and 2.76 m diameter, holding 200 t of grain), the

O_2 transmission rate (OTR) of the plastic liner should be 162 mL/(m² d) to maintain the O_2 ingress rate at 0.05% per day (specific calculations are not shown here). This OTR value is consistent with reported permeance values for silo bag liners (10.34×10^{-4} m³ d⁻¹ m⁻² atm⁻¹ (see Section 4.2.1.3). Arias Barreto et al. (2016) integrated a population growth rate model for *R. dominica* and *Sitophilus oryzae* (L.) with a heat and mass transfer model for silo bags, simulating wheat storage conditions in different regions of Argentina. They concluded that insect population growth would be limited by temperature in the central and southern regions of Argentina but not in the subtropical regions in the north of the country. Additionally, they found that in a structurally intact silo bag, for an initial density of 1 insect/kg, the O_2 concentration ranged from 2% to 5% v/v, effectively arresting insect development since the O_2 ingress rate remained below 0.05% per day.

4.7.3 Control of Insects in Silo Bags

In situations requiring insect control, the utilization of gaseous treatments appears to be the optimal approach, capitalizing on the hermetic nature of the silo bag (Cardoso et al., 2012).

4.7.3.1 Fumigation with Phosphine

There is limited research documenting phosphine fumigation trials specifically conducted in silo bags. Cardoso et al. (2009) conducted a fumigation study where 3 g of aluminum phosphide (equivalent to 1 g of phosphine per t of wheat) was applied every 5 m along a silo bag (Figure 4.11). They determined that this dosage effectively maintained a concentration of 200 ppm for 5 d, ensuring a satisfactory distribution of the gas throughout the entire silo bag. Similarly, Ridley et al. (2011) employed a phosphine dosage of 1.5 g per m³ and achieved concentrations ranging from 1,350 to 1,500 ppm after 8 d. They confirmed a well-dispersed gas distribution and observed complete control of all life stages of a resistant strain of *R. dominica* at all locations within the silo bag. Additionally, they implemented a phosphine venting procedure to ensure proper dissipation of the gas (98% of phosphine gas was vented within 1.5 h). Carpaneto et al. (2016) examined the relationship among phosphine concentration, applied dosage, and airtightness of the silo bag. They found that less than half of the tested bags in the field had a suitable PDT of 90 s for fumigation, primarily due to inadequate sealing of the bag ends. However, they demonstrated that heat sealing the bag ends provided sufficient airtightness for successful fumigation, resulting in a sustained concentration of 200 ppm for 5 d using a dosage of 1 g of phosphine per m³. Finally, Bartosik et al. (2016) developed an automated

Figure 4.11 Inserting aluminum phosphide tablets in a silo bag to fumigate the grain.

PDT device to assess the airtightness of silo bags, synthesized the available information, establishing a correlation between PDT results and phosphine dosage, and developed a phosphine dosage calculator.

4.7.3.2 Controlled Atmosphere

Carpaneto et al. (2016) conducted a study on the application of CA treatments using CO_2 in silo bags with varying levels of hermeticity (sealed and unsealed). The injection of CO_2 was performed at the center of the bag by means of a cylinder connected to a gasifier (heat exchanger) (Figure 4.12), aiming to achieve a minimum concentration of 80% CO_2 at any location within the bag. Air purging was accomplished through perforations at both ends of the bag, resulting in purging efficiencies ranging from 51% to 59% in both trials. In the unsealed silo bag, the CO_2 concentration dropped below 35% at several locations after 3 d. However, in the sealed silo bag (with a PDT of 45 min), the CO_2 concentration remained above 70% even after 18 d, and at 24 d after injection, the CO_2 concentration remained above 60% at all measurement points. According to Annis (1986), to ensure complete control of all insect species [except *Trogoderma granarium* (Everts)], an initial CO_2 concentration of 70%, declining to 35% over a 15 d period, is necessary. Based on this criterion, CA treatment in sealed silo bags would provide effective insect control. One of the main

Figure 4.12 Carbon dioxide cylinders and heat exchanger ready for injecting gas in a CA treatment in a silo bag with organic grain.

challenges in implementing CA treatments in stored grains is the cost and labor involved in achieving a suitable airtight condition in traditional storage systems, such as bins and warehouses. Considering that silo bags are a cost-effective storage solution and that achieving an airtight condition for CA treatment in silo bags does not require significant additional expenses or labor, silo bags could facilitate the implementation of CA treatments for specialty grains.

4.8 Monitoring

4.8.1 Airtightness Assessment

Evaluating the physical integrity of a silo bag in field conditions poses considerable challenges, especially when it comes to identifying small perforations, particularly those situated at the top or bottom of the bag. A practical approach to assess the airtightness of hermetic systems involves implementing a PDT. Navarro and Zettler (2001) conducted a study to determine the correlation between different levels of perforated areas and gas leakage. They concluded that for flexible structures with a capacity of less than 500 t and filled with grain, a PDT of 1.5 min, 3 min, and 5 min was required for fumigation (phosphine), controlled atmosphere treatments, and modified atmosphere treatments, respectively. In order to facilitate a rapid assessment of silo bag airtightness, Bartosik et al. (2016) developed a portable device for automatic pressure decay testing. This device comprises electrical fans, connecting hoses, a pressure gauge, valves, and Bluetooth connectivity to a cellphone (Figure 4.13).

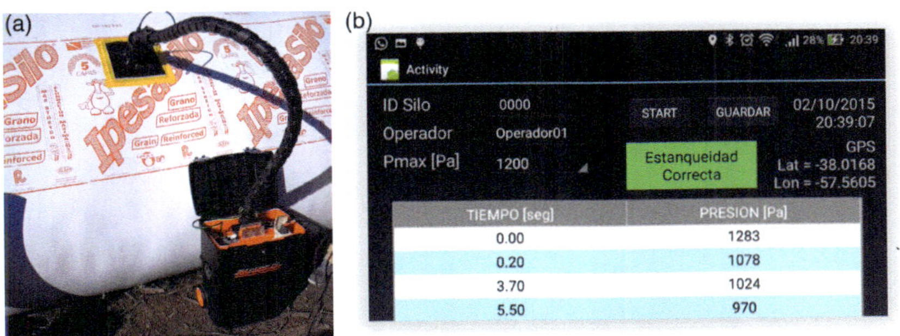

Figure 4.13 (a) Assessing airtightness in a silo bag with a pressure decay test (PDT) device. (b) Smartphone screen capture with the report of the test.

4.8.2 Biological Activity and Grain Quality Losses

Specialized monitoring strategies have been devised specifically for silo bag storage systems. The extent of modification in the internal gas composition of silo bags can be indicative of the activity of biotic components within the stored grain mass, allowing for the early detection of spoilage (Cardoso et al., 2008; Rodríguez et al., 2008). This concept has led to the development of a dedicated monitoring device (Silcheck, Junin, Argentina) that measures the internal CO_2 concentration at 6-m intervals along the bag using a portable meter [Figure 4.14(a)]. The collected data are then transmitted to a server and subjected to analysis utilizing a proprietary algorithm that evaluates the risk of storage issues (Bartosik et al., 2013). Alternative monitoring technologies involve the insertion of a probe equipped with a CO_2 sensor into various points within the bag (Smart Silobag, Wiagro, Argentina). Data from these probes are transmitted wirelessly to an online monitoring system [Figure 4.14(b)]. Unlike traditional methods, this approach eliminates the need for operators to be physically present at the silo bag. However, owing to cost constraints, a limited number of probes, about three, are typically deployed for each bag.

Subsequently, Taher et al. (2019) formulated a predictive model for estimating soybean losses (in terms of kilograms of spoiled, noncommercialized grain) in silo bag storage. The model relies on monitoring the CO_2 concentration and other relevant variables to provide accurate predictions. Carbon dioxide monitoring has the advantages of being of simple implementation (it is easy to train an operator to conduct the monitoring) and fast (it takes only a few minutes per bag) and providing a comprehensive traceability.

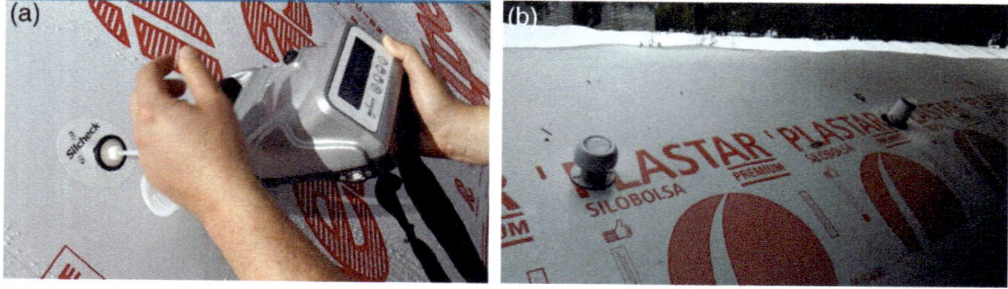

Figure 4.14 Monitoring silo bags by measuring the internal carbon dioxide concentration with a portable (a) and a stationary (b) sensing device.

References

Abadía B, Bartosik R (2013) Manual de buenas prácticas en poscosecha de granos. [Grain postharvest best management practices handbook], 1st ed. INTA Ediciones, Buenos Aires, Argentina.

Abadía B, San Martino S, Bartosik RE (2022) Can anoxic atmospheres protect the quality of maize seeds during storage? J Stored Prod Res 96, 1–8. doi: 10.1016/j.jspr.2021.101927

Abalone R, Gastón A, Bartosik R, Cardoso L, Rodríguez J (2011) Gas concentration in the interstitial atmosphere of a wheat silo-bag. Part II: Model sensitivity and effect of grain storage conditions. J Stored Prod Res 47, 276–283. doi: 10.1016/j.jspr.2011.05.003

Adler C, Athanassiou C, Carvalho MO, Emekci M, Gvozdenac S, Hamel D, Riudavets J, Stejskal V, Trdan S, Trematerra P (2022) Changes in the distribution and pest risk of stored product insects in Europe due to global warming: Need for pan-European pest monitoring and improved food-safety. J Stored Prod Res 97, 101977. doi: 10.1016/J.JSPR.2022.101977

Alvarez E, Cardoso L, Bartosik R, Depetris G, Castellari C, Montiel MD (2016) A preliminary comparative study of conventional and hermetic storage of wet distillers grains with solubles. pp. 324–328. In: Navarro S, Jayas DS, Alagusundaram K (eds), Proc 10th Intern Conf Controlled Atm Fumi Stored Prod (CAF2016), CAF Permanent Committee Secretariat, Winnipeg, Canada.

Al-Yahaya SA, Bern CJ, Misra MK, Bailey TB (1993) Carbon dioxide evolution of fungicide-treated high-moisture corn. Trans ASAE 36, 1417–1422.

Annis PC (1986) Towards rational controlled atmosphere dosage schedules: a review of current knowledge. pp. 128–148. In: Donahaye J, Navarro S (eds), Proc 4th Intern Conf in Stored-Product Protection. IWCSPP, Tel Aviv, Israel, September 1986.

Arias Barreto A, Abalone R, Gastón A (2013) Mathematical modelling of momentun, heat and mass transfer in grains stored in silos. Part I: Model development and validation. Lat Am Appl Res 43, 337–384.

Arias Barreto A, Abalone R, Gastón A (2016) Prediction of insect development in a wheat (Triticum aestivum) silo-bag by computer simulation. pp. 316–323. In: Navarro S, Jayas DS, Alagusundaram K (eds), Proc 10th Intern Conf Controlled Atm Fumi Stored Prod (CAF2016), CAF Permanent Committee Secretariat, Winnipeg, Canada.

Arias Barreto A, Abalone R, Ochandio D, Cardoso L, Bartosik R (2017) Validation of a heat, moisture and gas concentration transfer model for soybean (Glycine max) grains stored in plastic bags (silo bags). Biosyst Eng 8, 23–37. doi: 10.1016/j.biosystemseng.2017.03.009

ASAE (2001) ASAE D245.5 JAN01 - Moisture relationships of plant-based agricultural products.

Banks HJ, Annis PC (1990) Comparative advantages of high carbon dioxide and low oxygen types of controlled atmospheres for grain storage. In: Calderon M, Barkai Golan R (eds), Food Preservation by Modified Atmospheres. CRC Press, Boca Raton. pp. 93–122.

Banks HJ, Annis PC, Rigby GR (1990) Controlled atmosphere storage of grain: the known and the future. pp. 695–706. In: Fleurat-Lessard F, Ducom P (eds), Proc 5th Intern Working Conf Stored-Prod Prot, Bordeaux, France.

Bartosik R (2012) An inside look at the silo-bag system. pp. 117–128. In: Navarro S, Banks HJ, Jayas DS, Bell CG, Noyes RT, Ferizli AD, Emekci M, Isikber AA, Alagusundaram K (eds), Proc 9th Intern Conf Controlled Atm Fumi Stored Prod, Antalya, Turkey.

Bartosik R, Cardoso L, Albino J, Busato P (2013) CO_2 monitoring of grain stored in silobag through a web application. pp. 24–27. In: Proceedings of the EFITA, WCCA, CIGR 2013 Conference: Sustainable Agriculture through ICT Innovation. CIGR, Torino, Italy.

Bartosik R, Cardoso L, de la Torre D, Abadía B, Igartua M (2016) An approach for developing a phosphine dosage procedure and calculator for silo bags. pp. 371–376. In: Navarro S, Jayas DS, Alagusundaram K (eds), Proc 10th Intern Conf Controlled Atm Fumi Stored Prod (CAF2016), CAF Permanent Committee Secretariat, Winnipeg, Canada.

Bartosik R, Cardoso L, Rodríguez J (2008a) Early detection of spoiled grain stored in hermetic plastic bags (silo-bags) using CO_2 monitoring. pp. 550–554. In: Daolin G, Navarro S, Jian Y, Cheng T, Zuxun J, Yue L, Yanf L, Haipeng W (eds), Proc 8th Intern Conf Controlled Atm Fumi Stored Prod, Chengdu, China.

Bartosik R, Ochandio D, Cardoso L, de la Torre D (2012) Storage of malting barley with different moisture contents in hermetic silo bags. pp 549–554. In: Navarro S, Banks HJ, Jayas DS, Bell CG, Noyes RT, Ferizli AD, Emekci M, Isikber AA, Alagusundaram K (eds), Proc 9th Intern Conf Controlled Atm Fumi Stored Prod, Antalya, Turkey.

Bartosik R, Rodríguez J, Cardoso L (2008b) Storage of corn, wheat, soybean and sunflower in hermetic plastic bags. pp. 0–13. In: Proceeding of the 2008 International Grain Quality & Technology Congress. Chicago, Illinois, USA.

Bartosik R, Urcola H, Cardoso L, Maciel G, Busato P (2023) Silo-bag system for storage of grains, seeds and by-products: A review and research agenda. J Stored Prod Res 100, 1–11. doi: 10.1016/j.jspr.2022.102061

Behr E, Cardoso L, Bartosik R, Marcos Valle F, de la Torre D, Taher H, Maciel G (2021) Evaluation of storage of sunflower pellets in silo bags. pp. 123–130. In: Jayas DS, Jian F (eds), Proc 11th Intern Conf Controlled Atm Fumi Stored Prod, CAF Permanent Committee Secretariat, Winnipeg, Canada.

Bern CJ, Steele JL, Morey RV (2002) Shelled corn CO_2 evolution and storage time for 0.5% dry matter loss. Appl Eng Agric 18, 703–706.

Braga, C., 2023. Reporte Anual de la Cámara Argentina de Fabricantes de Maquinaria Agrícola (CAFMA) - Análisis Sector Externo [Anual Report of the Argentine Chamber of Agricultural Machinery

Manufacturers- Export Sector Analysis]. Buenos Aires, Argentina, URL HYPERLINK "https://linkprotect.cudasvc.com/url?a=https%3a%2f%2fcafma.org.ar%2fnovedades%2f200&c=E,1,eyWMr7nt LcApDhh1VmBC6Rn-cYYRyM0NPlPlDlWL_3zOlX-ikis1lIBW fmB3zfubuGqqmBeOCobRgrATRZ6XqD11A1XpBDc6XBIIE3CyrW U,&typo=1"https://cafma.org.ar/novedades/200

Butts CL, Ward JK (2016) Storing peanuts in grain bags. pp. 1–16. In: 2016 American Society of Agricultural and Biological Engineers Annual International Meeting, ASABE 2016. American Society of Agricultural and Biological Engineers, East Lansing, Michigan USA, Orlando, Florida, USA July 17–20, 2016. doi: 10.13031/aim.20162456481

Cardoso L, Bartosik R, Campabadal C, de la Torre D (2012) Air-tightness level in hermetic plastic bags (silo-bags) for different storage conditions. pp. 583–589. In: Navarro S, Banks HJ, Jayas DS, Bell CG, Noyes RT, Ferizli AD, Emekci M, Isikber AA, Alagusundaram K (eds), Proc 9th Intern Conf Controlled Atm Fumi Stored Prod, Antalya, Turkey.

Cardoso L, Bartosik R, de la Torre D, Abadía B, Santa Juliana M (2014) Almacenamiento de granos en silo bolsa. Resultados de investigación 2009–2013. [Storage of grains in silo-bag. Research report 2009–2013]. Ediciones INTA, Buenos Aires, Argentina.

Cardoso L, Bartosik R, Milanesio D (2009) Phosphine concentration change during fumigation in hermetic plastic bags (Silobags). In: CIGR (Ed.), Proceedings of the CIGR Section V International Symposium. Rosario, Argentina.

Cardoso L, Bartosik R, Rodríguez J, Ochandio D (2008) Factors affecting carbon dioxide concentration in interstitial air of soybean stored in hermetic plastic bags (silo-bag). pp. 565–568. In: Daolin G, Navarro S, Jian Y, Cheng T, Zuxun J, Yue L, Yanf L, Haipeng W (eds), Proc 8th Intern Conf Controlled Atm Fumi Stored Prod, Chengdu, China.

Cardoso L, Ochandio D, de la Torre D, Bartosik R, Rodriguez J (2010) Storage of quality malting barley in hermetic plastic bags. pp. 331–337. In: Carvalho MO, Fields PG, Adler CS, Arthur FH, Athanassiou CG, Campbell JF, Fleurat-Lessard F, Flinn PW, Hodges RJ, Isikber AA, Navarro S, Noyes RT, Riudavets J, Sinha KK, Thorpe GR, Timlick BH, Trematerra P, White NDG (eds), Proc 10th Intern Working Conf Stored-Prod Prot, Estoril, Portugal.

Carpaneto B, Bartosik R, Cardoso L, Manetti P (2016) Pest control treatments with phosphine and controlled atmospheres in silo bags with different airtightness conditions. J Stored Prod Res, 143–151. doi: 10.1016/j.jspr.2016.07.007

Casini C (2014) Technological evolution of grain storage in plastic bags. pp. 1–19. In: Gastón A, Abalone R, Bartosik R (eds), 1st International Silo Bag Conference. Instituto Nacional de Tecnologia Agropecuaria (INTA), Mar del Plata/Balcarce, Argentina. October 13–16.

Castellari C, Cendoya MG, Marcos Valle F, Barrera V, Pacin AM (2015) Factores extrínsecos e intrínsecos asociados a poblaciones fúngicas micotoxigénicas de granos de maíz (*Zea mays* L.) almacenados en silos bolsa en Argentina. [Extrinsic and intrinsic factors associated with mycotoxigenic fungal populations of corn stored in Argentina]. Rev Argent Microbiol 47, 350–359. doi: 10.1016/j.ram.2015.08.003

Castellari C, Marcos Valle F, Mutti J, Cardoso L, Bartosik R (2010) Toxigenic fungi in corn (maize) stored in hermetic plastic bags. pp. 501–504. In: Carvalho MO, Fields PG, Adler CS, Arthur FH, Athanassiou CG, Campbell JF, Fleurat-Lessard F, Flinn PW, Hodges RJ, Isikber AA, Navarro S, Noyes RT, Riudavets J, Sinha KK, Thorpe GR, Timlick BH, Trematerra P, White NDG (eds), Proc 10th Intern Working Conf Stored-Prod Prot, Estoril, Portugal.

Chelladurai V, Jian F, Jayas DS, White NDG (2016a) Permeability of silo bag material for carbon dioxide and oxygen. pp 377–381. In: Navarro S, Jayas DS, Alagusundaram K (eds), Proc 10th Intern Conf Controlled Atm Fumi Stored Prod (CAF2016), CAF Permanent Committee Secretariat, Winnipeg, Canada.

Chelladurai V, Jian F, Jayas DS, White NDG, Fields PG, Manickavasagan A (2016b) Feasibility of storing canola at different moisture contents in silo bags under Canadian Prairie conditions. Can Biosyst Eng 58, 309–320. https://library.csbe-scgab.ca/docs/journal/58/C16277.pdf

Chotikasatian C, Chayaprasert W, Pathaveerat S (2017) A study on the feasibility of quantifying the population density of stored product insects in air-tight grain storage using CO_2 concentration measurements. J Stored Prod Res 73, 21–29. doi: 10.1016/j.jspr.2017.05.005

Clemente G, Peppi B, Casini C, Pagliero M, Santa Juliana M (2009) Study of the effect of shade cloth cover on quality of soybean seeds (*Glycine max* (L.) Merr.) stored in plastic bags. pp. 1–4. II: CIGR Proceeding of Technology and Management to Increase the Efficiency in Sustainable Ag. Systems. CIGR, Rosario, Argentina September 2009.

Cofie-Agblor R, Muir WE, Sinha RN (1995) Comparative heat of respiration of five grain beetles in stored wheat. Postharvest Biol Technol 5, 167–175.

Darby J, Caddick LP (2007) Review of grain harvest bag technology under Australian conditions. CSIRO Entomology Technical Report N° 105. ISBN 0643 091130. Australia.

Diawara B, Cahagnier B, Richard-Molar D (1986) Oxygen consumption by a wet grain ecosystem in hermetic silos at various water activites. pp. 77–84. In: Donahaye J, Navarro S (eds), Proc 4th Intern Working Conf Stored-Prod Prot, Tel Aviv, Israel.

Faroni LRDA, de Alencar ER, Paes JL, da Costa AR, Roma RCC (2009) Armazenamento de soja em silos tipo bolsa. [Storage of soybean in silo-bags]. Eng Agric 29, 91–100. doi: 10.1590/S0100-69162009000100010

Fields PG (1992) The control of stored-product insects and mites with extreme temperatures. J Stored Prod Res 28, 89–118. doi: 10.1016/0022-474X(92)90018-L

Fleurat-Lessard F (2017) Integrated management of the risks of stored grain spoilage by seedborne fungi and contamination by storage mould mycotoxins – An update. J Stored Prod Res 71, 22–40. doi: 10.1016/j.jspr.2016.10.002

Gaikwad KK, Rubino M (2017) Water vapor permeability of packaging materials (LDPE) and fabricated package systems. Technical Report. School of Packaging, Michigan State University, East Lansing, Michigan, USA.

Garg A, Valavala P, Lokhande A, Ruiz E, Raimondi G (2019) Mechanical failure in agricultural silo bags, in: Annual Technical Conference - ANTEC, Conference ProcSociety Plastics Eng, March 2019, Detroit, USA.

Gastón A, Abalone R, Bartosik R, Rodríguez J (2008) Heat and mass transfer in soybean stored in hermetic plastic bags (silobags), in: CIGR (ed), Proc Int Conf Agri Eng, XXXVII Brazilian Congress Agri Eng, Foss do Iguazu, Brasil.

Gastón A, Abalone R, Bartosik R, Rodríguez J (2009) Mathematical modelling of heat and moisture transfer of wheat stored in plastic bags (silobags). Biosyst Eng 104, 72–85. doi: 10.1016/j.biosystemseng.2009.06.012

Giorni P, Battilani P, Pietri A, Magan N (2008) Effect of a_w and CO_2 level on *Aspergillus flavus* growth and aflatoxin production in high moisture maize post-harvest. Int J Food Microbiol 122, 109–113. doi: 10.1016/J.IJFOODMICRO.2007.11.051

Gregori R, Meriggi P, Pietri A, Formenti S, Baccarini G, Battilani P (2013) Dynamics of fungi and related mycotoxins during cereal storage in silo bags. Food Control 30, 280–287. doi: 10.1016/j.foodcont.2012.06.033

Hagstrum DW (2001) Immigration of insects into bins storing newly harvested wheat on 12 Kansas farms. J Stored Prod Res 37, 221–229. doi: 10.1016/S0022-474X(00)00023-0

Hagstrum DW, Phillips TW, Cuperus G (2012) Stored Product Protection, Kansas State University, Manhattan, Kansas, USA.

Howlader AJ, Matin ASMA (1988) Observations on the pre-harvest infestation of paddy by stored grain pests in Bangladesh. J Stored Prod Res 24, 229–231. doi: 10.1016/0022-474X(88)90024-0

Jian F, Chelladurai V, Jayas DS, Demianyk CJ, White NDG (2014) Interstitial concentrations of carbon dioxide and oxygen in stored canola, soybean, and wheat seeds under various conditions. J Stored Prod Res. 57, 63–72. doi: 10.1016/j.jspr.2013.12.002

Jian F, Chelladurai V, Jayas DS, White NDG (2015) Three-dimensional transient heat, mass, and momentum transfer model to predict conditions of canola stored inside silo bags under Canadian Prairie conditions: Part II. Model of canola bulk temperature and moisture content. Trans ASABE 58, 1135–1144. doi: 10.13031/trans.58.11053

Magan N, Lacey J (1984) Effects of gas composition and water activity on growth of field and storage fungi and their interactions. Trans Br Mycol Soc 82, 305–314. doi: 10.1016/s0007-1536(84)80074-1

Maier D, Bartosik R (2020) Grain aeration storage app released [WWW Document]. Iowa Grain Qual Initiat URL https://www.extension.iastate.edu/grain/grain-aeration-storage-app-released (accessed 3.22.23).

Marcos Valle F, Castellari C, Yommi A, Pereyra MA, Bartosik R (2021a) Evolution of grain microbiota during hermetic storage of corn (*Zea mays* L.). J Stored Prod Res 92, 101788. doi: 10.1016/j.jspr.2021.101788

Marcos Valle F, Gastón A, Abalone R, de la Torre D, Castellari C, Bartosik R (2021b) Study and modelling the respiration of corn seeds (*Zea mays* L.) during hermetic storage. Biosyst Eng 8, 45–57. doi: 10.1016/j.biosystemseng.2021.05.009

Marín S, Companys E, Sanchis V, Ramos AJ, Magan N (1998) Effect of water activity and temperature on competing abilities of common maize fungi. Mycol Res 102, 959–964. doi: 10.1017/S0953756297005613

Massigoge J, Cardoso L, Bartosik R, Rodríguez J, Ochandio D (2010) Almacenamiento de cebada cervecera en silos bolsa. [Storage of maltin barley in silo-bags], in: IX Cong Latinoam Caribe de Inge Agr - CLIA 2010 y XXXIX Cong Brasileiro Eng Agrícola - CONBEA. Asociación Brasileira de Engenharia Agrícola, Vitoria, Brasil, July 25–29 2010.

Navarro S, Donahaye JE, Fishman S (1994) The future of hermetic storage of dry grains in tropical and subtropical climates. pp. 130–138. In: Highley E, Wright EJ, Banks HJ, Champ BR (eds), Pro 6th Intern Working Conf Stored-Prod Prot, Canberra, Australia.

Navarro S, Timlick B, Demianyk CJ, White NDG (2012) Controlled or modified atmospheres, In: Hagstrum D, Phillips T, Cuperus G (eds), Stored Product Protection S156, Kansas State University, Manhattan, Kansas, pp. 191–202.

Navarro S, Zettler J (2001) Critical limits of degree for sealing for successful application of controlled atmosphere or fumigation. pp. 507–520. In: Donahaye EJ, Navarro S, Leesch JG (eds), Proc 6th Intern Conf Controlled Atm Fumi Stored Prod, Fresno, USA, 2000.

Ochandio D, Bartosik R, Gastón A, Abalone R, Arias Barreto A, Yommi A (2017) Modelling respiration rate of soybean seeds (*Glycine max* (L.) in hermetic storage. J Stored Prod Res 74, 36–45. doi: 10.1016/j.jspr.2017.09.001

Ochandio D, Cardoso L, Bartosik R, de la Torre D, Rodríguez J, Massigoge J (2010) Storage of canola in hermetic plastic bags. pp. 323–330. In: Carvalho MO, Fields PG, Adler CS, Arthur FH, Athanassiou CG, Campbell JF, Fleurat-Lessard F, Flinn PW, Hodges RJ, Isikber AA, Navarro S, Noyes RT, Riudavets J, Sinha KK, Thorpe GR, Timlick BH, Trematerra P, White NDG (eds), Proc 10th Intern Working Conf Stored-Prod Prot, Estoril, Portugal.

Ochandio D, Massigoge J, Castellari C (2014) Almacenamiento hermético in vitro de granos de avena, canola, girasol, maíz, soja y sorgo. [In-vitro hermetic storage of oat, canola, sunflower, corn soybean and sorghum]. In: Gastón A, Abalone R, Bartosik R (eds), Memorias Primer Cong Int Alm Granos Silo Bolsa. INTA Ediciones, Mar del Plata/Balcarce, Argentina. pp. 159–167.

Paster, N, Bullerman, LB., (1988). Mould spoilage and mycotoxin formation in grains as controlled by physical means. Int. J. Food Microbiol. 7, 257–265. https://doi.org/10.1016/0168-1605(88)90044-X

Pacin AM, Bovier EC, González HL, Whitechurch EM, Martínez EJ, Resnik SL (2009) Fungal and fumonisins contamination in argentine maize (*Zea mays* L.) silo bags. J Agric Food Chem 57, 2778–2781. doi: 10.1021/jf803609c

Ridley AW, Burrill PR, Cook CC, Daglish G (2011) Phosphine fumigation of silo bags. J Stored Prod Res 47, 349–356. doi: 10.1016/j.jspr.2011.06.001

Rodríguez J, Bartosik R, Cardoso L, Crocce D (2008) Factors affecting carbon dioxide concentration in interstitial air of wheat stored in hermetic plastic bags (silo-bag). pp. 589–592. In: Daolin G, Navarro S, Jian Y, Cheng T, Zuxun J, Yue L, Yanf L, Haipeng W (eds), Proc 8th Intern Conf Controlled Atm Fumi Stored Prod, Chengdu, China.

Santa Juliana M, Cardoso L (2014) Componentes del sistema de silo bolsa. [Silo-bag system components], in: Cardoso L, Bartosik R, de la Torre D, Abadía B, Santa Juliana M (eds.) 22. Instituto Nacional de Tecnologia Agropecuaria (INTA), Buenos Aires, Argentina Granos En Silo Bolsa: Resultados de Investigación 2009-2013, pp. 15–22.

Scarabino A (2019) Geometric shape and tensile forces on silo bags for grain storage. J Agric Eng 50, 143–149. doi: 10.4081/jae.2019.973

Sone J (1999) Carbon dioxide production in stored maize as affected by moisture content, level of broken corn and foreign materials and infestation by *Sitophilus zeamais* Motschulsky. J Asia Pac Entomol 2, 133–141. https://doi.org/10.1016/S1226-8615(08)60041-2

Steele JL, Saul RA, Hukill WV (1969) Deterioration of shelled corn as measured by carbon dioxide production Trans ASAE 12, 685–689.

Subramanyam B, Lakshmikantha C, Campabadal C, Lawrence J, Cardoso L, Maier DE (2012) Evaluation of silo bags for temporary storage of wheat. pp. 535–541. In: Navarro S, Banks HJ, Jayas DS, Bell CG, Noyes RT, Ferizli AD, Emekci M, Isikber AA, Alagusundaram K (eds), Proc 9th Intern Conf Controlled Atm Fumi Stored Prod, Antalya, Turkey.

Taher H, Urcola H, Cardoso L, Bartosik R (2014) Percepción de los productores sobre beneficios y problemáticas en el almacenamiento en silo bolsa. [Farmers perception about pros and cons of the silo-bag storage system]. In: Gastón A, Abalone R, Bartosik R. (eds), Memorias Primer Cong Int Almacen Granos Silo Bolsa. INTA, Mar del Plata, Argentina. pp. 35–42.

Taher H, Urcola HA, Cendoya MG, Bartosik R (2019) Predicting soybean losses using carbon dioxide monitoring during storage in silo bags. J Stored Prod Res 82, 1–8. doi: 10.1016/j.jspr.2019.03.002

Thompson TL (1972) Temporary storage of high-moisture shelled corn using continuous aeration. Trans ASAE 15, 333–337.

Trematerra P (2015) Adult dispersal of *Sitotroga cerealella* in a conventional small-farm in Southern Italy. Bull Insectology 68, 111–118.

Ward J, Davis J (2013) A system to assess grain bag storage internal environment. Trans ASABE 56, 1503–1509.

Weinberg ZG, Yan Y, Chen Y, Finkelman S, Ashbell G, Navarro S (2008) The effect of moisture level on high-moisture maize (*Zea mays* L.) under hermetic storage conditions-in vitro studies. J. Stored Prod Res 44, 136–144. doi: 10.1016/j.jspr.2007.08.006

Zufiaurre E, Abba A, Bilenca D (2018) Assessment of stakeholder perceptions of the damage to silo bags by vertebrate species in Argentina. Hum Dimens Wildl 24, 1–7. doi: 10.1080/10871209.2019.1538434

Zufiaurre E, Abba A, Bilenca D (2020) Damage to silo bags by mammals in agroecosystems: A contribution for mitigating human-wildlife conflicts. Wildl Res Early View. doi: 10.1071/WR20045

Chapter 5

High-Nitrogen and High-Carbon Dioxide Systems

Shlomo Navarro, Hagit Navarro, and Nadav Inbari

5.1 Introduction

An alternative description of high-nitrogen and high–carbon dioxide systems is modified or controlled atmosphere (MA/CA) storage. The objective of MA treatment is to attain a composition of atmospheric gases rich in CO_2 and low in O_2 or a combination of these two gases at normal or altered atmospheric pressure within the treatment enclosure, for the exposure time necessary to control the storage pests. The CA method is intended to compensate for possible small leakages of gases, which cause an increase of the O_2 or decrease of the CO_2 content in the enclosure and which are almost impossible to avoid. Thus, the term CA, although commonly used to describe the entire subject, actually has its own specific meaning. Such a treatment system is commonly applied for dry and fresh agricultural products. For fresh products the objective is to maintain quality and freshness and avoid microflora damage, whereas for dry products the main objective is control of storage pests. Quality preservation is also achieved under MA/CA storage systems.

DOI: 10.1201/9781003309888-5

Since ancient times, a kind of MA for the preservation of dry agricultural commodities was practiced in the form of hermetic storage in underground pits in dry and subtropical regions of the Middle East, North Africa, India, and possibly other dry-climate regions of the globe. Today, such storage of small lots of grain is still in use in some less developed agricultural communities in dry-climate areas of Africa. However, the application of MA/CA technology on a large commercial scale could receive sufficient attention only after the vast development work launched in the late 1980s, particularly in Australia and Italy. In spite of all the research and development invested in MA/CA technology, its full-scale application is still in its infancy, as it was almost 40 years ago. MA/CA technology could not fully replace the alternative technologies available for protecting agricultural commodities.

The two available technologies most competitive with MA/CA technologies are aeration and refrigeration for cooling grain for insect management and fumigation to control insects. Cooling grain and fumigation technologies complete each other by providing low-temperature grain storage in the globe's cool regions and full control of insects by fumigation in hot climates. Both technologies come at a competitive cost compared to MA/CA storage technologies. The cost of cooling by applying aeration is about 2 (kW/h)/t of grain, and the cost of fumigation is about USD0.2 (only the cost of the fumigant).

The advantage of applying MA/CA technology is that it is an environmentally sound and sustainable technology, free of chemicals and free of the danger of polluting the environment. Furthermore, for most products considered organic, application of MA/CA technology is preferred over cold storage because of the cost and the disadvantages of the low-temperature effects on the products. For large grain masses, cooling is effective for delaying the development of insects but not for controlling them. To achieve full control of insects, freezing temperatures should be attained, rendering the technology economically infeasible except in most northern latitudes for three or four months when ambient temperatures are below $-20°C$.

In addition to being used for controlling insects, MA/CA technology has been adapted for several niche applications. Among those are storage of all kinds of dry organic products such as dried fruits, nuts, and oilseeds; control of bed bugs; control of wood boring beetles in dwellings and museums; and control of narcissus flies.

However, the costs of applying MA/CA, including those related to rendering the structure gastight, the use of a pressure relief valve and respiration bag, and the acquisition and handling of the gas used (CO_2 or N_2), should be compared with those associated with the application of the fumigant, which has more modest requirements and is more tolerant with respect to the gastightness of the structure. For the application of MA/CA, an extremely gastight structure is needed, and precautions should be taken to prevent structural damage due to sudden changes in pressure.

In this chapter, the application technologies for fumigation of dry cereal grain stored in bulk in silos and ships is presented, along with some other applications to control such insect pests as museum pests and bedbugs. Other applications, such as that of high-pressure carbon dioxide or low-pressure (vacuum) systems, are not included in this chapter.

Along with all the advantages of applying MA/CA comes a disadvantage. The slow reaction of insects to the treatment, although comparable to the slow reaction of insects to phosphine gas, is a challenge of the treatment. This disadvantage is offset, however, by MA/CA's excellent reputation among those whose top priorities are the environment and not having toxic chemicals in food. Therefore, in this chapter, the latest available information related to the application of high-nitrogen and high–carbon dioxide systems is presented.

5.2 Technical Aspects Before Applying High Nitrogen and High Carbon Dioxide

5.2.1 Achieving Gastightness

The first challenge to the use of MA/CA is to have a sufficiently gastight structure to hold the atmospheric composition for sufficient time to achieve the mortality of insects. In practice, warehouses and silos are not designed as gastight structures. Therefore, unless the structure was purposely designed to be gastight, it needs to be sealed before MA/CA treatment. CAs have been used in a wide array of grain storage structures. The most important consideration is that they must be airtight for long-term storage or relatively airtight for CO_2 or N_2 fumigation.

A bolted, galvanized iron silo (21.5 t) was sealed using a polyvinyl resin formulation sprayed onto joints from the inside. The silo was loaded with wheat into which cages of insect-infested wheat were introduced. Conditions were monitored with thermocouples and gas sampling lines. Oxygen levels were reduced to <1% by purging with N_2, and similar levels were then maintained by a slow N_2 bleed for 35 d, after which the silo was emptied. All adult insects were dead, but, as expected, some immatures survived. This was because the exposure period was too short for ensuring a complete kill at the observed grain temperatures of <15°C (Williams et al., 1980).

Airtight, galvanized-steel silos have been manufactured in Australia for the past 30 years and are commercially available (Moylan Silos, 2023).

Welded-steel hopper silos can be modified for CO_2 fumigation for a few hundred dollars. Carbon dioxide from dry ice must be recirculated through the grain, and a pressure relief valve must be installed at the top of the silo. The top and bottom hatches must be gasket-sealed. After 10 d at 20°C, 75% of the applied CO_2 was retained, while 99% of the caged *Cryptolestes ferrugineus* (Stephens) were killed (Mann et al., 1999).

Carbon dioxide fumigation of grain has been successful in concrete elevators holding 209 t of wheat. The bottom hopper was sealed, and the grain was purged with CO_2 for 4 h (1 t of CO_2), with additional gas added as needed. All caged test insects were killed (White and Jayas, 2003). A large installation for the application of CO_2-based CA was installed to treat more than 200,000 t of rice annually in flat storage structures, each of 5,000 t capacity, in Mianyang, China (Figure 5.1).

To control stored-grain insects, nitrogen was applied in three concrete silos in Cyprus, each containing 2,400 t of grain (Navarro et al., 2012). Structural sealing was carried out, and airtight valves were installed to improve the airtightness of the silos.

Figure 5.1 Application of CO_2-based CA in a 5,000-concrete horizontal silo bin in Mianyang, Sichuan, China.

5.2.2 Testing Gastightness

Methods to determine gastightness have been investigated for different purposes. For the analysis of the energy requirements of buildings, air infiltration, which is a primary source of energy loss, can be measured experimentally. Carbon dioxide has also been used experimentally as a tracer gas (Navarro, 1997a; Navarro and Zettler, 2001). Hunt (1980) reviewed some tracer gas techniques to measure air infiltration into buildings and compared fan-pressurization-evacuation procedures to estimate the comparative tightness of those structures. The dynamic characteristics of air infiltration into buildings have been studied to predict the heating and cooling load of seasonal energy requirements (Hill and Kusuda, 1975). The *ASHRAE Handbook of Fundamentals* describes the air change method and the crack method for predicting air infiltration rates (Anonymous, 1972). The crack method or the constant pressure test is usually regarded as more accurate as long as the leak characteristics can be evaluated properly (Hill and Kusuda, 1975). With this method, airflow is expressed as

$$Q = b \times \Delta P^n$$

where Q = the volumetric flow rate of air, in m^3/s; b = the proportionate constant; n = the exponent, and ΔP = the pressure difference exerted on an enclosure, in Pa.

Meiering (1982) investigated a constant pressure test and a variable pressure test for measuring specific silo permeability in silage systems, where the sealed shell is designed to limit the entry of O_2 so as to minimize losses in quality. The effect of variations in environmental temperature and atmospheric pressure on the O_2 intake of silos was simulated, and permeability limits required for proper O_2 control in silos containing silage were defined. Gas interchange within freight containers and factors leading to gas interchange between containers and the external atmosphere were detailed by Banks et al. (1975). They found that the relationship between applied pressure and gas leak rate gave a useful measure of gastightness. Banks and Annis (1977) developed a practical guide for the storage of dry grain under MA and specified requirements for silo gastightness. Their specifications correspond to the pressure decay times needed to maintain the atmospheric composition in the silos. Sharp (1982), using the constant pressure test, measured the gastight level of sealed structures. These tests are designed to estimate the permissible limits for effectively maintaining the gas composition in the stores during the treatment.

A decision about the degree of gastightness that is satisfactory or what gas concentrations can be maintained under given environmental and structural conditions should be made before the gaseous treatment. This decision should be weighed against the investment involved in sealing a leaky structure to prevent excessive loss of the treatment gas. Navarro (1998) described the use of the variable and constant pressure tests and summarized provisional recommendations for pressure tests for the successful application of gaseous treatments to control storage insects.

5.2.2.1 Rigid Structures

Rigid structures (categorized below as small, medium, and large) can withstand the positive pressures exerted on them during the test without practical changes in volume. Rigid structures may be constructed of concrete, metal, or a combination of the two. For MA/CA treatments, the structures may be equipped with a pressure relief valve in order to avoid structural failure under extreme and sudden pressure variations. In conducting the pressure tests, care should be taken to carefully monitor the pressure applied, especially within the rigid structures, so as not to exceed the pressure limits the structure can withstand. A small blower used to pressurize the structure can eventually produce enough pressure to cause structural damage. This is particularly important in large stores. It is always advisable to seek the advice of a civil engineer regarding the structural soundness of a storage before conducting pressure tests. Most structures of medium-size warehouses and silos can withstand a pressure of 250 Pa, which may be taken as an upper limit.

1. Small-size structures

    These structures are conveniently used for the treatment of stacked commodities. They are often referred to as fumigation chambers. Their volumes can range from a few cubic meters to 500 m³.
2. Medium size warehouses and silos

    Structures in this category may be vertical or horizontal (longer than they are tall) and the volumes may range between 500 and 2,000 m³. The commodities in store may be stacked or in bulk.
3. Large warehouses and silos

    The warehouse and silo sizes in this category are commonly used as central stores and may range from 2,000 to 10,000 m³ in volume. Single bins of larger sizes are generally less common.

5.2.2.2 Flexible Structures

Flexible structures include plastic-sheeted structures, bagged stacks sealed in plastic enclosures, and bunker-type storage. Positive pressures exerted on them cause uncontrolled expansion in volume. Therefore, pressure tests in such structures should be performed by creating a negative pressure through evacuation until the sheeting adheres to the commodity or the packed material.

Flexible structures can also be categorized as small, medium, and large, as follows:

1. Small-size structures

    These are mostly destined for indoor gaseous treatments of stacked commodities. The dimensions of the structure are dictated by the manageability of the stack. Sizes of such structures for indoor gaseous treatments can range up to 500 m^3 in volume.

    A recent development is to store stacked commodities in fumigatable small flexible structures outdoors using heavy-duty liners. They are used for MA storage of stacks of 10–50 t capacity termed storage cubes or GrainPro Cocoons™, and designed for storage at the farmer-cooperative and small trader level (Donahaye et al., 1991).

2. Medium-sized flexible structure.

    In this category, structures are made from heavy-duty material. The flexible silo linings are contained within a circular wall consisting of metal-weld mesh. Such structures are fumigatable and suitable for the hermetic storage or MA storage of grain in bulk or in bags and have capacities of up to 1,500 m^3 (Calderon et al., 1989; Navarro et al., 1990, 1994; Navarro and Donahaye, 1993).

3. Large-size structures

    These are mostly bunker-type stores with capacities of 1,500 m^3 or more. These bunkers are usually used for bulk storage of commodities. Ramps of concrete, metal, or earth border the bunkers on three sides in order to raise the walls and increase capacity. Before loading the bunkers, the floor and the ramps are lined with plastic sheeting and then an over-liner is used as a cover. After loading, the over-liner is welded closed in order to form a continuous enclosure, and the two liners are overlapped and folded to form a hermetic seal. Bunker storage for hermetic storage of grain in large bulks of 10,000 to 15,000 t capacity have been in use in Cyprus and Israel (Navarro et al., 1984; 1993; Navarro and Donahaye, 1993).

5.2.3 Comparative Results with Variable Pressure Test

In a variable pressure test, the structure is pressurized to a value above atmospheric, using a fan. The air supply is then shut off and the pressure is allowed to fall by natural leakage to a new value. The time taken to fall from the high positive pressure serves as a measure of the degree of sealing.

Time elapsed to reach half the pressure is usually considered for comparisons of gastightness level. In this chapter, the time to half the pressure is also referred to as the "half-life" pressure decay time,. A similar test can also be conducted by lowering the pressure in the enclosure by vacuum and then allowing the pressure to rise. A constant static pressure test (various airflow rates are applied and correlated to the obtained static pressures within the tested container) is more accurate than a variable pressure decay test and has the advantage of being independent of bin volume. However, the variable pressure decay test is quicker and therefore the most practical means of measuring gastightness in grain storage (Zahradnik, 1968; Banks, 1984b; Anonymous, 1989; Navarro, 1997a).

To minimize the thermal influence, tests should be carried out preferably before sunrise and in still weather. A pressure of 250 Pa may be taken as an upper limit, but for some structures, even this pressure may cause poor seals to open. Welded-steel cells and concrete silos may be able to stand 500 Pa, but higher pressures are usually detrimental to the structure and unnecessary.

Comparative tests with variable pressure tests are scarce. Table 5.1 was prepared to provide provisional guidelines based on the best estimates available in the literature. The suggested times given in Table 5.1 were doubled for empty storages as an approximation to the intergranular airspace,

Table 5.1 Provisional Recommended Ranges for Variable Pressure Tests Carried Out in Structures Destined for Gaseous Treatments to Control Storage Insects

Type of Gaseous Treatment	Structure Volume (m^3)	Variable Pressure Test Decay Time (min) 250–125 Pa	
		Empty Structure	95% Full
Fumigants	Up to 500	3	1.5
	500 to 2,000	4	2
	2,000 to 15,000	6	3
CA	Up to 500	6	3
	500 to 2,000	7	4
	2,000 to 15,000	11	6
MA, including airtight storage	Up to 500	10	5
	500 to 2,000	12	6
	2,000 to 15,000	18	9

Source: From Navarro (1998).

since for barley, corn, rough rice, and wheat this free space is in the range of 35% to 65% of the total volume (Trisviaskii, 1966).

Banks and Ripp (1984) tested sealed flat storages of from 4,500 to 27,000 t capacity and compared their fumigant effectiveness using phosphine with pressure decay times from 150 to 75 Pa. Their tests resulted in successful control of insects when half-life pressure decay time was 3 min for storage capacities of 15,600 and 16,500 t, whereas full insect control could not be achieved when half-life pressure decay time was less than 1 min for capacities of 4,800 t. In Table 5.1, a minimum of 3 min for large-size structures and a minimum of 1.5 min for the small range should serve as the basis for half-life pressure decay times for full storages when using fumigants.

For the special case of hay fumigation by applying methyl bromide (MB) in containers 12.2 m (40 ft) long, pressures from 200 to 100 Pa for a decay time of >10 s were reported by Ball and van S Graver (1997). Although the minimum 200 g-h/m^3 Ct product was attained using this standard decay time of >10 s, gas loss was significant. For exposure times of 12–15 h, MB concentration dropped from 58 g/m^3 to the range of 12–18 g/m^3, and after 22–27 h exposure to the range of 10–18.5 g/m^3, pressure halving times ranged from 11 to 85 s.

With sorptive fumigants like methyl bromide (MB), a pressure decay test for the assessment of gas retention is not a precise concept since there are several forces that contribute to gas loss through the leaks in the structure. Sorption by the treated material is a major cause of the drop in gas concentration during the exposure time. Wheat flour treated using MB in experimental jars resulted in sorption of 70% of the initial gas concentration after 24 h exposure at 25°C (Navarro, 1977). Since the rate of gas loss is independent of leakage from the enclosure, in well-sealed structures the overall gas loss rate will approach that expected from sorption only.

For controlled atmosphere storage in Australia, with structures of 300 to 10,000 t capacity, a decay time of 5 min for an excess pressure drop of 2,500 to 1,500 Pa or 1,500 to 750 Pa or 500 to 250 Pa was regarded as satisfactory (Banks et al., 1980). According to Banks and Annis (1980), this range of pressures was chosen because it is the highest usable without unduly stressing the storage fabric of the store. They commented also that above 10,000-tonne capacity, pressure testing is difficult to carry out satisfactorily since it requires very stable atmospheric conditions. From analysis of the data presented by Banks et al. (1980) it would appear that for storages with capacities in the range of 1,600–1,900-tonne in CA with an initial CO_2 concentration of about 60–85% and an average decay time of 11 min, the daily decay rate was about 4% CO_2. With a similar range of initial CO_2 concentration in a structure of 150 m^3 capacity, daily gas loss was correlated to different levels of pressure decay times (Navarro et al., 1997a). Their comparison resulted in a pressure decay time of 3 min for a daily decay rate of about 4% CO_2.

The influence of hermetic storage on controlling insects was examined using small-scale sealed plastic structures of 15, 30, and 52 m^3 capacity for outdoor storage of wheat, paddy, and corn (Navarro et al., 1995). Pressure decay rates were compared with daily CO_2 decay rates. With these structures, successful insect controls were obtained with <1% CO_2 daily decay rate, which was found to be equivalent to a 5 min half-life pressure decay time. Similarly, comparative data were obtained using hermetic bunker storages of about 19,000 m^3 capacity, where successful results were obtained when the half-life pressure decay time was about 9 min (Navarro et al., 1984).

According to Banks and Annis (1984), daily ventilation rates tolerable in various insect control processes are estimated as 2.6% for hermetic storage, 5% for N_2-based CA, 7% for CO_2-based CA, and 10% for phosphine fumigation. Based on the proportions of ventilation rates, this would account for ventilation rates for fumigation using phosphine being twice those of N_2-based CA, and the latter being as much as twice those of hermetic storage. These values were also considered in extrapolating the different ranges given in Table 5.1.

In Table 5.2, infiltration of O_2 and the loss rate of CO_2, PH_3, and MB from exposed and shaded empty structure of 7.5 m^3 capacity in relation to

Table 5.2 Infiltration of O_2 and Loss Rate of CO_2, PH_3, and MB from Exposed (EX) and Shaded (SH) Empty Structure of 7.5 m^3 Capacity in Relation to Half-Life Pressure Decay Time (min)

Time for Half-Life Pressure Decay (min)	Infiltration Rate (% O_2/d)		Loss Rate (% CO_2/d)		PH_3 Loss (ppm/d)		MB Loss (g/m^3-d)	
	EX	SH	EX	SH	EX	SH	EX	SH
1.0	1.60	1.15	6.65	5.00	220	145	1.95	2.05
2.0	1.35	1.05	5.45	4.20	190	120	1.70	1.85
3.0	1.15	0.95	4.40	3.45	175	100	1.45	1.55
4.0	0.95	0.70	3.30	2.70	155	75	1.15	1.25
5.0	0.80	0.50	2.20	1.95	130	52	0.93	1.10

Source: Navarro and Zettler (2001).

Note: The suggested maximum infiltration rate for modified atmospheres is 0.5% O_2/d increase. The suggested maximum loss rate for carbon dioxide is 2%/d. PH3 loss should not exceed 100 ppm/d. MB loss of 1.0 g/m^3-d is considered acceptable. For MB, the shaded structure was tested in October and the exposed structure was tested in November, when ambient temperatures were lower than in October.

half-life pressure decay time (min) was demonstrated (Navarro and Zettler, 2001). Time in minutes for the half-life pressure decay was correlated with daily ventilation rates of O_2, CO_2, PH_3, and MB. The ventilation rates were tested using different sizes of cross-section leak areas with orifices of 1.6 mm, 3.2 mm, and 6.4 mm ID. To evaluate the influence of temperature on the rate of gas exchange, an empty structure of 7.5 m^3 was tested when it was under cover (by providing shading to minimize the direct solar heating effect) and when it was exposed to direct solar heating. The oxygen infiltration rate for modified atmospheres was 0.5% O_2/d at a 5 min half-life pressure decay time when the chamber was under shade. For the same level of gastightness, gas loss was 0.8% O_2/d when the cube was exposed. The CO_2 loss rate was 2%/d at a 5 min half-life pressure decay time when the structure was under shade and 3%/d when the cube was exposed. The loss rate of PH_3 from the cube for the shaded conditions was 100 ppm/d at a 3 min half-life pressure decay time. A sealing equivalent to a 4.5 min half-life pressure decay time was required when the cube was exposed. The MB loss rate of 1.0 $g/m^3 \cdot d$ was approximately equivalent to a 5 min half-life pressure decay time. At the time tests were carried out with MB, ambient temperature differences were not large enough to show different loss rates between the shaded and the exposed structure.

5.2.4 Use of Pressure Relief Valve

In structures with good gastightness, there is a danger of structural damage due to sudden changes in pressure with either barometric pressure or temperature variations, particularly when the structure is exposed to solar irradiation and made of metal. Even at the initial stages of fumigation or purging, atmospheric gases like N_2 or CO_2, are under the influence of temperature changes between day and night, and pressure changes must be compensated for to avoid structural damage.

Pressure relief valves play a crucial role in ensuring the safe operation of grain storage silos. These valves are designed to release excess pressure that can build up within the silo due to various factors, such as temperature changes, grain respiration, and fermentation processes. Given below is some information on the use of pressure relief valves in grain-stored silos.

Pressure Build-Up: Grain stored in silos can undergo physical processes mainly due to day and night temperature fluctuations, changes in barometric pressure, sorption of CO_2 by the commodity, and at times biological processes that generate gases, heat, and moisture. These factors contribute to pressure buildup inside the silo. The pressure can increase significantly, and if not released, it can lead to structural damage or even explosion.

Pressure relief valve operation: Pressure relief valves typically remain closed under normal operating conditions. In spring-loaded devices, when the pressure inside the silo exceeds the set point, the force exerted on the valve's spring overcomes its resistance, causing the valve to open and release the excess pressure. There are other types of pressure relief valves; another common type is made of a U-tube filled with transmission oil or liquid glycerin (Figure 5.2). Yet another common type is based on the weight of the plate that blocks the valve (Figure 5.3).

It is important to consult with experts in silo design, construction, and safety to determine the specific requirements for pressure relief valves in grain storage silos. They can provide guidance on the appropriate valve sizing, installation location, and compliance with relevant safety codes and regulations.

Figure 5.2 Pressure relief valve made of a U-tube filled with transmission oil or liquid glycerin. (Courtesy of Chris Newman, Australia.)

Figure 5.3 Mechanical pressure relief valve based on the weight of the plate that blocks the valve.

5.2.5 Use of a Breather Bag

Breather bags may be installed to avoid gas loss due to temperature and barometric changes. In a very good gastight structure, with any change in barometric pressure or in the temperature of the silo or warehouse, gas loss is inevitable.

Breather bags, also known as gas exchange bags or gas permeable membranes, are commonly used in controlled atmosphere chambers or rooms for the storage of fruits, vegetables, and other perishable products. Given below is some information on the use of breather bags in controlled atmosphere chambers.

Gas exchange: CA storage involves modifying the composition of gases (oxygen, carbon dioxide, and nitrogen) surrounding the stored products to slow down the aging and ripening processes during the storage of perishables. After the desired gas composition is achieved, the target is to maintain that composition for the desired time period. However, this desired period of time for CA treatment may be several days or, depending on the commodity temperature, even weeks. Day and night temperature differences, sorption of CO_2, and refrigeration cause expansion and contraction of the CA composition of the treated enclosure. To prevent structural damage due to such changes in gas volume, the pressure relief valve is activated. During contraction, a portion of the external atmosphere infiltrates the CA-treated enclosure, causing dilution of the CA composition, and during expansion, a portion of the CA composition is exhausted. Breather bags facilitate gas exchange between the stored produce and the external environment, allowing for the regulation and maintenance of desired gas concentrations.

Purpose of breather bags: The primary purpose of breather bags is to avoid dilution of the CA composition due to gas exchange with the ambient atmosphere. If a CO_2-based CA is targeted, the goal is to maintain the CO_2 concentration, and if an N_2-based CA is targeted, the goal is to maintain the low O_2 concentration. The gas exchange is enabled through the breather bag. In the expansion stage, the exhausted CA composition is kept in the breather bag, and in the contraction stage, the CA composition is sucked into the treated CA enclosure. The breather bag allows the controlled movement of gases while preventing the ingress of the ambient atmosphere. It enables the exchange of gases, specifically O_2 and CO_2, which are critical for maintaining the desired atmospheric composition within the treated storage chamber.

Installation and placement: It is not common to install breather bags in silos. They are more commonly installed in CA chambers for fresh fruit and vegetable quality preservation and in fumigation chambers. Breather bags are installed within the controlled atmosphere chamber in a manner that allows efficient gas exchange. They are often placed in strategic locations, such as near the product storage areas or on walls or ceilings, where they may ensure proper circulation and distribution of gases.

5.3 Generation of High-Nitrogen and High–Carbon Dioxide Concentrations

The objective of MA treatment is to attain atmospheric gases rich in CO_2 or low in O_2 or a combination of the two within the storage enclosure or treatment chamber for the time necessary to control the storage pests. At present, the most widely used source for the production of such atmospheric gas

Table 5.3 Rates of Gas Supply Requirements for Modified Atmosphere Application[a]

Selected Atmospheric Gas Concentration	Application Phase	Amount of Gas per Tonne Commodity	Supply Time (h)
<1% O_2 in N_2	Purge	1–2 m^3 N_2	<12
	Maintenance	0.01–0.06 m^3 N^2	*
>70% CO_2 in air	Purge	0.5–1.9 m^3 CO^2	<48
	Maintenance	0.02–0.04 m^3 CO_2	*
Gas burner <1% O_2 with >14% CO_2	Purge	17–66 g $C_3 H_8$	<48
	Maintenance	0.6–1.2 g $C_3 H_8$	*
>70% CO_2 in air	Single-shot	0.5–1.0 m^3 CO_2	<48
Pressurized CO_2 at >20 bar	Single-shot	>18 kg CO_2	<0.5

Source: Navarro and Donahaye (1989).
Note: [a]Compiled from Banks (1984a) (except data on pressurized CO_2). Only gas composition supported by field experience is presented in this table. Basic assumptions for above requirements are: that storage is filled with grain (minimum headspace) and pressure decay time is <5 min for decay from 500 to 250 Pa.
*According to the dosage regime

compositions is tanker-delivered liquefied CO_2 or N_2. The availability and suitability of this means of gas supply must be questioned when the gases have to be transported over long distances from an industrial production area to the storage site. Therefore, alternative potential methods of generating MAs should also be considered.

The gas supply rates required for the application of selected MAs are listed in Table 5.3. The proposed supply time at the purge phase for an MA of <1% is considerably shorter than for the other MAs. This shorter purge time derives from the physical characteristics of N_2 (Banks and Annis, 1977).

5.3.1 Supply from Tankers

When the target MA gas composition is <1% O_2 or a high CO_2 concentration, a commonly used method is to supply N_2 or CO_2 from pressurized tankers. The practical aspects of purging grain storages have been described by Guiffre and Segal (1984) for CO_2 and by Banks et al. (1980) for N_2 and CO_2.

Figure 5.4 Application of CO_2-based CA in a 6,800-metal silo bin in Kingston, North Carolina.

A significant portion of the cost of applying MAs generated from tankers is for transportation and on-site purging. Bulk liquid gas is transported in conventionally insulated road tankers (Figure 5.4).

For large-scale applications of N_2 or CO_2, vaporizers are essential. These vaporizers consist of a suitably designed receptacle with a heating medium (electricity, steam, diesel fuel, or propane), a hot-water-jacketed super-heated coil, and a forced or natural draft. A forced-draft type of vaporizer with electrical superheating was found to be convenient (Guiffre and Segal, 1984).

5.3.2 Supply from Exothermic Gas Generators

For on-site generation of MAs by combustion of hydrocarbon fuel to produce a low-O_2 atmosphere containing some CO_2, commercial installations (termed exothermic gas generators or gas burners) are available. Such equipment was originally designed for MA storage of fresh fruits. It allows the presence of ~2–3% O_2 and removes CO_2 through scrubbers. Therefore, its use in the grain industry requires several adaptations, such as adjusting the equipment to obtain an O_2 level of <1%, taking full advantage of the CO_2 generated, and removing excessive humidity from the generated atmosphere. Combustion of propane yields approximately 13% CO_2, and combustion of butane yields approximately 15% CO_2. The MA generated is more toxic than is an N_2 atmosphere deficient in O_2. This is because the presence of CO_2 in the MA causes hypercarbia; hypercarbia and hypoxia together are synergistic in their effect on insect mortality (Calderon and Navarro, 1979, 1980; Bell, 1984; Navarro and Jay, 1987). Equipment has been designed to operate with open-flame burners, with catalytic burners, and as an internal combustion system. Full-scale field trials using open-flame burners (Storey, 1973; Fleurat-Lessard and Le Torc'h, 1987) and catalytic burners (Navarro et al., 1979) to provide a low-O_2 gas mixture have proven successful. Open-flame burners are capable of producing high gas-flow rates at a low O_2 tension. Consequently, the generated MA can be applied directly to purge the treated enclosure. On the other hand, catalytic systems reduce the O_2 concentration in the atmosphere by a fixed fraction during passage through the catalyst and therefore preferably should be used in a recirculation system. The development of a modified internal combustion engine for MA generation has been reported (Banks and Annis, 1984). In spite of its advantages over the open flame and catalytic burners as an easily operated, transportable, and independent system, information on field application of such a combustion system is lacking.

5.3.3 Supply from On-Site N_2 Generators

Commercial equipment using the process of O_2 adsorption from compressed air passed through a molecular sieve bed, also known as a pressure-swing adsorption system, is available (Zanon, 1980). For continuous operation, a set of two adsorbers is provided; these operate sequentially for O_2 adsorption and regeneration. Nitrogen at a purity of 99.9% can be obtained through the regulation of inlet airflow. Equipment is now manufactured that is rated to supply an outlet flow rate of 120 m³/h at an outlet purity of 98% N_2. However, in view of the high capital investment involved, it would seem wise to undertake a long-term cost-benefit analysis to explore the justification for using these installations.

5.4 Recirculation of Carbon Dioxide

Recirculation is known as an effective method of application and distribution of fumigant compounds for the treatment of grain stored in bulk. Gaseous-type fumigants may be recirculated upward or downward through the grain bulk to obtain a uniform concentration that will effectively control insect infestations. One of the most important advantages of this method is that lengthy exposure periods are reduced (Monro, 1969). Recirculation is an effective method of returning the air/fumigant mixture that has passed through the grain bulk back into a fan so that continuous circulation is achieved (Brown and Heseltine, 1949; Howe and Klepser, 1958).

Monro (1969) recommends that the air and fumigant mixture be recirculated at least twice and preferably four or five times to obtain thorough blending of the air-fumigant mixture in 15 to 20 min; a flow rate of 1 L of air per min for each 50 L of grain [approximately 3 $(m^3/h)/m^3$ of void space if porosity is 40% or 1.54 $(m^3/h)/t$ of grain if bulk density is 0.78 t/m^3] is considered adequate in the design of fumigant recirculation systems. The airflow rate is the basic factor needed for the design of such a system. Using existing knowledge of the design of aeration systems, it is possible to calculate the resistance to flow through the commodity and the friction loss in ducts, elbows, and air distribution systems and to subsequently determine the fan capacity required in a given grain bin (Shedd, 1953; Holman, 1966; Navarro and Noyes, 2002).

Cook (1984) indicated that low airflow recirculation can be effectively used with less sorptive fumigants to obtain even penetration and distribution in grain bulks. This method has been patented (Cook, 1980) for use with aluminum or magnesium phosphide. With this method, one air change in 8 to 12 h or approximately 0.10 $(m^3/h)/m^3$ of void space or 0.05 $(m^3/h)/t$ was found to be satisfactory in recirculating phosphine gas to attain and maintain a uniform gas concentration.

Carbon dioxide can be used to modify atmospheres in grain storage facilities to control stored-product insects (Jay, 1980). Since CO_2 is 50% heavier than air, after a certain time sinking of CO_2 results in a higher CO_2, concentration in the bottom layers of the bin. Therefore, after the structure is purged with CO_2, it has been found necessary to recirculate the mixture in order to obtain an even CO_2 concentration for effective insect control (Wilson et al., 1980; Mann et al., 1999). Banks and Annis (1980) recommend recirculating the storage atmosphere from the base of the bin into the headspace via external pipework with a small sealed blower to maintain an adequate CO_2-air mixture. They concluded from unpublished data that a recirculation rate of about 0.1 bin volume per day [approximately 0.01 $(m^3/h)/m^3$ void space] is adequate for this purpose.

Navarro et al. (1986) tested a 665.7-L experimental bin containing wheat and equipped with a recirculation system to determine the relationship between the time needed to attain uniform distribution of CO_2 and the recirculation rate. A cyclic pressure increase and decrease was observed, which was caused by the adsorption and desorption processes taking place during gas recirculation in the bin (Figure 5.5). An index based on the ratio of lowest to highest concentration of CO_2 at a given time was used to determine the distribution of the CO_2. An equation based on this index is proposed as a means of predicting the time needed to attain uniform distribution based on the recirculation rates used. For the application of CO_2, design requirements for recirculation systems should be made at the stage when the silos are being converted to the application of CO_2-based CA technology.

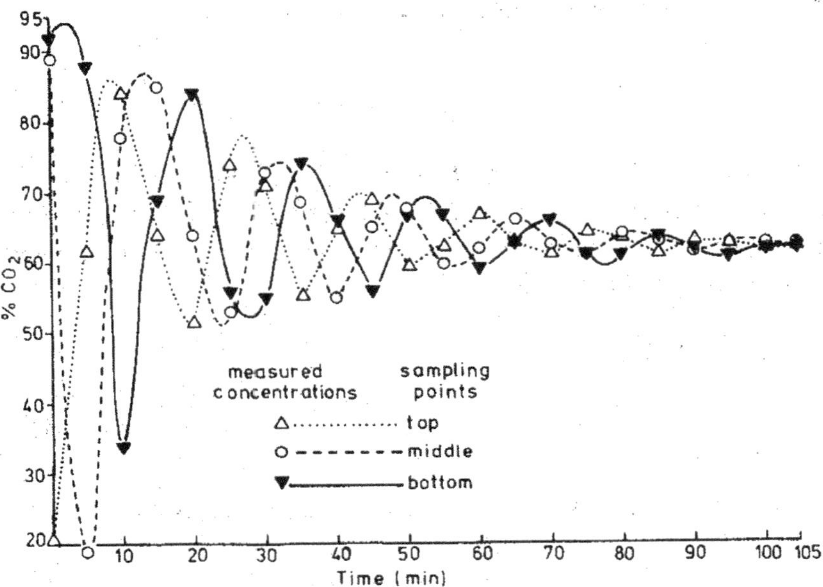

Figure 5.5 Cyclic CO_2 concentration changes observed with 3.737 $(m^3/h)/m^3$ recirculation rate, gas purge from the bottom, and the final concentration 62%. (Redrawn from Navarro et al., 1986.)

5.5 Sorption of Carbon Dioxide by the Commodity

Jay (1980) found that sorption of CO_2 by grain makes the gas effective against insect species whose immature stages feed inside the kernel. The adsorption mechanism of CO_2 into the grain has been found to be very similar to that observed in sorption of gases by charcoal and silica gel (Mitsuda et al., 1973). The sorption phenomenon causes the removal of some molecules of gas from the free space in the treated enclosure, which causes a progressive lowering of the concentration (partial pressure) of the gas in the free space of the treated enclosure.

5.5.1 Effect on Changes in Pressure of the Treated Container

Banks et al. (1980) observed an initial rapid decay of concentration shortly after CO_2 was purged into large bins containing wheat. This initial rapid decay seemed to be associated with sorption of the CO_2 by the grain.

Characteristics of CO_2 sorption by stored wheat were studied by Cofie-Agblor et al. (1993, 1995) and by Yamamoto and Mitsuda (1980). Typical reduced pressures created by sorption in the gastight containers were shown by Navarro et al. (1997). Since the amount of CO_2 sorbed is proportional to the amount of grain in the container at a given partial pressure of CO_2 and temperature, the resulting reduced pressure is also proportional to the void space of the system. Therefore, the drop in pressure can represent the experimental conditions only when the grain bulk volume occupies 93% of the total container capacity. Under these conditions, the lowest pressure recorded was 520 mm of Hg at 15°C (Navarro, 1997b) (Figure 5.6). The sorption rate changed inversely with the temperature of the wheat, and the highest absolute pressure of 606 mm of Hg was obtained at 30°C. Mitsuda and Yamamota (1980) reported that a 0.8-L container filled with grain (apparently rice) developed a negative pressure of 0.27 kg/cm² after 7 d. This negative pressure in terms of absolute pressure at standard temperature and pressure conditions was calculated to be about 555 mm of Hg.

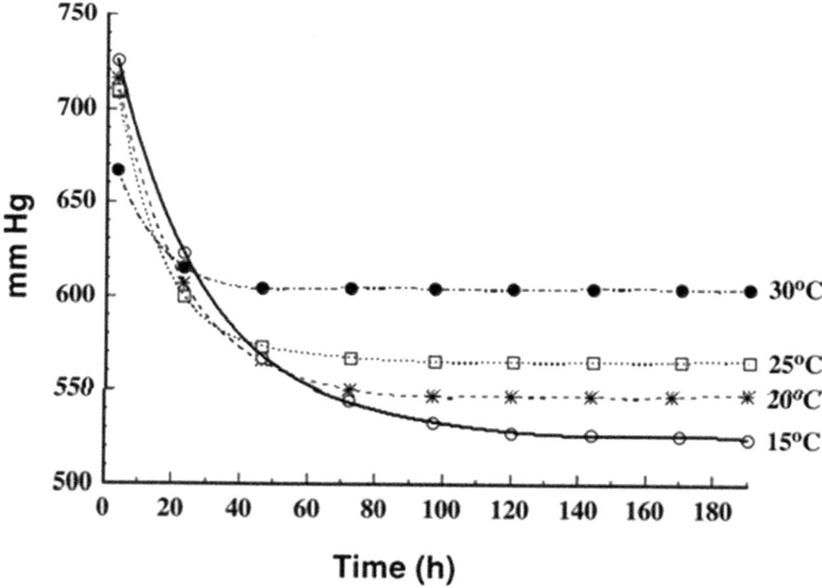

Figure 5.6 Pressure decay due to sorption of CO_2 in gastight containers filled to 93% capacity with wheat at four temperatures, with an initial CO_2 concentration of 99.8% and an initial pressure of 768 mm of Hg. (Redrawn from Navarro, 1997b.)

5.6 Experience Gained Applying High-Nitrogen Systems and High–Carbon Dioxide Systems

Reports on commercial-scale treatments of commodities using N_2 or CO_2 are scarce in the literature. According to Jonathan Banks (2023 personal communication), there are two large grain terminals in Australia that use N_2-based CA on export grain held at the terminal. One uses liquid N_2 as a CA source; the other uses an N_2 PSA generator. In addition, there is quite a lot of commercial use of CA with high-value products such as cocoa beans. However, there is limited use of CA in bulk grain with the up-country grain storage in Australia. Australian Biosecurity recently approved CA treatment of imported grain against khapra beetle at point of export. It is likely to be restricted to organic grains and similar products (personal communication).

According to Chris Newman (personal communication), N_2 is used routinely on bulk stored grain in Western Australia, but it appears there are no N_2 generators installed on the Eastern seaboard by the bulk handlers. There are some organic growers who have mobile PSA N_2 generators in Western Australia, but most organic growers use CO_2.

Newman was aware of one conventional grower in Western Australia who was treating his 14 silos, each of 90, using N_2 supplied from a mobile PSA unit. It appears that this grower was not able to secure a premium price for supplying N_2-treated grain, so he may have reverted to phosphine.

Australian grain handler CBH has a large-capacity PSA unit installed at the Port of Albany on the South Coast. This unit is destined to supply N_2 to 10 silos, each of 10,000 tonnes, of newly harvested canola as well as cereal grains. Newman reports that Kwinana grain terminal also has a high-capacity PSA unit to supply N_2 to the entire bank of 2,200-t concrete cells. Currently, CBH is in the process of installing large-capacity PSA units at the remaining export terminals to incorporate this highly effective and environmentally friendly insect control method. The system is highly effective because CBH can guarantee the gastightness of the storages for the very long fumigation period needed to kill all life stages of grain insects (Newman 2023).

As for Europe, Nico Vroom (personal communication) reported on the difficulty of estimating the amount of grain commodities treated using MA/CA on a yearly basis. His company is currently engaged in increasing the extent of use of MA/CA for stored products and artifacts in museums.

For chambers and silo treatments, most often N_2-based CA is used, as it is more affordable and easier to apply than CO_2. For stack treatment, both applications are used depending on the size and temperature of the product.

Vessel treatments are always done with CO_2-based MA. The largest MA treatment Vroom was involved in was treating a Panamax vessel with six holds and a capacity of 50.000 tonnes of grain.

According to Vroom, ECHA (European Chemical Agency) has placed N_2 and CO_2 under the biocide rule of the EU regulations. This means that N_2 and CO_2 now need a registration to be used, like PH_3. The European Union regulatory agency expects that in the coming years, 2024 or after, the control offices in each country will enforce the use of only registered N_2 and CO_2.

Vroom also reports that the most common products to be treated using MA/CA are cereal grains, pulses, nuts, sunflower seeds, dried currants, dried figs, cocoa beans, seeds, artifacts, and tobacco. In addition, space fumigation is performed using MA/CA in airplanes.

Vroom estimates that globally on an annual basis up to 6 Mt (million tonnes) of commodities are treated using MA/CA technologies. This excludes the emerging N_2-based vacuum packing and hermetic storage. What is mostly used is N_2, as CO_2 could have some effects on products. It is not allowed to be used on some products such as tobacco, as CORESTA's investigations indicate that CO_2 influences the color of the tobacco.

According to Vroom, there are industries other than cereal grains that are interested in applying MA/CA. Among them are the aviation industry (mostly CO_2-based MA against rodents in airplanes), the tobacco industry (N_2-based CA), and museums (N_2-based CA).

5.6.1 High-Nitrogen Systems

Haojie et al. (2014) reported that since the late 1960s China has carried out research on low-oxygen grain storage technology using either hermetic storage or the introduction of nitrogen into the grain mass. Beginning in 2004, some of China's grain storage depots began the large-scale application of nitrogen in stored grain, and results show success with insect control. Some of these results were presented at the 2008 International Conference on Controlled Atmosphere and Fumigation (Daolin et al., 2008) in Chengdu. China invested in nine depots for N_2-based CA treatment and gradually expanded the scale. By the end of 2012, the storage capacity that could be treated with N_2-CA was more than 10 Mt.

Low-oxygen control of stored-grain pests at 30°C was investigated in research by the Chinese Academy of State Administration of Grain. Experiments were done in the laboratory under conditions of 75 ± 5% RH and 100% high-purity nitrogen with different stored-product species as test insects. The LT values were determined for different exposure times. Previous studies have shown that the lethal effects of using nitrogen to replace O_2 were related to temperature. At moisture content below 12% at

25°C and a N_2 concentration of 98% to 100%, 28 d were needed for the complete kill. But at 18°C, 105 d were needed to achieve the same insecticidal effect. When the O_2 content was less than 1.0% and was maintained for at least 20 d, satisfactory control was achieved at storage above 25°C.

Oxygen has a significant effect on the metabolic activity of fungi. According to Haojie et al. (2014), when the oxygen concentration is < 2%, most fungi cannot reproduce. However, some species can tolerate a low-oxygen environment, and only pure nitrogen will eliminate those fungi. Reducing grain moisture content is necessary for the complete elimination of fungi.

Nitrogen production has changed considerably over the years. Pressure-swing absorption systems have proven successful, permitting a 13,660-m^3 bin to be purged to <1.0% O_2 in 7 d. Appropriate sealing allows for an accurate calculation of additional gas application required to compensate for gas loss due to sorption, as well as pressure cycling caused by pressure change (Cassels et al., 2000). It also ensures that gas concentration can be maintained for appropriate times. Liquid N_2 can be used for topping up the CA, but it may cost twice what other sources do. Although CA treatment of grain is an old and proven technology, its applications have remained limited. In a recent development reported by Clamp and Moore (2000), N_2 supplied as a bulk liquid under pressure was used to treat 1,800-tonne bins. Since the N_2 treatment was commissioned in 1993, more than 300,000 tonnes had been treated in the Newcastle facilities as of 2000 (Clamp and Moore, 2000).

Nitrogen also can be easily generated using molecular membrane generators. These are capable of purging vertical grain storages of 120 tonnes capacity within 3 h (Timlick et al., 2002). By maintaining a slight positive bin pressure, concentrations within sealed commercial storage could be maintained (compensation for leakage), and insect mortality was significant after 14 d at 17°C.

N_2 was applied for controlling stored-grain insects in three concrete silos in Cyprus, each containing 2,400 tonnes of grain. Structural sealing was carried out, and airtight valves were installed to improve the airtightness of the silos. Pressure decay tests (250 to 125 Pa) carried out in full silos showed decay times from 120 to 290 s. Using a PSA-N_2 generator, the O_2 concentration was reduced below 0.9% in 44–56 h, and after that, it was maintained between 0.2% and 0.9% for up to 23.8 d. Treatment for 18.7 and 23.8 d on grain at a temperature of 26° and 22°C, respectively, was effective for the control of the adults of important stored-grain insects *Oryzaephilus surinamensis* (L.), *T ribolium confusum* (J. duVal), and *Rhyzopertha dominica* (L.) placed above and inside the grain mass. In separate bioassays, which were placed in different zones right above the grain mass, complete (100%) mortality of *T. confusum* (all life stages), *O. surinamensis* (larvae and adults), *Sitophilus granarius* (L.) (adults), and *R. dominica* (adults) was achieved in 18.7 and 23.8 d as well.

5.6.2 High–Carbon Dioxide Systems

5.6.2.1 CO_2 Treatment of Cereal Grain

Navarro et al. (1985) reported the treatment with CO_2 of a welded-steel silo of 11,278 m^3 containing 6,881 tonnes of soft red winter wheat. The silo was purged from the aeration duct at the bottom of the silo for 39 h at a rate of 427 kg CO_2/h, and during the maintenance phase CO_2 was added from the top for an additional 106 h at a rate of 79 kg CO_2/h. The temperature of the grain ranged from 18° to 20°C on the surface and to 30–33°C in the center of the bulk (Figure 5.7). After 30 d, when grain samples were examined from the center midway to the bottom of the bulk, the percent reduction in insect population for *Tribolium castaneum* (Herbst) larvae, *R. dominica* adults, and *Cryptolestes ferrugineus* (Stephens) was 100%, while for *Sitophilus oryzae* (L.) it was 98.5%.

Alagusundaram et al. (1995) placed dry ice in insulated coolers under a CO_2-impervious plastic sheet above wheat 2.5 m deep in a 5.6-m-diameter silo. Carbon dioxide levels were 30% at 0.55 m above the floor, where 90% of rusty grain beetles, *C. ferrugineus*, were killed; CO_2 levels of 15% at 2.0 m above the floor resulted in 30% mortality.

Anonymous (1989) suggested recommendations for CO_2 fumigation of sealed stacks of cereal grains and related stored products in the ASEAN region. Annis and van S Graver (1990) discussed the uses and limitations of the process, equipment, workforce requirements, personnel protection and safety, preparation for fumigation using CO_2, pressure testing, the fumigation itself, long-term storage, and opening of the stack; they also provided a checklist of operations.

Mann et al. (1999) reported on CO_2-based MA fumigation in a welded-steel hopper bin to control the rusty grain beetle. Two welded-steel hopper bins were modified for fumigation with CO_2, and a method for efficiently purging the air from the bins was developed. Concentrations of CO_2 during experimental fumigations were less than the concentrations predicted theoretically but were high enough to kill more than 99% of caged adult rusty grain beetles in three separate experiments. Between 58% and 75% of the CO_2 initially added remained in the bin when the CO_2 concentrations peaked. The positive results from this research mean that stored-product insects in stored grain can be controlled using CO_2 rather than synthetic insecticides and fumigants that present health and environmental concerns.

5.6.2.2 Ship Fumigation Using CO_2

An evaluation of the application of an advanced insect control technology in preventing insect contamination in in-transit ship fumigation of organic commodities was reported by Navarro et al. (2022). It should be emphasized that such a gas monitoring study had not been reported in the

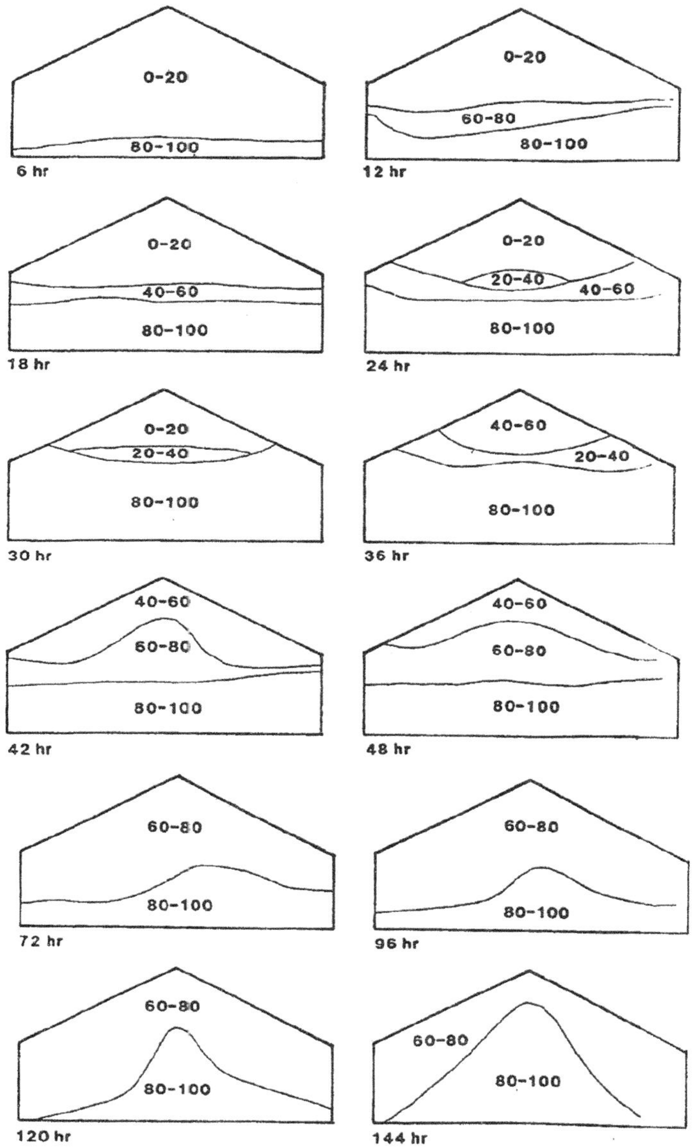

Figure 5.7 Iso-concentration lines recorded during the introduction of CO_2 into an 11,278-m^3 welded-steel bin partially filled with 6,881 tonnes of soft red winter wheat. The initial purge was through the main aeration duct and lasted 39 h. The maintenance of CO_2 was from the top of the bin and lasted 106 h. (Redrawn from Navarro et al., 1985.)

literature. Therefore, this investigation should be considered a pioneering study for in-transit ship fumigation using CO_2 gas. Van Someren (2004), to emphasize the importance of monitoring, made the statement "if you are not monitoring you are not fumigating." Reports on CO_2 distribution and retention in treated ship holds are not available in the literature. Without such information, the success of the application of CO_2 fumigation could not be estimated. In the report of Navarro et al. (2022), distribution and retention of CO_2 in the CO_2-fumigated five holds of a ship was investigated. The ship carried organic rapeseed meals, peas, cracked corn, corn bran pellets, and soybean meal. The total cargo was 27,553. The ship holds were fumigated using CO_2 immediately after loading the commodities. To take gas samples from the ship holds, gas sampling tubes were installed at the bottom, middle, and top layers in the center of the bulk of the commodities. The CO_2 gas was successfully applied. The attained gas concentrations were appropriate to achieve complete insect control. The data gathered enabled suggestion of improvements that might be considered by the company. To convey the large volume of CO_2 without pressure loss, an appropriate-size distribution duct should be provided. Also, gas distribution should be from more than one duct to enable more homogenous gas distribution.

Gas distribution was strongly affected by convection currents, gas sorption by the commodity, and leakage rate. Out of the 15 sampling tubes installed in five ship holds, two tubes were damaged at the initial phase of loading the commodities and two tubes were suspected of being blocked. The main leakage of the ship holds was recorded at the top layer. It is suspected that the hatches could not be completely sealed to create a gastight seal. However, the middle and bottom layers' gas readings were at a level to achieve complete insect control. Temperature sampling points should be installed inside the commodity, and sampling for insect presence should be carried out before and after treatment at the destination. The phytosanitary report at the discharge of the commodities was not received after the ship arrived at its destination. However, an investigation by the personnel responsible for quality control at the destination reported that no visible live insects were recorded. Improvements in sampling the commodities before the voyage and after arrival at the destination, in sampling tube placement, and in achieving gastightness of the hatches of the holds were recommended. The objective of the work carried out by Navarro et al. (2022) was to report on CO_2 fumigation carried out in a commercial vessel to assess the success of the treatment in controlling storage insects.

There were fluctuations in gas concentrations throughout the ship's voyage (Navarro et al., 2022). Some days, gas readings could not be taken due to high waves that sprayed the deck where the sampling tubes were bundled. In addition, it should be obvious that the measured gas concentrations might not be constant at the measured levels. Those levels (top, middle, and bottom) were selected to achieve minimum gas monitoring possibilities. As in other studies, gas distribution was strongly affected by convection currents, gas sorption by the commodity, and leakage rate (Navarro et al., 1985). In the following discussion, the authors assume that the temperature of the commodities in all ship holds throughout the journey was above 20°C. Based on this assumption, CO_2 levels were considered lethal when the concentration was at 40% for 17 d of exposure, at 60% for 11 d, at 80% for 8.5 d, or at 70% declining to 35% in 15 d at 20°C (Annis, 1987).

Although the five ship holds were monitored for gas concentration, Navarro et al. (2022) reported typical gas fluctuation during the monitoring of the CO_2 in the ship. Ship hold no. 2 (Figure 5.8) gas concentrations at the top layer immediately after the purge reached 100% CO_2. But within 24 h, they dropped to 58% and then remained for 11 d above 54%. This concentration has a marginal lethal effect on insect mortality. Therefore, the survival of insects at the top layer could be possible. Whereas the middle point, at the initial stage and immediately after the purge, reached 100% CO_2. That indicated that the purge point in ship hold no. 2 was close to the middle of the bulk. Four days after the initial purge, the middle layer concentrations were about 96%. At day 16, the gas concentration in the middle layer remained above 62%, indicating a lethal gas concentration for storage insects. The gradual decrease in gas concentration during the 16 d and 24 d after fumigation had little influence on the survival populations, since for the majority of the bulk the high CO_2 concentrations necessary to kill the insects were achieved (Figure 5.8) (Navarro et al., 2022).

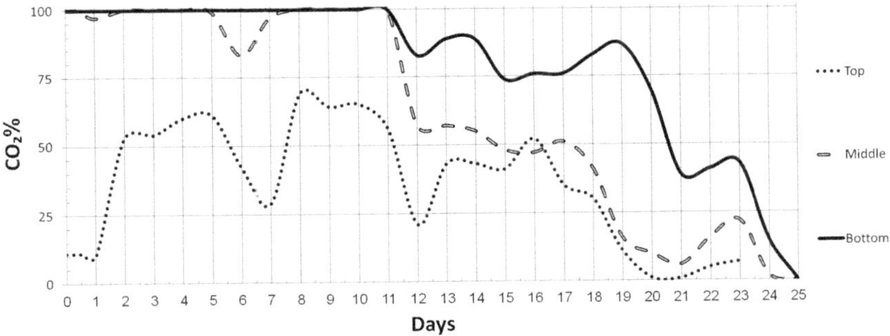

Figure 5.8 Carbon dioxide concentrations (%) recorded at three levels (top, middle, and bottom of the bulk) in the center of ship hold no. 2 during voyage of the ship. (Redrawn from Navarro et al., 2022.)

5.6.2.3 CO_2 Treatment of Museum Pests

Among stored-product insects, two common nonfood targets are museum artifacts and wood products. Since methyl bromide harms rubber products and due to its ozone-depleting properties has been phased out, nonchemical fumigations have been applied indoors to treat museums' valuables such as stuffed animals and woods from the Far East,. Treatments were applied under controlled conditions of 25°C for 12 d to control *Lyctus spp.* and *Antrenus verbessci* (L.) beetles. The museum artifacts consisted of various stuffed wild animals that had been collected by the Human and Animal Museum in Ramat Gan, Israel, and were on display inside the show window (Navarro and Navarro, 2018a). All items were placed inside a 13-m^3 gastight and flexible fumigation chamber on three levels of shelves. Carbon dioxide was applied from a pressurized cylinder using a high-pressure-resistant tube attached to the bottom of the fumigation chamber, while at the top an area of about 15 cm was left open. The purging of CO_2 was carried out directly using a siphon cylinder. The 2 kg/m^3 dosage was applied, and gas concentration was monitored during purging and until 100% concentration was reached. Measurements were carried out immediately after application and before opening. Fumigation was done using low-permeability flexible liners equipped with a zip-lock zipper. During the fumigation, the chamber was heated to maintain 25°C. Complete mortality of both species was obtained. Carbon dioxide fumigation applied indoors with controlled temperature is a suitable and comfortable method of application, with no need to transport the valuable and sometimes heavy artifacts to a special gastight chamber, normally located far from the museum.

5.6.2.4 CO_2 Treatment of Bed Bugs

As a result of withdrawal of residual insecticides such as organophosphates and carbamates throughout the world, infestations by *Cimex lectularius* (L.) have dramatically increased in recent years. The ability of this pest to starve for three months and the flattened shape of its body enable it to hide in tiny holes/cracks or folds, making it difficult to control. Therefore, as a complementary action, fumigating items like textile products and furniture within dwellings has become a common practice in Israel (Navarro and Navarro, 2018b). As an alternative to the conventional treatment of stripping the room, vacuuming, steaming the mattresses, and then spraying with a pyrethroid, a novel method was developed; instead of steaming the mattress and dry cleaning all clothes and other textile products, CO_2 based MA was applied to all movable furniture and textile products inside the dwelling. According to several scientific papers on the efficacy of CO_2 in controlling bed bugs, this method has been successfully implemented in Israel. All textile products are inserted into a sealed, low-permeability fumigation enclosure for 3 d of exposure time at room temperature to reach a calculated concentration of 100% CO_2. Since textile products absorb some of the CO_2, the concentration quickly drops to about 80%. During the summer of 2017, numerous treatments were carried out with 100% success; repeated treatments were not required. This treatment has proven to be a promising method of controlling *C. lectularius* with no need for evacuation of the residents; it also saves money and effort by eliminating the need to dry clean clothes and other textile products. It is highly effective against all life stages of the pest (Navarro and Navarro, 2018b).

5.6.2.5 CO_2 Treatment of Wax Moths

According to new Israeli legislation, farmers are not allowed to use phosphine any longer because control of *Galleria mellonella* (L.) and *Achroia grisella* (F.) are not included on its label. Therefore, bee keepers are looking for new solutions to treat their beehives. In Israel, *G. mellonella* especially is a major pest of beehives, causing monetary loss to the honey production industry if beehives are not treated during the storage period to control the pests.

Previous work carried out by Donahaye et al. (1998) and Yakobson et al. (1997) showed the possibility of controlling beeswax pests using varying CO_2 concentrations. In an unpublished work, Hagit Navarro (2018 personal communication) demonstrated to the Israel Agricultural Research Organization beekeepers' technical staff the use of a gastight flexible structure under CO_2 to control all stages of *G. mellonella*.

Heavily infested beehives were placed in a 1.75-m³ gastight flexible structure (GPC WB-3) (Figure 5.9). The gastight flexible structure consisted of a fumigation cube that was wrapped with a shrink tape to (a) prevent the effect of ballooning of the enclosure while the purged CO_2 was flowing out and (b) shrink the treated volume by using minimum CO_2. The total volume of the treated five beehives was 500 L, for which 1.35 kg of CO_2 was used to flush the air out to achieve 81.2% CO_2. After 20 d of exposure, the CO_2 concentration had dropped to 46.6% and the O_2 concentration had increased to 14.2%. The control beehives were stored in the same size flexible structure and were from the same infested source. During the trial, the temperature in the chamber where the trial was carried out fluctuated between 20° and 28°C. Unlike the control, the fumigation enclosure remained sound without holes and no larva activity was observed.

For a quality preservation test, the fumigation enclosure was left for another month. At the end of the treatment, the CO_2 concentration was

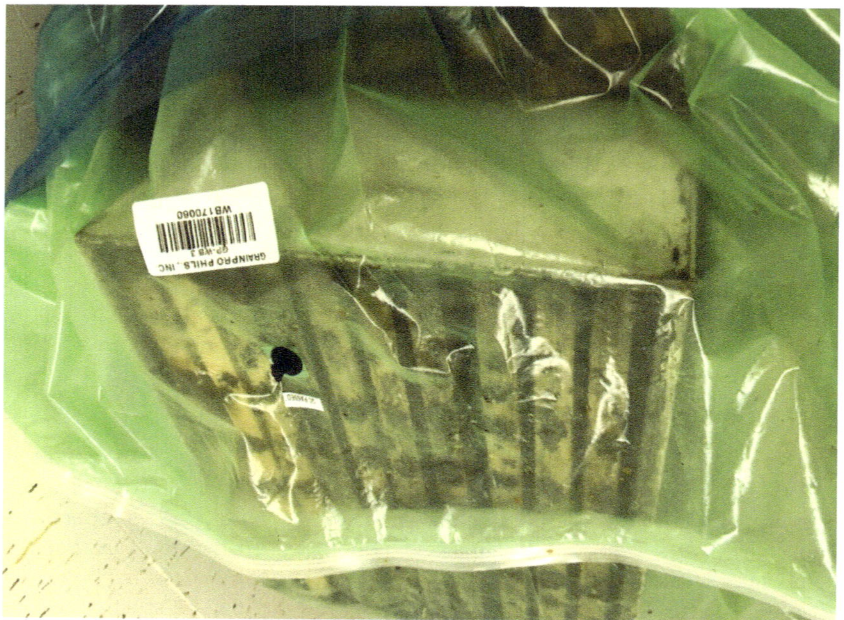

Figure 5.9 Heavily infested beehives placed in a 1.75 m³ gastight flexible structure consisting of a chamber for CO_2 fumigation. (From Hagit Navarro, 2018.)

20.6%. The beehives were observed to be in good condition (Figure 5.9). On average, CO_2 concentration decreased by 1.2%/d. The beekeepers were very satisfied with the results (Hagit Navarro, 2018).

5.7 Future Research Applying High-Nitrogen and High–Carbon Dioxide Concentrations

Sustainability and environmental impact: While CA storage can reduce reliance on chemical treatments, it is important to assess the overall environmental impact of this storage method. Further investigation might explore the energy requirements, carbon footprint, and potential emissions associated with controlled atmosphere storage systems. Since such information is lacking in the literature, finding ways to optimize energy efficiency and reduce environmental impacts will be crucial for sustainable implementation.

Economic viability and scalability: CA storage technology involves significant infrastructure and operational costs. In view of the current adverse effects of use of chemical pesticides, further investigation is needed to evaluate the economic viability of implementing CA storage systems, especially for small-scale farmers and regions with limited resources. Research could focus on developing cost-effective solutions, exploring alternative sealed storage structures, and assessing the scalability of the technology.

Integrated pest management strategies: While CA storage can help control pests, it is important to develop comprehensive integrated pest management strategies that combine multiple techniques for pest control. Further research could explore the integration of CA storage with other pest management approaches, such as temperature manipulation, biocontrol agents, and monitoring systems, to create more effective and sustainable pest management practices.

Consumer acceptance and sensory quality: Consumer acceptance plays a vital role in the adoption of CA storage. Further investigation is needed to assess the sensory quality, taste, texture, and overall consumer perception of cereals stored under CA. Understanding consumer preferences and conducting sensory evaluations can help ensure that the stored cereals meet the desired quality attributes and maintain consumer acceptance, particularly in the organic market.

By focusing on these areas of investigation, researchers and stakeholders can further enhance the effectiveness, efficiency, and practicality of CA storage for cereals and other commodities, thereby promoting its wider adoption and benefits across the agricultural and food sectors.

Acknowledgments

Authors thank Dr. Jonathan Banks, Scientist, Australia; Chris Newman, Grain Storage Specialist, Australia; and Nico Vroom, Consultant, No Infestation Consulting Organization, Netherlands, for the information they provided on the current use of MA/CA.

References

Alagusundaram K, Jayas DS, White NDG, Muir WE, Sinha RN (1995) Controlling *Cryptolestes ferrugineus* (Stephens) adults in wheat stored in bolted-metal bins using elevated carbon dioxide. Can Agric Eng 37, 217–223.

Annis PC (1987) Towards rational controlled atmosphere dosage schedules: a review of current knowledge. pp. 128–148. In: Donahaye J, Navarro S (eds), Proc 4th Intern Working Conf Stored-Prod Prot, Tel Aviv, Israel, 1986. Jerusalem, Israel: Caspit Press Ltd.

Annis PC, van S Graver J (1990) Suggested recommendations for the fumigation of grain in the ASEAN region. Part 2: carbon dioxide fumigation of bag-stacks sealed in plastic enclosures: an operations manual. ASEAN Food Handling Bureau.

Anonymous (1972) ASHRAE Handbook of Fundamentals. American Society of Heating, Refrigerating and Air-Conditioning Engineers, Inc. New York.

Anonymous (1989) Suggested recommendations for the fumigation of grain in the ASEAN region. 1. Principles and General Practice. ASEAN Food Handling Bureau, Kuala Lumpur and Australian Centre for International Agricultural Research, Canberra.

Ball S, van S Graver JE (1997) Pressure tests to determine need for sheeting loaded freight containers before fumigation. pp. 353–358. In: Donahaye ED, Navarro S, Varnava A (eds), Proc 5th Intern Conf Controlled Atm Fumi Stored Prod, Nicosia, Cyprus, 1996.

Banks HJ (1984a) Current methods and potential systems for production of controlled atmospheres for grain storage pp. 525–542. In: Ripp BE, Banks HJ, Calverley DJ, Jay EG, Navarro S (eds), Proc. Int Symp Practical Aspects of Controlled Atm and Fumi in Grain Storages, Amsterdam, Holland, 1983.

Banks HJ (1984b) Modified atmospheres—Testing of storage structures for gastightness. In: Champ BR Highley E (eds), Proc. Australian Dev. Asst. Course on the Preservation of Stored Cereals. Vol. 2. Commonwealth Scientific and Industrial Research Organization, Div. of Entomology, Canberra, Australia.

Banks HJ, Annis PC (1977) Suggested procedures for controlled atmosphere storage of dry grain. CSIRO Australia, Division of Entomology, Technical Paper No. 13, 23 p.

Banks HJ, Annis PC (1980) Conversion of existing grain storage structures for modified atmosphere use. pp. 461–473. In: Shejbal J (ed), Proc 1st Intern Conf Controlled Atm Storage of Grains, Castelgandolfo, Italy.

Banks HJ, Annis PC (1984) Importance of processes of natural ventilation to fumigation and controlled atmosphere storage. pp. 299–322. In: Ripp BE, Banks HJ, Calverley DJ, Jay EG, Navarro S (eds), Proc. Int Symp Practical Aspects of Controlled Atm and Fumi in Grain Storages, Amsterdam, Holland, 1983.

Banks HJ, Annis PC, Henning RC, Wilson AD (1980) Experimental and commercial modified atmosphere treatments of stored grain in Australia. pp. 207–224. In: Shejbal J (ed), Proc 1st Intern Conf Controlled Atm Storage of Grains, Castelgandolfo, Italy.

Banks HJ, Annis PC, Henning RC, Wilson AD (1980) Experimental and commercial modified atmosphere treatments of stored grain in Australia. pp. 207–224. In: Shejbal J (ed), Proc 1st Intern Conf Controlled Atm Storage of Grains, Castelgandolfo, Italy.

Banks HJ, Ripp BE (1984) Sealing of grain storages for use with fumigants and controlled atmospheres. pp. 375–390. In: Mills RB, Wright VF, Pedersen JR, McGaughey WH, Beeman RW, Kramer KJ, Speirs RD, Storey CL (eds), Proc 3rd Intern Working Conf Stored-Prod Ent, Manhattan, USA. 1983.

Banks HJ, Sharp AK, Irving AR (1975) Gas interchange in freight containers. pp. 519–531. In: Brady EU, Brower JH, Hunter PE, Jay EG, Lum PTM, Lund HO, Mullen MA, Davis R (eds), Proc 1st Intern Working Conf Stored-Prod Ent, Savanna, USA.

Bell CH (1984) Effects of oxygen on the toxicity of carbon dioxide to storage insects. pp. 67–73. In: Ripp BE, Banks HJ, Calverley DJ, Jay EG, Navarro S (eds), Proc. Int Symp Practical Aspects of Controlled Atm and Fumi in Grain Storages, Amsterdam, Holland, 1983.

Brown WB, Heseltine HK (1949) Fumigation of grain in silo bins: The provision of circulating systems. Milling 112, 229–230, 233.

Calderon M, Donahaye E, Navarro S, Davis R (1989) Wheat storage in a semi-desert region. Tropical Sci 29, 91–110.

Calderon M, Navarro S (1979) Increased toxicity of low oxygen atmospheres supplemented with carbon dioxide on *Tribolium castaneum* adults. Entomologia Experimentalis et Applicata 25, 39–44.

Calderon M, Navarro S (1980) Synergistic effect of CO_2 and O_2 mixture on stored grain insects. pp 79–84. In: Shejbal J (ed), Proc 1st Intern Conf Controlled Atm Storage of Grains, Castelgandolfo, Italy.

Cassels J, Banks J, Allanson R (2000) Applications of pressure swing absorption (PSA) and liquid nitrogen as methods for providing controlled atmospheres in grain terminals. In: Proc 6th Int Work Conf Stored-Product Protection, pp. 57–63.

Clamp P, Moore D (2000) Nitrogen treatment of grain, Newcastle Grain Terminal. Australian Postharvest Technical Conference, pp. 184–186.

Cofie-Agblor R, Muir WE, Cenkowski S, Jayas DS (1993) Carbon dioxide sorption in stored wheat. pp. 261–269. In: Navarro S, Donahaye E (eds), Proc 4th Intern Conf Controlled Atm Fumi Grain Storages, Winnipeg, Canada, 1992, Caspit Press Ltd., Jerusalem.

Cofie-Agblor R, Muir WE, Sinico R, Cenkowski S, Jayas DS (1995) Characteristics of carbon dioxide sorption by stored wheat. J Stored Prod Res 31(4), 317–324.

Cook JS (1980) U. S. Patent no. 4,200, 657.

Cook JS (1984) The use of controlled air to increase the effectiveness of fumigation of stationary grain storages. pp. 419–424. In: Ripp BE, Banks HJ, Calverley DJ, Jay EG, Navarro S (eds), Proc. Int Symp Practical Aspects of Controlled Atm and Fumi in Grain Storages, Amsterdam, Holland, 1983.

Daolin G, Jie T, Shengbin L, et al (2008) Evaluation of large, modern warehouse storages designed and constructed for application of carbon dioxide. pp. 8–13. In: Daolin G, Navarro S, Jian Y, Cheng T, Zuxun J, Yue L, Yanf L, Haipeng W (eds), Proc 8th Intern Conf Controlled Atm Fumi Stored Prod, Chengdu, China.

Donahaye EJ, Navarro S, Miriam R, Azrieli A (1998) Sensitivity of the greater wax moth *Galleria mellonella* to carbon dioxide enriched modified atmospheres. pp. 692–701 In: Zuxun J, Quan L, Yongsheng L, Xianchang T, Lianghua G (eds), Proc 7th Intern Working Conf Stored-Prod Prot, Beijing, China.

Donahaye E, Navarro S, Ziv A, Blauschild Y, Weerasinghe D (1991) Storage of paddy in hermetically sealed plastic liners in Sri Lanka. Tropical Sci 31, 109–121.

Fleurat-Lessard F, Le Torc'h JM (1987) Practical approach to purging grain with low Oxygen atmosphere for disinfestation of large wheat bins against the granary weevil, *Sitophilus granarius*. pp. 208–217. In: Donahaye J, Navarro S (eds), Proc 4th Intern Working Conf Stored-Prod Prot, Tel Aviv, Israel, 1986. Jerusalem, Israel: Caspit Press Ltd.

Guiffre V, Segal AI (1984) Practical approaches to purging grain storages with carbon dioxide in Australia. pp. 343–358. In: Ripp BE, Banks HJ, Calverley DJ, Jay EG, Navarro S (eds), Proc. Int Symp Practical Aspects of Controlled Atm and Fumi in Grain Storages, Amsterdam, Holland, 1983.

Haojie L, Jian Y, Pengcheng F, Xiaoping Y, Arthur FH, Kengkanpanich R, Suthisut D (2014). Application of nitrogen controlled atmosphere in grain storage in China. pp. pp. 24–28. In: Arthur FH, Kengkanpanich R, Chayaprasert W, Suthisut D (eds), Proc 11th Intern Working Conf Stored-Prod Prot, Chiang Mai, Thailand.

Hill JE, Kusuda T (1975) Dynamic characteristics of air infiltration. Trans Amer Soc Heat Refrig Air Cond Eng 81, 68–185.

Holman LE (1966) Aeration of grain in commercial storages Marketing Res. Rep. No. 178. USDA Agricultural Marketing Service, Research Division, 46 pp.

Howe RG, Klepser GE (1958) Forced recirculation control of stored-grain insects. Down to Earth 14(3), 6–16.

Hunt CM (1980) Air infiltration: A review of some existing measurement techniques and data. pp. 3–23. In: Hunt CM, King IC, Trechsel HR (eds), Building Air Change Rate and Infiltration Measurements. Amer. Soc. for Testing and Materials. ASTM STP 719.

Jay EG (1980) Methods of Applying Carbon Dioxide for Insect Control in Stored Grain. U.S. Department of Agriculture, Science Education Administration, AAT-S-13. 8 pp.

Mann DD, Jayas DS, Muir WE, White NDG (1999) Efficient carbon dioxide fumigation of wheat in welded-steel hopper bins. Appl Eng Agric 15, 57–63.

Meiering AG (1982) Oxygen control in sealed silos. Trans Amer Soc Agric Eng 25, 1349–1354.

Mitsuda H, Kawai F, Kuga M, Yamamota A (1973) Mechanism of carbon dioxide gas adsorption by -grains and its application to skin-packaging. J Nutr Sci Vitaminol 19, 71433.

Mitsuda H, Yamamota A (1980) Advances in grain storage in a CO_2 atmosphere in Japan. pp. 235–246. In: Shejbal J (ed) Proc 1st Intern Conf Controlled Atm Storage of Grains, Castelgandolfo, Italy.

Monro HAU (1969) Manual of fumigation for insect control FAO Agric Studies No. 79. Rome, 289 pp.

Moylan Silos (2023) Moylangrainsilos.com/moylanproducts.htm. July 22, 2023.

Navarro S (1977) A laboratory method for fumigating grain with methyl bromide. Rep Vis Scient Vic Plant Res Inst 1977, 54–62. Burnley, Victoria.

Navarro S (1997a) Sealing efficiency assessment in modified atmosphere storages. pp. 371–384. In: Donahaye ED, Navarro S, Varnava A (eds), Proc 5th Intern Conf Controlled Atm Fumi Stored Prod, Printco Ltd., Nicosia, Cyprus, 1996

Navarro S (1997b) Sorption of carbon dioxide by wheat. pp. 193–202. In: Donahaye ED, Navarro S, Varnava A (eds), Proc 5th Intern Conf Controlled Atm Fumi Stored Prod, Nicosia, Cyprus, Printco Ltd., Nicosia, Cyprus, 1996.

Navarro S (1998) Pressure tests for gaseous applications in sealed storages: Theory and practice. pp. 385–390. In: Zuxun J, Quan L, Yongsheng L, Xianchang T, Lianghua G (eds), Proc 7th Intern Working Conf Stored-Prod Prot, Beijing, China.

Navarro S, Athanassiou C, Varnava A, Vroom N, Yilasoumis D, Leandrou I, Hadjioannou S (2012) Control of stored grain insects by using nitrogen in large concrete silos in Cyprus. pp: 478–487, In: Navarro S, Banks HJ, Jayas DS, Bell CG, Noyes RT, Ferizli AD, Emekci M, Isikber AA, Alagusundaram K (eds), Proc 9th Intern Conf Controlled Atm Fumi Stored Prod, Antalya, Turkey.

Navarro S, Donahaye E (1989) Generation and application of modified atmospheres and fumigants for the control of storage insects. pp. 152–165. In: Champ BR, Highley E, Banks HJ (eds) Proc 3rd Intern Conf Fumi Controlled Atm Storage Grain, Singapore.

Navarro S, Donahaye E (1993) Preservation of grain by airtight storage. pp. 425–434. 5th Int. Cong. On Mechanization and Energy in Agriculture, 11–14 Oct. 1993, Kusadasi, Türkiye.

Navarro S, Donahaye E, Caliboso FM, Sabio GC (1995) Application of modified atmospheres under plastic covers for prevention of losses in stored grain. Final Report submitted to U.S. Agency for International Development, CDR Project No. C7-053, August 1990 - November 1995. 32 pp.

Navarro S, Donahaye E, Fishman Svetlana (1994) The future of hermetic storage of dry grains in tropical and subtropical climates. pp. 130–138. In: Highley E, Wright EJ, Banks HJ, Champ BR (eds), Pro 6th Intern Working Conf Stored-Prod Prot, Canberra, Australia.

Navarro S, Donahaye E, Jay E (1985) Modified Atmospheres for Controlling Stored-Product Insects. Final Scientific Rep. presented to Binational Agric. Res. Dev. Fund US-Israel (BARD) No. I-303-80, 184 p.

Navarro S, Donahaye E, Kashanchi Y, Pisarev V, Bulbul O (1984) Airtight storage of wheat in a P.V.C. covered bunker. pp. 601–614. In: Ripp BE, Banks HJ, Calverley DJ, Jay EG, Navarro S (eds), Proc. Int Symp Practical Aspects of Controlled Atm and Fumi in Grain Storages, Amsterdam, Holland, 1983.

Navarro S, Donahaye E, Rindner M, Azrieli A (1990) Airtight storage of grain in plastic structures. Hassadeh Quarterly 1(2), 85–88.

Navarro S, Donahaye E, Rindner M, Azrieli A (1999) Storage of dried fruits under controlled atmospheres for quality preservation and control of nitidulid beetles. Acta Hortic. 480, 221–226 DOI: 10.17660/ActaHortic.1998.480.38 https://doi.org/10.17660/ActaHortic.1998.480.38.

Navarro S, Gonen M, Schwartz A (1979) Large scale trials on the use of controlled atmospheres for the control of stored grain insects. pp. 260–270. In: Caswell GH, Boshoff WH, Daramola AM, Dina SO, Adesuyi SA, Singh SR, Ivbijaro MF, Adeyemi SAO, Taylor TA (eds), Proc 2nd Intern Working Conf Stored-Prod Ent, Ibaban, Nigeria, 1978.

Navarro S, Jay E (1987) Application of modified atmospheres for controlling stored grain insects. pp. 229–236. Proc. British Crop Protection Council, No. 37 Stored Products Pest Control. Univ. of Reading.

Navarro S, Jay EG, Leesch JG (1986) Recirculation rate requirements for adequate distribution of carbon dioxide in grain bins. Transactions of the ASAE 29(5), 1348–1354.

Navarro H, Navarro S (2018a) Commercial application of non-residual fumigation in museums. IOBC-WPRS Bulletin 130, 324–333.

Navarro H, Navarro S (2018b) A commercial method of controlling bedbugs (*Cimex lectularius*) using CO_2 in dwellings. Julius-Kühn-Archiv, (463), 1052–1058.

Navarro S, Navarro H, Inbari N (2022) Distribución y Retención de Dióxido de Carbono en Bodegas de Barcos Tratados Durante el Viaje. Granos. pp 8–14.

Navarro S, Noyes R (eds) (2002) The Mechanics and Physics of Modern Grain Aeration Management. CRC Press, Boca Raton, FL. 647 pp.

Navarro S, Varnava A, Donahaye E (1993) Preservation of grain in hermetically sealed plastic liners with particular reference to storage of barley in Cyprus. pp. 223–234. In: Navarro S, Donahaye E (eds), Proc 4th Intern Conf Controlled Atm Fumi Grain Storages, Winnipeg, Canada, 1992, Caspit Press Ltd., Jerusalem, Israel.

Navarro S, Zettler JL (2001). Critical limits of sealing for successful application of controlled atmosphere or fumigation. pp. 507–520. In: Donahaye EJ, Navarro S, Leesch JG (eds), Proc 6th Intern Conf Controlled Atm Fumi Stored Prod, Fresno, USA, 2000.

Sharp AK (1982) Measurement of gastightness with an automatic decay timer. Sci Tech Froid 1982-1, 361–367.

Shedd CK (1953) Resistance of grain and seeds to air flow. Agric Eng 34(9), 616–619.

Storey CL (1973) Exothermic inert atmosphere generators for control of insects in stored wheat. J Econ Entomol 66, 511–514.

Timlick B, Dickie G, McKinnon D (2002) Nitrogen as a major component of a controlled atmosphere to manage stored product insect pests in large vertical storage. In: Intergrated Protection of Stored Products, IOBC Bulletin 25(3), 193–197.

Trisviaskii LA (1966) Storage of Grain. Translated from Russian by DM Keane. Vol. 1–3, 1969, Natl. Lending Library Sci. Technol., Boston Spa, England, 845 pp.

Van Someren Graver JE (2004) Guide to Fumigation Under Gas-Proof Sheets, Food and Agricultural Organization of the United Nations, Rome, Italy.

White NDG, Jayas DS (2003) Controlled atmosphere storage of grain. In Chakraverty A, Majumdar AS, Raghavan GSW, Ramaswamy HS (eds), Handbook of Postharvest Technology. Cereals, Fruits, Vegetables, Tea, and Spices. Marcel Dekker Inc., New York. pp. 235–251.

Williams P, Minett W, Navarro S, Amos TG (1980) Sealing a farm silo for insect control by nitrogen swamping for fumigation. Aust J Exp Anim Husb 20, 108–114.

Wilson AD, Banks HJ, Annis PC, Guiffre V (1980) Pilot commercial treatment of bulk wheat with CO_2 for insect control: The need for gas recirculation. Aust J Exp Agric Anim Husb 20, 618–624.

Yakobson B, Navarro S, Donahaye EJ, Azrieli A, Slaveski Y, Ephrati H (1997) Control of beeswax moths using carbon dioxide in flexible plastic and metal structures. pp. 169–174. In: Donahaye ED, Navarro S, Varnava A (eds), Proc 5th Intern Conf Controlled Atm Fumi Stored Prod. Printco Ltd., Nicosia, Cyprus, 1996.

Yamamoto A, Mitsuda H (1980) Characteristics of carbon dioxide gas adsorption by grain and its components. In: Shejbal, J. (ed). Controlled Atmosphere Storage of Grains. Elsevier, Amsterdam, pp. 247–258.

Zahradnik JW (1968) Methods for evaluating controlled atmosphere (C-A) storage gas seals. American Society of Agricultural Engineers, Paper No. 68–887, 5 p.

Zanon K (1980) Systems of supply of nitrogen for the storage of grain. pp. 507–516. In: Shejbal J (ed.) Proc 1st Intern Conf Controlled Atm Storage of Grains, Castelgandolfo, Italy.

Chapter 6

Fumigation Process Using Chemicals and Controlled or Modified Atmospheres

Formulations, Introduction, and Monitoring

Watcharapol Chayaprasert and Dirk E. Maier

6.1 Introduction

Fumigation is a process in which a toxic gas is released in an enclosed structure to disinfest the structure or the commodity inside. Controlled atmosphere (CA) and modified atmosphere (MA) treatments are processes in which the atmospheric composition within the treated enclosure is manipulated. In CA, the atmospheric composition within the treated enclosure is controlled or maintained at a desired level and duration, while MA uses a gas composition that is initially modified and may change during the treatment period (Navarro et al., 2012). Although proper fumigation and CA/MA techniques have been covered in several publications (Bond, 1984; Annis and Graver, 1991; Graver, 2004; DAWE, 2018; Hopkins and Johnson,

DOI: 10.1201/9781003309888-6

2022), new regulations, technologies, and research findings have also been continually established. With this chapter, we intend to highlight the current technologies and practices used in conducting these treatments. Before proceeding further, it must be emphasized that pest disinfestation treatments must be conducted by trained (and often licensed) personnel, following specific safety and regulatory requirements.

6.2 General Treatment Procedure

Four primary types of fumigation are (1) quarantine and preshipment (QPS) application, (2) postharvest application, (3) structural fumigation, and (4) soil fumigation. QPS fumigation usually is conducted on import and export goods. During storage of agricultural commodities, fumigation is conducted to maintain low levels of postharvest pests. When a pest outbreak occurs in residential and commercial buildings, as well as food and feed processing facilities, fumigation is typically used as a disinfestation measure. To get rid of diseases and pests in soil, fumigation is performed before crops are planted. Controlled atmosphere/modified atmosphere treatments are typically used for the QPS and postharvest storage purposes.

The procedures of fumigation and CA/MA treatments are different in many aspects, but their fundamentals are mostly the same and can be divided into four phases (Table 6.1): preparation, gas introduction, exposure, and degassing. Once a fumigation or CA/MA treatment job is requested, the service provider must prepare for the job by first acquiring essential information, including the conditions of the commodity and/or structure to be fumigated (e.g., current infestation level, temperature and moisture content, size, gastightness, major infested areas, gas distribution paths), surrounding areas, target pests, weather conditions, customer's operating schedule, and records of previous fumigation jobs. Knowing this information enables the service provider to, among other things, better determine the sealing method, dose rate, gas introduction and monitor locations, and whether additional effort, such as preheating, is required. Depending on the enclosure type (e.g., grain bins, buildings, bag stacks, shipping containers), different sealing procedures could be used. In the case of structural fumigation, the fumigated building is cleaned to allow for greater gas penetration and distribution.

While the enclosure is being constructed, gas introduction and monitoring locations as well as circulation fan placement spots (if necessary) are selected. Plastic tubings with inner diameters between 3 and 6 mm are appropriate for monitoring lines. If the treatment chemical is in a pressurized gas formulation, it is typically released via high-pressure hoses with inner diameters between 8 and 12 mm. In order to quantitatively evaluate the treatment efficacy, bioassays are sometimes placed inside the

Table 6.1 Procedures During a Fumigation or CA/MA Treatment

Phase	Procedure
1. Preparation	1.1 Acquiring information
	1.2 Site inspection (if necessary)
	1.3 Constructing the enclosure
	1.4 Placing gas introduction and monitoring tubings
	1.5 Placing circulation fans and ducting (if necessary)
	1.6 Placing insect bioassays (if necessary)
	1.7 Pressure testing the enclosure (if necessary)
	1.8 Calculating the required dose rate
2. Gas introduction	2.1 Releasing the treatment gas
	A. Placing tablets, pallets, or sachets in case of solid metal phosphide formulation
	B. Injecting via hoses in case of pressurized gas formulation
	C. Utilizing a gas generator, reactor, or mixer
	2.2 Recirculating the treatment gas
3. Exposure	3.1 Monitoring gas concentrations
	3.2 Topping-up (if necessary)
	3.3 Disinfesting areas surrounding the enclosure with other treatments (if necessary)
4. Degassing	4.1 Opening and ventilating the enclosure
	4.2 Monitoring gas concentrations to ensure the threshold limit level (TLV)
	4.3 Retrieving equipment

enclosure. Once the enclosure has been properly sealed, its gastightness can be determined by the pressurization test (Navarro, 1998). Although pressure testing is routinely done on enclosures for CA/MA treatments, it is not usually performed in other applications due to additional costs. The proper dose rate (i.e., amount of treatment gas per unit volume) depends on several factors, especially the type of chemicals, available exposure time, weather conditions, and target insect species and life stages. Once the dose rate has been determined, the total amount (i.e., dose) of the treatment gas needed can be calculated based on the total volume of the enclosure. The combination of the gas concentration and exposure time [i.e., concentration × time (Ct product)] is sometimes referred to as the dosage and is used as the indicator for treatment efficacy (Annis, 1998; Gandy and Chanter, 1976).

The *Ct* product is essentially the area under the gas concentration curve plotted over elapsed exposure time—namely, the integration of the concentration curve (Chayaprasert and Maier, 2010).

Depending on the formulation, different methods are used to release the fumigant. During the exposure period, the gas concentrations are monitored to ensure efficacy. In cases where the concentration falls below the chemical-specific lethal level, an additional amount could be added or the exposure time could be extended. While the space or content inside the enclosure is being exposed to the gas, other treatments, such as fogging or spraying, can be applied to the area surrounding the enclosure. Once the target exposure time is reached, all temporary seals are taken off and the enclosure is opened to vent off the gas. Degassing or ventilation can be done naturally or assisted by electric fans, but it must be done until the gas concentration decreases below the threshold limit level (TLV). Once TLV is attained, any treatment equipment is retrieved from the enclosure and the commodity and/or structure are handed back to the customer.

6.3 Chemicals

Despite having the same general treatment procedure, each of the widely used fumigants and CA/MA gases has its own set of unique advantages and limitations. The following subsections highlight important details about these chemicals.

6.3.1 Methyl Bromide

Methyl bromide (MB) is an ozone-depleting substance whose use as a fumigant for commodities and soils has been declining since 1992, with the adoption of the Montreal Protocol (UNEP, 2020). Nearly all nations had successfully phased out MB by January 2015 (UNEP, 2016). As of 2021, only a few nations requested critical-use exemptions (CUE) and less than 100 metric tonnes of MB in total were permitted (UNEP, 2021b). However, the amounts of MB used for QPS purposes in many countries have remained consistently high for the past few years (Table 6.2). Currently, production of MB for QPS occurs in five countries (China, India, Israel, Japan, and the United States),and the total production in 2020 was 10,143 metric tonnes (UNEP, 2022). This shows that MB is still an essential fumigant for QPS even though alternative treatments are available. To reduce its impact on the environment, some countries are beginning to impose restrictions on the use of MB in QPS applications, such as recapture requirements and bans of

Table 6.2 Amounts in Metric Tonnes of Methyl Bromide Used for Quarantine and Preshipment Purposes in Some Countries

Year	Australia	China	Indonesia	India	Japan	Thailand	United States
2021	641.2	735.2	80.0	1,553.4	467.1	164.1	2,494.1
2020	695.1	797.6	72.0	2,104.1	450.7	156.1	1,688.8
2019	793.8	871.2	135.8	1,778.4	467.9	156.1	857.4
2018	682.0	1,227.8	89.5	1,551.7	483.1	188.1	2,254.0
2017	897.8	1,360.7	96.0	1,384.4	445.0	188.1	1,326.6
2016	708.5	1,141.6	128.0	1,180.9	434.1	157.4	739.6
2015	864.3	932.1	224.0	329.5	449.0	194.7	1,639.9

Source: UNEP (2021a).

certain applications. For example, starting January 1, 2023, New Zealand's government has banned MB in ship's hold fumigation and required that 80% of MB be recaptured from every container fumigation (EPANZ, 2021).

Methyl bromide fumigant products are usually stored in the form of pressurized gas in metal canisters or cylinders of various volumes at either 100% MB or 98% MB plus 2% chloropicrin concentrations. The QPS dose rate and exposure time for any exported commodity largely depend on the requirements of the importing country, as summarized in Table 6.3. It must be noted that MB can cause phytotoxic damage, especially in perishable commodities, and thus overdosing and extending the exposure time unnecessarily should be avoided (Kamel et al., 1970; Houck et al., 1994).

Some countries may implement a quarantine program preventing specific invasive species. For example, the governments of Australia and New Zealand have specific measures for safeguarding against *Halyomorpha halys* (Stål), the brown marmorated stink bug, which require that high-risk goods from target risk countries be treated before import. One of the approved treatments is MB fumigation at a dose rate of 24 g/m^3, at 10°C or above, for at least 12 h (but no more than 24 h), with a minimum end-point concentration of 12 g/m^3 (DAFF, 2022b; MPI, 2022a). Another example is the quarantine requirements against *Trogoderma granarium* (Everts), the khapra beetle, by the Australian government. According to the requirements, freight containers packed with high-risk plant products in a khapra beetle target risk country must be treated using an approved treatment option. One of the options is MB fumigation at a dose rate of 80 g/m^3, at 21°C or above, for 48 h, with a minimum end-point concentration of 20 g/m^3 (DAFF, 2022a).

Table 6.3 Methyl Bromide Dose Rates and Exposure Times for Some Commodities as Required by Different Importing Countries

Importing Country	Commodity	Dose Rate (g/m^3)	Exposure Time (h)	Temperature (°C)
Australia	Pineapple	32	2	> 21
Brunei Darussalam	Ornamental plants	32	2	-
	Groundnut	32	2	-
	Red bean	32	2	-
European Union	Orchid cut flowers	20–24	1.5	-
India	Coconut seed nut	16	12	-
	Tangerine	32	2	> 21
	Pomelo	32	2	> 21
Iran	Orchid cut flowers	24	1.5	> 21
	Pineapple	32	6	> 21
Israel	Ribbon dracaena	32	3	> 21
Malaysia	Rice and flour	32	24	-
	Water convovulus	16	12	26
New Zealand	Rice	48	24	10–15
	Orchid cut flowers	32	2	> 21
Papua New Guinea	Rice	80	48	> 21
Republic of Fiji	Mung bean	40	30	> 21
South Korea	Mangosteen	32	2	> 21
Taiwan	Betel nut	32	4	> 21
United States	Asparagus shoot	24	2.5	> 27
		32	2.5	21–26
	Blueberry	32	3.5	> 21
	Garlic	32	1.5	> 32
		32	2	27–32
	Kale	32	2	> 21
		40	2	16–21
	Yam	40	4	> 32
		48	4	27–32

Source: Data from OAR-DOA (2017) and APHIS (2016).

Table 6.4 Methyl Bromide Dose Rates and Exposure Times for Logs and Sawn Timber as Required by Different Importing Countries

Importing Country	Commodity	Dose Rate (g/m³)	Exposure Time (h)	Temperature (°C)
Brazil	Logs	48	16	> 21
	Sawn timber	48	16	> 21
China	Logs	80	16	> 15
European Union	Logs	Fumigated to ISPM15 requirements		
	Sawn timber	Fumigated to ISPM15 requirements		
Malaysia	Logs	64	16	> 21
	Sawn timber	64	16	> 21
Philippines	Sawn timber	48	24	> 21
United States of America	Logs	48	16	> 21
	Sawn timber	48	16	> 21

Source: MPI (2022b).

For wood packaging, under the International Standards for Phytosanitary Measures (ISPM) No. 15, one of the acceptable treatments is MB fumigation that achieves a *Ct* product of 650 g-h/m³ with a 24 h exposure time while the wood's temperature is maintained at at least 21°C (IPPC, 2018). This is equivalent to a dose rate of 48 g/m³ with a minimum end-point concentration of 24 g/m³ at 24 h. For logs, some countries do not have phytosanitary import requirements, some allow certification based only on visual pest inspection, and some may require a phytosanitary treatment (FAO, 2011). For logs and sawn timber that are fumigated with MB, some countries follow the ISPM15 standards and some mandate different dose rates and exposure times. Some examples of the required dose rates and exposure times are shown in Table 6.4.

6.3.2 Sulfuryl Fluoride

A gas fumigant in the form of 99.8% sulfuryl fluoride (SF) and 0.2% inert ingredients packaged as a liquid in pressurized steel cylinders was developed under the trade name Vikane® in the 1950s specifically for the control of drywood termites (Derrick et al., 1990). Sulfuryl fluoride has been registered for postharvest use in the United States since 2004 under the trade name ProFume® (Dow AgroSciences, 2004). In Europe, it was used for structural treatment of postharvest facilities starting in 2003 in Switzerland (Bell, 2004). In 2011, ProFume® was registered for use in 17 countries

(Barnekow, 2011). Both of these products are currently supplied by Douglas Products (2022). There are other suppliers of similar SF fumigant products, such as (Ensystex, 2022) Chengdu Taiyu Industrial Gases (2022) and Indo Global Trade (2021).

According to the product label (Douglas Products, 2016), ProFume® must be used with Fumiguide™, proprietary fumigation management software. Fumigators must use the software to determine the dose rate based on various input parameters, a few of which are the dimensions of the fumigated enclosure, the temperature inside the enclosure, and the estimated fumigant leakage rate (Schneider et al., 2003). In addition, based on monitored fumigant concentrations, the software is used to determine the completion of the fumigation and whether or not corrective actions such as extending the exposure time or adding more fumigant are needed. Similar SF fumigation management software developed by other companies (Barcan İlaçlama Fumigasyon Hizmetleri, 2021; Ensystex, 2023) can also be found. The leakage rate is represented by half-loss time (HLT), the time at which the fumigant concentration in the enclosure reduces by half, and the completion of the fumigation is indicated by the achieved *Ct* product. The calculation of the *Ct* product is described in Section 6.7.2. Wontner-Smith (2005) gathered data on the efficacy of SF on stored-product pests from various other studies (Table 6.5). It must be noted that the ProFume® applicator manual (Douglas Products, 2017) specifies that fumigations with SF concentrations above 128 g/m³ or *Ct* products above 1,500 g-h/m³ are not allowed.

The egg is generally the life stage most tolerant of SF (Bell et al., 1998; Bell et al., 2002; Kenaga, 1961; Williams and Sprenkel, 1990). Outram

Table 6.5 *Ct* Products of Sulfuryl Fluoride Required for 100% Control of the Egg Stage of Some Stored-Product Pests

	Ct (g-h/m³)	Temp (°C)
Cryptolestes turcicus (Grouvelle) (Turkish grain beetle)	784	25
Ephestia kuehniella (Zeller) (Mediterranean flour moth)	912	25
Oryzaephilus surinamensis (Linnaeus) (saw-toothed grain beetle)	957	20
Plodia interpunctella (Hübner) (Indian meal moth)	446	20
Sitophilus granarius (Linnaeus) (granary weevil)	1,339	25
Tribolium castaneum (Herbst) (red flour beetle)	1,669	25
Tribolium confusum (Jacquelin du Val) (confused flour beetle)	1,908	25

Source: Wontner-Smith (2005).

(1967a, 1967b) suggested that resistance is due mainly to the inability of SF to penetrate the chorion (the outer layer of the egg). Athanassiou et al. (2012) found that for stored-product psocids (Psocoptera) fumigated with SF for 48 h at 27.5°C, the complete mortality of adults and nymphs was recorded at concentrations between 4 and 24 g/m^3. However, some survival of eggs was found even at 96 g/m^3. Jagadeesan et al. (2015) fumigated adults and immature stages of *Tribolium castaneum*, with SF for 48 h and found that at LC50, eggs were 29 times more tolerant than the other stages. However, the *Ct* products required to control insect eggs can be substantially reduced by increasing the temperature. Bell (2006) reported that 95% mortality of eggs of *Rhyzopertha dominica* (Fabricius) could be achieved at *Ct* products of about 800, 600, and 300 $g\text{-}h/m^3$ at 20, 25, and 30°C, respectively. Bell and Drinkall (2000) suggested that increasing temperature from 20° to 30°C could potentially reduce the efficacy of SF up to five-fold, depending on the species.

Given that the global warming potential (GWP) for SF was calculated to be 4,780 over 100 years (GWP is 1 for carbon dioxide), there is a growing concern regarding its being released into the atmosphere (Papadimitriou et al., 2008; Mühle et al., 2009; Gressent et al., 2021). However, at the time of writing of this chapter, no requirement for recapturing SF after fumigation, to prevent it from being released into the atmosphere, had been established. Although the widely used MB and phosphine fumigants also leave residues, there seemed to be an elevated concern with regard to residues in SF-fumigated commodities, especially the fluoride ion residue (Pest Control Technology, 2011; Fluoride Action Network, 2013; Hwaidi et al., 2015). Sulfuryl fluoride residue in food contributes only a very small portion of total exposure to fluoride, but when it is combined with fluoride exposure from other channels, such as drinking water and toothpaste, the total exposure could potentially exceed the Federal Food, Drug, and Cosmetic Act safety limits (US-EPA, 2016). European Food Safety Authority et al. (2021) conducted an extensive review of the maximum residue levels (MRLs) of SF and fluoride ions according to the existing European Union (EU) regulation (Article 12 of Regulation No. 396/2005). Based on this review, a standard consumer risk assessment on fluoride ions could not be performed due to a lack of information on the toxicological reference values. However, an indicative estimate in the same review suggested that the contribution to overall consumer exposure to fluoride ions from authorized SF uses was low. In addition, these authors suggested MRLs of SF and fluoride ions of several commodities. For example, for cereal grains, while the existing EU MRLs of SF and fluoride ions are 0.05 and 2 mg/kg, respectively, MRLs of 0.01 and 2 mg/kg, respectively, were proposed by the authors.

Despite the challenges mentioned in the previous paragraph, with proper dose rates and management plans, SF is still considered safe and a viable alternative to MB in many applications. In addition to storage and processing facilities, SF has been registered for use in raw agricultural and processed food commodities such as cereal grains, flour, and dried fruits (EPPO, 2009; DAWE, 2022). Aside from MB, SF is the other fumigant approved for the control of *Halyomorpha halys* by the governments of Australia and New Zealand, allowing the same dose rates and minimum end-point concentrations as for MB (DAFF, 2022b). However, SF fumigation is not yet an approved treatment for *Trogoderma granarium* (DAFF, 2022a). It should be noted that while the ProFume® applicator manual does not allow for Ct products above 1,500 g-h/m^3, Ghimire et al. (2015) found the egg stage of *Trogoderma granarium* to be most tolerant of SF, requiring Ct products above 2,000 g-h/m^3 at 30°C to achieve quarantine level control. The ISPM15 standard for wood packaging approves SF fumigation with a minimum Ct product of 1,400 g-h/m^3 for a 24 h exposure time at 30°C (IPPC, 2018). If the wood's temperature is below 20°C, the minimum Ct product and required exposure time must not be less than 3,000 g-h/m^3 and 48 h, respectively.

Adam et al. (2010) used the economic engineering and partial budgeting approach to compare the costs of MB and SF fumigations. The data were obtained by interviewing six commercial fumigators. The costs of MB and SF as reported by the authors were 15 USD/kg and 11–15 USD/kg, respectively. They concluded that (1) the required amounts of SF and MB are the factor influencing the relative profitability of the two fumigants, (2) despite the fact that SF has a lower unit price than MB, fumigation of a 28,317-m^3 warehouse using SF is 28% to 55% more expensive than the same fumigation using MB, and (3) SF can be less expensive than MB for fumigating cocoa beans because the amount of SF required for cocoa beans is substantially lower. In addition, due to the lack of cross-resistance to phosphine (PH$_3$) (Jagadeesan and Nayak, 2017), SF has been suggested as a resistance breaker in managing phosphine-resistant insects (Nayak et al., 2014; Opit et al., 2016).

6.3.3 Phosphine

Phosphine, or hydrogen phosphide gas, is the most commonly used fumigant for bulk-stored cereal grains, oil seeds, and other bulk dried commodities due to its low cost and relative ease of use (Phillips et al., 2012). It can be obtained from both solid metal phosphide and cylinderized gas formulations. The two most common types of solid metal phosphide formulations are aluminium phosphide (AlP) and magnesium phosphide (MgP), which are available in the form of tablets, pellets, sachets, or plates. Phosphine gas is generated by the reaction between AlP or MgP with water molecules in the environment. The ready-to-use cylinderized PH$_3$ gas mixed with either carbon dioxide (CO$_2$) or nitrogen (N$_2$) is available under the trade name ECO$_2$FUME® (Solvay, 2021a), with 2% PH$_3$ + 98% CO$_2$, or Frisin® (S&A, 2022), with 1.7% PH$_3$ + 98.3% N$_2$, respectively. Another cylinderized gas formulation is cylinderized PH$_3$ gas with more than 90% purity, such as VAPORPH$_3$OS® (Solvay, 2021b) or UltraPhos® (Specialty Gases, 2022). Due to the ability of PH$_3$ gas to self-ignite at concentrations of more than 1.8% by volume, this high-purity PH$_3$ has to be diluted on site either with air or with other inert gases, using specifically designed equipment such as the Horn Diluphos System® (HDS) (Horn and Horn, 2006; Fosfoquim, 2022) or the UltraPhos® Air Mixing System (Ryan et al., 2014).

Despite its cost advantage, in many cases the use of PH$_3$ fumigation is limited due to the facts that (1) PH$_3$ is corrosive to certain metals, such as copper, silver, gold, and their alloys, (2) PH$_3$ fumigation requires substantially longer exposure time than other treatments, and (3) some species of stored-product insects have developed strong resistance to PH$_3$. The corrosive property and long exposure time restrict PH$_3$ fumigation primarily to bulk-stored commodities where delicate metals or electrical systems are not present and the commodity is planned for extended storage. Insect resistance to PH$_3$ is a well-known problem that has been aggravated over the past two decades (Nayak et al., 2020). Ineffective fumigation where PH$_3$ concentration and/or exposure time are insufficient allows insects to survive. These surviving insects then reproduce, passing on their genetic trait. Over time, this can result in a population of insects that is largely resistant to PH$_3$ (Friendship et al., 1986). PH$_3$ resistance in stored-product insects has been observed in Australia, Brazil, China, Greece, India, Morocco, Pakistan, South Korea, Turkey, the United States, Vietnam, and many other countries in Africa and Asia (Chayaprasert, 2022).

The test method for evaluating insect resistance to PH$_3$ was established by FAO (1975). Based on this method, the level of PH$_3$ resistance is expressed as the resistance factor, also referred to as the resistance ratio or resistance coefficient, which is calculated by dividing the lethal concentration (LC) of PH$_3$ achieved on the resistant population by the LC of PH$_3$ on a susceptible strain of the same species. Several studies have been conducted to investigate the severity of the PH$_3$ resistance problem in different stored-product insects, as shown in Table 6.6. This table shows that different strains of the same insect species can have vastly different PH$_3$ resistances. In addition, the resistance level in a geographical area may change over time due to changing pest management practices. Insects can become more tolerant of PH$_3$ because of poor pest management practices and less tolerant with good pest management practices. For example, Pengkum et al. (2018) found *Cryptolestes* spp. with resistance factors as low as 1.54 and as high as more than 13,577. As a simpler alternative to the FAO resistance test, a PH$_3$ tolerance test kit is commercially available (Cato et al., 2019; Afful et al., 2021). Unlike for MB and SF, for PH$_3$ insect mortality is

Table 6.6 Resistance Factors of Some Stored-Product Insects to PH_3 Reported by Various Studies

Insect Species	Author	Resistance Factor
Cryptolestes spp.	Pengkum et al. (2018)	1.54 – more than 13,577
	Kengkanpanich et al. (2019)	2 – more than 81
Cryptolestes ferrugineus (Stephens)	Ling (1998)	10.4–109.9
	Yuchi et al. (2008)	350.51–449.14
	Nayak et al. (2013)	7.3–1,457.8
Oryzaephilus surinamensis	Pimentel et al. (2007)	1.9–32.2
	Suthisut et al. (2020)	2–22
Tribolium castaneum	Pimentel et al. (2007)	1.0–186.2
	Uraichuen and Pengkum (2015)	23.55–62.63
	Wakil et al. (2021)	73.43–120.42
Rhyzopertha dominica	Ling (1998)	3.7–295
	Lorini et al. (2007)	3–549
	Pimentel et al. (2007)	1.0–71.0
	Yuchi et al. (2008)	5.39–32.64
	Song et al. (2011)	2.429–610.286
	Wakil et al. (2021)	74.02–126.67
Sitophilus oryzae (Linnaeus)	Ling (1998)	142.6–336.7
Sitophilus zeamais (Motschulsky)	Yuchi et al. (2008)	0.6–18.55
Sitophilus granarius	Wakil et al. (2021)	24.18–64.63
Trogoderma granarium	Wakil et al. (2021)	37.06–84.57

not well correlated with the area under the concentration curve. With PH_3, it has been found that given the same *Ct* product, a longer exposure time is more effective than a shorter one in achieving control (Winks, 1984; WHO, 1988; Tiongson, 1992). In other words, the *Ct* product required for a particular level of mortality decreased as exposure time was extended (Hole et al., 1976). This increase in required exposure time could be attributed to the narcosis effect by which treated insects are immobilized but after being exposed to fresh air become active again. As mentioned by Winks (1985), in several laboratory studies when insects were exposed to high levels of PH_3 concentration, the narcosis effect was observed. It has been hypothesized that during narcosis the PH_3 uptake rate of insects decreases, leading to insect survival and PH_3 resistance.

In order to prevent resistance, it is important that PH_3 fumigation be performed as effectively as possible. Fumigators must apply an optimal dose rate and exposure time that yields complete control of all insect species and life stages. Annis (2001) gathered PH_3 dosage response data on non-PH_3-resistant stored-product insects of most species from more than 70 studies and produced a 2D scatterplot, with data points showing complete and incomplete mortalities at various combinations of PH_3 concentrations and exposure times. Additional data points from PH_3-resistant *Rhyzopertha dominica* were added to the plot by Annis and Dowsett (2001). From the data points of these two review papers, a line was drawn dividing the plot into two areas in which the PH_3–exposure time combinations yielded complete and incomplete mortalities (Figure 6.1). From this plot, it can be deduced that in fumigation with an exposure time of less than 10 d, a PH_3 concentrations below 100 part per million (ppm) would most likely not produce successful control. In addition, for *Cryptolestes* spp., which is one of the most PH_3-resistant insects (Yuchi et al., 2008; Nayak et al., 2013), Liu et al. (2003), Reichmuth et al. (2004) suggested that 100 ppm at 30 d and 1,500 ppm at 60 h, respectively, would not produce successful control. Several other authors suggested combinations of PH_3 concentrations and exposure times for effective control of *Cryptolestes ferrugineus* (Table 6.7). Notice that these suggested combinations all lie within the "complete mortality" area in Figure 6.1.

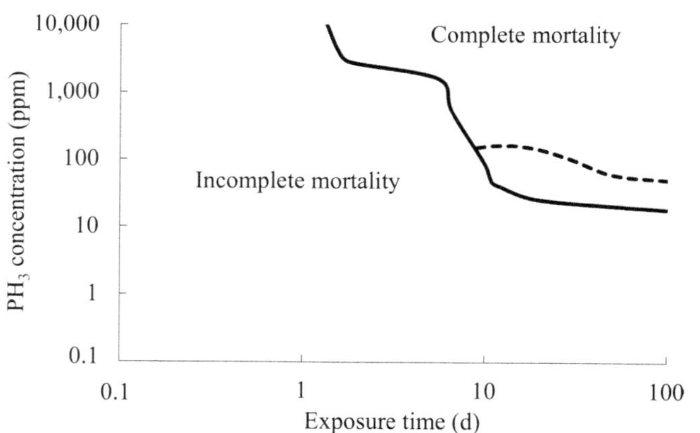

Figure 6.1 Phosphine dosage responses that produced complete and incomplete mortality, adapted from Annis (2001) (solid line) and Annis and Dowsett (2001) (dashed line).

Table 6.7 Recommended PH$_3$ Concentrations and Exposure Times for Effective Control of *Cryptolestes ferrugineus*

Author	Recommended Concentrations and Exposure Times
Li and Yan (2008)	200 ppm for 28 d
Wang et al. (2008)	300–500 ppm for 16–25 d
Collins (2009)	400 ppm for 14 d or 200 ppm for 25 d
Nayak et al. (2010)	720 ppm for 24 d or 360 ppm for 30 d
Kaur and Nayak (2015)	720 ppm for 15 d or 1,440 ppm for 6 d

6.3.4 Ethyl Formate

Another MB alternative that has been extensively studied (Ryan and De Lima, 2014; Ryan and Dominiak, 2021) and used as a fumigant in stored grains, nuts, and fresh or dried produce is ethyl formate (EF). It is also commonly used as a solvent in flavoring or fragrance production and rapidly breaks down into formic acid and ethanol. Ethyl formate is generally recognized as safe (GRAS) (FDA, 2022). It is a colorless liquid with a relatively low boiling point (52–54°C) and is highly volatile and flammable at normal ambient temperature (Lewis et al., 2016). Due to its high flammability, a diluted formulation or specifically designed dispensing system is needed.

A commercial fumigant under the trade name VAPORMATE™ is a liquid mixture of 16.7% EF and 83.3% CO$_2$ by weight, packaged in pressurized cylinders where CO$_2$ acts as a carrier gas and fire retardant (Linde, 2014). VAPORMATE™ has been registered in Australia, New Zealand, South Korea, Indonesia, Malaysia, Israel, and the Philippines (Wolmarans et al., 2017). Liquid EF with 98–99% concentration by weight, such as EMATE® (APVMA, 2022) and FUMATE® (SafeFume, 2022a), is also available. These EF liquid formulations must be used with specifically designed dispensers in which liquid EF is evaporated and mixed with CO$_2$ (Ryan and De Lima, 2014) or N$_2$ (SafeFume, 2022b). As an example, Table 6.8 shows the dose rates for some perishables and cereal grains given by the labels of these fumigants. Although EF is a safer fumigant than MB, SF, or PH$_3$ and has been reported to have significant toxicity against several insect species, its limitation seems to be higher sorption rates and lower penetrability (Phillips et al., 2012; Ramadan et al., 2022). However, when incorporated into appropriate cases, EF can be an effective pest management tool, especially in QPS situations.

Table 6.8 Examples of Dose Rates Given by the Labels of Commercial EF Fumigants

	VAPORMATE™	EMATE®	FUMATE®
Sweet potato	30 g/m³ at > 15°C for 2 h	-	-
Sweet pepper	-	12 g/m³ at > 15°C for 2 h	-
Banana	420 g/m³ at > 15°C for 6 h	70 g/m³ at > 15°C for 6 h	35 g/m³ at > 15°C for 4 h
Citrus	90 g/m³ at > 10°C for 1 h	62 g/m³ at > 10°C for 6 h	70 g/m³ at > 5°C for 4 h
Cereal grains	420 g/m³ at > 15°C for 24 h	70 g/m³ at > 15°C for 24 h	-

6.3.5 Ozone

Ozone (O_3) is formed naturally in the earth's atmosphere by ultraviolet (UV) radiation or electric discharges through oxygen (O_2) molecules. These processes break O_2 molecules into individual atoms, which then react with other O_2 molecules to form O_3. Commercial O_3 generators work based on the UV radiation or electric discharge principle. Ozone is a strong antimicrobial agent and considered as GRAS (Emekci, 2010). Ozone molecules are not stable and decay naturally back to O_2, leaving no residue. The O_3 decay rate depends on the surface characteristics of the materials with which O_3 comes into contact and temperature (Hardin et al., 2010). A primary concern when using O_3 as a fumigant is that it can oxidize a wide range of materials, including metals, rubber, and plastics. Due to this concern and O_3's rapid degradation in water, Armstrong et al. (2014) concluded that O_3 is not a potential MB alternative for QPS treatments of export logs. The U.S. Food and Drug Administration permitted the use of O_3 in gaseous and aqueous phases on meat, poultry, and agricultural commodities for commercial purposes (Armstrong et al., 2014).

Ozone treatment, also referred to as ozonisation, can be used to control pest insects and mold in stored products (Maier et al., 2006; White et al., 2010; Hansen et al., 2013; Changchai et al., 2016). Jian et al. (2013) summarized O_3 dosage applications from seven studies, ranging from 25 ppm for 5 d to 120 ppm for 1.2 d. At these O_3 dosages, the mortalities of the treated stored-product insects were at least 90%. Under laboratory conditions, McDonough et al. (2010) exposed adult stages of *Tribolium castaneum* and *Sitophilus zeamais* to O_3 and achieved 100% mortality at 1,800 ppm for 120 min. However, Kells et al. (2001) treated 8.9 tonnes of infested stored maize in a galvanized steel grain bin with O_3 and found that 100% insect mortality could not be reached at 50 ppm for 3 d. Although O_3 does not penetrate through barriers readily, Hansen et al. (2012) achieved full mortality of internal stages of *Sitophilus* spp. and *Rhyzopertha dominica* residing within kernels at 135 ppm for 8 d. While O_3 can control pests and mold in stored products, effective treatments require proper dosages.

6.3.6 Other Fumigants

Several other fumigants have been explored as alternatives to MB and commercially used in QPS, postharvest, and structural fumigations. These include ethanedinitrile (EDN), hydrogen cyanide (HCN), propylene oxide (PPO), and carbonyl sulphide (COS).

Ethanedinitrile, also known as cyanogen, is a colorless gas with an almond-like odor effective at killing a wide range of pests, including insects, mites, and fungi (Emekci, 2010). It is a highly flammable liquefied gas with a flammability range of 6–32% by volume (Ryan et al., 2006). It is marketed as EDN™, EDN Fumigas™, or Sterigas™ in the form of compressed gas containing the active ingredient EDN in CO_2 at a concentration of 200 or 1,000 g/kg (Ryan et al., 2006; Environmental Protection Authority, 2021b; Draslovka, 2023). Due to its phytotoxicity and ineffectiveness on *Sitophilus* spp, EDN is not typically used as a MB alternative in grain fumigation (Ducom, 2006). It is mostly used in timber and soil fumigation. According to Environmental Protection Authority (2021b), EDN is approved in Australia for fumigation in timber and strawberries and is in various stages of approval or evaluation for use as a fumigant in Canada, China, India, Malaysia, Russia, South Korea, and the United States.

Hydrogen cyanide was widely used as a fumigant in the early 1900s, but it is now only used in a few applications. In the European Union, HCN is registered as a biocide in some countries under the trade name BLUEFUME™, CYANOPUR®, or URAGAN D2 and is packaged in cylinders or cans containing liquid HCN (boiling point of 25.6°C) as the active ingredient, plus other components (Stejskal et al., 2012; Hnatek et al., 2018; ECHA, 2023). In New Zealand, BLUEFUME™ is registered for use as a fumigant for the control of various storage pests and rodents in mills, warehouses, food factories, and empty ship holds and for the treatment of surface pests on bananas (Environmental Protection Authority, 2021a). In Australia, BLUEFUME™ is currently undergoing the registration process (AgroPages, 2023).

Propylene oxide is a colorless liquid at room temperature. It has a boiling point of 34°C and is flammable at 3–37% by volume. In the United States, a formulation of PPO in CO_2 called Propoxide 892 (Aberco Inc., 2014) is approved for use to control microbial and insect contamination in structures and several dried herbs and spices (Maille et al., 2023). In Australia, it is registered under the trade name Dibbs Progas for the control of microbiological spoilage in shelled almonds (APVMA, 2012).

Carbonyl sulphide is a colorless gas at room temperature (boiling point of $-50.2°C$) and is flammable at $11.9–28.5\%$ by volume (BOC, 2010). It was developed as a stored-product fumigant and trademarked in Australia as COSMIC™ (Wright, 2003). However, according to Armstrong et al. (2014), the application for both COS and COSMIC™ as fumigants in Australia in 2007 was not approved, and the manufacturer did not intend to pursue further registrations because of the successful registration of EDN.

To expand the use of the aforementioned fumigants as MB alternatives, ongoing research has been conducted to improve their efficacy and safety. In addition, a variety of commercially available chemicals have been approved for soil fumigation, including 1-3 dichloropropene (Telone®), chloropicrin (Chlor-O-Pic®), sodium tetrathiocarbonate (Enzone®), metam-sodium (Vapam®), dazomet (Basamid®), and metam-potassium (K-Pam®). After being applied, soil fumigants rapidly form gas via either volatilization or chemical transformation (Ajwa et al., 2010). For instance, sodium tetrathiocarbonate breaks down in the soil to carbon disulfide gas. However, we will not discuss the details of these chemicals as soil fumigants as that is not the focus of this chapter.

6.4 Controlled or Modified Atmospheres

Most CA and MA processes manipulate the concentrations of CO_2 or O_2 gases within the storage environment, which can be achieved by adding gaseous or solid CO_2, adding a gas of low O_2 content (e.g., pure N_2 or output from a hydrocarbon burner), or allowing metabolic processes within an airtight storage to remove O_2 (Jayas and Jeyamkondan, 2002). The processes in which CO_2 or N_2 is added are sometimes referred to as CO_2 or N_2 fumigations, respectively. In some cases, the disinfestation efficacy of gas concentration manipulation is enhanced by an increase or decrease in the pressure inside the treated enclosure. Pressurized gas cylinders are commonly used as the source of CO_2 or N_2. Nitrogen can also be produced at the treatment site via an N_2 generator.

Carbon dioxide and N_2 treatments are considered residue-free treatments that can be used on organic commodities (Clamp and Moore, 2004; DAWE, 2016). The minimum required concentration and the exposure time for CO_2 fumigation are 35% and 15 d, respectively, at a temperature of at least $20°C$ (Annis and Graver, 1991). In addition to being conducted at atmospheric pressure, CO_2 fumigation is also conducted at high pressure for use on a commercial scale, mostly for dried fruits, nuts, herbs, and spices. High-pressure CO_2 treatments performed in specially designed metal chambers at 1,000–3,700 kPa can shorten the exposure duration from days to hours (Navarro, 2006; Navarro et al., 2012). When producing a low-O_2 atmosphere with N_2, it is recommended that the O_2 level be maintained below 1% for controlling most common insect species of stored grain (Navarro, 2010; Athanassiou and Sakka, 2022). For N_2 treatments in 40 1,800-tonne concrete grain silos, at <1% O_2 concentration and 30°C, Clamp and Moore (2004) recommended at least 9 d exposure for complete disinfestation. Sakka et al. (2020) conducted an N_2 treatment in a 195-m^3 chamber and found that after 9 d of exposure to 1% O_2 and 28°C, a small percentage of *Tribolium castaneum* adults survived and production of a small number of progeny was recorded. However, with 1–3% O_2 concentration and 20–29°C temperature, the exposure times required for 100% kill increased to more than 14 d and in some cases more than 21 d (Navarro, 2006). For example, Yonglin et al. (2021) reported complete disinfestation of 100-tonne steel and 2,200-tonne concrete silos after 1–3 weeks at 20–25°C with N_2 generators delivering 97–99% pure N_2. To increase the efficacy of CA/MA treatments, combinations of high CO_2 and low O_2 concentrations as well as manipulation of temperature have also been explored (Neven and Mitcham, 1996; Mann et al., 1999; Navarro, 2006). Research and innovation in the CA/MA area continue to improve the treatment efficacy and cost effectiveness.

6.5 Enclosures

Fumigation and CA/MA treatments are typically conducted in various types of enclosures (Table 6.9). For fumigating of granular commodities in bags or bulk, enclosures are often constructed by covering bag stacks or bulk piles with tarpaulins. When more gastightness is required, a tarpaulin is placed beneath the materials to be fumigated; however, this requires advance planning to ensure that a layer of tarpaulins is placed under the stack or pile at the filling stage so as to avoid another handling. Prefabricated flexible storages such as the GrainPro Cocoon (GrainPro, 2022) and the Silo bag (Silo Bags, 2022) do not have structural rigidity but typically provide better gastightness. In rigid structures (i.e., buildings, grain bins, or ship holds), fumigation is typically performed without the entire structure being completely covered. Fumigation chambers are specifically built to be gastight. In addition, they can be constructed to withstand high pressure (Navarro et al., 2012) or vacuum (Quality Foodstuff Factory, 2022). To maintain product quality and/or increase treatment efficacy, some treatment chambers are equipped with systems to release treatment gas, monitor gas concentration, and control temperature and/or relative humidity (Doernemann, 1991; Kostyukovsky et al., 2021; b-Cat, 2022).

Table 6.9 Types of Fumigation Enclosures

Type of Enclosure	Construction	Size
Ship holds (cargo holds)	Rigid	Medium to large
Shipping containers	Rigid	Small
	Flexible if covered with tarps	
Railcars	Rigid	Small
Prefabricated flexible storages	Flexible	Small to medium
Bag stacks	Flexible	Small to medium
Bulk piles	Flexible	Small to large
Grain bins	Rigid	Small to large
• Steel		
• Concrete		
Fumigation or CA/MA chambers	Rigid	Small to medium
• Atmospheric		
• High or vacuum pressure		
Building structures	Rigid	Small to large
• Residential	Flexible if covered with tarps	
• Commercial		
• Factory or warehouse		

Navarro (1998) classified structures for gaseous treatments into rigid and flexible in terms of construction. Structures with volumes less than 500 m^3, between 500 and 2,000 m^3, and greater than 2,000 m^3 were classified as small, medium, and large, respectively, by this author. Sealing can be done either permanently or temporarily. Permanent sealing is typically preferred, provided it is allowed. The methods for constructing and sealing enclosures have been described in more detail by other authors (Graver, 2004; Jones et al., 2017). These methods include

- covering entire structures, bag stacks, or grain piles with tarpaulins and weighing the edges with sand snakes or taping them to the floor;
- closing all doors, windows, hatches, manholes, slide gates, etc.; and
- sealing all seams and openings with plastic sheets, adhesive tape, expandable spray foam, glue, sealants, and/or heat.

Typically, gastightness of enclosures can be checked by visual inspection before fumigation or estimated either from historical fumigant concentration readings or from those measured during fumigation. A more quantitative and accurate method for determining gastightness is pressure testing the enclosure (Navarro, 1998). Pressure tests can be performed with either variable or constant pressure. In the variable pressure test, pressure inside an enclosure is increased or decreased to a certain level and allowed to decay naturally. The pressure decay time is then measured and used as the gastightness indicator. In the constant pressure test, the enclosure is pressurized at different constant pressure levels using a specially calibrated fan. At each pressure level, the rate of air flow that passes through the fan is measured. The pressure–flow rate measurements are then plotted to form the leakage characteristic curve of the enclosure. The constant pressure test, also known as the blower door test, is a standard method for determining a building's thermal performance (ASTM, 1996; ISO, 2006) and requires more sophisticated equipment and procedures. For fumigation purposes, the variable pressure test is often preferred.

For CO_2 fumigation in plastic enclosures, Annis and Graver (1991) recommended that it take at least 10 min for a pressure of 500 Pa to fall to 250 Pa. This measure is known as the pressure half-life. For rigid structures, such as sealed steel silos, a positive pressure is applied and a minimum pressure half-life of 5 min is required, according to Standards Australia (2010). For MB fumigation chambers, DAWE (2018) requires a minimum pressure half-life of 10 s. Based on information gathered from various publications, Navarro (1998) recommended pressure half-lives of 1–3 min, 3–6 min, and 5–9 min for fumigation, CA treatments, and MA treatments, respectively, in 95% filled enclosures.

6.6 Gas Introduction Methods

Several methods can be used to introduce treatment gases (i.e., fumigants or CA/MA gases) into enclosures, as summarized in Table 6.10. When selecting the introduction method, the primary factor is the source of treatment gas, which could be in the form of a solid, cylinderized gas, or a generator. Other factors such as the type of enclosure, type of fumigated goods and their packaging, time constraints, surrounding area, and environmental conditions must also be considered. Other details that must be decided include whether a gas recirculation system should be used, evaporation of gas should be assisted using a heat exchanger, or rapid gas distribution should be obtained via initial flushing or vacuuming. As an example, different gas introduction methods are graphically illustrated in Figure 6.2 as they are implemented in a grain bin treatment. It must be emphasized that each fumigant product has its unique set of requirements for safety and proper application. Therefore, fumigators must understand

Table 6.10 Summary of Methods for Introducing Fumigants or CA/MA Gases into Enclosures

Solid formulation	Fumigation	PH_3	**Method 1: Direct placement** • Solid formulation is placed directly inside an enclosure. • Solid formulation could be (a) placed directly on commodity or a suitable surface, (b) put in breathable bags or open containers and placed around commodity, or (c) mixed in with commodity.
			Method 2: Introduction via a reactor • Treatment gas is produced from a solid formulation in a reaction chamber and released into an enclosure. • In the reaction chamber, solid formulation is exposed to liquid water (wet process) or a high flow rate of moist air (dry process).
Cylinderized gas	Fumigation	PH_3, EF	**Method 3: Introduction via a mixer/blender** • Treatment gas with high purity and flammability is diluted with a carrier gas in a mixing machine to eliminate the risk of explosion. • Mixed gas is released into an enclosure. • To maintain lethal concentration, treatment gas can be released continuously or intermittently throughout the exposure period.
	Fumigation	PH_3, MB, SF, EF, EDN, HCN, PPO	**Method 4: Direct release** • Treatment gas is released into an enclosure directly from cylinder(s). • A heat exchanger is occasionally used to assist with evaporation. • Most of the time, gas is released via hoses, pipes, or tubes, but in some cases (for example, structural fumigation with multiple floors), gas cylinders are placed inside the enclosure. • To maintain lethal concentration, treatment gas can be released continuously or intermittently throughout the exposure period • For CA/MA, this method is typically used in small enclosures.
	CA/MA	CO_2, N_2	
Generator	Fumigation	O_3	**Method 5: Introduction via a generator** • Treatment gas is produced in a generator and released into an enclosure. • Treatment gas can be released continuously or intermittently throughout the exposure period.
	CA/MA	N_2	

Notes:

1. Variations of recirculation systems are often incorporated into the treatment gas introduction method.
2. Initial flushing or vacuuming could be applied to the enclosure to facilitate uniform gas distribution.
3. For Methods 3, 4, and 5, an automatic control system could be implemented to maintain gas concentration at a desired level.

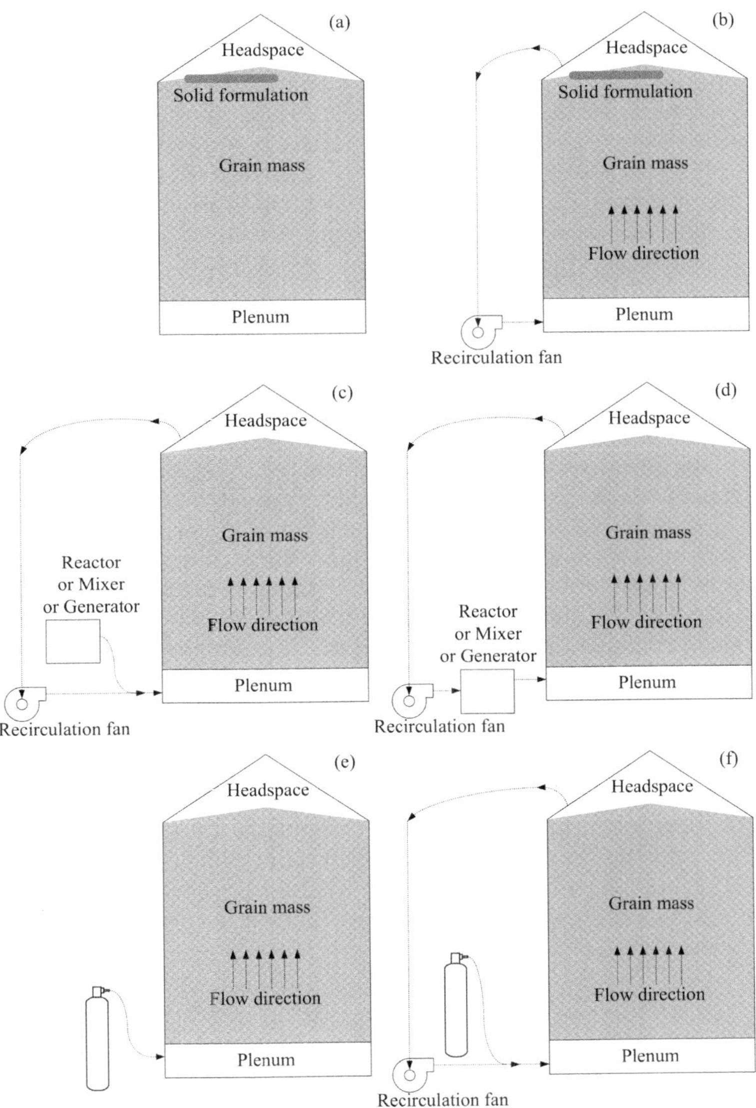

Figure 6.2 Graphical illustration of the fumigant and CA/MA gas introduction methods listed in Table 6.10. (a) Method 1 without recirculation, (b) Method 1 with recirculation, (c) Methods 2, 3, and 5 with recirculation fan in parallel, (d) Methods 2, 3, and 5 with recirculation fan in series, (e) Method 4 without recirculation, and (f) Method 4 with recirculation. Note that the point of gas entry can vary depending on several factors.

and operate according to the applicator's manual provided by the product's manufacturer. Although the machines used in Methods 2 and 3 are sometimes referred to by researchers and marketed by manufacturers as generators, for the purpose of the discussion in this chapter they are classified as reactors and mixers, respectively. Note that most fumigants used for soil fumigation are in the form of cylinderized gas and their introduction methods are typically shank injection or drip application (Klose et al., 2007). However, these soil fumigant introduction methods are not discussed in this chapter.

6.6.1 Direct Placement

For metal phosphide solid formulations that release PH_3, the simplest method is to place the tablets, pallets, sachets, or plates inside the enclosure. This direct placement method without recirculation in grain bin headspace is shown in Figure 6.2(a). It can also typically be seen in other types of enclosures. Tablets or pallets can be put in fire-resistant and breathable pouches or other containers for ease of residue collection. Although not a common practice in many countries, tablets or pellets are sometimes mixed in with the fumigated commodity. A typical recirculation system for a grain bin includes an electric fan and ducting, as shown in Figure 6.2(b). Jones et al. (2017) outline the process for designing such a recirculation system. Although the direct placement method is simple and requires no expensive equipment, the release of PH_3 gas is rather slow. Under favorable conditions (30°C and 90% relative humidity), an AlP tablet takes at least 35 h to react and release most of its PH_3 content (1 g of PH_3 gas) (Xianchang, 1994). In addition, when agricultural commodities are stored in enclosed spaces, condensation tends to occur, especially in metal grain bins. Thus, placing metal phosphide formulations in grain bin headspaces presents a risk of fire if condensation occurs and liquid water comes in contact with the formulations.

6.6.2 Introduction via a Reactor

To overcome the slow PH_3 release of the direct placement method, studies have been conducted to develop processes and equipment to speed up the reaction process of solid metal phosphide formulations. Since these formulations react with water molecules to produce PH_3 gas, the reaction can be accelerated by increasing the amount of water that comes in contact with the formulations. This can be done by combining them with liquid water (wet process) (Asher and Devidas, 2009) or maximizing the amount of air flowing through them (dry process) (Waterford and Winks, 2002). Based

Table 6.11 Some Commercial Models of Wet- and Dry-Type PH_3 Reactors

Wet-type	Model: Degesch Phosphine Generator Manufacturer: Degesch America (2002) Max capacity: 8.51 kg of PH_3 gas
	Model: QuickPHlo-R Phosphine Generator Manufacturer: UPL (2023) Max capacity: 25 kg of PH_3 gas
	Model: GFSL8-2A Manufacturer: Beijing Liangmao Technology Development (2010) Max capacity: 2.7 kg of PH_3 gas
Dry-type	Model: Speedbox® Manufacturer: Degesch America (2022) Max capacity: 396 g of PH_3 gas
	Model: Phosphine Fumigation Box Manufacturer: Kotzur (2021) Max capacity: 7,000 metric tonne silo

on these principles, a few PH_3 reactors of various designs have been successfully developed and commercialized (Table 6.11).

Figure 6.3(a) illustrates the basic working principle of the wet-type PH_3 reactor. The primary component of most wet-type PH_3 reactors is a reaction chamber made of stainless steel. In the chamber, a metal phosphide formulation is mixed with liquid water by either adding the formulation into water or vice versa. The formulation is generally in the form of either tablets or pallets. Some PH_3 reactors, such as the Degesch and QuickPHlo-R models, must be used with specialized formulations of magnesium and aluminium phosphide granules, respectively (D & D Holdings Inc., 2006; United Phosphorus Inc., 2008). The PH_3 gas produced is carried out of the chamber by forced air flow or injection of inert gas such as CO_2.

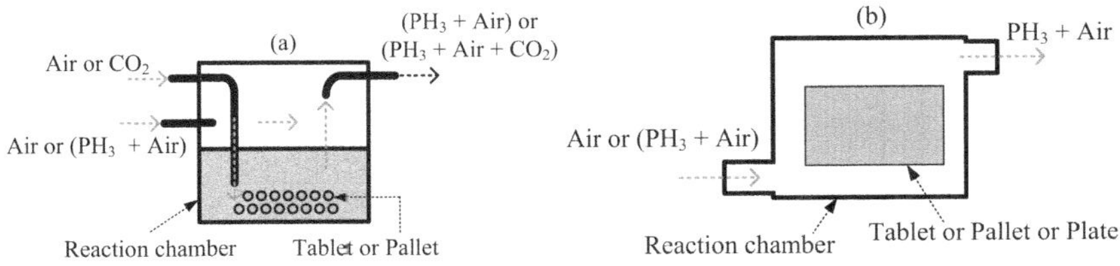

Figure 6.3 (a) Wet-type and (b) dry-type PH_3 reaction chambers.

To prevent self-ignition of PH_3 gas, the PH_3 concentration in the reaction chamber must be maintained below 1.8% by volume, which can be achieved by supplying a sufficient flow rate of the carrying air or CO_2. Furthermore, the CO_2 gas retards the flammability of PH_3. The forced air flow or injected CO_2 is also used to agitate the metal phosphide–water slurry. The PH_3 gas is then transferred to the fumigated enclosure typically via ducting and an electric fan, as shown in Figure 6.2(c) and (d), depending on the reactor design. A wet-type PH_3 reactor is typically used in conjunction with a gas recirculation system. It can be installed in parallel or in series with the recirculation fan. In addition to the reaction chamber, these PH_3 reactors consist of other components such as a control unit, an input/output user interface, a heating element, and a deactivation mechanism or a scrubber. These components add safety features and ease of use to the reactors. For example, in case of power loss during operation, the machine must be safely shut down.

In principle, the dry-type PH_3 generator is a chamber containing a metal phosphide formulation, as shown in Figure 6.3(b). The Speedbox® is designed to be used exclusively with magnesium phosphide plates (Degesh America, 2023), while the Phosphine Fumigation Box can be used with metal phosphide tablets or pallets. This chamber is installed in series in the recirculation flow circuit. The recirculation fan pulls moist air from the enclosure and pushes it through the chamber [Figure 6.2(d)]. The formulation reacts with the water vapor in the passing air, generating PH_3 gas. The PH_3 gas is then expelled back into the enclosure.

The advantages of using the wet- or dry-type reactor rather than the direct placement method include the shorter PH_3 release time, the ability to apply formulations at ground level, less risk of working in confined spaces, less risk of fire or combustion in enclosures, and easier residue collection. If the reaction rate is accelerated with liquid water, the entire required amount of PH_3 gas could be released within 1.5–6 h (Steuerwald et al., 2006; Asher, 2008; Ryan et al., 2010), depending on the reactor design and the size of the enclosure. In the case of the Speedbox®, in one study (Kostyukovsky et al., 2010) PH_3 concentration peaked after 15–17 h.

The main downside of using reactors is that they require capital investment, logistic and maintenance costs, and increased expenses for the applicators. Some models have rather low PH_3-producing capacity, and some require CO_2 gas to operate. Furthermore, as PH_3 can self-ignite at 18,000 ppm concentration, generating PH_3 gas in a small volume presents a severe explosion hazard. The risk of explosion substantially increases in circumstances where PH_3 gas accumulates in the reaction chamber. This can happen in various accidental scenarios. Primarily, PH_3 accumulation occurs when the flow rate of the carrier gas passing through the reaction chamber is below the desired level. In the case of the wet-type reactor, for example, the flow rate of injected CO_2 decreases when the CO_2 cylinder is nearly empty or the injection nozzle is obstructed. For the dry-type reactor, while the recirculation fan is still running, the concentration in the chamber can be maintained below the self-ignition level. The fan's operation can be interrupted, for example, due to an electricity power outage or a motor failure. Therefore, systems must be in place to prevent and manage these accidental scenarios. Firstly, the machine should be designed to prevent and detect such scenarios. Secondly, the personnel operating the machine must be properly trained. Thirdly, an uninterrupted power supply (UPS) should be used as back-up. Moreover, in the event of an unexpected fire or explosion, mitigating measures such as installation of a pressure relief panel, fire-extinguishing equipment, and evacuation plans should be implemented.

6.6.3 Introduction via a Mixer/Blender

As previously mentioned, before being released into an enclosure, high-purity cylinderized PH_3 (i.e., VAPORPH$_3$OS® and UltraPhos®) needs to be mixed with a carrier gas, diluting its concentration. The carrier gas could be CO_2, N_2, or air, depending on the specifications of the mixer. Various mixer models have been developed with a primary purpose to ensure that the concentration of mixed gas is below the self-ignition point. As an example, one of the simpler design is illustrated in Figure 6.4. Pressure regulators and flow control orifices are used to maintain the proper fixed flow rates of PH_3 and CO_2. More sophisticated designs can incorporate mass meters, control valves, and the electronics to allow for an adjustable flow rate while maintaining the proper PH_3-carrier gas ratio. Fosfoquim SA (2022) offers five models of the HDS PH_3 mixers, ranging in PH_3-releasing capacity from

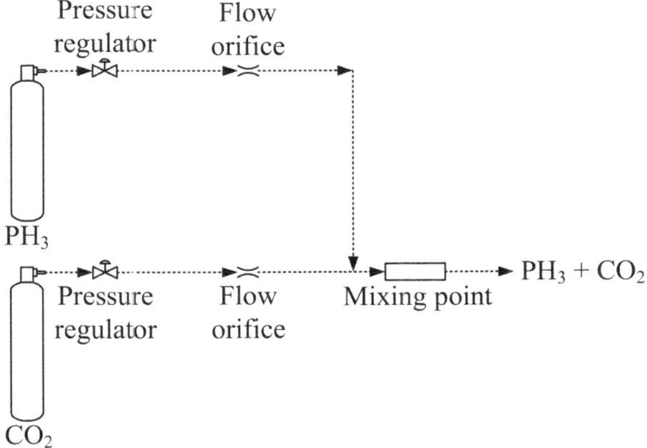

Figure 6.4 PH_3 and CO_2 blending equipment as simplified. (From Cytec Industries Inc., 2013.)

0.2 to 190 g/min. A different PH_3 mixer design with 1.4 g/min capacity was developed by GasApps Australia P/L (Ryan et al., 2014).

The Commonwealth Scientific and Industrial Research Organisation (CSIRO) developed cylinderized PH_3-dispensing processes for fumigating grain silos called SIROFLO®, SIROCIRC®, and SIROFUME® (Tumambing et al., 2012). SIROFLO® is a flowthrough process in which a 90-60 ppm PH_3 mixture is continuously released at the bottom of a silo and pushed through the grain mass, exiting at the top, similar to Figure 6.2(e), during the entire exposure period. The SIROCIRC® process includes a recirculation system similar to Figure 6.2(d) or (f). While in the SIROFLO® process the PH_3-carrier gas release rate is kept relatively constant, in the SIROCIRC® process as soon as PH_3 begins to recycle back from the top of the silo the release rate is lowered. The lowering of the PH_3-carrier gas release rate can be done manually or automatically, with the aim of maintaining the PH_3 concentration in the silo at a desired level. With SIROFUME®, all the required amount of PH_3 is released into the silo headspace at the beginning of the exposure period. Note that the fumigant gas for these processes could be obtained from an on-site PH_3 mixer or a premixed PH_3 formulation. The HDS PH_3 mixers in combination with any of the dispensing processes have been used to successfully fumigate grain and oilseeds in silos, bunkers, and flat storages ranging in size from 1,000 to 400,000 tonnes (Newman et al., 2001; Horn and Horn, 2006; Tumambing et al., 2012). Compared to the direct placement method and the use of solid metal phosphide reactors, on-site mixing of high-purity PH_3 gas can provide the fumigant at a higher rate without residue dust or slurry to be disposed of. A full cylinder of VAPORPH$_3$OS® contains 22 kg of PH_3 (Cytec Industries Inc., 2013), which is equivalent to 22,000 AlP tablets. The fumigator can benefit from ease of operation and a safer working environment.

A commercial EF-dispensing machine called SFM-1 is offered by SafeFume (2022b). In this machine, liquid EF (i.e., EMATE® and FUMATE®) is evaporated by and blended with heated N_2 gas, yielding a nonflammable EF-N_2 mixture (Yang et al., 2017). It has been mostly used in QPS of fresh produce (SafeFume, 2022b).

6.6.4 Direct Release

Most chemicals for fumigation and CA/MA are packaged in the form of ready-to-use cylinderized gas. In some cases, CO_2 or N_2 could be in on-site bulk storages or CO_2 could be used in solid form (dry ice). The usual gas introduction method for the cylinderized gas is simply releasing the contents of the cylinder directly into the enclosure, as in Figure 6.2(e) and (f). Usually, gas cylinders are located outside the enclosure and gas introduction is made through hoses, pipes, or tubes. In addition to being leakproof, the materials and installation of these hoses must be able to withstand high pressure, chemical corrosion, and temperature changes. Often, a pressure regulator that has been specially approved for the product must be used. For large enclosures, multiple cylinders can be released at multiple locations at the same time. For small and medium enclosures, the gas cylinder is placed on a scale to ensure that the required dose is accurately applied. In the case of MB, the fumigant can be released from a dispenser into which the target amount has been prefilled from the gas cylinder. To assist the evaporation of the product, the introduction hose can be routed through a heat exchanger such as an electric heater or a mass of boiling water. The outlet end of the introduction hose must be placed so that the gas injection is not directed toward commodities or any obstructions. When there is no risk of explosion, corrosion, or leakage, one or more recirculation fans is placed inside the enclosure and the outlet is placed in the direction of the air flow from the fan(s).

For premixed cylinderized PH_3, SIROFLO®, SIROCIRC®, and SIROFUME®, similar processes could be implemented. For CA/MA with CO_2 or N_2, in order to rapidly achieve the desired concentration (i.e., > 35% of CO_2 or < 1% of O_2, respectively), it is sometimes necessary to alternate between releasing the treatment gas and vacuuming the enclosure during the initial phase of gas introduction. To maintain a lethal concentration, the treatment gas can be released continuously or intermittently throughout the exposure period. This process can be controlled manually or automatically. The direct release method does not require capital investment in expensive equipment such as a reactor, mixer/blender, or generator. On the other hand, in large fumigation or CA/MA jobs where a substantial amount of treatment gas needs to be released, many gas cylinders must be transported and handled. Dry ice can be spread directly on grain without any quality deterioration, and gas can be recirculated to enhance uniformity of the CO_2 concentration throughout the bulk (Mann et al., 1999).

6.6.5 Introduction via a Generator

Currently, O_3 and N_2 can be generated on site based on processes that do not require cylinderized gas. Typical setups of O_3 and N_2 generators are shown in Figure 6.2(c) and (d). Note that setups without the recirculation loop can also be implemented. Ozone can be generated by passing atmospheric air through a high-voltage metallic grid or an ultraviolet (UV) light source. The former method, known as corona discharge, can in general provide higher O_3-generation capacity than the latter. Hardin et al. (2010) treated three galvanized steel grain bins, each containing 13.6 tonnes of hard red winter wheat, using an O_3 generator with a 120 g/h capacity. These researchers found that the generator could sustain O_3 concentrations between

60 and 100 ppm. Treating two grain bins storing 644 tonnes of popcorn with a 260 g/h O_3 generator, Maier et al. (2006) successfully maintained O_3 concentrations of at least 80 ppm. Several manufacturers offer O_3 generators with capacities of up to 1,500–3,000 g/h (Oxidation Technologies, 2017; Blue Planet, 2022; Oxyzone International, 2023).

Two commercially available N_2-generation technologies are pressure swing adsorption (PSA) and membrane generators, with the latter generally providing higher capacity but slightly lower purity. According to GRDC (2018), grain silos suitable for an N_2 treatment must be gastight with a pressure half-life of at least 5 min, complying with Standards Australia (2010). Furthermore, 99% N_2 gas should be delivered at the silo bottom until the O_2 concentration in the headspace reaches 1%, which typically requires at least two air changes. As an example, a 60 m^3/h N_2 generator would take about 4 h 20 min to achieve two air changes in a 130-m^3 silo. An N_2 generator with the size of a 12.2-m (40-ft) shipping container and a capacity of up to 6,800 m^3/h is available (Generon, 2023).

6.7 Concentration Monitoring

6.7.1 Conversion of Concentration Units

Fumigant and CA/MA gas concentrations can be expressed in either mass or volumetric concentrations—namely, g/m^3 or parts per million by volume (ppm), respectively. The exact conversion between the two types of concentration units can be calculated using the following equation:

$$ppm = \frac{c \cdot R \cdot T}{M \cdot P} \cdot 10^6 \tag{6.1}$$

where C is the concentration of the gas (g/m^3), R is the ideal gas constant (82.05×10^{-6} $m^3 \cdot$atm/mol·K), T is the temperature (K), M is the molecular mass (g/mol), and P is the pressure (atm). As an example, at 25°C and 1 atm, a concentration of 1 g/m^3 of MB, of which the molecular mass is 94.944 g/mol, is equivalent to

$$ppm = \frac{1 \cdot (82.05 \times 10^{-6}) \cdot (25 + 273.15)}{(94.944) \cdot 1} \cdot 10^6 = 258$$

Another commonly used mass concentration unit is ounces per 1,000 cubic feet (oz/Mcf). Note that the conversion factor from oz/Mcf to g/m^3 is 1.0012 and thus 1 oz/Mcf of MB is also approximately 258 ppm. The g/m^3-to-ppm conversion for the other treatment gases can be done in a similar fashion. For instance, for PH_3 (molecular mass = 34 g/mol) and

SF (molecular mass = 102.06 g/mol) at 25°C and 1 atm, a concentration of 1 g/m³ is equal to 720 and 240 ppm, respectively.

6.7.2 Concentration × Time (Ct) Product

As previously mentioned, the *Ct* product is the integration of the concentration curve. Chayaprasert (2007) and Cryer (2008) modeled gas concentrations during structural fumigation based on first-order kinetic approximations:

$$C_t = \frac{C_i}{2^{\frac{t}{\mathrm{HLT}}}} \tag{6.2}$$

where C_t is the current concentration (g/m³) at time t and C_i is the initial concentration (g/m³). The elapsed exposure time t and HLT are in units of hours. With an assumption that HLT is a constant, integration of equation 6.2 yields the corresponding *Ct* product in units of g-h/m³:

$$Ct = \frac{-C_i \cdot \mathrm{HLT} \cdot \left(2^{-\frac{t}{\mathrm{HLT}}} - 1 \right)}{\ln(2)} \tag{6.3}$$

However, the *Ct* product can also be approximated by summing the areas of trapezoids under the concentration curve (Technical Panel on Forest Quarantine, 2010). Note that for some gases such as PH_3 or CO_2 the *Ct* product model does not very well explain insect mortality.

6.7.3 Monitoring Method and Equipment

Gas monitors used in fumigation and CA/MA function based on various sensing principles, including colorimetry, electrochemistry (EC), thermal conductivity (TC), photoionization (PID), optical interferometry (OI), and near-infrared spectroscopy (NIR). For both fumigation and CA/MA, gas concentration monitoring is performed primarily for two purposes: efficacy and safety. Typically, gas monitors designed for one purpose could not be used for the other, as they operate in different measurement ranges. Efficacy monitoring to evaluate the disinfestation effectiveness of the treatment is done at gas concentrations in the higher ranges. Safety monitoring is done at lower concentrations close to the TLV levels to ensure that the enclosure is safe for reentry. For example, while PH_3 concentrations during a typical fumigation can vary from a few hundred to more than 1,500 ppm, the TLV of PH_3 is only 0.3 ppm (Graver, 2004).

Traditionally, a colorimetric tube such as that offered by Dräger (2022) is used for measuring gas concentrations; it is a glass tube containing a chemical reagent. When a specified volume of sample gas is drawn through the tube, the regent changes color. The gas concentration is indicated by the length of the stain or the intensity of the color. Except for colorimetric tubes, most gas monitors are equipped with electronic components that are battery operated or require plugging-in to an electrical outlet. Although several manufacturers produce low-range gas detectors for safety purposes, fumigation service providers must verify that the specifications of a particular detector (e.g., measuring range and sensitivity) meet the requirements of the fumigant label to be used as a clearance detection device. For instance, the ProFume® product label (Douglas Products, 2016) specifies that only a detector with a limit of detection of ≤1 ppm, such as the Interscan GF1900 (Interscan Corporation, 2023), can be used. Low-range electronic monitoring devices for fumigant and CA/MA gases are available from various manufacturers (Riken Keiki, 2019; Dräger, 2023; JJS Technical Services, 2023; Spectros Instruments, 2023b). For efficacy monitoring, traditionally gas samples are drawn through the plastic tubings placed into the monitoring device during filling and construction of the enclosure. Concentrations are usually measured at multiple locations to ensure disinfestation throughout the enclosure. For QPS fumigation with MB in enclosures larger than 30 m^3, the Australian government requires at least three gas monitoring locations (DAWE, 2018): at a top corner, in the middle, and at the bottom corner diagonally opposite the top corner (Figure 6.5). The sampled gas is passed

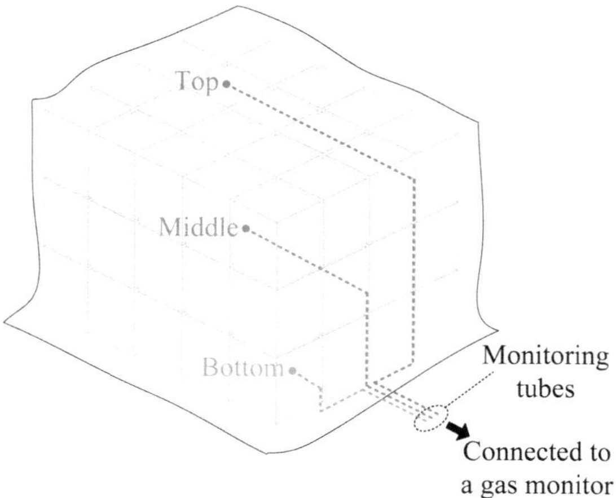

Figure 6.5 Example of top, middle, and bottom gas monitoring locations in an enclosure.

Table 6.12 Some Commercial High-Range Gas Monitors

Model	Type	Gas
Fumiscope (Key Chemical & Equiment Co, 2022)	TC	MB, SF
SF-ReportIR (Spectros Instruments, 2023a)	NIR	SF
PM200 (Spectros Instruments, 2023a)	NIR	PH_3
SiloChek (The Canary Company, 2022)	EC	PH_3
FI-21 (RKI, 2002)	OI	MB, PH_3, CO_2
MiniRAE 3000 (Honeywell International, 2023)	PID	MB, PH_3
G110 (AFC International, 2022)	NIR	CO_2

through a monitoring device and then either returned to the enclosure or vented to a safe unoccupied area. In addition to colorimetric tubes, some of the commercially available high-range monitors for various fumigation and CA/MA gases are given in Table 6.12. When selecting the monitoring device, factors such as price, accuracy, cross-sensitivity, ease of use, life span, durability, maintenance needs, and governmental regulations should be considered. For instance, a set of colorimetric tubes with an accompanying hand pump costs less than most electronic monitors, and there is no need for regular calibration. Meanwhile, electronic monitors require regular calibration, typically every six to twelve months. On the other hand, a colorimetric tube can give only one measurement while an electronic monitor can be repeatedly used, with an expected lifespan in the range of years.

Recent advancements in fumigation and CA/MA gas monitoring have led to the development of dosimeter tubes, automated monitors, and wireless sensors. The dosimeter tube, which is currently available for PH_3, is a colorimetric tube designed to measure the Ct product in units of ppm·h (Uniphos Envirotronic, 2013). During a fumigation, the tube is placed inside an enclosure where the reagent inside the tube reacts to the fumigant gas. At the end of the exposure period, the reagent's discoloration is proportional to the Ct product. Brabec et al. (2021) fumigated hard red winter wheat in 0.21-m³ (55-gallon) barrels using MgP pallets and compared the Ct product readings from dosimeter tubes against those calculated from concentration measurements at 1-h intervals. When Ct products were below 70,000 ppm·h, the dosimeter readings were within ±25% of the calculated Ct products. At levels above 70,000 ppm·h, the average difference between the dosimeter readings and the calculated Ct products was 44%. However, given the fact that the efficacy of PH_3 does not exactly follow the Ct product model, further studies might be needed to access the correlation between dosimeter readings and insect mortality.

The frequency of gas concentration measurements may vary depending on the length of the exposure period, ranging from measurements taken every couple of hours to those taken every one to three days. Working on treatments with long exposure periods at remote sites can be particularly challenging for service providers, as it can significantly increase their operating costs. This is often due to the fact that they must travel to the job site just to take readings. To address this issue, the pest-management industry is adopting automation and internet of things (IOT) technologies, and the monitoring process is one of the first areas where these technologies are being implemented. Manufacturers like Spectros Instruments (2022) and Uniphos Envirotronic (2022) offer commercial automatic monitoring systems that feature mobile internet integration and web or mobile application interfaces. A conceptual diagram of such systems is shown in Figure 6.6. The gas monitors of these systems have internet connectivity, which allows for receiving control commands and sending resulting measurements to the cloud server. This enables authorized personnel, including service providers, customers, and regulators, to remotely control the systems and access real-time gas concentrations from multiple job sites. Furthermore, additional functionalities can technically also be implemented. The system can alert relevant parties in real time in the event of excessive gas leakage, enabling prompt corrective actions to be taken. Other important parameters, such as temperature inside the enclosure and surrounding weather conditions, can also be tracked. With proper data security management, prompt presentation of treatment data to government officials can ideally facilitate the phytosanitary certification process.

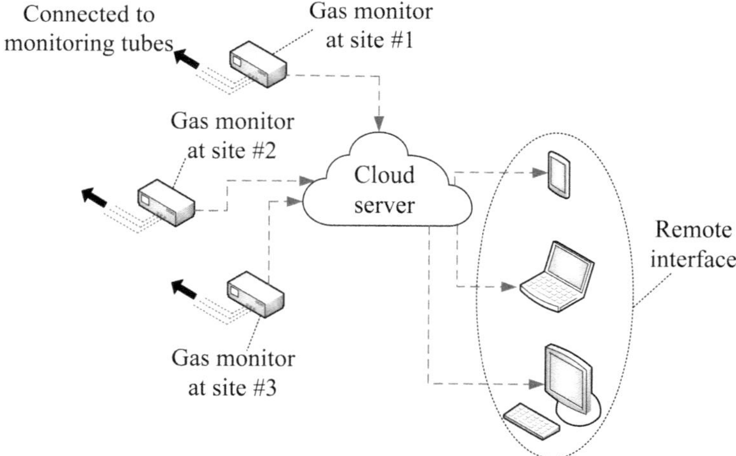

Figure 6.6 Conceptual diagram of an automatic gas monitoring system with remote interface.

Centaur Analytics (2023) and GrainPro (2023) are two companies that have recently introduced wireless monitoring systems for PH_3 and CO_2, respectively. These systems essentially comprise wireless sensors, a communication hub, and control software. The PH_3 sensor has a cylindrical shape with a diameter of 8 cm and a length of 17 cm, while the CO_2 sensor has a smaller size. The effectiveness of the PH_3 wireless monitoring system has been tested during fumigations in on-farm steel bins (Brabec et al., 2019) and in hopper-bottom railcars (Brabec et al., 2022). In the former study, fumigation trials were conducted on two bins, each storing 120 tonnes of wheat. Two wireless PH_3 sensors were placed about 10 cm below the surface of the grain mass in the headspace of each bin. The system was configured to record concentration measurements every 2 h, which was unlikely to be done with manual monitoring. In the latter study, four wireless sensors were placed on the grain surface in each railcar, which was loaded with 95 tonnes of corn grit. The railcars were transporting corn grits on an 8-d trip during which the fumigation took place. Unlike traditional monitoring systems, wireless gas sensors can be placed directly at desired locations inside the enclosure, eliminating the need to draw gas samples through plastic tubing to a monitor outside the enclosure. This can significantly reduce the preparation time of each treatment. Furthermore, for an in-transit fumigation scenario, continuous gas monitoring might not even be possible without this technology. However, while conventional monitoring can be done with one gas monitor and a network of tubings, wireless systems require a number of sensors to sufficiently capture gas distribution profiles in a large fumigated area. Thus, in order for the systems to be widely adopted, the overall cost of such systems will need to be made reasonable.

References

Aberco Inc. (2014) Propoxide 892, Aberco Inc, New Hampton, New York.

Adam BD, Bonjour EL, Criswell JT (2010) Cost comparison of Methyl Bromide and Sulfuryl Fluoride (ProFume®) for fumigating food processing facilities, warehouses, and cocoa beans. pp. 314–321. In: Carvalho MO, Fields PG, Adler CS, Arthur FH, Athanassiou CG, Campbell JF, Fleurat-Lessard F, Flinn PW, Hodges RJ, Isikber AA, Navarro S, Noyes RT, Riudavets J, Sinha KK, Thorpe GR, Timlick BH, Trematerra P, White NDG (eds), Proc 10th Intern Working Conf Stored-Prod Prot. Estoril, Portugal.

AFC International (2022) G110 CO_2 Analyzer with 0-100% CO_2 Carbon Dioxide Range. AFC International Inc. https://afcintl.com/p/g110-co2-analyzer-with-0-100-co2-range/. Accessed 10 March 2022.

Afful E, Cato A, Nayak MK, Phillips TW (2021) A rapid assay for the detection of resistance to phosphine in the lesser grain borer, *Rhyzopertha dominica* (F.) (Coleoptera: Bostrichidae). J Stored Prod Res 91, 101776.

AgroPages (2023) Australia to Approve Hydrogen Cyanide in the Products Bluefume Fumigant and Bluefume-D Fumigant. Chongqing Stanley Info-tech Co., Ltd. https://news.agropages.com/News/NewsDetail---45390.htm. Accessed 13 February 2023.

Ajwa H, Ntow WJ, Qin R, Gao S (2010) Properties of soil fumigants and their fate in the environment. In: Krieger R (ed), Hayes' Handbook of Pesticide Toxicology, 3rd ed. Academic Press, New York, pp 315–330.

Annis PC (1998) The relative effects of concentration, time, temperature and other factors in fumigant treatments. pp. 331–337. In: Zuxun J, Quan L, Yongsheng L, Xianchang T, Lianghua G (eds), Proc 7th Intern Working Conf Stored-Prod Prot, Beijing, China.

Annis PC (2001) Phosphine dosage regimes required for high mortality a data-base approach. pp. 45–55. In: Donahaye EJ, Navarro S, Leesch JG (eds), Proc 6th Intern Conf Controlled Atm Fumi Stored Prod, Fresno, USA, 2000.

Annis PC, Dowsett HA (2001) Time to kill phosphine resistant Rhyzopertha dominica using a concentration profile typical of field exposure. pp. 781–787. In: Donahaye EJ, Navarro S, Leesch JG (eds), Proc 6th Intern Conf Controlled Atm Fumi Stored Prod, Fresno, USA, 2000.

Annis PC, Graver JvS (1991) Suggested Recommendations for the Fumigation of Grain in the ASEAN Region: Part 2 Carbon Dioxide Fumigation of Bag-Stacks Sealed in Plastic Enclosures: An Operations manual. Australian Centre for International Agricultural Research. https://www.aciar.gov.au/publication/books-and-manuals/suggested-recommendations-fumigation-grain-asean-region. Accessed 5 January 2022.

APHIS (2016) Treatment Manual. Animal and Plant Health Inspection Service, United States Department of Agriculture, Riverdale, Maryland.

APVMA (2012) Public release summary on the evaluation of the propylene oxide in the product Dibbs Progas Fumigant. Australian Pesticides and Veterinary Medicines Authority, Australian Capital Territory, Australia.

APVMA (2022) VAPORFAZE EMATE INSECTICIDE: Label No. 64944. Australian Pesticides and Veterinary Medicines Authority, Sydney, Australia.

Armstrong JW, Brash DW, Waddell BC (2014) Comprehensive Literature Review of Fumigants and Disinfestation Strategies, Methods and Techniques Pertinent to Potential Use as Quarantine Treatments for New Zealand Export Logs, The New Zealand Institute for Plant & Food Research Ltd, Auckland, New Zealand.

Asher PP, Devidas SR (2009) Apparatus and method for rapid and continuous generation of phosphine gas. Patent Number US7556785 B2. United States of America.

Asher PP (2008) QuickPHlo-R formulation fumigant generators: a new safer and environmentally friendly phosphine fumigation process. pp. 402–405. In: Daolin G, Navarro S, Jian Y, Cheng T, Zuxun J, Yue L, Yanf L, Haipeng W (eds), Proc 8th Intern Conf Controlled Atm Fumi Stored Prod, Chengdu, China.

ASTM (1996) Standard E 1827-96: Standard Test Methods for Determining Airtightness of Buildings Using an Orifice Blower Door, ASTM International, West Conshohocken, Pennsylvania.

Athanassiou CG, Phillips TW, Aikins MJ, Hasan MM, Throne JE (2012) Effectiveness of sulfuryl fluoride for control of different life stages of stored-product psocids (Psocoptera). J Econ Entomol 105, 282–287.

Athanassiou CG, Sakka MK (2022) Using nitrogen for the control of stored product insects: One single application for multiple purposes. Agrochemicals 1, 22–28.

Barcan İlaçlama Fumigasyon Hizmetleri (2021) FumiTrack. 1.5 edn. Barcan İlaçlama Fumigasyon Hizmetleri, İzmir, Turkey.

Barnekow D (2011) Regulatory update for ProFume® gas fumigant. 2011 Annual International Research Conference on Methyl Bromide Alternatives and Emissions Reductions.

b-Cat (2022) The solutions for controlled atmosphere. b-Cat B.V. https://www.insect-treatment.com/products/. Accessed 31 December 2022.

Beijing Liangmao Technology Development (2010) User Manual for Phosphine Generator - Model: GFSL8-2A, Beijing Liangmao Technology Development Co. Ltd, Beijing, China.

Bell CH (2004) The use of sulphuryl fluoride in europe for structure and commodity disinfestation. In: Batchelor T, Alfarroba F (eds) Proceedings of International Conference on Alternatives to Methyl Bromide. European Commission, Brussels, Belgium, pp 237–240.

Bell CH (2006) Factors affecting the efficacy of sulphuryl fluoride as a fumigant. pp. 519–526. In: Lorini L, Bacaltchuk B, Beckel H, Deckers D, Sundfeld E, Santos JP, Biagi JD, Celaro JC, D'A LR, Faroni, Bortolini LOF, Sartori MR, Elias MC, Guedes RNC, Fonseca RG, Scussel VM (eds), Proc 9th Intern Working Conf Stored-Prod Prot, Campinas, São Paulo, Brazil.

Bell CH, Drinkall MJ (2000) Sulfuryl fluoride to replace methyl bromide use in the flour milling industry. 2000 Annual International Research Conference on Methyl Bromide Alternatives and Emissions Reductions. Methyl Bromide Alternatives Outreach.

Bell CH, Savvidou N, Wontner-Smith TJ (1998) The toxicity of sulfuryl fluoride (Vikane®) to eggs of insect pests of flour mills. pp. 345–350. In: Zuxun J, Quan L, Yongsheng L, Xianchang T, Lianghua G (eds), Proc 7th Intern Working Conf Stored-Prod Prot, Beijing, China.

Bell CH, Wontner-Smith TJ, Savvidou N (2002) Some properties of sulphuryl fluoride in relation to its use as a fumigant in the cereals industry. pp. 910–915. In: Credland PF, Armitage DM, Bell CH, Cogan PM, Highley E (eds), Proc 8th Intern Working Conf Stored-Prod Prot, York, UK.

Blue Planet (2022) Atom 4 ozone generator. Blue Planet. https://eblue-planet.eu/en/main/214-atom-4-ozone-generator.html. Accessed 17 December 2022.

BOC (2010) Safety Data Sheet: COSMIC 1000 Fumigant, BOC Limited, New South Wales, Australia.

Bond EJ (1984) Manual of fumigation for insect control. FAO Plant Production and Protection Papers - 54. Food and Agriculture Organization of the United Nations, Rome, Italy.

Brabec DL, Campbell J, Arthur F, Casada M, Tilley D, Bantas S (2019) Evaluation of wireless phosphine sensors for monitoring fumigation gas in wheat stored in farm bins. Insects 10, 121.

Brabec D, Kaloudis E, Athanassiou CG, Campbell J, Agrafioti P, Scheff DS, Bantas S, Sotiroudas V (2022) Fumigation monitoring and modeling of Hopper-Bottom railcars loaded with corn grits. J Biosyst Eng 47, 358–369.

Brabec D, Morrison W, Campbell JF, Arthur FH, Bruce A, Yeater K (2021) Evaluation of dosimeter tubes for monitoring phosphine fumigations. J Stored Prod Res 91, 101762.

Cato A, Afful E, Nayak MK, Phillips TW (2019) Evaluation of knockdown bioassay methods to assess phosphine resistance in the red flour beetle, *Tribolium castaneum* (Herbst) (Coleoptera: Tenebrionidae). Insects 10, 140.

Centaur Analytics (2023) Intelligent pest management and fumigation. Centaur Analytics Inc. https://centaur.ag/pest-control-and-fumigations/. Accessed 19 March 2023.

Changchai S, Varith J, Jaturonglumlert S, Narkprasom K, Theanworrakant N, Klinkajorn P (2016) Application of ozone gas to disinfest fruit fly (*Bactrocera latifrons*) in chilli. J Agric Res Ext 33, 13–23.

Chayaprasert W (2007) Development of CFD Models and an Automatic Monitoring and Decision Support System for Precision Structural Fumigation. Department of Agricultural and Biological Engineering. Purdue University, West Lafayette, Indiana, p 227.

Chayaprasert W (2022) A review of current fumigation practices in Thailand. Agric Nat Resour 56, 1233–1248.

Chayaprasert W, Maier DE (2010) Evaluating the effects of sealing quality on gas leakage rates during structural fumigation by pressurization testing and computational fluid dynamics simulations. Trans ASABE 53, 853–861.

Chengdu Taiyu Industrial Gases (2022) Sulfuryl Fluoride, SO2F2 Industrial Gas. Chengdu Taiyu Industrial Gases Co., Ltd. https://www.taiyugas.com/industrial-gases/sulfuryl-fluoride-so2f2-industrial-gas.html. Accessed 19 November 2022.

Clamp P, Moore D (2004) Nitrogen treatment of grain, Newcastle Grain Terminal. In: Wright EJ, Banks HJ, Highley E (eds) Proceedings of the 2nd Australian Postharvest Technical Conference. Stored Grain Research Laboratory, CSIRO, pp 184–186.

Collins P (2009) High Resistance to Phosphine. Grains Research and Development Corporation. https://grdc.com.au/resources-and-publications/grdc-update-papers/tab-content/grdc-update-papers/2009/09/high-resistance-to-phosphine. Accessed 3 January 2022.

Cryer SA (2008) Predicted gas loss of sulfuryl fluoride and methyl bromide during structural fumigation. J Stored Prod Res 44, 1–10.

Cytec Industries Inc. (2013) Application Manual for VAPORPH3OS® Phosphine Fumigant, CYTEC Industries Inc, Woodland Park, New Jersey.

D & D Holdings Inc. (2006) Degesch Magtoxin® Granules: Applicator's Manual, D & D Holdings Inc, Weyers Cave, Virginia.

DAFF (2022a) Requirements for sea containers to protect against khapra beetle. Department of Agriculture, Water and the Environment, Australian Government. https://www.awe.gov.au/biosecurity-trade/pests-diseases-weeds/plant/khapra-beetle/sea-container-measures#methyl-bromide-fumigation. Accessed 15 November 2022.

DAFF (2022b) Seasonal measures for Brown marmorated stink bug (BMSB). Department of Agriculture, Water and the Environment, Australian Government. https://www.agriculture.gov.au/biosecurity-trade/import/before/brown-marmorated-stink-bugs. Accessed 15 November 2022.

DAWE (2016) National Standard for Organic and Bio-Dynamic Produce. Department of Agriculture, Water and the Environment, Australian Government. https://www.awe.gov.au/sites/default/files/sitecollectiondocuments/aqis/exporting/food/organic/national-standard-edition-3-7.pdf. Accessed 28 February 2022.

DAWE (2018) Methyl bromide fumigation methodology - Version 2.0. Department of Agriculture, Water and the Environment, Australian Government. https://www.awe.gov.au/sites/default/files/sitecollectiondocuments/biosecurity/import/general-info/qtfp/aust-methyl-bromide-fumigation.pdf. Accessed 24 December 2021.

DAWE (2022) Quarantine and pre-shipment uses of methyl bromide, 2017–2020. Department of Agriculture, Water and the Environment, Australian Government, Canberra, Australia.

Degesch America (2002) Degesch Phosphine Generator, Degesch America Inc., Weyers Cave, Virginia.

Degesch America (2022) Degesch Speedbox®. Degesch America Inc. https://www.degeschamerica.com/wp-content/uploads/2016/05/Degesch-Speedbox.pdf. Accessed 1 February 2022.

Degesh America (2023) Applicator's manual for Degesch Fumi-Cel® and Fumi-Strip®. Degesh America Inc. http://www.degeschamerica.com/wp-content/uploads/2021/01/Degesch-17949-REV-Nov-2020-3-1.pdf. Accessed 1 February 2023.

Derrick MR, Burgess HD, Baker MT, Binnie NE (1990) Sulfuryl fluoride (Vikane): A review of its use as a fumigant. J Am Inst Conserv 29, 77–90.

Doernemann M (1991) Bulk material treatment and apparatus. Patent Number US4989363. United States of America.

Douglas Products (2016) ProFume® Gas Fumigant: Label (U.S. EPA Reg. No. 1015-79), Douglas Products and Packaging Co LLC, Liberty, Missouri.

Douglas Products (2017) Applicator Manual for ProFume® Gas Fumigant, Douglas Products and Packaging Co LLC, Liberty, Missouri.

Douglas Products (2022) Pest management. Douglas Products and Packaging Co LLC. https://douglasproducts.com/pest-management/.

Dow AgroSciences (2004) ProFume®, Product Label (U.S. EPA Reg. No. 62719-376). Dow AgroSciences LLC, Indianapolis, Indiana.

Dräger (2022) Dräger short-term tubes. Drägerwerk AG & Co. KGaA. https://www.draeger.com/en_seeur/Products/Short-term-Tubes?s=354. Accessed 22 February 2022.

Dräger (2023) Multi gas detectors. Drägerwerk AG & Co. KGaA. https://www.draeger.com/en-us_us/Productselector/Portable-Gas-Detection/Multi-Gas-Detectors?page=1&i=safety. Accessed 18 March 2023.

Draslovka (2023) EDN®. Draslovka Holding a.s. https://www.draslovka.com/edn. Accessed 8 January 2023.

Ducom PJF (2006) The return of the fumigants. pp. 510–516. In: Lorini L, Bacaltchuk B, Beckel H, Deckers D, Sundfeld E, Santos JP, Biagi JD, Celaro JC, D'A LR, Faroni, Bortolini LOF, Sartori MR, Elias MC, Guedes RNC, Fonseca RG, Scussel VM (eds), Proc 9th Intern Working Conf Stored-Prod Prot, Campinas, São Paulo, Brazil.

ECHA (2023) Biocidal product factsheet. European Chemicals Agency. https://echa.europa.eu/information-on-chemicals/biocidal-products/-/disbp/factsheet/CZ-0008969-0000/authorisationid. Accessed 10 February 2023.

Emekci M (2010) Quo Vadis the fumigants? pp. 303–313. In: Carvalho MO, Fields PG, Adler CS, Arthur FH, Athanassiou CG, Campbell JF, Fleurat-Lessard F, Flinn PW, Hodges RJ, Isikber AA, Navarro S, Noyes RT, Riudavets J, Sinha KK, Thorpe GR, Timlick BH, Trematerra P, White NDG (eds), Proc 10th Intern Working Conf Stored-Prod Prot, Estoril, Portugal.

Ensystex (2022) Zythor. Ensystex Inc. https://www.ensystex.com/zythor. Accessed 28 November 2022.

Ensystex (2023) Fumicalc. 1.6.3 edn. Ensystex Inc., Fayetteville, North Carolina, United States of America.

Environmental Protection Authority (2021a) Decision (APP203876). Environmental Protection Authority, New Zealand Government, Wellington, New Zealand.

Environmental Protection Authority (2021b) Updated science memo (APP202804-EDN). Environmental Protection Authority, New Zealand Government, Wellington, New Zealand.

EPANZ (2021) Reassessment of methyl bromide. Environmental Protection Authority, New Zealand Government. https://www.epa. govt.nz/public-consultations/decided/reassessment-of-methyl-bromide/#:~:text=In%20general%20terms%2C%20recapture%20 of,methyl%20bromide%20and%20associated%20controls. Accessed 28 October 2022.

EPPO (2009) Sulfuryl Fluoride Fumigation of Dried Fruits and Nuts to Control Various Stored Product Insects, European and Mediterranean Plant Protection Organization, Paris, France.

European Food Safety Authority, Anastassiadou M, Bernasconi G, Brancato A, Carrasco Cabrera L, Ferreira L, Greco L, Jarrah S, Kazocina A, Leuschner R, Magrans JO, Miron I, Nave S, Pedersen R, Reich H, Rojas A, Sacchi A, Santos M, Scarlato AP, Theobald A, Vagenende B, Verani A (2021) Review of the existing maximum residue levels for sulfuryl fluoride according to Article 12 of Regulation (EC) No 396/2005. EFSA Journal 19, e06390.

FAO (1975) Recommended methods for the detection and measurement of resistance of agricultural pests to pesticides. 16: Tentative method for adults of some stored cereals, with methyl bromide and phosphine. FAO Plant Protection Bulletin. Food and Agriculture Organization of the United Nations, Rome, Italy.

FAO (2011) Guide to Implementation of Phytosanitary Standards in Forestry, Food and Agriculture Organization of the United Nations, Rome, Italy.

FDA (2022) CFR - Code of Federal Regulations Title 21. U.S. Food and Drug Administration. https://www.accessdata.fda.gov/scripts/cdrh/ cfdocs/cfcfr/CFRSearch.cfm?fr=184.1295. Accessed 31 December 2022.

Fluoride Action Network (2013) The universal consensus on sulfuryl fluoride = keep it away from food. Fluoride Action Network. https:// fluoridealert.org/content/sf_consensus/. Accessed 4 December 2022.

Fosfoquim SA (2022) HDS Phosphine Fumigation Equipment. Fosfoquim SA. https://www.fosfoquim.cl/index.php?op=detalle-servicios&lang= ing&sub_area=4&id=7. Accessed 1 February 2022.

Friendship CAR, Halliday D, Harris AH (1986) Factors causing resistance to phosphine in insect pests of stored produce. In: Howe V (ed) Proceedings of GASGA Seminar on Fumigation Technology in Developing Countries. Tropical Development and Research Institute, pp 141–149.

Gandy DG, Chanter DO (1976) Some effects of time, temperature of treatment and fumigant concentration on the fungicidal properties of methyl bromide. Ann Appl Biol 82, 279–290.

Generon (2023) Generon family of nitrogen generators. Generon Inc. https:// www.generon.com/product_type/membrane-nitrogen-generator/.

Ghimire MN, Myers SW, Phillips TW (2015) Sulfuryl fluoride as a quarantine treatment for khapra beetle. 2015 Annual International Research Conference on Methyl Bromide Alternatives and Emissions Reductions.

GrainPro (2022) GrainPro Cocoon Cargo. GrainPro Inc. Ltd. https://www.grainpro.com/grainpro-cocoon-cargo. Accessed 31 December 2022.

GrainPro (2023) Ecowise standard. GrainPro Inc. Ltd. https://help.gpecowise.com/index.php/ecowise-standard.html. Accessed 19 March 2023.

Graver JvS (2004) Guide to Fumigation Under Gas-Proof Sheets, Food and Agriculture Organization of the United Nations, Rome, Italy.

GRDC (2018) Nitrogen- Notes for use in stored grain. Grains Research and Development Corporation. https://storedgrain.com.au/nitrogen-notes-use-stored-grain/. Accessed 6 March 2023.

Gressent A, Rigby M, Ganesan AL, Prinn RG, Manning AJ, Mühle J, Salameh PK, Krummel PB, Fraser PJ, Steele LP, Mitrevski B, Weiss RF, Harth CM, Wang RH, O'Doherty S, Young D, Park S, Li S, Yao B, Reimann S, Vollmer MK, Maione M, Arduini J, Lunder CR (2021) Growing atmospheric emissions of sulfuryl fluoride. J Geophys Res Atmos 126, e2020JD034327.

Hansen LS, Hansen P, Jensen KM (2012) Lethal doses of ozone for control of all stages of internal and external feeders in stored products. Pest Manag Sci 68, 1311–1316.

Hansen LS, Hansen P, Jensen K-MV (2013) Effect of gaseous ozone for control of stored product pests at low and high temperature. J Stored Prod Res 54, 59–63.

Hardin JA, Jones CL, Bonjour EL, Noyes RT, Beeby RL, Eltiste DA, Decker S (2010) Ozone fumigation of stored grain; closed-loop recirculation and the rate of ozone consumption. J Stored Prod Res 46, 149–154.

Hnatek J, Stejskal V, Jonas A, Malkova J, Aulicky R, Weiss V (2018) Two new fumigation preparations (EDN® and Bluefume™) to control soil, wood, timber, structural and stored product pest arthropods - an overview. The Kharkov Entomological Society Gazette 26, 115–118.

Hole BD, Bell CH, Mills KA, Goodship G (1976) The toxicity of phosphine to all developmental stages of thirteen species of stored product beetles. J Stored Prod Res 12, 235–244.

Honeywell International (2023) MiniRAE 3000. Honeywell International Inc. https://sps.honeywell.com/us/en/products/safety/gas-and-flame-detection/portables/minirae-3000. Accessed 19 March 2023

Hopkins JD, Johnson DR (2022) General fumigation - Classification 3: Training manual. Cooperative Extension Service, University of Arkansas. https://www.uaex.uada.edu/farm-ranch/pest-management/docs/training-manuals/AG1161.pdf. Accessed 23 November 2022.

Horn P, Horn F (2006) Large scale grain fumigations using pure cylinderized phosphine together with the HORN DILUPHOS SYSTEM. pp. 527–533. In: Lorini L, Bacaltchuk B, Beckel H, Deckers D, Sundfeld E, Santos JP, Biagi JD, Celaro JC, D'A LR, Faroni, Bortolini LOF, Sartori MR, Elias MC, Guedes RNC, Fonseca RG, Scussel VM (eds), Proc 9th Intern Working Conf Stored-Prod Prot, Campinas, São Paulo, Brazil.

Houck LG, Jenner JF, Mackey BE (1994) Phytotoxicity of methyl bromide fumigation to citrus fruit packed in shipping boxes. HortSci 29, 536a–536.

Hwaidi M, Collins PJ, Sissons M, Pavic H, Nayak MK (2015) Sorption and desorption of sulfuryl fluoride by wheat, flour and semolina. J Stored Prod Res 62, 65–73.

Indo Global Trade (2021) Indofum 99 GA. PT Indo Global Trade. https://indo-gt.com/product/fumigant/indofum.html. Accessed 28 December 2022.

Interscan Corporation (2023) GF1900 sulfuryl fluoride [Vikane®/ProFume®] monitor. Interscan Corporation. https://www.vikane-monitor.com/vikane-monitoring-products/gf1900-sulfuryl-fluoride-vikane-profume-monitor/. Accessed 18 March 2023.

IPPC (2018) Regulation of wood packaging material in international trade. International Plant Protection Convention, Food and Agriculture Organization of the United Nations, Rome, Italy.

ISO (2006) Standard 9972: Thermal Performance of Buildings - Determination of Air Permeability of Buildings - Fan Pressurization Method, International Organization for Standardization, Geneva, Switzerland.

Jagadeesan R, Nayak MK (2017) Phosphine resistance does not confer cross-resistance to sulfuryl fluoride in four major stored grain insect pests. Pest Manag Sci 73, 1391–1401.

Jagadeesan R, Nayak MK, Pavic H, Chandra K, Collins PJ (2015) Susceptibility to sulfuryl fluoride and lack of cross-resistance to phosphine in developmental stages of the red flour beetle, *Tribolium castaneum* (Coleoptera: Tenebrionidae). Pest Manag Sci 71, 1379–1386.

Jayas DS, Jeyamkondan S (2002) Modified atmosphere storage of grains meats fruits and vegetables. Biosyst Eng 82, 235–251.

Jian F, Jayas DS, White NDG (2013) Can ozone be a new control strategy for pests of stored grain? Agr Res 2, 1–8.

JJS Technical Services (2023) ToxiRAE Pro. JJS Technical Services. https://www.rae-gasmonitors.com/rae-systems-toxirae-pro.html. Accessed 18 March 2023.

Jones C, Hardin J, Bonjour E (2017) Design of closed-loop fumigation systems for grain storage structures. Division of Agricultural Sciences and Natural Resources, Oklahoma State University Extension. https://extension.okstate.edu/fact-sheets/design-of-closed-loop-fumigation-systems-for-grain-storage-structures.html. Accessed 3 January 2022.

Kamel AH, Fam EZ, Mahdy MT, Lotfy A, Sheltawy EM (1970) The phytotoxicity of methyl bromide fumigation on the germination of some seeds of certain crops. Bull Entomol Soc Egypt 4, 1–6.

Kaur R, Nayak MK (2015) Developing effective fumigation protocols to manage strongly phosphine-resistant *Cryptolestes ferrugineus* (Stephens) (Coleoptera: Laemophloeidae). Pest Manag Sci 71, 1297–1302.

Kells SA, Mason LJ, Maier DE, Woloshuk CP (2001) Efficacy and fumigation characteristics of ozone in stored maize. J Stored Prod Res 37, 371–382.

Kenaga EE (1961) Time, temperature and dosage relationships of several insecticidal fumigants. J Econ Entomol 54, 537–542.

Kengkanpanich R, Suthisut D, Noochanapai P, Sitthichaiyakul S (2019) Efficacy Studies on ECO2FUME Phosphine Fumigant for Control the Phosphine Resistant *Cryptolestes ferrugineus* Strains. Postharvest and Processing Research and Development Division, Department of Agriculture, Bangkok, Thailand.

Key Chemical & Equiment Co (2022) Fumiscope Version 5.1 Manual. Key Chemical & Equiment Co. http://www.fumiscope.com/pdf/51english languagefumiscopemanual.pdf. Accessed 25 February 2022.

Klose S, Ajwa HA, ShemTov S, Fennimore SA, Subbarao KV, MacDonald JD, Ferris H, Martin F, Gerik J, Mellano MA, Greene I (2007) Shank and drip applied soil fumigants as potential alternative to methyl bromide in California-grown cut flowers. 2007 Annual International Research Conference on Methyl Bromide Alternatives and Emissions Reductions.

Kostyukovsky M, Hazan T, Abu Alhija A, Shaikh Ibrahim A (2021) The use of FumBox® - mobile fumigation container for date treatment with phosphine. In: Zhu J (ed) Proceedings of the 1st International Electronic Conference on Entomology. MDPI, Basel, Switzerland.

Kostyukovsky M, Trostanetsky A, Yasinov G, Menasherov M, Hazan T (2010) Improvement of phosphine fumigation by the use of Speedbox. pp. 377–380. In: Carvalho MO, Fields PG, Adler CS, Arthur FH, Athanassiou CG, Campbell JF, Fleurat-Lessard F, Flinn PW, Hodges RJ, Isikber AA, Navarro S, Noyes RT, Riudavets J, Sinha KK, Thorpe GR, Timlick BH, Trematerra P, White NDG (eds), Proc 10th Intern Working Conf Stored-Prod Prot, Estoril, Portugal.

Kotzur (2021) Kotzur Phosphine Fumigation Boxes. Kotzur Pty. Ltd. https://kotzur.com/wp-content/uploads/2021/07/Fumigation-box-2021-07-01.pdf. Accessed 3 January 2022.

Lewis KA, Tzilivakis J, Warner DJ, Green A (2016) An international database for pesticide risk assessments and management. Hum Ecol Risk Assess 22, 1050–1064.

Linde (2014) VAPORMATE™ Manual for Fumigation, Linde AG, Pullach, Germany.

Ling Z (1998) Development and countermeasures of phoshine resistance in stored grain insects in Guangdong of China. pp. 642–647. In: Zuxun J, Quan L, Yongsheng L, Xianchang T, Lianghua G (eds), Proc 7th Intern Working Conf Stored-Prod Prot, Beijing, China.

Liu C, Xie W, Leng Y (2003) Test on recirculation fumigation in warehouse. Grain Storage 32, 11–14.

Li L, Yan Z (2008) Studies on prevention of resistance in Cryptolestes ferrugineus. pp. 616–619. In: Daolin G, Navarro S, Jian Y, Cheng T, Zuxun J, Yue L, Yanf L, Haipeng W (eds), Proc 8th Intern Conf Controlled Atm Fumi Stored Prod, Chengdu, China.

Lorini I, Collins PJ, Daglish GJ, Nayak MK, Pavic H (2007) Detection and characterisation of strong resistance to phosphine in Brazilian *Rhyzopertha dominica* (F.) (Coleoptera:Bostrychidae). Pest Manag Sci 63, 358–364.

Maier DE, Hulasare R, Campabadal CA, Woloshuk CP, Mason L (2006) Ozonation as a non-chemical stored product protection technology. pp. 773–777. In: Lorini L, Bacaltchuk B, Beckel H, Deckers D, Sundfeld E, Santos JP, Biagi JD, Celaro JC, D'A LR, Faroni, Bortolini LOF, Sartori MR, Elias MC, Guedes RNC, Fonseca RG, Scussel VM (eds), Proc 9th Intern Working Conf Stored-Prod Prot, Campinas, São Paulo, Brazil.

Maille JM, Edde PA, Phillips TW (2023) Efficacy of propylene oxide and ethyl formate as fumigants to control *Lasioderma serricorne* (F.) (Coleoptera: Ptinidae). J Stored Prod Res 100, 102047.

Mann DD, Jayas DS, White NDG, Muir WE (1999) Mortality of adult *Cryptolestes ferrugineus* (Stephens) exposed to changing CO_2 concentrations. J Stored Prod Res 35, 385–395.

McDonough MX, Mason LJ, Woloshuk CP, Campabadal CA (2010) Ozone technology in the post-harvest storage environment – a comparison of efficacy of high doses of ozone to insects treated under laboratory conditions and field conditions. pp. 386–388. In: Carvalho MO, Fields PG, Adler CS, Arthur FH, Athanassiou CG, Campbell JF, Fleurat-Lessard F, Flinn PW, Hodges RJ, Isikber AA, Navarro S, Noyes RT, Riudavets J, Sinha KK, Thorpe GR, Timlick BH, Trematerra P, White NDG (eds), Proc 10th Intern Working Conf Stored-Prod Prot, Estoril, Portugal.

MPI, (2022a) Brown marmorated stink bug: requirements for importers. Ministry for Primary Industries, New Zealand Government. https://www.mpi.govt.nz/import/vehicles-machinery-parts/brown-marmorated-stink-bug-requirements-for-importers/. Accessed 15 November 2022.

MPI (2022b) Forestry importing countries phytosanitary requirements. Ministry for Primary Industries, New Zealand Government. https://www.mpi.govt.nz/export/export-requirements/icpr-importing-countries-phytosanitary-requirements/forestry-icprs/. Accessed 18 November 2022.

Mühle J, Huang J, Weiss RF, Prinn RG, Miller BR, Salameh PK, Harth CM, Fraser PJ, Porter LW, Greally BR, O'Doherty S, Simmonds PG, Krummel PB, Steele LP (2009) Sulfuryl fluoride in the global atmosphere. J Geophys Res Atmos 114, D05306.

Navarro S (1998) Pressure tests for gaseous applications in sealed storages: theory and practice. pp. 385–390. In: Zuxun J, Quan L, Yongsheng L, Xianchang T, Lianghua G (eds), Proc 7th Intern Working Conf Stored-Prod Prot, Beijing, China.

Navarro S (2006) Modified atmospheres for the control of stored-product insects and mites. In: Heaps JW (ed), Insect Management for Food Storage and Processing, 2nd ed. AACC International, St. Paul, Minnesota, pp 105–146.

Navarro S (2010) Commercial applications of oxygen depleted atmospheres for the preservation of food commodities. In: Doona CJ, Kustin K, Feeherry FE (eds), Case Studies in Novel Food Processing Technologies. Woodhead Publishing Limited, Cambridge, United Kingdom, pp 321–350.

Navarro S, Timlick B, Demianyk CJ, White NDG (2012) Controlled or modified atmospheres. In: Hagstrum DW, Phillips TW, Cuperus G (eds), Stored Product Protection. Kansas State Research and Extension, Manhattan, Kansas, pp 191–202.

Nayak MK, Daglish GJ, Phillips TW, Ebert PR (2020) Resistance to the fumigant phosphine and its management in insect pests of stored products: A global perspective. Annu Rev Entomol 65, 333–350.

Nayak MK, Holloway JC, Emery RN, Pavic H, Bartlet J, Collins PJ (2013) Strong resistance to phosphine in the rusty grain beetle, *Cryptolestes ferrugineus* (Stephens) (Coleoptera: Laemophloeidae): Its characterisation, a rapid assay for diagnosis and its distribution in Australia. Pest Manag Sci 69, 48–53.

Nayak M, Holloway J, Pavic H, Head M, Reid R, Patrick C (2010) Developing strategies to manage highly phosphine resistant populations of flat grain beetles in large bulk storages in Australia. pp. 396–401. In: Carvalho MO, Fields PG, Adler CS, Arthur FH, Athanassiou CG, Campbell JF, Fleurat-Lessard F, Flinn PW, Hodges RJ, Isikber AA, Navarro S, Noyes RT, Riudavets J, Sinha KK, Thorpe GR, Timlick BH, Trematerra P, White NDG (eds), Proc 10th Intern Working Conf Stored-Prod Prot, Estoril, Portugal.

Nayak MK, Jagadeesan R, Kaur R, J DG, Reid R, Pavic H, Smith L, Collins PJ (2014) Developing sulfuryl fluoride as a 'phosphine resistance breaker' – the Australian experience. pp. 475–479. In: Arthur FH, Kengkanpanich R, Chayaprasert W, Suthisut D (eds), Proc 11th Intern Working Conf Stored-Prod Prot, Chiang Mai, Thailand.

Neven LG, Mitcham EJ (1996) CATTS (controlled atmosphere temperature treatment system): A novel tool for the development of quarantine treatments. Am Entomol 42, 56–59.

Newman C, Russell G, Shore W, Gock D, Ryan R (2001) Australian SIROCIRC® recirculatory phosphine fumigation systems at Xizui grain terminal and inland depots in China. pp. 297–306. In: Donahaye EJ, Navarro S, Leesch JG (eds), Proc 6th Intern Conf Controlled Atm Fumi Stored Prod, Fresno, USA, 2000.

OAR-DOA (2017) Training Manual on Pests, Regulations and Phytosanitary Requirements of Import Countries, Office of Agricultural Regulation, Department of Agriculture, Bangkok, Thailand.

Opit GP, Thoms E, Phillips TW, Payton ME (2016) Effectiveness of sulfuryl fluoride fumigation for the control of phosphine-resistant grain insects infesting stored wheat. J Econ Entomol 109, 930–941.

Outram I (1967a) Factors affecting the resistance of insect eggs to sulphuryl fluoride-I: The uptake of sulphuryl-^{35}S fluoride by insect eggs. J Stored Prod Res 3, 255–260.

Outram I (1967b) Factors affecting the resistance of insect eggs to sulphuryl fluoride-II: The distribution of sulphuryl-^{35}S fluoride in insect eggs after fumigation. J Stored Prod Res 3, 353–358.

Oxidation Technologies (2017) Products. Oxidation Technologies LLC. https://www.oxidationtech.com/products.html. Accessed 17 December 2022.

Oxyzone International (2023) Products. Oxyzone International Pty. Ltd. https://www.oxyzone.com.au/products/ozone/. Accessed 17 February 2023.

Papadimitriou VC, Portmann RW, Fahey DW, Mühle J, Weiss RF, Burkholder JB (2008) Experimental and theoretical study of the atmospheric chemistry and global warming potential of SO_2F_2. J Phys Chem A 112, 12657–12666.

Pengkum K, Uraichuen J, Suthisut D (2018) The Proper Phosphine Fumigation to Control Flat Grain Beetle Resistance Strains. Postharvest and Processing Research and Development Division, Department of Agriculture, Bangkok, Thailand.

Pest Control Technology (2011) EPA Proposes to Withdraw Use of Pesticide Sulfuryl Fluoride on Food. GIE Media Inc. https://www.pctonline.com/article/pct-011311-epa-proposes-withdraw-use-pesticide-sulfuryl-fluoride-food/. Accessed 5 December 2022.

Phillips TW, Thoms EM, DeMark J, Walse S (2012) Fumigation. In: Hagstrum DW, Phillips TW, Cuperus G (eds), Stored Product Protection. Kansas State Research and Extension, Manhattan, Kansas, pp 157–177.

Pimentel MA, Faroni LR, Tótola MR, Guedes RN (2007) Phosphine resistance, respiration rate and fitness consequences in stored-product insects. Pest Manag Sci 63, 876–881.

Quality Foodstuff Factory (2022) QFS - Vacuum fumigation chamber -by-CO_2. Quality Foodstuff Factory LLC. http://www.quality-ind.com/services.php?id=5. Accessed 2 January 2023.

Ramadan GRM, Maille JM, Phillips TW (2022) Sorption and desorption dynamics of ethyl formate and propylene oxide as fumigants in durable agricultural commodities. J Stored Prod Res 99, 102007.

Reichmuth CH, Klementz D, Rassmann W, Stenglein M, Münzel M (2004) Phosphine resistance in the stored product insect pests the rusty grain beetle Crytolestes ferrugineus and the granary weevil Sitophilus granarius. P, 613. In: Donahaye EJ, Navarro S, Bell C, Jayas D, Noyes R, Phillips TW (eds), Proceedings of the International Conference on Controlled Atmosphere and Fumigation in Stored Products. FTIC Ltd. Publishing, Israel.

Reichmuth CH, Klementz D, Rassmann W, Stenglein M, Münzel M (2011) Phosphine resistance in the stored product insect pests the rusty grain beetle *Crytolestes ferrugineus* and the granary weevil *Sitophilus granarius*. pp. 613. In: Donahaye EJ, Navarro S, Bell C, Jayas D, Noyes R, Phillips TW, Daolin G (eds), Proc 7th Intern Conf Controlled Atm Fumi Stored Prod,. Gold-Coast Australia 8–13th August 2004. Sichuan Publishing Group, Sichuan, China.

Riken Keiki (2019) Portable Gas Leak Detector: SP-220 Series. Riken Keiki Co. Ltd. https://www.rkiinstruments.com/pdf/SP-220_Manual_FUM_SC_Types.pdf. Accessed 24 February 2022.

RKI (2002) Optical Gas Monitor Model FI-21: Instruction Manual. RKI Instruments Inc. https://www.rkiinstruments.com/pdf/mFI-21.pdf. Accessed 25 February 2022.

Ryan R, Dominiak BC (2021) Ethyl formate: review of a rapid acting fumigant. pp. 269–275. In: Jayas DS, Jian F (eds), Proc 11th Intern Conf Controlled Atm Fumi Stored Prod (CAF2020), CAF Permanent Committee Secretariat, Winnipeg, Canada.

Ryan RF, Grant N, Nicolson J, Beven D, Harvey A (2006) Sterigas™ and Cosmic™: update on proposed new fumigants. pp. 624–629. In: Lorini L, Bacaltchuk B, Beckel H, Deckers D, Sundfeld E, Santos JP, Biagi JD, Celaro JC, D'A LR, Faroni, Bortolini LOF, Sartori MR, Elias MC, Guedes RNC, Fonseca RG, Scussel VM (eds), Proc 9th Intern Working Conf Stored-Prod Prot, Campinas, São Paulo, Brazil.

Ryan RF, De Lima CPF (2014) Ethyl formate fumigation an overview update. pp. 969–978. In: Arthur FH, Kengkanpanich R, Chayaprasert W, Suthisut D (eds), Proc 11th Intern Working Conf Stored-Prod Prot, Chiang Mai, Thailand.

Ryan RF, Shore WP, NewMan CJE (2010) Phosphine generator trial using external air dilution. pp. 430–432. In: Carvalho MO, Fields PG, Adler CS, Arthur FH, Athanassiou CG, Campbell JF, Fleurat-Lessard F, Flinn PW, Hodges RJ, Isikber AA, Navarro S, Noyes RT, Riudavets J, Sinha KK, Thorpe GR, Timlick BH, Trematerra P, White NDG (eds), Proc 10th Intern Working Conf Stored-Prod Prot, Estoril, Portugal.

Ryan RF, Shore WP, Nicolson J (2014) On-site mixing of phosphine – continuous improvements. pp. 523–530. In: Arthur FH, Kengkanpanich R, Chayaprasert W, Suthisut D (eds), Proc 11th Intern Working Conf Stored-Prod Prot, Chiang Mai, Thailand.

S&A (2022) Frisin®. S&A Service und Anwendungstechnik GmbH. https://www.s-und-a.de/geschaeftsfelder/begasungen-spezialloesungen/frisin. Accessed 17 December 2022.

SafeFume (2022a) FUMATE®. SafeFume Inc. https://en.safefume.com/fumate. Accessed 31 December 2022.

SafeFume (2022b) SFM (Fumate exclusive fumication chamber). SafeFume Inc. https://en.safefume.com/sfm. Accessed 31 December 2022.

Sakka MK, Gatzali F, Karathanos VT, Athanassiou CG (2020) Effect of nitrogen on phosphine-susceptible and -resistant populations of stored product insects. Insects 11, 885.

Schneider B, Voglewede C, Shodrock D (2003) Developing precision fumigation plans using the ProFume Fumiguide. 2003 Annual International Research Conference on Methyl Bromide Alternatives and Emissions Reductions.

Silo Bags (2022) A New System for Storing Grains. Silo Bags International Ltd. https://silobags.com/. Accessed 31 December 2022.

Solvay SA (2021a) ECO2FUME®. Solvay S.A. https://www.solvay.com/en/brands/eco2fume. Accessed 29 December 2021.

Solvay SA (2021b) VAPORPH3OS® Phosphine Fumigant. Solvay S.A. https://www.solvay.com/en/product/vaporph3os-phosphine-fumigant. Accessed 24 December 2021.

Song X-h, Wang P-p, Zhang H-y (2011) Phosphine resistance in *Rhyzopertha dominica* (Fabricius) (Coleoptera: Bostrichidae) from different geographical populations in China. Afr J Biotechnol 10, 9911–9917.

Specialty Gases (2022) UltraPhos Phosphine Fumigant. Specialty Gases Pty Ltd. https://specialtygases.com.au/products/ultraphos-phosphine-fumigant/. Accessed 17 December 2022.

Spectros Instruments (2022) PM400 phosphine monitor. Spectros Instruments Inc. https://www.spectrosinstruments.com/prod_pm400.html. Accessed 14 February 2022.

Spectros Instruments (2023a) Fumigation monitor. Spectros Instruments Inc. https://www.spectrosinstruments.com/products.html. Accessed 19 March 2023.

Spectros Instruments (2023b) SFExplorIR - sulfuryl fluoride clearance monitor. Spectros Instruments Inc. https://www.spectrosinstruments.com/prod_sf-explor-ir.html. Accessed 18 March 2023.

Standards Australia (2010) Australian Standard AS2628-2010: Sealed grain-storage silos - sealing requirements for insect control. Standards Australia, Sydney, Australia.

Stejskal V, Liskova J, Ptacek P, Kucerova Z, Aulicky R (2012) Hydrogen cyanide for insecticide phytoquarantine treatment of package wood. pp. 687–693. In: Navarro S, Banks HJ, Jayas DS, Bell CG, Noyes RT, Ferizli AD, Emekci M, Isikber AA, Alagusundaram K (eds), Proc 9th Intern Conf Controlled Atm Fumi Stored Prod, Antalya, Turkey.

Steuerwald R, Dierks-Lange H, Schmitt S (2006) The Degesch phosphine generator – a fast phosphine application. pp. 559–563. In: Lorini L, Bacaltchuk B, Beckel H, Deckers D, Sundfeld E, Santos JP, Biagi JD, Celaro JC, D'A LR, Faroni, Bortolini LOF, Sartori MR, Elias MC, Guedes RNC, Fonseca RG, Scussel VM (eds), Proc 9th Intern Working Conf Stored-Prod Prot, Campinas, São Paulo, Brazil.

Suthisut D, Kengkanpanich R, Noochanapai P, Pobsok P, Sitthichaiyakul S (2020) Efficacy of phosphine against resistant strains of *Oryzaephilus surinamensis* L. (Coleoptera: Silvanidae). Entomology and Zoology Gazette 38, 12–22.

Technical Panel on Forest Quarantine (2010) Calculating CT values – Problems and possible solutions TPFQ paper for discussion. International Plant Protection Convention, Food and Agriculture Organization of the United Nations. https://www.ippc.int/static/media/files/publications/en/2013/06/05/1284067679_2010-TPFQ-37_-_Calculating_CT_Pr.pdf. Accessed 26 February 2022.

The Canary Company (2022) Silo-Chek. The Canary Company Pty. Ltd. https://canaryco.com.au/product/canary-silo-chek/. Accessed 14 February 2023.

Tiongson R (1992) Fumigation for the control of insect pests in storage. In: Semple RL, Hicks PA, Lozare JV, Castermans A (eds) Towards Integrated Commodity and Pest Management in Grain Storage. Food and Agriculture Organization of the United Nations, Rome, Italy, p 526.

Tumambing J, Depalo M, Garnier JP, Mallari R (2012) ECO$_2$FUME and VAPORPH$_3$OS phosphine fumigants - global application updates. pp. 363–373. In: Navarro S, Banks HJ, Jayas DS, Bell CG, Noyes RT, Ferizli AD, Emekci M, Isikber AA, Alagusundaram K (eds), Proc 9th Intern Conf Controlled Atm Fumi Stored Prod, Antalya, Turkey.

UNEP (2016) Minimising quarantine and pre-shipment (QPS) uses of methyl bromide: Tools for controlling, monitoring and reporting. United Nations Environment Programme, Nairobi, Kenya.

UNEP (2020) The Montreal protocol on substances that deplete the ozone layer. United Nations Environment Programme. https://ozone.unep.org/treaties/montreal-protocol. Accessed 7 November 2022.

UNEP (2021a) Country Data. United Nations Environment Programme. https://ozone.unep.org/countries. Accessed 7 November 2022.

UNEP (2021b) Critical-use exemptions approved by Meetings of the Parties. United Nations Environment Programme. https://ozone.unep.org/treaties/montreal-protocol/critical-use-exemptions-approved-meetings-parties. Accessed 4 November 2022.

UNEP (2022) Montreal protocol on substances that deplete the ozone layer - Report of the technology and economic assessment panel (Volume 1: Progress report). United Nations Environment Programme, Nairobi, Kenya.

Uniphos Envirotronic (2013) Uniphos Dosimeter Tubes for Fumigation Monitoring. Uniphos Envirotronic Inc. https://www.uniphosamericas.com/wp-content/uploads/2015/06/Technical-Note-07.pdf. Accessed 3 March 2022.

Uniphos Envirotronic (2022) FumiTrack Multiport with Expansion Module. Uniphos Envirotronic Pvt. Ltd. https://uniphos-envirotronic.com/fumigation-monitoring-equipment/phosphine/multi-port-with-expansion-module/. Accessed 14 February 2022.

United Phosphorus Inc (2008) Applicator's Manual for QuickPHlo-R® Granules, United Phosphorus Inc, King of Prussia, Pennsylvania.

UPL (2023) QuickPHlo-R Phosphine Generator. UPL Limited. https://www.upl-ltd.com/in/crop-protection/fumigants/quickphlo-r-phosphine-generator. Accessed 14 January 2023.

Uraichuen J, Pengkum K (2015) Phosphine resistance detection of red flour beetle (*Tribolium castaneum*). Postharvest and Processing Research and Development Division, Department of Agriculture, Bangkok, Thailand.

US-EPA (2016) EPA proposes to withdraw sulfuryl fluoride tolerances. United States Environmental Protection Agency. https://archive.epa.gov/oppsrrd1/registration_review/web/html/evaluations.html. Accessed 5 December 2022.

Wakil W, Kavallieratos NG, Usman M, Gulzar S, El-Shafie HAF (2021) Detection of phosphine resistance in field populations of four key stored-grain insect pests in Pakistan. Insects 12, 288.

Wang Y, Li J, He F (2008) Study on the control of flat grain beetle (*Cryptolestes ferrugineus*, Stephens) effectively with multi-fumigation technology and multi pesticide. pp. 507–510. In: Daolin G, Navarro S, Jian Y, Cheng T, Zuxun J, Yue L, Yanf L, Haipeng W (eds), Proc 8th Intern Conf Controlled Atm Fumi Stored Prod, Chengdu, China.

Waterford CJ, Winks RG (2002) Formulations containing a phosphinde for use in the controlled generation of phosphine. Patent Number US6440390 B2. United States of America.

White SD, Murphy PT, Bern CJ, van Leeuwen J (2010) Controlling deterioration of high-moisture maize with ozone treatment. J Stored Prod Res 46, 7–12.

WHO (1988) Phosphine and Selected Metal Phosphides - Environmental Health Criteria 73, World Health Organization, Geneva, Switzerland.

Williams LH, Sprenkel RJ (1990) Ovicidal activity of sulfuryl fluoride to anobiid and lyctid beetle eggs of various ages. J Entomol Sci 25, 366–375.

Winks RG (1984) The toxicity of phosphine to adults of *Tribolium castaneum* (Herbst): Time as a dosage factor. J Stored Prod Res 20, 45–56.

Winks RG (1985) The toxicity of phosphine to adults of *Tribolium castaneum* (Herbst): Phosphine-induced narcosis. J Stored Prod Res 21, 25–29.

Wolmarans F, Bikoba V, Mitcham EJ (2017) A review of VaporMate™ (ethyl formate & CO2): an effective post-harvest fumigant. 2017 Annual International Research Conference on Methyl Bromide Alternatives and Emissions Reductions.

Wontner-Smith TJ (2005) Evaluation of the Use of Sulphuryl Fluoride (Profume) in the Malting Industry in the United Kingdom. Home-Grown Cereals Authority, London, UK, p 11.

Wright EJ (2003) Carbonyl sulfide (COS) as a fumigant for stored products: progress in research and commercialisation. In: Wright EJ, Webb MC, Highley E (eds) Proceedings of the Australian Postharvest Technical Conference. Commonwealth Scientific and Industrial Research Organisation, Acton ACT, Australia.

Xianchang T (1994) Evolution of phosphine from aluminium phosphide formulations at various temperatures and humidities. pp. 201–203. In: Highley E, Wright EJ, Banks HJ, Champ BR (eds), Pro 6th Intern Working Conf Stored-Prod Prot, Canberra, Australia.

Yang JO, Kim HM, Park YJ, Park MG, Ren YL, Lee BH (2017) New quarantine trials for using liquid ethyl formate with nitrogen application on imported citrus fruits_cost effectiveness and worker safety. Proceedings of the International Symposium and Annual Meeting of the Korean Society of Pesticide Science (KSPS), p 245.

Yonglin R, Agarwal M, Newman J, Pan S, Li J, Xiao Y (2021) Delivery and adoption of nitrogen technology for the management of grain storage pests and grain qualtiy. pp. 160. In: Jayas DS, Jian F (eds), Proc 11th Intern Conf Controlled Atm Fumi Stored Prod (CAF2020), CAF Permanent Committee Secretariat, Winnipeg, Canada.

Yuchi C, Hui W, Jinhuo L, Qiang Z (2008) How to effectively control phosphine resistance development in stored-grain insects by integrated pest management. pp. 610–615. In: Daolin G, Navarro S, Jian Y, Cheng T, Zuxun J, Yue L, Yanf L, Haipeng W (eds), Proc 8th Intern Conf Controlled Atm Fumi Stored Prod, Chengdu, China.

Chapter 7

Extreme Temperatures for Insect Control

*Christos G. Athanassiou, Maria K. Sakka,
Marina Gourgouta, and Paraskevi Agrafioti*

7.1 Introduction

The utilization of extreme temperature is probably one of the most classical methods for the control of insect infestations in different kinds of food, both animal and plant materials and has been applied for centuries in many parts of the globe. Nevertheless, the application of extreme temperatures gained ground only during the last decades, especially after the withdrawal of methyl bromide and other active ingredients that were used extensively in different types of facilities. The phase-out of methyl bromide left an irreplaceable gap in processing facilities that were subjected to structural treatments at regular intervals. An additional parameter, which contributed further to the investigation of the feasibility of extreme temperatures at the industrial scale, is the development of resistance of key insect species to several active ingredients, particularly phosphine.

The term *extreme temperatures* does not correspond to specific application scenario and includes the application of both heat and cold, either on empty spaces or in the commodity. In terms of the duration of the treatment, extreme temperatures can range from short applications (i.e., usually less than 24 h), which often apply for heat, to longer intervals (i.e., 7 d or more), which are typically the case for cold. However, both heat and cold are also

DOI: 10.1201/9781003309888-7

used for longer or shorter intervals, respectively, depending on the application scenario. For instance, short exposures to elevated temperatures can be implemented through infrared or similar technologies that are applied directly on the commodity (Athanassiou et al., 2016). In the following sections, we describe the concept of the application of extreme temperatures.

7.2 Heat Treatment

The use of extreme temperatures is a promising alternative method to the use of chemicals for the control of stored-product insects. This method aims at increasing and maintaining the temperature (usually ranging between 50° and 60°C) in the treated area to control stored-product insects. Lethality in stored-product insects at elevated temperatures depends on temperature and exposure time. Heat treatment is based on the application of high temperatures for a relatively short time for the control of various stored-product insect species (Mahroof et al., 2003a; Wang et al., 2006). At high temperatures, insects die through desiccation and cell/enzymatic disruption (Hepburn, 1985; Hulasare et al., 2010). It involves an increase in the ambient air temperature inside the treated facility to between 50° and 60°C. In theory, the target temperature should be at least 50°C, in which case only some minutes are needed to kill all life stages (Dowdy and Fields, 2002; Wright et al., 2002; Yu et al., 2011).

Heat treatment has been long regarded as a viable means for stored-product insect control (Mahroof et al., 2003a,b; Agrafioti et al., 2019; Sakka et al., 2022b). Heat can be applied with different techniques through electric heaters that are placed inside the facility or steam heaters that are placed outside of the facility, with hot air forced to circulate through ducts within the facility (Agrafioti et al., 2019; Sakka et al., 2022b). Heaters, electric or steam, are used to gradually increase the air temperature to the desired level (Mahroof and Subramanyam, 2006). There are cases; however, where both techniques are applied simultaneously (Opit et al., 2011; Brijwani et al., 2012a; Agrafioti et al., 2019; Sakka et al., 2022b). The target temperature of heat treatment is typically 50°C, where all life stages of most stored-product insects die (Tables 7.1, 7.2) (Dean, 1911; Fields, 1992; Dowdy and Fields, 2002; Yu et al., 2011). Temperatures as high as 60°C may damage sensitive equipment inside the facility (Mahroof et al., 2003a; Campolo et al., 2013). The duration of the treatment is usually 24–36 h, depending on various characteristics, such as the size of the facility, its structural complexity, and the target species (Roesli et al., 2003; Mahroof et al., 2003a; Yu et al., 2011). Several studies have been published on the efficacy of the use of elevated temperatures against stored-product insects (Tables 7.1 and 7.2) (Mahroof et al., 2003a; Boina and Subramanyam, 2004; Yu et al., 2011).

Table 7.1 Summary of Studies Dealing with Use of Heat Treatment with Different Stored-Product Insects under Field Conditions, Based on Research That Was Published after Fields (1992)

Order	Target Insect Pest	Life Stage	Temperature (°C)	Reference
Coleoptera	Tribolium castaneum Sitophilus oryzae	adults	>50	Sakka et al. (2022b)
	Rhyzopertha dominica Oryzaephilus surinamensis	adults	>50	Agrafioti et al. (2019)
	Oryzaephilus surinamensis	adults	45, 50, 55, 60, 65, 70, 75	Athanassiou et al. (2016)
	Tribolium confusum	adults		
	Tribolium castaneum	larvae, adults	50–60	Jian et al. (2013)
	Tribolium confusum	eggs, larvae, adults	>50	Campolo et al. (2013)
	Gnatocerus cornutus	larvae		
	Sitophilus oryzae	adults		
	Rhyzopertha dominica	adults		
	Tribolium castaneum	eggs, young larvae, old larvae, pupae, adults	>50	Brijwani et al. (2012b)
	Rhyzopertha dominica Tribolium castaneum	adults	>50	Opit et al. (2011)
	Lasioderma serriconrne	eggs, young larvae, old larvae, adults	50–60	Yu et al. (2011)
	Tribolium castaneum	young larvae	50–60	Hulasare et al. (2010)
	Tribolium confusum	old larvae		
	Lasioderma serricorne	eggs		
	Stegobium paniceum	old larvae		
	Carpophilus hemipterus	larvae	50	Finkelman et al. (2006)

(Continued)

Table 7.1 (Continued) Summary of Studies Dealing with Use of Heat Treatment with Different Stored-Product Insects under Field Conditions, Based on Research That Was Published after Fields (1992)

Order	Target Insect Pest	Life Stage	Temperature (°C)	Reference
	Tribolium castaneum	eggs, young larvae, old larvae, pupae, adults		Mahroof et al. (2003b)
	Tribolium confusum	eggs, young larvae, old larvae, pupae, adults	50–60	Boina and Subramanyam (2004)
Psocoptera	Liposcelis corrodens Liposcelis decolor	adults	>50	Opit et al. (2011)
Lepidoptera	Ephestia elutella	larvae	45, 50, 55, 60, 65, 70, 75	Athanassiou et al. (2016)
	Plodia interpunctella	old larvae	50–60	Hulasare et al. (2010)
	Plodia interpunctella	eggs, fifth instars, pupae, adults	44–52	Mahroof and Subramanyam (2006)

7.2.1 Studies of Various Species and Life Stages

There are species and life stages that are considered more tolerant than others of heat treatment (Tables 7.1, 7.2). For instance, Boina and Subramanyam (2004) reported that, among the species and life stages tested, young larvae of the confused flour beetle, *Tribolium confusum* (Jacquelin du Val) (Coleoptera: Tenebrionidae), were most susceptible to heat, with old larvae being the least susceptible. In that study, the authors indicated that the old larvae of *T. confusum* were the least susceptible at 46, 48, 50, 58, and 60°C, while the adult stage was the most susceptible at 54°C. For the cigarette beetle, *Lasioderma serricorne* (F.) (Coleoptera: Anobiidae), Yu et al. (2011) tested all life stages under laboratory conditions and found that the egg was the most heat-tolerant

Table 7.2 Summary of Studies Dealing with Use of Heat Treatment with Different Stored-Product Insects under Laboratory Conditions, Based on Research That Was Published after Fields (1992)

Order	Target Insect Pest	Life Stage	Temperature (°C)	Reference
Coleoptera	*Tribolium confusum*	adults	30, 35, 37.5, 40, 45, 50	Lü et al. (2021)
	Lasioderma serricorne	adults, pupae, larvae, eggs	50	Li et al. (2018)
	Tribolium confusum Gnatocerus cornutus Sitophilus oryzae Rhyzopertha dominica	eggs, larvae, adults	>50	Campolo et al. (2013)
	Sitophilus granarius	larvae	50	Kljajić et al. (2009)
	Carpophilus hemipterus	eggs, larvae, pupae, adults	50 and 55	Sen and Emekci (2007)
	Carpophilus hemipterus	larvae	40, 45, 50, 55	Finkelman et al. (2006)
	Tribolium castaneum	larvae, pupae, adults	36–54	Arthur (2006)
	Rhyzopertha dominica	adults, pupae and prepupae	50–60	Beckett and Morton (2003)
	Trogoderma variabile	eggs, nondiapausing larvae, pupae, adults	50–60	Wright et al. (2002)
Lepidoptera	*Plodia interpunctella*	0–24 h eggs	42 or 46	Lewthwaite et al. (1998)

life stage, but the results were inconclusive in real-world heat treatments. According to Mahroof and Subramanyam (2006), old larvae of the Indian meal moth, *Plodia interpunctella* (Hübner) (Lepidoptera: Pyralidae), were more tolerant to heat than eggs, young larvae, pupae, and adults. In commercial applications in different types of facilities, Agrafioti et al. (2019) noted that there were no significant trends among the species tested, and survival was mostly due to the conditions of the application, rather than a species-mediated trend. Campolo et al. (2013) tested heat treatment in a flour mill against the rice weevil, *Sitophilus oryzae* (L.) (Coleoptera: Curculionidae), and found complete mortality after exposure for 24 and 36 h at 50–60°C. Moreover, Sakka et al. (2022b) evaluated the effectiveness of heat treatment against phosphine-susceptible and -resistant populations of the red flour beetle, *Tribolium castaneum* (Herbst) (Coleoptera: Tenebrionidae), and *S. oryzae* in commercial flour and rice mills and found no significant differences among the different populations. Probably the stored-product insect species most tolerant to heat is the khapra beetle, *Trogoderma granarium* Everts (Coleoptera: Dermestidae), which, as diapausing and acclimated larvae, can survive for 36–48 h at 45°C (Wilches et al., 2019).

In contrast with other control methods, heat treatment is highly effective against eggs (Brijwani et al., 2012b; Wilches et al., 2019). In general, eggs and adults are considered more susceptible to heat than old larvae for *T. confusum* and *T. castaneum* (Mahroof et al., 2003a, 2003b; Boina and Subramanyam, 2004). Moreover, Beckett et al. (2007) found that eggs and pupae were more tolerant of heat than other life stages for the lesser grain borer, *Rhyzopertha dominica* (F.) (Coleoptera: Bostrychidae). Agrafioti et al. (2019) found that eggs and newly hatched larvae of *R. dominica* and the saw-toothed grain beetle, *Oryzaephilus surinamensis* (L.) (Coleoptera: Silvanidae), were particularly susceptible to heat at different types of facilities and under different conditions. Burges (2008) reported that exposure to 60°C for 30 min is sufficient to control all life stages of *T. granarium*.

7.2.2 Acclimation

Few studies have examined the acclimation to high temperatures of stored-product insects. Indeed, Gonen (1977) showed that acclimated granary weevils, *Sitophilus granarius* (L.) (Coleoptera: Curculionidae), were more tolerant of 40°C than the nonacclimated insects. More recently, Lü and Zhang (2016) tested the effect of acclimation of the maize weevil, *Sitophilus zeamais* Motschulsky (Coleoptera: Curculionidae), to high temperatures (i.e., 36, 43, 47, 51, and 55°C) and different exposure times (i.e., 0, 1, 3, and 5 h) and found that acclimation to sublethal temperatures can enhance the heat tolerance of *S. zeamais* adults and decrease their mortality.

7.2.3 Mechanisms

As insects are poikilotherms, temperature is an important factor that affects growth, reproduction, and development (Wang et al., 2022). Physiological changes can be developed in response to heat treatment. These changes are related to biochemical changes in the production of heat shock proteins (HSPs) (Guedes et al., 2008). The vast majority of the pests have many genes encoding the HSP family (Mahroof et al., 2005). These genes have been examined for different pests including stored-product insects and psocids such as the bamboo borer, *Omphisa fuscidentalis* (Hampson) (Lepidoptera: Crambidae); *P. interpunctella*; *T. castaneum*; the book lice, *Leptinotus reticulatus* (Enderlein) (Psocoptera: Trogiidae); and the grain psocid, *Liposcelis entomophila* (Enderlein) (Psocoptera: Liposcelididae) (Lewthwaite et al., 1998; Mahroof et al., 2003, 2005; Guedes et al., 2008; Shim et al., 2009; Tungjitwitayakul et al., 2015). For instance, Guedes et al. (2008) tested the tolerance to heat shock stress of different stored-product psocids and found high variability and tolerance in *L. entomophila* and *L. teticulatus*. Moreover, the effect of heat stress on enzymes that are related to high temperatures have been evaluated for stored-product pests (Miao et al., 2020; Farahani et al., 2020). In a recent study, Wang et al. (2022) tested the effect of metabolic enzyme (superoxide dismutase [SOD] and peroxidase [POD]) and detoxification enzyme (acetylcholinesterase [AChE] and glutathione-S-transferase [GST]) activity on *S. oryzae* and *L. serricorne* and found that under high temperatures the metabolic enzyme activities of all enzymes except AChE were higher than those in the control.

7.2.4 Direct Application of Heat

One additional way to apply heat for the disinfestation of stored products is directly to the commodity. Athanassiou et al. (2016) found that the application of elevated temperatures through an infrared heater on Corinthian currants was able to control *O. surinamensis* provided that the application was 30 to 35 s at different temperatures (i.e., 45, 50, 55, 60, 65, 70, and 75°C). At the same time, these conditions reduced molds in the product (Athanassiou et al., 2016). In an earlier study, Khamis et al. (2010) found that a flameless catalytic infrared unit was able to completely control *R. dominica* on winter wheat. Navarro et al. (2004) found that heat treatment can be used with success to control the infestation of dried dates by the dried-fruit beetle, *Carpophilus hemipterus* (L.) (Coleoptera: Nitidulidae). In that work, the authors underlined that heat forced insects to go out of the fruit, an additional benefit of the method (Navarro et al., 2004). Similar studies have been carried out for different types of products, such as nuts and cereals (Wang and Tang, 2004; Beckett et al., 2007). Although heating large product quantities, such as bulked grains in silos, may not be economically feasible, the application of heat directly on the product through specific flow techniques (e.g., the conveyor belt) can be effective. In all cases, the quality of the final product should be preserved at the desired levels. Athanassiou et al. (2016) found that the use of infrared technology directly on currants and sultanas had a negative effect on the quality of the product when the temperature was above 60°C.

7.2.5 Combining Heat with Other Methods

One of the most important positive characteristics of the application of elevated temperatures is its compatibility with other methods (Table 7.3). For instance, heat can notably shorten the duration of the application of modified atmospheres (Athanassiou et al., 2017; Sakka et al., 2020, 2022a). Agrafioti et al. (2022) found that in chambers, the application of nitrogen to reduce the oxygen level to less than 1% was able to provide 100% mortality in all insect species tested at 3 d when the application was carried out at 28°C, while the same control was achieved in only 2.5 d when the temperature was greater than 28°C. Similar results have been reported in a series of studies (Athanassiou et al., 2016; Sakka et al., 2020; 2022a). For instance, Sakka et al. (2020) tested two different temperatures (28° and 40°C) with three exposure intervals (2.5, 3, and 9 d) and found that complete parental mortality was recorded for all combinations, with the exception of some survival at 28°C after 3 d of exposure. In those studies, no adverse effects were recorded on the product, even when temperatures were as high as 38°C (Athanassiou et al., 2016). One additional combination that has been evaluated with success is that of heat with diatomaceous earth (Dowdy, 1999; Dowdy and Fields, 2002). In field trials with heat, Dowdy and Fields (2002) noted that the simultaneous application of diatomaceous earth resulted in high mortality levels of *T. confusum*, due to the increased insect stress that had been caused at elevated temperatures. In principle, the increase in temperature increased the efficacy of diatomaceous earth (Dowdy, 1999; Dowdy and Fields, 2002). Moreover, Kljajić et al. (2009) reported that a pretreatment of insects at 50°C (for less than 60 min) increased the insecticidal effect of contact insecticides. Bandara et al. (2021) combined heat treatment with kairomones and essential oils and found that the attraction for adults exposed in heat was notably increased. Moreover, heat treatment was found to be effective for the control of different populations of stored-product beetle species, regardless of the resistance of these populations to phosphine (Agrafioti et al., 2019; Sakka et al., 2022b), underlining the importance of heat as a tool to combat resistance.

7.2.6 Applications

In commercial applications, heat treatments should be carefully designed and executed in accordance with the structural complexity of the target

Table 7.3 Summary of Studies Dealing with Use of Heat Treatment in Combination with Other Methods

Order	Target Insect Pest	Life Stage	Combined Method	Reference
Coleoptera	*Trogoderma granarium*	eggs, diapausing larvae, nondiapausing larvae, adults	low oxygen	Sakka et al. (2022a)
	Tribolium castaneum	adults		
	Callosobruchus maculatus	adults		
	Rhyzopertha dominica *Oryzaephilus surinamensis*	adults	low oxygen	Agrafioti et al. (2022)
	Tribolium castcneum	adults	trapping, orientation	Bandara et al. (2021)
	Oryzaephilus surinamensis *Tribolium castaneum* *Sitophilus oryzae*	adults	nitrogen	Sakka et al. (2020)
	Sitophilus orryzae *Rhyzopertha dominica*	adults	contact insecticides	Wijayaratne and Rajapakse (2018)
	Tribolium confusum	eggs, larvae, pupae, adults	nitrogen	Athanassiou et al. (2017)
	Oryzaephilus surinamensis	adults		
	Tribolium castaneum	adults	diatomaceous earth	Frederick and Subramanyam (2016)
	Sitophilus granarius	adults	contact insecticides	Kljajić et al. (2009)
	Sitophilus oryzae *Tribolium confusum*	adults	low O_2 and/or high CO_2	Downes et al. (2008)
	Tribolium castaneum	adults	contact insecticides	Arthur and Dowdy (2003)
	Ahasverus advena *Cryptolestes. pusillus* *Lasioderma serricorne* *Rhyzopertha dominica* *Sitophilus granarius* *Sitophilus oryzae* *Tribolium castaneum* *Tribolium confusum*	adults	trapping	Roesli et al. (2003)

(Continued)

Table 7.3 (Continued) Summary of Studies Dealing with Use of Heat Treatment in Combination with Other Methods

Order	Target Insect Pest	Life Stage	Combined Method	Reference
	Tribolium confusum	1st, 2nd, 3rd, 4th, 5th instar larvae, pupae, adults	diatomaceous earth	Dowdy and Fields (2002)
	Tribolium castaneum	adults	diatomaceous earth	Dowdy (1999)
	Tribolium confusum	adults	diatomaceous earth	Fields et al. (1997)
Lepidoptera	*Ephestia elutella*	eggs, larvae	nitrogen	Athanassiou et al. (2017)
	Plodia interpunctella	larvae	diatomaceous earth, phosphine, CO_2	Kim et al. (2015)
	Plodia interpunctella	adults	diatomaceous earth	Kim et al. (2014)
	Cadra cautella	adults	trapping	Roesli et al. (2003)

Note: *Species that were analyzed for differences in trap captures immediately before and immediately after heat treatment.

facility and its sealing capacities. It should be taken into account that in some areas heat does not reach 50°C, given that insect survival may be high in these areas (Dowdy, 1999; Mahroof et al., 2003b). This can be monitored from outside with wireless temperature sensors (Agrafioti et al., 2019; Sakka et al., 2022b), while proper hot air distribution can be achieved through fans placed in critical locations (Mahroof et al., 2003b). The wireless temperature sensors can be placed in the treated facilities at different points. However, even if certain insect individuals survive the heat, there is a considerable level of postexposure (delayed) mortality or reduced progeny production capacity (Mahroof et al., 2005; Agrafioti et al., 2019). Cleaning and sanitation measures before the application are of outmost importance, since heat does not penetrate deeply into products such as flour, so any infestation foci within residues may remain unaffected (Brijwani et al., 2012b). It should be noted that heat is applied in empty storage and processing facilities such as machineries, but not in the commodity. Finally, food products should be removed before the initiation of the treatment along with other materials that are sensitive to increased temperatures (e.g., packaging material).

Extreme temperatures have been considered a major issue in wheat quality characteristics. According to Ghaly and Taylor (1982), elevated temperatures during thermal treatments can damage the quality of the grain. In the same study, the authors mentioned that heat treatment can affect the germination and turbidity of the wheat grains and the effect depends on the variety. Athanassiou et al. (2017) tested nitrogen treatment in combination with elevated temperatures against different life stages and species of stored-product insects in stored dried currants and found no effect on microbial, yeast, and mold counts when the temperature was 38–43°C. Lee et al. (2003) noted that the application of far infrared radiation (FIR) may activate different phenolic compounds that are natural antioxidants on rice. For pasta products, Giannetti et al. (2021) noted that heat treatment can cause Maillard reaction and changes in the organoleptic and nutrient characteristics of the pasta. Heat treatment was also tested in sorghum flour by Meera et al. (2011). The use of dry heat treatment on wheat flour at 50–200°C can increase the pseudo-crystalline structure of the amylopectin starch molecules.

7.3 Cold Treatment

7.3.1 Mechanisms

The mode of action of low temperatures on stored-product insects is not fully understood, but several mechanisms have been suggested. Three of the most widely proposed mechanisms are cold-induced membrane phase changes, protein denaturation, and loss of transmembrane ion balance, as well as a combination of more than one of these mechanisms (MacMillan et al., 2015). An additional approach, referred to as "supercooling," relies on the capability of the pests to be chilled prior to the onset of spontaneous ice formation in their bodily fluids. The temperature at which body water naturally solidifies is denoted as the supercooling threshold or supercooling point (SCP) (Zachariassen, 1985; Johnston and Lee, 1990).

Cooling can cause changes in the phase and fluidity of the cell membranes of insects, compromising their physical integrity (Quinn, 1985; Drobnis et al., 1993; Hazel, 1995). In addition, cold temperatures can induce the denaturation of proteins. Increased expression of heat shock proteins has been observed in relation to chill injury, indicating protein denaturation as a potential mechanism (Goto and Kimura, 1998; Yocum, 2001; Nielsen et al., 2005; Sinclair et al., 2007; Colinet et al., 2010; Teets and Denlinger, 2013). Finally, chilling injury in insects is closely associated with a loss of ion balance, particularly an increase in extracellular potassium ions (K+). Low temperatures reduce active ion transport, leading to membrane depolarization and a gradual loss of ion and water balance. This depolarization, combined with high extracellular K+ levels, initiates a cascade of events that can ultimately lead to cell destruction (Hoyle, 1954; Rheuben, 1972; Wareham et al., 1974; Kostal et al., 2004; Koštál et al., 2006; MacMillan and Sinclair, 2011; MacMillan et al., 2014). In a study conducted by MacMillan et al. (2015) on locusts, the researchers investigated the primary cause of chill injury using in vivo and in situ experiments. They examined the progressive development of injury in locusts during chronic cold exposure and recovery. They also measured the concentration of K+ in the haemolymph of the insect and quantified muscle damage. Cold temperatures can disrupt the physiological processes and homeostasis of insects, leading to paralysis, injury, and death. The mechanisms underlying chilling injury include changes in cell membrane properties, protein denaturation, and disturbances in ion balance.

7.3.2 Studies of Various Species and Life Stages

As noted above, there are few studies that have examined cold in "space" treatments. In the United States, the use of cold is applied to small structures, within a facility (Johnson and Valero, 2003; Arthur et al., 2015). In many parts of the world where there is an extensive winter season, it may be possible to use cold as a treatment. Most available data are based on freezers, in which insects are placed to estimate their susceptibility at different combinations of temperature and exposure interval (Table 7.4) (Arthur et al., 2015; Athanassiou et al., 2018a; 2018b). Many studies have focused on the susceptibility or tolerance of different insect life stages to cold treatment (Tables 7.4, 7.5) (Arthur et al., 2015; Flinn et al., 2015; Andreadis and Athanassiou, 2017). Specifically, Andrea is and Athanassiou (2017) reviewed the cold tolerance of different stored-product insects and reported that the cold tolerance varies according to age, life stage, species, and exposure intervals to low temperatures. However, low temperatures can be also a viable solution for insect control in large product quantities, such as in grain bulks through chilling (Maier and Rulon, 1996; Rulon et al., 1999; Nayyar and Kaushal, 2002). Although aeration can also be used to cool the grain bulk, the success of such an application largely depends on the outside temperatures, and thus it cannot be fully effective in all geographical locations (Flinn et al., 1997; Arthur et al., 1998, 2001, 2011, 2020; Arthur and Flinn, 2000). On the other hand, grain chilling (often also called chilled aeration) can be very effective in maintaining grain quality, along with providing insect and mold control (Maier, 1994). Often chilling is combined with drying; grains that have high moisture content can remain stored for longer intervals when they are chilled than when they are stored at higher

Table 7.4 Studies Reporting Exposure of Different Stored-Product Insect Species at Low Temperatures Based on Research That Was Published after Fields (1992)

Order	Species	Life stage	Temperature (°C)	Reference
Coleoptera	*Alphitobius diaperirus*	adults	−2, 0, 4, 8, 10, 12, 36	Renault (2011)
		adults	3	Colinet (2011)
		adults	0, 5, 10, 15	Colinet et al. (2011)
		adults	10	Renault et al. (2006)
		adults	0, 5, 10, 15, 20, 25, 30	Renault et al. (2004)
		adults	6, 10	Renault et al. (2003a)
		adults	10	Renault et al. (2003b)
		adults	6, 10	Renault et al. (1999)
	Callosobruchus maculatus	adults	0, −5, −7.5, 10, −12.5	Masoumi et al. (2021)
		eggs, larvae, pupae, adults	−5, −10, −15	Loganathan et al. (2011)
		eggs, larvae, pupae, adults	−18	Johnson and Valero (2003)
	Callosobruchus rhodesicnus	eggs, larvae, pupae, adults	−18	Dohino et al. (1999)
	Cryptolestes turcicous	eggs, young larvae, old larvae, pupae, adults	−10	Ganesan et al. (2021)
	Cryptolestes pusillus	eggs, young larvae, old larvae, pupae, adults	−10	Ganesan et al. (2021)

(Continued)

Table 7.4 (Continued) Studies Reporting Exposure of Different Stored-Product Insect Species at Low Temperatures Based on Research That Was Published after Fields (1992)

Order	Species	Life stage	Temperature (°C)	Reference
	Cryptolestes ferrugineus	eggs, young larvae, old larvae, pupae, adults	−10	Ganesan et al. (2021)
		adults	−14 to −7	Burks and Hagstrum (1999)
		adults	5, 10, 15	Fields et al. (1998)
	Lasioderma serricorne	larvae	7, 10, 13	Yang et al. (2022)
		adults	−3, 0, 3, 6	Wang et al. (2022)
		eggs, larvae, pupae, adults	5, 0, −5, −10, −15, −20	Imai and Harada (2006)
		egg, larvae, pupae, adults	−15, −10, −5, 0	Imai and Harada (2006)
	Oryzaephilus surinamensis	adults	−15, −10, −5, 0	Athanassiou et al. (2021)
		eggs	−18	Karimzadeh et al. (2021)
		eggs, larvae, adults	−15, −10, −5, 0	Athanassiou et al. (2019)
		eggs, larvae, adults	−15, −10, −5, 0	Athanassiou et al. (2018a)
		larvae, adults	−16	Eliopoulos et al. (2011)
		adults	−14 to −7	Burks and Hagstrum (1999)
		adults	−5, 5	Mignon et al. (1996)
		adults	0, 5, 10	Mignon et al. (1995)

(Continued)

Table 7.4 (Continued) Studies Reporting Exposure of Different Stored-Product Insect Species at Low Temperatures Based on Research That Was Published after Fields (1992)

Order	Species	Life stage	Temperature (°C)	Reference
		eggs, larvae, pupae, adults	−18, −10, −5, 0	Donahaye et al. (1995)
	Rhyzopertha dominica	adults	−14 to −7	Burks and Hagstrum (1999)
	Sitophilus granarius	eggs, larvae, pupae, adults	−18	Dohino et al. (1999)
		adults	5, 10, 15	Fields et al. (1998)
		adults	5, −5	Mignon et al. (1996)
		adults	−10	Mignon et al. (1995)
	Sitophilus oryzae	adults	6, 3, 0, −3	Wang et al. (2022)
		adults	6 ± 1	Ghasemzadeh et al. (2011)
		adults	6 ± 1	Gasemzadeh et al. (2010)
		adults	−14 to −7	Burks and Hagstrum (1999)
	Stegobium paniceum	eggs, larvae, pupae, adults	−10, −14, −18	Adler (2010)
		eggs, young larvae, old larvae, pupae, adults	5, −5, −10, −15	Abdelghany et al. (2010)
	Tenebrio molitor	larvae	−40	Pedersen et al. (2006)
	Tribolium castaneum	adults	−15, −10, −5, 0	Izadi et al. (2019)
		adults	4	Jiang et al. (2019)
		adults	0	Scharf et al. (2016)

(Continued)

Table 7.4 (Continued) Studies Reporting Exposure of Different Stored-Product Insect Species at Low Temperatures Based on Research That Was Published after Fields (1992)

Order	Species	Life stage	Temperature (°C)	Reference
		eggs, larvae, pupae, adults	−18	Arthur et al. (2015)
		eggs	4, 0, −2, −4, −8, −10, −12	Flinn et al. (2015)
		larvae, adults	0	Wijayaratne and Fields (2010)
		adults	6 ± 1	Gasemzadeh et al. (2010)
		adults	−14 to −7	Burks and Hagstrum (1999)
		eggs, larvae, pupae, adults	−18, −10, −5, 0	Donahaye et al. (1995)
	Trobolium confusum	eggs, larvae, pupae, adults	−15, −10, −5, 0	Athanassiou et al. (2021)
		eggs, larvae, pupae, adults	−15, −10, −5, 0	Athanassiou et al. (2019)
		eggs, larvae, pupae, adults	−15, −10, −5, 0	Athanassiou et al. (2019)
		eggs, larvae, pupae, adults	−15, −10, −5, 0	Athanassiou et al. (2018a)
		Larvae, adults	−16	Eliopoulos et al. (2011)
	Trogoderma granarium	larvae, adults	−16	Eliopoulos et al. (2011)
		eggs, larvae, pupae, adults	−18	Dohino et al. (1999)
	Trogoderma inclusum	eggs, larvae, pupae, adults	−18	Arthur et al. (2015)
	Trogoderma variabile	larvae	−10, −5, 0	Gerken et al. (2020)
		larvae	−20, −15, −10, −5	Mohammadzadeh and Izadi (2018a)

(*Continued*)

Table 7.4 (Continued) Studies Reporting Exposure of Different Stored-Product Insect Species at Low Temperatures Based on Research That Was Published after Fields (1992)

Order	Species	Life stage	Temperature (°C)	Reference
		larvae	−20, −15, −10, −5	Mohammadzadeh and Izadi (2018b)
		eggs, larvae, pupae, adults	−19, −16, −10, −5, 0	Abdelghany et al. (2015)
Lepidoptera	*Amyelois transitella*	eggs, larvae, pupae	0, 5, 10	Johnson (2007)
	Sitotroga cerealella	eggs	6, 9, 12, 15, 24	Pitcher et al. (2002)
	Plodia interpunctella	eggs	9, 4	Haque et al. (2021)
		larvae, pupae, adults	0, 5, 10, 15, 20, 25, 30,	Carrillo et al. (2015)
		eggs, larvae, pupae, adults	−10, −14, −18	Adler (2010)
		eggs, larvae, pupae	0, 5, 10	Johnson (2007)
		eggs, larvae, pupae, adults	−18	Dohino et al. (1999)
		non diapausing larvae, daupausing larvae	−20, −10	Naeemullah et al. (1999)
	Ephestia cautella	eggs, larvae, pupae, adults	0, −5, −10, −18	Donahaye et al. (1995)
	Ephestia kuehniella	eggs, larvae, pupae, adults	−5, −7.5, −10, −12.5	Andreadis et al. (2012)
		eggs, larvae, pupae, adults	−18	Dohino et al. (1999)

Table 7.5 The Supercooling Points of Some Stored-Product Insects.

Order	Species	Life Stage	Supercooling Point (°C)	References
Coleoptera	*Alphitobius diaperinus*	adults (nonacclimated)	−8.9 ±1.6	Renault et al. (2004)
		adults (acclimated 1 month at 15° C)	−11.8 ± 2.1	Renault et al. (2004)
		adults (males)	−14.5 ± 2.4	Salin et al. (2000)
		adults (females)	−10.3 ± 1.3	Salin et al. (2000)
		adults (males starved for 1 month at 20°C)	−17 ± 2.6	Salin et al. (2000)
		adults (females starved for 1 month at 20°C)	−11.2 ± 1.8	Salin et al. (2000)
	Callosobruchus maculatus	adults	−18.0	Masoumi et al. (2021)
	Cryptolestes ferrugineus	eggs	−25.4 ± 0.5	Ganesan et al. (2021)
		young larvae	−23.4 ± 0.4	Ganesan et al. (2021)
		old larvae	−22.4 ± 0.4	Ganesan et al. (2021)
		pupae	−24.6 ± 0.3	Ganesan et al. (2021)
		adults	−21.4 ± 0.5	Ganesan et al. (2021)
		eggs (acclimated*)	−25.1 ± 0.2	Ganesan et al. (2021)
		young larvae (acclimated*)	−25.5 ± 0.5	Ganesan et al. (2021)
		old larvae (acclimated*)	−22.2 ± 0.4	Ganesan et al. (2021)
		pupae (acclimated*)	−25.1 ± 0.4	Ganesan et al. (2021)
		adults (acclimated*)	−23.7 ± 0.4	Ganesan et al. (2021)

(Continued)

Table 7.5 (Continued) The Supercooling Points of Some Stored-Product Insects.

Order	Species	Life Stage	Supercooling Point (°C)	References
	Cryptolestes turcicus	eggs	−25.8 ± 0.1	Ganesan et al. (2021)
		young larvae	−23.9 ± 0.4	Ganesan et al. (2021)
		old larvae	−26.1 ± 0.2	Ganesan et al. (2021)
		pupae	−26.1 ± 0.4	Ganesan et al. (2021)
		adults	−21.3 ± 0.5	Ganesan et al. (2021)
		eggs (acclimated*)	−25.4 ± 0.3	Ganesan et al. (2021)
		young larvae (acclimated*)	−24.4 ± 0.7	Ganesan et al. (2021)
		old larvae (acclimated*)	−20.5 ± 0.7	Ganesan et al. (2021)
		pupae (acclimated*)	−26.7 ± 0.4	Ganesan et al. (2021)
		adults (acclimated*)	−23.4 ± 0.4	Ganesan et al. (2021)
	Crytolestes pusillus	eggs	−26.2 ± 0.2	Ganesan et al. (2021)
		young larvae	−23.1 ± 0.3	Ganesan et al. (2021)
		old larvae	−23.0 ± 0.6	Ganesan et al. (2021)
		pupae	−26.7 ± 0.1	Ganesan et al. (2021)
		adults	−20.1 ± 0.5	Ganesan et al. (2021)
		eggs (acclimated*)	−22.4 ± 0.4	Ganesan et al. (2021)
		young larvae (acclimated*)	−23.0 ± 0.5	Ganesan et al. (2021)
		old larvae (acclimated*)	−24.4 ± 0.7	Ganesan et al. (2021)

(Continued)

Table 7.5 (Continued) The Supercooling Points of Some Stored-Product Insects.

Order	Species	Life Stage	Supercooling Point (°C)	References
		pupae (acclimated*)	−25.7 ± 0.4	Ganesan et al. (2021)
		adults (acclimated*)	−22.2 ± 0.4	Ganesan et al. (2021)
	Tribolium castaneum	adults (28 ± 1°C)	−10.7 ± 0.4	Izadi et al. (2019)
		adults (cold acclimation, 12°C for 10 d)	−18.4 ± 0.7	Izadi et al. (2019)
		adults (rapid cold hardening, 0 °C for 4 h)	−16.1 ± 0.9	Izadi et al. (2019)
		adults (fluctuating acclimation**)	−14.2 ± 0.5	Izadi et al. (2019)
	Trogoderma variabile	eggs	−25.3 ± 0.3	Abdelghany et al. (2015)
		young larvae	−16.1 ± 0.8	Abdelghany et al. (2015)
		old larvae	−16.4 ± 0.6	Abdelghany et al. (2015)
		pupae	−19.1 ± 0.8	Abdelghany et al. (2015)
		adults	−20.6 ± 0.8	Abdelghany et al. (2015)
Lepidoptera	*Ephestia kuehniella*	young larvae	−16.1	Andreadis et al. (2012)
		old larvae	−19.5	Andreadis et al. (2021)
		pupae	−23.3	Andreadis et al. (2021)
		adults	−21.6	Andreadis et al. (2021)
	Plodia interpunctella	larvae (nonacclimated, 25°C)	−7.4 + 0.6	Fields and Timlick (2010)

(Continued)

Table 7.5 (Continued) The Supercooling Points of Some Stored-Product Insects.

Order	Species	Life Stage	Supercooling Point (°C)	References
		larvae (acclimated, 10°C for 4 weeks)	−9.8 + 0.6	Fields and Timlick (2010)
		larvae (diapausing, 20°C, 12L:12D; nonacclimated, 25°C)	−0.7 + 0.4	Fields and Timlick (2010)
		larvae (diapausing, 20°C, 12L:12D; acclimated, 10°C for 4 weeks)	−13.4 + 0.5	Fields and Timlick (2010)
		eggs (1 d old)	−24.4 ± 0.3	Carillo and Cannon (2005)
		eggs (4 d old)	−24.2 ± 0.6	Carillo and Cannon (2005)
		larvae (1st instar)	−23.5 ± 0.9	Carillo and Cannon (2005)
		larvae (2nd instar)	−12.9 ± 0.7	Carillo and Cannon (2005)
		larvae (3rd instar)	−14.1 ± 0.8	Carillo and Cannon (2005)
		larvae (4th instar)	−11.8 ± 0.6	Carillo and Cannon (2005)
		larvae (5th instar)	−14.4 ± 0.3	Carillo and Cannon (2005)
		larvae (5th instar, 2 d old)	−11.8 ± 0.6	Carillo and Cannon (2005)
		larvae (5th instar, 4 d old)	−16.2 ± 0.5	Carillo and Cannon (2005)

(Continued)

Table 7.5 (Continued) The Supercooling Points of Some Stored-Product Insects.

Order	Species	Life Stage	Supercooling Point (°C)	References
		larvae (5th instar, 7 d old)	-15.2 ± 0.6	Carillo and Cannon (2005)
		larvae (5th instar, 9 d old)	-14.8 ± 0.6	Carillo and Cannon (2005)
		larvae (5th instar, 12 d old)	-15.1 ± 0.7	Carillo and Cannon (2005)
		pupae	-22.2 ± 0.5	Carillo and Cannon (2005)
		adults	-22.4 ± 0.4	Carillo and Cannon (2005)
		pupae	$-5.0 + 0.2$	Carillo and Cannon (2005)
	Tineola bisselliella	eggs	-22.6 ± 2.2	Chauvin and Vannier (1997)
		larvae (2nd instar)	-13.0 ± 3.6	Chauvin and Vannier (1997)
		larvae (last instar)	-16.2 ± 5.1	Chauvin and Vannier (1997)
		pupae	-16.9 ± 3.4	Chauvin and Vannier (1997)
		adults (females)	-18.9 ± 1.1	Chauvin and Vannier (1997)
		adults (males)	-18.7 ± 1.3	Chauvin and Vannier (1997)

Notes:

* Held at 18, 10, and 5°C, for 1 wk at each temperature.

** In a cycle of 240 min at 5°C followed by 20 min at −10°C followed by 240 min at 5°C followed by 940 min at 33°C.

temperatures (Tompsett and Pritchard, 1998; Beckett, 2011). Building cooling has also been performed with good results, especially in countries where outside temperatures are extremely low for a long period within the year, but certain parameters should be considered, such as the removal of water from the equipment to avoid freezing, which can cause irreversible damage. Moreover, care should be taken to avoid detrimental effects on the different products that may exist in the area that is to be treated. The concepts of aeration and chilled aeration are presented in detail in Chapter 10.

7.3.3 Impact of Different Temperatures

There are several studies that have illustrated the utilization of different degrees of cold on different types of commodities. Flinn et al. (2015) exposed bags of flour to nine internal target temperatures ranging from $-12°C$ to $8°C$ and found that when they aimed for temperatures between $-12°C$ and $4°C$, all eggs of *T. castaneum* in bags positioned at both the periphery and the center of the pallet died. In that study, the authors modeled with temperature loggers the reduction of temperature through time within the flour mass and indicated that, when exposed to the dynamic temperature curve spanning 24 d, which includes both cooling and warming periods to reach the goal temperature of $0°C$, all *T. castaneum* eggs died, whether placed in peripheral or central bag locations (Flinn et al., 2015). In contrast with heat, where target temperatures and target treatment intervals are $50°C$ for 24–36 h for buildings, the application of cold ranges remarkably and depends largely on the equipment available for this treatment. Most data available examine the susceptibility of stored product-insects to temperatures ranging between $0°C$ and $0°F$ $(-17.8°C)$ and for exposure intervals that range from few hours to several weeks (Fields, 1992; Abdelghany et al., 2010; Arthur et al., 2015; Flinn et al., 2015; Athanassiou et al., 2018a). At $0°C$, complete mortality of most stored-product insects can be achieved after exposure for long intervals. For instance, at this temperature, all life stages of *P. interpunctella* died after 7 d (Athanassiou et al., 2018a). In principle, exposure at $0°C$ for some weeks may not drastically affect insect survival and may just cause immobilization and then quick recovery when the temperatures prevailing are suitable for development again (Evans, 1986). In this context, Athanassiou and Arthur (2020) found that for a series of stored-product beetle species, gradual cooling before exposure to $0°C$ for 7 d and gradual warming after that exposure did not have any effect on insect survival.

At subzero temperatures, the interval that is needed is drastically shortened. Indicatively, Athanassiou et al. (2018a) evaluated *P. interpunctella* and the psocid *Liposcelis bostrychophila* (Badonnel) (Psocoptera: Liposcelididae) under controlled conditions. Specifically, the temperatures tested were 0, –5, –10, and –15°C, and the insects were exposed for 2, 4, and 8 h and also for 1, 2, 3, and 7 d. For *P. interpunctella*, larvae were the most cold-tolerant life stage, given that after 3 d of exposure, complete control was noted at –10°C (Athanassiou et al., 2018a). In the case of *L. bostrychophila*, the most cold-tolerant life stage was eggs, as egg survival was recorded after 3 d at –15°C (Athanassiou et al., 2018a). *Tribolium confusum* and *O. surinamensis* were generally considered as moderately susceptible to cold, given that all life stages of these species died after exposure for 1 d at –10°C (Athanassiou et al., 2018b). Moreover, *T. granarium* is particularly tolerant to low temperatures and can survive for 15 d at –20°C, a combination that is lethal for most major stored-product insect species (Fields, 1992 Wilches et al., 2017; Athanassiou et al., 2019b). Arthur et al. (2015) tested the susceptibility of *T. castaneum* and the larger cabinet beetle, *Trogoderma inclusum* (LeConte) (Coleoptera: Dermestidae), to –18°C for different time intervals and showed that *T. inclusum* was more tolerant than *T. castaneum*. The results of this study indicated that the most tolerant life stages of *T. castaneum* were the eggs and the larvae, whereas for *T. inclusum* the most tolerant life stage was the larvae (Arthur et al., 2015).

Among the life stages, eggs seem to be the one most tolerant of cold for the vast majority of the species tested so far (Fields, 1992; Arthur et al., 2015; Athanassiou et al., 2018a, b, 2020). Arthur et al. (2015) noted that 8 h of exposure was required to control *T. castaneum* eggs at –18°C and this life stage was by far more cold tolerant than adults or larvae. In contrast, for some species, eggs are more susceptible to cold than other life stages, as in the case of tests conducted by Athanassiou et al. (2018b) where *O. surinamensis* adults were found to be more tolerant than the eggs but the effect was reversed with a decrease of temperature, suggesting that short exposures to cold may have a stronger inhibitory effect on eggs than on adults. Moreover, eggs of *P. interpunctella* were far more susceptible to cold than were pupae or larvae (Athanassiou et al., 2018a). Regarding *T. granarium*, the most cold-tolerant stage was diapausing larvae, which could survive at –10°C for 87 d, far longer than the interval required for the eggs of this species (Wilches et al., 2017). As these findings suggest, previous knowledge of the target species that is to be treated with cold may lead to species-specific cold treatment protocols.

7.3.4 Acclimation

Acclimation may play an important role in cold tolerance (Andreadis and Athanassiou, 2017). The induction of cold hardiness, also known as acclimation, is strongly influenced by exposure to low temperature levels (Salt, 1961). While acclimation is indeed an important factor in cold hardiness, however, it is not the sole determinant. Numerous biotic and abiotic factors contribute to the complexity of this phenomenon (Andreadis and Athanassiou, 2017). Acclimated larvae of *T. granarium* were more cold tolerant than nonacclimated ones (Wilches et al., 2017). However, in another study, acclimation of *O. surinamensis* and *T. confusum* had dissimilar results, often increasing the susceptibility to cold (Athanassiou et al., 2019a). Athanassiou et al. (2019a) highlighted that the adults of both species reacted the same way to acclimation. However, the nonacclimated and acclimated pupae of *T. confusum* were equally susceptible to cold. For the larval stage of both species, *O. surinamensis* gave mixed results on their cold tolerance, whereas results for *T. confusum* indicated that the nonacclimated larvae were more tolerant of cold than the acclimated ones. The acclimation reduced egg survival for *O. surinamensis*, while eggs of *T. confusum* that had not been previously exposed to cold were not affected by acclimation (Athanassiou et al., 2019). Considering the above series of data, a previous exposure to low temperatures is likely to increase cold tolerance in certain species, which should be considered in adjusting the pretreatment storage conditions (Andreadis and Athanassiou, 2017).

7.3.5 Temperature Stratification

Cold gradually penetrates deeper in the mass of several products, such as flour. Flinn et al. (2015) highlighted that the desired temperature level has a direct effect on the time that cold requires to reach the center of the treated bags. In addition, Arthur et al. (2015) found that in bags with flour that were exposed to $-17.8°C$, a temperature decrease was detected immediately after initiation of the treatment at the external zone of the product, but it took 5.5 d to reach the "heart" of the product within the bag. Apparently, the same trend continues after termination of the treatment, when the increase in temperature is more gradual in the "heart" than in the peripheral zones, which means that insects can still be negatively affected after termination of the treatment through delayed mortality. In the study of Flinn et al. (2015), for treatments where the center of the pallet was $-12°$ to $4°C$, egg mortality was 100% in bags located in the periphery and in the center of the pallet. In a large bulk mass, the time required for the total cooling to a target temperature is dependent on the actual size of that bulk mass (Arthur et al., 2020). For instance, Athanassiou and Buchelos (2020) indicated that the temperature stratification in large structures like silos varies remarkably among different zones of the bulk; the edge zone close to the walls cools down more rapidly than the central zone. On the other hand, the presence of low quantities of product did not have much effect on the efficacy of the cold, suggesting that cold treatments may easily kill the small quantities of insects that remain in the area due to poor sanitation measures (Athanassiou et al., 2021). However, the "insulating effect" of cold in terms of reduced insect mortality has not been investigated in detail.

7.4 Future Research

Although both techniques, heat and cold, are considered more expensive than conventional methods, their actual cost has to be updated based on the current cost of energy as well as other parameters that underline the importance of individual applications, such as the value of the product or the area that is to be treated, and the potential withdrawal of certain conventional insecticides. In addition, there are some negative effects on the commodity that have been poorly investigated and must be examined more thoroughly (Athanassiou et al., 2021). It has been reported that heat can cause some structural damage to buildings and to some materials (e.g., plastics) if not done properly (Boina and Subramanyam 2001; Agrafioti et al., 2019), while cold can affect the product quality, causing certain changes such as drying (Harke et al., 2022). All of the above must be examined in detail, on the basis of standardized management practices. Despite the fact that either heat or cold can be used as a standalone control method, in principle they should both be part of integrated pest management protocols that include other methods as well, such as sanitation and chemical insecticides. This is essential when an extreme temperature-based strategy is planned, given that insects are able to develop mechanisms to increase their tolerance to extreme temperatures, reducing the efficacy of these applications (Andreadis and Athanassiou, 2017; Wilches et al., 2017). Nevertheless, the development of this tolerance is likely to be more gradual than that recorded in the case of conventional methods, such as fumigants and contact insecticides,

Acknowledgments

The chapter was supported by the General Secretariat for Research and Innovation of the Ministry of Development and Investments under the PRIMA Programme with the title "GreenDriedFruits." PRIMA is an Art.185 initiative supported and co-funded under Horizon 2020, the European Union's Programme for Research and Innovation. Also, it was partially supported of the project «PrecisionFEEDProtect: Precision protection of stored feed from entomological infestations using innovative technologies» (Project code: KMP6-0077613) under the framework of the Action «Investment Plans of Innovation» of the Operational Program «Central Macedonia 2021-2027», which is co-funded by the European Regional Development Fund and Greece.

References

Abdelghany AY, Awadalla SS, Abdel-Baky NF, El-Syrafi HA, Fields PG (2010) Effect of high and low temperatures on the drugstore beetle (Coleoptera: Anobiidae). J Econ Entomol 103: 1909–1914.

Abdelghany AY, Suthisut D, Fields PG (2015) The effect of diapause and cold acclimation on the cold-hardiness of the warehouse beetle, *Trogoderma variabile* (Coleoptera: Dermestidae). Can Entomol 147: 158–168.

Adler C (2010) Low temperature to control Plodia interpunctella and Stegobium paniceum. pp. 608–613. In: M.O. Carvalho, P.G. Fields, C .S. Adler, F.H. Arthur, C.G. Athanassiou, J.F. Campbell, F. FleuratL essard, P.W. Flinn, R.J. Hodges, A.A. Isikber, S. Navarro, R.T. Noye s, J. Riudavets, K.K. Sinha, G.R. Thorpe, B.H. Timlick, P. Trematerra , N.D.G. White (eds.), Proceedings of the 10th International Working Conference on Stored Product Protection, 27 June-2 July, 2010. Estoril, Portugal, Julius Kuhn-Institut, Berlin, Germany (2010),

Agrafioti P, Athanassiou CG, Subramanyam B (2019) Efficacy of heat treatment on phosphine resistant and susceptible populations of stored product insects. J Stored Prod Res 81: 100–106.

Agrafioti P, Kaloudis E, Athanassiou CG (2022) Utilizing low oxygen to mitigate resistance of stored product insects to phosphine. J Sci Food Agric 102: 6080–6087.

Andreadis SS, Athanassiou CG (2017) A review of insect cold hardiness and its potential in stored product insect control. Crop Protect 91: 93–99.

Andreadis SS, Eliopoulos PA, Savopoulou-Soultani M (2012) Cold hardiness of immature and adult stages of the Mediterranean flour moth, *Ephestia kuehniella*. J Stored Prod Res 48: 132–136.

Arthur FH (2006) Initial and delayed mortality of late-instar larvae, pupae, and adults of *Tribolium castaneum* and *Tribolium confusum* (Coleoptera: Tenebrionidae) exposed at variable temperatures and time intervals. J Stored Prod Res 42: 1–7.

Arthur FH, Dowdy AK (2003) Impact of high temperatures on efficacy of cyfluthrin and hydroprene applied to concrete to control *Tribolium castaneum* (Herbst). J Stored Prod Res 39: 193–204.

Arthur, FH, Flinn, PW (2000) Aeration management for stored hard red winter wheat: simulated impact on rusty grain beetle (Coleoptera: Cucujidae) populations. J of econ entomol 93: 1364–1372.

Arthur FH, Hartzer KL, Throne JE, Flinn PW (2015) Susceptibility of *Tribolium castaneum* (Coleoptera: Tenebrionidae) and *Trogoderma inclusum* (Coleoptera: Dermestidae) to cold temperatures. J Stored Prod Res 64: 45–53.

Arthur FH, Morrison WR, Trdan S (2020) Feasibility of using aeration to cool wheat stored in Slovenia: A predictive modeling approach using historical weather data. Appl Sci 10: 6066.

Arthur FH, Throne JE, Maier DE, Montross MD (1998) Feasibility of aeration for management of maize weevil populations in corn stored in the southern United States: Model simulations based on recorded weather data. Am Entomol 44: 118–123.

Arthur FH, Throne JE, Maier DE, Montross MD (2001) Impact of aeration on maize weevil (Coleoptera: Curculionidae) populations in corn stored in the northern United States: Simulation studies. Am Entomol 47: 104–110.

Arthur FH, Yang Y, Wilson LT (2011) Use of a web-based model for aeration management in stored rough rice. J Econ Entomol 104: 702–708.

Athanassiou CG, Arthur FH (2020) Cool down–warm up: Differential responses of stored product insects after gradual temperature changes. Insects 11: 158.

Athanassiou GG, Arthur FH, Hartzer KL (2018a) Efficacy of low temperatures for the control of all life stages of *Plodia interpunctella* and *Liposcelis bostrychophila*. J Pest Sci 91: 1363–1369.

Athanassiou CG, Arthur FH, Kavallieratos NG, Hartzer KL (2018b) Susceptibility of different life stages of *Tribolium confusum* (Coleoptera: Tenebrionidae) and *Oryzaephilus surinamensis* (Coleoptera: Silvanidae) to cold treatment. J Econ Entomol 111: 1481–1485.

Athanassiou CG, Arthur FH, Kavallieratos NG, Hartzer KL (2019a) To acclimate or not to acclimate? Simultaneous positive and negative effects of acclimation on susceptibility of *Tribolium confusum* (Coleoptera: Tenebrionidae) and *Oryzaephilus surinamensis* (Coleoptera: Silvanidae) to low temperatures. J Econ Entomol 112: 2441–2449.

Athanassiou CG, Arthur FH, Kavallieratos NG, Hartzer KL (2021) Influence of the presence of flour on the efficacy of low temperatures against stored product insects. Crop Protect 144: 105514.

Athanassiou CG, Buchelos CT (2020) Grain properties and insect distribution trends in silos of wheat. J Stored Prod Res 88: 101632.

Athanassiou CG, Chiou A, Rumbos CI, Karagiannis A, Nikolidaki EK, Panagopoulou EA, Kouvelas A, Karathanos VT (2016) Effects of electric infrared heating with light source penetration on microbial and entomological loads of dried currants and their organoleptic characteristics. J Pest Sci 89: 931–943.

Athanassiou CG, Chiou A, Rumbos CI, Sotiroudas V, Sakka M, Nikolidaki EK, Panagopoulou EA, Karathanos VT (2017) Effect of nitrogen in combination with elevated temperatures on insects, microbes and organoleptic characteristics of stored currants. J Pest Sci 90: 557–567.

Athanassiou CG, Phillips TW, Wakil W (2019b) Biology and control of the khapra beetle, *Trogoderma granarium*, a major quarantine threat to global food security. Ann Rev Entomol 64: 131–148.

Bandara HMDS, Wijayaratne LKW, Egodawatta WCP, Morrison Iii WR (2021) Orientation of *Tribolium castaneum* (Coleoptera: Tenebrionidae) adults to 4, 8-dimethyldecanal, kairomone and botanical oils following ambient, low, or high temperature exposure. J Stored Prod Res 94: 101893.

Beckett SJ (2011) Insect and mite control by manipulating temperature and moisture before and during chemical-free storage. J Stored Prod Res 47: 284–292.

Beckett SJ, Fields PG, Subramanyam B (2007) Disinfestation of stored products and associated structures using heat. pp. 182–237. In: Tang, J, Mitcham, E, Wang, S and Lurie, S (eds), Heat Treatments for Postharvest Pest Control: Theory and Practice. CABI, Wallingford UK..

Beckett SJ, Morton R (2003) Mortality of *Rhyzopertha dominica* (F.) (Coleoptera: Bostrychidae) at grain temperatures ranging from 50 C to 60 C obtained at different rates of heating in a spouted bed. J Stored Prod Res 39: 313–332.

Boina D, Subramanyam B (2004) Relative susceptibility of *Tribolium confusum* life stages exposed to elevated temperatures. J Econ Entomol 97: 2168–2173.

Brijwani M, Subramanyam B, Flinn PW (2012a) Impact of varying levels of sanitation on mortality of *Tribolium castaneum* eggs and adults during heat treatment of a pilot flour mill. J Econ Entomol 105: 703–708.

Brijwani M, Subramanyam B, Flinn PW, Langemeier MR, Hartzer M, Hulasare R (2012b) Susceptibility of *Tribolium castaneum* life stages exposed to elevated temperatures during heat treatments of a pilot flour mill: Influence of sanitation, temperatures attained among mills floors, and costs. J Econ Entomol 105: 709–717.

Burges HD (2008) Development of the khapra beetle, *T. granarium*, in the low part of its temperature range. J Stored Prod Res 44: 32–35.

Burks CS, Hagstrum DW (1999) Rapid cold hardening capacity in five species of coleopteran pests of stored grain. J Stored Prod Res 35: 65–75.

Campolo O, Verdone M, Laudani F, Malacrinò A, Chiera E, Palmeri V (2013) Response of four stored products insects to a structural heat treatment in a flour mill. J Stored Prod Res 54: 54–58.

Carrillo, MA, Cannor, CA, Wilcke, WF, Morey, RV, Kaliyan, N, Hutchison, WD (2005). Relationship between supercooling point and mortality at low temperatures in Indianmeal moth (Lepidoptera: Pyralidae). J Econ Entomol 98: 618–625.

Carillo MA, Cannon CA (2005) Supercooling point variability in the Indian meal moth, *Plodia interpunctella* (Hübner) (Lepidoptera: Pyralidae). J Stored Prod Res 41: 556–564.

Chauvin G, Vannier G (1997) Supercooling capacity of *Tineola bisselliella* (Hummel) (Lepidoptera: Tineidae): Its implication for disinfestation. J Stored Prod Res 33: 283–287.

Colinet H (2011) Disruption of ATP homeostasis during chronic cold stress and recovery in the chill susceptible beetle (*Alphitobius diaperinus*). Comparative Comp Biochem Physiol A Mol Integr Physiol 160: 63–67.

Colinet H, Lalouette L, Renault D (2011) A model for the time–temperature–mortality relationship in the chill-susceptible beetle, *Alphitobius diaperinus*, exposed to fluctuating thermal regimes. J Therm Biol 36: 403–408.

Colinet H, Lee SF, Hoffmann A (2010) Knocking down expression of Hsp22 and Hsp23 by RNA interference affects recovery from chill coma in *Drosophila melanogaster.* J Exp Biol 213: 4146–4150.

Dean GA (1911) Heat as a means of controlling mill insects. J Econ Entomol 4: 142–161.

Dohino T, Masaki A, Matsuoka I, Tanno M, Takano T (1999) Low temperature as an alternative to fumigation for disinfecting stored products. Res Bull Plant Prot Serv Jpn 35: 5–15.

Donahaye EJ, Navarro S, Rindner M (1995) Low temperature as an alternative to fumigation for disinfecting dried fruit from three insect species. J Stored Prod Res 31: 63–70.

Dowdy AK (1999) Heat sterilization as an alternative to methyl bromide fumigation in cereal processing plants. pp. 1089–1095. In: Zuxun, J., Quan, L., Yongsheng, L., Xianchang, T., Lianghua, G. (eds), Proceedings of the Seventh International Working Conference on Stored Product Protection, 2, 14–19 October 1998. Sichuan Publishing House of Science & Technology, Chengdu, Sichuan Province, People's Republic of China.

Dowdy AK (1999) Mortality of red flour beetle, *Tribolium castaneum* (Coleoptera: Tenebrionidae) exposed to high temperature and diatomaceous earth combinations. J Stored Prod Res 35: 175–182.

Dowdy AK, Fields PG (2002) Heat combined with diatomaceous earth to control the confused flour beetle (Coleoptera: Tenebrionidae) in a flour mill. J Stored Prod Res 38: 11–22.

Downes CJ, Van Epenhuijsen CW, Lill RE, Downes JE, Carpenter A, Brash D (2008) Calorimetric evaluation of responses of *Sitophilus oryzae* and *Tribolium confusum* to elevated temperatures and controlled atmospheres. J Stored Prod Res 44: 295–303.

Drobnis EZ, Crowe LM, Berger T, Anchordoguy TJ, Overstreet JW, Crowe JH (1993) Cold shock damage is due to lipid phase transitions in cell membranes: A demonstration using sperm as a model. J Exp Zool 265: 432–437.

Eliopoulos PA, Prasodimou GZ, Pouliou AV (2011) Time–mortality relationships of larvae and adults of grain beetles exposed to extreme cold. Crop Protect 30: 1097–1102.

Evans DE (1986) Some biological and physical constraints to the use of heat and cold for disinfesting and preserving stored products. pp. 149–164. In: Donahaye E, Navarro S (eds), Proceedings of the 4th International Working Conference on Stored-Product Protection, Tel-Aviv, Israel.

Farahani S, Bandani AR, Alizadeh H, Goldansaz SH, Whyard S (2020) Differential expression of heat shock proteins and antioxidant enzymes in response to temperature, starvation, and parasitism in the Carob moth larvae, *Ectomyelois ceratoniae* (Lepidoptera: Pyralidae). PloS One 15: e0228104.

Fields PG (1992) The control of stored-product insects and mites with extreme temperatures. J Stored Prod Res 28: 89–118.

Fields PG, Dowdy A, Marcotte M (1997) Diatomaceous earth combined with heat to control insects in structures. In: 1997 Proceedings of the Annual International Research Conference on Methyl Bromide Alternatives and Emissions Reductions. San Diego, CA.

Fields PG, Fleurat-Lessard F, Lavenseau L, Febvay G, Peypelut L, Bonnot G (1998) The effect of cold acclimation and deacclimation on cold tolerance, trehalose and free amino acid levels in *Sitophilus granarius* and *Cryptolestes ferrugineus* (Coleoptera). J Insect Physiol 44: 955–965.

Fields PG, Timlick B (2010) The effect of diapause, cold acclimation and icenucleating bacteria on the cold-hardiness of *Plodia interpunctella*. In: Carvalho, M.O., Fields, P.G., Adler, C.S., Arthur, F.H., Athanassiou, C.G., Campbell, J.F., Fleurat-Lessard, F., Flinn, P.W., Hodges, R.J., Isikber, A.A., Navarro, S., Noyes, R.T., Riudavets, J., Sinha, K.K., Thorpe, G.R., Timlick, B.H., Trematerra, P., White, N.D.G. (eds), Proceedings of the 10th International Working Conference on Stored Product Protection, 27 June–2 July 2010, Estoril, Portugal. Julius Kühn-Archiv, Berlin, Germany.

Finkelman S, Navarro S, Rindner M, Dias R (2006) Use of heat for disinfestation and control of insects in dates: Laboratory and field trials. Phytoparasitica 34: 37–48.

Flinn PW, Arthur FH, Throne JE, Friesen KS, Hartzer KL (2015) Cold temperature disinfestation of bagged flour. J Stored Prod Res 63: 42–46.

Flinn PW, Hagstrum DW, Muir WE (1997) Effects of time of aeration, bin size, and latitude on insect populations in stored wheat: A simulation study. J Econ Entomol 90: 646–651.

Frederick JL, Subramanyam B (2016) Influence of temperature and application rate on efficacy of a diatomaceous earth formulation against *Tribolium castaneum* adults. J Stored Prod Res 69: 86–90.

Ganesan L, Fields PG, Jayas DS, Jian F (2021) Effects of developmental stage, cold acclimation and diet on the cold tolerance of three species of *Cryptolestes* (Coleoptera: Laemophloeidae). J Stored Prod Res 91: 101773.

Gasemzadeh S, Pourmirza AA, Safaralizadeh MH, Maroufpoor M (2010) Effect of microwave radiation and cold storage on *Tribolium castaneum* Herbst (Coleoptera: Tenebrionidae) and *Sitophilus oryzae* L. (Coleoptera: Curculionidae). J Plant Prot Res 2: 140–145.

Gerken AR, Abts SR, Scully ED, Campbell JF (2020) Artificial selection to a nonlethal cold stress in *Trogoderma variabile* shows associations with chronic cold stress and body size. Envi Entomol 49: 422–434.

Ghaly TF, Taylor PA (1982) Quality effects of heat treatment of two wheat varieties. J Agric Eng Res 27: 227–234.

Ghasemzadeh S, Pourmirza AA, Safaralizadeh MH, Maroufpoor M, Aramideh S (2011) Combination of microwave radiation and cold storage for control of *Oryzaephilus surinamensis* (Col.: Silvanidae) and *Sitophilus oryzae* (Col.: Curculionidae). Egypt Acad J Biol Sci F Toxicol 3: 43–50.

Giannetti V, Mariani MB, Colicchia S (2021) Furosine as marker of quality in dried durum wheat pasta: Impact of heat treatment on food quality and security–A review. Food Control 125: 108036.

Gonen M (1977) Survival and reproduction of heat-acclimated *Sitophilus granarius* (Coleoptera: Curculionidae) at a moderately high temperature. Entomol Exp Appl 21: 249–253.

Goto SG, Kimura MT (1998) Heat-and cold-shock responses and temperature adaptations in subtropical and temperate species of Drosophila. J Insect Physiol 44: 1233–1239.

Guedes RNC, Zhu KY, Opit GP, Throne JE (2008) Differential heat shock tolerance and expression of heat-inducible proteins in two stored-product psocids. J Econ Entomol 101: 1974–1982.

Haque A, Islam S, Bari A, Hossain A, Athanassiou CG, Hasan M (2021) Cold storage-mediated rearing of *Trichogramma evanescens* Westwood on eggs of *Plodia interpunctella* (Hübner) and *Galleria mellonella* L. Plos One 16: e0253287.

Harke S, Megeri GB, Hadimani G, Megeri T (2022) Effect of storage on quality characteristics of wheat flour. Int J Res Engin Sci 10(10): 12–33.

Hazel JR (1995) Thermal adaptation in biological membranes: Is homeoviscous adaptation the explanation? Annu Rev Physiol 57: 19–42.

Hepburn HR (1985) Structure of the integument. Comprehensive Insect Physiology, Biochemistry and Pharmacology, Vol. 3, Pergamon Press, New York, pp. 1–58

Hoyle G (1954) Changes in the blood potassium concentration of the African migratory locust (*Locusta migratoria migratorioides* R. & F.) during food deprivation, and the effect on neuromuscular activity. J Exp Biol 31: 260–270.

Hulasare R, Subramanyam B, Fields PG, Abdelghany AY (2010) Heat treatment: A viable methyl bromide alternative for managing stored-product insects in food-processing facilities. pp. 661–667. In: Carvalho MO, Fields PG, Adler CS, Arthur FH, Athanassiou CG, Campbell JF, Fleurat-Lessard FP, Flinn PW, Hodges RJ, Isikber AA, Navarro S, Noyes RT, Riudavets J, Sinha KK, Thorpe GR, Timlick BH, Trematerra P, White NDG (eds), Stored Product Protection. Proceedings of the tenth International Working Conference on Stored Product Protection, 27 June–2 July 2010, Estoril, Portugal. Julius-Kühn-Archiv.

Imai T, Harada H (2006) Low-temperature as an alternative to fumigation to disinfest stored tobacco of the cigarette beetle, *Lasioderma serricorne* (F.) (Coleoptera: Anobiidae). Appl Entomol Zool 41: 87–91.

Izadi H, Mohammadzadeh M, Mehrabian M (2019) Cold tolerance of the *Tribolium castaneum* (Coleoptera: Tenebrionidae), under different thermal regimes: Impact of cold acclimation. J Econ Entomol 112: 1983–1988.

Jiang H, Zhang N, Chen M, Meng X, Ji C, Ge H, Dong F, Miao L, Yang X, Xu X, Qian K, Wang J (2019) Transcriptional and post-translational activation of AMPKα by oxidative, heat, and cold stresses in the red flour beetle, *Tribolium castaneum*. Cell Stress Chaperones 24: 1079–1089.

Jian F, Subramanyam B, Jayas DS, White ND (2013) Models to predict mortality of *Tribolium castaneum* (Coleoptera: Tenebrionidae) exposed to elevated temperatures during structural heat treatments. J Econ Entomol 106: 2247–2258.

Johnson JA (2007) Survival of Indian meal moth and navel orangeworm (Lepidoptera: Pyralidae) at low temperatures. J Econ Entomol 100: 1482–1488.

Johnson JA, Valero KA (2003) Use of commercial freezers to control cowpea weevil, *Callosobruchus maculatus* (Coleoptera: Bruchidae), in organic garbanzo beans. J Econ Entomol 96: 1952–1957.

Johnston SL, Lee RE Jr (1990) Regulation of supercooling and nucleation in a freeze intolerant beetle (*Tenebrio molitor*). Cryobiology 27: 562–568.

Karimzadeh R, Salehpoor M, Saber M (2021) Initial efficacy of pyrethroids, inert dusts, their low-dose combinations and low temperature on *Oryzaephilus surinamensis* and *Sitophilus granarius*. J Stored Prod Res 91: 101780.

Khamis M, Subramanyam B, Flinn PW, Dogan H, Jager A, Gwirtz JA (2010) Susceptibility of various life stages of *Rhyzopertha dominica* (Coleoptera: Bostrichidae) to flameless catalytic infrared radiation. J Econ Entomol 103: 1508–1516.

Kim H, Yu YS, Lee KY (2014) Synergistic effects of heat and diatomaceous earth treatment for the control of *Plodia interpunctella* (L epidoptera: P yralidae). Entomol Res 44: 130–136.

Kim H, Yu YS, Lee KY (2015) Differential induction of heat shock protein genes to the combined treatments of heat with diatomaceous earth, phosphine or carbon dioxide on *P lodia interpunctella*. Entomol Res 45: 332–338.

Kljajić P, Andrić G, Perić I (2009) Impact of short-term heat pre-treatment at 50 C on the toxicity of contact insecticides to adults of three *Sitophilus granarius* (L.) populations. J Stored Prod Res 45: 272–278.

Kostal V, Vambera J, Bastl J (2004) On the nature of pre-freeze mortality in insects: Water balance, ion homeostasis and energy charge in the adults of *Pyrrhocoris apterus*. J Exp Biol 207: 1509–1521.

Koštál V, Yanagimoto M, Bastl J (2006) Chilling-injury and disturbance of ion homeostasis in the coxal muscle of the tropical cockroach (*Nauphoeta cinerea*). Comparative Biochemistry and Physiology Part B: Biochem Mol Biol 143: 171–179.

Lee SC, Kim JH, Jeong SM, Kim DR, Ha JU, Nam KC, Ahn DU (2003) Effect of far-infrared radiation on the antioxidant activity of rice hulls. J Agric Food Chem 51: 4400–4403.

Lewthwaite SE, Dentener PR, Alexander SM, Bennett KV, Rogers DJ, Maindonald JH, Connolly PG (1998) High temperature and cold storage treatments to control Indian meal moth, *Plodia interpunctella* (Hübner). J Stored Prod Res 34: 141–150.

Li M, Li XJ, Lü JH, Huo MF (2018) The effect of acclimation on heat tolerance of *Lasioderma serricorne* (Fabricius) (Coleoptera: Anobiidae). J Therm Biol 71: 153–157.

Loganathan M, Jayas DS, Fields PG, White NDG (2011) Low and high temperatures for the control of cowpea beetle, *Callosobruchus maculatus* (F.) (Coleoptera: Bruchidae) in chickpeas. J Stored Prod Res 47: 244–248.

Lü J, Jian F, Fields P (2021) Preference of *Tribolium confusum* (Coleoptera: Tenebrionidae) adults to grain patches under constant and rising temperatures. J Stored Prod Res 90: 101756.

Lü J, Zhang H (2016) The effect of acclimation to sublethal temperature on subsequent susceptibility of *Sitophilus zeamais* Mostchulsky (Coleoptera: Curculionidae) to high temperatures. PLoS One 11: e0159400.

MacMillan HA, Baatrup E, Overgaard J (2015) Concurrent effects of cold and hyperkalaemia cause insect chilling injury. Proceedings of the Royal Society B: Biological Sciences 282(1817): 20151483.

MacMillan HA, Findsen A, Pedersen TH, Overgaard J (2014) Cold-induced depolarization of insect muscle: Differing roles of extracellular K+ during acute and chronic chilling. J Exp Biol 217: 2930–2938.

MacMillan HA, Sinclair BJ (2011) Mechanisms underlying insect chill-coma. J Insect Physiol 57: 12–20.

Mahroof R, Subramanyam B (2006) Susceptibility of *Plodia interpunctella* (Lepidoptera: Pyralidae) developmental stages to high temperatures used during structural heat treatments. Bull Entomol Res 96: 539–545.

Mahroof R, Subramanyam B, Eustace D (2003b) Temperature and relative humidity profiles during heat treatment of mills and its efficacy against *Tribolium castaneum* (Herbst) life stages. J Stored Prod Res 39: 555–569.

Mahroof R, Subramanyam B, Throne JE, Menon A (2003a) Time-mortality relationships for *Tribolium castaneum* (Coleoptera: Tenebrionidae) life stages exposed to elevated temperatures. J Econ Entomol 96: 1345–1351.

Mahroof R, Zhu KY, Subramanyam B (2005) Changes in expression of heat shock proteins in *Tribolium castaneum* (Coleoptera: Tenebrionidae) in relation to developmental stage, exposure time, and temperature. Ann Entomol Soc Am 98: 100–107.

Maier DE (1994) Chilled aeration and storage of US crops-a review. In Proceedings. 6th International Working Conference on Stored-Product Protection (Vol. 1, pp. 300–311). Wallingford: CAB International.

Maier DE, Rulon RA (1996) Evaluation and optimization of a new commercial grain chiller. Appl Eng Agric 12: 725–730.

Masoumi Z, Noghabi SS, Izadi H (2021) Trehalose and proline failed to enhance cold tolerance of the cowpea weevil, *Callosobruchus maculatus* (F.) (Col.: Bruchidae). J Stored Prod Res 93: 101853.

Meera MS, Bhashyam MK, Ali SZ (2011) Effect of heat treatment of sorghum grains on storage stability of flour. LWT-Food Science and Technology 44: 2199–2204.

Miao ZQ, Tu YQ, Guo PY, He W, Jing TX, Wang JJ, Wei DD (2020) Antioxidant enzymes and heat shock protein genes from *Liposcelis bostrychophila* are involved in stress defense upon heat shock. Insects 11: 839.

Mignon J, Haubruge E, Gaspar C (1995) The use of low temperatures and ice-nucleating bacteria against stored product insects pests. Mededelingen van de Faculteit Landbouwkundige en Toegepaste Biologische Wetenschappen, 60(3b).

Mignon J, Haubruge E, Gaspar C (1996) Influence of thermal acclimation on the survival of *Sitophilus granarius* (L) and *Oryzaephilus surinamensis* (L) at low temperatures. Neth J Zool 46: 3–4.

Mohammadzadeh M, Izadi H (2018a) Different diets affecting biology, physiology and cold tolerance of *Trogoderma granarium* Everts (Coleoptera: Dermestidae). J Stored Prod Res 76: 58–65.

Mohammadzadeh M, Izadi H (2018b) Cooling rate and starvation affect supercooling point and cold tolerance of the Khapra beetle, *Trogoderma granarium* Everts fourth instar larvae (Coleoptera: Dermestidae). J Therm Biol 71: 24–31.

Naeemullah M, Tanaka K, Tsumuki H, Takeda M (1999) Relationship of cold tolerance to developmental determination in the Indian meal moth, *Plodia interpunctella* (Lepidoptera: Phycitidae). Appl Entomol Zool 34: 267–276.

Navarro S, Finkelman S, Rindner M, Dias R. (2004) Heat treatment for disinfestation of nitidulid beetles from dates. pp. 297–300. In: Batchelor T, Alfarroba F (eds), Proceedings of Fifth International Conference on Alternatives to Methyl Bromide, 27–30 September, 2004, Lisbon..

Nayyar H, Kaushal SK (2002) Chilling induced oxidative stress in germinating wheat grains as affected by water stress and calcium. Biol Plant 45: 601–604.

Nielsen MM, Overgaard J, Sørensen JG, Holmstrup M, Justesen J, Loeschcke V (2005) Role of the heat-shock factor during heat and cold hardening and for the resistance to severe heat and cold stress. J Insect Physiol 51: 1320–1329.

Opit GP, Arthur FH, Bonjour EL, Jones CL, Phillips TW (2011) Efficacy of heat treatment for disinfestation of concrete grain silos. J Econ Entomol 104: 1415–1422.

Pedersen SA, Kristiansen E, Hansen BH, Andersen RA, Zachariassen KE (2006) Cold hardiness in relation to trace metal stress in the freeze-avoiding beetle *Tenebrio molitor*. J Insect Physiol 52: 846–853.

Pitcher SA, Hoffmann MP, Gardner J, Wright MG, Kuhar TP (2002) Cold storage of *Trichogramma ostriniae* reared on *Sitotroga cerealella* eggs. BioControl 47: 525–535.

Quinn PJ (1985) A lipid-phase separation model of low-temperature damage to biological membranes. Cryobiology 22: 128–146.

Renault D (2011) Long-term after-effects of cold exposure in adult *Alphitobius diaperinus* (Tenebrionidae): The need to link survival ability with subsequent reproductive success. Ecol Entomol 36: 36–42.

Renault, D., Salin, C., Vannier, G., Vernon, P. (1999) Survival and chill-coma in the adult lesser mealworm, Alphitobius diaperinus (Coleoptera: Tenebrionidae), exposed to low temperatures. J of Therm Biol 24, 229–236.

Renault D, Bouchereau A, Delettre YR, Hervant F, Vernon P (2006) Changes in free amino acids in *Alphitobius diaperinus* (Coleoptera: Tenebrionidae) during thermal and food stress. Comp Biochem Physiol A Mol Integr Physiol 143: 279–285.

Renault D, Hance T, Vannier G, Vernon P (2003a) Is body size an influential parameter in determining the duration of survival at low temperatures in *Alphitobius diaperinus* Panzer (Coleoptera: Tenebrionidae)? J Zool 259: 381–388.

Renault D, Hervant F, Vernon P (2003b) Effect of food shortage and temperature on oxygen consumption in the lesser mealworm, *Alphitobius diaperinus* (Panzer) (Coleoptera: Tenebrionidae). Physiol Entomol 28: 261–267.

Renault D, Nedved O, Hervant F, Vernon P (2004) The importance of fluctuating thermal regimes for repairing chill injuries in the tropical beetle *Alphitobius diaperinus* (Coleoptera: Tenebrionidae) during exposure to low temperature. Physiol Entomol 29: 139–145.

Rheuben MB (1972) The resting potential of moth muscle fibre. Physiol J 225: 529–554.

Roesli R, Subramanyam B, Fairchild FJ, Behnke KC (2003) Trap catches of stored-product insects before and after heat treatment in a pilot feed mill. J Stored Prod Res 39: 521–540.

Rulon RA, Maier DE, Boehlje MD (1999) A post-harvest economic model to evaluate grain chilling as an IPM technology. J Stored Prod Res 35: 369–383.

Sakka MK, Gatzali F, Karathanos VT, Athanassiou CG (2020) Effect of nitrogen on phosphine-susceptible and-resistant populations of stored product insects. Insects 11: 885.

Sakka MK, Gatzali F, Karathanos V, Athanassiou CG (2022a) Efficacy of low oxygen against *Trogoderma granarium* Everts, *Tribolium castaneum* (Herbst) and *Callosobruchus maculatus* (F.) in commercial applications. J Stored Prod Res 97: 101968.

Sakka MK, Jagadeesan R, Nayak MK, Athanassiou CG (2022b) Insecticidal effect of heat treatment in commercial flour and rice mills for the control of phosphine-resistant insect pests. J Stored Prod Res 99: 102023.

Salin C, Renault D, Vannier G, Vernon P (2000) A sexually dimorphic response in supercooling temperature, enhanced by starvation, in the lesser mealworm *Alphitobius diaperinus* (Coleoptera: Tenebrionidae). J Therm Biol 25: 411–418.

Salt RW (1961) Principles of insect cold-hardiness. Annu Rev Entomol 6: 55–74.

Scharf I, Wexler Y, MacMillan HA, Presman S, Simson E, Rosenstein S (2016) The negative effect of starvation and the positive effect of mild thermal stress on thermal tolerance of the red flour beetle, *Tribolium castaneum*. Sci Nat 103: 1–10.

Sen GG, Emekci M (2007) Laboratory evaluation of the effectiveness of heat treatment against the dried fruit beetles. pp. 138–1. In: 2007 Annual International Research Conference on Methyl Bromide Alternatives and Emissions Reductions, San Diego, California, USA, 29th October–1st November. United States Environmental Protection Agency.

Shim JK, Aye TT, Kim DW, Kwon YJ, Kwon JH, Lee KY (2009) Gamma irradiation effects on the induction of three heat shock protein genes (piac25, hsc70 and hsp90) in the Indian meal moth, *Plodia interpunctella*. J Stored Prod Res 45: 75–81.

Sinclair BJ, Gibbs AG, Roberts SP (2007) Gene transcription during exposure to, and recovery from, cold and desiccation stress in *Drosophila melanogaster*. Insect Mol Biol 16: 435–443.

Teets NM, Denlinger DL (2013) Physiological mechanisms of seasonal and rapid cold-hardening in insects. Physiol Entomol 38: 105–116.

Tompsett PB, Pritchard HW (1998) The effect of chilling and moisture status on the germination, desiccation tolerance and longevity of *Aesculus hippocastanum* L. seed. Ann Bot 82: 249–261.

Tungjitwitayakul J, Tatun N, Vajarasathira B, Sakurai S (2015) Expression of heat shock protein genes in different developmental stages and after temperature stress in the maize weevil (Coleoptera: Curculionidae). J Econ Entomol 108: 1313–1323.

Wang D, Collins PJ, Gao X (2006) Optimising indoor phosphine fumigation of paddy rice bag-stacks under sheeting for control of resistant insects. J Stored Prod Res 42: 207–217.

Wang J, Germinara GS, Feng Z, Luo S, Yang S, Xu S, Li C, Cao Y (2022) Comparative effects of heat and cold stress on physiological enzymes in *Sitophilus oryzae* and *Lasioderma serricorne*. J Stored Prod Res 96: 101949.

Wang SJ, Tang JM (2004) Radio frequency heating: A potential method for post-harvest pest control in nuts and dry products. J Zhejiang Univ Sci Ang A 5: 1169–1174.

Wareham AC, Duncan CJ, Bowler K (1974) The resting potential of cockroach muscle membrane. Comp Biochem Physiol A Mol Integr Physiol 48: 765–797.

Wijayaratne LKW, Fields PG (2010) Effect of methoprene on the heat tolerance and cold tolerance of *Tribolium castaneum* (Herbst) (Coleoptera: Tenebrionidae). J Stored Prod Res 46: 166–173.

Wijayaratne LKW, Rajapakse RHS (2018) Effects of spinosad on the heat tolerance and cold tolerance of *Sitophilus oryzae* L. (Coleoptera: Curculionidae) and *Rhyzopertha dominica* F.(Coleoptera: Bostrichidae). J Stored Prod Res 77: 84–88.

Wilches DM, Laird RA, Floate KD, Fields PG (2017) Effects of acclimation and diapause on the cold tolerance of *Trogoderma granarium*. Entomol Exp Appl 165: 169–178.

Wilches DM, Laird RA, Floate KD, Fields PG (2019) Control of *Trogoderma granarium* (Coleoptera: Dermestidae) using high temperatures. J Econ Entomol 112: 963–968.

Wright EJ, Sinclair EA, Annis PC (2002) Laboratory determination of the requirements for control of *Trogoderma variabile* (Coleoptera: Dermestidae) by heat. J Stored Prod Res 38: 147–155.

Yang JP, Zhang Y, Yu F, Dai RH, Yang H, Hu DM, Zhang XM, Wang Y (2022) Biological quality of *Anisopteromalus calandrae* (Hymenoptera: Pteromalidae) reared with cold-stored larvae of *Lasioderma serricorne* (Coleoptera: Anobiidae). J Stored Prod Res 97: 101974.

Yocum GD (2001) Differential expression of two HSP70 transcripts in response to cold shock, thermoperiod, and adult diapause in the Colorado potato beetle. Insect Physiol 47: 1139–1145.

Yu C, Subramanyam B, Flinn PW, Gwirtz JA (2011) Susceptibility of *Lasioderma serricorne* (Coleoptera: Anobiidae) life stages to elevated temperatures used during structural heat treatments. J Econ Entomol 104: 317–324.

Zachariassen KE (1985) Physiology of cold tolerance in insects. Physiol Rev 65: 799–832.

Chlorine Dioxide for Insect and Pathogen Control in Stored Commodities

Rania Marie Buenavista, Selladurai Manivannan, Xinyi E, Kaliramesh Siliveru, and Bhadriraju Subramanyam

8.1 Introduction

Microorganisms can easily contaminate food at various stages of crop production including planting, harvest, processing, transportation, and storage. Foodborne pathogen contamination contributes an estimated 48 million incidents of foodborne illness annually in the United States (Scallan et al., 2011; Whitham et al., 2022). Several studies have detected populations of bacterial species, such as *Listeria* sp., *Salmonella* sp., *Escherichia coli*, *Cyclospora* sp., *Clostridium* sp., *Vibrio* sp., *Campylobacter* sp., *Pseudomonas* sp., *Acinetobacter* sp., and *Shigella* sp., in contaminated food, making it unsafe for human consumption (Panwar et al., 2022). Symptoms of foodborne illness may vary among individuals based on its severity and may include headache, nausea, stomach cramps, diarrhea, vomiting, and

DOI: 10.1201/9781003309888-8

279

fatigue. In worst-case scenarios, the illness can be life-threating and may require medical treatment (Nazir et al., 2023). Despite continuous preventive measures, preventing foodborne pathogen outbreaks is still a critical public health challenge. Microbial safety of stored commodities has been a growing concern over the years due to numerous recalls associated with foodborne illnesses caused by consumption of pathogen-contaminated food products (Han et al., 2004).

Chlorine dioxide (ClO_2), a strong oxidation agent, has been widely used for pathogen control in a large range of stored commodities due to its antimicrobial effects. Chlorine dioxide has been demonstrated to be effective against bacteria, fungi, and viruses on fresh fruits and vegetables (Praeger et al., 2018; Sun et al., 2019). Notably, it does not produce any toxic byproducts or compromise the sensory qualities of treated produce (Malka and Park, 2022). Treatment of edible fungi with ClO_2 reduced the microbial load and decreased the activity of the enzyme, polyphenol oxidase (PPO), that is responsible for browning (Ku et al., 2006).

Chlorine dioxide is applied in gaseous and aqueous forms. Gaseous ClO_2 is advantageous over the aqueous form in terms of efficacy, as it can permeate protective sites within the commodities that are responsible for inhibiting microbial activity (Han et al., 2001). Given its cost-effectiveness, ClO_2 has emerged as a viable alternative to ozone and is predominantly (> 95%) employed in the wood pulp industry for bleaching purposes. Recently, studies have shown that ClO_2 is also effective in controlling major stored-product insects (E et al., 2018). Nevertheless, the practical application of ClO_2 poses a minor challenge due to its explosive nature at high concentrations and pressures above 10 kPa. As a result, storing ClO_2 is impractical, leading to the need for on-site generation methods.

Chlorine dioxide was introduced in the 1800s in the paper industry as a bleaching agent during pulp processing (Simpson, 2005). In the late 1960s, it was approved as an aqueous solution for disinfecting fresh fruits and vegetables, as an odor-controlling agent, and for sanitizing food processing equipment (Smith et al., 2014; Ahmed et al., 2017). Over the past few decades, ClO_2 has completely replaced chlorine in pulp mills (Gordon and Rosenblatt, 2005). In addition, the use of ClO_2 in drinking water systems has seen a considerable increase due to increasing regulation of chlorinated byproducts. Because ClO_2 gas has a high oxidation property, more than 2.5 times that of chlorine in hypochlorous acid, the exposure time and quantity of ClO_2 required for the same bactericidal action as is obtained with chlorine is significantly smaller (Benarde et al., 1965; Huang et al., 1997; Beuchat et al., 2004). Beyond its use in aqueous solutions for treatment of water and odor elimination in manufacturing plants, the use of ClO_2 in gaseous form has seen a reasonable increase (Gordon and Rosenblatt, 2005). The gaseous form of ClO_2 came into existence in 1984 for use as a sterilizing agent. Since 2006, the United States Food and Drug Administration (U.S. FDA) has granted approval for the use of ClO_2 on agricultural postharvest products (Gómez-López et al., 2009). The use of ClO_2 both in aqueous and in gaseous forms has enabled a considerable increase in microbial decontamination. The application of ClO_2 for insect control has potential, with its efficacy as a fumigant for stored-product insects (E et al., 2017).

In terms of chemical methods for stored-product pest management, the registered fumigants for postharvest disinfestation include phosphine, methyl bromide, sulfuryl fluoride, and ethyl formate (Phillips et al., 2012). Methyl bromide has been phased out following the Montreal Protocol due to its ozone-depleting potential (MBTOC, 2014). Phosphine has been widely used for controlling many stored-product insects, as it is cost-effective and versatile in application (Manivannan, 2015; Nayak et al., 2020). However, overexploitation of phosphine, combined with substandard fumigation practices and genetic factors, has led to developed resistance to phosphine among various stored-product insect species (Nayak et al., 2020; Lampiri et al., 2021). Managing phosphine-resistant insect strains has become a major concern in many developing and developed countries (Opit et al., 2012; Nayak et al., 2020). Although sulfuryl fluoride has demonstrated insecticidal efficacy against many insect pests, concerns regarding fluoride residues (U.S. EPA, 2016) and its greenhouse status (Gressent et al., 2021) undermine its wide acceptability. Chlorine dioxide gas is a prospective alternative to commonly used fumigants, as it has powerful oxidizing properties with high penetration capability (Han et al., 2016). Several studies have shown the insecticidal potency of ClO_2 on stored-product insects and some household pests (Kim et al., 2015; E et al., 2018).

This chapter provides an overview of the use of ClO_2 as an antimicrobial agent and as a fumigant for pathogen and insect control in stored commodities. It also presents properties, the mechanism of action against microbial and insect populations, and relevant research results on these applications.

8.2 Chlorine Dioxide

8.2.1 Properties

Chlorine dioxide is a yellow to reddish-yellow color gas, which readily dissolves in water at high concentrations. At 0°C, anhydrous ClO_2 is present in liquid form; it has a deep red color but becomes colorless when diluted with water (Young, 2016). Chlorine dioxide has an irritating nature and emits a pungent odor like chlorine or bleach (NIOSH, 2010).

Five electrons are accepted by ClO_2 when it is reduced to chloride ions. The oxidation capacity of chloride ions is greater than 2.5 times that of chlorine in hypochlorous acid. Chlorine dioxide has lower oxidation strength, which makes it more selective in its reactions. Also, oxidation and chlorination of organic compounds occur at a slower rate in reaction to ClO_2 than in reaction to chlorine. In most cases, the oxidation reaction predominates over the chlorination action (Fukayama et al., 1986). Low pH and organic matter decrease the efficacy of ClO_2 in terms of oxidation (Beuchat, 1998).

To assess the suitability of ClO_2 for large-scale fumigations, it is crucial to compare its key properties related to fumigant action with those of other well-known fumigants. Here we compare the important attributes of ClO_2 as a fumigant with those of phosphine and methyl bromide to understand its suitability as a methyl bromide alternative. Chlorine dioxide has a molecular weight (67.45 g/mol), which is approximately twice that of phosphine (34 g/mol) but lower than that of methyl bromide (94.94 g/mol). The boiling point of ClO_2 (11°C) is higher than those of phosphine (–87.7°C) and methyl bromide (3.56°C). Another important factor to consider for the fumigant effectiveness of ClO_2 in large-scale applications is vapor pressure. The higher the vapor pressure of a fumigant, the better its ability to penetrate into the grain mass. Chlorine dioxide gas has a vapor pressure of 1 atm, compared to 1.87 atm for methyl bromide and 41.3 atm for phosphine. The vapor pressure of ClO_2 fumigant is more than 40 times lower than that of phosphine fumigant, indicating that the penetration capability of ClO_2 cannot be comparable to that of phosphine, and this should be considered during large-scale trials. However, maintaining a balance among temperature, reactivity, and decomposition tendency is crucial for ClO_2. Another factor to consider is vapor density, which shows how much heavier the gas is in relation to air at a given temperature and pressure. The vapor density of ClO_2 (2.33) is lower than that of methyl bromide (3.3) but higher than that of phosphine (1.2) (NIOSH, 1999). This means that chlorine dioxide gas is heavier than air and tends to settle or accumulate in lower areas or confined spaces. Considering these properties, ClO_2 appears to be a potential alternative to methyl bromide and phosphine fumigants. However, careful evaluation and balancing of these variables are essential to ensure the effectiveness of ClO_2 as a fumigant.

8.2.2 Generation

Due to its explosive nature, shipment of ClO_2 gas is dangerous; hence, it is mostly generated on site or transported as a diluted solution stored in sealed containers. It can be generated through various methods using aqueous solutions of sodium chlorite or sodium chlorate as the precursor, along with specific chemical reactions or by electrolysis. The system for generating ClO_2 is decided based on the quantity and form of ClO_2 needed for a specific application (Knapp and Battisti, 2001). For large-scale applications, the most economical method of generation is through chlorate ion reduction, while smaller-scale generation systems use sodium chlorite solution in producing ClO_2 (Simpson, 2005).

Sodium chlorite solution undergoes an acidification process to generate an aqueous solution of ClO_2. Activation of sodium chlorite is done with the use of reducing agents such as hydrochloric acid or sulfuric acid. The chlorine-chlorite solution method generates higher ClO_2 yields, wherein reaction between chlorite ions and chlorine of hydrochloric acid (HCl) results in formation of ClO_2. However, the drawback of this method is its slow production rate. A faster method with high ClO_2 yield is through vaporization of sodium chlorite solution, which then reacts with chlorine gas under vacuum conditions. In an industrial setting, three main feedstocks such as hydrochloric acid, sodium chlorite, and sodium hypochlorite are used in generators to avoid the restricted storage of liquid chlorine (Trinetta et al., 2012). On the other hand, ClO_2 gas is usually generated using 4% chlorine (Cl_2) that undergoes bubbling in a nitrogen gas carrier and successively flows through solid sodium chlorite cartridges. Storing solid sodium chlorite as tablets, sachets, or pouches allows controlled amounts to be transported; they then are usually activated using water or a humid environment to generate ClO_2 gas (White, 2010).

The U.S. FDA provided the following guidelines on ClO_2 gas generation:

i. Reaction involving aqueous sodium chlorite solution and chlorine gas or sodium hypochlorite and hydrochloric acid solution.

$$2NaClO_2 + Cl_2 \rightarrow 2ClO_2 + 2NaCl$$

$$2NaClO_2 + 2NaOCl + 2HCl \rightarrow 2ClO_2 + 2NaCl + 2H_2O$$

ii. Reaction involving aqueous sodium chlorate and hydrogen peroxide (an oxidizer) with sulfuric acid.

$$NaClO_3 + H_2O_2 + 2H_2SO_4 \rightarrow 2\ ClO_2 + 2\ H_2O + 2NaHSO_4 + O_2$$

iii. Generating chlorine dioxide gas through electrolysis of an aqueous solution of sodium chlorite.

$$ClO_2^- \rightarrow 2ClO_2 + e^-$$

Shirasaki et al. (2016) reported using 3.35% sodium chlorite solution and 85% phosphoric acid solution to generate ClO_2 gas as follows:

$$15NaClO_2 + 4H_3PO_4 \rightarrow 12ClO_2 + 6H_2O + 3NaCl + 4Na_3PO_4$$

8.3 Insect Control

Addressing the significant concern about postharvest losses attributed to stored-product insects has become imperative in light of the escalating demand for food due to the growing global population. Fumigation plays a pivotal role in integrated pest management programs for controlling stored-product insects (Phillips et al., 2012). The utilization of fumigants offers significant advantages over that of residual insecticides. Gaseous fumigants have the ability to deeply penetrate the grain and ensure more uniform distribution of the gas, without leaving behind significant residues on treated commodities (Manivannan, 2015). This addresses a major concern associated with residual treatments using insecticides: they can contaminate the food with toxic residues affecting the health of the consumer and the environment. Phosphine has been the predominant fumigant globally accepted since the 1960s (Chaudhry, 2000; Manivannan, 2015). However, the development of phosphine resistance among stored-product insects is the major factor responsible for control failures in large-scale fumigations. Phosphine resistance has been reported in a range of stored-product insect species (Champ and Dyte, 1976; Price and Mills, 1988; Rajendran and Gunasekaran, 2002; Lorini et al., 2007; Jagadeesan et al., 2012; Kaur et al., 2015; Gautam et al., 2020). Due to the phase-out of methyl bromide following the Montreal Protocol, finding viable alternatives to methyl bromide is crucial for effective management of stored-product insects, especially for commodities and grain processing facilities. Among potential alternatives, ClO_2 gas has shown promising insecticidal potency on stored-product insects.

8.3.1 Mechanism of Action

Chlorine dioxide gas enters the respiratory system of insects through spiracles (small openings on the insect's body for gas exchange). Once it enters the respiratory system, ClO_2 reacts with respiratory tissues, disrupting the insect's ability to breathe and causing suffocation. The primary mechanism responsible for mortality of insects upon treatment with ClO_2 is oxidative stress as a result of production of reactive oxygen species (ROS) (Kim et al., 2015; Young, 2016). Although ClO_2 treatment induced the production of antioxidant enzymes such as superoxide dismutase (SOD) and thioredoxin peroxidase (Tpx), the overflow of ROS production following ClO_2 treatment exerted an oxidative stress in insects, causing a lethal effect (Kim et al., 2015). Supporting this mechanism is the finding that silencing the SOD and Tpx genes in the red flour beetle, *Tribolium castaneum* (Herbst), through the introduction of dsRNAs significantly enhanced the insecticidal activity of ClO_2. Another mechanism by which ClO_2 induces mortality in treated insects is through the reduction of hemocytes (blood cells) and impairing hemocyte spreading behavior. Impaired hemocyte spreading affects the insect's immune system, rendering it more susceptible to bacterial infections, as explained by Kumar et al. (2015).

8.3.2 Efficacy on Various Insect Species

Chlorine dioxide gas exhibits significant potential for controlling stored-product insects. However, while numerous studies have reported the microbial decontamination efficacy of ClO_2 gas, only a limited number of studies have explored its impact on stored-product insects (E et al., 2018). Of the various methods of generating ClO_2 gas listed in Section 8.2.2, most of the studies have utilized the electrolysis method. Research indicates that ClO_2 gas is effective against all stages of stored-product insects. The susceptibility of insect stages to ClO_2, as to other fumigants, is associated with their metabolic activity, as presented in Table 8.1. Typically, larva and adult stages, being active, tend to be more susceptible to fumigants, whereas egg and pupae, with their dormant state, pose greater challenges for eradication (Price and Mills, 1988). Studies of the effect of phosphine fumigation on the immature stages of *T. castaneum* demonstrated that the order of susceptibility of immature stages to phosphine was second instars > fourth instars > sixth instars > pupae > eggs (Manivannan, 2015). Similarly, Channaiah et al. (2012) found that the order of susceptibility of *T. castaneum* to ClO_2 was adults > young larvae > old larvae > eggs. These findings indicate that insect susceptibility is influenced by variations in ClO_2 uptake by active and dormant insect stages. However, the concentrations they reported are significantly higher than those reported in the other studies in the literature. Interestingly, studies by Han et al. (2016) involving tests with ClO_2 against the different life stages of the Indian meal moth, *Plodia interpunctella* (Hübner), indicated that eggs were more susceptible than the other stages tested; the order of susceptibility to ClO_2 was eggs > adults > fifth instars > pupae. Since the action of a fumigant is influenced by the metabolic rate of the insect stage and the eggs are dormant, this result is quite surprising. Hence, the mode of action of ClO_2 on insects could influence this variation in responses between stages of different insect species. Furthermore, the dosage required for complete mortality, provided in Table 8.1, shows that the susceptibility of insects to ClO_2 is also related to the insect species dealt with (E et al., 2017). For instance, E et al. (2017) observed differences in susceptibility

Table 8.1 Effect of ClO_2 on Stored-Product Insect Species and Their Life Stages

Insect Species	Insect Stage Tested	Method of Chlorine Dioxide Generation	Dosage Required for Complete Mortality	Results	Order of Susceptibility	Limitation/Comments	Reference
T. castaneum	Eggs, young larvae, old larvae, adults	Sodium hypochlorite and hydrochloric acid mixture added to water	834.4 g-h/m^3	Mortality in eggs and old larvae was 10% and 20%, respectively.	Adults > young larvae > old larvae > eggs	The presence of food (wheat flour) decreased the mortality considerably in all stages except adult stage due to ClO_2 reaction with organic matter.	Channaiah et al. (2012)
Tribolium confusum (Jacquelin du Val)	Eggs, young larvae, old larvae, adults	Sodium hypochlorite and hydrochloric acid mixture added to water	834.4 g-h/m^3	Mortality in eggs and old larvae was 11% and 31%, respectively.	Adults > young larvae > old larvae > eggs	The presence of food (wheat flour) decreased the mortality considerably in all stages except adult stage due to ClO_2 reaction with organic matter. All stages were susceptible compared to T. castaneum.	
T. castaneum	Larvae, female adults	Electrolysis of aqueous sodium chlorite solution	12.96 g-h/m^3	Complete mortality of both the insect stages was achieved.	Both stages succumbed to the tested dosage.	Tests were performed with food source, and in tests with flour insects burrowed and escaped fumigant. Tests with brown rice showed better effect.	Kim et al. (2015)

(Continued)

Table 8.1 (Continued) Effect of ClO_2 on Stored-Product Insect Species and Their Life Stages

Insect Species	Insect Stage Tested	Method of Chlorine Dioxide Generation	Dosage Required for Complete Mortality	Results	Order of Susceptibility	Limitation/Comments	Reference
P. interpunctella	Fifth instars	Electrolysis of aqueous sodium chlorite solution	12.96 g-h/m³	Complete mortality was achieved.	NA	Tests only involved fifth instars with brown rice granules as food source.	Kumar et al. (2015)
P. interpunctella	Eggs Fifth instars PupaeAdults	Electrolysis of aqueous sodium chlorite solution	3.24 g-h/m³ 12.96 g-h/m³ 12.96 g-h/m³ 3.24 g-h/m³	Eggs were more susceptible.	Adults > eggs > pupae > fifth instars	Tests were performed with food source.	Han et al. (2016)
T. castaneum O.surinamensis R. dominica S. zeamais S. oryzae	Adults Adults Adults Adults Adults	Electrolysis of aqueous sodium chlorite solution	14.79 g-h/m³ 8.41 g-h/m³ 15.79 g-h/m³ 14.53 g-h/m³ 7.67 g-h/m³ (after 5 days with food)	Complete reduction in progeny production of R. dominica was observed at 10.8 g-h/m³ dosage, while no progeny were observed after treatment with ClO_2 at least dosage of 2.7 g-h/m³ for T. castaneum, S. oryzae, and S. zeamais.	S. oryzae > O. surinamensis > S. zeamais T. castaneum > R. dominica>	Compared to tests without food, tests performed with food source required higher dosages for complete mortality.	E et al. (2017)

(Continued)

Table 8.1 (Continued) Effect of ClO_2 on Stored-Product Insect Species and Their Life Stages

Insect Species	Insect Stage Tested	Method of Chlorine Dioxide Generation	Dosage Required for Complete Mortality	Results	Order of Susceptibility	Limitation/Comments	Reference
T. castaneum	Adults	Electrolysis of aqueous sodium chlorite solution	16.92 g-h/m³	Complete mortality was achieved only for T. castaneum and O. surinamensis. No progeny production was observed in T. castaneum and O. surinamensis for any of the tested concentrations. Complete production was observed in R. dominica, S. zeamais, and S. oryzae at 8.46 g-h/m³ and above.	O. surinamensis > T. castaneum > S. oryzae > S. zeamais > R. dominica	Compared to tests without food, tests performed with food source required higher dosages for complete mortality. Delayed mortality was observed.	E et al. (2018)
O. surinamensis	Adults		5.64 g-h/m³				
R. dominica	Adults		16.92 g-h/m³				
S. zeamais	Adults		16.92 g-h/m³				
S. oryzae	Adults		16.92 g-h/m³				
			16.92 g-h/m³ (after 5 days with food)				

(Continued)

Table 8.1 (Continued) Effect of ClO_2 on Stored-Product Insect Species and Their Life Stages

Insect Species	Insect Stage Tested	Method of Chlorine Dioxide Generation	Dosage Required for Complete Mortality	Results	Order of Susceptibility	Limitation/Comments	Reference
P. interpunctella S. zeamais	Fifth instar, adults Adults	Electrolysis of aqueous sodium chlorite solution	12.96 g-h/m³ 12.96 g-h/m³	Outer space had better larvae mortality (66.67%) than inner bags. Adult mortality was 100% at all dosages. Complete mortality was not achieved at any tested concentration.	Adults > fifth instars n/a	P. interpunctella larvae were more tolerant and resulted in poor mortality when tested with grains. Lower temperature and humidity contributed to poor mortality. S. zeamais was more tolerant than P. interpunctella adults.	Kim et al. (2019)
R. dominica	Adults	Electrolysis of aqueous sodium chlorite solution	32.4 g-h/m³	Complete mortality was achieved at 25.92 g-h/m³ at all locations of sample placement except for middle location.	100% reduction in adult progeny was not achieved at any tested concentration.	Treatment was performed with 10 g of wheat.	Buenavista et al. (2023)

to ClO_2 among the different species tested; they found *Oryzaephilus suri-namensis* (L.) to be the most susceptible to ClO_2 gas and the adults of *T. castaneum* and *Rhyzopertha dominica* (F.) to be the most tolerant insect species. Kim et al. (2019) observed the adults of *P. interpunctella* to be susceptible to ClO_2, while the larvae of *P. interpunctella* and the adults of *Sitophilus zeamais* (Motschulsky) were tolerant. The differences in susceptibility among species to ClO_2 gas could be due to variations in the physiology such as metabolism and respiration, similar to those reported earlier for phosphine fumigant. (Pimentel et al., 2007).

Most of the studies compared ClO_2 efficacy on stored-product insects with and without food and concluded that the presence of food during fumigation influenced the efficacy of ClO_2 on insects, with insects treated with food requiring longer treatment times for complete mortality (Channaiah et al., 2012; E et al., 2017, 2018). Another important aspect along these lines was observed by Kim et al. (2015), who reported that the burrowing nature of *T. castaneum* larvae and adults in wheat flour during tests with ClO_2 was responsible for their reduced ClO_2 uptake. This refuge-seeking behavior was also reported by Kim et al. (2019) in their studies with ClO_2 tested against the fifth instars of *P. interpunctella*. E et al. (2018) showed the influence of temperature on the efficacy of ClO_2 with five stored-product insects. The efficacy of fumigants in general tends to increase with temperature (Phillips et al., 2012). Furthermore, treatment of grains with ClO_2 at lower temperatures can increase the adsorption of gas molecules on the surface of the grains, reducing the amount of free gas molecules available for insecticidal efficacy. Hence, during ClO_2 treatments, it is imperative to maintain warm temperatures (>30°C) so as to increase the metabolic rate of insects, thereby increasing the uptake of fumigant by insects, resulting in their death. For most of the treatments discussed in this chapter, the exposure time required for complete mortality of the insect stages was within 48 h, which is shorter than the periods required for phosphine. Delayed mortality of insects was observed following ClO_2 treatment. For instance, the adults of *T. castaneum, O. surinamensis, Sitophilus oryzae* (L.), *S. zeamais,* and *Rhyzopertha dominica* (F.) and *P. interpunctella* fifth instars showed delayed mortality after exposure to ClO_2 (Kim et al., 2015; Kumar et al., 2015; E et al., 2017, 2018).

A recent study conducted by E and Subramanyam (unpublished work) found that sorption of ClO_2 gas into wheat and corn kernels was inevitable during fumigation, hindering the efficacy of ClO_2 against *T. castaneum, O. surinamensis, R. dominica,* and *S. zeamais.* Three identical cylindrical columns containing grains (11 kg of wheat or 11.3 kg of corn) were fumigated with 350 ppm ClO_2 gas. Figures 8.1a and 8.2a show the results of this study, wherein the second application of ClO_2 gas [Figures 8.1b and 8.2b] resulted in a higher ClO_2 headspace concentration through time than

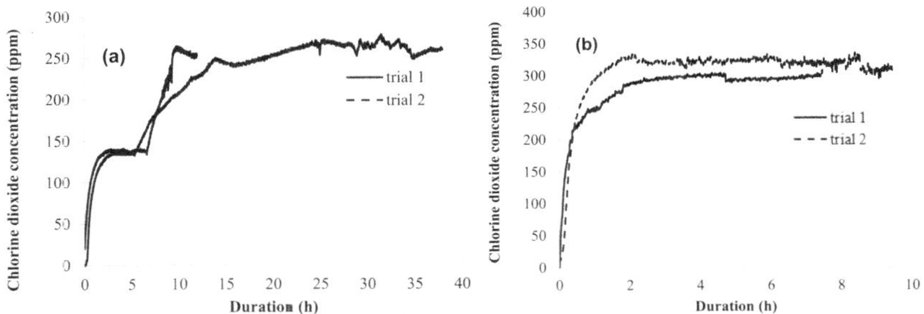

Figure 8.1 (a) Chlorine dioxide concentrations (ppm) in the headspace of the columns during the first wheat fumigation. (b) Chlorine dioxide concentrations (ppm) in the headspace of the columns during the second wheat fumigation.

the first application, using the same batch of wheat or corn. In both initial and repeated fumigations, wheat (Figure 8.1) adsorbed a greater quantity of ClO_2 gas than did corn (Figure 8.2), which was attributed to the surface binding to active sites on kernels. The difference in ClO_2 sorption quantity between wheat and corn can be attributed to their specific surface area. Wheat has a larger specific surface area than corn, as more wheat kernels can fit in the same volume. Larger specific surface area (m² surface area per m³ volume) results in higher fumigant sorption, as more active sites are accessible for fumigant binding (Hwaidi et al., 2015).

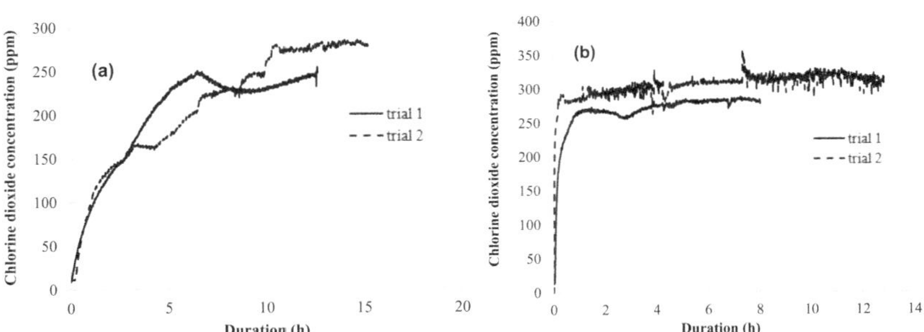

Figure 8.2 (a) Chlorine dioxide concentrations (ppm) in the headspace of the columns during the first corn fumigation. (b) Chlorine dioxide concentrations (ppm) in the headspace of the columns during the second corn fumigation.

8.3.3 Status on Cross Resistance

Cross resistance refers to the phenomenon whereby an organism develops resistance to one or more substances or agents as a result of being exposed to another substance or agent with a similar or dissimilar mode of action or chemical structure. To date, no instances of cross resistance to ClO_2 have been reported in any stored-product insect pests. In studies conducted by E et al. (2017, 2018), where phosphine-resistant strains of five stored-product insects (*T. castaneum, O. surinamensis, S. oryzae, S. zeamais,* and *R. dominica*) were exposed to ClO_2, no significant differences in the mortality response were observed between the phosphine-susceptible and -resistant insect strains. This indicated that phosphine-resistant insects did not exhibit resistance to ClO_2, and therefore ClO_2 can be a viable option for managing resistant strains. Likewise, in a study by Jagadeesan and Nayak (2017), it was observed that phosphine-resistant strains did not exhibit cross resistance when subjected to sulfuryl fluoride treatment.

8.4 Pathogen Control

Pathogen contamination of stored commodities may occur due to improper handling, unsanitary surfaces and equipment, or lack of regular sanitary practices during processing, storage, or transportation (Yeni et al., 2016). Minimizing levels of microorganisms during postharvest operations is important in increasing the stability and shelf-life of these commodities and decreasing the incidence of illness related to foodborne pathogens (Goodburn and Wallace, 2013). Chlorine dioxide has been proven to be effective against various pathogenic organisms such as *Legionella, Listeria, E. coli, Salmonella,* bacterial spores, *Cryptosporidium,* and amoebal cysts. Another advantage of ClO_2 is its effectiveness against biofilm, which retards bacterial regrowth (Joshi et al., 2013).

8.4.1 Mechanism of Action

Chlorine dioxide penetrates through the cell wall of microorganisms and significantly destroys metabolic function. This ability of ClO_2 is observed under a wide range of pH conditions, thus making it suitable and effective against a wide range of microorganisms. Four mechanisms of action are attributed to the antimicrobial action of ClO_2: interruption of protein synthesis, damage to the oxidative cell membrane, oxidation of amino acids or proteins, and oxidation of deoxyribonucleic acid (DNA) and ribonucleic acid (RNA). Various cellular processes are disrupted upon exposure to chlorine dioxide due to the irreversible damage it does to proteins. Chlorine dioxide has the ability to denature proteins that are important for cellular function and integrity. As ClO_2 reacts to bacterial cells, a decrease in α-helical content and increase in oxygen atoms in proteins are observed (Ogata, 2007). Its antimicrobial efficiency has a broad range, making it effective against various spoilage and pathogenic microorganisms such as *Salmonella* spp., *Listeria monocytogenes*, and *Escherichia coli* O157:H7 (Sun et al., 2014). In terms of its mode of action against *E. coli*, increased permeability of cells was indicated by the outflow of potassium, leading to damage in the transmembrane ionic gradient (Berg et al., 1986). Upon oxidation, both DNA and RNA form 8-oxoguanine from its base guanine, which results in inhibition of nucleic acid replication via base pairing. Thus, the inactivation of both DNA and RNA leads to a lack of coded information for translation of protein and ribosome functions. Another study, done by Bakhmutova-Albert et al. (2008), attributed the microbial inactivation of ClO_2 to the oxidation of dihydronicotinamide adenine dinucleotide (NADH), which is an important co-enzyme for numerous biological redox reactions such as ATP and electron-transport chain synthesis. Bakhmutova-Albert et al. (2008) concluded that 2 mol of ClO_2 can convert 1 mol of NADH into NAD+ in 30 min at temperatures ranging from 4°C to room temperature.

In a study by Yu et al. (2020), reduction of aflatoxin B_1 (AFB_1) in corn was attributed to the alteration of the furan ring, cyclopentenone, the methoxyl structure, or the lactone ring. After ClO_2 treatment on corn kernels, the C8-C9 double bond of the furan ring was damaged, resulting in the disappearance of toxic sites of AFB_1. In almonds, reduction of *Salmonella enterica* populations was explained to be caused by lipid oxidation as exhibited by increased levels of peroxide, thiobarbituric acid–reactive substances, free fatty acids, and conjugated dienes (Wang et al., 2023). It also induced formation of volatile compounds, including aldehydes, carboxylic acids, and alcohols. Kinetic models for the bacterial inactivation were established using first-order kinetics on strawberries (Mahmoud et al., 2007) and lettuce leaves against *E. coli* O157:H7 and *Salmonella enterica* (Mahmoud and Linton, 2008) and using the Weibull model for whole cantaloupe against *Salmonella* (Poona), *L. monocytogenes*, and *E. coli* O157:H7 (Mahmoud et al., 2008).

8.4.2 Efficacy Against Microbial Populations

Table 8.2 lists the log reductions associated with microbial populations on stored commodities after ClO_2 treatment. The cited studies concluded that ClO_2 was effective in reducing the levels of microbial populations under investigation. Chlorine dioxide concentrations and environmental

Table 8.2 Effect of Aqueous and Gaseous ClO_2 on Microbial Populations Associated with Stored Commodities

Commodity	Microorganism	Type of ClO_2	ClO_2 Concentration	Time	Temperature	Log Cycle Reduction	Reference
Alfalfa seeds	Salmonella L. monocytogenes Shiga toxin–producing E. coli	1.6% $NaClO_2$	1.6%	15 min 15 min 15 min	Room temperature	5.4 5.1 7.3	Tan et al. (2020)
Almond	Aspergillus flavus	Gas	5.67 mg/L	24 h	Room temperature	2.4	Rane et al. (2023)
Cabbage	Salmonella E. coli O157:H7 L. monocytogenes	Gas	4.1 mg/L	30.8 min 20.5 min 29.3 min	22°C	4.42 3.13 3.6	Sy et al. (2005)
Carrot	Salmonella E. coli O157:H7	Gas	4.1 mg/L	30.8 min 20.5 min 29.3 min	22°C	5.15 5.62 5.88	Sy et al. (2005)
Carrot	Total aerobic bacteria Psychrotrophs Lactic acid bacteria Yeasts	Gas	1.33 mg/L	6 min	28°C	1.88 1.71 2.6 0.66	Gómez-López et al. (2007)
Green pepper	E. coli O157:H7	Gas	0.62 mg/L 1.24 mg/L	30 min	22°C	3.03 6.45	Han et al. (2001)
Lettuce	Salmonella E. coli O157:H7 L. monocytogenes	Gas	4.1 mg/L	30.8 min 20.5 min 29.3 min	Room temperature	1.58 1.57 1.53	Sy et al. (2005)
Lettuce	E. coli O157:H7	Gas	1.00 mg/L	15 min	Room temperature	2.31	Singh et al. (2002)

(Continued)

Table 8.2 (Continued) Effect of Aqueous and Gaseous ClO_2 on Microbial Populations Associated with Stored Commodities

Commodity	Microorganism	Type of ClO_2	ClO_2 Concentration	Time	Temperature	Log Cycle Reduction	Reference
Lettuce	E. coli O157:H7 S. enterica	Gas	5 mg/L	10 min	Room temperature	3.9 2.8	
Lettuce seeds	E. coli O157:H7	Gas	10 mg/L	3 min	Room temperature	1.8	Trinetta et al. (2011)
Mungbean	Salmonella L. monocytogenes Shiga toxin–producing E. coli	Gas	1.6%	15 min 30 min 15 min 30 min 15 min 30 min	Room temperature	4.3 4.7 5.0 5.5 5.0 6.7	Tan et al. (2020)
Mungbean sprouts	S. enterica	Gas (with mechanical mixing) Gas (without mechanical mixing)	0.5 mg/L of air 0.5 mg/L of air	15 min 30 min 60 min 15 min 30 min 60 min	21°C	3.0 4.0 5.5 3.0 3.0 4.0	Prodduk et al. (2014)
Onion	Salmonella Yeasts and molds	Gas	1.4–4.1 mg/L	6–25 min	22°C	0.83–1.94 0.22–0.36	Sy et al. (2005)
Potato	P. aeruginosa Total aerobic bacteria Yeasts and molds	Gas	40 mg/L 16 mg/L 16 mg/L	5 h 2.5 h 2.5 h	22°C	6 1.7 1.9	Wu and Rioux (2010)

(Continued)

Table 8.2 (Continued) Effect of Aqueous and Gaseous ClO_2 on Microbial Populations Associated with Stored Commodities

Commodity	Microorganism	Type of ClO_2	ClO_2 Concentration	Time	Temperature	Log Cycle Reduction	Reference
Potato	*Erwinia carotovora*	Gas	0.02 mg/L	10 min	22°C	7.4	Tsai et al. (2001)
Radish seeds	*E. coli* O157:H7	Aqueous	25 mg/L	5 min	Room temperature	0	Singh et al. (2002)
Radish seeds	*E. coli* O157:H7 Aerobic bacteria Yeast and molds	Aqueous	500 mg/L	5 min	Room temperature	>5 >5 ~1	Bang et al. (2011)
Spinach leaves	*E. coli* O157:H7 *Salmonella* *L. monocytogenes*	Gas	0.13 mg/L	20 min	21°C	1.25–2.54	Park and Kang (2015a)
Spinach leaves	*E. coli* O157:H7 *Salmonella* *L. monocytogenes*	Gas	0.03 mg/L	20 min	22°C	4.6 4.7 4.3	Mahmoud et al. (2007)
Tomato	*S. enterica*	Gas	0.15–0.85 mg/L	58 min	4°C	3.95	Netramai et al. (2016)
Tomato	*E. coli* O157:H7 *Salmonella* *L. monocytogenes*	Gas	0.08 mg/L	20 min	22°C	> 1.39 > 1.22 > 1.52	Park et al. (2018)

(Continued)

Table 8.2 (Continued) Effect of Aqueous and Gaseous ClO_2 on Microbial Populations Associated with Stored Commodities

Commodity	Microorganism	Type of ClO_2	ClO_2 Concentration	Time	Temperature	Log Cycle Reduction	Reference
Tomato	E. coli O157:H7 Salmonella L. monocytogenes	Gas	0.05 mg/L	30 min	25°C	1.53–1.88	Park and Kang (2018)
Tomato	L. monocytogenes Salmonella	Gas	0.05 mg/L	12 min	22°C	> 5	Bhagat et al. (2010)
Tomato	S. enterica	Gas	8 mg/L 10 mg/L 10 mg/L	1 min 2 min 3 min	25°C	2.94 3.86 4.87	Trinetta et al. (2010)
Tomato	A. alternata	Gas	10 mg/L	1 min	23°C	2.71 2.63	Trinetta et al. (2013)
Tomato	S. enterica	Gas	0.4 mg/L	4 h	13°C	4.6–5	Olanya et al. (2015)
Tomato	E. coli O157:H7 Salmonella	Gas	0.03 mg/L	20 min	22°C	3.9 3.5	Park and Kang (2015b)
Tomato	A. alternata	Gas	0.01 mg/L	14 d	20°C	1.6–4.0 2.9–4.7	Sun et al. (2019)
Tomato	Salmonella Yeasts and molds	Gas	1.4–4.1 mg/L	6–25 min	22°C	1.11–4.33 0.87–1.61	Sy et al. (2005)

conditions in these studies differed depending on the tested microorganism and type of stored commodity. In summary, ClO_2 was investigated for reducing microbial populations in potable drinking water, solid surfaces, fruits and vegetables, seeds, sprouts, and meat and seafood products (Trinetta et al., 2012).

In general, Gram-positive (G+) bacteria have higher tolerance against ClO_2 than Gram-negative (G−) bacteria, which may be attributed to the presence of a thin plate mesh peptidoglycan layer in G− bacteria. Chlorine dioxide encounters a thicker barrier to penetrate with the peptidoglycan layer of G+ bacteria (Vandekinderen et al., 2009a). Lower concentrations of ClO_2 than of other disinfectants were needed to reach a reduction effect or complete removal of microorganisms (Anfruns-Estrada et al., 2019; Xue et al., 2013). Anfruns-Estrada et al. (2019) stated that about a two to three times higher concentration of hypochlorite and peracetic acid was needed to give the same effect as that of ClO_2 in reducing microbial populations. However, research on resistance status and development against ClO_2 in various microorganisms is necessary to determine the overall long-term efficacy of ClO_2. Valderrama et al. (2009) found that 3 mg/L ClO_2 can result in a 4-log unit reduction of *L. monocytogenes* in 90 s. A ClO_2-based disinfectant powder diluted in water at 100 mg/L can effectively reduce populations of *Staphylococcus aureus* and *Escherichia coli* (Zhu et al., 2008). Although the results for this disinfectant powder are promising, further research is needed with regard to its potential use in sanitizing surfaces of stored commodities. Among numerous studies investigating the efficacy on microbial populations on stored commodities, results highlighted significant log reduction at low concentration levels. Some of these findings include the reduction of 3.13–4.42 log CFU/g for cabbage, 1.53–1.58 log CFU/g for lettuce, 4.33 log CFU/g for tomato, 5.15–5.88 log CFU/g for carrots, and 1.94 log CFU/g for onion. The relative microbial inactivation action increases with more acidic environments and higher temperatures (Ridenour and Ingols, 1947; Joshi et al., 2013).

8.5 Quality and Safety Aspects of Chlorine Dioxide Treatments

In deciding whether a technology is applicable for insect and pathogen control, several criteria must be considered, including its ability to effectively reduce pathogenic and spoilage microorganisms, maintain food quality at the level required safe consumption and consumer acceptance, increase product shelf life, and avoid negative effects on the health of consumers. Hence, various studies have investigated the effect of ClO_2 on product quality and safety.

8.5.1 Product Quality

Changes in product quality after ClO_2 treatment are variable among different stored commodities. In terms of respiration rate, no significant change was observed in the respiration rate of shredded iceberg lettuce (Lopez-Galvez et al., 2010b) and grated carrots (Gómez-López et al., 2007), while the respiration rate of shredded white cabbage increased by about 22% (Gómez-López et al., 2007). Meanwhile, carbon dioxide production was reduced in ClO_2-treated whole green bell peppers (Jin-Hua et al., 2007). Decreased peroxidase activity in lettuce (Chen et al., 2010) and polyphenol oxidase activity of asparagus lettuce increased the stability of these produce (Fu et al., 2007).

Chlorine dioxide treatment may affect the phytochemical composition of stored commodities, as it can react with phenols (Napolitano et al., 2005). Moreover, ClO_2 can easily oxidize nutrients such as ascorbic acid. Although changes in composition are recorded, it is claimed that they are not so big as to negatively impact the entire commodity, as limited exposure to ClO_2 occurs, mainly at the surface of the stored commodities (Gómez-López, 2012). Results in Du et al. (2007) showed that the rate of vitamin C loss in ClO_2-treated whole green peppers was observed to accelerate for the first 10 d in storage under ambient conditions but retarded after 20 d in storage. After 40 d, the end result was higher concentrations of vitamin C in ClO_2-treated whole green peppers. In iceberg letture, ClO_2 treatment of 3.7 mg/L for 30 min did not alter the phenolic and vitamin C content (Lopez-Galvez et al., 2010b). Vandekinderen et al. (2008) observed a decrease in total phenol and β-carotene content and no change in antioxidant capacity of ClO_2-treated grated carrots. Decreased total phenols, antioxidant capacity, and β-carotene and no effect on vitamin C content were observed in white cabbage (Vandekinderen et al., 2009b). In iceberg lettuce, no significant changes in carotenes, vitamin C, antioxidant capacity, or total phenols were observed (Vandekinderen et al., 2009c). In leek, vitamin C decreased, but carotenoids, antioxidant capacity, and total phenols were not significantly altered by ClO_2 (Vandekinderen et al., 2009d).

Wang et al. (2019) investigated the effect of ClO_2 gas treatment on the sensory and nutritional qualities of tomatoes inoculated with *S. enterica*. Chlorine dioxide treatments at 1.9 mg/L for 5 h and 4.3 mg/L for 2.5 and 5 h did not adversely affect the off-odor, color, firmness, lycopene content, appearance, or vitamin C content of grape tomatoes, while reducing the *S. enterica* population from 5.4 log CFU/fruit to a nondetectable level (the limit of detection is 1.70 log CFU/fruit). In peanuts, Liu et al. (2022) found that ClO_2 treatment (0.45%, w/w relative to dry matter) did not significantly alter acid and iodine values or fatty acid composition of the crude oil extracted from ClO_2-treated peanuts. The changes observed in ClO_2-treated peanuts were increased hydroperoxides content, moisture content, and oil yield, suggesting that the process might be a useful detoxification method for oil processing or the oilseed storage industry. In almonds, increased levels of aldehydes and lipid oxidation due to ClO_2 gas exposure may alter the favor and consumer acceptance. Further sensory evaluation is necessary to determine the extent of changes in flavor and taste of treated almonds.

In terms of visual quality, higher ClO_2 concentrations may cause darkening due to oxidation of phytochemicals (Du et al., 2007). Similar findings were reported in Singh et al. (2002), wherein discoloration, bleaching, and browning of food products were attributed to oxidation of chlorophyll and phenols. In the study of Mahmoud and Linton (2008), lettuce leaves changed color from green to white. Variation in effects on visual attributes suggests that concentration and length of treatments must be established specific to the stored commodity, as reaction rates differ. Moreover, further research is needed to determine the extent of this visual degradation for repeated exposure to ClO_2.

Since many of the studies evaluating the insecticidal efficacy of ClO_2 on stored-product insects are laboratory-oriented, not much information is currently available on the quality of treated grains. Han et al. (2018) evaluated the effect of ClO_2 treatment on the viability of rice and wheat grains exposed to 50, 100, and 200 ppm concentrations for 0, 6, 12, 24, and 48 h. Their results showed that the viability of rice seeds was not significantly affected when they were treated at 50 and 100 ppm concentrations until the length of treatment reached 48 h. On the other hand, they observed that the viability of wheat grains was significantly affected when they were treated at concentrations of 50 and 100 ppm. Exposing the rice and wheat grains to a 200-ppm concentration of ClO_2 significantly affected the seed viability, most notably after 48 h of treatment. Similarly, the germination rate of rice was not significantly affected when it was treated at 200 ppm concentration of ClO_2 for 24 h; however, the germination ability of wheat grains was reduced from 82% to 72% following ClO_2 treatment under the same conditions (Han et al., 2018). Buenavista et al. (2023) observed more than a 50% reduction in the germination rate of wheat grains fumigated with the lethal ClO_2 concentration of 32.4 g/m^3 required for achieving complete mortality of *R. dominica* adults. Moreover, they observed increased lightness, reduced pH, and reduced peak and final viscosities of flour obtained from wheat kernels treated to a dosage of 12.96 $g\text{-}h/m^3$ or more. In addition, they observed that the functionality parameters of flour obtained from ClO_2-treated wheat kernels were comparable to those of the control samples.

8.5.2 Safety of Foods Exposed to Chlorine Dioxide

No chlorinated toxic byproducts are formed after ClO_2 treatment, which is its major advantage over chlorine for treatment of stored commodities. The main byproducts of ClO_2 include chlorate and chlorite upon its decomposition. It does not form any organo-halogen byproducts, as its reaction to food products is only through oxidation; no electrophilic substitution takes place. However, chlorinated byproducts may form due to occurrence of impurities, such as chloramines and chlorine, in the production of the ClO_2. Hence, generation of ClO_2 must be done with proper measures in place (Trinetta et al., 2011). In the study of Trinetta et al. (2010), detected levels of ClO_2, chlorite, and chlorate were reportedly undetected after a day. In contrast, ClO_2, chlorite, and chlorate have been detected in lettuce. Aqueous ClO_2 at 3.7 mg/L did not leave any trihalomethanes at detectable levels (<5 ppb) on lettuce after washing (López-Gálvez et al., 2010a).

During the fumigation process, ClO_2 readily engages in oxidation reactions and byproduct rapidly decomposes, giving rise to byproducts such as chlorite (ClO_2-) and chlorate (ClO_3-) ions, which can further convert into chloride ($Cl-$) ions. Research conducted by Han et al. (2018) demonstrated that the treatment of rice grains with a concentration of 12.96 g/m^3 of ClO_2 resulted in an average ClO_2 residue of 1.86 mg/kg of grain immediately after exposure. However, this residue decreased to 1.39 mg/kg of grain after one day and completely dissipated after 10 days, indicating the propensity of ClO_2 gas to react with organic matter and form other byproducts (Simpson, 2005). As for chlorine, no residues were detected either immediately or after 10 d of treatment. Immediately after treatment, chlorite residues in treated rice grains were assessed to be 0.44 mg/kg of grain, significantly decreasing to 0.22 mg/kg of grain after 10 d. On the other hand, both chlorine and chlorite residues were detected both immediately and after 1 d of treatment in wheat grains. This suggests that ClO_2 reacted with wheat grains to generate byproducts during and after fumigation, with the adsorbed residues dissipating after a 9-d waiting period. However, the study did not measure the levels of chlorates, which can potentially form as byproducts of ClO_2 breakdown. Since there are no residue standards specifically for ClO_2 and its byproducts, the current standards set by the EPA for drinking water were used as a basis for comparison with the reported results. The observed levels in the grains were found to be below the EPA standards established for drinking water. It is important to note that the composition of the product and the conditions of treatment can significantly influence the formation of byproducts in organic matter such as grains. Therefore, thorough investigations are necessary to determine the suitability of ClO_2 for insect control in food grains.

8.6 Future Directions

Given the growing demand for food and the necessity of sustainable agricultural production, it is vital to address the issue of postharvest losses caused by stored-product insects. Although fumigation has proven effective in controlling these pests, the banning of numerous fumigants and the emergence of resistance to existing fumigants present challenges. In this regard, ClO_2 gas exhibits potential as an alternative fumigant for managing stored-product insects. To further investigate the efficacy of ClO_2 as a fumigant, it is essential to conduct large-scale trials. These trials would enable assessment of the effectiveness of ClO_2 in controlling stored-product insects and its impact on treated grains, particularly regarding the formation of byproducts and biochemical changes in the treated commodities. Additionally, research efforts should focus on understanding the gas distribution patterns of ClO_2 at various temperature and humidity conditions, considering properties such as vapor pressure and vapor density, which are inferior in ClO_2 relative to what they are in phosphine fumigants, known for their better penetration and diffusivity. One possible solution could be to implement closed circulation systems that facilitate circulation to achieve a uniform distribution during fumigation treatments (Chapter 16). Furthermore, it is crucial to conduct extensive studies that encompass different insect species and stages, since the uptake of the fumigant can vary significantly. Such investigations would shed light on the feasibility of ClO_2 for large-scale commodity fumigations. Moreover, extensive research is necessary for understanding ClO_2 byproducts chemistry and safety issues related to pre- and post-treatment of the stored commodities. Scientific knowledge on these matters will help in seeking regulatory approval of ClO_2 for a wider range of applications in insect and pathogen control. Both the aqueous and the gaseous ClO_2 should be evaluated on grains to determine their effectiveness against pathogenic microorganisms.

8.7 Summary and Conclusion

Chlorine dioxide possesses a notable advantage by exhibiting dual efficacy as both an insecticide and an antimicrobial agent, which distinguishes it from other fumigants. While studies have demonstrated the effectiveness of ClO_2 as a fumigant, further comprehensive research is required to assess its performance in diverse storage systems. This research is essential to ascertain the optimal treatment conditions and dosage requirements tailored to specific insect species and pathogens. Meticulous investigation is necessary to understand the sorption of ClO_2 by commodities and the potential impact of treatment on the byproducts and nutritional composition of the treated commodities. Furthermore, research studies must be conducted in relation to repeated applications of ClO_2 for both insect and pathogen control on the quality and safety of stored commodities. These investigations are crucial to ensuring the efficacy and safety of ClO_2 in practical applications.

Acknowledgments

We thank PureLine Systems, Inc. (Bensenville, Illinois) for donating the trailer with the capability of producing chlorine dioxide gas and for help in ensuring that the unit worked properly. We acknowledge the support of the Plant Biosecurity Cooperative Research Centre (Project No: PBCRC 3038), established and supported under the Australian government's Cooperative Research Centre Program. This chapter is contribution number 24-084-B of the Kansas Agricultural Experiment Station.

References

Ahmed ST, Bostami AR, Mun HS, Yang CJ (2017) Efficacy of chlorine dioxide gas in reducing *Escherichia coli* and *Salmonella* from broiler house environments. J Appl Poult Res 26: 84–88.

Anfruns-Estrada E, Bottaro M, Pintó RM, Guix S, Bosch A (2019) Effectiveness of consumers washing with sanitizers to reduce human norovirus on mixed salad. Foods 8: 637.

Bakhmutova-Albert EV, Margerum DW, Auer JG, Applegate BM (2008) Chlorine dioxide oxidation of dihydronicotinamide adenine dinucleotide (NADH). Inorg Chem 47: 2205–2211.

Bang J, Kim H, Kim H, Beuchat LR, Ryu JH (2011) Combined effects of chlorine dioxide, drying, and dry heat treatments in inactivating microorganisms on radish seeds. Food Microbiol 28: 114–118.

Benarde MA, Israel BM, Olivieri VP, Granstrom ML (1965) Efficiency of chlorine dioxide as a bactericide. Appl Microbiol 13: 776–780.

Berg JD, Roberts PV, Matin A (1986) Effect of chlorine dioxide on selected membrane functions of *Escherichia coli*. J Appl Microbiol 60: 213–220.

Beuchat L (1998) Surface decontamination of fruits and vegetables eaten raw: a review. World Health Organization. Food Safety Unit. Available from https://www.who.int/publications/i/item/WHO-FSF-FOS-98.2.

Beuchat LR, Pettigrew CA, Tremblay ME, Roselle BJ, Scouten AJ (2004) Lethality of chlorine, chlorine dioxide, and commercial fruit and vegetable sanitizer to vegetative cells and spores of *Bacillus cereus* and spores of *Bacillus thuringiensis*. J Food Protect 67: 1702–1708.

Bhagat A, Mahmoud BS, Linton RH (2010) Inactivation of *Salmonella enterica* and *Listeria monocytogenes* inoculated on hydroponic tomatoes using chlorine dioxide gas. Foodborne Pathog Dis 7: 677–685.

Buenavista RM, Xinyi E, Subramanyam B, Rivera JL, Casada M, Siliveru K (2023) Evaluation of wheat kernel and flour quality as influenced by chlorine dioxide gas treatment. J Stored Prod Res 102: 102127.

Champ BR, Dyte CE (1976) Report of the FAO Global Survey of Pesticide Susceptibility of Stored Grain Pests, Food and Agricultural Organization, Rome, Italy.

Channaiah L, Wright C, Subramanyam B, Maier DE (2012) Evaluation of chlorine dioxide gas against eggs, larvae, and adults of *Tribolium castaneum* and *Tribolium confusum*. In Proceedings, 9th International Conference on Controlled Atmosphere and Fumigation in Stored Products. pp. 15–19.

Chen Z, Zhu C, Zhang Y, Niu D, Du J (2010) Effects of aqueous chlorine dioxide treatment on enzymatic browning and shelf-life of fresh-cut asparagus lettuce (*Lactuca sativa* L.). Postharvest Biol Technol 58: 232–238.

Du JH, Fu MR, Li MM, Xia W (2007) Effects of chlorine dioxide gas on postharvest physiology and storage quality of green bell pepper (*Capsicum frutescens* L. var. Longrum). Agri Sci China 6: 14–219.

E X, Li B, Subramanyam B (2018) Toxicity of chlorine dioxide gas to phosphine-susceptible and-resistant adults of five stored-product insect species: Influence of temperature and food during gas exposure. J Econ Entomol 111: 1947–1957.

E X, Subramanyam B, Li B (2017) Responses of phosphine susceptible and resistant strains of five stored-product insect species to chlorine dioxide. J Stored Prod Res 72: 21–27.

Fukayama MY, Tan H, Wheeler WB, Wei CI (1986) Reactions of aqueous chlorine and chlorine dioxide with model food compounds. Environ Health Perspect 69: 267–274.

Fu Y, Zhang K, Wang N, Du J (2007) Effects of aqueous chlorine dioxide treatment on polyphenol oxidases from golden delicious apple. LWT-Food Sci Technol 40: 1362–1368.

Gautam SG, Opit GP, Konemann C, Shakya K, Hosoda E (2020) Phosphine resistance in saw-toothed grain beetle, *Oryzaephilus surinamensis* in the United States. J Stored Prod Res 89: 101690.

Gómez-López VM (2012) Decontamination of Fresh and Minimally Processed Produce. John Wiley & Sons, New York, NY. 576 pp.

Gómez-López VM, Devlieghere F, Ragaert P, Debevere J (2007) Shelf-life extension of minimally processed carrots by gaseous chlorine dioxide. Int J Food Microbiol 116: 221–227.

Gómez-López VM, Rajkovic A, Ragaert P, Smigic N, Devlieghere F (2009) Chlorine dioxide for minimally processed produce preservation: A review. Trends Food Sci Technol 20: 17–26.

Goodburn C, Wallace CA (2013) The microbiological efficacy of decontamination methodologies for fresh produce: A review. Food Cont 32: 418–427.

Gordon G, Rosenblatt AA (2005) Chlorine dioxide: The current state of the art. Ozone Sci Eng 27: 203–207.

Gressent A, Rigby M, Ganesan AL, Prinn RG, Manning AJ, Mühle J, Salameh PK, Krummel PB, Fraser PJ, Steele LP, Lunder CR (2021) Growing atmospheric emissions of sulfuryl fluoride. J Geophy Res Atmos 126: e2020JD034327.

Han Y, Floros JD, Linton RH, Nielsen SS, Nelson PE (2001) Response surface modeling for the inactivation of *Escherichia coli* O157: H7 on green peppers (*Capsicum annuum* L.) by chlorine dioxide gas treatments. J Food Prot 64: 1128–1133.

Han GD, Kwon H, Kim BH, Kum HJ, Kwon K, Kim W (2018) Effect of gaseous chlorine dioxide treatment on the quality of rice and wheat grain. J Stored Prod Res 76: 66–70.

Han GD, Kwon H, Na J, Kim YH, Kim W (2016) Sensitivity of different life stages of Indian meal moth *Plodia interpunctella* to gaseous chlorine dioxide. J Stored Prod Res 69: 217–220.

Han Y, Linton RH, Nielsen SS, Nelson PE (2001) Reduction of *Listeria monocytogenes* on green peppers (*Capsicum annuum* L.) by gaseous and aqueous chlorine dioxide and water washing and its growth at 7°C. J Food Prot 64: 730–1738.

Han Y, Selby TL, Schultze KK, Nelson PE, Linton RH (2004) Decontamination of strawberries using batch and continuous chlorine dioxide gas treatments. J Food Prot 67: 2450–2455.

Huang J, Wang L, Ren N, Ma F (1997) Disinfection effect of chlorine dioxide on bacteria in water. Water Res 31: 607–613.

Hwaidi M, Collins PJ, Sissons M, Pavic H, Nayak MK (2015) Sorption and desorption of sulfuryl fluoride by wheat, flour, and semolina. J Stored Prod Res 62: 65–73.

Jagadeesan R, Collins PJ, Daglish GJ, Ebert PR, Schlipalius DI (2012) Phosphine resistance in the rust red flour beetle, *Tribolium castaneum* (Coleoptera: Tenebrionidae): Inheritance, gene interactions and fitness costs. PLosone 7: e31582.

Jagadeesan R, Nayak MK (2017) Phosphine resistance does not confer cross-resistance to sulfuryl fluoride in four major stored grain insect pests. Pest Manag Sci 73: 1391–1401.

Jin-Hua DU, Mao-Run FU, Miao-Miao LI, Wei XIA (2007) Effects of chlorine dioxide gas on postharvest physiology and storage quality of green bell pepper (Capsicum frutescens L. var Longrum). *Agricultural Sciences in China* 6:214–219.

Joshi K, Mahendran R, Alagusundaram K, Norton T, Tiwari BK (2013) Novel disinfectants for fresh produce. Trends Food Sci Technol 34: 54–61.

Kaur R, Subbarayalu M, Jagadeesan R, Daglish GJ, Nayak MK, Naik HR, Schlipalius DI (2015) Phosphine resistance in India is characterised by a dihydrolipoamide dehydrogenase variant that is otherwise unobserved in eukaryotes. Heredity 115: 188–194.

Kim BH, Han GD, Kwon H, Chun YS, Na J, Kim W (2019) Chlorine dioxide fumigation to control stored product insects in rice stored in a room. J Stored Prod Res 84: 101527.

Kim Y, Park J, Kumar S, Kwon H, Na J, Chun Y, Kim W (2015) Insecticidal activity of chlorine dioxide gas by inducing an oxidative stress to the red flour beetle, *Tribolium castaneum*. J Stored Prod Res 64: 88–96.

Knapp JE, Battisti DL (2001) Chlorine dioxide. In: Block S (ed) Disinfection, Sterilization and Preservation, 5th ed. Lippincott, Williams, & Wilkins, New York. pp. 215–227.

Kumar S, Park J, Kim E, Na J, Chun YS, Kwon H, Kim Y (2015) Oxidative stress induced by chlorine dioxide as an insecticidal factor to the Indian meal moth, *Plodia interpunctella*. Pestic Biochem Physiol 124: 48–59.

Ku KJ, Ma YH, Shin HY, Lee SH, Park JH, Kim LH, Song KB (2006) Effects of chlorine dioxide treatment on quality and microbial change of *Agaricus bisporus* Sing during storage. J Korean Soc Food Sci Nutri 35: 955–959.

Lampiri E, Agrafioti P, Athanassiou CG (2021) Delayed mortality, resistance, and the sweet spot, as the good, the bad and the ugly in phosphine use. Sci Rep 11: 933.

Liu Z, Cao Z, Wang J, Sun B (2022) Chlorine dioxide fumigation: An effective technology with industrial application potential for lowering aflatoxin content in peanuts and peanut products. Food Cont 136: 08847.

López-Gálvez F, Allende A, Truchado P, Martínez-Sánchez A, Tudela JA, Selma MV, Gil MI (2010a) Suitability of aqueous chlorine dioxide versus sodium hypochlorite as an effective sanitizer for preserving quality of fresh-cut lettuce while avoiding by-product formation. Postharvest Biol Technol 55: 53–60.

Lopez-Galvez F, Gil MI, Truchado P, Selma MV, Allende A (2010b) Cross-contamination of fresh-cut lettuce after a short-term exposure during pre-washing cannot be controlled after subsequent washing with chlorine dioxide or sodium hypochlorite. Food Microbiol 7: 199–204.

Lorini I, Collins PJ, Daglish GJ, Nayak MK, Pavic H (2007) Detection and characterisation of strong resistance to phosphine in Brazilian *Rhyzopertha dominica* (F.) (Coleoptera: Bostrychidae). Pest Manag Sci 63: 58–364.

Mahmoud BS, Bhagat AR, Linton RH (2007) Inactivation kinetics of inoculated *Escherichia coli* O157: H7, *Listeria monocytogenes* and *Salmonella enterica* on strawberries by chlorine dioxide gas. Food Microbiol 24: 736–744.

Mahmoud BSM, Linton RH (2008) Inactivation kinetics of inoculated *Escherichia coli* O157: H7 and *Salmonella enterica* on lettuce by chlorine dioxide gas. Food Microbiol 25: 244–252.

Mahmoud BSM, Vaidya NA, Corvalan CM, Linton RH (2008) Inactivation kinetics of inoculated *Escherichia coli* O157: H7, *Listeria monocytogenes* and *Salmonella* Poona on whole cantaloupe by chlorine dioxide gas. Food Microbiol 25: 857–865.

Malka SK, Park MH (2022) Fresh produce safety and quality: Chlorine dioxide's role. Frontiers Plant Sci 12: 775629.

Manivannan S (2015) Toxicity of phosphine on the developmental stages of rust-red flour beetle, *Tribolium castaneum* Herbst over a range of concentrations and exposures. J Food Sci Technol 52: 6810–6815.

MBTOC (2014) Report of the methyl bromide technical options committee. 2006 assessment (p. 277). Nairobi: UNEP. ozone.unep.org

Napolitano MJ, Greer BJ, Nicoson JS, Margerum DW (2005) Chlorine dioxide oxidations of tyrosine, N-acetyltyrosine, and dopa. Chem Res Toxicol 18: 501–508.

Nayak MK, Daglish GJ, Phillips TW, Ebert PR (2020) Resistance to the fumigant phosphine and its management in insect pests of stored products: A global perspective. Annual Review of Entomology 65: 333–350.

Nazir A, Ochani S, Nazir A, Fatima B, Ochani K, Al Hasibuzzaman M, Ullah K (2023) Rising trends of foodborne illnesses in the US. Annals Med Surg 85: 2280.

Netramai S, Kijchavengkul T, Sakulchuthathip V, Rubino M (2016) Antimicrobial efficacy of gaseous chlorine dioxide against *Salmonella enterica* Typhimurium on grape tomato (*Lycopersicon esculentum*). Int J Food Sci Technol 51: 2225–2232.

NIOSH (1999) International Chemical Safety Cards. Chlorine dioxide. 10049-04-4. ICSC 0127. Oct 20, 1999. Washington, DC: Center Disease Control, NIOSH. Available from, as of Dec 12, 2005: http://niosh. dnacih.com/nioshdbs/ipcsneng/neng0127.htm

NIOSH (2010) NIOSH Pocket Guide to Chemical Hazards. Department of Health & Human Services, Centers for Disease Control & Prevention. National Institute for Occupational Safety & Health. DHHS (NIOSH) Publication No. 2010-168 (2010) Available from: https://www.cdc.gov/ niosh/npg

Ogata N (2007) Denaturation of protein by chlorine dioxide: Oxidative modification of tryptophan and tyrosine residues. Biochem 46: 4898–4911.

Olanya OM, Annous BA, Taylor J (2015) Effects of *Pseudomonas chlororaphis* and gaseous chlorine dioxide on the survival of *Salmonella enterica* on tomatoes. Int J Food Sci Technol 50: 1102–1108.

Opit GP, Phillips TW, Aikins MJ, Hasan MM (2012) Phosphine resistance in *Tribolium castaneum* and *Rhyzopertha dominica* from stored wheat in Oklahoma. J Econ Entomol 105: 1107–1114.

Panwar S, Duggirala KS, Yadav P, Debnath N, Yadav AK, Kumar A (2022) Advanced diagnostic methods for identification of bacterial foodborne pathogens: Contemporary and upcoming challenges. Crit Rev Biotech 43: 1–19.

Park SH, Kang DH (2015a) Antimicrobial effect of chlorine dioxide gas against foodborne pathogens under differing conditions of relative humidity. LWT-Food Sci Technol 60: 186–191.

Park SH, Kang DH (2015b) Combination treatment of chlorine dioxide gas and aerosolized sanitizer for inactivating foodborne pathogens on spinach leaves and tomatoes. Int J Food Microbiol 207: 103–108.

Park SH, Kang DH (2018) Effect of temperature on chlorine dioxide inactivation of *Escherichia coli* O157: H7, *Salmonella* Typhimurium, and *Listeria monocytogenes* on spinach, tomatoes, stainless steel, and glass surfaces. Int J Food Microbiol 275: 39–45.

Park SH, Kim WJ, Kang DH (2018) Effect of relative humidity on inactivation of foodborne pathogens using chlorine dioxide gas and its residues on tomatoes. Letters Appl Microbiol 67: 154–160.

Phillips TW, Thoms EM, DeMark J, Walse S (2012) Fumigation. In: Stored Product Protection. K-State Research and Extension. Manhattan, Kansas. pp. 157–177. https://www.postharvestinfo.com/_files/ugd/9125d4_bff86d8deada492fbdffcf3f858fd873.pdf#page=163.

Pimentel MAG, Faroni LRDA, Tótola MR, Guedes RNC (2007) Phosphine resistance, respiration rate and fitness consequences in stored-product insects. Pest Manag Sci 9: 876–881.

Praeger U, Herppich WB, Hassenberg K (2018) Aqueous chlorine dioxide treatment of horticultural produce: Effects on microbial safety and produce quality–a review. Crit Rev Food Sci Nutri 58: 18–333.

Price LA, Mills KA (1988) The toxicity of phosphine to the immature stages of resistant and susceptible strains of some common stored product beetles, and implications for their control. J Stored Prod Res 24: 51–59.

Prodduk V, Annous BA, Liu L, Yam KL (2014) Evaluation of chlorine dioxide gas treatment to inactivate *Salmonella enterica* on mungbean sprouts. J Food Prot 77: 1876–1881.

Rajendran S, Gunasekaran N (2002) The response of phosphine-resistant lesser grain borer *Rhyzopertha dominica* and rice weevil *Sitophilus oryzae* in mixed-age cultures to varying concentrations of phosphine. Pest Manag Sci 58: 277–281.

Ridenour GM, Ingols RS (1947) Bactericidal properties of chlorine dioxide. J Amer Water Works Assoc 39: 561–567.

Scallan E, Hoekstra RM, Angulo FJ, Tauxe RV, Widdowson MA, Roy SL, Jones JL, Griffin PM (2011) Foodborne illness acquired in the United States—Major pathogens. Emerg Infect Dis 17: 7.

Shirasaki Y, Matsuura A, Uekusa M, Ito Y, Hayashi T (2016) A study of the properties of chlorine dioxide gas as a fumigant. Exp Anim 65: 303–310.

Simpson GD (2005) Practical Chlorine Dioxide: Foundations. Simpson & Associates, Colleyville, Texas. 288 pp.

Singh N, Singh RK, Bhunia AK, Stroshine RL (2002) Efficacy of chlorine dioxide, ozone, and thyme essential oil or a sequential washing in killing *Escherichia coli* O157: H7 on lettuce and baby carrots. LWT - Food Sci Technol 35: 720–729.

Singh N, Singh RK, Bhunia AK, Stroshine RL (2002) Effect of inoculation and washing methods on the efficacy of different sanitizers against *Escherichia coli* O157: H7 on lettuce. Food Microbiol 19: 183–193.

Smith DJ, Ernst W, Giddings JM (2014) Distribution and chemical fate of 36Cl-chlorine dioxide gas during the fumigation of tomatoes and cantaloupe. J Agri Food Chem 62: 11756–11766.

Sun X, Bai J, Ference C, Wang ZHE, Zhang Y, Narciso JAN, Zhou K (2014) Antimicrobial activity of controlled-release chlorine dioxide gas on fresh blueberries. J Food Prot 77: 127–1132.

Sun X, Baldwin E, Bai J (2019) Applications of gaseous chlorine dioxide on postharvest handling and storage of fruits and vegetables–a review. Food Cont 95: 18–26.

Sun X, Zhou B, Luo Y, Ference C, Baldwin E, Harrison K, Bai J (2017). Effect of controlled-release chlorine dioxide on the quality and safety of cherry/grape tomatoes. Food Cont 82: 26–30.

Sy KV, McWatters KH, Beuchat LR (2005) Efficacy of gaseous chlorine dioxide as a sanitizer for killing *Salmonella,* yeasts, and molds on blueberries, strawberries, and raspberries. J Food Prot 68: 1165–1175.

Sy KV, Murray MB, Harrison MD, Beuchat LR (2005) Evaluation of gaseous chlorine dioxide as a sanitizer for killing *Salmonella, Escherichia coli* O157: H7, *Listeria monocytogenes,* and yeasts and molds on fresh and fresh-cut produce. J Food Prot 68: 1176–1187.

Tan JN, Hwang CA, Huang L, Wu VC, Hsiao HI (2020) In situ generation of chlorine dioxide for decontamination of *Salmonella, Listeria monocytogenes,* and pathogenic *Escherichia coli* on cantaloupes, mung beans, and alfalfa seeds. J Food Prot 83: 287–294.

Trinetta V, Linton RH, Morgan MT (2013) The application of high-concentration short-time chlorine dioxide treatment for selected specialty crops including Roma tomatoes (*Lycopersicon esculentum*), cantaloupes (*Cucumis melo* ssp. melo var. cantaloupensis) and strawberries (*Fragaria ananassa*). Food Microbiol 34: 296–302.

Trinetta V, Linton RH, Morgan MT (2013) Use of chlorine dioxide gas for the postharvest control of *Alternaria alternata* and *Stemphylium vesicarium* on Roma tomatoes. J Sci Food Agri 93: 3330–3333.

Trinetta V, Morgan MT, Linton RH (2010) Use of high-concentration-short-time chlorine dioxide gas treatments for the inactivation of *Salmonella enterica* inoculated onto Roma tomatoes. Food Microbiol 27: 1009–1015.

Trinetta V, Morgan M, Linton R (2012) Chlorine dioxide for microbial decontamination of food. In: Microbial Decontamination in the Food Industry. Woodhead Publishing Limited, Sawston, Cambridge, United Kingdom. pp. 533–562.

Trinetta V, Vaidya N, Linton R, Morgan M (2011) Evaluation of chlorine dioxide gas residues on selected food produce. J Food Sci 76: T11–T15.

Tsai LS, Huxsoll CC, Robertson G (2001) Prevention of potato spoilage during storage by chlorine dioxide. J Food Sci 66: 472–477.

US Environmental Protection Agency (US EPA) (2016) About pesticide tolerances. Retrieved from https://www.epa.gov/pesticide-tolerances/about-pesticide-tolerances. Accessed 21 Aug 2023.

Valderrama WB, Mills EW, Cutter CN (2009) Efficacy of chlorine dioxide against *Listeria monocytogenes* in brine chilling solutions. J Food Prot 72: 2272–2277.

Vandekinderen I, Devlieghere F, Van Camp J, Kerkaert B, Cucu T, Ragaert P, De Bruyne J, De Meulenaer B (2009a) Effects of food composition on the inactivation of foodborne microorganisms by chlorine dioxide. Int J Food Microbiol 131: 138–144.

Vandekinderen I, Van Camp J, De Meulenaer B, Veramme K, Bernaert N, Denon Q, Ragaert P, Devlieghere F (2009c) Moderate and high doses of sodium hypochlorite, neutral electrolyzed oxidizing water, peroxyacetic acid, and gaseous chlorine dioxide did not affect the nutritional and sensory qualities of fresh-cut iceberg lettuce (*Lactuca sativa* Var. capitata L.) after washing. J Agr Food Chem 57: 4195–4203.

Vandekinderen I, Van Camp J, Devlieghere F, Veramme K, Denon Q, Ragaert P, De Meulenaer B (2008) Effect of decontamination agents on the microbial population, sensorial quality, and nutrient content of grated carrots (*Daucus carota* L.). J Agri Food Chem 56: 5723–5731.

Vandekinderen I, Van Camp J, Devlieghere F, Veramme K, Bernaert N, Denon Q, Ragert P, De Meulenaer B (2009b) Effect of decontamination on the microbial load, the sensory quality and the nutrient retention of ready-to-eat white cabbage. Eur Food Res Technol 229: 443–455.

Vandekinderen I, Van Camp J, Devlieghere F, Ragaert P, Veramme K, Bernaert N, Denon Q, Meulenaer De2009d) Evaluation of the use of decontamination agents during fresh-cut leek processing and quantification of their effect on its total quality by means of a multidisciplinary approach. Inn Food Sci Emerg Technol 10: 63–373.

Wang L, Sokorai K, Wu VC, Fan X (2019) Gaseous chlorine dioxide maintained the sensory and nutritional quality of grape tomatoes and reduced populations of *Salmonella enterica* serovar Typhimurium. Food Cont 96: 99–309.

Wang W, Smith DJ, Ngo H, Jin ZT, Mitchell AE, Fan X (2023) Lipid oxidation and volatile compounds of almonds as affected by gaseous chlorine dioxide treatment to reduce *Salmonella* populations. J Agri Food Chem 71: 5345–5357.

White GC (2010) Chlorine dioxide. pp. 700–757. In: White's Handbook of Chlorination and Alternative Disinfectants, 5th ed. Wiley: New York.

Whitham HK, Gilliland AE, Collier SA, Scallan Walter E, Hoffmann S (2022) Direct outpatient health care costs among commercially insured persons for common foodborne pathogens and acute gastroenteritis, 2012–2015. Foodborne Pathog Dis 9: 558–568.

Wu VC, Rioux A (2010) A simple instrument-free gaseous chlorine dioxide method for microbial decontamination of potatoes during storage. Food Microbiol 27: 179–184.

Xue B, Jin M, Yang D, Guo X, Chen Z, Shen Z, Wang X, Qiu Z, Wang J, Zhang B, Li J (2013) Effects of chlorine and chlorine dioxide on human rotavirus infectivity and genome stability. Water Res 47: 3329–3338.

Yeni F, Yavaş S, Alpas H, Soyer Y (2016) Most common foodborne pathogens and mycotoxins on fresh produce: A review of recent outbreaks. Crit Rev Food Sci Nutri 56: 1532–1544.

Young RO (2016) Chlorine dioxide (ClO$_2$) as a non-toxic antimicrobial agent for virus, bacteria, and yeast (*Candida albicans*). Int J Vaccines Vaccin 2: 00052.

Yu Y, Shi J, Xie B, He Y, Qin Y, Wang D, Shi H, Ke Y, Sun Q (2020) Detoxification of aflatoxin B1 in corn by chlorine dioxide gas. Food Chem 328: 127121.

Zhu M, Zhang LS, Pei XF, Xu X (2008) Preparation and evaluation of novel solid chlorine dioxide-based disinfectant powder in single-pack. Biomed Environ Sci 21: 57–162.

Chapter 9

Ethanedinitrile (EDN)

A New Broad-Spectrum Fumigant for Forest Products

Katie Stevens and Swaminathan Thalavaisundaram

9.1 Introduction

Fumigation has become the accepted practice to treat timber and wooden structures (ISPM, 2019). Fumigants can reach the surface and penetrate into timber and logs that other pesticides do not easily reach. Methyl bromide is widely used as an effective means of controlling a range of insect pests found in exported and imported commodities. The Montreal Protocol on Substances That Deplete the Ozone Layer is a multilateral treaty to reduce the use and production of human-made substances that damage the stratospheric ozone layer. The widely used fumigant methyl bromide (MB) is an ozone-depleting substance whose use was unanimously banned in many countries. Following the growing awareness of the dangers associated with the use of methyl bromide, Australia's Commonwealth Scientific and Industrial Research Organization (CSIRO) identified ethanedinitrile (EDN) as a fumigant in 1996 and advocated its use as a viable alternative to methyl bromide use in biosecurity and phytosanitary treatments.

Ethanedinitrile is a highly volatile fumigant compound that has shown excellent potential as an MB alternative. It was first synthesized in

DOI: 10.1201/9781003309888-9

1815 by Joseph Louis Gay-Lussac, who prepared a small amount of EDN through the thermal decomposition of silver cyanide (Brotherton and Lynn, 1959). At the time, there were no viable commercial applications; however, its future use enabled the expansion of the synthetic fertilizer industry. Ethanedinitrile was patented in 1960 for nitrate fertilizer production and remains a vital intermediate in the production of many fertilizers. It is also used as a stabilizer in the production of nitrocellulose, which has applications in a wide range of industries, including medicine, cosmetics, and molecular biology.

In January 2011, Phytos [previously known as Stakeholders in Methyl Bromide Reduction Inc. (STIMBR)] in New Zealand commenced a comprehensive research program, exploring a range of approaches including management of the log supply chain, physical treatments such as debarking, the use of electricity to heat logs, chemical treatments, and methods that would permit use of MB such as destroying or recapturing the gas or reducing its concentration (STIMBR, 2018). In extensive laboratory and commercial-scale confirmatory tests, Phytos research demonstrated the efficacy of Probit 9. Phytos concluded that EDN is the only alternative fumigant treatment available to replace MB as a phytosanitary treatment for exported pine logs and timber.

9.2 Properties of EDN

Ethanedinitrile, otherwise known as cyanogen, carbon nitride, dicyan, oxalonitrile, or dicyanogen, is a colorless, flammable gas at atmospheric pressure, with an almond-like odor. It is naturally present in the air when organic matter is decomposing or being burned. Ethanedinitrile has a low boiling point ($-21.1°C$) and high vapor pressure (3800 mmHg), compared to 3.6°C and 1,420 mmHg, respectively, for MB.

Ethanedinitrile is not an ozone-depleting substance or greenhouse gas. It degrades to ammonium, and nitrates occur naturally in the atmosphere. As EDN and its degradants exist naturally, its use is safe for the environment. In 2020, the Solar Impulse Foundation, an organization that independently assesses products and programs to address environmental challenges without compromising economic growth (https://solarimpulse. com/solutions-explorer/edn), recognized EDN as an alternative to help protect the environment.

Ethanedinitrile enters the target organism through insect spiracles' inhalation. Upon contact with moisture within the organism, EDN breaks down, forming hydrogen cyanide (HCN) and a cyanate ion. Cyanide targets cytochrome C oxidase, preventing the use of oxygen within the cells. Consequently, reduction in available oxygen causes tissue damage throughout the body, with those organs with high oxygen demands being the most vulnerable. Oxygen tension rises in peripheral tissues, causing a decrease in the unloading gradient for oxyhemoglobin and the death of the target organism.

9.3 Applications of EDN: Penetration and Sorption of EDN into Logs and Timber

Ethanedinitrile can be applied to logs and other timber products that are in an enclosed space, including in a chamber, shipping container, or ship or under a sheet (tarpaulin). The existing application setup available for MB can be used for EDN treatments with minimal changes. However, due to the volatility of EDN, it can be applied directly to the treatment area via tubing from the cylinder. In contrast, MB requires the use of heaters to volatilize the compound.

Chemical and physical characteristics of fumigants impact sorption, penetration, and therefore efficacy. The low boiling point of EDN means it is very volatile at ambient atmospheric pressure. Additionally, large molecules such as those of MB (molecular weight 94.94 g) are less volatile than molecules of EDN (molecular weight 52.04 g) (Cichowlaz, 2005). This is evident during the fumigant application process. These properties allow EDN to penetrate logs and timber across and along the grain better than MB (Viljoen and Ren, 2001; Park et al., 2021).

Ren et al. (2011) studied the penetration and sorption rates of MB, sulfuryl fluoride (SF), EDN, and phosphine (PH_3) into Douglas fir (*Pseudotsuga menziesii*) wood blocks (10 cm × 10 cm × 30 cm) in a 30-L stainless steel chamber. The treatment doses used were 48 g/m^3 (MB, SF, and EDN) and 1 g/m^3 (PH_3) with a treatment time of 48 h. The rate of timber absorption was the highest for MB (70%), followed by EDN (63%), SF (35%), and PH_3 (25%). Ethanedinitrile penetrated the timber blocks quickly compared to SF and PH_3. This study found that within 1 h, the concentration of EDN at 15 cm depth (timber block) was half that in the chamber; after 6 h of fumigation, the EDN concentration in the chamber and at 5, 10, and 15 cm depth was 25 g/m^3. At the end of the fumigation, the concentration of EDN at all sampling points was 22 g/m^3.

In comparison, the MB concentration was not equal in the chamber or at 5, 10, and 15 cm depth over the 48-h fumigation. The maximum variation between the chamber and 15 cm depth for EDN and MB concentrations were 81.3% and 1.5% at 24 h and 76.8% and 0.5% at 48 h of fumigation, respectively. The study demonstrated that EDN penetrates both along and across the grain of the timber, while MB penetrates only along the grain. This demonstrates that EDN has an advantage over MB due to its high sorption.

Ren et al. (2014) measured the penetration of EDN and MB into *Pseudotsuga menziesii* (Douglas fir), *Pinus radiata* (Monterey pine), *Intsia bijuga* (Colebr.) Kuntze (Merbau), and *Eucalyptus marginata* (Jarrah). Two sizes of timber blocks were used (10 cm × 10 cm × 30 cm and 20 cm × 20 cm × 30 cm). The moisture content of the timber blocks ranged between 9.9% and 13.5% for dry timber and 14.3% and 19.9% (dry basis) for high-moisture-content sawn wood. A set of monitoring holes was placed at 5, 10, 15, 20, and 25 cm from one end of each timber block along the middle to test the movement of fumigants. Both ends of the timber block were sealed with silicone to test across the grain movement in the timber block. Two loading factors (20% and 40%) were used to reflect commercial pallets and saw timber fumigation. The application was conducted in 15- and 60-L chambers fitted with gas sampling ports. The treatment dose was 48 g/m^3 for 24 h.

Both MB and EDN penetrated *P. menziesii*, *P. radiata*, and *I. bijuga* up to 15 cm at both 20% and 40% load factors after 24 h; however, penetration varied with timber thickness. For 10 cm × 10 cm × 30 cm timber, EDN penetration decreased between a depth of 5 and 15 cm by 14% with a fill rate of 20% and by 10% when the fill rate was 40%. For the same timber size, MB had an 85% drop between 5 and 15 cm depth with a 20% load factor and an 89% drop between 5 and 15 cm with a 40% load factor. For 20 cm × 20 cm × 30 cm timber, EDN penetration decreased by 40% between 5 and 15 cm for both 20% and 40% load factors. For the same type of timber, MB penetration decreased by 85% and 89% for 20% and 40% fill rates, respectively. For *E. marginata* (10 × 10 × 30 cm), EDN penetration dropped to 70% between 5 and 15 cm depth at a 20% load factor and 75% at a 40% load factor. Penetration of EDN decreased when *E. marginata* thickness increased to 20 cm × 20 cm × 30 cm at a 20% or 40% load factor; however, it did penetrate up to 10 cm. Methyl bromide did not penetrate *E. marginata* at either timber thickness (10 cm × 10 cm × 30 cm or 20 cm × 20 cm × 30 cm). Ethanedinitrile penetrated across the timber grain for all timber species (10 cm × 10 cm × 30 cm) and *P. menziesii*, *P. radiata*, and *I. bijuga* (20 cm × 20 cm × 30 cm) at both a 20% and a 40% loading factor but failed for *E. marginata* (20 cm × 20 cm × 30 cm) at a 40% loading factor. Methyl bromide only penetrated across the timber grain (10 cm × 10 cm × 30 cm) for a 20% and a 40% loading factor. It did not penetrate *P. menziesii*, *P. radiata*, and *I. bijuga* (20 cm × 20 cm × 30 cm) at a 20% and a 40% loading factor or *E. marginata* at both sizes and loading factors. This study shows that EDN's penetration is superior to MB's and EDN can be used to control both surface and internal insect pests.

Hall et al. (2018) found that there is a temperature-dependent response during the treatment of *Pinus radiata* logs with EDN, where sorption of EDN increased as the fumigation temperature increased. Due to the highly volatile and sorptive nature of fumigants, the concentration of fumigant left at the end of the treatment (end-point concentration) is an important factor to consider. In the case of EDN, Najar-Rodriguez et al. (2020) found that less than 1% of the applied concentration of EDN remained in the enclosed atmosphere at the conclusion of 16-, 20-, and 24-h treatment times. Ethanedinitrile can, therefore, safely be released into the atmosphere during ventilation without the use of forced extraction and/or recapture of the fumigant from the headspace before venting.

9.4 Efficacy with Wood and Timber Pests

Research regarding the efficacy of EDN has shown that it is an excellent alternative to MB (Ren et al., 2014; Seabright et al., 2020). Ethanedinitrile has been evaluated for efficacy against a wide range of pests in timber and timber products (Park et al., 2014; Lee et al., 2017b). Based on the results of these studies, many scientists and organizations have identified EDN as the best option for providing control of quarantine pests (Seabright et al., 2020). A summary of the presented data can be found in Table 9.2 at the end of this chapter.

The international trade of nursery stock, logs, sawn timber, and solid wood products, including pallets, provides pathways for the unintended introduction of these insect pests and pathogens. The pests can be grouped into generic clusters as follows.

9.4.1 Wood Borers

9.4.1.1 Burnt-Pine Longhorn Beetle [*Arhopalus ferus* (Mulsant, 1839)]

The burnt-pine longhorn beetle, *Arhopalus ferus* (Mulsant, 1839), is considered a key quarantine pest in many countries such as New Zealand, China, and Australia (Najar-Rodriguez et al., 2016). As the common name suggests, *A. ferus* adults are strongly attracted to recently burned trees, but they are also known to infect stumps, logs, and dead or dying pine (*Pinus* spp.) trees. As *A. ferus* primarily attack burned and dead trees, their impact on forests is minimal; however, they are considered a major pest due to their rapid infestation of salvageable timber. In some cases, adult *A. ferus* will travel 3 km to find a new host and can infest trees within 24 h of burning.

Damage caused by *A. ferus* includes tunnels created by larval feeding, which greatly reduce wood quality and marketability. When populations are especially high, the outer 50 mm of the sapwood may be heavily damaged. In addition, adult *A. ferus* vector fungi that cause discoloration of the internal timber, also known as sapstain. The presence of sapstain also reduces the quality and marketability of timber.

Ethanedinitrile has been shown to effectively move throughout the timber profile and provide control of *A. ferus*. Pranamornkith et al. (2014) reported 99% control of adult *A. ferus* when treated with an average of 12.6 g/m^3 for 3 h. Najar-Rodriguez et al. (2015) compared the efficacy of EDN and MB on *A. ferus* eggs, larvae, and adults for 3 and 4 h, respectively. Results show that the overall average concentration × time (*Ct*) product required to control 99% of the *A. ferus* population for EDN and MB is 14.1 and 52.4, respectively, with eggs being especially susceptible to EDN. These results suggest that EDN is more effective than MB against *A. ferus*.

9.4.1.2 Japanese Pine Sawyer (Monochamus alternatus)

The Japanese pine sawyer beetle, *Monochamus alternatus*, is native to not only Japan, but also China, Taiwan, Vietnam, Laos, and Korea. This insect pest is known to infest several species of *Pinus*, *Picea*, and *Cedrus* trees when they are stressed or have recently been felled. *Monochamus alternatus* lays its eggs into these trees, which reduces their value as timber; however, this is not the primary concern about this pest. Like other *Monochamus* species, *M. alternatus* is also a vector for the highly damaging pine wilt nematode, *Bursaphelenchus xylophilus*.

To reduce the risk of introducing these pests to susceptible locations, treatment of forest products is mandated by several countries; however, the use of MB is highly restricted or banned and heat treatment is not economically viable for large-scale operations. Preliminary laboratory and field studies have shown that EDN is highly efficacious against *M. alternatus* larvae when treated for 6 h at a range of temperatures (5, 12, and 20°C) (Lee et al., 2017a). Park et al. (2014) reported higher toxicity of EDN than of metam sodium to *M. alternatus*, particularly at the low temperatures (4.4–6.1°C) that are typical for fumigation of timber in some locations.

9.4.1.3 European House Borer (Hylotrupes Bajulus L.)

Originating from North Africa, the house longhorn beetle, old house borer, or European house borer (*Hylotrupes bajulus* L.) can now be found worldwide due to the international trade of infested timber, wood products, and solid wood packaging material timber products. It is believed that *H. bajulus* was the first longhorn beetle species to be unintentionally introduced to a nonnative country, in 1758. Since the initial introduction, *H. bajulus* has often been referred to as one of the most dangerous wood-boring insects, second only to termites. In Turkey, *H. bajulus* was found to have a higher damage index than several other wood-destroying insects (Yalcin *et al.*, 2020).

Hylotrupes bajulus is a major pest of untreated seasoned coniferous wood, particularly that used for structural frameworks. Species of pine (*Pinus* spp.), fir (*Abies* spp.), Douglas fir (*Pseudotsuga* spp.), and spruce (*Picea* spp.) are the main hosts for *H. bajulus*. Despite its moniker, these beetles prefer young wood over old, as protein content declines with timber age. Subsequently, infestations tend to occur in structures less than ~10 yr old with high moisture areas (Puckett and Tomberlin, 2019).

Treatment for *H. bajulus* often includes the application of heat or methyl bromide. Studies have shown, however, that MB provides marginal control of *H. bajulus* larvae (Emery et al., 2014). When timber blocks artificially infested with larval *H. bajulus* were treated with 40 mg/L EDN and 48 mg/L MB for 24 h, the result was 100% and 97% control, respectively (Emery et al., 2014). Additionally, *H. bajulus* larvae were completely controlled when treated with 50 g/m³ EDN for 3 h in wooden blocks (Malkova et al., 2016). These results indicate that EDN is more efficacious against *H. bajulus* and is a practical alternative to MB.

9.4.1.4 Asian Longhorn Beetle [*Anoplophora glabripennis* (Motschulsky, 1853)]

The Asian longhorn beetle, *Anoplophora glabripennis* (Motschulsky, 1853), is an invasive insect that has the unique ability to infest a wide range of host tree species, including maple (*Acer* spp.), birch (*Betula* spp.), and elm (*Ulmus* spp.). These wood-boring pests are native to East Asia, where their damage causes an estimated 1.5 billion USD annual loss (Meng et al., 2015). In 1996, *A. glabripennis* was detected in the United States, with subsequent discoveries in Canada and several European countries (Javal et al., 2019). The introduction of *A. glabripennis* is believed to have occurred via infested wood pallets or wood packaging materials from China (Carter et al., 2010). There is no known biological or chemical control method for *A. glabripennis* larvae; therefore, infected trees must be destroyed (Ren et al., 2006). In the United States, eradication costs from 1998 to 2006 were estimated at 249 million USD (Meng et al., 2015).

To reduce the risk of further introduction of *A. glabripennis*, many countries require the treatment of forest products with heat or fumigation with MB. Research has shown that the MB alternative, EDN, is highly efficacious against *A. glabripennis*, especially at low temperatures. Ren et al. (2006) determined the CT products required for EDN to control 99% of *A. glabripennis* at 4.4, 10.0, 15.6, and 21.1°C for 6 h and compared values

Table 9.1 Dose Estimates for Exposure of Naked *A. glabripennis* Larvae to Ethanedinitrile (EDN) for 6 h and Methyl Bromide (MB) for 24 h at Different Temperatures

Temperature (°C)	$L(CT)_{99.5}$ (mg-h/L)	
	EDN	MB
4.4	353.38	1,000
10.0	221.82	826
15.6	126.77	786
21.1	56.62	729

with those obtained for 24-h treatment with MB (Table 9.1). These studies show that EDN can provide control against *A. glabripennis* at lower application doses for a shorter application period than MB. In another study, sulfuryl fluoride (SF) was ineffective against *A. glabripennis* when applied at 4.4° and 10.0°C (Barak et al., 2005). Methyl bromide requires a dose 3.5 times the size of the EDN dose to achieve 99% control at 10°C (Ren et al., 2006). In addition, EDN can control *A. glabripennis* under low temperatures, while MB must be applied when conditions are more favorable (Ren et al., 2006).

9.4.2 Bark Beetles and Weevils

In Europe, the European spruce bark beetle (*Ips typographus* L.), spruce wood engraver (*Pityogenes chalcographus*), bark beetle (*Dryocoetes hectographus*), and lesser spruce shoot beetle [*Hylurgops palliates* (Gyllenhal, 1813)] share Norway spruce (*Picea abies*) as their preferred host. These beetles are all native to parts of Europe where the Norway spruce is widely distributed and has been extensively planted for harvest (Johann et al., 2004). Öhrn et al. (2021) reported 200 million seedlings being planted in Norway annually. Schroeder and Cocoş (2018) observed that the host range of these insects may also include firs (*Abies* spp.), larches (*Larix* spp.), pines (*Pinus* spp.), and other spruce species (*Picea* spp.).

While most bark beetle species are considered to be secondary colonizers (i.e., they attack stressed trees), some insect species such as *I. typographus* are primary colonizers, attacking healthy trees. These attacking species may have a symbiotic relationship with pathogenic fungi belonging to the *Ophistoma* and *Diplodia* genera (Davydenko, 2021). Members of both genera stain the wood. As infected trees weaken, other bark beetle species such as *P. chalcographus*, *H. palliates*, and *D. hectographus* are attracted to the stressed trees, increasing the burden on the trees. Additionally, drought adds to tree stress, attracting these secondary colonizers (Öhrn et al., 2021).

In the United States, Ward et al. (2023) reported that *I. typographus* was intercepted 505 times between 1914 and 2008. Despite reports of this species being found during surveys in Indiana and Maryland in 1995 and 2002, respectively, later surveys have not reported the presence of this species (Liebhold et al., 2017). While *I. typographus* has not yet established itself in the United States, some models expect it to become well established due to climate change (Jönsson et al., 2011; Bentz et al., 2019). In addition, *H. palliates* is one of the most frequently intercepted species at U.S. ports (Haack, 2001).

Due to the internal feeding habit of these pests, control measures must be able to effectively penetrate timber. Stejskal et al. (2017) reported complete control of all life stages of *I. typographus*, *P. chalcographus*, *H. palliates*, and *D. hectographus* when treated in logs with EDN for 8 h. Alternatively, the use of phosphine requires longer treatments spanning days to achieve an acceptable level of control (Chaudhry, 1997).

The black pine bark beetle, *Hylastes ater* (Paykull), and the golden-haired pine bark beetle, *Hylurgus ligniperda* (F.), are both native to Europe but have been unintentionally introduced to other locations including China, Australia, New Zealand, and the Americas. The native range of *H. ligniperda* includes southern Europe, Asia Minor, Caucasus, Algeria, and Ukraine (Pfeffer, 1995). While these pests have a relatively wide host range, they primarily infest pines (*Pinus* spp.).

Hylastes ater is known to feed on the bark and phloem of host tree seedlings, which has caused 70% and 90% seedling loss in Australia and Chile, respectively (Reay et al., 2001). Both pests also infest tree stumps, logs, and fallen trees. Direct damage caused by these pests is due to their feeding and breeding habits under the bark. Feeding also introduces decay and saps stain fungi, greatly reducing the quality of the wood. Its status as a quarantine pest necessitates the use of phytosanitary treatment prior to shipping.

Najar-Rodriguez et al. (2020) demonstrated the efficacy of EDN against *H. ater* and *H. ligniperda* in laboratory- and field-based trials. Laboratory-based studies show that *H. ater* was more tolerant to EDN than *H. ligniperda* at all life stages and temperatures. Commercial-sized field studies show that EDN provided 100% mortality of all pest life stages, including *H. ater* pupae, and is a viable alternative to MB. Additionally, EDN has been proven to provide complete control of *H. ligniperda* larvae at lower doses than MB (15 and 48 g/m^3, respectively) (Brash et al., 2007).

9.4.3 Termites

9.4.3.1 West Indian Drywood Termite [Cryptotermes brevis (Walker)]

The West Indian drywood termite, *Cryptotermes brevis* (Walker), is inadvertently introduced to new locations more frequently than any other termite species (Evans et al., 2013). Endemic to the coastal desert in Chile and Peru, *C. brevis* is a major pest throughout tropical and subtropical regions worldwide, excluding Asia (Bahder et al., 2009). Control efforts for *C. brevis* in the United States, Australia, New Zealand, South America, and South Africa have been costly (Gay and Watson, 1982; Su and Scheffrahn, 1990; Rust and Su, 2012; Constantino, 2022). In Australia alone, it is estimated that in excess of 10 million AUD has been spent to control *C. brevis* (Peters and Fitzgerald, 2007; Horwood, 2008).

To prevent the further spread of *C. brevis*, forest products are often required to undergo treatment prior to shipment or upon receipt. Currently, the use of heat or fumigation with methyl bromide (MB) is required for treatment; however, the use of MB is being restricted worldwide due to its ozone-depleting status. Ethanedinitrile has no impact on the atmosphere and has been recognized as an environmentally friendly alternative to MB. Research has also shown that EDN is highly effective against *C. brevis* at a low application rate and more efficacious than MB and sulfuryl fluoride (Dowsett and Ren, 2004).

9.4.4 Terrestrial Invertebrates

9.4.4.1 Pinewood Nematode [*Bursaphelenchus xylophilus* (Steiner & Buhrer) Nickle]

The pinewood nematode, *Bursaphelenchus xylophilus* (Steiner & Buhrer) Nickle, is the causal agent of pine wilt disease, a devastating disease that primarily affects trees of the *Pinus* genus. Host trees in parts of North America make up the native range of *B. xylophilus* and are naturally resistant to the disease it vectors; however, it is a serious threat to trees in Europe and East Asia, particularly China, Japan, South Korea, and Taiwan. In 1979, Japan lost 2.4 million m³ of timber due to pine wilt disease, the heaviest loss since the initial infection in 1925 (Mamiya, 1988). In Europe, concern about *B. xylophilus* has increased, as the nematode was detected in imported timber in Portugal, Finland, Sweden, and France (Bonifácio et al., 2014; Lee et al., 2017b). Modeling of an unregulated *B. xylophilus* infestation in Europe estimated that the loss of timber from 2008 to 2030 would cost 22 billion euros (Soliman et al., 2012).

The spread of *B. xylophilus* is predominantly facilitated by beetles belonging to the *Monochamus* genus; therefore, both pests are considered quarantine pests in many countries. The risk of pest introduction is high due to the international trade of fresh and minimally processed logs, so phytosanitary treatment is often required to mitigate this risk in these commodities (Seabright et al., 2020). The International Plant Protection Convention (IPPC) developed the International Standards for Phytosanitary Measures No. 15 (ISPM-15) partly in response to the spread of *B. xylophilus* in wood packaging material.

Control of *B. xylophilus* can be extremely difficult, as they spend most of their life inside timber, where many control methods cannot reach (Park et al., 2014). ISPM-15 mandates the use of MB or SF as chemical treatment options; however, studies show that these chemicals may not completely control *B. xylophilus* under some conditions. Seabright et al. (2020) tested the efficacy of MB, SF, and EDN against *B. xylophilus* in wood chips (25 mm × 38 mm × 6 mm), blocks (75 mm × 75 mm × 150 mm), and 1.5-m-long logs (19.1–39.1 cm diameter) at 20°C. Results from this study show that SF is ineffective for blocks and effective in wood chips only when treated for 48 h at higher doses (70, 80, and 90 mg/L). Blocks treated for 16 h with 120 mg/L MB had live *B. xylophilus* in all blocks; however, when the treatment time was increased to 48 h, complete mortality was achieved. Ethanedinitrile was applied to wood chips, blocks, and logs at a range of 30 –100 mg/L for 24 h and provided complete control in all treatments. This study concluded that EDN represents the best option for providing a level of control suitable for quarantine pests.

9.4.4.2 Sirex Woodwasp (*Sirex noctilio*)

Native to Europe, Asia, and northern Africa, the sirex woodwasp, *Sirex noctilio* (F.), is a serious threat to forests worldwide. Extensive losses to nonnative plantations in Australia, New Zealand, Chile, and South Africa have caused an estimated 16 to 60 million USD in annual damages due to the international trade of timber products. This pest primarily affects *Pinus* spp. but can also infest species in the genera *Abies*, *Larix*, *Picea*, and *Pseudotsuga*. Unlike other Siricidae species, *S. noctilio* not only damage stressed trees, but can also heavily damage healthy trees.

Sirex noctilio are obligately associated with the white rot fungus (*Amylostereum areolatum*), which females deposit into trees along with a phytotoxic mucus as they oviposit. This fungus disrupts vascular tissues in trees, while the mucus aids in fungal survival (Hajek and van Frankenhuyzen, 2017). As *A. areolatum* spreads throughout the tree, *S. noctilio* larvae feed on the fungus for 10–11 mo, creating tunnels which reduce timber quality. Pupation occurs near the bark surface and lasts 3 wk until adults emerge from the tree.

322

A field experiment was conducted to test the toxicity of EDN against *S. noctilio* (M. K. Hall unpublished data). Ethanedinitrile was applied at 100 g/m³ for 24 h, providing control that satisfies the quarantine requirement for approval. While methyl bromide has been approved for *S. noctilio* without any data, this new efficacy study supports the use of EDN for controlling *S. noctilio*.

9.5 Regulatory Status

As mentioned above, the use of EDN as a fumigant was discovered by CSIRO, which is a pioneering Australian government scientific research organization. As it was discovered locally, Australia was the first country in the world to register this fumigant and it satisfies the country-of-origin requirements. Ethanedinitrile is a broad-spectrum fumigant that is highly effective against insect pests, nematodes, soil-borne pathogens, and weeds. Ethanedinitrile is also used as a soil fumigant in some countries.

Ethanedinitrile was first approved in Australia for timber and logs in 2013. At that time, the previous registrant provided limited data to the Australian Pesticide and Veterinary Medicine Authority (APVMA), which restricted its use pattern, including loading factor, scrubbing requirements, and ventilation period in Australia. These restrictions made EDN commercially unviable compared with ozone-depleting MB for timber and log treatment. As a result, the replacement of MB was not progressing and the Australian commitment to addressing the Montreal Protocol requirement has not been achieved. In 2015, the Draslovka Group, a family-owned entity based in the Czech Republic and the sole global manufacturer and supplier of EDN, made a significant investment both in Australia and in other countries to address the restriction and to make the product commercially viable. As a result, Draslovka received approval for soil fumigation with EDN in Australia in 2018. New efficacy, human health, and environmental data were provided to the APVMA to remove the prior restrictions. Following the submission, APVMA approved the label changes for timber and logs in 2023.

Ethanedinitrile was approved in South Korea for timber and logs in 2019, followed by Malaysia for timbers, logs, grains, empty storage structures, and soil treatment in 2020. Ethanedinitrile was approved in Russia and New Zealand for timber and logs treatment in 2021 and 2022, respectively. EDN was approved in Turkey for timber and logs treatment recently and as soil fumigant in 2023. Also in 2023, EDN was approved for timber and log treatment in Uruguay.

Ethanedinitrile registration is currently in the approval process in the United States (approval expected for timber and logs treatment in 2024), Canada (timber and logs treatment in 2024), South Africa (soil and timber treatment 2024), India (timber and logs treatment in 2025), and China (timber and logs treatment in 2026).

9.5.1 EDN Phytosanitary Approvals

For the international trade of timber products to occur, each country must establish the phytosanitary standards required to be performed by the exporting country or upon arrival to reduce the risk of introducing alien pests. Countries that currently accept or are in the approval process for EDN as a phytosanitary treatment include Australia, South Korea, Malaysia, Russia, New Zealand, and Turkey.

9.5.2 ISPM-28 and -15 Status

The International Plant Protection Convention (IPPC) has developed and adopted the International Standards for Phytosanitary Measures to prevent the introduction and spread of regulated plant pests and to define the appropriate measures for their control. These ISPMs cover a wide range of concerns including pest surveillance and reporting, phytosanitary certifications, and treatments. ISPM-28 details the accepted phytosanitary treatments for regulated pests. As EDN is highly efficacious against many forest-product pests, the New Zealand Ministry for Primary Industries has proposed EDN as a treatment option to be included in ISPM-28.

While the trade of forest products is a major pathway for invasive species to travel, the use of wood packaging material (WPM) such as crates, dunnage, and pallets can also result in the transport and spread of invasive pests. For this reason, ISPM-15 was developed to provide treatment options for WPM used in international trade. In the United States alone, implementation of ISPM-15 reduced the rate of pest interception by 36–52% (Haack et al., 2014). With the high efficacy of EDN in wood products, it is projected to be included as a treatment option for ISPM-15 as well.

9.6 Conclusion

Due to its high vapor pressure and low boiling point, EDN can effectively penetrate forest products while also being safe for bystanders and the atmosphere. In addition, many studies have confirmed the efficacy of EDN against a wide range of timber pests under laboratory and commercial field settings (Table 9.2). Registration of EDN has been achieved thus far in several countries, with many more to come.

Table 9.2 Summary of Published Studies Conducted with Ethanedinitrile Against Target Pests and Presented in This Chapter

Order	Species	Life Stage	Treatment Conditions	Mortality	Reference
Coleoptera	*Arhopalus ferus*	Adult	12.6 g EDN/m^3 for 3 h	99%	Pranamornkith et al., 2014
		Egg, larvae, adult	1.5–22.0 g-h/m^3	99%	Najar-Rodriguez et al., 2015
Coleoptera	*Monochamus alternatus*	Larvae	194.9 g-h/m^3	99%	Lee et al., 2017a
		Larvae	329.4 g-h/m^3	100%	Park et al., 2014
Coleoptera	*Hylotrupes bajulus*	Larvae, adult	40 mg/L for 24 h	100%	Emery et al., 2014
		Larvae	50 g/m^3 for 3 and 6 h	100%	Malkova et al., 2016
Coleoptera	*Anoplophora glabripennis*	Larvae	10–95 g-h/m^3	99%	Ren et al., 2006
Coleoptera	*Ips typographus*	Egg, larvae, adult	50 g/m^3 for 24 h	100%	Stejskal et al., 2017
Coleoptera	*Hylastes ater*	Larvae, pupae, adult	45.21–135.10 g-h/m^3	100%	Najar-Rodriguez et al., 2020
Coleoptera	*Hylurgus ligniperda*	Larvae, pupae, adult	8.0–62.64 g-h/m^3	100%	Najar-Rodriguez et al., 2020
		Larvae	15 g/m^3 for 24 h	100%	Brash et al., 2007
Coleoptera	*Pityogenes chalcographus*	Egg, larvae, adult	50 g/m^3 for 24 h	100%	Stejskal et al., 2017
Coleoptera	*Dryocoetes hectographus*	Egg, larvae, adult	50 g/m^3 for 24 h	100%	Stejskal et al., 2017
Coleoptera	*Hylurgops palliates*	Egg, larvae, adult	50 g/m^3 for 24 h	100%	Stejskal et al., 2017
Blattodea	*Cryptotermes brevis*	Worker	4.0 g-h/m^3 for 6 h	100%	Dowsett and Ren, 2004
Aphelenchida	*Bursaphelenchus xylophilus*	n/a	329.4 g-h/m^3	100%	Park et al., 2014
		n/a	30–100 g/m^3 for 24 h	100%	Seabright et al., 2020
		n/a	100, 120, and 150 g/m^3 for 24 h	100%	Lee et al., 2017b

References

Bahder BW, Scheffrahn RH, Křeček J, Keil C, Whitney-King S (2009) Termites (isoptera: Kalotermitidae, rhinotermitidae, termitidae) of Ecuador. Annales de la Societe Entomologique de France 45, 529–536. doi: 10.1080/00379271.2009.10697634

Barak A, Wang Y, Zhan G, Wang X, Xu L, Huang Q, Zhu Y, Wu Y (eds), 2002 Pp. 1–2. In: Proc 2002 Annu Intern Res Conf Methyl Bromide Alternatives, Orlando, USA.

Barak A, Wang Y, Xu L, Rong Z, Hang X, Zhan G (2005) Methyl bromide as a quarantine treatment for *Anoplophora glabripennis* (Coleoptera: Cerambycidae) in regulated wood packing material. J Econ Entomol 98, 1911–1916.

Bentz BJ, Jönsson AM, Schroeder M, Weed A, Wilcke RAI, Larsson K (2019) *Ips typographus* and *Dendroctonus ponderosae* models project thermal suitability for intra- and inter-continental establishment in a changing climate. Front For Glob Change 2. doi: 10.3389/ffgc.2019.00001

Bonifácio LF, Sousa E, Naves P, Inácio ML, Henriques J, Mota M, Barbosa P, Drinkall MJ, Buckley S (2014) Efficacy of sulfuryl fluoride against the pinewood nematode, *Bursaphelenchus xylophilus* (Nematoda: Aphelenchidae), in *Pinus pinaster* boards. Pest Manag Sci 76, 10–17. doi: 10.1002/ps.3507

Brash D, Epenhuijusen K, Zhang Z, Somerfield K (2007) Ethanedinitrile for Control of Quarantine Pests on Export Timber and Logs. New Zealand Institute for Crop & Feed Research Limited, Palmerston North, New Zealand.

Brotherton TK, Lynn JW (1959) The synthesis and chemistry of cyanogen. Chem Rev 59, 841–883. doi: 10.1021/cr50029a003

Carter M, Smith M, Harrison R (2010) Genetic analyses of the Asian long-horned beetle (Coleoptera, Cerambycidae, *Anoplophora glabripennis*), in North America, Europe and Asia. Biol Invasions 12, 1165–1182. doi: 10.1007/s10530-009-9538-9

Cichowlaz SD (2005) Fumigation: The use of poisonous and lethal fumigants. Nevada State Department of Agriculture. https://agri.nv.gov/uploadedFiles/agrinvgov/Content/Plant/PEST/Study_Manuals/New%20C4%20fumigation%20Manual-full.pdf. Accessed 25 February 2023.

Chaudhry M (1997) A review of the mechanisms involved in the action of phosphine as an insecticide and phosphine resistance in stored-product insects. Pestic Sci 49, 213–228.

Constantino R (2022) The pest termites of South America: Taxonomy, distribution and status. J Appl Entomol 126, 355–365.

Davydenko K (2021) New insights into the role of phytopathogenic fungi vectored by pine bark beetles in pine decline, in: Forestry Academy of Sciences Ukraine. doi: 10.15421/412122

Dowsett HA, Ren Y (eds), 2004 Pp. 577. In: Proc 7th Intern Conf Controlled Atm Fumi Stored Prod, Brisbane, Australia.

Emery RN, Ren YL, Newman J, Thalavaisundaram S (eds), 2014 Pp. 1–8. In: Proc 11th Intern Working Conf Stored-Prod Prot, Chiang Mai, Thailand.

Evans TA, Forschler BT, Kenneth J (2013) Biology of invasive termites: A worldwide review. Annu Rev Entomol 58, 455–474. doi: 10.1146/annurev-ento-120811-153554

Gay F, Watson J (1982) The genus Cryptotermes in Australia (Isoptera: Kalotermitidae). Aust J Zool Supplementary Series 30, 1–64.

Haack RA (2001) Intercepted Scolytidae (Coleoptera) at U.S. ports of entry: 1985–2000, Integr Pest Manag Rev 6, 253–282. https://doi.org/10.1023/A:1025715200538

Haack RA, Britton KO, Brockerhoff EG, Cavey JF, Garrett LJ, Kimberley M, Lowenstein F, Nuding A, Olson LJ, Turner J, Vasilaky KN (2014) Effectiveness of the international phytosanitary standard ISPM No. 15 on reducing wood borer infestation rates in wood packaging material entering the United States. PLoS One 9, e96611. doi: 10.1371/journal.pone.0096611

Hajek a, van Frankenhuyzen K (2017) Use of entomopathogens against forest pests. In: Lacey L (ed) Microbial Control of Insect and Mite Pests. Academic Press, Massachusetts, pp. 313–330.

Hall M, Adlam A, Matich A, Najar-Rodriguez A, Pal P, Brash D (2018) Quantification of hydrogen cyanide as a potential decomposition product of ethanedinitrile during pine log fumigation. NZ J For Sci 48, 7. doi: 10.1186/s40490-018-0114-x

Horwood M (2008) West Indian drywood termite. New South Wales Department of Primary Industries, Primefact. 826(2). Sydney (Australia): New South Wales Department of Primary Industries. https://fumapest.com.au/pdf/West%20Indian%20Drywood%20Termite.pdf. Accessed 02 March 2023.

ISPM 15 (2019) ISPM 15 Regulation of Wood Packaging Material in International Trade. Food and Agriculture Organization of the United Nations. https://www.fao.org/3/mb160e/mb160e.pdf. Accessed 05 March 2023.

Javal M, Lombaert E, Tsykun T, Courtin C, Kerdelhué C, Prospero S, Roques A, Roux G (2019) Deciphering the worldwide invasion of the Asian long-horned beetle: A recurrent invasion process from the native area together with a bridgehead effect. Mol Ecol 28, 951–967. doi: 10.1111/mec.15030

Johann E, Agnoletti M, Axelsson AL, Burgi M, Ostlund L, Rochel X, Schmidt U, Schular A, Skovsgaard J, Winiwarter V (2004) History of secondary Norway spruce forests in Europe, in: Norway Spruce Conversion. Brill, pp. 25–62.

Jönsson AM, Harding S, Krokene P, Lange H, Lindelöw Å, Økland B, Ravn HP, Schroeder LM (2011) Modelling the potential impact of global warming on *Ips typographus* voltinism and reproductive diapause. Clim Change 109, 695–718. doi: 10.1007/s10584-011-0038-4

Lee BH, Park CG, Ren Y (2017a) Evaluation of different applications of ethanedinitrile (C2N2) in various fumigation chambers for control of *Monochamus alternatus* (Coleoptera: Cerambycidae) in naturally infested logs. J Econ Entomol 110, 502–506. doi: 10.1093/jee/tox052

Lee BH, Yang JO, Beckett S, Ren Y (2017b) Preliminary trials of the ethanedinitrile fumigation of logs for eradication of *Bursaphelenchus xylophilus* and its vector insect *Monochamus alternatus*. Pest Manag Sci 75(2), 556–563. doi: 10.1002/ps.4476

Liebhold AM, Brockerhoff EG, Kimberley M (2017) Depletion of heterogeneous source species pools predicts future invasion rates. J Appl Ecol 54, 1968–1977. doi: 10.1111/1365-2664.12895

Malkova J, Aulicky R, Dlouhy M, Stejskal V, Hnatek J, Hampl J, Trocha A (eds), 2016 Pp. 477–478. In: Proc 10th Intern Conf Controlled Atm Fumi Stored Prod, New Delhi, India.

Mamiya Y (1988) History of pine wilt disease in Japan. J Nematol 20, 219–226.

Meng PS, Hoover K, Keena MA (2015) Asian longhorned beetle (Coleoptera: Cerambycidae), an introduced pest of maple and other hardwood trees in North America and Europe. J Integr Pest Manag 6, 4. doi: 10.1093/jipm/pmv003

Najar-Rodriguez A, Hall M, Adlam A, Hall A, Burgess S, Somerfield K, Page B, Brash D (2015) Developing new fumigation schedules for the phytosanitary treatment of New Zealand export logs: Comparative toxicity of two fumigants to the burnt pine longhorn beetle, *Arhopalus ferus*. N Z Plant Prot 68, 19–25.

Najar-Rodriguez A, Hall M, Jakoubek P, Thalavaisundaram S (2016) Comparative toxicity of ethanedinitrile and methyl bromide to a key quarantine pest-the Burnt Pine longhorn Beetle, *Arhopalus ferus*. N Z Plant Prot. doi: 10.13140/RG.2.1.2891.4328

Najar-Rodriguez AJ, Afsar S, Esfandi K, Hall M, Adlam AR, Wilks C, Noakes E, Richards K (2020) Laboratory toxicity and large-scale commercial validation of the efficacy of ethanedinitrile, a potential alternative fumigant to methyl bromide, to disinfest New Zealand Pinus radiata export logs. J Stored Prod Res 88, 101671. doi: 10.1016/j.jspr.2020.101671

Öhrn P, Berlin M, Elfstrand M, Krokene P, Jönsson AM (2021) Seasonal variation in Norway spruce response to inoculation with bark beetle-associated bluestain fungi one year after a severe drought. For Ecol Manag 496, 119443. https://doi.org/10.1016/j.foreco.2021.119443

Park CG, Son J, Lee B, Cho JINH, Ren Y (2014) Comparison of EDN and metam sodium for control of *Bursaphelenchus xylophilus* (Nematoda:Aphelenchidae) and *Monochamus alternatus* in naturally infested logs at low temperatures. J Econ Entomol 107, 2055–2060.

Park M, Ren Y, Lee B (2021) Preliminary study to evaluate ethanedinitrile (C2N2) for quarantine treatment of four wood destroying pests. Pest Manag Sci 77, 5213–5219. doi: 10.1002/ps.6562

Peters B, Fitzgerald C (2007) Developments in termite management in Queensland, Australia: Life after Cyclodienes (Isoptera). Sociobiology 49, 231–250.

Pfeffer A (1995) Bark and Ambrosia beetles from the central and west palaearctic region (Coleoptera, Scolytidae, Platypodidae). Entomol Basil 17, 5–310.

Pranamornkith T, Hall MKD, Adlam AR, Somerfield KG, Page BBC, Hall AJ, Brash DW (2014) Effect of fumigant dose, timber moisture content, end-grain sealing, and chamber load factor on sorption by sawn timber fumigated with ethanedinitrile. N Z Plant Prot 67, 66–74. doi: 10.30843/nzpp.2014.67.5753

Puckett R, Tomberlin J (2019) Urban entomology, in: Byrd J, Tomberlin J (eds), Forensic Entomology, 3rd ed. CRC Press, Florida, pp. 507–519.

Reay SD, Thwaites JM, Farrell RL, Walsh PJ (2001) The role of the bark beetle, *Hylastes ater* (Coleoptera: Scolytidae), as a sapstain fungi vector to *Pinus radiata* seedlings: A crisis for the New Zealand forestry industry?. Integr Pest Manag Rev 6, 283–291.

Ren Y, Agarwal M, Newman J, Du B (2014) A report prepared for BOC Limited: Comparison of ethanedinitrile (EDN) with methyl bromide (MB) as a biosecurity fumigant for timber and log. Plant Biosec Coop Res Centre.

Ren Y, Wang Y, Barak AV, Wang X, Liu Y, Dowsett HA (2006) Toxicity of ethanedinitrile to *Anoplophora glabripennis* (Coleoptera: Cerambycidae) larvae. J Econ Entomol 99, 308–312. https://doi.org/10.1603/0022-0493-99.2.308

Ren YL, Lee BH, Padovan B (2011) Penetration of methyl bromide, sulfuryl fluoride, ethanedinitrile, and phosphine into timber blocks and the sorption rate of the fumigants. J Stored Prod Res 47, 63–68.

Ren YL, Sarwar M, Wright EJ (eds), 2002 Pp. 1–4. In: Proc 2002 Annu Intern Res Conf Methyl Bromide Alternatives, Orlando, Florida.

Rust MK, Su NY (2012) Managing social insects of urban importance. Annu Rev Entomol 57, 355–375. doi: 10.1146/annurev-ento-120710-100634

Schroeder M, Cocoş D (2018) Performance of the tree-killing bark beetles *Ips typographus* and *Pityogenes chalcographus* in non-indigenous lodgepole pine and their historical host Norway spruce. Agric For Entomol 20, 347–357. doi: 10.1111/afe.12267

Seabright K, Davila-Flores A, Myers S, Taylor A (2020) Efficacy of methyl bromide and alternative fumigants against pinewood nematode in pine wood samples. J Plant Dis Prot 127, 393–400. doi: 10.1007/s41348-019-00297-7

Soliman T, Mourits MCM, van der Werf W, Hengeveld GM, Robinet C, Lansink AGJMO (2012) Framework for modelling economic impacts of invasive species, applied to pine wood nematode in Europe. PLoS One 7, e45505. doi: 10.1371/journal.pone.0045505

Stejskal V, Jonáš A, Hnátek J, Aulický R, Mochán M, Vybíral O (2017) New technology of wood fumigation against bark beetle first trial results-EDN product used for forest protection in Czech Republic. Forest Work Magazine 11, 19–21.

STIMBR (2018) Ethanedinitrile (EDN): An alternative to methyl bromide. Stakeholders in Methyl Bromide Reduction. http://www.stimbr.org. nz/uploads/1/4/1/0/14100200/edn_faqs_2018__2018-02-28_.pdf. Accessed 15 March 2023.

Su NY, Scheffrahn RH (1990) Efficacy of sulfuryl fluoride against four beetle pests of museums (Coleoptera: Dermestidae, Anobiidae). J Econ Entomol 83, 879–882. doi: 10.1093/jee/83.3.879

Viljoen JH, Ren YL (eds), 2001 Pp. 1–2. In: Proc 2001 Annu Intern Res Conf Methyl Bromide Alternatives, San Diego, California.

Ward SF, Brockerhoff EG, Turner RM, Yamanaka T, Marini L, Fei S, Liebhold AM (2023) Prevalence and drivers of a tree-killing bark beetle, *Ips typographus* (Coleoptera, Scolytinae), in international invasion pathways into the USA. J Pest Sci 96, 845–856. doi: 10.1007/ s10340-022-01559-4

Yalcin M, Akcay C, Tascioglu C, Yuksel B, Ozbayram AK (2020) Damage severity of wood-destroying insects according to the Bevan damage classification system in log depots of Northwest Turkey. Sci Rep 10, 13705. doi: 10.1038/s41598-020-70696-6

Aeration and Chilled Aeration for Insect, Mite, and Microflora Management and Grain Quality Preservation

Fuji Jian and Digvir S. Jayas

10.1 Introduction

Aeration is a process of forcing low flow rate ambient air through a grain bulk in a storage structure mainly to change grain temperature. Aeration is also called "active," "mechanical," "low volume," or "forced" ventilation because air movement is produced by a fan or blower. This chapter uses the most common term—aeration. In storage grain ecosystems (Jayas et al., 1995), aeration can modify one or more abiotic factors (temperature, humidity, atmospheric composition) that influence the development and multiplication of biotic factors such as insects, mites, and microflora and thereby maintain the stored grain under a safe storage condition. The beneficial effect of this modification can be optimized by monitoring

DOI: 10.1201/9781003309888-10

331

both ambient and grain conditions and controlling the aeration process so that the effect of maintaining favorable storage conditions for the safe preservation of grain quality can be maintained. If low ambient air temperature is not available, chilled or refrigerated air (referred to as chilled aeration) can be used. Chilled aeration is conducted by cooling ambient air through a bank of refrigeration coils. The relative humidity (RH) of the cooled air will increase but not the specific humidity. If the relative humidity of the cooled air is higher than 60 to 75%, the ambient air is cooled to lower its dew point temperature and water is condensed out. The cooled air is reheated a few degrees to achieve the desired RH and temperature of the chilled air. The amount of reheating and the final air temperature are adjustable by the operator. The reason chilled aeration is more expensive than aeration is that aeration uses only ambient air. However, because chilled aeration is required infrequently (once every two or three months, depending on the size of the storage structure) and chilling units are typically mobile, a single chilling unit can be used for multiple bins in a complex or in multiple complexes, thus reducing the cost of chilled aeration per bin. The basic principle and application of chilled aeration are the same as for aeration.

Aeration is an energy-efficient and economical method to reduce grain temperatures. As expected, the total aeration cost increases with increases in initial grain temperature and decreases with later harvest dates (Bridges et al., 2005). Aeration generally costs less than using grain fumigants or protectants. Even though the chilled aeration of grain has a higher cost than the aeration, it was applied commercially in over 50 countries even before the 1990s (Maier, 1994) due to the benefit of maintaining grain quality under low temperatures. At present, grain aeration is one of the most effective, sustainable, and nonchemical methods in use for the management of stored-grain conditions, biological activity, and grain quality losses. Throughout the world, low-temperature storage through aeration, with or without chilling, is widely used, and this storage technology can be part of an integrated pest management paradigm for reducing damage from insect and mite pests and mold and for reducing the application of fumigants and other treatments. Grain chillers are primarily used as a means of preserving high-value products (e.g., popcorn, polished rice, processed food), while aeration is widely used to remove the heat in stored grain after harvest. For example, rice storage at low temperature has been widely adopted in Japan, and over 4 million tonnes of brown rice is commercially stored in low-temperature warehouses during summer each year (Nakakita and Ikenaga, 1997). Grain chilling has been applied commercially in over 50 countries during the past 70 years and gained recognition in 1994 in the United States (Maier, 1994). Maier (1994) estimated that over 20 million tonnes of grains are cooled with chilling systems. Chilling of high-moisture-content grain for animal feeds is widely used in the world (Maier, 1994). Now a primary application of grain chilling is to maintain the quality of dry grain without using chemicals. The storage building used for grain chilling is usually insulated because not all grain in a uninsulated structure (i.e., grain along the periphery of the structure) can be maintained at the low temperatures required to limit insect and mold multiplication (Hunter and Taylor, 1980). In China, almost all warehouses in state grain depots are aerated every year so that the grain can be safely stored for more than five years. Different control strategies of aeration have been developed for wide applications in the world. These strategies have not been reviewed and optimal strategies have not been outlined.

Unlike thermal treatments, which must be precise because of the narrow margin between efficacy and commodity tolerance of the elevated temperature, aeration for cooling grain does not need precision because the population of insects, mites, and microflora will not increase when the temperature is lower than their minimum development temperature, and stored grain will not be damaged or has a slow deterioration rate under low temperature. Aeration also takes advantage of two important physical properties of grain bulks: porosity (percentage of the intergranular pore space over the total volume of the grain bulk) and low thermal diffusivity of the stored grain. Stored grain has a porosity of 27% to 59% (Jian and Jayas, 2022). The porous nature of the bulk grain permits forced air to pass through and come into contact with grain kernels along the air pathway. After the grain is cooled by cold air supplied by the fan during aeration, the low thermal diffusivity can slow down the warming up process of the grain by ambient weather. This enables maintenance of a "modified microclimate" for a few weeks or months after the grain bulk has been aerated. This principle can also apply to grain warming. Different aeration strategies (e.g., warming and cooling) could result in different modified microclimates, and there are different aeration processes to modify grain temperature and moisture content. Even though several studies have been published on the temperature and moisture content of grain modified by aeration (Reed et al., 1998; Reed and Harner, 1998; Arthur and Casada, 2005, 2010), the optimal range of temperature and moisture content in practical grain management has not been outlined.

Scientists have known for many years that dry and cold conditions will increase the shelf life of biological materials. The knowledge accumulated over the last seven decades has formed the basis for the present aeration technology. Using aeration techniques to manage grain temperature was pioneered in the 1950s by U.S. researchers such as Foster (1951), Robinson et al. (1951), Foster (1953), Shedd (1953), and Hukill (1953). A commercial cold-air chilling system for aeration was sold in Germany in 1958 (Maier, 1994). After these pioneering studies and this commercialization, the research was expanded to related fields such as insect control

(Cuperus et al., 1986; Arthur and Flinn, 2000), maintaining grain quality (Vertucci and Roos, 1990), chilled aeration (Navarro et al., 1973a, 1973b; Hunter and Taylor, 1980; Maier et al., 1992), and aeration control techniques (Reed and Harner, 1998; Arthur and Casada, 2010). Several books have chapters on this topic (Reed and Arthur, 2000; Navarro et al., 2012). In 2002, a book entitled *The Mechanics and Physics of Modern Grain Aeration Management* was published (Navarro and Noyes, 2002). These publications reviewed aeration from physical and engineering design aspects. However, a detailed review on insect and microflora management through aeration is needed. Also, existing reviews do not cover optimal aeration strategies developed in the last 20 years, and different researchers in different times made different recommendations. Therefore, this chapter reviews (1) grain temperature and moisture content management in practical aeration; (2) aeration for insect, mite, and microflora management and grain quality preservation; and (3) the advanced strategies of aeration developed through mathematical modeling in recent years.

10.2 Grain Temperature and Moisture Content Management During Practical Aeration

10.2.1 Grain Temperature During Practical Aeration

Temperatures higher than 42°C can be used for most insect and mite disinfestation because most insects and mites will be killed when the temperature is higher than 42°C over a long period (e.g., a month) and 60°C temperature kills in a few minutes. Even though heat treatment of storage structures is a common practice of grain storage management (Fields, 1992), warming up of the entire stored grain for insect, mite, and mold control is rarely used, for three reasons. First is the high energy cost. For example, warming up 100 t of wheat from 20° to 45°C needs at least 2.5×10^6 kJ, which requires running a 10-kW heater for at least 70 h. Second is the overdrying effect. For example, a fan with a 10-kW heater will significantly overdry the grain. Third is the fact that the grain should be cooled after heat treatment because storing grain at the high treatment temperature will significantly damage its quality, and conducting this cooling will increase the operational complexity and cost. Therefore, most research on warming grain for insect control was conducted before the 1990s (Hansen et al., 2011), and the process is occasionally applied in warmer regions such as Australia (Dermott and Evans, 1978; Evans et al., 1983; Sutherland et al., 1986) by using a grain dryer and not aeration. In many countries, ambient temperatures reach up to 49°–50°C but do not stay at this level long enough to warm up the entire grain mass, and thus insects are not controlled. Some small landholders in

Asia and Africa do spread their grain in the sun to warm it to high enough temperatures to kill insects and then cool it by moving grain to a shaded area or keeping it outside overnight when outside temperatures drop to 30°–35°C (Fawki et al., 2022).

During aeration when ambient air is forced by a fan or blower, the air temperature might increase about 1° to 5°C due to frictional heat produced by the fan against high pressure in plenum (a rise in static pressure of 850 Pa will generate a 1°C increase in temperature, and pressure during aeration can be about 2 to 8 kPa, depending on the grain type and the depth of grain in the bin). According to the ideal gas law, this high pressure will produce heat and decrease air volume. When the forced air temperature in plenum is higher than the grain temperature or higher than the preset temperature, the ambient air can be cooled by chilling or the aeration should be stopped (unless the purpose of the aeration is to warm up grain, as is done during spring in Canada).

Temperatures lower than 0°C can be used to kill insects and delay mold multiplication and development. Insects will die in seconds when the temperature is lower than their supercooling point (SCP). Even though different stages and species have different SCPs, the SCP for different insects is usually lower than –15°C. Old larvae of *Trogoderma variabile* (Ballion) have an SCP of –16.4°C when not acclimated and –25.1°C when acclimated (Abdelghany et al., 2015). The SCP of *Cryptolestes* ranges from –20.6° to –26.7°C, and acclimated insects have a slightly lower SCP than nonacclimated insects (Ganesan et al., 2021). *Trogoderma granarium* (Everts) larvae have an SCP of –14.4°C when not acclimated and –24.3°C when acclimated in diapause (Wilches et al., 2017). Therefore, to quickly kill insects, the grain temperature should be lower than –18°C (Arthur et al., 2017). However, if grain is frozen to lower than 0°C, water condensed on the surface of grain kernels might become ice, which could block the airway during aeration. To prevent this problem, the frozen grain should be warmed up in spring, and condensation might occur when the cold grain is warmed up. Therefore, even though temperatures lower than 0°C could effectively disinfest insects over a long period and effectively delay mold development, 5°–10°C of grain temperature after aeration is recommended for sound stored-grain management.

For the above-mentioned reasons, aeration in grain storage management is of only one type: forcing cold air through grain and reducing the grain temperature to about 5° to 10°C. Therefore, this chapter refers to the temperature from 5° to 10°C as aeration temperature. Aeration to cool grain is recommended for managing stored grains in temperate climates like those of northern Europe (Armitage, 1987; Lasseran and Fleurat-Lessard, 1990), Canada (Metzger and Muir, 1983), and north and central United States (Gardner et al., 1988). Chilled aeration is mostly conducted in warmer regions such as South America and South Asia.

10.2.2 Grain Moisture Content During Practical Aeration

The main objective of aeration is to move the cold temperature front through the entire grain bulk. The low airflow of less than 0.31 m^3 min^{-1} t^{-1} used for aeration will rarely reduce average moisture content more than one percentage point, even when operated for long periods (Harner and Hagstrum, 1990). It is not recommended to store high-moisture-content grain when the grain is cooled using aeration because mold can multiply even at aeration temperature if the grain moisture content is higher than the recommended safe-storage moisture content. Burrell and Laundon (1967) reported that at points where the temperature on the completion of cooling was 10°C or more, there was a subsequent rise of some 2° to 5°C during the storage of damp barley. Infestation of insects and mold at these warming locations could continue even during winter.

Even though rewetting of grain by aeration with humid air is a very slow process, grain rewetting occurs when aeration is conducted during rainy days or days with high relative humidity. Although the rewetted grain is mostly at the air entrance location, mold can develop in the wetted grain. Therefore, it is advisable to wait for low ambient humidity air (e.g., stop aeration during rainy or high-humidity days) to continue the aeration (Harner and Hagstrum, 1990). The relative humidity of air used for aeration is usually less than 70%. The drying effect of aeration at lower than 70% RH is minimal, and only a small amount of grain usually at the air entrance location can be dried. If forced air has a much lower RH (e.g., lower than 60%), part of the grain at the air entrance location might be overdried. This overdried grain can be rewetted later, after the ambient air changes to higher RH. Thus, some grain near the air entrance could go through multiple rewettings and dryings if the aeration period is too long; however, aeration rarely takes more than a few days to pass the cold front through the entire grain bulk. Therefore, usually aeration is not stopped even at low ambient RH, but aeration may be stopped on rainy days.

10.3 Effects of Aeration

10.3.1 Preventing Moisture Migration

Warm grain is usually loaded into storage structures after harvest due to high daytime temperatures and solar radiation on the heads of crop plants (Prasad et al., 1978) or the warm temperature of grain dried by a high-temperature drier (Jian and Jayas, 2022). Ambient temperature gradually decreases in fall and winter, and the grain near the bin walls is gradually cooled in the temperate climates. If this grain is not aerated, temperature gradients develop, with a higher temperature at the center of the storage bin (Figures 10.1 and 10.2) because of the low thermal diffusivity of stored

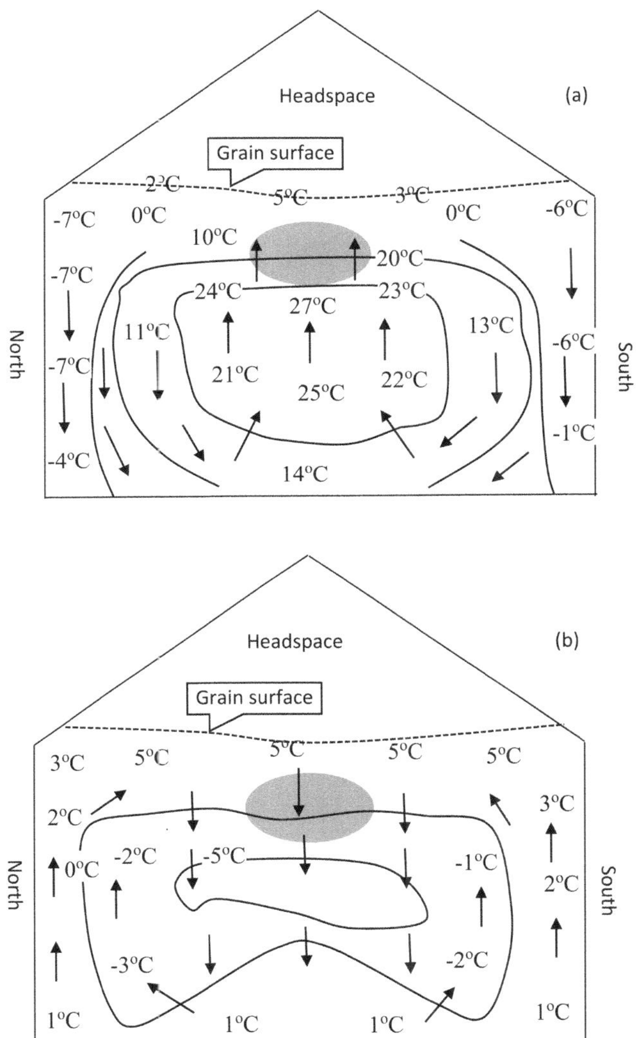

Figure 10.1 Measured temperatures and approximate isotherms of 20 t of wheat in (a) late November (after 3 months storage) and (b) late April (after 8 months storage) in a galvanized-steel bin near Winnipeg, Manitoba, Canada. The wheat ranged in temperature from 25° to 30°C and 13.6 ± 0.1% moisture content when harvested and stored to a depth of 2.6 m in the 3.7-m-diameter cylindrical bin in late August. Arrows show the directions of convection currents, and shaded areas show the cumulated moisture areas. (Reprinted with permission from Jian and Jayas, 2022).

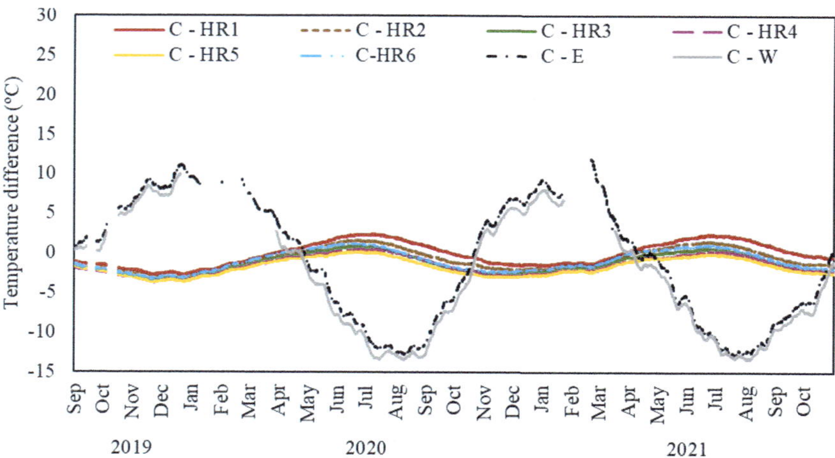

Figure 10.2 Temperature differences between center and other locations at 3.5 m from the surface of the grain in a 10 m diameter flat bottom bin filled with 300 t of wheat with 12.5% moisture content and stored for three years. Temperature difference > 0 implies that the temperature at center is higher than those at other locations. In the legend, `C' is the temperature cable located at 0.85 m from the centre of the bin along the north direction, and `HR' represents the cables located at half radius and numbered from 1 to 6, in order, from north in clockwise direction. `C-W' and `C-E' are the cable located at 0.6 m away from the walls in the west and east direction, respectively. Figure is reprinted with permission from Bharathi et al. (2023).

grain (Montross et al., 2002). Jian et al. (2009) reported that the average temperature gradient in a 20-t wheat bin was $5.1 \pm 1.0°C/m$, with the highest temperature gradient of $32.4°C/m$ at the center and 1.6 m height from the bottom of the bin. The intergranular air in the cold grain is denser than that in the warm grain. The force of gravity on denser air produces a downward pressure, and buoyancy in warm air produces an upward pressure. These pressure differences at different locations help each other to create continuous air circulation in the storage grain bulk, with denser air moving down and warmer air moving up. This air movement is termed free convection currents. The convection currents might pass through locations without a temperature gradient because air might be pushed or pulled by the moving air at adjacent locations. The air from the headspace might or might not join the circulated convection currents. The air speed of the free convection currents is very slow (lower than 6 to 15 m/d) (Gough et al., 1990; Jian and Jayas, 2022). The grain temperature is not influenced by the convection currents because (1) the moving air equilibrates with the grain in its path due to the low movement speed; (2) the energy difference of the slow-moving air between adjacent locations is negligible; and (3) the Rayleigh number is small enough not to influence grain temperature (Smith and Sokhansanj, 1990).

The warmer air has a higher water-holding capacity (Jian and Jayas, 2022) and gains more water from the warmer grain through a dynamic equilibrium process between the grain and the moving air. As this warm air moves through the top layer of cooler grain, the air is cooled and loses some of its water to the grain because (1) low-temperature air has a lower water-holding capacity than higher-temperature air; (2) the dynamic equilibrium process between the grain and the moving air occurs as in the warm grain; and (3) grain absorbs water through this equilibrium process. No condensation occurs because of the slow air movement and equilibrium process. The result of the free convection currents is moisture migration from the warmer locations to the locations with a rapid temperature decrease. The grain at the center of the bin and about 1 m down from the grain surface usually has a larger temperature decrease (Schmidt, 1955; Jian et al., 2009) because of the combined effect of the headspace and the cylindrical bin structure. Therefore, the migrated water accumulates at this location (Figure 10.1).

In spring and summer, if the grain is continually stored in the same bin without aeration, the grain near the walls warms up gradually, producing a reversed temperature gradient like that in fall and winter (Figures 10.1-10.2). This reversed temperature gradient generates reversed free convection currents. The moisture migration occurs due to the above-mentioned mechanism. The migrated moisture accumulates at a location similar to that in fall and winter because the location has the most rapid temperature decrease. Also, as the air temperature rises along the walls, it creates a downward pulling effect on the air in the center, thus bringing warm air from the top center to cooler grain in the center, resulting in moisture accumulation at a slightly lower location than that in fall and winter and sometimes near the bottom of the bin. Typically, this moisture migration can make the moisture increase at these locations by up to 1.5 percentage points (Muir et al., 1980; Jian and Jayas, 2022).

When aeration is completed in the desired time, there is either no temperature gradient or only a negligible temperature gradient, which cannot induce big enough free convection currents. Moisture migration due to free convection currents cannot be limited in bins without aeration. Therefore, aeration should be completed by cooling the grain and creating a uniform temperature distribution throughout the grain mass as soon as possible after the grain is binned.

10.3.2 Insect Control

10.3.2.1 Temperature Effect on Insect Population

Many factors influence the multiplication and development of stored-grain insects and mites. These topics have been reviewed by several research groups (Howe, 1965; Imura, 1990; Strang, 1992; Kiritani, 1997; Beckett, 2011; Collins, 2012; Kiritani, 2012; Fields and Abdelghany, 2014; Stejskal et al., 2019), and this chapter will only summarize the conclusions related to aeration for management of stored-product insects.

Stored-grain insects are tropical or subtropical in origin and require 24° to 32°C for development (White, 1995). Closely related species (e.g., in the same family) have similar thermal requirements (Jarosik et al., 2011). Hagstrum and Milliken (1988) found that the relative influence of environmental factors on development in order from high to low was temperature > moisture > diet. However, near the optimal temperature, moisture and diet influence larval development more than temperature. Insects are sensitive to low temperatures, and the development of most insect species is generally stopped below 20°C (Table 10.1). The stored-product insects most commonly found in the United States and Canada stop development at 7° to 17.6°C. Therefore, grain temperatures of 5° to 15°C are considered "safe" for insect management, because feeding and breeding are low at low temperature and most species do not multiply fast enough to cause significant damage to stored grain until the temperature is about 3° to 5°C above the minimum and the RH is at least 10% above the minimum. The suboptimum (below the optimum) zone for temperatures is between 13° and 25°C (Fields, 1992). For example, *Cryptolestes pusillus* (Schonherr) survives less than 24 wk at 11°C and stops egg laying at 14°C (Fields and White, 1997). *Cryptolestes ferrugineus* (Stephens) and C. *pusillus* exposed to decreased temperature from 25° to 0°C for up to 40 wk stops egg laying at 16°C (Fields and Abdelghany, 2014). If insects are at chill-coma, they cannot hatch and lay eggs (Nakakita and Ikenaga, 1997). Hatching of *Sitophilus zeamais* (Motschulsky) and *Sitophilus oryzae* (L.) is greatly reduced at 15°C, while pupation and eclosion proceed successfully. Both hatching and metamorphosis of each species are completely inhibited at 10°C. Population increase of *S. oryzae* is completely suppressed at 15°C, while a small number of F1 of *S. zeamais* emerge, resulting in a marginal increase in population (Nakakita and Ikenaga, 1997). Therefore, the practical aeration temperature can slow down or stop insect development and multiplication.

10.3.2.2 Time Required to Control Insects

Controlling (killing) insects through aeration usually requires a few months because a complete life cycle at these temperatures takes three months or more (White, 1995). Temperatures below 20°C can significantly suppress insect population (Tables 10.1 and 10.2). The temperature of 6°C is progressively lethal (100% chill-coma after 12 d and mortality after 22 d) to *Alphitobius diaperinus* (Panzer), while at 10°C, 70% of the insects were alive after one month (Renault et al., 2003). At 9°C and 45% RH, 15.2 wk is required to kill all *C. ferrugineus*, while 23.6 wk is required to kill all *Sitophilus granarius* (Evans, 1983). Maintaining a low temperature (9° to 13.5°C) in dry grain for 3 to 6 months causes 99% mortality of several species of Coleopteran pests [e.g., *C. ferrugineus*, *Oryzaephilus surinamensis* (L.), *Rhyzopertha dominica* (Fabricius), *S. granarius*, *S. oryzae*, and *Tribolium castaneum* (Herbst)] (Evans, 1983; Beckett, 2010).

Table 10.1 Minimum Temperature and Relative Humidity for Growth of Individual Stored-Grain Insects (Beetles, Moths, and Psocids) and Mites

Insect or Mite	T^a	RH^b	Source
Mites	5–15	>62.5	Stejskal et al. (2019), Collins (2012)
Acanthoscelides obtectus	17	30	Howe and Currie (1964)
Ahasverus advena[d]	14.1		Kiritani (1997)
Alphitobius diaperinus	17.3		Kiritani (1997)
Apomyelois ceratoniae	10.6		Kiritani (1997), Imura (1990)
Anthrenus verbasci	9.6–11.3		Kiritani (1997)
Anagasta kuehniella[d]	10	30 [c]	Howe (1965)
Araecerus fasciculatus	10.6	65	Howe (1965), Li and Li (2009)
Cadra calidella	9.8–10.5	25	Burks and Johnson (2012), Jarosik et al. (2011)
C. cautella	11.5	60	Kiritani (1997), Imura (1990)
C. figuliella	10.4		Jarosik et al. (2011)
Callosobruchus analis	16.6	30[c]	Howe and Currie (1964), Kiritani (1997)
C. chinensis	10.1	30	Howe and Currie (1964), Imura (1990)
C. maculatus	14.5	10	Howe and Currie (1964), Imura (1990)
C. rhodesianus	13.7	30[c]	Howe and Currie (1964), Imura (1990)
Caryedon gonagra	22	30	Howe (1965)
C. hemipterus	8.6	50	Lindgren and Vincent (1959), Imura (1990)
Corcyra cephalonica[d]	13.5	30	Howe and Currie (1964), Jarosik et al. (2011)
Cryptolestes capensis	11.1	30	Howe (1965), Imura (1990)
C. ferrugineu[d]	15.7	10	Smith (1965), Imura (1990)

(Continued)

Table 10.1 (Continued) Minimum Temperature and Relative Humidity for Growth of Individual Stored-Grain Insects (Beetles, Moths, and Psocids) and Mites

Insect or Mite	T^a	RH^b	Source
C. pusillus[d]	14.6	40	Bishop (1959), Kiritani (1997
C. pussilloides	11		Beckett (2011)
C. turcicus	10.2	50	Bishop (1959), Lefkovitch (1962), Kiritani (1997)
C. Uganda	14.2		Kiritani (1997), Imura (1990)
Dermestes frischii	15.2		Kiritani (1997), Imura (1990)
D. haemorrhoidalis	13.2		Jarosik et al. (2011)
D. Lardarius	15.0		Strang (1992)
D. macrdatus	15.0	30	Howe (1965), Subramanyam et al. (1990)
Dienerella argus	9.1–10.7		Imai and Miyamoto (2019)
Ectomyelois ceratoniae	13.0		Burks and Johnson (2012)
Endrosis sarcitrella	8.5		Woodroffe (1951b), Kiritani (1997), Imura (1990)
E. kuehniella	7.5		Kiritani (1997), Imura (1990)
Ephestia eulutella	15	30	Burks and Johnson (2012)
Gibbium psylloides	11.8	30	Howe (1953), Kiritani (1997)
Gnathocerus cornutus	16	40	Howe (1965)
G. maxillosus	13.3–16.1		Jarosik et al. (2011)
Hofmannophila pseudospretella	6.9		Woodroffe (1951a), Imura (1990)
Lasioderma serricorne[d]	15.7	30	Howe (1957), Howe (1962b)
Liposcelis badia[d]	10.0		Jiang et al. (2008)
L. bostrychophila[d]	8.2–9.2		Jarosik et al. (2011)
L. decolor[d]	13.0		Tang et al. (2008)
L. entomophila[d]	15.2–16.6		Jarosik et al. (2011)
L. tricolor[d]	11.3		Dong et al. (2007)
L. paeta	18.1–20.9		Dong et al. (2007)
L. yunnaniensis	14.8		Hassan et al. (2011)
Mezium affine	11.2		Kiritani (1997), Imura (1990)

(Continued)

Table 10.1 (Continued) Minimum Temperature and Relative Humidity for Growth of Individual Stored-Grain Insects (Beetles, Moths, and Psocids) and Mites

Insect or Mite	T[a]	RH[b]	Source
Nemapogon granella[d]	7.0	65	Trematerra and Lucchi (2014)
Necrobia rufipes	22	50	Howe (1965)
Nitpus hololeucus	10	50	Howe (1953)
Ophiomyia simplex	12.1		Morrison III et al. (2014)
Oryzaephilus mercator[d]	16.6	10	Howe (1956a), Imura (1990)
O. surinamensis[d]	8.8	10	Howe (1956b), Subramanyam et al. (1990)
Palorus ratzeburgii	15.4		Jarosik et al. (2011)
P. subdepressus	15.6		Jarosik et al. (2011), Imura (1990)
Phthorimaea operculella	10.6–18	60	Jarosik et al. (2011), Subramanyam et al. (1990)
Plodia interpunctella[d]	12.1	40	Tzanakakis (1959), Williams (1964)
Prostephanus truncates[d]	15.1–16.2		Kiritani (1997), Imura (1990)
Pseudeurostus hilleri	15	50	Howe (1953), Howe (1956c)
Ptinus clavipes	18	50	Howe (1959)
P. pusillus	19	50	Howe (1959)
P. fur[d]	10	50	Howe (1953)
P. sexpunctatus	19	50	Howe (1953)
P. tectus	7.1–7.5	50	(Howe, 1962a), Imura (1990)
Pyralis farinalis[d]	14.8		Shang et al. (2013)
Rhyzopertha dominica[d]	13.2	30	Birch (1953), Subramanyam et al. (1990)
Stegobium paniceum[d]	12.1–13.9	60	(Howe, 1965), Kiritani (1997)
Stethomezium sguamosum	19	50	Howe (1953)
Sitotroga cerealella[d]	16	30	Howe (1965)
S. granarius[d]	10	50	Howe (1953), Subramanyam et al. (1990)

(*Continued*)

Table 10.1 (Continued) Minimum Temperature and Relative Humidity for Growth of Individual Stored-Grain Insects (Beetles, Moths, and Psocids) and Mites

Insect or Mite	T^a	RH^b	Source
S. oryzae[d]	12.6	60	Birch (1953), (Hardman, 1978)
S. zeamais[d]	11.3–15.1	45	Sinha and Watters (1985), Kiritani (1997)
Tinea translucens	10.0–15.0		Kiritani (2012)
T. bisselliella	9.0		Strang (1992)
Tinas unicolor	12	60	Howe (1955)
Tribolium castaneum[d]	17.6	30[c]	Howe (1962b), Subramanyam et al. (1990)
T. confusum	15.7–16.7	30[c]	Howe (1965), Howe (1960), Kiritani (1997)
Trichomyrmex destructor	12.6		Imura (1990)
Tribolium madens	17.4		Kiritani (1997), Imura (1990)
Trigonogenius globulus	18	50	Howe (1953)
T. glabrum	10.2–10.8		Kiritani (1997), Jarosik et al. (2011)
Trogoderma anthrenoides	15.5–18.4		Kiritani (1997)
T. granarium	13.3–13.9	30[c]	Howe (1965), Imura (1990)
T. inclusum[d]	11.4–12.4		Kiritani (1997)
T. variabile[d]	12.1	30	Partida and Strong (1975)
T. versicolor	17.0		Kiritani (1997), Jarosik et al. (2011)
Typhaea stercorea[d]	14.5		Kiritani (1997), Imura (1990)
Zabrotes subfasciatus	14.3–15.0	30	Howe and Currie (1964), Imura (1990)

Notes:

[a] T = Minimum growth temperature (°C). At minimum growth temperature, insects and mites require higher than the minimum RH.

[b] RH = minimum growth relative humidity required (%). At minimum growth RH, insects and mites require higher than the minimum temperature.

[c] Estimated from the resource.

[d] Most common stored-product insects found in the United States (United States Department of Agriculture, 2016) and Canada.

Table 10.2 Survival Time of Insect Adults or Mites under Aeration Temperatures in Dry Grain

Insect/Mite	S^a	$T \& RH^b$	$LT_{50}{}^c$	$LT_{99}{}^c$	Source
Acarus siro	E	0, 20	3.5	10	Cunnington (1984)
	E	0, 90	10	21.4	Cunnington (1984)
Alphitobius diaperinus	A	6, 65	1.5		Renault et al. (1999)
Callosobruchus maculatus	E	5, ?	0.6	1.1	Mullen and Arbogast (1979)
Cryptolestes ferrugineus	A	9, 45	7.4	15.2	Evans (1983)
C. pusilus	E	2, 35	0.1	0.6	Williams (1954)
	-	2, 70	0.2	0.3	Williams (1954)
Ephestia cautella	E	5, ?	0.7	1.1	Mullen and Arbogast (1979)
E. kuehniella	E	4.5, ?	0.1		Daumal et al. (1974)
	-	2, 80	1.0	7.2	Le Torc'h (1977)
Lasioderma serricorne	E	5, ?	0.5	2.0	Mullen and Arbogast (1979)
	-	4.4, 75	1.0	3.0	Childs et al. (1970)
Oryraephilus surinamensis	A	9, 45	7.8	14.4	Evans (1983)
	E	5, 70	0.2	0.5	Jacob and Fleming (1986)
Plodia interpunctella	A	10, 60	7.0	14	Johnson et al. (1997)
	E	2.4, 85	0	0.5	Cline (1970)
	E	10, 60	1.0	2.0	Johnson et al. (1997)
	L	2, 80	2.9	5.7	Le Torc'h (1977)
Rhyzopertha dominica	A	9, 45	7.7	11.3	Evans (1983)
	E	4.4, 67		2.0	David et al. (1977)
Sitophilus granaries	A	9, 45	7.4	23.6	Evans (1983)
	L	4.5, 67	2.0	6.0	David et al. (1977)
	E, L, P	5, 67		14.3	Mathlein (1961)
S. oryzae	A	9, 45	5.5	8.7	Evans (1983)
Tribolium castaneum	A	9, 45	6.1	8.9	Evans (1983)
	E, L, P	13.5, 60		13	Evans (1987)
T. confusum	A	7, 74	2.0		Nagel and Shepard (1934)
	E	7, 60	0.7		Nagel and Shepard (1934)
	A	7, 74	2		Nagel and Shepard (1934)

Notes:

[a] S = insect stage. A = adult, E = egg, L = larva, P = pupa.

[b] The number before the comma is the temperature (°C), and the one after the comma is the relative humidity (%). ? indicates that the moisture content and relative humidity of the diet were not reported in the source.

[c] LT_{50} and LT_{99} = lethal time of 50% and 99% of the population in days, respectively.

Even though there is a wide range of temperatures that can result in insect death, temperatures below 5° to 15°C will result in the death of most species of stored-product insects after a long time (Table 10.1 and 10.2), especially in the immature stages (Banks and Fields, 1994). At conditions roughly 6°C below the threshold for population growth, >99% mortality of major Coleopteran species is possible after 9 months at 45% RH (Beckett, 2011). A grain temperature of 0°C is sufficient to kill all insects except *C. ferrugineus* in 3 months, but 5°C is insufficient to achieve 100% mortality (Bareil et al., 2018). Therefore, longer than half a year is required to disinfest dry grain at lower than 9°C.

10.3.2.3 Acclimation During Aeration

Insects can acclimate to aeration temperature because aeration is not usually completed in one day. Exposure of insects to temperatures from 10° to 20°C for some days might result in their getting acclimatized to the conditions, in which case their cold hardiness may increase two- to tenfold (Evans, 1983; Fields, 1992). Ganesan et al. (2021) found acclimated insects were anywhere from having no increase in cold tolerance to being 14-fold more cold tolerant than the corresponding nonacclimated species at that stage. Therefore, different acclimation processes might influence insect tolerance to low temperature, explaining disagreements among published results. For example, there is a disagreement on what stage of *C. ferrugineus* has the most cold tolerance. Smith (1970) found that nonacclimated *C. ferrugineus* adults and old larvae were more cold hardy than pupae, young larvae, and eggs, while Fields (1992) found that adults were the most cold hardy. However, in 300 kg of wheat, Abdelghany and Fields (2017) found immature stages were more cold tolerant than adults. Therefore, different aeration temperatures could result in different insect tolerances, and their resistance to cold could be reduced by means of rapid cooling of the stored grain because completing an aeration cycle in a short period could minimize insect acclimation to the low temperature.

Grain not only provides an environment that prevents fast cooling but also provides a diet for insects. *Cryptolestes* adults reared on a grain diet (whole wheat kernels, cracked wheat kernels, and wheat germ, 90:5:5 mass ratio) are the most cold tolerant, while those reared on a white-wheat flour and brewer's yeast diet (95:5 mass ratio) have the next highest cold tolerance, followed by the adults reared on 100% white-wheat flour (Ganesan et al., 2021). Therefore, insect tolerance to low temperature in stored grain might have a wide variation due to the variation of grain temperature during aeration and the variation in grains.

10.3.2.4 Effect of Temperature Combined with Moisture Content

The effect of low temperature on insect survival could be influenced by grain moisture content (intergranular RH) (Imura, 1990) because low RH (low grain moisture content) could decrease insect survival at low temperatures (Table 10.2). The effect of moisture is most obvious at the lowest conditions (temperature and RH), which are severely detrimental to the physiological function of insects (Beckett, 2011). Insects live longer at 70% RH than at 45% RH (Evans, 1983). Aeration is mostly conducted on grain with a safe storage moisture content, and aeration usually does not change grain moisture content by much. Therefore, even though the effect of grain moisture on insect survival during aeration is usually not considered (because the effect is usually significant only when storage time is a few months), the effect of low moisture content under aeration conditions can be significant for the total insect population because at low grain moisture content adults do not live as long, the number of eggs laid per female is reduced, and immature stages take longer to develop and struggle to survive.

Regardless of insect stage, grain moisture content, and degree of acclimation, if temperature is lower than 10°C, all insects except some cold-hardy species will be killed in one month (Table 10.2). To control the cold-hardy insects, more than five weeks are required (Fields, 1992). Therefore, aeration is mostly used to suppress insect population and seldom used to kill insects. Due to the wide variation of insect species and stages, aeration strategies, weather conditions, and daily weather fluctuations, using aeration to control insects might have different results.

Mites can multiply when the temperature is higher than 4° to 7°C (Solomon, 1962). The crucial control parameter for mite pests is the RH because mites require an RH higher than 65% for their development—similar to the requirement for storage-grain fungi. Wheat with 12.5% moisture content at 25°C (the equilibrium RH is less than 65%) can effectively suppresses mite (Cunnington, 1984; Navarro et al., 2002) and mold development. Therefore, it is not recommended to control mites in damp grain (14% to 16% moisture content wet-basis cereal) or wet grain (higher than 16% moisture content cereal) by using aeration because controlling grain moisture is cheaper than maintaining low grain temperature in high-moisture grain for a long storage time.

10.3.3 Controlling Mold Multiplication

Most fungi involved in the deterioration of stored grain thrive at 10° to 40°C, with the optimal temperature being in the range of 25° to 35°C and different species having different optimal ranges. The number of fungal species that can be found colonizing stored grain is wide, and there are variations of dominant species on different crop grains. The most common molds of stored grain are *Penicillium* spp. and A*spergillus* spp. (Frisvad, 1995). The following species can also be found: *Eurotium, Fusarium, Rhizopus*, and thermophilics like *Talaromyces thermophilus* Stolk, *Rhizomucor pusillus* (Lindt) Schipper, and *Thermomyces lanuginosus* Tsiklinsky. The most dominant species of fungi on canola are *Penicillium* spp., *A. glaucus* group, and *A. candidus* Link, while a small number of seeds maybe infected by *Alternaria alternata* (Fr.) Keissler and *Cladosporium* spp. (Sun et al., 2014). The most dominant species on hemp seeds are *Alternaria alternata* (Fr.) Keisler, *Aspergillus flavus* Link, *A. candidus, Aspergillus ochraceus* Wilhelm, *Aspergillus wentii* Wehmer, and P*enicillium* spp. (Jian et al., 2019b). The most dominant species on rye are *Penicillium* spp. and *Aspergillus* (Sathya et al., 2008). *Penicillium* spp. and *A. glaucus* group are most dominant on durum wheat (Nithya et al., 2011).

The temperature and water activity (a_w) are the main environmental factors influencing fungi growth and multiplication. Germination of *Fusarium moniliforme* (Sheldon) and *Fusarium proliferatum* (Nirenberg) is very rapid at $a_w > 0.94$, with an almost linear increase with time. Germination rates of *F. moniliforme* are slower than those of *F. proliferatum* at marginal a_w (0.88) levels and 5° to 25°C, while at higher temperature (30° to 37°C) the former germinates more rapidly than the latter (Marin et al., 1996). Even though storage fungi do not need higher water activity for their growth and multiplication, they still require an a_w of at least 0.65 for their slow growth (Magan and Lacey, 1984b). The difference between the minimum a_w requirement for spore germination and growth of some field and storage fungi on wheat is negligible (the a_w difference between spore germination and growth is less than 0.01 to 0.03) (Marin et al., 1996). Even though the minimum a_w for storage fungi growth and multiplication changes with temperature (Magan and Lacey, 1988), when water activity is higher than 0.7 their growth and multiplication (germinating, growing, and sporulating in stored grain) can be continued and initiate grain spoilage. Successful spore germination results in the formation of extending hyphae that can colonize the substrate. When water activity is lower than 0.7, storage fungi will not cause storage loss for up to one year of storage.

Water activity may interact with gas composition especially in increasing the lag phase before spore germination occurs. Magan and Lacey (1984a) reported that the lag phase for some *Aspergillus* and *Penicillium* spp. at an a_w of 0.90 was 16 to 18 d with 15% CO_2 and 21% O_2, respectively, compared to 4 d with 0.03% CO_2 and 21% O_2. Even though low O_2 concentration can hinder the growth and multiplication of fungi, there are difficulties to conducting aeration with low-O_2 gas (e.g., high N_2 or CO_2) due to its cost.

Temperatures lower than 5°C are needed for the suppression of most mold development. To completely control fungi, a temperature lower than 0°C is required (Table 10.3). Therefore, it is not recommended to cure mold

Table 10.3 Minimum Temperature (T) and Water Activity (Ma_w) for Growth of Storage Fungi in Stored Grain

Fungi	T^a	$Ma_w{}^b$	Reference
Alternaria alternata	5	0.88	Sanchis and Magan (2004), Sun et al. (2014)
A. tenuissima	10	0.85	Sanchis and Magan (2004), Young et al. (1980)
Aspergillus amstelodami	6	0.70	Northolt and Bullerman (1982)
A. candidus	10	0.75	Northolt and Bullerman (1982), Lacey et al. (1980)
A. carbonarius	8	0.80	Sanchis and Magan (2004)
A. chevalieri	10	0.71	Northolt and Bullerman (1982)
A. flavus	8	0.75	Sanchis and Magan (2004), Sorenson et al. (1967)
A. fumigatus	12	0.82	Northolt and Bullerman (1982), Magan and Lacey (1984b)
A. nidulans	12	0.75	Northolt and Bullerman (1982), Lacey et al. (1980)
A. niger	12	0.85	Northolt and Bullerman (1982), Marín et al. (1998)
A. ochraceus		0.77	Ramos et al. (1998), Northolt et al. (1979)
A. oryzae	12	0.86	Northolt and Bullerman (1982)
A. parasiticus		0.81	(Sanchis and Magan (2004)
A. repens	8	0.75	Magan and Lacey (1984b)
A. restrictus	4	0.68	Lacey et al. (1980)
A. versicolor	6	0.83	Northolt and Bullerman (1982), Lacey et al. (1980)
Cladosporium cladosporiodes	8	0.88	Magan and Lacey (1984b)
C. herbarum	7	0.88	Northolt and Bullerman (1982), Magan and Lacey (1984b)
Epicocum nigrum	5	0.95	Magan and Lacey (1984b)
Eurotium Spp.	4	0.68	Lacey et al. (1980)
Fusarium culmorum	7	0.90	Magan and Lacey (1984b)
F. graminearum		0.90	Sanchis and Magan (2004)
F. moniliforme	5	0.88	Marin et al. (1996)
F. proliferatum	5	0.88	Marin et al. (1996)

(*Continued*)

Table 10.3 (Continued) Minimum Temperature (T) and Water Activity (Ma_w) for Growth of Storage Fungi in Stored Grain

Fungi	T^a	$Ma_w{}^b$	Reference
Penicillium aurantiogriseum	−8	0.82	Lacey et al. (1980)
P. brevicompactum	2	0.81	Northolt and Bullerman (1982), Magan and Lacey (1984b)
P. capsulatum	18	0.80	Lacey et al. (1980)
P. citrinum	12	0.80	Northolt and Bullerman (1982)
P. corylophilum	4	0.68	Lacey et al. (1980)
P. galbrum	2	0.82	Lacey et al. (1980)
P. expansum	1	0.82	Sanchis and Magan (2004)
P. hordei	5	0.82	Magan and Lacey (1984b)
P. islandicum	10	0.83	Northolt and Bullerman (1982)
P. patulum	4	0.81	Sanchis and Magan (2004)
P. piceum	18	0.80	Lacey et al. (1980)
P. roquefortii	5	0.85	Magan and Lacey (1984b)
P. rugulosum	4	0.68	Lacey et al. (1980)
P. verrucosum	5	0.80	Sanchis and Magan (2004), (Magan and Lacey, 1984b)
Paecilomyces variotii	18	0.80	Lacey et al. (1980)
Stachybotrys atra	2	0.94	Northolt and Bullerman (1982)
Verticillium lecanii	4	0.95	Magan and Lacey (1984b)
Wallemia sebi	4	0.68	Lacey et al. (1980)

Notes:

[a] T = minimum temperature for growth (°C).

[b] Ma_w = minimum water activity for growth (decimal).

in high-moisture grain by conducting aeration because of (1) the high cost to maintain the grain at low temperature during the entire storage period; (2) the condensation issue when the grain temperature is lower than 0°C, and (3) the fact that most fungi grow at higher than 65% RH (Table 10.3), which is equivalent to roughly 13.5% to 14.5% moisture content for cereal grains and 8.5% to 9% for canola at 20° to 25°C (Jian and Jayas, 2022). These moisture contents are the recommended safe storage moisture contents in temperate climates. In tropical and subtropical climates, safe moisture contents for cereal grains should be 11% or less. However, mold multiplication in grain can be reduced by aeration because aeration can reduce grain temperature and move the heat produced by mold respiration. Therefore, grain with a safe storage moisture content can be stored for a few years at 5° to 15°C. Lasseran et al. (1994) reported a decrease in the total number of bacteria and molds in the aerated bin.

10.3.4 Minimizing Hotspot Development

The moisture content and temperature of stored grain in a bin always have variation due to moisture and temperature variation at harvest (Prasad et al., 1978), at bin loading (Loschiavo, 1985), and after drying (Sinha et al., 1985) or aeration (Ghaly, 1984; Sinha et al., 1985). These variations might provide suitable environmental conditions at some places in a bin for multiplication and development of insects, mites, and mold (Figure 10.3). Both micro-organisms and insects can initiate a hotspot in stored grain bulk (Sinha and Wallace, 1965). Conditions for initializing a hotspot are complex, and hotspot development is a chain reaction of insect and mold multiplication, grain sprouting, and chemical reaction (Jian and Jayas, 2014) because the heat and water produced in this chain reaction create a more suitable environment for the development and multiplication of the insects and microorganisms. The heat and water produced at the end of this chain reaction can

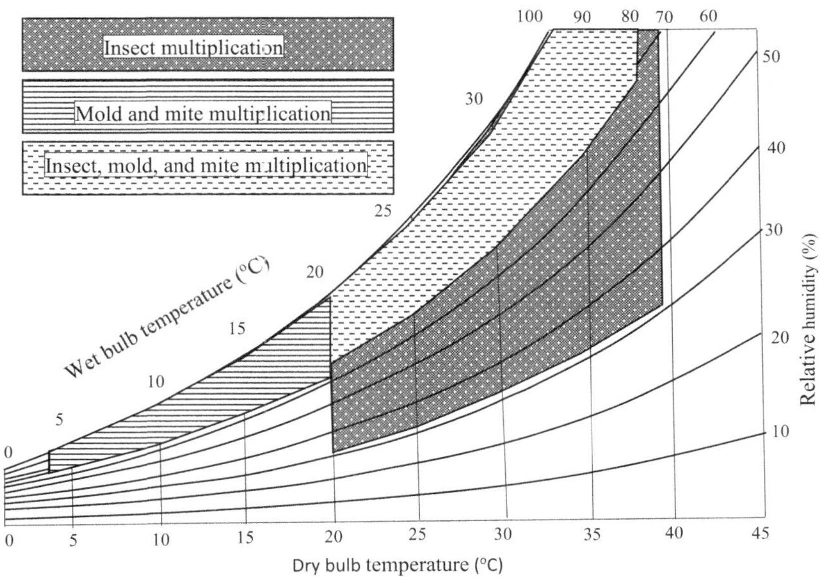

Figure 10.3 Multiplication conditions of insect, mite, and mold shown on a psychromatic chart.

induce a self-accelerating heating process (Sinha, 1961; Sinha and Wallace, 1965) because the low thermal conductivity and diffusivity of the stored grain and low free convective currents mean that the heat produced in these processes cannot be rapidly dissipated out of the bin. The temperature in a developing hotspot in the center of a wheat bulk increases by 15°C, from 5° to 20°C, in 2 wk and then further increases by 45°C to reach a maximum of about 65°C in only one more week (Sinha and Wallace, 1965). Aerating the grain before a hotspot develops further is critical because aeration can (1) prevent and delay the development and multiplication of insects and microorganisms due to the low temperature effect, (2) move the heat and water produced by these agents out of the grain (Jian et al., 2014a), and (3) stop the chain reactions before extensive chemical oxidization starts (Jian and Jayas, 2014). It might be too late to conduct aeration when the temperature is higher than 80°C in the core of the hotspot because aeration can supply enough O_2 for the oxidation process. Armitage and Burrell (1978) used aeration and a perforated duct to cool warmer grain (a potential hotspot).

10.3.5 Improving Distribution of Fumigants and Removing Fumigant Residues and Odors

It is possible to take advantage of existing aeration systems during and after fumigation because aeration systems can be used to successfully distribute fumigants such as phosphine and hydrogen cyanide (Phillips, 1957). The desorption of fumigants at the end of a fumigation can be achieved with relatively low airflow rates (lower than or similar to the flow rate for aeration). The same process can be conducted to remove the musty odor produced by insects and mold. The aeration system can be operated intermittently (e.g., 10 to 15 min every 2 to 3 h) to flush the grain bulk. Even though this strategy is conducted as part of grain storage management, the removal effect is not evaluated in the literature. Commercial applications using aeration combined with ozone can also be used to reduce odors in grain (Liu et al., 2020). Thorpe and Elder (1982) found that in the temperate and subtropical wheat-growing regions of Australia, aeration could reduce usage of the pesticide methacrifos (O-2-methoxy carbonylprop-1-enyl O.O-dimethyl phosphorothioate) by factors of 7 and 4, respectively.

10.3.6 Maintaining Grain Quality

Low grain temperatures are desirable for better maintenance of grain quality. Within the range of 0° to 50°C, the lower the temperature, the longer the seeds maintain full viability. Within 5% to 14% grain moisture content, a seed's life span in storage is doubled for each 5°C decrease and for each

1 percentage point decrease in seed moisture content (Harrington, 1973). Equilibrating seeds between 19% and 27% RH (about 6% moisture content of soybean seeds) provides the optimal moisture level for maintaining seed longevity during long-term storage (such as seed storage for several decades in a seed bank) (Vertucci and Roos, 1990).

There are a few studies on the effect of aeration on grain quality because grain quality will not significantly change if the grain is maintained at 5° to 15°C and at lower than the recommended safe storage moisture content. Calderwood et al. (1984) stored rice with 11.1% to 11.8% moisture content for 54 months in the warm and humid climate of Texas by controlling the rice temperature at lower than 15°C through aeration. They also stored another batch of the same rice without aeration but with PH_3 fumigation to control insects. The rice in the aerated bins did not change in quality during this long storage period, while the rice in the fumigated bins without aeration increased in moisture content to 14% in some locations due to moisture migration, decreased in grade from no. 1 to no. 5 due to decreased test weight, increased in heated grain kernels due to microorganism multiplication, decreased in germination from about 92% to about 20%, and increased in FAV (from about 14 to 55 mg KOH per 100 g dry mass) and undesired taste and appearance. Even though fumigation can prevent insect infestation, the rice stored in the phosphine-fumigated bin without aeration also downgraded due to high temperature and high moisture content at some locations. Navarro et al. (1969) found that aerated grain maintained at 10.5° to 14.3°C was in excellent condition throughout a two-year storage period.

Srzednicki et al. (2006) compared the milling yield, cooking properties, pasting properties, and germination of corn and rice after the grain was dried and then aerated. Rice with an initial moisture content of 18.3% was dried and aerated to the range of 12.6% to 13.4% using chilled air (6°C and less than 70% RH) or ambient air (28°C and less than 70% RH). A similar treatment was conducted for corn (initial moisture content was 19.3%, and it was dried and aerated to the range of 12.4% to 11.5%). They found that the milling yield of the rice was not significantly affected by the drying and aeration treatments. Significant differences in the pasting properties of rice and corn were observed between the two different treatments. The peak viscosity of corn varied by as much as 40 Rapid-Visco Analyser units between the two treatments. A highly significant difference in germination was observed. A significantly reduced cooking time was observed for rice treated with chilled air. Lower gruel solid loss was observed for the ambient air treatment. They concluded that chilled aeration could better preserve grain viability. Shafiekhani and Atungulu (2020) found chilled/cold storage of rough rice (16% to 21% moisture content) at 10°, 15°, and 20°C retarded discoloration of the milled rice. Calderwood et al. (1984) found that the cooler average rice temperatures in aerated bins compared with rice temperatures in nonaerated bins kept germination of seed at a higher level and slowed the buildup of fat acidity for long-term rice storage (54 mo). There was no insect infestation, and the rice remained at U.S. no. 1 grade and was more than marginally acceptable to most taste panel participants.

10.4 Aeration System Configuration and Control

10.4.1 Aeration System Configuration

Even though an aeration system is designed based on the local climate and grain storage requirements in different countries and geographical locations, it is usually equipped with one or more fans and ducts. A fully perforated floor is usually not needed for aeration, and a larger perforated floor area is desired to minimize the dead aeration locations. Permanent or removable floor ducts with different configurations are widely used around the world. Jayas and Muir (2002) described the design requirements of an aeration system for different aeration configurations, including vertical upward, vertical downward, and horizontal airflow. Combinations of fan location at the bottom or top of the bin and different configurations of ducts can make different airflow directions. Navarro et al. (2012) outlined the minimum area of aeration ducts. Interested readers can consult their work for design details. Silva et al. (2014) found that using an exhauster system in a rice storage silo reduced the loss of dry matter, the energy required for aeration maintenance over a 9-mo period, grain hardening, development of grain staining defects and darkening of the grain, while preserving the conclusion temperature of gelatinization and the enthalpy, the final viscosity and the head rice yield, and not affecting the onset temperature or the peak temperature of gelatinization.

A galvanized-steel cylindrical bin with a flat or hopper bottom is the most common storage structure in the world. In round mud, mat, or steel bins, either radial flow distribution from a central ducting or in-floor ducting is used (Figure 10.4). A warehouse is another storage structure widely used in the world, and it is usually equipped with on-floor or in-floor ducting aeration systems. Only the duct covered by grain is perforated, and air is distributed to the grain through the perforated area of the air distribution ducts (Figure 10.).

Upward airflow is preferred by many grain storage managers over downward airflow for four reasons. First, the fan can deliver maximum airflow against pressure. (A fan can push more mass of cool dense air than of warm light air, and there is less pressure resistance for the upward airflow than for the downward flow.) Second, the upward system generates more uniform airflow through the grain mass than does a suction system. Third, condensation on the ceiling of the bin is visible and usually dries

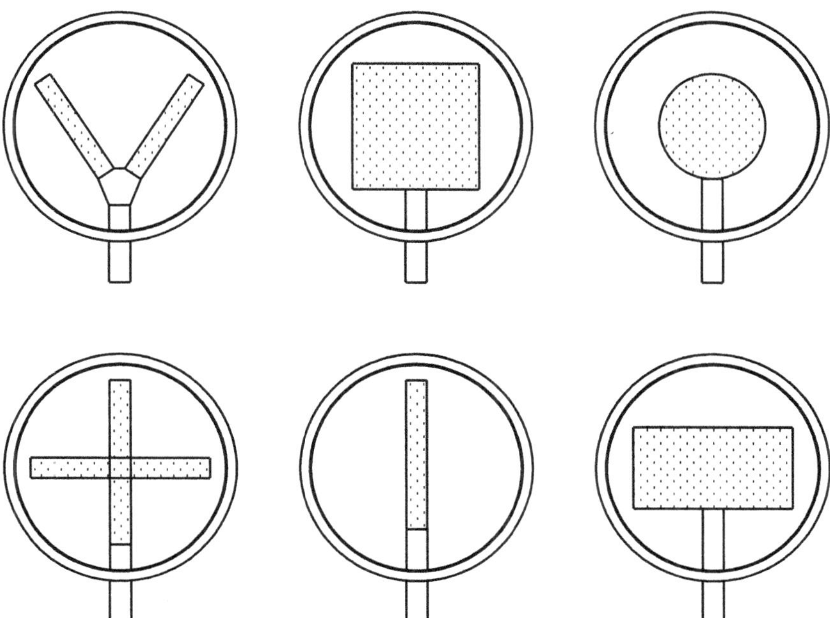

Figure 10.4 Variations in the configuration of partially perforated floors for aeration. (From Jayas and Muir, 2002, with permission.)

up by the end of the aeration cycle, whereas bottom condensation damage caused by suction fans cannot be seen until the bin is unloaded. Fourth, the upward system eliminates winter bin roof collapse. (Roof vents may freeze over during suction cooling.) Arthur and Casada (2016) pointed out that headspace temperature (especially during the afternoon) should be considered when downward aeration is used, and downward aeration would be more beneficial than upward aeration for controlling insect pests in wheat stored in the southern plains of the United States (Arthur and Casada, 2010) because most insects prefer the center location near the grain surface, given the warmer environment in the headspace. Therefore, suction cooling at night during summer can effectively cool the top 1 to 3 m of grain by 5°C to 10°C within 3 to 5 wk of harvest in most regions of United States, thus dramatically slowing insect population buildup at the top of the grain bulk. However, downward aeration requires more aeration time. Bartlett et al. (2002) found that the volume of grain cooled by upward aeration was 20% more than that cooled by downward aeration using the same aeration system. The duct wall resistance was 16% lower when blowing wheat than when sucking.

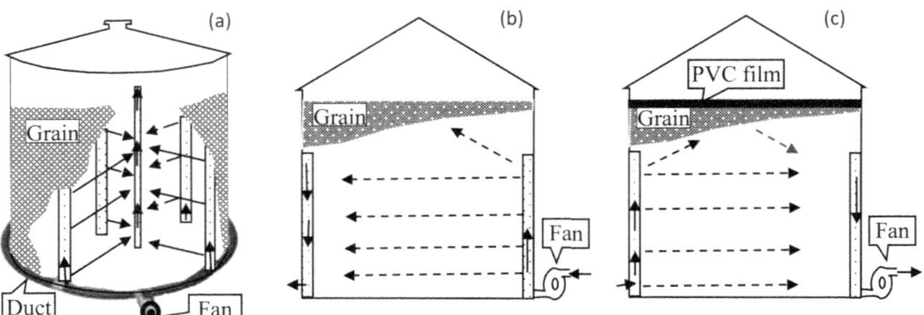

Figure 10.5 Horizontal aeration system in a round bin (a) and a warehouse (b and c). In the graph, vertical columns show the air distribution and collection ducts. The arrows show air movement directions. Positive (push, b) and negative (pull, c) aeration can be conducted in both the round bin and the warehouse. When negative aeration is conducted, a PVC film should be used to cover the grain surface to control air distribution (c). Note that (c) shows only one pair of the air distribution ducts.

In recent years, horizontal aeration has been widely adopted to decrease fan power (Figure 10.5) because of the low airflow resistance in the horizontal direction (Jayas and Sokhansanj, 1989) and short air travel distance. In a round bin, several vertical air distribution ducts (usually four or six) are fixed on the bin walls, and one air collection duct is installed in the center of the bin. The system can also conduct negative (pull) aeration through sucking air with the fan. In a warehouse, many pairs of vertical aeration ducts (the distance between two adjacent pairs is 2 to 3 m, and the total number of pairs depends on the length of the warehouse) are fixed on warehouse walls to conduct positive (pushing of air by fan) or negative (sucking of air by fan) aeration (Figure 10.5). Two ducts of each pair are facing each other so that the air travel distance is the shortest. When negative aeration is conducted, a PVC film can be used to cover the grain surface so that the air distribution in the grain bulk can be manipulated, and application of PVC film can minimize the aeration dead areas (Figure 10.5).

Grain in big bins or warehouses usually has a wide variation of temperatures. Aeration can be conducted to cool only the warmer grain if the air duct is introduced at the bottom of the warmer grain (in a process termed partial aeration). The introduced air can move upward, downward, or both, depending on the configuration of the system. This partial aeration is occasionally conducted by farmers and grain managers. Uneven grain temperature at the end of aeration should be avoided because uneven airflow distribution might result from this not well-designed aeration.

Air distribution in bins with a longer air travel distance is usually more uniform than that in bins with a shorter air travel distance because the airflow resistance produced by the grain will balance the airflow distribution. However, longer air travel distance produces higher airflow resistance, which results in more pressure drop. Therefore, air distribution in upright bins is more uniform than in flat storage for an upward airflow. This might be true for horizontal aeration also. Panigrahi et al. (2020) simulated the airflow distribution inside a round silo equipped with a V-shaped aeration floor. The simulation showed that 14.3% of the total grain volume exhibited a low airflow rate below the recommended value of 0.002 $m^3s^{-1}t^{-1}$ when upward aeration was conducted. The most significant portion with low airflow rates was the top grain volume above the eave height. Due to 85% of the nonperforated section on the silo floor, 12.6% of the lower grain volume up to a maximum height of 4.4 m above the floor exhibited a low airflow rate. To compensate for uneven airflow distribution in a bin, a higher airflow rate should be supplied to the bin.

10.4.2 Aeration Airflow Rate

Aeration flow rate should be selected based on the available aeration time and the minimum time required to complete an aeration cycle at a geographical location. The process by which a temperature front is completely moved through the entire grain bulk is an aeration cycle. The minimum time (h) is the time required to complete an aeration cycle. Airflow rate influences the minimum time, with a higher airflow rate resulting in a lower minimum time. Harner and Hagstrum (1990) found that an airflow rate of 1.7 $m^3min^{-1}t^{-1}$ (1.5 cfm/bu) would cool wheat by an average of 6°C with approximately 9 h of fan operation, which had the potential to reduce insect population growth by 80% to 97%. A longer minimum time might result in a larger temperature gradient in the grain bulk after the cycle has been completed. Temperature gradients exist in a bin if the aeration is completed with more than 10 h of fan operation (Harner and Hagstrum, 1990). The available aeration time is the time in which one or several cycles can be completed to maintain grain quality and seed vigor without infestation of insects, mites, and mould. The aeration flow rate can be as small as the available aeration time permits. The flow rate should be increased if the available aeration time is smaller than the minimum time needed.

A rule of thumb for aeration is that about 1,000 volumes of air are required to cool one volume of grain. Based on this rule, the minimum time can be estimated as

$$Q_t = \frac{30 + W_T}{500Q_r} \tag{10.1}$$

where W_T is the test weight of the grain (kg/m³) and Q_t and Q_r are the aeration time of an aeration cycle (h) and the airflow rate ($m^3min^{-1}t^{-1}$), respectively.

Equation 10.1 was developed by running the simulations using the mathematical model developed by Jian et al. (2019a). Lawrence and Maier (2011) estimated the aeration time of an aeration cycle (h) as 16.5 divided by the airflow rate in $m^3min^{-1}t^{-1}$. These estimated minimum times are consistent with measured values and have a variation (about 15 h) due to different grain types, initial grain temperature and moisture content, aeration system configuration, and variations in ambient air. Reducing the airflow rate beyond the recommended rate for aeration may cause the estimated time needed to cool grain satisfactorily exceeds practical requirements. Reed and Harner (1998) reported that at an airflow rate of 0.11 $m^3min^{-1}t^{-1}$, a minimum of 120 h was required to cool wheat from a high temperature to a desired threshold, whereas only 40 h was required with an airflow rate of 0.33 $m^3min^{-1}t^{-1}$; tripling the airflow rate required reduced the time to one-third. Harner and Hagstrum (1990) studied the effect of high airflow rates (>1.65 $m^3min^{-1}t^{-1}$) for aerating wheat during warm summer months, and they reported that wheat was cooled by an average of 6°C with approximately 9 h of fan operation. Harner and Hagstrum (1990) evaluated the airflow rate from 0.02 to 0.042 $m^3s^{-1}m^{-3}$ (1.7 to 3.6 $m^3min^{-1}t^{-1}$, or 1.8 to 3.1 cfm/bu) and concluded that a smaller than 10°C difference between average ambient air temperature and initial wheat temperature justified a cooling cycle, and no adverse effects on test weight or moisture content were noticed with the airflow rates used. If the airflow rate is doubled, it takes approximately half as long to cool the grain (Harner and Hagstrum, 1990), the same conclusion as had been arrived at by Reed and Harner (1998).

Airflow rates used for wheat aeration in the United States range from 0.05 to 0.31 $m^3min^{-1}t^{-1}$ (Arthur and Casada, 2016). When cereal grain has a lower than 12% moisture content, the airflow rate can be lower than the recommended rate. Reed and Arthur (2000) recommended 0.08 $m^3min^{-1}m^{-3}$ (0.1 $m^3min^{-1}t^{-1}$, or 0.1 cfm/bu) or less for stored wheat in the middle of the United States. Akdogan and Casada (2006) concluded that airflow rates higher than 0.1 $m^3min^{-1}t^{-1}$ were necessary to achieve sufficient grain cooling in the summer in Texas, Oklahoma, and southeastern Kansas. However, Montross et al. (2004) found that airflow rates of at least 0.6 $m^3min^{-1}t^{-1}$ (0.5 cfm/bu) would be more adequate to complete an aeration cycle for summer-harvested grain in southern regions of the United States, while during July and August only the northern half of the United States would have sufficient time available for cooling grain below 17°C using an airflow rate of 0.1 $m^3min^{-1} t^{-1}$ (0.1 cfm/bu).

Even though the recommended airflow rate used for aeration is 0.22 to 0.31 $m^3min^{-1}t^{-1}$ (0.10 to 0.25 cfm/bu) (Navarro et al., 2012), this range can be extended from 0.05 to 1 cfm/bu depending on crop type, minimum time required to complete an aeration cycle, and available aeration time.

Each type of grain has different storage requirements for temperature and moisture and thus requires a unique aeration protocol. Early-harvested crops should be cooled during relatively frequent cool nights soon after harvest. Crops harvested later may not need aeration or may require fewer cycles of aeration. Oil seeds generally require lower temperatures to prevent development of rancidity and spoilage. Therefore, the grain storage manager should consider all the factors outlined above in selecting the most suitable airflow rate, which might be lower or higher than the recommended rate for aeration.

10.4.3 Aeration Control

10.4.3.1 Methods of Aeration Control

Aeration fans can be controlled by manual, semiautomatic, or automatic controllers. These different control systems are all commercially available. Manual control requires only a single on-off switch. In Texas, farm owners manually activate a single on-off switch in mid-October until the grain is cooled to 5° to 10°C. Semiautomatic control is completed by using either a time clock or an electric eye for night and day identification. Systems with an electric eye are generally operated at night. Automatic controls use thermostat or humidistat control or a combination of the two. Thermostat control may be based on ambient temperature only or consider both grain temperature (average grain temperature or temperature in the top grain layer) and ambient temperature and their difference. Humidistat control may be based on the ambient RH only or consider both the ambient RH and the grain equilibrium RH and their difference. More complex controllers measure internal temperature at specified locations in the grain mass and ambient temperature outside the bin and are activated by comparing ambient and internal grain temperature. After the activation temperature and/or RH is set, fans will operate only when the ambient temperature and/or RH is within these set points, thereby cooling the grain through a progressive cooling front.

Automatic aeration is a low-cost management option that can cool stored grains more effectively than manual aeration. Temperature and RH thresholds can be specified for each cycle; the temperature thresholds are gradually decreased, with the last threshold temperature being the target grain temperature. In the United States, it is recommended that aeration be started immediately so that grain will be cooled to lower than 23.7°C during the summer cycle, followed by an early autumn cooling cycle to 15°C and then a late autumn cooling cycle to 7.2°C (Arthur and Casada, 2005). Reed and Harner (1998) set the temperature threshold of the thermostatically controlled aeration at 21°, 15°, and 7°C, respectively, for the three individual cycles.

Due to the weather RH effect, using wet-bulb temperature to control aeration (CWBT) was investigated by several researchers (Wilson and

Desmarchelier, 1994; Akdogan and Casada, 2006). Noyes and Navarro (2002) provided the details of the CWBT method. Wilson and Desmarchelier (1994) used seed wet-bulb temperature to identify available cooling aeration periods and found that the CWBT method had an advantage in warm climates with low wet-bulb temperatures at night that can be used to cool dry grain. During summer aeration, however, Akdogan and Casada (2006) found that grain cooling was highly influenced by humidity, and actual available aeration hours were approximately 78% less during the periods than was indicated by the calculations based on dry-bulb temperature alone because of the heat of condensation released during adsorption-reduced aeration cooling.

Since the aeration to cool grain lasts many days and several cooling fronts might be produced because of variation in ambient temperature, monitoring grain temperature during and after aeration is critical. The monitored grain temperature can be coded in the automatic control system. Once the aeration is completed, the fan can be automatically turned off. Sensor location should be carefully selected. Temperature cables near the center or near the edges of silos or at randomized locations do not represent the entire grain mass (Plumier and Maier, 2021); a location that has a high chance of insect and mold multiplication should be used. In the silos evaluated by Plumier and Maier (2021), the optimal location for aeration control and grain quality monitoring was about one-third to two-thirds of the distance between the core and the periphery of the grain mass. Arthur and Casada (2016) found that installing a thermostat for controlling aeration utilizing the headspace temperature as a set point, instead of relying on outside ambient temperature, might increase the efficiency of downward aeration.

10.4.3.2 Optimal Aeration Strategy

Brooker et al. (1974) reported that aeration cooling of grain could be started whenever the ambient air temperature was 5.6° to 8.3°C lower than the grain temperature. This recommendation has been modified in different geographical regions. In humid tropical areas, the unavailability of dry cold air demands more careful management than in other areas. Kanujoso et al. (1995) reported that using 80% RH air did not rewet grain as long as the air temperature was 5.6°C below the grain temperature. Using 10 years of weather data, Sinicio and Muir (1998) evaluated the aeration strategies for preventing spoilage of wheat stored in tropical and subtropical climates (Brazil). They reported that the best aeration conditions for grain at 13% initial moisture content were a differential thermostat setting between 5° and 7°C, a linear airflow rate between 0.06 and 0.18 $m^3min^{-1}m^{-3}$, and an air temperature increment between 1° and 5°C. Lopes and Neto (2019) projected that climate change would tend to reduce the cooling potential while increasing fan operation time and electrical energy requirements in tropical and subtropical climates (Brazil).

In hot subtropical summer, few or no fan run hours are available to cool grain. The maintenance aeration concept was adopted by Lawrence and Maier (2011). Maintenance aeration utilizes ambient air to maintain the grain temperature and moisture content in equilibrium with the average ambient conditions. That way, hotspot development can be prevented because the aeration can remove heat produced by molds, insects, and grain and prevent moisture migration and surface crusting. Lawrence and Maier (2011) simulated maintenance aeration by using 5-year weather data in Ambala, Haryana State, India, and compared 15 different aeration strategies at the aeration flow rate of 0.11, 0.34, and 0.67 $m^3min^{-1}t^{-1}$. They concluded that the best strategy was to operate fans 4 h during the morning and/or evening with an airflow of 0.11 $m^3min^{-1}t^{-1}$, with equilibrium moisture content (EMC) control but without temperature control during the maintenance period and with both EMC and temperature control during the cooling period.

In states in the southwest of the United States, such as Oklahoma, automatically controlled aeration has been used as an insect and mold management tool (Noyes et al., 1994). An extension fact sheet on the use of automatically controlled aeration for insect and mold management was developed in 1987 and revised in 1990. This fact sheet is widely used by farmers (Noyes et al., 1994). In the southern state of Arkansas, three bins filled with 18% rice were near ambient air dried to 14% and then were aerated at approximately 1.3 $m^3min^{-1}t^{-1}$ (1.2 cfm/bu). The aeration was controlled by a thermostatic controller to cool the rice to 24°, 15.5°, and 7.2°C in three cycles. A humidistat was used to prevent the rewetting of the rice. Both live insects and total emerged adult insects were significantly lower within the aerated bins (Ranalli et al., 2002). Arthur et al. (2008) compared the automatic aeration in Texas with manual aeration and found that total heat units at temperatures above 15°C were 150 to 300 degree days lower in the bins with automatic aeration. Temperatures from September 17 through the middle of October were 8° to 10°C less in bins with automatic aeration than in those with manual aeration and 3° to 6°C lower during the remainder of the year. Insect populations in the automatic aeration bins were significantly lower than in the manual aeration bins. In a region with a short available aeration time, Yang et al. (2017) reported that an increase in aeration airflow rate from 0.69 to 1.38 $m^3min^{-1}t^{-1}$ had a much greater impact in reducing insect populations than did a decrease in the aeration target temperature.

In the middle of the United States, in Kansas, Reed and Harner (1998) conducted a comparison study between automatically controlled aeration and manual aeration. They concluded that automated intermittent fan operation cooled the wheat one to two months faster than fans controlled manually according to standard recommendations and several months faster than natural cooling. Grain moisture was conserved using a three-cycle approach rather than a single aeration cycle late in the fall. They also found that thermostatically controlled aeration immediately after summer harvest could control insects. Arthur and Casada (2005) compared temperature profiles and insect populations in wheat that had been aerated at 0.11 m³min⁻¹t⁻¹ (0.1 cfm/bu) during the summer in addition to two autumn aeration cycles, wheat aerated in autumn only, and unaerated wheat in Manhattan, Kansas. They found distinct declines in temperature for each aeration cycle during the first two years of the study but not in the third year. At the end of the summer aeration cycle, the numbers of lesser grain borers, red flour beetles, and rice weevils were consistently higher in bins that had not been aerated during the summer. Casada et al. (2002) monitored wheat aerated in summer immediately after harvest in Kansas. They found that there were sufficient hours with air temperatures below 24°C to cool the grain with an airflow rate of 0.11 m³min⁻¹t⁻¹ (0.1 cfm/bu). However, high humidities during these nighttime periods of low temperatures in one year resulted in final temperatures higher than 24°C. Therefore, both temperature and humidity should be evaluated for adequate cooling potential when air temperatures are near the upper acceptable limit.

In temperate regions (e.g., Britain, the north of the United States, and Canada), there is enough available aeration time for the aeration to cool grain after harvest. Most farmers take advantage of this and manually start aeration after grain is binned. The aeration fan is not stopped during rainy days, which results in some rewetting of grain. So most farmers manually run the fan more than required and try to dry the rewetted grain. Armitage (1987) examined the effectiveness of manually controlled aeration versus automatically controlled aeration (using a thermostat) to control *O. surinamensis* and *S. granarius* in wheat during British winters and concluded that the automatically aerated bins were cooled more quickly, minimizing the time that the grain was exposed to the optimal temperatures for insect growth. Lasseran et al. (1994) concluded that aeration in temperate climates is the best technique for preserving grain whatever the quality criterion might be. Kaliyan et al. (2007) compared the following aeration strategies in north- and east-central regions of the United States: no aeration, continuous aeration, and intermittent aeration at 0.11 m³min⁻¹t⁻¹ (0.1 cfm/bu) downward. They found that a fan control strategy that involved turning the aeration fan on when the grain temperature at 0.4 m depth was greater than the head-space air temperature was the best strategy for managing *P. interpunctella* larvae. Metzger and Muir (1983) had a similar recommendation for western Canadian farmers.

10.4.4 Mathematical Modeling and Simulation of Aeration

Automatic aeration control should include a fan control strategy to accommodate local weather conditions. This strategy should be developed by using a verified mathematical model and local weather data to guarantees sufficient fan operation to achieve the desired control objective. Ten years of historic weather data are usually a minimum for evaluation due to variation of weather during different years, and 20 to 30 years of data are recommended (Arthur et al., 1998; Arthur and Casada, 2005).

Many mathematical models for simulating natural air drying and aeration have been developed and reviewed by several research groups (Sharp, 1982; Sokhansanj and Bruce, 1987; Jayas et al., 1991; Cenkowski et al., 1993; Maier and Montross, 1998; Parde et al., 2003; Jian et al., 2019a). These models are classified as nonequilibrium (partial differential equations), equilibrium, or logarithmic. A logarithmic model is rarely used due to its low prediction accuracy (Lopes et al., 2006; Jian et al., 2019a). Partial differential models of aeration were developed by Thorpe (2001) and later modified by Lopes et al. (2006). However, these models were not verified and are not widely used because their computer calculation times are longer than those of equilibrium models. For three reasons, widely used aeration models are the equilibrium and near equilibrium models. First, the assumption of equilibrium conditions between grain and air has been shown to provide a reasonable prediction of moisture content in a deep bed (bin) for aeration (Jian et al., 2019a). Second, equilibrium models are a simplification of partial differential models. Third, the equilibrium or near equilibrium model does not need a drying rate equation for a product, and it is easy to develop a three-dimensional model. Therefore, the equilibrium model uses significantly less computer time than partial differential models. Different groups developed equilibrium models to simulate aeration and natural air drying (Thompson et al., 1968; Sun and Woods, 1997; Jian et al., 2019a). The most often used equilibrium model was developed by Thompson et al. (1968) and further modified by Bowden et al. (1983) and Morey et al. (1979). A weakness of these developed models is their inadequacy at making realistic predictions near the bottom of bins (Morey et al., 1979; Bowden et al., 1983). Jian et al. (2019a) further modified the equilibrium model by considering the desorption, sorption, and hysteresis of grain during aeration and natural air drying. They claim that the prediction accuracy of their model is higher than that of the previously published equilibrium models.

To simulate aeration effect on the management of insects and microorganism, models of insect and microorganism development are needed. Mathematical models to simulate population dynamics of the following insects have been developed: *P. interpunctella* (Throne and Arbogast, 2010), *S. oryzae* (Hardman, 1978; Cuff and Hardman, 1980; Thorpe et al.,

1982; Driscoll et al., 2000; Yang et al., 2017; Bareil et al., 2018), *R. dominica* (Driscoll et al., 2000; Yang et al., 2017; Bareil et al., 2018), *S. granarius, C. ferrugineus, O. surinamensis* (Driscoll et al., 2000; Bareil et al., 2018), *T. castaneum* (Driscoll et al., 2000), *T. confusum* (Bareil et al., 2018), *Ophiomyia simplex* (Morrison III et al., 2014), and *T. granarium* (Rajendran et al., 2022). Several models have been developed to predict grain germination of canola (Jian et al., 2014b), hemp seeds (Jian et al., 2019b), and wheat at 19% moisture content (Karunakaran et al., 2001). Safe storage times for many crop seeds have been measured, and charts of their safe storage times based on germination drop have been developed. However, these safe storage charts and germination models have not been used in aeration simulations.

Population models of the following insect species have been combined with aeration models to simulate the effect of aeration: *S. oryzae* (Hardman, 1978; Thorpe et al., 1982; Driscoll et al., 2000; Yang et al., 2017; Bareil et al., 2018), *R. dominica* (Driscoll et al., 2000; Yang et al., 2017; Bareil et al., 2018), *S. granarius, C. ferrugineus, O. surinamensis* (Driscoll et al., 2000; Bareil et al., 2018), *T. castaneum* (Bareil et al., 2018), and *T. confusum* (Bareil et al., 2018). The most often simulated insects are *S. oryzae* (Hardman, 1978; Thorpe et al., 1982; Driscoll et al., 2000; Yang et al., 2017; Bareil et al., 2018) and *R. dominica* (Driscoll et al., 2000; Yang et al., 2017; Bareil et al., 2018). A web-based expert system that integrates a bin cooling model with population models for *R. dominica* and *S. oryzae* has been developed and used (Arthur et al., 2011).

10.5 Future Research

Aeration cooling has not generally been considered as a process that produces high levels of insect control quickly, easily, and reliably. However, responses of stored-product insect and mite species to the right strategy of aeration control can be substantial and catastrophic for some species, especially in grain bulk when the insects have not had enough time to acclimate to low temperatures. The combined effect of the factors low temperature and low RH is also substantial because this environment can not only minimize multiplication of insects and mites but also limit the multiplication of microflora and maintain grain quality. However, there is no comprehensive study on all these effects under different airflow rates and different aeration strategies.

Mathematical models for a few insect species have been developed; the model of *S. oryzae* has been developed by six research groups. Mathematical models of population dynamics for each insect species should be developed. To develop a model of a stored grain ecosystem, a mathematical model should be created to simulate various bio-factors including multiple insect species, fungi, and mites under optimal conditions for their growth but also realistic conditions expected to occur in storage. Such insect development models could be used to formulate a practical strategy of aeration under different climatic conditions.

Guidelines have been developed on safe storage times for many crop seeds: barley (Jacobsen and Fleurat-Lessard, 2002), canola (Sathya et al., 2009; Sun et al., 2014), corn (Williams et al., 2017), durum wheat (Nithya et al., 2011), pinto bean (Rani et al., 2013), rye (Sathya et al., 2008), soybean (Maciel et al., 2020), and wheat (Wilkins, 1983). These guidelines are based on 5 to 20 percentage point drops in germination from its initial level. They are mostly used to predict the grain quality under storage conditions including the aeration condition. However, these guidelines are for the most part presented in chart format. Jian (2022) developed a general model to predict germination and safe storage time of crop seeds, including cereal grain (durum wheat, wheat, and rye) and oilseed (canola and hemp seeds). More mathematical models of safe storage time are needed for the development of an optimal aeration strategy.

References

Abdelghany AY, Fields PG (2017) Mortality and movement of *Cryptolestes ferrugineus* and *Rhyzopertha dominica* in response to cooling in 300-kg grain bulks. J Stored Prod Res 71: 119–124.

Abdelghany AY, Suthisut D, Fields PG (2015) The effect of diapause and cold acclimation on the cold-hardiness of the warehouse beetle, *Trogoderma variabile* (Coleoptera: Dermestidae). Can Entom 147: 158–168.

Akdogan H, Casada ME (2006) Climatic humidity effects on controlled summer aeration in the hard red winter wheat belt. Transac of the ASAB E49: 1077–1087.

Armitage DM (1987) Controlling insects by cooling grain. British Columbia Provi Commi 37: 218–227.

Armitage DM, Burrell NJ (1978) The use of aeration spears for cooling infested grain. J Stored Prod Res 14: 223–226.

Arthur FH, Casada ME (2016) Temperature stratification and insect pest populations in stored wheat with suction versus pressure aeration. Appl Eng Agric 32: 849–860.

Arthur FH, Flinn PW (2000) Aeration management for stored hard red winter wheat: Simulated impact on rusty grain beetle (Coleoptera: Cucujidae) populations. J Econ Ento 93: 1364–1372.

Arthur FH, Casada ME (2005) Feasibility of summer aeration to control insects in stored wheat. Appl Eng Agric 21: 1027–1038.

Arthur FH, Casada ME (2010) Directional flow of summer aeration to manage insect pests in stored wheat. Appl Eng Agric 26: 115–122.

Arthur FH, Hartzer KL, Throne JE, Flinn PW (2017) Freezing for control of stored-product psocids. J Stored Prod Res 72: 166–172.

Arthur FH, Throne JE, Maier DE, Montross MD (1998) Feasibility of aeration for management of maize weevil populations in corn stored in the Southern United States: Model simulations based on recorded weather data. American Entom 44: 118–123.

Arthur FH, Yang Y, Wilson LT (2011) Use of a web-based model for aeration management in stored rough rice. J Econ Ento 104: 702–708.

Arthur FH, Yang Y, Wilson LT, Siebenmorgen TJ (2008) Feasibility of automatic aeration for insect pest management for rice stored in east Texas. Appl Eng Agric 24: 345–350.

Banks H, Fields PG (1994) Physical methods for insect control in stored-grain ecosystems. In: Jayas D, White N, Muir W, Sinha R (eds.) Stored-Grain Ecosystems, Marcel Dekker, Inc., New York.

Bareil N, Crepon K, Piraux F (2018) Prediction of insect mortality in cooled stored grain. J Stored Prod Res 78: 110–117.

Bartlett D, Armitage DM, Harral B (2002) Optimising the performance of vertical aeration systems. pp. 946–955. In: Credland PF, Armitage DM, Bell CH, Cogan PM, Highley E (eds.), Proc 8th Intern Working Conf Stored-Prod Prot, York, UK.

Beckett SJ (2010) Protecting and disinfesting stored products by drying and cooling, and disinfesting stored products during handling by mechanical treatments. pp. 219–228. In: Carvalho MO, Fields PG, Adler CS, Arthur FH, Athanassiou CG, Campbell JF, Fleurat-Lessard F, Flinn PW, Hodges RJ, Isikber AA, Navarro S, Noyes RT, Riudavets J, Sinha KK, Thorpe GR, Timlick BH, Trematerra P, White NDG (eds.), Proc 10th Intern Working Conf Stored-Prod Prot, Estoril, Portugal.

Beckett SJ (2011) Insect and mite control by manipulating temperature and moisture before and during chemical-free storage. J Stored Prod Res 47: 284–292.

Bharathi VSK, Jian F, Jayas DS (2023) Study on 300 t of wheat stored in corrugated steel bin for two years in Canada. Part I – Temperature and moisture profiles of the grain. J Stored Prod Res 100: 102057.

Birch LC (1953) Experimental background to the study of the distribution and abundance of insects. Ecology 34: 698–711.

Bishop GW (1959) The comparative bionomics of American *Cryptolestes* (Coleoptera-Cucujidae) that infest stored grain. Anna Entom Soci Amer 52: 657–665.

Bowden PJ, Lamond WJ, Smith EA (1983) Simulation of near-ambient grain drying: I. comparison of simulations with experimental results. J Agric Eng Res 28: 279–300.

Bridges TC, Montross MD, McNeill SG (2005) Aeration strategies and fan cost comparisons for wheat in mid-south production regions. Appl Eng Agric 21: 115–124.

Brooker DB, Bakker-Arkema FW, Hall CW (1974) Drying Cereal Grains, AVI Publishing Company, Westport, Conn.

Burks CS, Johnson JA (2012) Biology, behavior and ecology of stored fruit and nut insects. In: Hagstrum DW, Phillips TW, Cuperus G (eds.) Stored Product Protection. Kansas State Research and Extension, Kansas.

Burrell NJ, Laundon JHJ (1967) Grain cooling studies–I: Observations during a large scale refrigeration test on damp grain. J Stored Prod Res 3: 125–126, IN123, 127–144.

Calderwood DL, Cogburn RR, Webb BD, Marchetti MA (1984) Aeration of rough rice in long-term storage. Transac of the ASABE 27: 1579–1585.

Casada ME, Arthur FH, Akdogan H (2002) Temperature monitoring and aeration strategies for stored wheat in the central plains. ASAE Annual International Meeting, Chicago, Illinois, USA.

Cenkowski S, Jayas DS, Pabis S (1993) Deep-bed grain drying - a review of particular theories. Drying Tec 11: 1553–1582.

Childs DP, Overby JE, Watkins PJ, Niffenegger D (1970) Low temperature effect upon third-and fourth-instar cigarette beetle larvae. J Econ Ento 63: 1860–1864.

Cline DL (1970) Indian meal moth egg hatch and subsequent larval survival after short exposures to low temperature. J Econ Ento 63: 1081–1082.

Collins DA (2012) A review on the factors affecting mite growth in stored grain commodities. Exp Appl Acar 56: 191–208.

Cuff WR, Hardman JM (1980) A development of the Leslie matrix formulation for restructuring and extending an ecosystem model: The infestation of stored wheat by *Sitophilus oryzae*. Eco Model 9: 281–305.

Cunnington AM (1984) Resistance of the grain mite *Acarus siro* L. (Acarina, Acaridae) to unfavourable physical conditions beyond the limits of its development. Agric Ecosys Envir 11: 319–339.

Cuperus GW, Prickett CK, Bloome PD, Pitts JT (1986) Insect populations in aerated and unaerated stored wheat in Oklahoma. J Kansas Entom Soci 59: 620–627.

Daumal J, Jourdheuil P, Tomassone R (1974) Variaibite des effects letaux des basses temperatures en fonction du stade de developpement embryonnaire chez la pyrale de la farine (*Anagasta kuhniella* Zell., Lepid., Pyralidae). Annales de Zoologie/Ecologie Animal 6: 229–243.

David MH, Mills RB, White NDG (1977) Effects of low temperature acclimation on developmental stages of stored-product insects. Envir Entom 6: 181–184.

Dermott T, Evans DE (1978) An evaluation of fluidized-bed heating as a means of disinfesting wheat. J Stored Prod Res 14: 1–12.

Dong P, Wang JJ, Jia FX, Hu F (2007) Development and reproduction of the psocid *Liposcelis tricolor* (Psocoptera: Liposcelididae) as a function of temperature. Anna Entom Soci Amer 100: 228–235.

Driscoll R, Longstaff BC, Beckett S (2000) Prediction of insect populations in grain storage. J Stored Prod Res 36: 131–151.

Evans DE (1983) The influence of relative humidity and thermal acclimation on the survival of adult grain beetles in cooled grain. J Stored Prod Res 19: 173–180.

Evans DE (1987) The survival of immature grain beetles at low temperatures. J Stored Prod Res 23: 79–83.

Evans DE, Thorpe GR, Sutherland JW (1983) Large scale evaluation of fluid-bed heating as a means of disinfesting grain. pp. 523–530. In: Mills RB, Wright VF, Pedersen JR, McGaughey WH, Beeman RW, Kramer KJ, Speirs RD, Storey CL (eds.), Proc 3rd Intern Working Conf Stored-Prod Ent, Manhattan, USA.

Fawki S, Fields PG, Jian F, Yousery A (2022) Control of *Sitophilus oryzae* (Coleoptera: Curculionidae) in bags of wheat using solar radiation. J Stored Prod Res 96: 101941.

Fields PG (1992) The control of stored-product insects and mites with extreme temperatures. J Stored Prod Res 28: 89–118.

Fields PG, Abdelghany A (2014) Low temperature to control insects in grain bins and flour mills. p. 842. In: Arthur FH, Kengkanpanich R, Chayaprasert W, Suthisut D (eds.), Proc 11th Intern Working Conf Stored-Prod Prot, Chiang Mai, Thailand.

Fields PG, White NDG (1997) Survival and multiplication of stored-product beetles at simulated and actual winter temperatures. Can. Entom 129: 887–898.

Foster GH (1951) Mechanical ventilation of stored grain. Agric Eng 32: 606–608.

Foster GW (1953) Minimum airflow requirements for drying grain with unheated air. Agric Eng 34: 681–684.

Frisvad JC (1995) Mycotoxins and mycotoxigenic fungi in storage. In: Jayas DS, White NDG, Muir WE (eds), Stored-Grain Ecosystems, Marcel Dekker, Inc, NY, Basel, Hong Kong, pp. 251–288.

Ganesan L, Fields PG, Jayas DS, Jian F (2021) Effects of developmental stage, cold acclimation and diet on the cold tolerance of three species of *Cryptolestes* (Coleoptera: Laemophloeidae). J Stored Prod Res 91: 101773.

Gardner RD, Harein PK, Subramanyam B (1988) Management of stored barley in Minnesota: practices versus recommendations. Am Entomol 34: 22–26.

Ghaly TF (1984) Aeration trial of farm-stored wheat for the control of insect infestation and quality loss. J Stored Prod Res 20: 125–131.

Gough MC, Uiso CBS, Stigter CJ (1990) Air convection current in metal silos storing maize grain. Tropi Sci 30: 217–222.

Hagstrum DW, Milliken GA (1988) Quantitative analysis of temperature, moisture, and diet factors affecting insect development. Anna Entom Soci Amer 81: 539–546.

Hansen JD, Johnson JA, Winter DA (2011) History and use of heat in pest control: A review. Int J Pest Management 57: 267–289.

Hardman JM (1978) A logistic model simulating environmental changes associated with the growth of populations of rice weevils, *Sitophilus oryzae*, reared in small cells of wheat. J Appl Eco 15: 65–87.

Harner JP, Hagstrum DW (1990) Utilizing high airflow rates for aerating wheat. Appl Eng Agric 6: 315–321.

Harrington JF (1973) Biochemical basis of seed longevity. Seed Sci Tech 1: 453–461.

Hassan MW, Dou W, Chen L, Jiang H, Wang J (2011) Development, survival, and reproduction of the psocid *liposcelis yunnaniensis* (Psocoptera: Liposcelididae) at constant temperatures. J Econ Ento 104: 1436–1444.

Howe RW (1953) Studies on beetles of the family *Ptinidae*. VIII. the intrinsic rate of increase of some *Ptinid* beetles. Ann Appl Bio 40: 121–133.

Howe RW (1955) Studies on beetles of the family *Ptinidae*. XII. The biology of *Tipnus unicolor* Pill. and Mitt. Entomol's Mon Mag 91: 253–257.

Howe RW (1956a) The biology of the two common storage species of *Oryzaephilus* (Coleoptera, Cucujidae). Ann Appl Bio 44: 341–355.

Howe RW (1956b) The effect of temperature and humidity on the rate of development and mortality of *Tribolium castaneum* (Herbst) (Coleoptera: Tenebrionidae). Ann Appl Bio 44: 356–368.

Howe RW (1956c) Studies on beetles of the family *Ptinidae*. XIV. The biology of *Ptinus pusillus* Sturm. Entomol's Mon Mag XCII: 331–333.

Howe RW (1957) A Laboratory study of the cigarette beetle, *Lasioderma serricorne* (F.) (Col., Anobiidae) with a critical review of the literature on its biology. Bull Ent Res 48: 9–56.

Howe RW (1959) Studies on beetles of the family *Ptinidae*. XVII. conclusions and additional remarks. Bull Ent Res 50: 287–326.

Howe RW (1960) The effects of temperature and humidity on the rate of development and mortality of *Tribolium confusum* Duval (Coleoptera: Tenebrionidae). Ann Appl Bio 48: 363–376.

Howe RW (1962a) The effect of temperature and relative humidity on the rate of development and the mortality of *Tribolium madens* (Charp.) (Coleoptera, Tenebrionidae). Ann Appl Bio 50: 649–660.

Howe RW (1962b) The effects of temperature and humidity on the oviposition rate of *Tribolium castaneum* (Hbst.) (Colemptera: Tenebrionidae). Bull Ent Res 53: 301–310.

Howe RW (1965) A summary of estimates of optimal and minimal conditions for population increase of some stored products insects. J Stored Prod Res 1: 177–184.

Howe RW, Currie JE (1964) Some laboratory observations on the rates of development, mortality and oviposition of several species of Bruchidae breeding in stored pulses. Bull Ent Res 55: 437–477.

Hukill WV (1953) Grain cooling by air. Agric Eng 34: 456–458.

Hunter AJ, Taylor PA (1980) Refrigerated aeration for the preservation of bulk grain. J Stored Prod Res 16: 123–131.

Imai T, Miyamoto Y (2019) The developmental parameters of the minute brown scavenger beetle *Dienerella argus* (Coleoptera: Latridiidae). App Ent Zool 54: 75–78.

Imura O (1990) Thermal requirements for development of stored-product insects. Tribolium Infor Bull 30: 58–68.

Jacob TA, Fleming DA (1986) The effect of temporary exposure to low temperature on the viability of eggs of *Oryzaephilus surinamensis* (L.) (Col., Silvanidae). Entomol's Mon Mag 122: 117–119.

Jacobsen EE, Fleurat-Lessard F (2002) Estimation of safe storage periods for malting barley using a model of heat production based on respiration experiments. pp. 456–463. In: Credland PF, Armitage DM, Bell CH, Cogan PM, Highley E (eds.), Proc 8th Intern Working Conf Stored-Prod Prot, York, UK.

Jarosik V, Honek A, Magarey RD, Skuhrovec J (2011) Developmental database for phenology models: Related insect and mite species have similar thermal requirements. J Econ Ento 104: 1870–1876.

Jayas DS, Cenkowski S, Pabis S (1991) Review of thin-layer drying and wetting equations. Drying Tec 9: 551–588.

Jayas DS, Muir WE (2002) Aeration systems design. In: Navarro S, Noyes R (eds), The Mechanics and Physics of Modern Grain Aeration Management, CRC Press, Boca Raton London New York Washington, D. C, pp. 195–249.

Jayas DS, Sokhansanj S (1989) Design data on resistance of airflow through canola (rapeseed). Transac of the ASAE 32: 295–296.

Jayas DS, White NDG, Muir WE (1995) Stored-Grain Ecosystems, Marcel Dekker Inc., New York.

Jian F (2022) A general model to predict germination and safe storage time of crop seeds. J Stored Prod Res 99: 102041.

Jian F, Jayas DS (2014) Understanding the initiation and development of hotspots in storage-grain ecosystems. J Appl Zool Res 25: 1–10.

Jian F, Jayas DS (2022) Grain: Engineering Fundamentals of Drying and Storage, CRC Press, Boca Raton, London, New York.

Jian F, Jayas DS, White NDG (2009) Temperature fluctuations and moisture migration in wheat stored for 15 months in a metal silo in Canada. J Stored Prod Res 45: 82–90.

Jian F, Jayas DS, White NDG (2014a) Heat production of stored canola seeds under airtight and non-airtight conditions. Transac of the ASABE 57: 1151–1162.

Jian F, Liu J, Jayas DS (2019a) A new mathematical model to simulate sorption, desorption and hysteresis of stored canola during aeration. Drying Tec 38: 2190–2201.

Jian F, Mamun MAA, White NDG, Jayas DS, Fields PG, McCombe J (2019b) Safe storage times of FINOLA® hemp (*Cannabis sativa*) seeds with dockage. J Stored Prod Res 83: 34–43.

Jian F, Sun K, Vellaichamy C, Jayas DS, White NDG (2014b) Quality changes in high and low oil content canola during storage: Part II: Mathematical models to predict germination. J Stored Prod Res 59: 328–337.

Jiang HB, Liu JC, Wang ZY, Wang JJ (2008) Temperature-dependent development and reproduction of a novel stored product psocid, *Liposcelis badia* (Psocoptera: Liposcelididae). Envir Entom 37: 1105–1112.

Johnson JA, Valero KA, Hannel MM (1997) Effect of low temperature storage on survival and reproduction of Indianmeal moth (Lepidoptera: Pyralidae). Crop Prot 16: 519–523.

Kaliyan N, Morey RV, Wilcke WF, Carrillo MA, Cannon CA (2007) Low-temperature aeration to control Indianmeal moth, *Plodia interpunctella* (Hubner), in stored grain in twelve locations in the United States: A simulation study. J Stored Prod Res 43: 177–192.

Kanujoso B, Chung DS, Song A, Erickson LE (1995) Study of desorption and adsorption during grain aeration. Drying Tec 13: 183–196.

Karunakaran C, Muir WE, Jayas DS, White NDG, Abramson D (2001) Safe storage time of high moisture wheat. J Stored Prod Res 37: 303–312.

Kiritani K (2012) The low development threshold temperature and the thermal constant in insects and mites in Japan. Bull Nation Insti Agro-Envir Sci 31: 1–74.

Kiritani KT (1997) The low development threshold temperature and the thermal constant in insects, mites and nematodes in Japan. Miscell Pub Nation Insti Agro-Envir Sci 21: 1–72.

Lacey J, Hill ST, Edwards MA (1980) Micro-organisms in stored grains: Their enumeration and significance. Trop Stored Prod Inf 39: 19–32.

Lasseran JC, Fleurat-Lessard F (1990) Aeration of grain with ambient or artificially cooled air: A technique to control weevils in temperate climates. pp. 1221–1232. In: Fleurat-Lessard F, Ducom P (eds.), Proc 5th Intern Working Conf Stored-Prod Prot, Bordeaux, France.

Lasseran JC, Niquet G, FleuratLessard F (1994) Quality enhancement of grain by improved design and management of aeration. pp. 296–299. In: Highley E, Wright EJ, Banks HJ, Champ BR (eds.), Proc 6th Intern Working Conf Stored-Prod Prot, Canberra, Australia.

Lawrence J, Maier DE (2011) Aeration strategy simulations for wheat storage in the sub-tropical region of north India. Transac of the ASABE 54: 1395–1405.

LeTorc'h JM (1977) Le froid: Moyen de protection contre les ravageurs des denrees stockees. Essais de laboratoire sur les insectes des pruneaux. Revue de Zoologie Agricole et de Pathologie Vegetale 76: 109–117.

Lefkovitch LP (1962) The biology of *Cryptolestes turcicus* (Grouvelle) (Coleoptera: Cucujidae), a pest of stored and processed cereals. Proc Zool Soc Lond 138: 23–35.

Li C, Li Z (2009) Influence of temperature on development and reproduction of experimental populations of *Araecerus fasciculatus* (Coleoptera: Anthribidae). Acta Entom Sin 52: 1385–1389.

Lindgren DL, Vincent LE (1959) Biology and Control of *Trogoderma granarium* Everts. J Econ Ento 52: 312–319.

Liu C, Zhang Y, Li H, Li L, Zheng X (2020) Effect of ozone treatment on processing properties of wheat bran and shelf life characteristics of noodles fortified with wheat bran. J Food Sci Tech 57: 3893–3902.

Lopes DC, Neto AJS (2019) Effects of climate change on the aeration of stored beans in Minas Gerais State, Brazil. Biosys Eng 188: 155–164.

Lopes DDC, Martins JH, Melo EDC, Monteiro PMDB (2006) Aeration simulation of stored grain under variable air ambient conditions. Postharvest Bio Tech 42: 115–120.

Loschiavo SR (1985) Post-harvest grain temperature, moisture, and insect infestation in steel granaries in Manitoba. Can. Entom 117: 7–14.

Maciel G, Torre DDI, Cardoso LM, Cendoya MG, Wagner JR (2020) Determination of safe storage moisture content of soybean expeller by means of sorption isotherms and product respiration. J Stored Prod Res 86: 1–7.

Magan N, Lacey J (1984a) The effects of gas composition and water activity on growth of field and storage fungi and their interactions. Trans Br Mycol Soc 82: 305–314.

Magan N, Lacey J (1984b) Effects of temperature and pH on water relations of field and storage fungi. Trans Br Mycol Soc 82: 71–81.

Magan N, Lacey J (1988) Ecological determinants of mould growth in stored grain. Int J Food Micr 7: 245–256.

Maier DE (1994) Chilled aeration and storage of U.S. crops -a review. pp. 300–311. In: Highley E, Wright EJ, Banks HJ, Champ BR (eds.), Proc 6th Intern Working Conf Stored-Prod Prot, Canberra, Australia.

Maier DE, Montross MD (1998) Modeling aeration and storage managemnet strategies. pp. 1279–1300. In: Zuxun J, Quan L, Yongsheng L, Xianchang T, Lianghua G (eds.), Proc 7th Intern Working Conf Stored-Prod Prot, Beijing, China.

Maier DE, Moreira RG, Bakker-Arkema FW (1992) Comparison of conventional and chilled aeration of grains under Texas conditions. Appl Eng Agric 8: 661–667.

Marín S, Sanchis V, Sáenz R, Ramos AJ, Vinas I, Magan N (1998) Ecological determinants for germination and growth of some *Aspergillus* and *Penicillium* spp. from maize grain. J Appl Microbio 84: 25–36.

Marin S, Sanchis V, Teixido A, Saenz R, Ramos AJ, Vinas I, Magan N (1996) Water and temperature relations and microconidial germination of *Fusarium moniliforme* and *F. proliferatum* from maize. Can J Microbio 42: 1045–1050.

Mathlein R (1961) Experiments with fresh-air treatment for the control of grain storage pests. National Insti Plant Prot Contribution 12: 99–125.

Metzger JF, Muir WE (1983) Aeration of stored wheat in the Canadian Prairies. Can Agric Eng 25(1): 127–137.

Montross MD, Maier DE, Haghighi K (2002) Development of a finite-element stored grain ecosystem model. Transac of the ASAE 45: 1455–1464.

Montross MD, McNeill SG, Bridges TC (2004) Seasonal aeration rates for the Eastern United States based on long-term weather patterns. Appl Eng Agric 20: 665–669.

Morey RV, Cloud HA, Gustafson RJ, Petersen DW (1979) Evaluation of the feasibility of solar energy grain drying. Transac of the ASAE 22: 409–417.

Morrison III WR, Andresen J, Szendrei Z (2014) The development of the asparagus miner (*Ophiomyia simplex* Loew; Diptera: Agromyzidae) in temperate zones: A degree-day model. Pest Management Sci 70: 1105–1113.

Muir WE, Fraser BM, Sinha RN, Shejbal J (1980) Simulation model of two-dimensional heat transfer in controlled-atmosphere grain bins. Controlled Atmosphere Storage of Grains, Elsevier Scientific Plub. Co, Amsterdam, pp. 385–398.

Mullen MA, Arbogast RT (1979) Time-temperature-mortality relationship for various stored-product insect eggs and chilling times for selected commodities. J Econ Ento 72: 476–478.

Nagel RH, Shepard HH (1934) The lethal effect of low temperatures on the various stages of the confused flour beetle. J Agric Res 48: 1009–1016.

Nakakita H, Ikenaga H (1997) Action of low temperature on physiology of *Sitophilus zeamais* motschulsky and *Sitophilus oryzae* (L) (Coleoptera: Curculionidae) in rice storage. J Stored Prod Res 33: 31–38.

Navarro S, Donahaye E, Calderon M (1969) Observations on prolonged grain storage with forced aeration in Israel. J Stored Prod Res 5: 73–81.

Navarro S, Donahaye E, Calderon M (1973a) Studies on aeration with refrigerated air–II. chilling of soybeans undergoing spontaneous heating. J Stored Prod Res 9: 261–268.

Navarro S, Donahaye E, Calderon M (1973b) Studies on aeration with refrigerated air–I. chilling of wheat in a concrete elevator. J Stored Prod Res 9: 253–259.

Navarro S, Noyes R (2002) The Mechanics and Physics of Modern Grain Aeration Management, CRC Press, Boca Roton, London, New York, Washington, DC.

Navarro S, Noyes R, Armitage D, Maier DE (2002) Objectives of aeration. In: Navarro S, Noyes R (eds), The Mechanics and Physics of Modern Grain Aeration Management, CRC Press, Boca Raton, London, New York, Washinton D.C, pp. 1–34.

Navarro S, Noyes RT, Casada M, Arthur FH (2012) Grain aeration. In Hagstrum DW, Phillips TW, Cuperus G (eds.) Stored Product Protection. Kansas State Research Extension, Kansas.

Nithya U, Chelladurai V, Jayas DS, White NDG (2011) Safe storage guidelines for durum wheat. J Stored Prod Res 47: 328–333.

Northolt MD, Bullerman LB (1982) Prevention of mold growth and toxin production through control of environmental conditions. J Food Prot 45: 519–526.

Northolt MD, Van EHP, Paulsch WE (1979) Ochratoxin a production by some fungal species in relation to water activity and temperature. J Food Prot 42: 485–490.

Noyes R, Cuperus GW, Kenkel P (1994) Using controlled aeration for insect and mould management in the south-western United States. pp. 323–334. In: Highley E, Wright EJ, Banks HJ, Champ BR (eds.), Proc 6th Intern Working Conf Stored-Prod Prot, Canberra, Australia.

Noyes R, Navarro S (2002) Operating aeration systems. In: Navarro S, Noyes R (eds), The Mechanics and Physics of Modern Grain Aeration Management. CRC Press, Boca Raton, London, New York, Washinton, D.C, pp. 315–412.

Panigrahi SS, Singh CB, Fielke J (2020) CFD modelling of physical velocity and anisotropic resistance components in a peaked stored grain with aeration ducting systems. Comp Elect Agric 179: 1–10.

Parde SR, Jayas DS, White NDG (2003) Grain drying: A review. Sciences Des Aliments: 23(5): 589–622.

Partida GJ, Strong RG (1975) Comparative studies on the biologies of six species of Trogoderma: *T. variabile*. Anna Entom Soci Amer 68: 115–125.

Phillips GL (1957) Experiments on distributing methyl bromide in bulk grains with aeration systems. In: Agricultural Marketing Service, United States Department of Agricultute, Manhattan, Kans.

Plumier B, Maier DE (2021) Effect of temperature sensor numbers and placement on aeration cooling of a stored grain mass using a 3d finite element model. Agriculture 11: 1–14. doi:10.3390/agriculture 11030231

Prasad DC, Muir WE, Wallace HAH (1978) Characteristics of freshly harvested wheat and rapeseed. Transac of the ASAE 21: 782–784.

Rajendran V, Jian F, Fields PG, Jayas DS (2022) Mathematical modeling of population dynamics of *Trogoderma granarium* (Coleoptera: Dermistidae). J Econ Ento 115: 353–364.

Ramos AJ, Labernia N, MarÃ-n S, Sanchis V, Magan N (1998) Effect of water activity and temperature on growth and ochratoxin production by three strains of *Aspergillus ochraceus* on a barley extract medium and on barley grains. Int J Food Microbio 44: 133–140.

Ranalli RP, Howell TA Jr, Arthur FH, Gardisser DR (2002) Controlled ambient aeration during rice storage for temperature and insect control. Appl Eng Agric 18: 485–490.

Rani PR, Chelladurai V, Jayas DS, White NDG, Kavitha-Abirami CV (2013) Storage studies on pinto beans under different moisture contents and temperature regimes. J Stored Prod Res 52: 78–85.

Reed C, Arthur FH (2000) Aeration. In: Subramanyam B, Hagstrum BW (eds), Alternatives to Pesticides in Stored-Product IPM, Kluwer Academic Publishers, Boston, Massachusetts, USA, pp. 51–72.

Reed C, Arthur FH, Trigo-Stockli D (1998) Conditioning practices and their effects on infestation and quality of corn stored on Kansas farms. Appl Eng Agric 14: 623–630.

Reed C, Harner J (1998) Thermostatically controlled aeration for insect control in stored hard red winter wheat. Appl Eng Agric 14: 501–505.

Renault D, Salin C, Vannier G, Vernon P (1999) Survival and chill-coma in the adult lesser mealworm, *Alphitobius diaperinus* (Coleoptera: Tenebrionidae), exposed to low temperatures. J Thermal Bio 24: 229–236.

Renault D, Hance T, Vannier G, Vernon P (2003) Is body size an influential parameter in determining the duration of survival at low temperatures in *Alphitobius Diaperinus* Panzer (Coleoptera: Tenebrionidae)? J Zool 259: 381–388.

Robinson RN, Hukill W V, Foster GH (1951) Mechanical ventilation of stored grain. Agric Eng 32: 606–608.

Sanchis V, Magan NE (2004) Environmental conditions affecting mycotoxins. In: Magan N, Olsen M (eds), Mycotoxins in Food: Detection and Control, Woodhead Publishing Ltd, Cambridge, pp. 174–189.

Sathya G, Jayas DS, White NDG (2008) Safe storage guidelines for rye. Can Biosys Eng 50: 1–8.

Sathya G, Jayas DS, White NDG (2009) Safe storage guidelines for canola as the seeds slowly dry. Can Biosys Eng 51: 29–38.

Schmidt JL (1955) Wheat Storage Research at Hutchinson, Kansas, and Jamestown. U.S. Department of Agriculture, Washington D. C, p. 98.

Shafiekhani S, Atungulu GG (2020) Effect of rice chilling on drying, milling, and quality characteristics. Appl Eng Agric 36: 767–776.

Shang X, Yang M, Zhang C, Cai L, Feng Y, Qiu T (2013) Eects of temperature on the growth and development of *Pyralis farinalis* (Lepidoptera: Pyralidae), one insect used for producing insect tea in China. Acta Entom Sci 56: 671–679.

Sharp JR (1982) A review of low temperature drying simulation models. J Agric Eng Res 27: 169–190.

Shedd CK (1953) Resistance of grains and seeds to air flow. Agric Eng 34: 616–619.

Silva WSV, Vanier NL, Ziegler V, Oliveira M, Guerra AR, Elias MC (2014) Effects of using eolic exhausters as a complement to conventional aeration on the quality of rice stored in metal silos. J Stored Prod Res 59: 76–81.

Sinha RN (1961) Insects and mites associated with hot spots in farm stored grain. Can Entom 93: 609–621.

Sinha RN, Muir WE, Sanderson DB (1985) Quality assessment of stored wheat during drying with near-ambient temperature air. Can Plant Sci 65: 849–866.

Sinha RN, Wallace HAH (1965) Ecology of a fungus-induced hot spot in stored grain. Can J Plant Sci 45: 48–59.

Sinha RN, Watters FL (1985) Insect Pests of Flour Mills, Grain Elevators, and Feed Mills and Their Control. Research Branch, Agriculture Canada Publication 1776, Ottawa, Ontario.

Sinicio R, Muir WE (1998) Aeration strategies for preventing spoilage of wheat stored in tropical and subtropical climates. Appl Eng Agric 14: 517–527.

Smith EA, Sokhansanj S (1990) Moisture transport caused by natural convection in grain stores. J Agric Eng Res 47: 23–34.

Smith LB (1965) The intrinsic rate of natural increase of *Cryptolestes ferrugineus* (Stephens) (Coleoptera, Cucujidae). J Stored Prod Res 1: 35–49.

Smith LB (1970) Effects of cold-acclimation on supercooling and survival of the rusty grain beetle, *Cryptolestes ferrugineus* (Stephens) (Col.: Cucujidae) at subzero temperatures. Can Zool 48: 853–858.

Sokhansanj S, Bruce DM (1987) A conduction model to predict grain temperatures in grain drying simulation. Transac of the ASAE 30: 1181–1184.

Solomon ME (1962) Ecology of the flour mite, *Acarus siro L.* (Tyroglyphus farinae DeG). Ann App Bio 50: 178–184.

Sorenson WG, Hesseltine CW, Shotwell OL (1967) Effect of temperature on production of aflatoxin on rice by *Aspergillus flavus*. Mycopathol Mycol Appl 33: 49–55.

Srzednicki G, Singh M, Driscoll RH (2006) Effects of chilled aeration on grain quality. pp. 985–993. In: Lorini L, Bacaltchuk B, Beckel H, Deckers D, Sundfeld E, Santos JP, Biagi JD, Celaro JC, D'A FDF, Bortolini LOF, Sartori MR, Elias MC, Guedes RNC, Fonseca RG, Scussel VM (eds.), Proc 9th Intern Working Conf Stored-Prod Prot, Campinas, São Paulo, Brazil.

Stejskal V, Vendl T, Li Z, Aulicky R (2019) Minimal thermal requirements for development and activity of stored product and food industry pests (Acari, Coleoptera, Lepidoptera, Psocoptera, Diptera and Blattodea): A review. Insects 10: 1–26.

Strang TJK (1992) A review of published temperatures for the control of pest insects in museums. Collection Forum 8: 41–67.

Subramanyam BH, Hagstrum DW, Harein PK (1990) Upper and lower temperature thresholds for development of six stored-product beetles. pp. 2029–2038. In: Fleurat-Lessard F, Ducom P (eds.), Proc 5th Intern Working Conf Stored-Prod Prot, Bordeaux, France.

Sun DW, Woods JL (1997) Deep-bed simulation of the cooling of stored grain with ambient air: A test bed for ventilation control strategies. J Stored Prod Res 33: 299–312.

Sun K, Jian F, Jayas DS, White NDG (2014) Quality changes in high and low oil content canola during storage: Part I – Safe storage time under constant temperatures. J Stored Prod Res 59: 320–327.

Sutherland JW, Evans DE, Fane AG, Thorpe GR (1986) Disinfestation of grain with heated air. pp. 261–274. In: Donahaye J, Navarro S (eds.), International Working Conference on Stored-Prod Product Protection, Tel Aviv, Israel.

Tang PA, Wang JJ, He Y, Jiang HB, Wang ZY (2008) Development, survival, and reproduction of the psocid *Liposcelis decolor* (Psocoptera: Liposcelididae) at constant temperatures. Ann Entom Soc Amer 101: 1017–1025.

Thompson TL, Peart RM, Foster GH (1968) Mathematical simulation of corn drying — A new model, Transac of the ASAE 11: 582–586.

Thorpe G (2001) Physical basis of aeration. In: Navarro S, Noyes RT (eds.) The Mechanics and Physics of Modern Grain Aeration Management. CRC Press.

Thorpe GR, Cuff WR, Longstaff BC (1982) Control of *Sitophilus oryzae* infestation of stored wheat: An ecosystem model of the use of aeration. Eco Model 15: 331–351.

Thorpe GR, Elder WE (1982) Modelling the effects of aeration on the persistence of chemical pesticides applied to stored bulk grain. J Stored Prod Res 18: 103–114.

Throne JE, Arbogast RT (2010) A computer model for simulating population development of the indianmeal moth (Lepidoptera: Pyralidae) in stored corn. J Econ Ento 103: 1503–1507.

Trematerra P, Lucchi A (2014) *Nemapogon granellus* (L.) pest on corks of wine bottles stored for aging. pp. 885–893. In: Arthur FH, Kengkanpanich R, Chayaprasert W, Suthisut D (eds.), Proc 11th Intern Working Conf Stored-Prod Prot, Chiang Mai, Thailand.

Tzanakakis ME (1959) An ecological study of the Indian-meal moth *Plodia interpunctella* (Hubner) with emphasis on diapause. Hilgardia 29: 205–245.

United States Department of Agriculture (2016). Stored-Graingrain insect reference. In: A. M. Service (ed.). Federal Grain Inspection Service, United State Department of Agriculture, Washington, D. C, p. 63.

Vertucci CW, Roos EE (1990) Theoretical basis of protocols for seed storage. Plant Physio 94: 1019–1023.

White NDG (1995) Insects, mites, and insecticides in stored-grain ecosystems. In: Jayas DS, White NDG, Muir WE (eds), Stored-Grain Ecosystems, Marcel Dekker, New York, pp. 527–567.

Wilches DM, Laird RA, Floate KD, Fields PG (2017) Effects of acclimation and diapause on the cold tolerance of *Trogoderma granarium*. Entom Exp Appl 165: 169–178.

Wilkins D (1983) Safe storage for wheat. Country Guide, Sept.

Williams GC (1954) Observations on the life history of *Laemophloeus minutus* (Ol.) (Col. Cucujidae) when bred on various stored cereals and cereal products. Bull Ent Res 45: 341–350.

Williams GC (1964) The life-history of the Indian meal-moth, *Plodia interpunctella* (Hübner) (Lep. Phycitidae) in a warehouse in Britain and on different foods. Ann Appl Bio 53: 459–475.

Williams SB, Murdock LL, Baributsa D (2017) Safe storage of maize in alternative hermetic containers. J Stored Prod Res 71: 125–129.

Wilson SG, Desmarchelier JM (1994) Aeration according to seed wet-bulb temperature. J Stored Prod Res 30: 45–60.

Woodroffe GE (1951a) A life-history study of the brown house moth, *Hofmannophila pseudospretella* (Staint.) (Lep., Cecophoridae). Bull Ent Res 41: 529–553.

Woodroffe GE (1951b) A life-history study of *Endrosis lactella* (Schiff.) (Lep., Cecophoridae). Bull Ent Res 41(4): 749–760.

Yang Y, Wilson LT, Arthur FH, Wang J, Jia C (2017) Regional analysis of bin aeration as an alternative to insecticidal control for post-harvest management of *Sitophilus oryzae* (L.) and *Rhyzopertha dominica* (F.). Eco Model 359: 165–181.

Young AB, Davis ND, Diener UL (1980) Effect of temperature and moisture on tenuazonic acid production by *Alternaria tenuissima*. Phytopathology 70: 607–609.

Chapter 11

Sulfuryl Fluoride and Propylene Oxide as Fumigants for Stored-Product Pest Control

Vimala S. K. Bharathi and Digvir S. Jayas

11.1 Introduction

Fumigation is one of the effective methods adopted worldwide for managing pests in stored products. Previously, fumigants such as phosphine and methyl bromide (MB) were extensively employed for this purpose. However, the ozone-depleting characteristics of MB have led to restrictions on its use in numerous countries, as mandated by the Montreal Protocol. While phosphine is economical and readily available and does not leave any residues in the grains, overuse and improper use of phosphine have resulted in the development of resistance in stored-grain insects. Consequently, researchers are exploring alternative fumigants such as ethyl formate, carbonyl sulfide, carbon disulfide, carbon dioxide, hydrogen cyanide, sulphur dioxide, sulfuryl fluoride, propylene oxide, methyl isothiocyanate, methyl iodide, acetaldehyde, and ozone (Fields and White, 2002; Navarro and Navarro, 2016a; Saini et al., 2022). Among these alternatives, carbon disulfide has been deregistered for

DOI: 10.1201/9781003309888-11

use in a lot of countries due to concerns about potential toxic residues (Fields and White, 2002; Ren and Mahon, 2006). As for carbon dioxide, a reasonably airtight storage structure is required for its application, leading to higher capital costs, and the process is relatively slow (Annis and Graver, 1986). A comprehensive review on ethyl formate (Ryan and Lima, 2014) and ozone (Boopathy et al., 2022) already exists. Thus, this chapter focuses specifically on two fumigants: sulfuryl fluoride and propylene oxide.

11.2 Sulfuryl Fluoride

Sulfuryl fluoride (also spelled as sulphuryl fluoride) or sulfuryl difluoride (SO_2O_2) is a colorless, odorless, and nonflammable gas. On hydrolysis, the compound breaks down and releases fluoride and fluorosulphate ions. Originally, sulfuryl fluoride was primarily used to control termites, including but not limited to *Cryptotermes cavifrons* (Banks), *Incisitermes schwarzi* (Banks), and *Coptotermes formosanus* (Shiraki) (Su and Scheffrahn, 1986). As the need arose to find an alternative to the ozone-depleting MB, sulfuryl fluoride received attention and was subsequently investigated for its efficacy for controlling pests in stored products. This transition not only addressed environmental concerns but also broadened the scope of sulfuryl fluoride's utility in pest management for stored commodities.

The tolerable daily intake of sulfuryl fluoride recommended by Health Canada is 0.122 (mg/kg bw)/d, and the acceptable worker exposure level over an 8 h workday is 0.13 ppm (PMRA, 2006). Sulfuryl fluoride was approved by the United States Environmental Protection Agency (U.S. EPA) for application in cereal grains and their products, cocoa and coffee beans, and dried fruits and nuts, as well as in food handling and processing facilities (EPA, 2016).

11.2.1 Trade Names

Sulfuryl fluoride has been used under trade names such as Vikane, ProFume, and Xun Mie Jin. Among these, Vikane has been popular in the United States to control termites and other pests that damage wooden structures (Williams and Sprenkel, 1990). ProFume (99.8% sulfuryl fluoride) has been approved by the U.S. EPA (EPA, 2016) and Canada's PMRA (PMRA, 2006) for use in food commodities. It is also registered in Mexico and Australia and across Europe in countries including Switzerland, Italy,

France, Belgium, Germany, and the United Kingdom (Prabhakaran, 2006; Fields, 2012). Xun Mie Jin is the trade name of sulfuryl fluoride in China (Guogan et al., 1998).

11.2.2 Fumigant Characteristics

Because of the higher vapor pressure of sulfuryl fluoride (13,442 mm of Hg at 25°C) compared to that of MB (1,824 mm of Hg), sulfuryl fluoride exhibits greater penetrative capability than MB (Rajendran, 2009). More precisely, sulfuryl fluoride was reported to penetrate approximately 10 times more rapidly than MB (Bell, 2006). Moreover, sulfuryl fluoride exhibits no reactivity with structural materials or components of electronic equipment and other machinery (Drinkall et al., 2002; Bell, 2006). It does not cause any adverse effects on a range of materials including but not limited to stainless steel, silver, aluminum, and butyl rubber (Kenaga, 1957; Ohh et al., 2001). The boiling point of sulfuryl fluoride (–55.2°C at standard atmospheric pressure) is lower than that of a lot of alternative fumigants [ethyl formate (54.1°C), methyl iodide (42°C), carbonyl sulfide (–50.2°C), and methyl isothiocyanate (117°C)] used to treat stored-product pests (Zettler et al., 1999; Zettler and Arthur, 2000; Rajendran, 2009, 2016; PubChem, 2023). Thus, in practical fumigation environments, it exists as a gas, leaving almost no residue and requiring a shorter aeration period. The fumigant is stable up to 400°C and can be safely used in the presence of heat. In addition, low-density polyethylene (LDPE) sheeting or oil-based painting effectively contains the fumigant, minimizing the risk of leakage during fumigation (Bell, 2006). According to Fields and White (2002), phosphine and sulfuryl fluoride are highly promising substitutes to MB, and currently, considering the insect resistance to phosphine, sulfuryl fluoride stands out as an effective fumigant for controlling stored-product insects. The U.S. EPA also reports that sulfuryl fluoride serves as a substitute for MB and is exclusively approved as a structural fumigant for residential applications (EPA, 2023). In addition, lack of cross resistance of phosphine-resistant strains to sulfuryl fluoride makes it a potential alternative to phosphine (Bell et al., 2002) and has led to its being recognized as a "phosphine resistance breaker" (Jagadeesan and Nayak, 2023). Moreover, the likelihood of insects developing resistance to sulfuryl fluoride fumigant is reported to be minimal (PMRA, 2006). Prabhakaran et al. (2001) utilized a sequential quantitative insecticide resistance model to simulate the potential development of insecticide resistance in *Tribolium castaneum* (Herbst), a significant pest known for its tendency to become resistant to established fumigants. They concluded that *T. castaneum* exhibits a very low potential for developing resistance to sulfuryl fluoride.

11.2.3 Mode of Action

The fumigant infiltrates actively respiring life stages through respiration or penetrates egg stages through diffusion. Upon entry, the fumigant breaks down and releases fluoride anion, which is an insecticide. The compound disrupts both glycolysis and citric acid cycles in the insects, ultimately depriving the insects of vital energy for survival. The inhibition of these cycles forces insects to utilize protein and amino acids for maintaining energy. However, these alternative processes cannot help maintain the minimum metabolic rate required for the survival of insects (Meikle et al., 1963). The efficacy of the fumigant increases with an increase in temperature due to higher insect metabolism at higher temperatures (Muhareb, 2010).

In the case of eggs, different species have different respiratory openings on the chorion, which could influence the diffusion of fumigant into the egg and hence the efficacy of the fumigant. For instance, eggs of *Ephestia elutella* (Hubner) have a lot of aeropyles and only one micropyle, whereas *Carpophilus hemipterus* (L.) eggs contain just two aeropyles, without any micropyle. On the other hand, the eggs of *Amyelois transitella* (Walker), *Plodia interpunctella* (Hubner), and *Lasioderma serricorne* (F.) consist of several aeropyles and micropyles, while the eggs of *T. castaneum* do not have any aeropyles or micropyles (Gautam et al., 2014a, 2015). In addition, the chorionic ultrastructure, types of layers, and relative thicknesses of various chorion layers differ depending on the species (Gautam et al., 2015). These factors could help us understand the species-specific impact of sulfuryl fluoride on the eggs of stored-product insects. After exposure to sulfuryl fluoride, the eggs of *Araecerus fasciculatus* (De Geer) were observed to undergo the following changes: (1) destroyed chorion (no clear difference or boundary was observed between exochorion and endochorion), (2) coagulated yolk, and (3) decrease in size (probably due to damage of membrane) (Salbiah et al., 2023).

11.3 Sulfuryl Fluoride Treatment for Control of Stored-Product Pests

The efficacy of sulfuryl fluoride depends on factors such as treatment temperature, exposure time, initial fumigant concentration, type and volume of stored commodity, and characteristics of the targeted insect species, including life stage, age, and type of strain (Table 11.1). For instance, the complete mortality of *Ephestia cautella* (Walker) eggs at the treatment temperatures of 15°, 20°, 25°, 30°, and 35°C was achieved at the sulfuryl fluoride fumigant concentrations of 190, 140, 90, 60, and 30 g/m^3 with a 24 h exposure period at 75% RH (Akan and Ferizli, 2010). According to Bell and Drinkall (2000),

Table 11.1 Effect of Sulfuryl Fluoride on Various Stored-Product Pests

Insect	Life Stage	Commodity	Concentration	Exposure Time	Temperature (°C)	Mortality (%)	References
Acanthosce-lides obtectus (Say)	Eggs	-	1,070 g h/m³	24 h	20	100	Bell (2006)
			605 g h/m³	96 h		100	
			763 g h/m³	24 h	25	100	
			379 g h/m³	48 h		100	
			480 g h/m³	24 h	30	100	
			259 g h/m³	48 h		100	
Ahasverus advena (Waltl)	Eggs	-	4,656 g h/m³	40 h	20	100	Bell (2006)
			3,072 g h/m³	96 h		100	
			1,966 g h/m³	168 h		100	
Araecerus fasciculatus (De Geer)	Adults	500 g of dried cassava	12 g/m³	24 h	30	100*	Salbiah et al. (2023)
	Pupae		12 g/m³	24 h		64.16*	
			24 g/m³			84.14*	
			36 g/m³			93.18*	
			48 g/m³			97.06*	
			60 g/m³			100*	
			12 g/m³	48 h		81.46*	
			24 g/m³			91.77*	
			36 g/m³			100*	
	Larvae		12 g/m³	24 h		92.63*	
			24 g/m³			95.39*	
			36 g/m³			100*	
			12 g/m³	48 h		100*	
	Eggs		12 g/m³	24 h		27.39*	
			24 g/m³			60.57*	
			36 g/m³			73.67*	
			48 g/m³			93.27*	
			60 g/m³			94.45*	
			72 g/m³			96.46*	
			12 g/m³	48 h		58.18*	
			24 g/m³			78.86*	
			36 g/m³			89.37*	
			48 g/m³			97.78*	
			60 g/m³			100*	

(Continued)

Table 11.1 (Continued) Effect of Sulfuryl Fluoride on Various Stored-Product Pests

Insect	Life Stage	Commodity	Concentration	Exposure Time	Temperature ($^\circ$C)	Mortality (%)	References
Cryptolestes ferrugineus (Stephens)	Adults	1 kg of wheat flour	36 g/m^3	24 h	-	100	Widayanti et al. (2016)
	Eggs					100	
Ephestia kuehniella (Zeller)	Eggs	Flour	1,353 g h/m^3 (mean in a 9-floor mill)**	22 h 15 min	27	100	Drinkall et al. (2002)
	Pupae	Food made of bran, wheat, yeast, corn flour, white flour, honey, and glycerol				100	
	Eggs, larvae, pupae	-	35 g/m^3	48 h	20	100	Drinkall et al. (1996)
			11.7 g/m^3	72 h	20	100	
Lasioderma serricorne (F.)	Adults	1 kg of tobacco	36 g/m^3	24 h	-	100	Widayanti et al. (2016)
	Eggs					100	
Oryzaephilus surinamensis (L.)	Mixed	-	13.3 g/m^3	24 h	20	100	Drinkall et al. (1996)
Plodia interpunctella (Hubner)	Eggs	Flour	1353 g h/m^3 (mean in a 9-floor mill)**	22 h 15 min	27	100	Drinkall et al. (2002)
	Pupae	Food made of bran, wheat, yeast, corn flour, white flour, honey, and glycerol				100	
	1-d-old eggs	A few g of flour	5 mg/L	48 h	30	100	Schneider and Hartsell (1998)
	2-d-old eggs					98.6	
	3-d-old eggs					73.7	
	1-d-old eggs		10 mg/L			98.7	
	2-d-old eggs					97.8	

(Continued)

Table 11.1 (Continued) Effect of Sulfuryl Fluoride on Various Stored-Product Pests

Insect	Life Stage	Commodity	Concentration	Exposure Time	Temperature (°C)	Mortality (%)	References
	3-d-old eggs					86.9	
	1-d-old eggs		15 mg/L			100	
	2-d-old eggs					100	
	3-d-old eggs					97.5	
	1-d-old eggs		20 mg/L			100	
	2-d-old eggs					100	
	3-d-old eggs					98	
	1-d-old eggs					100	
	2-d-old eggs		25 mg/L			100	
	3-d-old eggs					98	
	1- to 3-d-old eggs		115 mg h/L**		25	17	
			213 mg h/L**			50	
			238 mg h/L**			98.4	
			374 mg h/L**			99	
			453 mg h/L**			98	
			538 mg h/L**			9.3	
			653 mg h/L**			99.7	
			814 mg h/L**			100	
			854 mg h/L**			99	
	Eggs, larvae, pupae	-	11.7 g/m³	24 h	20	100	Drinkall et al. (1996)
Rhyzopertha dominica (F.)	Mixed	5 g of maize	1,353 g h/m³ (mean in a 9-floor mill)**	22 h 15 min	27	100	Drinkall et al. (2002)
	Eggs	-	912 g h/m³**	20 h	20	100	Bell (2006)
			762 g h/m³**	58 h		100	
			939 g h/m³**	120 h		100	
			656 g h/m³**	20 h	25	100	
			525 g h/m³**	58 h		100	
			638 g h/m³**	120 h		100	

(Continued)

Table 11.1 (Continued) Effect of Sulfuryl Fluoride on Various Stored-Product Pests

Insect	Life Stage	Commodity	Concentration	Exposure Time	Temperature (°C)	Mortality (%)	References
			415 g h/m³**	20 h	30	100	
			304 g h/m³**	58 h		100	
			155 g h/m³**	120 h		100	
Sitophilus granarius (L.)	Eggs	-	35 g/m³	24 h	20	100	Drinkall et al. (1996)
			18.6 g/m³	72 h		100	
	Larvae		13.3 g/m³	24 h		100	
	Pupae		13.3 g/m³			100	
	Adults		13.3 g/m³			100	
	Adults	5 g of brown rice	3.5 mg/L	3 h	15	43.7	Misumi et al. (2010)
			4 mg/L			72.4	
			4.5 mg/L			84	
			5 mg/L			86	
			5.5 mg/L			82	
Sitophilus oryzae (L.)	Mixed	5 g of maize	1,353 g h/m³ (mean in a 9-floor mill) **	22 h 15 min	27	100	Drinkall et al. (2002)
Sitophilus zeamais (Motschulsky)	Adults	1 kg of milled rice	36 g/m³	24 h	-	100	Widayanti et al. (2016)
	Eggs					100	
Stegobium paniceum (L.)	Mixed	15 g of dog food pellets	1,353 g h/m³ (mean in a 9-floor mill)**	22 h 15 min	27	100	Drinkall et al. (2002)
	Mixed	-	18.2 g/m³	24 h	20	100	Drinkall et al. (1996)
Tenebrio molitor (L.)	Larvae	-	11.7 g/m³	24 h	20	100	Drinkall et al. (1996)
	Pupae		11.7 g/m³			100	
	Adults		11.7 g/m³			100	
Tribolium castaneum (Herbst)	Eggs	Flour	1,353 g h/m³ (mean in a 9-floor mill)**	22 h 15 min	27	100	Drinkall et al. (2002)
	Adults	Flour				100	
	Adults	1 kg of maize	36 g/m³	24 h	-	100	Widayanti et al. (2016)
	Eggs		72 g/m³			100	
	1- to 3-d-old eggs	A few g of flour	233 mg h/L**	48 h	25	83.9	Schneider and Hartsell (1998)
			484 mg h/L**			95.1	
			621 mg h/L**			98.9	
			758 mg h/L**			98.9	
			910 mg h/L**			99.8	
			1,157 mg h/L**			99.8	

(Continued)

Table 11.1 (Continued) Effect of Sulfuryl Fluoride on Various Stored-Product Pests

Insect	Life Stage	Commodity	Concentration	Exposure Time	Temperature (°C)	Mortality (%)	References
Tribolium confusum (Jacquelin du Val)	Eggs	Flour	1,353 g h/m³ (mean in a 9-floor mill)**	22 h	27	100	Drinkall et al. (2002)
	Adults	Flour		15 min		100	
	1- to 3-d-old eggs	A few g of flour	233 mg h/L**	48 h	25	60.6	Schneider and Hartsell (1998)
			490 mg h/L**			99.5	
			624 mg h/L**			99.5	
			792 mg h/L**			99.7	
			941 mg h/L**			100	
			1,178 mg h/L**			100	
	Mixed	-	13.3 g/m³	24 h	20	100	Drinkall et al. (1996)
Trogoderma granarium (Everts)	Eggs	-	16 mg/L	72 h	25	99.2	Myers et al. (2021)
			24 mg/L			99.6	
			32 mg/L			100	
			16 mg/L	48 h	30	67.2	
			20 mg/L			90.7	
			24 mg/L			94.2	
			28 mg/L			95.4	
			32 mg/L			97.9	
Trogoderma inclusum (LeConte)	Mixed	-	18.2 g/m³	24 h	20	100	Drinkall et al. (1996)
	Mixed	-	11.7 g/m³	48 h	20	100	
Trogoderma variabile (Ballion)	1- to 3-d-old eggs	A few g of flour	238 mg h/L**	48 h	25	13.8	Schneider and Hartsell (1998)
			514 mg h/L**			31.3	
			619 mg h/L**			96.9	
			814 mg h/L**			96.9	
			950 mg h/L**			97.4	
			1,195 mg h/L**			98.0	

* Expressed in terms of the suppression rate for adult emergence.
** *Ct* product (concentration × time).

a rise in temperature from 20° to 30°C can potentially reduce the concentration time (*Ct*) products by about five times, depending on the targeted insect species. Baltaci et al. (2008) reported that *Oryzaephilus mercator* (Fauvel) and phosphine-resistant strains of *Cryptolestes ferrugineus* (Stephens) were completely controlled at a sulfuryl fluoride concentration of 10 g/m³ with an exposure period of 72 h at 15°C; however, the phosphine-susceptible strains of *C. ferrugineus* were controlled only at 20 g/m^3 with an exposure period of 48 h. On the other hand, *Sitophilus zeamais* (Motschulsky) adults were more tolerant of sulfuryl fluoride treatment than *Sitophilus granarius* (L.) adults (Misumi et al., 2010). Myers et al. (2021) reported that diapausing larvae were more tolerant of sulfuryl fluoride fumigation in the case of recently collected field strains, whereas this pattern was not observed in a strain cultured in the laboratory for more than 30 yr. Among the four major stored-grain insects tested, *T. castaneum* exhibited the highest tolerance toward sulfuryl fluoride, followed by *Sitophilus oryzae* (L.), *C. ferrugineus*, and *Rhyzopertha dominica* (F.) (Jagadeesan and Nayak, 2016). The sulfuryl fluoride *Ct* product requirement at 30°C to completely eliminate *R. dominica* eggs is 300 g h/m^3 for an exposure period of 20 h, whereas for a 120 h exposure period the required *Ct* product is just 155 g h/m^3 (Bell, 2006). Furthermore, the adult, larval, and pupal stages of *A. fasciculatus* reached 100% mortality at the fumigation concentrations of 12, 36, and 60 g/m^3, respectively, at a 24 h exposure period. However, the eggs reached 100% mortality only at a 48 h exposure period at 60 g/m^3 (Salbiah et al., 2023). Eggs are the most tolerant stage in the case of insects, while postembryonic stages are the most tolerant in the case of mites (Bell et al., 2002). Bell and Savvidou (1999) examined the toxicity of various concentrations of sulfuryl fluoride at different exposure periods and temperatures on *Ephestia kuehniella* (Zeller) eggs in different age groups and concluded that 1- to 2-d-old eggs exhibited the highest tolerance, followed by 2- to 3-d, 0- to 1-d, and 3- to 4-d-old eggs. Contrarily, Karakoyun and Emekci (2010) reported that 0- to 1-d-old eggs of *C. hemipterus* exhibited higher tolerance than 1- to 2-d-old eggs at the tested temperatures (15°, 20°, and 25°C). Moreover, 180 g/m^3 of sulfuryl fluoride for a 24 h exposure period at 25°C was required to achieve the complete mortality of *C. hemipterus* eggs (Karakoyun and Emekci, 2010), whereas a fumigant concentration of 21.3 g/m^3 for 48 h at 20° and 25°C was sufficient to achieve complete mortality of *E. elutella* eggs 1- to 4-d old (Baltaci et al., 2009). A treatment temperature range of 25° to 30°C and/ or longer exposure times could enhance the efficacy of the fumigant and reduce the concentrations used against egg stages of most stored-product insect species. This effect is associated with the reduced development time of eggs at those temperatures (Bell, 2006). Su and Scheffrahn (1990) also reported delayed hatching of *L. serricorne* eggs exposed to sulfuryl fluoride fumigation. The maximum registered rate of sulfuryl fluoride (1,500 g h/m^3 *Ct* product) in Australia (APVMA, 2014) is ineffective in controlling the egg stages of *T. castaneum* at a exposure period of 48 h (Jagadeesan et al., 2015)

The fumigant concentrations of 200 g/m^3 for 72 h, 214 g/m^3 for 72 h, and 40 g/m^3 for 24 h for allyl acetate, ethyl formate, and sulfuryl fluoride, respectively, are recommended by Rajendran (2009), to control all life stages of insect species such as *T. castaneum*, *S. oryzae*, and *Trogoderma granarium* (Everts) in bag stacks of wheat at 25°C. Sulfuryl fluoride demonstrated efficacy at a lower concentration and with a shorter exposure time than the other two fumigants. Similarly, Leesch and Zettler (2000) compared the efficacy of fumigants carbonyl sulfide, methyl iodide, and sulfuryl fluoride against diapausing larvae of *Cydia pomonella* (L.) at normal atmospheric pressure, at 15.6°C, and for a 24 h exposure period, in the presence of 33% walnuts load. The study's findings revealed that sulfuryl fluoride was the most effective fumigant of the three tested. On comparing the efficacy of sulfuryl fluoride and phosphine against warehouse insects such as *Alphitobius diaperinus* (Panzer) and *T. castaneum*, Subekti and Syahadan (2021) reported that sulfuryl fluoride is more effective with shorter exposure time. In addition to being more effective than phosphine, sulfuryl fluoride demonstrated efficacy against phosphine-resistant strains of *C. ferrugineus* (Nayak et al., 2014), *R. dominica*, and *T. castaneum* (Opit et al., 2016). Small (2009) studied the effect of sulfuryl fluoride on *T. castaneum* and *Tribolium confusum* (Jacquelin du Val) and the effect of heat treatment on *T. confusum* and *E. kuehniella* insect populations in flour mills in the United Kingdom. The researchers revealed that sulfuryl fluoride was effective in controlling the tested insects, whereas the heat treatment was not found to be effective against the tested insect species due to uneven distribution of heat.

Small (2007) compared the efficacy of methyl bromide and sulfuryl fluoride against stored-product insects and concluded that sulfuryl fluoride was effective against *T. confusum* and *E. kuehniella*, and its efficacy was comparable to that of MB. Hartzer et al. (2010) documented that the efficacies of sulfuryl fluoride and MB against eggs and large larvae of *T. castaneum* were not significantly different. Tsai et al. (2006) also confirmed that sulfuryl fluoride can be used as an alternative to MB in flour mills to control *P. interpunctella* and *T. castaneum*. Vinghes and Ducom (2001) reported that MB requires about 12 h of exposure at the concentration of 50 g/m^3 to completely treat *Curculio elephas* (Gyllenhal); while sulfuryl fluoride requires only about 4 h of exposure at 40 g/m^3. This could be associated with the higher toxicity or lower sorption of sulfuryl fluoride. Contrarily, MB was reported to be more effective than sulfuryl fluoride to control psocids including *Liposcelis paeta* (Pearman), *Liposcelis decolor* (Pearman), *Liposcelis bostrychophila* (Badonnel), *Liposcelis entomophila* (Enderlein), and *Lepinotus reticulatus* (Enderlein) (Athanassiou et al., 2012, 2015). However, by increasing the concentration or exposure time, the complete mortality of psocids could be achieved using sulfuryl fluoride (Athanassiou et al., 2012).

Using sulfuryl fluoride, Rajendran et al. (2008) fumigated grain stacks (milled raw rice of 150 t each and wheat of 170 t each) containing mixed-age cultures or eggs of *Oryzaephilus surinamensis* (L.), *R. dominica*, *S. oryzae*, *T. castaneum*, and *T. granarium* and concluded that the fumigant concentration of 40 g/m^3 for 24 h of exposure at 32°C could effectively control all the tested insect species. Recently, Jagadeesan and Nayak (2023) evaluated the dose mortality response of seven phosphine-resistant reference strains and eighteen field populations of *T. castaneum* from diverse regions in Australia and reported that a sulfuryl fluoride concentration of 1.7 mg/L applied over a 48 h period serves as an effective discriminating dose. This concentration can be utilized to screen for the presence of potentially sulfuryl fluoride–resistant insects. Sulfuryl fluoride treatments could help in ensuring adherence to quarantine regulations, as they could be effectively used to target and mitigate the risk of quarantine pest infestation in stored commodities. Cottrell et al. (2020) reported that 1,300 g h/m³ could be used as a *Ct* dosage for quarantine treatment against fourth instar larvae of *Curculio caryae* (Horn). Thus, sulfuryl fluoride not only supports compliance with quarantine regulations but also contributes to the overall protection of agricultural products.

Apart from controlling stored-product insects, sulfuryl fluoride treatment was also explored to control the mites *Tyrophagus putrescentiae* (Schrank) (Mueller et al., 2006; Abbar et al., 2018). However, at preferred commercial sulfuryl fluoride application rates (300 or 1,000 g h/m³), complete control of mites was not possible at 40°C within an exposure period of 48 h (Abbar et al., 2018). Hasan et al. (2021) also confirmed that sulfuryl fluoride is not suitable to control *T. putrescentiae* but can be used against *Necrobia rufipes* (DeGeer), a stored-product beetle.

Thus, sulfuryl fluoride has demonstrated variable effectiveness across different life stages and strains of insects, with eggs being the most tolerant stage. The fumigant exhibits advantages over other alternatives, such as MB, phosphine, methyl iodide, carbonyl sulfide, ethyl formate, and ally acetate, for achieving control against certain insect species, with lower concentrations and shorter exposure times. Overall, sulfuryl fluoride represents a valuable tool in integrated pest management strategies for stored commodities.

11.4 Effect of Sulfuryl Fluoride Treatment on the Stored Commodity

Sorption of the fumigant by the stored product is one of the crucial factors to be considered when implementing pest control measures. Hwaidi et al. (2017) reported that sulfuryl fluoride sorption by wheat reduced the efficacy rate of the fumigant on adults and eggs of *R. dominica*. This is because through sorption, the grain reduces the biologically active fumigant available for insects. Moreover, the sorption of fumigant by the wheat followed an exponential decay trend. The authors concluded that optimizing both exposure time and fumigant concentration, while taking sorption into account, is essential to improve the overall effectiveness of pest control measures. Moreover, on analyzing the effectiveness of sulfuryl fluoride sorption and desorption by bread and durum wheat, flour, and semolina as a function of temperature (15°, 25°, and 35°C), MC (12% and 15%), and *Ct* product combination (1,500 mg h/L obtained by 4.167 mg/L for 360 h, 8.928 mg/L for 168 h, and 31.25 mg/L for 48 h), Hwaidi et al. (2015) concluded that the sorption rate at each dosage was directly proportional to commodity temperature and MC. More precisely, the sorption was the highest at 35°C and 15% MC and the lowest at 15°C and 12% MC. Moreover, initial concentration did not affect the rate of sorption. Thus, sorption rates mainly depend on the particle size of the fumigated commodity, temperature, MC, and exposure time.

The sorption of sulfuryl fluoride (25% to 37.5%) by walnuts was low compared to that of methyl iodide (88.1% to 91.9%), carbonyl sulfide (52.2% to 61.3%), and MB (about 80%) (Leesch and Zettler, 2000). Sulfuryl fluoride sorption into chestnuts was also reported to be lower than that of MB (Vinghes and Ducom, 2001). Similarly, the sorption of sulfuryl fluoride into flour was less than 75 mg/kg, whereas that of MB was 750 mg/kg (Bell, 2006). Guogan et al. (1998) also documented about 30% sorption of sulfuryl fluoride by wheat, corn, and soybean, along with rapid desorption of sulfuryl fluoride compared to methyl bromide. These observations were further confirmed by Hwaidi et al. (2015), who reported that the sulfuryl fluoride sorption rate by food products is slow compared to those of MB and carbonyl sulfide, whereas it is about four to five times faster than that of phosphine. Phillips et al. (2007) reported that the sorption of sulfuryl fluoride by rough rice was higher than that of polished rice. Sriranjini and Rajendran (2008) studied the sulfuryl fluoride sorption of 68 food commodities including 15 cereals and cereal products, 16 pulses and pulse products, 8 oil seeds, 12 spices and spice products, 11 dried fruits and tree nuts, and 3 beverage crops, as well as jaggery, white sugar, and sago. Among these, 54%, 34%, and 12% of the commodities exhibited low (≤25%), medium (26% to 50%), and high (>50%) sorption of sulfuryl fluoride, respectively. Only white oats showed a very high level (>75%) of sorption.

Sulfuryl fluoride is not known to cause adverse effects on the sensory characteristics of the treated commodity (Vinghes and Ducom, 2001). Sen et al. (2012) reported that sulfuryl fluoride fumigation on dried apricot and raisins did not affect their surface color, total soluble solids, water content, or titratable acidity. In addition, the sulfuryl fluoride–fumigated hazelnut kernels showed no adverse effects on their color or fatty acid composition even after 7 mo of storage.

The residues of sulfuryl fluoride in whole grains such as wheat, corn, rice, brown rice, Chinese sorghum, and millets fumigated for 24 h were about 2.2%, 1.1%, 1.7%, 5.3%, 5.3%, and 2%, respectively, whereas the residues in maize flour and Chinese sorghum flour were about 31.6% and 66.5%, respectively. This implies that the sulfuryl fluoride residue in whole grains is very low compared to that in the processed products; also, there was more residue in brown rice than in regular rice (Guogan et al., 1998). The sulfuryl fluoride–fumigated raisins analyzed one week after the treatment showed no residues of sulphur dioxide (Sen et al., 2012). On fumigating wheat samples with sulfuryl fluoride at 45 mg/L for 1 wk and performing the multi-fumigations for the second and third time after 4 wk of aeration in between the fumigations, Zhang (2021) observed the residues of fluoride ions increasing by 2.7, 2.8, and 3.8 mg/kg after the first, second, and third fumigation treatments, respectively. Thus, multiple sulfuryl fluoride treatments could lead to accumulation of fluoride ions, depending on the time between the fumigation treatments and the concentration.

11.5 Sulfuryl Fluoride in Combination with Other Treatments

Co-fumigation entails the utilization of a blend of two or more fumigants to mitigate the limitations associated with their individual use. For instance, MB is effective in controlling egg stages of *T. castaneum* and *P. interpunctella*, which are more tolerant of sulfuryl fluoride. On the other hand, sulfuryl fluoride has better efficacy against *T. castaneum* pupae than does MB (Tsai, 2008). Thus, co-fumigation of MB with sulfuryl fluoride can provide efficacy against all stages of both *T. castaneum* and *P. interpunctella*. Since MB usage has been restricted in many countries, this approach of co-fumigation is being adopted for various other fumigants. Thus, to enhance the effectiveness and reduce the concentration of sulfuryl fluoride, it is used in combination with other treatments such as carbon dioxide, phosphine, propylene oxide, and low pressure. Table 11.2 compares the combined effects of sulfuryl fluoride and various other treatments to the effects of sulfuryl fluoride alone on a range of stored-product insects.

11.5.1 Carbon Dioxide

Carbon dioxide is often used in combination with other fumigants to enhance the penetration of a fumigant. Xi et al. (2021) used a combination of sulfuryl fluoride and CO_2, in the weight ratio of 7:3, against all four stages of *L. serricorne* at a concentration of 30 g/m³ at different treatment times (48, 72, 96, and 120 h) at 28°C and 60% RH in the presence of tobacco leaves. Complete mortality of all life stages occurred after 96 h of fumigation. Comparatively, the egg stage exhibited the highest tolerance, followed by the larval stage. No notable change was reported in chemical composition of the treated tobacco leaves except a slight reduction in total plant alkaloid content. In addition, no significant effect on sensory quality and no residue was observed. However, the authors reported the requirement of a highly airtight chamber to achieve maximum efficacy in the treatment. If not, optimization in concentration or time is required to achieve complete control of the insects.

Table 11.2 Comparison of the Effectiveness of Sulfuryl Fluoride Alone Versus Its Combination with Other Treatments for Controlling Stored-Product Pests

| Insect Species | Life Stage | Temperature (°C) | SF Alone | | | Combined Treatment | | | | | References |
			Conc. (mg/L)	Time (h)	Efficacy (%)	SF Conc. (mg/L)	Other Treatment	Other Treatment Conc.	Time (h)	Efficacy (%)	
Cryptolestes ferrugineus (Stephens)	Adults and 0- to 3-d-old eggs	25	1.1	168	100 (adults) 83.6 (eggs)	1.1	Phosphine	0.5 mg/L	168	100 (adults) 98.9 (eggs)	Jagadeesan et al. (2018a)
						2.2		0.5 mg/L		100 (adults and eggs)	
			2.2		100 (adults) 98.8 (eggs)	1.1		1 mg/L		100 (adults and eggs)	
						2.2		1 mg/L		100 (adults and eggs)	
Sitophilus zeamais (Motschulsky)	Adults	15	20	3	29.2	20	Phosphine	0.05 mg/L	3	70.9	Misumi et al. (2011a)
								0.10 mg/L		82.7	
	Larvae	15	0.7	24	30.2	0.7		0.1 mg/L	24	83.2	
								0.5 mg/L		94.7	
								1 mg/L		96	
	Pupae	15	0.9		43.2	0.9		0.1 mg/L		9.4	
								0.5 mg/L		14.2	
								1 mg/L		16.5	
Cydia pomonella (L.)	Larvae	20	37.3	2	50	21.3	Vacuum	100 mm Hg	2	50	Zettler et al. (1999)
			51.5		95	28.0				95	
	0- to 1-d-old eggs		65.8		50	67				50	
			173.8		95	187.9				95	
	1- to 2-d-old eggs		127.1		50	160.9				50	
			339.4		95	393.8				95	
	2- to 3-d-old eggs		131.3		50	160.0				50	
			347.3		95	352.6				95	

(Continued)

Table 11.2 (Continued) Comparison of the Effectiveness of Sulfuryl Fluoride Alone Versus Its Combination with Other Treatments for Controlling Stored-Product Pests

| Insect Species | Life Stage | Temperature (°C) | SF Alone | | | Combined Treatment | | | | | References |
			Conc. (mg/L)	Time (h)	Efficacy (%)	SF Conc. (mg/L)	Other Treatment	Other Treatment Conc.	Time (h)	Efficacy (%)	
Plodia interpunctella (Hubner)	Eggs	15.6	18.5	24	50	85.2	Vacuum	100 mm Hg	4	50	Walse et al. (2009)
			36.8		95	181.2				95	
			48.9		99	247.7				99	
		21.1	11.1		50	60.8				50	
			25.7		95	104.9				95	
			36.4		99	131.5				99	
		26.7	5.4		50	21.9				50	
			14.2		95	62.3				95	
			21.3		99	96.1				99	
Tribolium castaneum (Herbst)		15.6	31.7		50	141.9				50	
			81.5		95	396.7				95	
			120.6		99	607.3				99	
		21.1	29.6		50	162.5				50	
			69.2		95	291.7				95	
			98.2		99	371.8				99	
		26.7	27.6		50	45.3				50	
			43.7		95	204.9				95	
			52.9		99	383.1				99	
Carpophilus hemipterus (L.)		15.6	214.7		50	846.8				50	
			444		95	1,092.7				95	
			599.9		99	1,214.5				99	

Note: SF represents sulfuryl fluoride; Conc. represents concentration.

11.5.2 Phosphine

Even though phosphine is still one of the fumigants most commonly used to treat stored grains, the development of phosphine resistance in stored-grain insects poses a significant challenge to its effectiveness. As stated earlier, the egg stages are the most tolerant of sulfuryl fluoride treatment. Thus, combining phosphine with sulfuryl fluoride could help overcome this challenge. For instance, Misumi et al. (2011a) reported that a combination treatment of sulfuryl fluoride and phosphine provides synergistic effect and has the potential to lower the concentration of each fumigant used. The researchers concluded that combining sulfuryl fluoride with phosphine could potentially help treat all stages of *Sitophilus* species (*S. zeamais* and *S. granarius*). The synergistic effect of phosphine and sulfuryl fluoride treatment was true for adult, larval, and egg stages, whereas in the case of pupae, addition of a very small amount of phosphine (0.1 mg/L) decreased the efficacy of sulfuryl fluoride (Table 11.2). Nevertheless, the efficacy gradually increased with increase in the concentration of phosphine. The reason for the reduction in efficacy for pupae could be related to the difference in the respiration mechanism of pupae, as compared to larvae and adults (Misumi et al., 2011a). While Misumi et al. (2011a) reported the synergistic effect of the combination against *S. zeamais* and *S. granarius*, Jagadeesan et al. (2018b), who tested the co-fumigation of a phosphine and sulfuryl fluoride mixture on *R. dominica*, *T. castaneum*, *S. oryzae*, and *C. ferrugineus*, reported that the mixture followed an additive model. Moreover, the application of sulfuryl fluoride with phosphine reduced the phosphine requirement by 50% to achieve 99.9% mortality in *C. ferrugineus* and *S. oryzae*.

11.5.3 Vacuum

Even though sulfuryl fluoride treatment at a vacuum of 100 mm Hg reduced the concentration of fumigant required to control the diapausing larvae of *C. pomonella*, it did not influence the fumigant efficacy on the most tolerant stage (eggs) (Zettler et al., 1999). Similarly, Walse et al. (2009) reported that atmospheric fumigation of sulfuryl fluoride is better suited than fumigation under vacuum to controlling eggs of stored-product insects such as *P. interpunctella*, *T. castaneum*, and *C. hemipterus*.

11.6 Propylene Oxide

Propylene oxide (C_3H_6O) is a liquid fumigant with a boiling point of 35°C and a vapor pressure of 445 mm Hg at 20°C. It possesses a sweet, alcoholic, and ether-like odor and is soluble in water. It is used as an insecticidal fumigant and sterilant to control bacterial and mold contamination, microbial spoilage, and insect infestation and is registered in the United States for food items such as figs, raisins, prunes, cacao beans, and cocoa powder (EPA, 2006). It is also used as a food emulsifier, for cosmetics, as a surfactant, and as a starch modifier and does not have any negative impact on the flavor of the treated commodity (Navarro, 2006). Propylene oxide undergoes degradation into propylene glycol in both soil and the human stomach. Importantly, propylene glycol is inherently nontoxic and biodegradable (Rajendran, 2001) and is used as a food additive (Griffith and Warren, 2001).

Propylene oxide acts as a fumigant and sterilant by interfering with the essential cellular functions of insects and microorganisms. It is known to cause mutagenic effects in microorganisms and insects. Additionally, in vitro studies have reported that propylene oxide can cause mutations and damage to DNA, as well as chromosomal effects in mammalian cells (WHO, 1985). In contrast to the case of sulfuryl fluoride, egg stages of major stored-product insects were more susceptible to propylene oxide treatment, whereas the pupal stage was the most tolerant (Navarro et al., 2004; Isikber et al., 2017; Myers et al., 2021).

11.7 Propylene Oxide in Combination with Other Treatments

Despite its efficacy, a notable drawback of propylene oxide lies in its flammability, ranging from 3% to 37% when mixed with air. So, to address this limitation, it is administered under low pressure or high CO_2 atmosphere.

11.7.1 Low Pressure

Isikber et al. (2004a) examined the effectiveness of propylene oxide under low-pressure conditions of 100 mm Hg against all the life stages of *T. castaneum*, *P. interpunctella*, *E. cautella*, and *O. surinamensis* in the absence of food product at 30°C and reported that propylene oxide treatment at low pressure was effective against the tested stored-product insect species. On comparing the tolerances of pupal stages, the researchers reported that *O. surinamensis* exhibited the highest tolerance, followed by *T. castaneum*, *E. cautella*, and *P. interpunctella*. Later, Işıkber et al. (2007) investigated the efficacy of the same treatment against *P. interpunctella*, both in the absence and in the presence of nuts such as peanuts, almonds, and walnuts. Eggs and larvae were reported to be the most tolerant at tested conditions. The dosage requirement varied depending on the food commodity present. More precisely, a concentration of 13.9 mg/L was sufficient to achieve 99% mortality of *P. interpunctella* larvae for an exposure period of 4 h in the absence of nuts, whereas in the presence of peanuts, about 60.4 mg/L was required; 72.1 and 93.1 mg/L fumigation concentrations were required in the presence of almonds and walnuts, respectively, for the same 4 h treatment time. In a similar study, the effect of low pressure of 100 mm Hg against *P. interpuntella* and *E. cautella* in the absence and presence of figs was investigated (Isikber et al., 2012). After 4 h of exposure, a fumigation concentration of 32.4 mg/L was required to achieve 99% mortality of *P. interpunctella* larvae in the presence of figs. Thus, for *P. interpunctella* larvae, the LC_{99} values increased by about 6.5 times in the presence of walnuts, from four- to five-fold in the presence of peanuts and almonds, and 2.85 times in the presence of figs. Isikber et al. (2004b) tested the efficacy of propylene oxide at 100 mm Hg against *T. castaneum* at various temperatures (16°, 22°, and 30°C) and reported that the efficacy decreased with decrease in temperature. Isikber et al. (2004a) reported that the toxicity of propylene oxide at low pressure against *T. castaneum* pupae was lower than that of phosphine, but higher than those of MB, carbon tetrachloride, and ethylene dichloride.

11.7.2 Carbon Dioxide

Since the application of vacuum-based propylene oxide treatment is cost ineffective, propylene oxide in combination with CO_2 has been explored to control stored-product pests. However, this treatment requires high levels of airtightness. Navarro and Navarro (2016b) used co-fumigation of 92% CO_2 and 8% propylene oxide against adults of *O. surinamensis, R. dominica*, and *T. castaneum* in three different experimental setups: (1) 2.85 L desiccators containing 500 g of polished rice (laboratory scale), (2) 1.5 m³ polyethylene (PE)–laminated envelopes with 1 t of polished rice (semi-commercial scale), and (3) 9.8 m³ PE-laminated envelopes or woven polyvinyl chloride (PVC) cocoons containing 8 t of polished rice (commercial scale). The fumigation was performed for 24 h at 30°C with the propylene oxide concentration of 35 g/m³ in 437.5 g/m³ of CO_2. Complete mortality of the tested species was reported in all trials. Similarly, Gautam et al. (2014b) studied the effect of propylene oxide and CO_2 fumigation in the ratio of 8:92 against eggs of *C. hemipterus, T. castaneum, L. serricorne, A. transitella, E. elutella*, and *P. interpunctella*. The *Ct* products of the tested insect species needed to achieve 99% control were 414.8, 348.6, 113.1, 133.1, 104.9, and 95.3 mg h/L, respectively, at 25°C for the exposure period of 24 h. This implies that *C. hemipterus* eggs were the most tolerant, followed by *T. castaneum, A. transitella, L. serricorne, E. elutella*, and *P. interpunctella* eggs. Misumi et al. (2011b) compared the efficacy of propylene oxide fumigation in the presence of inert gas (N_2) with that in the presence of CO_2 against diapause larvae of *P. interpunctella* and pupae of *S. granarius* and concluded that the CO_2 combination is better suited to controlling the tested insect species than the N_2 combination. To further improve the efficacy, a mixture of propylene oxide and CO_2 along with sulfuryl fluoride has been reported to provide a synergistic effect against stored-product insects (Ryan and Shore, 2010, reciting Muhareb et al., 2009),

Propylene oxide fumigation in the presence of carbon dioxide atmosphere was explored not only for stored-product pests but also as a postharvest treatment for fresh fruits, targeting pests such as *Caliothrips fasciatus* (Pergande), *Aonidiella aurantii* (Maskell), *Naupactus godmanni* (Crotch), *Brevipalpus lewisi* (McGregor), *Brevipalpus californicus* (Banks), and *Lorryia formosa* (Cooreman) (Gautam et al., 2021).

Combining propylene oxide with low pressure or CO_2 not only reduces the flammability risk of propylene oxide but also enhances the efficacy of propylene oxide fumigation. Navarro et al. (2004) studied the efficacy of propylene oxide treatment as well as treatment at low-pressure and high-CO_2 environments against *T. castaneum*. They concluded that propylene oxide treatment in combination with low pressure or CO_2 is more effective than propylene oxide alone to control stored-product insects. Similarly, Isikber et al. (2017) tested the efficacy of propylene oxide alone and the propylene oxide treatment at low pressure (100 mm Hg) and at high levels of CO_2 (92%) against all the life stages of *E. cautella*. The results of the study revealed that the LC_{50} and LC_{99} values of all the life stages were significantly lower when propylene oxide was used in combination with a low-pressure or high-CO_2 environment. Moreover, the researchers concluded that the propylene oxide treatment combined with a low-pressure environment was more effective than the treatment combined with a high-CO_2 environment.

11.7.3 Other Fumigants

Lasioderma serricorne larvae are more tolerant of propylene oxide fumigation, while *L. serricorne* pupae are more tolerant of ethyl formate fumigation. However, eggs and adults stages are susceptible to both the fumigants (Maille, 2019). Thus, combining propylene oxide with ethyl formate could provide effective control against all the life stages of *L. serricorne*. Similarly, *T. granarium* eggs and diapausing larvae are tolerant of sulfuryl fluoride treatment. Thus, in order to overcome the species-specific ovicidal and larvicidal limitations of sulfuryl fluoride, the fumigant is combined with propylene oxide. The combination treatment at 10°C with a sulfuryl fluoride concentration of 96 mg/L and a propylene oxide concentration of 40 mg/L, as well as the treatment at 20°C with 80 mg/L sulfuryl fluoride and 40 mg/L propylene oxide, was found to be effective in completely controlling the eggs and diapausing larvae of *T. granarium*.

11.8 Sorption, Desorption, and Residues in the Treated Commodity

The sorption of propylene oxide by the treated commodity could result in reduced efficacy of the fumigant. For instance, during a study conducted to control stored-product pests such as *L. serricorne, T. putrescentiae,* and *N. rufipes,* sorption of propylene oxide by pet food, flour, and tobacco led to reductions in mortality by 20%, 30%, and 80%, respectively, during a 12 h treatment period; during a 24 h treatment period, reductions in mortality of 10%, 30%, and 40%, respectively, were observed (Maille, 2019). Ramadan et al. (2022) reported that the propylene oxide sorption by tested food products such as wheat flour, wheat kernels, corn, pinto beans, and tobacco leaves followed a logarithmic function. After 24 h, those food products sorbed about 90.6%, 42.8%, 76.5%, 97.6%, and 77% of the initial concentration, respectively. Thus, among the tested products, pinto beans absorbed the most fumigant, whereas wheat kernels absorbed the least fumigant. Moreover, when they compared the sorption of ethyl formate and propylene oxide by those commodities, they reported ethyl formate to be highly sorptive and propylene oxide to be moderately sorptive.

Isikber et al. (2005) studied the sorption of propylene oxide by commodities such as wheat, corn, cocoa beans, and narciccus bulbs. After the first hour of fumigation, a drop in fumigation concentration from 40% to 76% of the initial concentration was observed for corn and cocoa beans, whereas a reduction in concentration from 25% to 41% was observed for narcissus bulbs and wheat. In the case of figs, the reduction in concentration during the first hour was about 50% of the initial concentration (Isikber et al., 2012). In a similar study, Işıkber et al. (2007) reported high sorption of propylene oxide by nuts such as peanuts, almonds, and walnuts. More precisely, after the first hour of exposure, the concentrations were reduced by about 86% in the presence of walnuts and about 77% to 82% in the presence of peanuts and almonds. However, the residue levels after 3 d of aeration were below the maximum allowed residual limit (300 ppm) in all the cases. Similarly, the residue levels were reported to be below the tolerance of 300 ppm immediately after 4 h of fumigation in wheat (173 ppm), corn (157 ppm), and cocoa beans (117 ppm), and there was almost no residue (14, 6, and 8 ppm for wheat, corn, and cocoa beans, respectively) after 3 d of aeration (Isikber et al., 2005); there was about 85 ppm of residue in the case of fumigated figs after 3 d of aeration (Isikber et al., 2012). This implies that, although the commodities quickly absorb the fumigant, there is a rapid desorption of the propylene oxide as well. Moreover, Isikber et al. (2005) reported an increase in the average sorption rate with increases in initial concentration.

Propylene oxide–fumigated food products could also contain residues of propylene chlorohydrin (PCH) (1-chloropropen-2-ol and 2-chloropropen-1-ol), as well as propylene bromohydrin (PBH) (1-bromopropene-2-ol and 2-bromopropene-1-ol) (Jimenez et al., 2014). PBH residues are not a concern for the EPA, since they are minimal compared to those of PCH and PPO in treated food products (Ji, 2014).

11.9 Summary and Conclusion

High toxicity at a shorter exposure time with minimal effect on sensory attributes of the treated commodity, as well as lower environmental risks, make sulfuryl fluoride and propylene oxide promising alternatives to phosphine and MB. The limited ovicidal properties of sulfuryl fluoride and the flammability of propylene oxide are the major drawbacks. Co-fumigation of two or more fumigants or sequential fumigation, in which one fumigant is followed by another, could increase the efficacy and reduce the concentration of fumigant used, thus reducing the environmental toxicity (Misumi et al., 2011a). With the strategic utilization of alternative fumigants such as sulfuryl fluoride and propylene oxide, the development of resistance in insects to phosphine can be mitigated. This approach not only diversifies pest control methods but also helps in overcoming challenges associated with insect resistance, ensuring more effective and sustainable pest management practices. Additionally, the use of alternative fumigants contributes to a comprehensive and integrated pest management strategy, reducing reliance on a single chemical and enhancing the overall resilience of pest control measures. Future works should address (1) developing holistic integrated pest management strategies focusing on the integration of several alternative fumigants, including sulfuryl fluoride and propylene oxide, along with other physical, chemical, and biological control methods; (2) developing sustainable fumigation protocols by considering factors such as resistance management, ecological impact, and environmental feasibility; (3) optimizing the integrated fumigant concentration, exposure time, and environmental conditions for a broader spectrum of stored-product pests, and (4) exploring and adopting emerging technologies, including controlled atmospheric storage, advanced data analytics, and sensor-based real-time monitoring systems, to enhance the fumigation process.

References

Abbar S, Sağlam Ö, Schilling MW, Phillips TW (2018) Efficacy of combining sulfuryl fluoride fumigation with heat to control the ham mite, *Tyrophagus putrescentiae* (Schrank) (Sarcoptiformes: Acaridae). J Stored Prod Res 76: 7–13. doi:10.1016/j.jspr.2017.11.008

Akan K, Ferizli AG (2010) Does sulfuryl fluoride and heat combination overcome the egg-weakness of almond moth? pp. 56-1-4. In: Obenauf GL (ed.), Annual International Research Conference on Methyl Bromide Alternatives and Emissions Reductions, Nov 2-5, 2010, Orlando, USA.

Annis PC, Graver JS (1986) Use of carbon dioxide and sealed storage to control insects in bagged grain and similar commodities. In: Champ BR, Highley E (eds) Pesticides and Humid Tropical Grain Storage Systems: Proceedings of the International Seminar Manila, Philippines (Vol. 364, Issue 14). ACIAR Proceedings, pp. 313–321.

APVMA (2014) Trade advice notice on sulfuryl fluoride in the product ProFume gas fumigant. Australian Pesticides and Veterinary Medicines Authority. https://www.apvma.gov.au/sites/default/files/publication/13476-tan-sulfuryl-fluoride.pdf. Accessed 28 November 2023.

Athanassiou CG, Hasan MM, Phillips TW, Aikins MJ, Throne JE (2015) Efficacy of methyl bromide for control of different life stages of stored-product psocids. J Econ Entomol 108:1422–1428. doi:10.1093/jee/tov069

Athanassiou CG, Phillips TW, Aikins MJ, Hasan MM, Throne JE (2012) Effectiveness of sulfuryl fluoride for control of different life stages of stored-product psocids (Psocoptera). J Econ Entomol 105:282–287. doi:10.1603/EC11209

Baltaci D, Klementz D, Gerowitt B, Drinkall MJ, Reichmuth C (2008) Sulfuryl fluoride against all life stages of rust-red grain beetle (*Cryptolestes ferrugineus*) and merchant grain beetle (*Oryzaephilus mercator*). pp. 87-1-3. In: Proc Annual Intern Res Conf MB Alts Emissions Reducs, Orlando, USA.

Baltaci D, Klementz D, Gerowitt B, Drinkall MJ, Reichmuth C (2009) Lethal effects of sulfuryl fluoride on eggs of different ages and other life stages of the warehouse moth *Ephestia elutella* (Hübner). J Stored Prod Res 45:19–23. https://doi.org/10.1016/j.jspr.2008.06.006

Bell CH (2006) Factors affecting the efficacy of sulphuryl fluoride as a fumigant. pp. 519–526. In: Lorini L, Bacaltchuk B, Beckel H, Deckers D, Sundfeld E, Santos JP, Biagi JD, Celaro JC, Faroni LRD, Bortolini LOF, Sartori MR, Elias MC, Guedes RNC, Fonseca RG, Scussel VM (eds.), Proc 9th Intern Working Conf Stored-Prod Prot, Campinas, São Paulo, Brazil.

Bell CH, Drinkall MJ (2000) Sulfuryl fluoride to replace methyl bromide use in the flour milling industry. Paper 75, In: Proc Annual Intern Res Conf MB Alts Emissions Reduc, Orlando, USA.

Bell CH, Savvidou N (1999) The toxicity of Vikane (sulfuryl fluoride) to age groups of eggs of the Mediterranean flour moth (*Ephestia kuehniella*). J Stored Prod Res 35:233–247. doi:10.1016/S0022-474X(99)00008-9

Bell CH, Wontner-Smith TJ, Savvidou N (2002) Some properties of sulphuryl fluoride in relation to its use as a fumigant in the cereals industry. pp. 910–915. In: Credland PF, Armitage DM, Bell CH, Cogan CM, Highley E. (eds), Proc 8th Intern Working Conf Stored-Prod Prot, York, UK.

Boopathy B, Rajan A, Radhakrishnan M (2022) Ozone: An alternative fumigant in controlling the stored product insects and pests: A status report. Ozone Sci Eng 44:79–95. doi:10.1080/01919512.2021.1933899

Cottrell TE, Aikins MJ, Thoms EM, Phillips TW (2020) Efficacy of sulfuryl fluoride against fourth-instar pecan weevil (Coleoptera: Curculionidae) in pecans for quarantine security. J Econ Entomol 113:1152–1157. doi:10.1093/jee/toaa021

Drinkall MJ, Dugast JF, Reichmuth UK, Scholler M (1996) The activity of the fumigant sulfuryl fluoride on stored product insect pests. pp. 525–528. Proc 2nd Intern Conf Urban Pests.

Drinkall MJ, Zaffagnini V, Suss L, Locatelli DP (2002) Efficacy of sulfuryl fluoride on stored-product insects in a semolina mill trial in Italy. pp. 884–887. In: Credland PF, Armitage DM, Bell CH, Cogan PM, Highley E (eds.), Proc 8th Intern Working Conf Stored-Prod Prot, York, UK.

EPA (2006) Reregistration eligibility decision for propylene oxide. https://archive.epa.gov/pesticides/reregistration/web/pdf/propylene_oxide_red.pdf. Accessed 26 November 2023.

EPA (2016) ProFume®. United States Environmental Protection Agency, Washington, DC. https://www3.epa.gov/pesticides/chem_search/ppls/001015-00079-20160831.pdf. Accessed 27 November 2023.

EPA (2023). Sulfuryl Fluoride https://www.epa.gov/ingredients-used-pesticide-products/sulfuryl-fluoride. Accessed 22 November 2023.

Fields PG (2012) Comparison of efficacy of methyl bromide and sulfuryl fluoride fumigations in Canadian pasta plants. pp. 215–221. In: Navarro S, Banks HJ, Jayas DS, Bell CG, Noyes RT, Ferizli AD, Emekci M, Isikber AA, Alagusundaram K (eds.), Proc 9th Intern Conf Controlled Atm Fumi Stored Prod, Antalya, Turkey.

Fields PG, White NDG (2002) Alternatives to methyl bromide treatments for stored-product and quarantine insects. Annu Rev Entomol 47:331–359. doi:10.1146/annurev.ento.47.091201.145217

Gautam SG, Obenland D, Walse S, Grafton-Cardwell E (2021) Efficacy of propylene oxide as a postharvest fumigant for fresh citrus pest disinfestation and its impact on citrus fruit quality. pp. 193–200. In: Jayas DS, Jian F (eds.), Proc 11th Intern Conf Controlled Atm Fumi Stored Prod (CAF2020), CAF Permanent Committee Secretariat, Winnipeg, Canada.

Gautam SG, Opit GP, Margosan D, Hoffmann D, Tebbets JS, Walse S (2015) Comparative egg morphology and chorionic ultrastructure of key stored-product insect pests. Ann Entomol Soc Am 108:43–56. doi:10.1093/aesa/sau001

Gautam SG, Opit GP, Margosan D, Tebbets JS, Walse S (2014a) Egg morphology of key stored-product insect pests of the United States. Ann Entomol Soc Am 107:1–10. doi:10.1603/AN13103

Gautam SG, Walse S, Tebbets JS, Opit GP (2014b) Efficacy of propylene oxide in combination with carbon dioxide against eggs of postharvest insect pests at normal atmospheric pressure. pp. 398–406. In: Arthur FH, Kengkanpanich R, Chayaprasert W, Suthisut D (eds.), Proc 11th Intern Working Conf Stored-Prod Prot, Chiang Mai, Thailand.

Griffith T, Warren M (2001) Propylene oxide. A registered fumigant, a proven insecticide. p. 763. In: Donahaye EJ, Navarro S, Leesch JG (eds.), Proc 6th Intern Conf Controlled Atm Fumi Stored Prod, Fresno, USA.

Guogan X, Zhongmer C, Zhao S, Nengzhi Q (1998) The development of sulphuryl fluoride (S02F2), in China-a brief introduction. pp. 562–566. In: Zuxun J, Quan L, Yongsheng L, Xianchang T, Lianghua G (eds.), Proc 7th Intern Working Conf Stored-Prod Prot, Beijing, China.

Hartzer M, Subramanyam B, Chayaprasert W, Maier DE, Savoldelli S, Campbell JF, Flinn PW (2010) Methyl bromide and sulfuryl fluoride effectiveness against red flour beetle life stages. pp. 365–370. In: Carvalho MO, Fields PG, Adler CS, Arthur FH, Athanassiou CG, Campbell JF, Fleurat-Lessard F, Flinn PW, Hodges RJ, Isikber AA, Navarro S, Noyes RT, Riudavets J, Sinha KK, Thorpe GR, Timlick BH, Trematerra P, White NDG (eds.), Proc 10th Intern Working Conf Stored-Prod Prot, Estoril, Portugal.

Hasan MM, Aikins MJ, Schilling MW, Phillips TW (2021) Sulfuryl fluoride as a methyl bromide alternative for fumigation of *Necrobia rufipes* (Coleoptera: Cleridae) and *Tyrophagous putrescentiae* (Sarcoptiformes: Acaridae), a major pests of animal-based stored products. J Stored Prod Res 91:101769. doi:10.1016/j.jspr.2021.101769

Hwaidi M, Collins PJ, Sissons M, Pavic H, Nayak MK (2015) Sorption and desorption of sulfuryl fluoride by wheat, flour and semolina. J Stored Prod Res 62:65–73. doi:10.1016/j.jspr.2015.04.005

Hwaidi MI, Collins PJ, Sissons M (2017) Does sorption of sulfuryl fluoride by wheat reduce its efficacy against adults and eggs of *Rhyzopertha dominica*? J Stored Prod Res 74:91–97. doi:10.1016/j.jspr.2017.10.010

Isikber AA, Navarro S, Finkelman S, Rindner M, Azrieli A, Dias R (2004a) Toxicity of propylene oxide at low pressure against life stages of four species of stored product insects. J Econ Entomol 97:281–285. doi:10.1603/0022-0493-97.2.281

Isikber AA, Navarro S, Finkelman S, Rindner M, Dias R (2004b) Influence of temperature on toxicity of propylene oxide at low pressure against *Tribolium castaneum*. Phytoparasitica 32:451–458. doi:10.1007/BF02980439

Isikber AA, Navarro S, Finkelman S, Rindner M, Dias R (2005) Sorption of propylene oxide by various commodities. J Stored Prod Res 41:311–321. doi:10.1016/j.jspr.2004.04.003

Işıkber AA, Navarro S, Finkelman S, Rindner M, Dias R (2007) Propylene oxide as an alternative to methyl bromide for quarantine purposes (8-13 August 2004). pp. 185–191. In: Donahaye EJ, Navarro S, Bell C, Jayas D, Noyes R, Phillips TW, Daolin G (eds.) Proc 7th Intern Conf Controlled Atm Fumi Stored Prod, Gold-Coast Australia 8-13th August 2004. Sichuan Publishing Group, Sichuan, China.

Isikber AA, Oztekin S, Dayisoylu KS, Duman AD (2012) Propylene oxide as potential quarantine fumigant for insect disinfestation of dried figs. pp. 345–352. In: Navarro S, Banks HJ, Jayas DS, Bell CG, Noyes RT, Ferizli AD, Emekci M, Isikber AA, Alagusundaram K (eds.), Proc 9th Intern Conf Controlled Atm Fumi Stored Prod, Antalya, Turkey.

Isikber AA, Tunaz H, Athanassiou CG, Bilgili Y, Er MK (2017) Toxicity of propylene oxide alone and in combination with low pressure or carbon dioxide against life stages of *Ephestia cautella* (Walker) (Lepidoptera: Pyralidae) under laboratory conditions. Crop Prot 98:56–60. doi:10.1016/j.cropro.2017.01.015

Jagadeesan R, Nayak MK (2016) Phosphine resistance does not confer cross-resistance to sulfuryl fluoride in four major stored grain insect pests. Pest Manag Sci 73:1391–1401. doi:10.1002/ps.4468

Jagadeesan R, Nayak MK (2023) A two-step approach to monitoring changes in susceptibility to sulfuryl fluoride in red flour beetle, *Tribolium castaneum* (Herbst) (Coleoptera: Tenebrionidae) and its implication for the industry. J Stored Prod Res 104: 102181. doi:10.1016/j.jspr.2023.102181

Jagadeesan R, Nayak MK, Pavic H, Chandra K, Collins PJ (2015) Susceptibility to sulfuryl fluoride and lack of cross-resistance to phosphine in developmental stages of the red flour beetle, *Tribolium castaneum* (Coleoptera: Tenebrionidae). Pest Manag Sci 71:1379–1386. doi:10.1002/ps.3940

Jagadeesan R, Nayak MK, Singarayan V, Ebert PR (2018a) Co-fumigation with phosphine and sulfuryl fluoride: Potential for managing strongly phosphine-resistant rusty grain beetle, *Cryptolestes ferrugineus* (Stephens). pp. 1021–1024. In: Adler CS, Opit G, Fürstenau B, Müller-Blenkle C, Kern P, Arthur FH, Athanassiou CG, Bartosik R, Campbell J, Carvalho MO, Chayaprasert W, Fields P, Li Z, Maier D, Nayak M, Nukenine E, Obeng-Ofori O, Phillips T, Riudavets J, Throne J, Schöller M, Stejskal V, Talwana H, Timlick B, Trematerra P (eds.), Proc 12th Intern Working Conf Stored-Prod Prot, Berlin, Germany.

Jagadeesan R, Singarayan VT, Chandra K, Ebert PR, Nayak MK (2018b) Potential of co-fumigation with phosphine (PH_3) and sulfuryl fluoride (SO_2F_2) for the management of strongly phosphine-resistant insect pests of stored grain. J Econ Entomol 111:2956–2965. doi:10.1093/jee/toy269

Ji Y (2014). Development and validation of a purge & trap - thermal desorption - gas chromatography - mass spectrometry method for the determination of propylene chrolohydrins (PCH) in hydroxyropyl starch and/or propylene oxide fumigated food products. MSc Thesis, The State University of New Jersey, New Brunswick, USA.

Jimenez RL, Tebbets JS, Opit GP, Gautam SG, Walse SS (2014) Commercial fumigation of stored products with a propylene oxide, carbon dioxide, and sulfuryl fluoride mixture. pp. 408–415. In: Arthur FH, Kengkanpanich R, Chayaprasert W, Suthisut D (eds.), Proc 11th Intern Working Conf Stored-Prod Prot, Chiang Mai, Thailand.

Karakoyun NS, Emekci M (2010) The efficacy of sulfuryl fluoride against egg stage of the dried fruit beetle. pp. 55–1–4. In: Obenauf GL (ed.), Proc Annual Intern Res Conf MB Alts Emissions Reducs, Orlando, USA.

Kenaga EE (1957) Some biological, chemical, and physical properties of sulfuryl fluoride as an insecticidal fumigant. J Econ Entomol 50:1–6. doi:10.1093/jee/50.1.1

Leesch JG, Zettler JL (2000) The effectiveness of fumigating walnuts with carbonyl sulfide, methyl iodide, or sulfuryl fluoride in controlling stored product pests. pp. 227–232. In: Walnut Research Reports, January 2000. Walnut Marketing Board, Sacramento, CA.

Maille JM (2019) The efficacy of propylene oxide, ethyl formate, and ethanedinitrile as fumigants to control *Lasioderma serricorne* (Fabricius, 1792) (Coleoptera: Anobiidae), *Tyrophagus putrescentiae* (Schrank, 1781) (Sarcoptiformes: Acaridae), and *Necrobia rufipes* (De Geer, 1775) (Coleoptera: Cleridae) Dissertation, Kansas State University, Manhattan, USA.

Meikle RW, Stewart D, Globus OA (1963) Drywood termite metabolism of Vikane fumigant as shown by labeled pool technique. J Agric Food Chem 11: 226–230. doi:10.1021/jf60127a022

Misumi T, Aoki M, Kitamura H (2011a) Synergistic and suffocative effects of fumigation with a lower-concentration phosphine and sulfuryl fluoride gas mixture on mortality of *Sitophilus species* (Coleoptera: Dryophthoridae), a stored-product pest part 2: Susceptibility test on *Sitophilus zeamais* for fumigation with a gas mixture, and verification test of a sequential fumigation with two fumigants. Res Bull Pl Prot Japan 47: 1–10.

Misumi T, Aoki M, Tanigawa N, Kitamura H, Suzuki N (2010) Synergistic and suffocative effects of fumigation with a lower-concentration phosphine and sulfuryl fluoride gas mixture on mortality of *Sitophilus* species (Coleoptera: Dryophthoridae), a stored product pest part 1: Dose-response test on *Sitophilus granarius* and *Sitophilus zeamais* for fumigation with sulfuryl fluoride. Res Bull Pl Prot Japan 46: 1–7.

Misumi T, Kitamura H, Suzuki N, Aoki M, Shukuya T, Yamada K, Nobuaki T (2011b) Susceptibility of stored-product insect pests (Indian meal moth, *Plodia interpunctella* [Lepidoptera: Pyralidae]; granary weevil, *Sitophilus granarius*; and maize weevil, *S. zeamais* [Coleoptera: Dryophthoridae]) to propylene oxide fumigation at normal atmospheric pressure. Res Bull Pl Prot Japan 47: 25–32.

Mueller DK, Kelley PJ, Vanryckeghem AR (2006) Mold mites *Tyrophagus putrescentiae* (Shrank) in stored products. pp. 1117–1122. In: Lorini L, Bacaltchuk B, Beckel H, Deckers D, Sundfeld E, Santos JP, Biagi JD, Celaro JC, Faroni LRD, Bortolini LOF, Sartori MR, Elias MC, Guedes RNC, Fonseca RG, Scussel VM (eds.), Proc 9th Intern Working Conf Stored-Prod Prot, Campinas, São Paulo, Brazil.

Muhareb JS (2010) A comparison to methyl bromide with two alternatives treatments; Sulfuryl fluoride and heat to control stored product insects. MAB Thesis, Kansas State University, Manhattan, USA.

Muhareb JS, Hartsell P, Pena B, Hurley JM (2009) Evaluation of combining sulfuryl fluoride, propylene oxide & CO2 for stored product insect control. pp. 64–1–64–2. In 16th Annual International Research Conference on Methyl Bromide Alternatives and Emissions Reduction, 10-13 November 2009, San Diego, CA, USA.

Myers SW, Ghimire MN, Arthur FH, Phillips TW (2021) A combination sulfuryl fluoride and propylene oxide treatment for *Trogoderma granarium* (Coleoptera: Dermestidae). J Econ Entomol 114:1489–1495. doi:10.1093/jee/toab124

Navarro S (2006) New global challenges to the use of gaseous treatments in stored products. pp. 126–133. In: Lorini L, Bacaltchuk B, Beckel H, Deckers D, Sundfeld E, Santos JP, Biagi JD, Celaro JC, Faroni LRD, Bortolini LOF, Sartori MR, Elias MC, Guedes RNC, Fonseca RG, Scussel VM (eds.), Proc 9th Intern Working Conf Stored-Prod Prot, Campinas, São Paulo, Brazil.

Navarro H, Navarro S (2016b) Comparative practical application of two novel candidate fumigants. pp. 147–152. In: Navarro S, Jayas DS, Alagusundaram K (eds.), Proc 10th Intern Conf Controlled Atm Fumi Stored Prod (CAF2016), CAF Permanent Committee Secretariat, Winnipeg, Canada.

Navarro S, Isikber AA, Finkelman S, Rindner M, Azrieli A, Dias R (2004) Effectiveness of short exposures of propylene oxide alone and in combination with low pressure or carbon dioxide against *Tribolium castaneum* (Herbst) (Coleoptera: Tenebrionidae). J Stored Prod Res 40:197–205. doi:10.1016/S0022-474X(02)00097-8

Navarro S, Navarro H (2016a) Current global challenges to the use of fumigants. pp. 126–133. In: Navarro S, Jayas DS, Alagusundaram K (eds.), Proc 10th Intern Conf Controlled Atm Fumi Stored Prod (CAF2016), CAF Permanent Committee Secretariat, Winnipeg, Canada.

Nayak MK, Jagadeesan R, Kaur R, Daglish GJ, Reid R, Pavic H, Smith L, Collins PJ (2014) Developing sulfuryl fluoride as a 'phosphine resistance breaker'-the Australian experience. pp. 24–28. In: Arthur FH, Kengkanpanich R, Chayaprasert W, Suthisut D (eds.), Proc 11th Intern Working Conf Stored-Prod Prot, Chiang Mai, Thailand.

Ohh MH, Sone J, Chung KH (2001) Mortality of cigarette beetle, *Lasioderma Serricorne,* Fabricius (Coleoptera: Anobiidae) and corrosion of metals as influenced sulfuryl fluoride (SO_2F_2). J Korean Soc Tob Sci 23: 141–148.

Opit GP, Thoms E, Phillips TW, Payton ME (2016) Effectiveness of sulfuryl fluoride fumigation for the control of phosphine-resistant grain insects infesting stored wheat. J Econ Entomol 109:930–941. doi:10.1093/jee/tov395

Phillips TW, Konemann CE, Atkinson TH, Schilling WA (2007) Sulfuryl fluoride for post harvest fumigation of arthropod pests in agricultural products. pp. 98:1-2. In: Proc Annual Intern Res Conf MB Alts Emissions Reducs, Orlando, USA.

PMRA (2006) Sulfuryl fluoride. Pest Management Regulatory Agency, Health Canada. https://publications.gc.ca/collections/Collection/H113-7-2006-15E.pdf. Accessed 26 November 2023

Prabhakaran S (2006) Commercial performance and global development status of ProFume® gas fumigant. pp. 15–18. In: Lorini L, Bacaltchuk B, Beckel H, Deckers D, Sundfeld E, Santos JP, Biagi JD, Celaro JC, Faroni LRD, Bortolini LOF, Sartori MR, Elias MC, Guedes RNC, Fonseca RG, Scussel VM (eds.), Proc 9th Intern Working Conf Stored-Prod Prot, Campinas, São Paulo, Brazil.

Prabhakaran S, Schneider B, Williams R, Ray S (2001) Computer modeling for simulating stored product insect pest population dynamics and insecticide resistance potential in food processing plants. pp. 523–529. In: Donahaye EJ, Navarro S, Leesch JG (eds.), Proc 6th Intern Conf Controlled Atm Fumi Stored Prod, Fresno, USA.

PubChem (2023) Methyl isothiocyanate. National Center for Biotechnology Information. https://pubchem.ncbi.nlm.nih.gov/compound/Methyl-isothiocyanate. Accessed 24 November 2023

Rajendran S (2001) Alternatives to methyl bromide as fumigants for stored food commodities. Pestic Outlook, 12:249–253. doi:10.1039/b110550g

Rajendran S (2009) Alternative fumigants for stored grain protection in India. In: Narayanasamy P, Mohan S, Awaknavar JS (eds), Pest Management in Store Grains, Satish Serial Publishing House, Delhi, India, pp. 83–91.

Rajendran S (2016) Status of fumigation in stored grains in India. Indian J Entomol 78:28–38. doi:10.5958/0974-8172.2016.00022.5

Rajendran S, Kumar VL, Sriranjini V (2008) Fumigation of grain stacks with sulfuryl fluoride. Int J Pest Manag 50: 192–195.

Ramadan GRM, Maille JM, Phillips TW (2022) Sorption and desorption dynamics of ethyl formate and propylene oxide as fumigants in durable agricultural commodities. J Stored Prod Res 99:102007. doi:10.1016/j.jspr.2022.102007

Ren YL, Mahon D (2006) Fumigation trials on the application of ethyl formate to wheat, split faba beans and sorghum in small metal bins. J Stored Prod Res 42:277–289. doi:10.1016/j.jspr.2005.04.002

Ryan RF, Lima DCPF (2014) Ethyl formate fumigation an overview update. pp. 969–978. In: Arthur FH, Kengkanpanich R, Chayaprasert W, Suthisut D (eds.), Proc 11th Intern Working Conf Stored-Prod Prot, Chiang Mai, Thailand.

Ryan RF, Shore WP (2010) Pre-mix and on-site mixing of fumigants. pp. 410–422. In: Carvalho MO, Fields PG, Adler CS, Arthur FH, Athanassiou CG, Campbell JF, Fleurat-Lessard F, Flinn PW, Hodges RJ, Isikber AA, Navarro S, Noyes RT, Riudavets J, Sinha KK, Thorpe GR, Timlick BH, Trematerra P, White NDG (eds.), Proc 10th Intern Working Conf Stored-Prod Prot, Estoril, Portugal. doi:10.5073/jka.2010.425.113

Saini MK, Singh S, Sharma DK (2022) Alternatives to phosphine fumigation in managing stored grain insect pests. Int J Agric Sci 18:522–534. doi:10.15740/has/ijas/18.1/522-534

Salbiah, Hidayat Y, Sudarjat (2023) Sulfuryl fluoride fumigation to eliminate all life stages of *Araecerus fasciculatus* (Coleoptera: Anthribidae). Stored Prod Res 103:102152. doi:10.1016/j.jspr.2023.102152

Schneider BM, Hartsell PL (1998) Control of stored product pests with Vikane gas fumigant (sulfuryl fluoride). pp. 406–408. In: Zuxun J, Quan L, Yongsheng L, Xianchang T, Lianghua G (eds.), Proc 7th Intern Working Conf Stored-Prod Prot, Beijing, China.

Sen F, Aksoy U, Tan G (2012) Determining effect of sulfuryl fluoride fumigation on dried apricots, raisins, and hazelnuts quality. pp. 315–320. In: Navarro S, Banks HJ, Jayas DS, Bell CG, Noyes RT, Ferizli AD, Emekci M, Isikber AA, Alagusundaram K (eds.), Proc 9th Intern Conf Controlled Atm Fumi Stored Prod, Antalya, Turkey.

Small GJ (2007) A comparison between the impact of sulfuryl fluoride and methyl bromide fumigations on stored-product insect populations in UK flour mills. J. Stored Prod. Res. 43:410–416. doi:10.1016/j.jspr.2006.11.003

Small GJ (2009) Evaluation of the impact of sulfuryl fluoride fumigation and heat treatment on stored-product insect populations in UK flour mills. Int Pest Control 51: 43–46.

Sriranjini VR, Rajendran S (2008) Sorption of sulfuryl fluoride by food commodities. Pest Manag Sci 64:873–879. doi:10.1002/ps.1577

Su NY, Scheffrahn RH (1986) Field comparison of sulfuryl fluoride susceptibility among three termite species (Isoptera: Kalotermitidae, Rhinotermitidae) during structural fumigation. J Econ Entomol 79:903–908. https://doi.org/10.1093/jee/79.4.903

Su NY, Scheffrahn RH (1990) Efficacy of sulfuryl fluoride against four beetle pests of museums (Coleoptera: Dermestidae, Anobiidae). J Econ Entomol 83:879–882. doi:10.1093/jee/83.3.879

Subekti N, Syahadan MA (2021) Comparison the effectiveness of the fumigants sulfuryl fluoride and phosphine in controlling warehouse pest insects. J Phys Conf Ser 1918:05021. doi:10.1088/1742-6596/1918/5/052021

Tsai WT (2008) Examination of insect management practices in food processing plants prior to, during, and post sulfuryl fluoride or methyl bromide fumigation. PhD Dissertation, Purdue University, West Lafayette, USA.

Tsai WT, Mason LJ, Ileleji KE (2006) A preliminary report of sulfuryl fluoride and methyl bromide fumigation of flour mills. pp. 595–599. In: Lorini L, Bacaltchuk B, Beckel H, Deckers D, Sundfeld E, Santos JP, Biagi JD, Celaro JC, Faroni LRD, Bortolini LOF, Sartori MR, Elias MC, Guedes RNC, Fonseca RG, Scussel VM (eds.), Proc 9th Intern Working Conf Stored-Prod Prot, Campinas, São Paulo, Brazil.

Vinghes C, Ducom P (2001) Preliminary study on chestnut insect disinfestation with sulfuryl fluoride. pp. 149–153. In: Donahaye EJ, Navarro S, Leesch JG (eds.), Proc 6th Intern Conf Controlled Atm Fumi Stored Prod, Fresno, USA.

Walse SS, Leesch JG, Tebbets S (2009) Ovicidal efficacy of sulfuryl fluoride to stored-product pests of dried fruit. pp. 60.1–60–2. In: Obenauf (ed.), Proc Annual Intern Res Conf MB Alts Emissions Reducs, San Diego, USA.

WHO (1985) Propylene oxide. World Health Organization. https://wedocs. unep.org/bitstream/handle/20.500.11822/29328/EHC56ProylnOxd. pdf?sequence=1&isAllowed=y. Accessed 25 November 2023

Widayanti S, Harahap IS, Rivai M, Asnan TAW, Wibawa KA (2016) Efficacy of sulfuryl fluoride against major insect pests of stored commodities in Indonesia. pp. 188–191. In: Navarro S, Jayas DS, Alagusundaram K (eds.), Proc 10th Intern Conf Controlled Atm Fumi Stored Prod (CAF2016), CAF Permanent Committee Secretariat, Winnipeg, Canada.

Williams LH, Sprenkel RJ (1990) Ovicidal activity of sulfuryl fluoride to anobiid and lyctid beetle eggs of various ages. J Entomol Sci 25:366–375. doi:10.18474/0749-8004-25.3.366

Xi J, Zhao Y, Li Y, Dong H, Wang J, Gu X, Song J (2021) Mixed fumigation of sulfuryl fluoride and carbon dioxide to tobacco lamina. pp. 297–300. In: Jayas DS, Jian F (eds.), Proc 11th Intern Conf Controlled Atm Fumi Stored Prod (CAF2020), CAF Permanent Committee Secretariat, Winnipeg, Canada.

Zettler JL, Arthur FH (2000) Chemical control of stored product insects with fumigants and residual treatments. Crop Prot 19:577–582. doi:10.1016/ S0261-2194(00)00075-2

Zettler JL, Leesch JG, Gill RF, Tebbets' JG (1999) Chemical alternatives for methyl bromide and phosphine treatments for dried fruits and nuts. pp. 554–561. In: Zuxun J, Quan L, Yongsheng L, Xianchang T, Lianghua G (eds.), Proc 7th Intern Working Conf Stored-Prod Prot, Beijing, China.

Zhang W (2021) Evaluation of fluoride ion residue in sulfuryl fluoride (SF) fumigated Australian wheat. MRes Thesis, Murdoch University, Perth, Australia.

Chapter 12

Ethyl Formate for the Treatment of Invertebrate Pests

YongLin Ren and Hagit Navarro

12.1 Introduction

Ethyl formate, with scientific synonyms ethyl methanoate, formic acid ethyl ester, formic ether, and ethyl formic ester, is a low-molecular-weight volatile compound produced by many fruits and vegetables. It has an important role as a flavor and aroma component. It is also used as a fumigant, possessing insecticidal and fungicidal properties.

The compound can potentially degrade to biogenic levels in the tissues of treated commodities, in contrast to conventional chemicals, which can persist as residues in food products. Ethyl formate was evaluated as a fumigant for stored grain as a potential alternative to the ozone-depleting fumigant methyl bromide (MB) and to phosphine, which is under pressure owing to the development of strong resistance in stored-grain insects.

Ethyl formate is an old fumigant, with several properties suggesting that it could be useful as a rapid fumigant for grains and similar durable commodities (Simmons and Fisher, 1946), and was evaluated for grain protection in the 1980s (Muthu et al., 1984). Ethyl formate is used as a food additive worldwide (FDA, 1979) and is currently registered as a fumigant on dried fruits in Australia (Banks and Hilton, 1996). It has valuable attributes

DOI: 10.1201/9781003309888-12

for durable commodities, such as rapid action, strong sorption, and rapid decay to natural products (Muthu et al., 1984; Ren and Desmarchelier, 2001), and has potential not only as a relatively safe grain fumigant but also as a grain protectant (Desmarchelier et al., 1998; Annis, 2000; Mahon et al., 2003; Ren et al., 2003; Ren and Mahon, 2006).

The U.S. Food and Drug Administration (FDA) has reviewed the use of ethyl formate as a flavoring agent and has characterized this compound as generally recognized as safe (FDA, 2004). Ethyl formate has been widely used as a fumigant for pests associated with dried fruits (Vincent and Lindgren, 1971; Desmarchelier et al., 1999). Ethyl formate was previously registered in the United States for control of several stored-product pests, including the confused flour beetle, *Tribolium confusum* Jacqueline du Val, and the raisin moth, *Ephestia figulilella* Gregson. Registrations for the use of ethyl formate at both federal and state levels have expired in the United States and in Israel.

Ethyl formate is currently registered as a fumigant for dried fruit treatment in Australia and has a previous history of safe use as a food additive. Ethyl formate naturally occurs in soil, water, vegetation, and a range of raw and processed foods, including vegetables, fruits, grain, beer, and animal products such as milk and cheese (Collin et al. 1993). Unlike phosphine, ethyl formate kills insects rapidly and its residues break down to naturally occurring products: formic acid and ethanol.

Commonwealth Scientific and Industrial Research Organization (CSIRO) Stored Grain Research Laboratory has produced information on the natural occurrence of ethyl formate in stored products and the behavior of residues on different grains. Safe practice in handling and application is an important aspect of the field evaluation. Trials (50–140 t) using ethyl formate have been undertaken on wheat, sorghum, and navy beans stored in unsealed farm bins. These trials have shown that ethyl formate, unlike phosphine, which takes days and requires a well-sealed store, kills insects within hours with a pulsed application dose of 2×90 g/m^3 to avoid reaching LEL. Ethyl formate was, however, less toxic at low grain temperatures (less than 15°C). At low temperatures, liquid ethyl formate does not readily vaporize, and distribution of the fumigant in grain is slower and less uniform.

The challenges are that ethyl formate is potentially flammable when the liquid formulation is used at high concentrations and is rapidly absorbed by grain and broken down to inert compounds. The fumigant therefore needs to be applied to grain at high dosage rates and in a quick-release form.

The toxicity of ethyl formate as a fumigant for stored-product insects has been evaluated and reviewed (Muthu et al., 1984; Banks and Hilton, 1996). The egg stages of both *Sitophilus oryzae* (L.) and *Tribolium castaneum* (Herbst) were the most susceptible and the pupae stages the most tolerant to ethyl formate fumigation (Muthu et al., 1984). Damcevski and

Annis (2002) reported that at the application rates of 109, 130, and 155 mg/L of ethyl formate for 48 h, all *S. oryzae* larvae in 2.7-L glass desiccators containing 500, 1,000, and 1,500 g of wheat, respectively, were killed.

An ethyl formate application rate of 90 g/m³ for a 24 h exposure in an 868-L-capacity fumigation chamber (without grain) is efficacious to control all the life stages of *S. oryzae*, *T. castaneum*, and *Rhyzopertha dominica* (F.) (Damcevski and Annis, 2002). Results from bin trials with wheat and peas (50–55 t) showed that an ethyl formate application rate of 90 g/m³ for a 48 h exposure is efficacious for all the life stages of *T. castaneum* and *R. dominica* but is marginal for immature *S. oryzae* (Desmarchelier et al., 1998; Mahon et al., 2003).

Ren and Mahon (2006) reported that the *Ct* products for *S. oryzae* were at least 2,200 mg h/L in 75–145 t farm bins. In all these examples, the true *Ct* product experienced by the target insects is not well defined, as continuous measurements of prevailing concentrations were not taken. The results were confounded by sorption effects that led to rapid loss of the fumigant. Calculating the *Ct* product on the basis of applied concentration and exposure period thus leads to an overestimation of the true *Ct* product needed for control. It also does not take into account that the required *Ct* products are probably strongly time dependent, with lower values at short exposures having the same effect as higher values at longer exposures (e.g., see Banks and Hilton, 1996).

In most previously reported bioassays (Ren and Mahon, 2006), carbon dioxide (CO_2) resulting from insect respiration and varying exposure concentrations due to sorption of ethyl formate on the culture medium were not adequately considered. When CO_2 is useds with ethyl formate, the toxicity to adults of *S. oryzae*, *T. castaneum*, and *R. dominica* can be greatly increased (Damcevski et al., 2003). Damcevski and Annis (2006), working with *S. oryzae* adults, reported that the presence or absence of grain had the largest influence on mortality, where the larger the grain quantity, the higher the ethyl formate dose required to achieve 99% mortality. They also reported that the toxicity of ethyl formate was quite strongly dependent on relative humidity, with toxicity being higher at higher humidities.

12.2 Physicochemical Properties of Ethyl Formate

12.2.1 Physio-chemical Properties of Ethyl Formate

Ethyl formate is a volatile liquid ester (CH_3CH_2OCHO) with a low molecular weight of 74.08 g/mol and a boiling point of 54°C (ECHA, 2022). Its general physicochemical properties are listed in Table 12.1. It has a pleasant odor (Ryan and Dominiak, 2021). Ethyl formate is naturally present at low levels

Table 12.1 General Physicochemical Properties

Cas number	09-94-4
Molecular formula	$C_3H_6O_2$
Molecular weight	74.08
Melting point	$-80.5°C$
Boiling point	$54.3°C$
Specific gravity (water = 1)	0.924 at 25°C
Vapor density (air = 1)	2.55
Lower explosive limit – upper explosive limit	2.8 (3.1–3.6)–16.5 (15.0–15.7)%
Flammability limit	84 mg /L (practical data) Max. 12.5% v/v of ethyl formate vapor in CO_2
Flash point	$-19.44°C$
Vapor pressure	192 mmHg (25.5 kPa) at 20°C, 300 mmHg (40 kPa) at 30°C
Conversion factor	1 ppm = 3.02 mg/m^3 at 25°C and 1 mg/m^3 = 0.331 ppm at 25°C
Solubility in other solvents	Ethanol, ether, benzene, acetone

(0.5–2 mg/kg) in certain fruits, vegetables, cheese, fish, meat, and wine (Muthu et al., 1984; Collin et al. 1993; Ren and Desmarchelier, 2001) and serves as a flavoring agent (FDA, 1979) with an acceptable daily intake of 3 (mg/kg body weight)/day (JECFA, 1997). Therefore, it is characterized as a generally recognized as safe (GRAS) compound (FDA, 1979).

Due to its vapor pressure of 25.5 kPa at 20°C (ECHA, 2022), it was found effective for grain fumigation in Australia (Neifert et al., 1925) and was reported in studies to control insect pests in dried fruits and stored wheat (Neifert et al., 1925; Cotton and Roark, 1928; Roark and Cotton, 1929; Simmons and Fisher, 1946). It was then evaluated for grain protection in the 1980s (Muthu et al., 1984).

Since ethyl formate is flammable at levels of 2.8–16.5% (v/v), to minimize potential for dangerous explosions in air, Jones (1933) established the specific nonflammable mixture range of ethyl formate in CO_2 as 1:6 by volume. Hazelhoff (1928) discovered that CO_2 also accelerates the penetration of insecticides into insects' spiracles. Moreover, it has a synergistic effect with an optimal range of 5%–20% (Haritos et al., 2006; Ryan and Shore, 2010) that rapidly kills storage insects (Haritos et al., 2003). Ryan and De Lima (2014) found that ethyl formate does not affect the quality or flavor of treated commodities, and it does not affect the viability or germination of seeds.

Compared to MB and phosphine, today's primary fumigant competitors (Agarwal et al., 2015), ethyl formate has a high TLV, indicating that it is 100 and 1,000 times less toxic, respectively (Anonymous, 1975, 1989).

12.3 Toxicity of Ethyl Formate to Organisms

Humans are continually exposed to naturally occurring ethyl formate from a wide range of foods (Desmarchelier et al., 1999). The most common pathway in the breakdown of ethyl formate is hydrolysis to formic acid and ethanol. Hydrolysis can be catalyzed in acidic or basic conditions or by enzymes such as esterase that are present in plants and animals, facilitating the breakdown (Haritos and Dojchinov, 2003; Haritos et al., 2003). In 1979, the federal register proposed to affirm the status of formic acid as a generally recognized as safe product due to its natural occurrence in many agricultural and processed products such as cheese, gelatins, bread, and beer (FDA, 1979).

Ethyl formate has low mammalian toxicity and causes no adverse effect and thus was assigned a high acceptable daily intake (ADI) of (3 mg/kg body weight)/day by regulatory authorities for the World Health Organization (WHO) (Haritos et al., 2000; Haritos et al., 2003). It has oral acute toxicity with an LD_{50} (rat) (lethal dose to kill 50% of experimental animals) of 1,850 mg/kg body weight (bw) (Lewis, 2016). There are no set values of LC_{50} (concentration of vapor that results in the death of 50% of test animals after 4 h of inhalation exposure). The reported LCs range from an LC_{83} (rat) of 24 g/m^3 (Smyth, 1954) to an LC_{100} (dog) of 30 g/m^3 (Opdyke, 1980), with an average of 20 g/m^3 (Haritos et al., 2003). The dermal LD_{50} (rabbit) is greater than 18 g/kg bw. Formic acid, the breakdown product, is of low acute toxicity via oral, dermal, and inhalation routes of exposure. It is not a dermal or eye irritant in rabbits or a dermal sensitizer in guinea pigs. It is not expected to be carcinogenic or mutagenic, based on negative results in the bacterial reverse mutation assay, mammalian cell gene mutation assay, and mammalian cell cytogenetics assays. Neurotoxicity and immunotoxicity studies are not available for review (EPA, 2020).

It is considered a safe fumigant due to its considerably high TLV of 100 ppm, compared to 1 ppm for MB and 0.3 ppm for phosphine.

12.3.1 Dangers to Humans from Ethyl Formate

The Standard for the Uniform Scheduling of Medicines and Poisons (SUSMP) specifies that ethyl formate be labeled S6 when it is packed for use as a fumigant. The maximum residue limit (MRL) for ethyl formate in the dried fruit industry is 1 mg/kg (ppm) (APVMA, 2023).

Currently, the APVMA lists three products containing ethyl formate: Orica Eranol insecticide, Vapormate fumigant, and Vaporfaze Emate insecticide. The National Institute for Occupational Safety and Health (NIOSH) specifies the TLV as 303 mg/m³ (8 h working day, 40 h week). NIOSH also specifies acute LD_{50} oral at 1,850 mg/kg (rat) and LD_{50} dermal at 2,000 mL/kg (rabbit).

It can be an irritant to mucous membranes, eyes, nose, and skin. Slight eye irritation in human subjects and strong persistent nasal irritation were noted on exposure to 330 ppm in the air. The irritating effects on the eyes and respiratory tract are probably due to rapid hydrolysis of the ester on contact with water, with the formation of alcohol and formic acid.

In persons with impaired pulmonary function, especially those with obstructive airway diseases, the breathing of ethyl formate might cause exacerbation of symptoms due to its irritant properties. Persons with pre-existing skin disorders may be more susceptible to the effects of this agent. Although ethyl formate is not known as a liver toxin in humans, the importance of this organ in the biotransformation and detoxification of foreign substances should be considered before exposing persons with impaired liver function. Although ethyl formate is not known as a kidney toxin in humans, the importance of this organ in the elimination of toxic substances justifies special consideration in those with possible impairment of renal function.

Ethyl formate is absorbed through the lungs, from the gastrointestinal tract, and to a small extent through the skin. This ester is hydrolyzed into ethyl alcohol and formic acid with subsequent metabolism via well-known pathways, primarily to CO_2 in the case of ethanol, while formic acid is reduced to biologically active methyl or excreted as the free acid.

The International Agency for Research on Cancer (IARC) lists ethyl formate as noncarcinogenic—that is, it has little or no tumorigenic potential. At the time of this review, no data were available to assess the mutagenic or genotoxic potential of this agent.

12.3.2 Material Data Sheet

A material data sheet (MDS) was reviewed by Desmarchelier et al. (1999). The TLV is 100 ppm (v/v) = 302 mg/m³ (303 µg/L) for ethyl formate, 1,000 ppm for EtOH, and 5 ppm for formic acid (NOHSC, 1995). The TLV of ethyl formate is much higher than those of CS_2 (10 ppm), MB (5 ppm), and PH_3 (0.3 ppm) (Harritos et al., 1999). Oral LD_{50} values for rats are 1,850 mg/kg for ethyl formate (Sax and Lewis, 1989), 1,100 mg/kg for EtOH (Sax and Lewis, 1989), and 1,210 mg/kg for formic acid (Weiss, 1986). The LC_{50} of ethyl formate in rats was 8,000 ppm after a 4 h exposure (Anonymous, 1975, 1989).

The acute, short-term, and subchronic toxicity of ethyl formate has been reviewed by Haritos (2000). Ethyl formate at high concentration is acutely toxic to animals, although compared with other chemicals it is considered of low toxicity. There are no available data on the long-term toxicity of ethyl formate, but ethyl formate is rapidly hydrolyzed to ethanol and formate and this process can partially occur prior to adsorption from the gastrointestinal tract in the human body. Ethyl formate was not carcinogenic in animal studies (Rády et al., 1981), nor was it mutagenic (U.S. Department of Commerce, 1976). Toxicology data are considered adequate for registration of ethyl formate as a grain fumigant (Haritos et al., 2000).

12.4 Insect Mortality in Laboratory Bioassays

12.4.1 Toxicity of Ethyl Formate to Stored-Product Insect Pests

The toxicity of ethyl formate to insects has been reviewed (Muthu et al., 1984; Adu and Muthu, 1985; Hilton and Banks, 1997). The LD_{95} was 10.00 for eggs, 21.13 for larvae, 47.32 for pupae, and 9.42 mg/L for adults of *Callobruchus chinensis* (L) (Adu and Muthu, 1985). This result showed pupae were most tolerant, followed by larvae, eggs, and adults. Muthu et al (1984) reported that eggs of both *S. oryzae* and *T. castaneum* were found to be the most susceptible stage and pupae the most tolerant stage for ethyl formate fumigation (Table 12.2).

Ethyl formate can have very rapid action against *S. oryzae* adults. With a 50% fill rate of wheat; exposure times of 12 min and 2 h resulted in 100% mortality for concentrations of 340 and 210 mg/L, respectively. Also, an exposure time of 3 h achieved 94% mortality for a concentration of 130 mg/L. All *S. oryzae* larvae were killed when exposed to application of 109, 130, and 155 mg/L of ethyl formate on 500, 1,000, and 1,500 g of wheat, respectively, for 48 h (Damcevski and Annis, 2001). Ethyl formate at 25 mg/L showed significant progeny reduction of psocids at 15°, 25°, and 30°C with 12, 24, and 48 h exposure time (Riudavets et al., 2000). Ethyl formate was also found to be highly efficient in controlling all life stages of *Trogoderma granarium* Everts at a dosage of 93.5 g/m^3 within 6 h of exposure time at 26°C and 55% RH. The first instar larva is the most susceptible stage, and it requires 30.7 g/m^3 to obtain LD_{99} mortality. The adult stage is quite susceptible, with 45.5 g/m^3 to obtain LD_{99} mortality. The egg stage follows in susceptibility, with 55.34 mg/L to obtain LD_{99} mortality (Table 12.2). The most tolerant life stage is the pupa, with 93.5 g/m^3 required to obtain LD_{99} mortality.

Table 12.2 Toxicity of Ethyl Formate Against Stored-Grain Insects

Target Insect	Life Stage	Type of Commodity and FR (%)	MC or RH (%)	Exposure Time	Temp. (°C)	Dose (mg/L)	Ct Product (mg h/L)	Mortality (%)	Reference
C. chinensis	Eggs			24 h		10.0		95	Adu and Muthu (1985)
C. chinensis	Larvae			24 h		21.1		95	
C. chinensis	Pupae			24 h		47.3		95	
C. chinensis	Adults			24 h		9.4		95	
S. oryzae	Eggs		60–80	24 h	27.4 ± 2		283	95	Muthu et al. (1984)
S. oryzae	Larvae		60–80	24 h	27.4 ± 2	14.8–27.1	384–727	95	
S. oryzae	Pupae		60–80	24 h	27.4 ± 2	44.0	1110	95	
S. oryzae	Adults		60–80	24 h	27.4 ± 2	19.2	588	95	
T. castaneum	Eggs		60–80	24 h	27.4 ± 2	5.3	154	95	
T. castaneum	Larvae		60–80	24 h	27.4 ± 2	17.4	590	95	
T. castaneum	Pupae		60–80	24 h	27.4 ± 2	27.9	669	95	
T. castaneum	Adults		60–80	24 h	27.4 ± 2	22.3	643	95	
S. oryzae	Adults	Wh. 50	60	12 min	25	340		100	Damcevski and Annis (2001)
S. oryzae	Adults	Wh. 50	60	2 h	25	210		100	
S. oryzae	Adults	Wh. 50	60	3 h	25	130		94	
S. oryzae	Larvae	Wh. 20	60	48 h	25	109		100	
S. oryzae	Larvae	Wh. 50	60	48 h	25	130		100	
S. oryzae	Larvae	Wh. 80	60	48 h	25	155		100	

(Continued)

Table 12.2 (Continued) Toxicity of Ethyl Formate Against Stored-Grain Insects

Target Insect	Life Stage	Type of Commodity and FR (%)	MC or RH (%)	Exposure Time	Temp. (°C)	Dose (mg/L)	Ct Product (mg h/L)	Mortality (%)	Reference
O. surinamensis	Eggs		60	8 h	25	37	248–249	78–99	Hilton and
O. mercator	Larvae					73	469–541	99.8	Banks
P. interpunctella	Pupae					95	566–586	100	(1997)
C. hemipterus	Adults					142	901–920	100	
T. castaneum						(Initial Con.)			
T. confusum									
O. surinamensis	Eggs		60	24 h	25	14	321	45–95	
O. mercator	Larvae					27	572–628	46–100	
P. interpunctella	Pupae					35	765–783	100	
C. hemipterus	Adults					52	911–1158	92–100	
T. castaneum						(Initial Con.)			
T. confusum									
S. oryzae	Adults		100	24 h	25	14	304	99.9	Damcevski
S. oryzae	Adults		30–50	24 h	25	30	660	99.4	and Annis
	Adults					36.4	790	99.9	(2002)
S. oryzae	Larvae + early pupae	0.3	30–50		25		660	100	Wright et al. (2001)
S. oryzae	Larvae + early pupae	0.3	100		25		300	99.9	

(Continued)

Table 12.2   (Continued) Toxicity of Ethyl Formate Against Stored-Grain Insects

Target Insect	Life Stage	Type of Commodity and FR (%)	MC or RH (%)	Exposure Time	Temp. (°C)	Dose (mg/L)	Ct Product (mg h/L)	Mortality (%)	Reference
S. oryzae	Adults	0.3	55	24 h	25–30		1,500	100	Damcevski and Annis (2002)
S. oryzae	Mid pupae	0.3	55	24 h	15		>2,200	100	
R. dominica	All	0.3	55	24 h	25 30		1,300 2,200	100	
T. castaneum	Larvae	0.3	55	24 h	15		1,300	100	
					25		1,500	99	
					30		1,700	99	
R. dominica	All	?	55	24 h	29	70		99	Damcevski and Annis (2002)
T. castaneum					24	66			
S. oryzae					16	60			
S. oryzae	Pupae	48	12% (MC)	48 h	25	300		99	Damcevski and Annis (2002)
						77		80	
S. oryzae	Pupae	48	12% (MC)	48 h	25	270		70	
S. oryzae	Pupae	48	12% (MC)	24 h	25	77		60	

Note: Wh. is wheat and FR is filling ratio. Based on these data, it can be concluded that at 25°C the pupae stage in 31 d is most tolerant (230 mg/L ethyl formate, 48 h, 84% killed) and at 30°C the pupae stage in 26 d is most tolerant (230 mg/L ethyl formate, 48 h, 92% killed).

12.4.2 Toxicity of Ethyl Formate for Field Insect Pests

A variety of arthropods infest California table grapes. Although they may be present at or below economic threshold levels in table grapes, many export markets have designated several of these arthropods as actionable pests, demanding quarantine levels of control. Here, we report the effectiveness of various concentrations of ethyl formate alone and in combination with elevated CO_2 against selected arthropods commonly associated with California table grapes. Western flower thrips, *Frankliniella occidentalis* (Pergande), are a common pest of many agricultural and horticultural crops in California, and both their feeding and oviposition on berries can cause damage (Table 12.2).

Thrips also can cause foliar damage but rarely to a level requiring chemical control (Flaherty et al., 1992). The grape mealybug, *Pseudococcus maritimus* (Ehrhorn), produces two broods per year and overwinters as eggs or crawlers. In the spring, crawlers move onto expanding green tissue, and they mature during May and June. Females then return to old wood and lay eggs. The eggs hatch during June and July, and most crawlers move to feed on fruit and foliage. It is the second brood that causes most of the fruit damage. During August and September, females are often found in the fruit and can lay eggs in berry clusters (Flaherty et al., 1992).

The Pacific spider mite, *Tetranychus pacificus* McGregor, feeds on both the foliage and the berries of table grapes and can cause significant damage when found in high numbers (Flaherty et al., 1992). The omnivorous leafroller, *Platynota stultana* Wasingham, is a major vineyard pest in California. It feeds on flowers and developing berries, and feeding damage provides entry for secondary rot organisms that further damage clusters (Flaherty et al., 1992).

Susceptibility to many types of control measures varies greatly with target pest age; therefore, a range of life stages for each target pest was tested with ethyl formate. Preliminary arthropod tests indicated that LC_{99} concentrations of ethyl formate would be in a range well tolerated by table grapes. Additionally, arthropod tests indicated that longer exposure times might be necessary to achieve complete mortality for the omnivorous leafroller and certain other pest life stages (Yonglin Ren, unpublished data).

12.4.3 Toxicity of Ethyl Formate to Snails in Stored Grain

Ren et al. (2019) reported that snails exposed to ethyl formate at levels of 30–45 g/m^3 were completely controlled after 3 h without grain. However, after 7 and 14 d of exposure at 23°C, all treated snails survived under 100 ppm phosphine in 95% N_2 + 5% O_2 treatment without grain. All snails survived after 7 d under 350 ppm phosphine in 90% N_2 + 10% O_2 treatment without grain. The first signs of mortality were achieved at 350 ppm phosphine in 90% N_2 + 10% O_2 over 14 and 28 d at 5–10% and 85–98%, respectively. Mortality was also achieved at 5–30% and 80–99%, respectively, for 14 and 28 d under 350 ppm phosphine in 90% N_2 + 10% O_2 treatment with grain (canola and barley). Consistently high phosphine levels (>350 ppm) for a longer fumigation holding period (>28 d) are required to effectively control snails.

12.4.4 Concentration × Time (Ct) Product

For sultanas fumigation with ethyl formate at typical commercial dose rates, a *Ct* of 1,493 mg h/L at 8 h and 3,876 mg h/L at 24 h were obtained at 25°C and 60% RH. A mixed-age culture of six pests showed that 8 h exposure was more effective than 24 h exposure, with the same *Ct* product. At 24 h, *Ct* products for *O. surinamensis, O. mercator, Plodia interpunctella,* and *Carpophilus hemipterus* were 765 mg h/L and for *T. confusum* was 1,158 mg h/L for 100% mortality and for *T. castaneum* was 1,158 mg h/L for 94% mortality. At 8 h, all of those pests were controlled in 541 mg h/L (Hilton and Banks, 1997).

For wheat fumigation with ethyl formate, *S. oryzae* adults with a short time exposure (<3 h) showed a wide range of response for a given *Ct,* but a high concentration appears much more effective than a low one (Annis, 1999). The concentration × time product for *S. oryzae* (late larva/early pupa stage) was at least 660 mg h/L at 25°C, 30%–50% RH (Wright et al., 2001) and 300 mg h/L for 99.9% mortality at 25°C, 100% RH. The concentration × time product for *S. oryzae* was 1,500 mg h/L on 24 h at 25° and 30°C; it was >2,000 mg h/L at 15°C. The young pupa stage of *S. oryzae* was the most tolerant stage (>2,200 mg h/L). The concentration × time product for *R. dominca* was 1,300–2,200 mg h/L. The concentration × time product for *T. castaneum* was 1,500 and 1,700 mg h/L at 25° and 30°C for survival of 0.1% in larvae; there was no survival at 15°C for 1,300 mg h/L. The 350–410 mg h/L of ethyl formate gave 100% mortality of adults and nymphs of psocids (Damcevski, personal communication).

12.4.5 Temperature and Moisture Content

Ethyl formate used in a fumigation chamber (868-L capacity) showed complete disinfestation of all stages of *R. dominica, T. castaneum,* and *S. oryzae* adults at 29°, 24°, and 16°C at a concentration of 70, 66, and 60 g/t, respectively, for 24 h. A slightly higher concentration of 75–80 g/t would possibly control all stage of *S. oryzae* at 24° and 16°C (Damcevski and Annis, 2000). This result showed toxicity of ethyl formate increased when the temperature was raised.

12.4.6 Applications

Dried fruit moths can infest fruit and other foods in storage from eggs laid during drying. The grubs and moths of this insect are destroyed by ethyl formate, but repeated applications may be necessary as new eggs hatch.

For control, place a few drops of the ethyl formate per kg of dried fruit in the sealed container with the fruit.

12.4.6.1 Ethyl Formate Mixed with Carbon Dioxide (CO_2)

Carbon dioxide (5%) was found to enhance the toxicity of ethyl formate against adult insects of the species *R. dominica, T. castaneum,* and *S. oryzae.* Although CO_2 showed no significant synergism with ethyl formate against larvae and pupae of *S. oryzae,* addition of CO_2 improved the kill rate of other stages of *S. oryzae.* Future application of ethyl formate with CO_2 could include less ethyl formate, reducing flammability risk (Haritos et al., 2003).

12.4.6.2 Ethyl Formate Mixed with Nitrogen (N_2)

Coetzee et al. (2021) reported on a study with mixed-age cultures of four species of stored-product insects: the cigarette beetle, *Lasioderma serricorne* (F.); the rice weevil, *S. oryzae* (L.); the warehouse beetle, *Trogoderma variabile* (Ballion); and the lesser grain borer, *R. dominica* (F.). They were fumigated with 90 g/m³ ethyl formate and purged with nitrogen into a 20-ft shipping container (6.1 m × 2.44 m × 2.59 m). Ethyl formate concentrations inside the container and the surrounding environment were monitored at timed intervals. The fumigation achieved *Ct* products in the range of 437.54–449.19 g h/m³ in the container, which was sufficient to exterminate all stages of most common insect pests (Table 12.3). This study demonstrated that on-site

Table 12.3 Insects Emerging from Mixed-Age Cultures of Treated and Untreated Insects after a 6 h Fumigation Period with 90 g/m³ Ethyl Formate + Nitrogen

Species	Treatment	Before Fumigation Live	After 6 h Fumigation Live	After 6 h Fumigation Dead	Live Insects Emerging after 7, 14, or 28 d 7	Live Insects Emerging after 7, 14, or 28 d 14	Live Insects Emerging after 7, 14, or 28 d 28
Lasioderma	Control	315	315	0	61	104	157
serricorne	Treated	1,218	0	1,218	0	0	0
Trogoderma	Control	308	308	0	82	71	40
variabile	Treated	1.176	0	1,176	0	0	0
Sitophilus	Control	319	319	0	49	57	129
oryzae	Treated	1,208	0	1,208	0	0	0
Rhyzopertha	Control	327	327	0	66	78	114
dominica	Treated	1,259	0	1,259	0	0	0

Source: Coetzee et al., 2021.

generation of a nonflammable ethyl formate and nitrogen fumigant can be achieved, and this new application technology ensures that ethyl formate is distributed evenly in the container within 0.5 h after application, with a variation of <3%. The research further demonstrated that an ethyl formate + nitrogen application can be used as a bio-secure preshipment treatment for controlling insect pests in shipping containers, with all specimens of the four stored-product insect species being killed. After a fumigation holding period, ventilation of the shipping container for 15 min allowed the ethyl formate to be successfully removed from the container at 0.5–35 ppm in different locations. The levels of ethyl formate in the workspace were <0.5 ppm during application, fumigation, and aeration, which is 200 times below the TLV level of 100 ppm (Coetzee et al., 2021). A low concentration of O_2 (2%) slightly increased the toxicity of ethyl formate against adult insects (*R. dominica, T. castaneum,* and *S. oryzae*), but it had no effect on immature larva stages. Also, the low concentration of O_2 may have antagonized the toxicity of ethyl formate in the pupa stage of *S. oryzae* (Haritos, 2000).

Coetzee et al. (2019) reported that in-transit fumigation trials were conducted in two 6.1-m (20-ft) shipping containers during a 2 d journey in both September and December 2017. Ethyl formate (90 g/m³) was purged with N_2 into the containers. Ethyl formate concentrations inside containers and in the surrounding environment were monitored at timed intervals throughout the journey. Fumigation achieved sufficient *Ct* products in the containers during the journey to exterminate all stages of most common insect pests (Table 12.4). The *Ct* products in transit were greater than those in a shipping container being fumigated in a stationary position at a dose rate of 90 g/m³ for 24 h exposure. Levels of ethyl formate in the environment between 1 and 15 m downwind from the containers and in the driver's cabin were less than 0.5 ppm at each of the timed intervals, 200 times below the ethyl formate TLV of 100 ppm. Their study indicated that in-transit ethyl formate + N_2 technology has the potential to deliver cost savings in the fumigation process through reduction of the labor cost, elimination of the time a container and cargo must remain stationary in a fumigation yard, and a significant decrease in total supply chain time (between container packing and receival) (Coetzee et al., 2019).

Coetzee et al. (2020) reported that ethyl formate + N_2 could be safely applied as an in-transit fumigant for containers transported on land and then by sea. In-transit fumigation trials were conducted in four 6.1-m (20-ft) shipping containers during a 4 d journey in December 2019 in Western Australia. Ethyl formate (90 g/m³) was released with N_2 into the containers. Ethyl formate concentrations inside the containers and in the surrounding environment on the barge were monitored at timed intervals throughout the overnight voyage. This study added new data on in-transit fumigation with ethyl formate + N_2 via road and successfully demonstrated the safety of in-transit fumigation with ethyl formate + N_2 via the marine sector. There was no detectable risk to the public, crew members on the barge, or workers throughout the journey. In

Table 12.4 Comparison of Achieved Ct Product (g h/m^3) for Different Exposure Times and Mortality of Common Stored-Product Insect Pests

Trials	Achieved Concentration × Time Product (g h/m^3) for Different Exposure Times (h)				
	3	4	6	10	48
Container 1, summer	261.20	323.52	420.12	504.51	598.4
Container 1, winter	267.26	356.20	516.17	765.66	2,197.8
Container 2, summer	257.11	332.78	471.74	723.51	2,313.1
Container 2, winter	257.07	330.81	475.02	761.07	2,935.9
Comparison mortality comments	LD$_{99.5}$ 207.4 (adult *S. oryzae*) LD$_{99.5}$ 167.1 (adult *T. castaneum*)[a] LD$_{99.5}$ 122.2 (adult *R. dominica*)[a] LD$_{99.5}$ 98.7 (adult Indian meal moth)[b]	100% kill of adults of common stored-product insect pests with 300[b,c,d,e,f]	100% kill of eggs of common stored-product insect pests with 450[b,c,d,e,f]	100% kill of larvae of common stored-product insect pests with 50[b,c,d,e,f]	100% kill of all stages of all common stored-product insect pests with 2,200[d]

Source: Coetzee et al., 2021.

Note:
[a] Ni et al. (2008).
[b] Muthu et al. (1984).
[c] Banks and Hilton (1996).
[d] Ren and Mahon (2006).
[e] Desmarchelier et al. (1998).
[f] Mahon et al. (2002).

addition, all tested containers were ready to be opened and unloaded without aeration or with 5–10 min aeration upon arrival (Coetzee et al., 2020).

12.4.7 Formulations

Aqueous formulation of the inert gum (methyl cellulose, carboxymethyl cellulose, and hydroxy propyl methyl cellulose), polyethylene glycol (PEG), glycerol, and oils (Kwickin®, Biotrol®, and soybean) all required the addition of an emulsifier (Docusate sodium, Tween 80) to make a homogeneous formulation. All-natural gums, such as acacia, guar, xanthan, and pectin, were not suitable.

The most effective formulations were 83% ethyl formate v/v% in PEG including 2.5% water without emulsifier, ethyl formate (80%) with formic acid (5%) and ethanol (5%) with 10% water, and ethyl formate (40–60%) with vegetable oil (less than 10%) and water with the required emulsifier. As a nonaqueous formulation, ethyl formate was fully soluble in any volume of vegetable oil and stable.

The PEG (lower MW such as 200, 600, and 1,000) were fully soluble in ethyl formate and produced a slightly viscous liquid. Inert substances such as Cab-O-Sil (fine silica) and Dryacide® formed good slurries that were stable and homogeneous. Kwickin® and Biotrol® were fully soluble in ethyl formate. Cab-O-Sil followed by Dryacide, PEG1000, PEG200, and formic acid/ethanol showed much less vaporization than neat ethyl formate. This formulation could be helpful to slow release of ethyl formate (Haritos, 1999).

12.5 Interactions

12.5.1 Interaction with Commodities

Ethyl formate reacts under neutral conditions (nonacidic or nonalkaline) with water to form formic acid and ethanol, a well-known hydrolysis reaction (Morrison and Boyd, 1959; Mata-Segreda, 1999). Mata-Segreda (1999) illustrated that this reaction proceeds to an unstated equilibrium ratio in excess liquid water at 30°C in approximately 4–5 h or at 50°C in 1 h, the reaction being very temperature dependent. Ethyl formate is water soluble (105 g/L at 20°C) and undergoes slow hydrolysis to ethanol and formic acid (Budavari, 1996). This reaction is reported to proceed much faster when catalyzed by acidic or alkaline conditions (Newling and Hinshelwood, 1936; Mai, 2006; Shah and Amis, 1954).

Other workers have claimed that ethyl formate is very stable in distilled water (Ghittori et al., 1984). The reaction equilibrium constant (K_c) for hydrolysis of ethyl formate at 25°C was found to be 0.03 for a dilute solution of ethyl formate in water. With a solution of 0.009 mol ethyl formate/mol H_2O, the reaction rate was measured as 0.38, and with 0.025 mol/L of HCl added to the solution as a catalyst to speed up the reaction, the rate was measured as 0.40 (Mai, 2006). These values were confirmed by measuring the reaction equilibrium constant of the reverse esterification reaction of ethanol with formic acid, which was found to be $K_c = 1/2.6 = 0.38$ (Mai, 2006). A reaction equilibrium constant of this size indicates that the reaction of ethyl formate in water will proceed such that less than half the ethyl formate available will react to form ethanol and formic acid (roughly 30 mol % for equivalent molar concentrations of reactants ethyl formate and H_2O).

This hydrolysis reaction is "thought to occur for ethyl formate adsorbed by grain although more rapidly" (Haritos et al., 2003); however, specific evidence demonstrating that grain-based hydrolysis occurs was not provided. Ethyl formate "freshly added" to samples of wheat to create a grain concentration of 100 mg/kg and left to "temper" for 24 to 72 h in sealed flasks was completely recovered using 24 h liquid methanol extraction to give an estimated 98–102 mg/kg concentration (Desmarchelier et al., 1999), indicating that the reaction of ethyl formate with grain constituents was minor for periods of 24 to 72 h.

Haritos et al. (1999) found that the levels of ethyl formate directly "dripped" onto wheat, when entering a sealed 50-tonne silo at 90 g ethyl formate per tonne of wheat, were not distinguishable from wheat at levels above control wheat samples of 1–3 mg/kg when out-turned, after a 4 wk "withholding" period in the unaerated silo. Reuss and Annis (2002) discuss the formation of ethanol during the adsorption of ethyl formate by a range of rice fractions but conclude that the formation of ethanol in rice products could not be distinguished from that occurring in pure air (i.e., the control experimental run).

Natural levels of ethyl formate have been measured in wheat (Desmarchelier et al., 1998) at ≤ 1 mg/kg; in wheat and barley at 0.02 and 1.0 mg/kg, respectively (Desmarchelier et al., 1999); and in grains (wheat, barley, oats, and canola) at 1–3 mg/kg (Ren and Desmarchelier, 2001). Natural levels of the ethyl formate hydrolysis reaction product formic acid were measured at relatively high rates of 197 to 243 ± 22 mg/kg for wheat and 237 ± 48 mg/kg for barley (Desmarchelier et al., 1999).

The hydrolysis reaction product ethanol occurs naturally at much lower levels of 0.006 ± 0.009 mg/kg for wheat and <0.01 mg/kg for barley (Desmarchelier et al., 1999). Fruits naturally produce volatile compounds that are important for aromatic and flavor characteristics (Nursten, 1970). Plant volatiles such as ethyl formate have been shown to have insecticidal properties (Vincent and Lindgren, 1971; Aharoni and Stewart, 1987; Rohitha et al., 1993).

It is expected that the hydrolysis of ethyl formate with water held in wheat kernels will be substantially slower than that of pure water due to the physical restriction of the available water reactant. Furthermore, the hydrolysis reaction equilibrium characteristic indicates that hydrolysis will not proceed to completion such that all the ethyl formate is reacted to hydrolysis products. But the hydrolysis reaction rate in grain cannot be estimated or predicted based on available information, nor has it been measured.

The fact that relatively large quantities of ethyl formate are adsorbed from air but the presence of pure ethyl formate residues in "aged" fumigated grain has been found to be negligible implies that the ethyl formate reacts in the grain or desorbs for reaction or dissipation in air. However, most methods used to extract ethyl formate from grain that had been dosed with ethyl formate by gaseous or liquid means and aged for many weeks (ethyl formate residues present in grain for many weeks) usually involved aqueous alcohol or pure water extraction solvents for up to 48 h. So any ethyl formate extracted is expected to hydrolyze in the excess water during this extraction period, preventing the measurement of pure ethyl formate present. The only analytical technique that extracted ethyl formate present in wheat with pure methanol (nonaqueous) recovered 100% of the ethyl formate dosed, but the wheat was only exposed to ethyl formate for 48 h.

12.5.2 Sorption of Ethyl Formate

Upon ethyl formate treatment, various stored cereals (wheat, rice, barley, and oats) and oilseeds, oaten hay, and field peas showed less sorption than oilseed canola, making them appear to be more difficult to treat. Paddy rice, sunflower seed, and kidney bean were unsuitable for treatment with ethyl formate because of rapid loss of the ethyl formate (Reuss and Annis, 2002). Paddy rice, brown rice, white rice, and rice flour were treated for 2 d with ethyl formate (60 g/t) at 10° and 25°C. The loss of ethyl formate was more rapid at 25°C than at 10°C in all treated rice types. However, ethyl formate treatment at 60 g/t had no negative effect on germination of paddy rice (Reuss et al., 2002). The rate of sorption of ethyl formate on sultanas was found to be independent of concentration but greatly increased with increased filling ratio and moisture content; there was a slight temperature effect (Hilton and Banks, 1997).

12.5.3 Desorption of Ethyl Formate after Fumigation

Desorption of ethyl formate depends on the moisture content of the wheat. Fumigated wheat with high moisture content (14.0%) released a small amount of ethyl formate, whereas low-moisture wheat (9.7%) released a remarkable amount of ethyl formate within 48 h. After 48 h, gradual desorption occurred.

12.6 Mode of Action

The mechanisms of toxicity of many fumigants are still in research around the world. The toxicity of ethyl formate for insects may be due to its inhibiting the respiratory electron transport chain enzyme cytochrome-c-oxidase, resulting in chemical asphyxiation of the insect's cells and ultimately death (Haritos and Dojchinov, 2003). When ethyl formate goes into the body of insects, it gets broken down to ethanol and formic acid (Haritos et al., 2000). Previously, Haritos and Dojchinov (2003) determined that formate esters, particularly ethyl formate, exert toxicity after they are rapidly metabolized to formic acid in vivo in the stored-product pest *S. oryzae*. The work of Haritos and Dojchinov (2003) additionally showed substantial formic acid liberation from ethyl formate in vitro by beetle homogenates. The insect's rate of respiration is a key factor in determining ethyl formate's toxicity, with a faster rate increasing susceptibility and a slower rate decreasing susceptibility (Haritos, 2005).

12.7 Field Trials

12.7.1 Field Trials on Stored Grain

Welded-steel cylindrical farm silos (50-t capacity), filled with 40 t of wheat (Desmarchelier et al., 1998), were treated with 90 g/t of ethyl formate as a 4% solution in water for 4 wk. The withholding period for falling below the experimental MRL 1 mg/kg was used. The researchers found that *R. dominica* and *T. castaneum* were fully controlled but there were survivors of *S. oryzae* adults (0.26% of total *S. oryzae*) and progeny (31 insects incubated 28 d at 30°C). A welded-steel silo (50 t), filled with barley (33 t), canola (33 t), and oats (25 t) (Ren and Mahon, 2003; Ren et al., 2023) and treated with 90 g/t of ethyl formate mixed with 4% solution for 7 d, gave a high level of control of external stages but incomplete control of immature stages, especially of *S. oryzae*. A Kotzer®-type (50 t) unsealed silo, filled with field pea (40 t) at relative low temperature (10–12°C) (Ren and Mahon 2006) and treated with ethyl formate at 90 g/t for 5 wk at low temperature, needed an extended withholding period to reduce the residue to the required level. Mortality of adult *T. castaneum* and *C. maculatus* was 100%, *S. oryzae* had an average mortality of 98%, and there was incomplete control of immature-stage *S. oryzae* and *C. maculatus*.

In unsealed farm bins (for which the two numbers in parentheses after each commodity give bin capacity/amount of grain) treated with ethyl formate of 85 g/t × 2, applied on top of the grain through a PVC probe (4 cm diameter, 1.2 m long), an effective concentration of ethyl formate was maintained for >10 h (Wright et al., in press). Treatments were conducted on wheat (125 t/125 t), MC 11.1%, grain temp (32°C) (Harden, New South Wales); on sorghum (145 t/140 t and 140 t/135 t), MC 11.8%, grain temp (27°, 27°C) (Warwick, Queensland); and on split fava beans (75 t/75 t), MC

15.2%, grain temp (20°, 10°C) (Two Wells, South Australia) and resulted in control of all insects (*S. oryzae, T. castaneum, R. dominica,* and *C. phaseoli*) except *S. oryzae* on sorghum. Residue on fava bean persisted longest (34.6–41.4 mg/kg).

12.7.2 Field Trials on Stored Fruit and Vegetables

Export of Pink Lady apples from Australia has been significantly affected by infestations of adult eucalyptus weevils (*Gonipterus platensis* Marelli). These weevils cling tenaciously to the pedicel of apple fruit when selecting overwintering sites. As a result, apples infested with live *G. platensis* adults are rejected for export. Since the Montreal Protocol restricted use of MB as postharvest treatment, it was necessary to consider alternative safer fumigants for disinfestation of the eucalyptus weevil. Agarwal et al. (2015) conducted laboratory experiments using concentrations of 5, 10, 15, 20, 25, 30, 40, and 80 mg/L of ethyl formate. Complete control (100% mortality) was achieved at 25–30 mg/L of ethyl formate at 22°–24°C for 24 h exposure without apples. However, with 90%–95% of the volume full of apples, complete control was achieved at 40 mg/L of ethyl formate at 22°–24°C for 24 h exposure. No phytotoxicity was observed, and after 1 d aeration, the residue of ethyl formate had declined to natural levels (0.05–0.2 mg/kg). Five ethyl formate field trials were conducted in cool storages (capacity from 250 to 900 t), and 100% kill of eucalyptus weevils was achieved at 50–55 mg/L at 7°–10°C for 24 h. Ethyl formate has great potential for preshipment treatment of apples. Its use is considerably cheaper and safer than that of already existing fumigants such as MB and phosphine (Agarwal et al., 2015).

Export of celery from Australia has been affected by a natural infestation of springtails, *Hypogastrura vernalis* (Collembola: Hypogstruridae). These insects contaminate fresh celery by living inside the celery head but do not damage it. As a result, springtail-infested celery has been rejected at export, with a dramatic impact on the market value of the fresh produce. Ahmed et al. (2018) evaluated the effects of fumigation of celery with ethyl formate, PH_3, and their combination on the mortality of springtails in naturally infested celery. Laboratory experiments were conducted using concentrations of 50, 60, and 90 mg/L of ethyl formate for 1, 2, and 4 h; 1, 1.5, 2, and 2.5 mg/L of PH_3 for 2, 4, and 6 h; 20, 30, and 40 mg/L of ethyl formate mixed with 1 mg/L PH_3 for 2 and 4 h. Complete control was achieved at 90 mg/L of ethyl formate for 2 h. However, phytotoxicity was observed in celery treated by ethyl formate at all concentrations. At 2.5 mg/L of PH_3, 100% mortality was achieved within 6 h, and no phytotoxicity was evident.

Mortality of 100% was achieved at 30 and 40 mg/L ethyl formate combined with 1 mg/L of PH_3 for 2 and 4 h exposure time with phytotoxicity. From the data, we can say that PH_3 alone has some potential as a fumigant for the preshipment treatment of celery (Ahmed et al., 2018).

The cotton aphid, *Aphids gossypii* (Glover), is known as a quarantine pest that is hard to control with short periods of fumigation with PH_3 or low concentrations of ethyl formate. Moreover, low-temperature fumigation with ethyl formate can lead to phytotoxic damage of some perishable commodities. Lee et al. (2016) reported that a laboratory study was conducted to evaluate the synergistic effect of mixing ethyl formate and PH_3 for the treatment of adults and nymphs of *A. gossypii*. Combined toxicity was observed and compared with that of a single dose of either ethyl formate or PH_3. When insects were exposed to 0.5 g/m^3 of PH_3 combined with different levels of ethyl formate from 1.6 to 16.3 g/m^3 at 5° and 20°C for 2 h, $L(Ct)_{50}$ and $L(Ct)_{99}$ values were significantly reduced below their values with a single dose of either ethyl formate or PH_3. The synergistic ratio (SR) is described as $L(Ct)$ of ethyl formate alone/$L(Ct)$ of ethyl formate + PH_3. The SR values of $L(Ct)_{50}$ and $L(Ct)_{99}$ for adult *A. gossypii* at 5°C were 4.55 and 2.33. At 20°C, the SR levels of $L(Ct)_{50}$ and $L(Ct)_{99}$ were 2.22 and 1.45 but still showed significant synergism (significant difference, $P < 0.5$). This new technology could meet quarantine preshipment (QPS) requirements for shorter exposure times and less damage of perishable commodities and could also be extended for controlling other quarantine pests and thereby be a useful alternative to MB for fruit and vegetable applications (Lee et al., 2016).

Navarro et al. (2022) report that adults and larvae of the western flower thrips were controlled on various edible flower species treated with ethyl formate mixed with CO_2 (1:6 ratio) at a dosage of 30 g/m^3 for 2 h exposure at 15°C, with no progeny after 6 d incubation at 3°C (Navarro et al., 2022).

12.7.3 Field Trials for Structural Treatment

For fumigation of mill machinery with ethyl formate (Wagga Wagga, New South Wales), Desmarchelier et al. (1998) applied 80 g/t of ethyl formate without dilution for 4–5 h and obtained 100% mortality for *T. castaneum*. Fumigation of empty cells (Port Giles, South Australia) with approximately 89 g/t of ethyl formate, applied at 4% and 8% or 2% (5 repeats) water solution on the structure (cell) and on the surface of the wheat, gave almost immediate control of the psocid (*Liposcelis decolor*) but not complete control for longer periods (1 wk or more).

For disinfestation of semi-sealed sampling equipment used for quarantine sampling (Fisherman Island, Newcastle), Allen and Desmarchelier (2000) applied 70 g/t of ethyl formate for 30 min and achieved 100% mortality for all species expect *T. castaneum* in 7% CO_2, which had 99% mortality, and *T. variabile,* which had 0% mortality. Ethyl formate was carried by a CO_2 stream into a >30-m-long sampling system.

12.7.4 Field Trials to Control the Brown Marmorated Stink Bug

Halyomorpha halys (Stål) (Hemiptera: Pentatomidae), commonly referred to as the brown marmorated stink bug (BMSB), is a polyphagous insect capable of feeding upon over 300 species of plants (Lee et al., 2013), ranging from fruit and vegetables to native vegetation. The insect's most likely route of entry into new countries/territories/regions is via the shipping of vehicles and other machinery from infested regions (Hoebeke and Carter, 2003). McKirdy (2022) has reported that ethyl formate was effective at applied concentrations above 12 mg/L for the control of adult BMSB irrespective of dormancy state and temperature conditions tested. The laboratory bioassay found that diapausing insects, as expected, were the most difficult to kill, with the lower temperature condition (10°C) requiring a higher ethyl formate concentration than at the 25°C temperature condition to achieve the desired mortality level (LD_{50}, LD_{99}, and Probit 9). The results from the laboratory bioassays are consistent with existing literature on the fumigation of insects, where there is an inverse relationship between temperature and the fumigant concentration required to achieve mortality (Bond and Monro, 1984).

In laboratory bioassays of McKirdy (2022), 10°C nondiapause treatments had an LD_{99} of 8.87 mg/L and a Probit 9 of 15.98 mg/L; Kawagoe et al. (2017) had a Probit 9 of 16.5 mg/L for a 2 h exposure and a Probit 9 of 10.5 mg/L for a 4 h exposure. Both of these studies contrast with research evaluating the efficacy of ethyl formate in treating grain insects (Coetzee et al., 2021), where the concentration used to achieve LD_{99} was 90 mg/L, approximately nine times greater, which indicates that BMSB is more susceptible to ethyl formate than are *L. serricorne, S. oryzae, Trogoderma v*ariabile (Ballion), and *R. dominica.* The Australian government has approved ethyl formate for minor use as a biosecurity treatment for Barrow Island at a *Ct* product nine times higher than what was required to achieve complete BMSB control in this study (Australian Pesticides and Veterinary Medicine Authority, 2021).

Work described in Abrams et al. (2020), using sulfuryl fluoride, found that diapausing BMSB required a gas concentration 2.1 times higher than was required by nondiapausing insects at 10°C to achieve Probit 9. In comparison, this study found that diapausing BMSB at 10°C required an ethyl formate concentration 1.3 times greater than did nondiapausing insects to achieve Probit 9. As a result, it could be assumed that ethyl formate is approximately twice as effective as sulfuryl fluoride at controlling diapausing BMSB (McKirdy, 2022).

McKirdy (2022) reported that commercial-scale trials using a 6.1-m (20-ft) shipping container validated the laboratory trials and showed that concentrations above 10 mg/L were sufficient to achieve 100% adult insect mortality within a day with each of the three treatments. These results, and the large number of insects tested (n = 37,380), enhance the findings from the laboratory bioassays. While nondiapausing adult insects were used in the commercial trials, it is possible to identify a required ethyl formate concentration to achieve LD_{99} for diapausing insects based on the laboratory-scale trials. A control condition was not used due to the availability of only one container and limited equipment and because laboratory trials indicated that the short exposure time (3 h) did not result in any control mortality.

References

Abrams AE, Kawagoe JC, Najar-Rodriguez A, Walse SS (2020) Sulfuryl fluoride fumigation to control brown marmorated stinkbug (Hempitera: Pentatomidae). Postharvest Biol Technol 163: 111111.

Adu OO, Muthu M (1985) The relative toxicity of seven fumigants to life cycle stages of Callosobruchus chinensis (L). Int J Trop Insect Sci 6: 75–78.

Agarwal M, Ren YL, Newman J, Learmonth S (2015) Ethyl formate: A potential disinfestation treatment for eucalyptus weevil (*Gonipterus platensis*) (Coleoptera: Curculionidae) in Apples. J Econ Entomol 108(6): 2566–2571.

Aharoni Y, Stewart JK (1987). Thrips mortality and strawberry quality after vacuum fumigation with acetaldehyde or ethyl formate. J Am Soc Hortic Sci 105: 926–929.

Ahmed Q, Ren Y, Emery R, Newman J, Agarwal M (2018) Evaluation of ethyl formate, phosphine, and their combination to disinfest harvested celery against purple scum springtails. HortTechnology 28(4): 492–501. https://doi.org/10.21273/HORTTECH04030-18

Allen SE, Desmarchelier JM (2000) Ethyl formate as a fast fumigant for disinfestations of sampling equipment at grain export terminals. In: Wright EJ, Banks HJ, Highley E (eds), Proceedings of the Australian Postharvest Technical Conference 2000, CSIRO Entomology, Canberra, ACT, Australia.

Annis P (1999) The relative effects of concentration, time, temperature and other factors in fumigant treatments. Pp. 331–337. In: Zuxun J, Quan L, Liang Y, Tan X, Guan L (ed), Proceedings of the 7th International Working Conference on Stored-Product Protection, Beijing, P.R. China, 14–19 October 1998, Sichuan Publishing House of Science and Technology, Chengdu.

Annis PC (2000) Ethyl formate—Where are we up to? In: Wright EJ, Banks HJ, Highley E (eds), Proceedings of the Australian Postharvest Technical Conference 2000, CSIRO Entomology, Canberra, ACT, Australia.

Anonymous (1975) Recommended methods for the detection and measurement of resistance of agricultural pests to pesticides 16. FAO Plant Protection Bulletin 23: 12–25.

Anonymous (1989) Suggested recommendations for the fumigation of grain in the ASEAN Region, Part 1, principles and general practice, ASEAN Food Handling Bureau, Kuala Lumpur.

APVMA (2023) Australian pesticides and veterinary medicines authority: Strategic review report - July 2023. https://www.agriculture. gov.au/sites/default/files/documents/APVMA%20-%20Strategic%20 Review%20Report.PDF

Australian Pesticides and Veterinary Medicines Authority (2021) Permit to allow supply and minor use of an unregistered agvet chemical product for in-transit disinfestation of foodstuffs and general goods transported to barrow Island. (PERMIT NUMBER—PER90604). Australian Government.

Banks HJ, Hilton SJ (1996) Ethyl formate as a fumigant of sultanas: Sorption and efficacy against six pest species. Pp. 409–422. In: Donahaye EJ, Navarro S., Varnava A (eds), Proceedings of the International Conference on Controlled Atmosphere and Fumigation in Grain Storage, Printco Ltd, Nicosia, Cyprus.

Bevilacqua AE, Califano AN (192) Changes in organic acids during ripening of Port Salut Argentino cheese. Food Chem 43: 345–349.

Bond EJ, Monro HAU (1984) Manual of fumigation for insect control (Vol. 54), FAO Rome.

Budavari S (1996) The Merck Index (12th Edition, p. 648). Merck & Co. Whitehouse Station.

Coetzee M, Du X, Thomas M, Ren YL, McKirdy SJ (2020) In-transit fumigation of shipping containers with ethyl formate+nitrogen on road and continued journey on sea. J Environ Sci Health, Part B-Pesticides, Food Contaminants, and Agricultural Wastes. https://doi.org/10.1080 /03601234.2020.1786328

Coetzee M, McKirdy H, Du X, McKirdy MJ, Ren YL (2021) Ethyl formate + nitrogen fumigant: A new, safe and environmentally friendly option for treating a 20ft shipping container loaded with general freight. J Environ Sci Health, Part B-Pesticides, Food Contaminants, and Agricultural Wastes 56(7): 650–657. https://doi.org/10.1080/036012 34.2021.1936850

Coetzee M, Newman J, Coupland G, Thomas M, van der Merwe J, Ren YL, McKirdy SJ (2019) Commercial trials evaluating the novel use of ethyl formate for in-transit fumigation of shipping containers. J Environ Sci Health, Part B-Pesticides, Food Contaminants, and Agricultural Wastes. ISSN: 0360-1234 (Print) 1532–4109 (Online) Journal homepage: https://www.tandfonline.com/loi/lesb20.

Collin S, Osman M, Delcambre S, El-Zayat AI, Dufour JP (1993) Investigation of volatile flavor components in fresh and ripened Domiati cheeses. J Agric Food Chem 41: 1659–1663.

Cotton RT, Roark RC (1928) Fumigation of stored products with certain alkyl and alkylene formates. Ind Eng Chem 20: 380.

Damcevski K, Annis P (2006) Influence of grain and relative humidity on the mortality of Sitophilus oryzae (L.) adults exposed to ethyl formate vapour. J Stored Prod Res 42(1): 61–74.

Damcevski KA, Annis PC (2000). The response of three stored product insect species to ethyl formate vapour at different temperatures. In: Wright, EJ, Banks, HJ, Highley, E (eds), Proceedings of the Australian Postharvest Technical Conference 2000, CSIRO Entomology, Canberra, ACT, Australia.

Damcevski KA, Annis PC (2002) The response of three stored product insect species to ethyl formate vapour at different temperatures. Pp. 78–81. In: Wright EJ, Banks HJ, Highley E (eds), Stored Grain in Australia: Proc. of the Australian Postharvest Tech. Conf., Adelaide, Australia.

Damcevski KA, Annis P (2001) Does ethyl formate have a role as a rapid grain fumigant? Preliminary findings. In: Paper presented at the Proceedings of the International Conference on Controlled Atmosphere and Fumigation in Stored Products.

Damcevski KA, Dojchinov G, Haritos VS (25–27 June, 2003) VAPORMATE™, a formulation of ethyl formate with CO2, for disinfestation of grain. Pp. 199–204. In: Wright EJ, Webb MC, Highley E (ed), Stored grain in Australia 2003: Proceedings of the Australian Postharvest Technical Conference, Canberra.

Desmarchelier JM, Allen SE, Ren YL, Moss R, Vu LT (1998) Commercial scale trials on the application of ethyl formate, carbonyl sulphide and carbon disulphide to wheat. CSIRO Entomology Technical Report No. 75, p. 63.

Desmarchelier JM, Johnston FM, Vu LT (1999) Ethyl formate, formic acid, and ethanol in air, wheat, barley and sultanas: Analysis of natural levels and fumigant residues. Pest Sci 55: 815–824.

ECHA (2022) The European Chemicals Agency. reports on the progress made in 2022 in dossier and substance evaluation in line with Article 54 of REACH. Progress in evaluation in 2022 - ECHA (europa.eu). Accessed 26 September 2023.

EPA (2020) Environmental protection authority. Annual report. ISSN 2203–2568.

FDA (1979) US food and drug administration, requirements on content and format for labeling for human prescription drugs rule.

FDA (2004) US food and drug administration, good manufacturing practices for the 21st century for food processing.

Flaherty DL, Christensen LP, Lanini WT, Marois JJ, Phillips PA, Wilson LT (1992) Grape pest management. University of California, Division of Agri-culture and Natural Resources, Davis, CA.

Ghittori S, Imbriani M, Borlini F, Pezzagno G, Zadra P (1984) Studies on the stability of esters in blood in vitro. Bollettino Società Italiana di Biologia Sperimentale 60: 2207–2213.

Haritos V (21–24 July-August, 2005) A new fumigant for stored grain. Australian Grain.

Haritos V, Dojchinov G (2003) Cytochrome c oxidase inhibition in the rice weevil Sitophilus oryzae (L.) by formate, the toxic metabolite of volatile alkyl formates. Comp Biochem Physiol C Toxicol Pharmacol 136(2): 135–143.

Haritos V, Damcevski K, Dojchinov G (2003) Toxicological and regulatory information supporting the registration of VapormateTM as a grain fumigant for farm storages. In: Paper presented at the Proc. Austr. Postharvest Techn. Conf., CSIRO Entomology, Canberra.

Haritos V, Ghabrial H, Ahokas J, Ching M (2000) Role of cytochrome P450 2D6 (CYP2D6) in the stereospecific metabolism of E-and Z-doxepin. Pharmacogenetics Genom 10(7): 591–603.

Haritos VS (August 2000) Review of toxicological studies of carbonyl sulphide, ethyl formate and carbon disulphide. CSIRO Entomology Technical Report No. 87, p. 36.

Haritos VS, Damcevski KA, Dojchinov G (2003) Toxicological and regulatory information supporting the registration of VAPOURMATE as a grain fumigant for farm storages. In: Wright EJ, Webb MC, Highley E (eds), Stored Grain in Australia 2003: Proceedings of the Australian Postharvest Technical Conference, Canberra, CSIRO Stored Grain Research Laboratory, Canberra.

Haritos VS, Damcevski KA, Dojchinov G (2006) Improved efficacy of ethyl formate against stored grain insects by combination with carbon dioxide in a 'dynamic'application. Pest Manag Sci Formerly Pesticide Sci 62(4): 325–333.

Haritos VS, Ren YL, Desmarchelier JM (1999) Regulatory toxicology of alternative fumigants. Pp. 356–363. In: Zuxun J, Quan L, Yongsheng L, Xianchang T, Lianghua G (eds), Proc 7th Intern Working Conf Stored-Prod Prot, Beijing, China.

Hazelhoff EH (1928) Carbon dioxide a chemical accelerating the penetration of respiratory insecticides into the tracheal system by keeping open the tracheal valves. J Econ Entomol 21: 790.

Hilton S, Banks H (1997) Ethyl formate as a fumigant of sultanas: Sorption and efficacy against six pest species.

Hoebeke ER, Carter ME (2003) Halyomorpha halys (Stål)(Heteroptera: Pentatomidae): A polyphagous plant pest from Asia newly detected in North America. Proc Entomol Soc Wash 105(1): 225–237.

JECFA (1997) 97th Joint FAO/WHO expert committee on food additives. Food additives: Ninety-seventh meeting - Joint FAO/WHO Expert Committee on Food Additives (JECFA). Accessed 26 September 2023.

Jones RM (1933) Reducing inflammability of fumigants with carbon dioxide. Ind Eng Chem 25: 394–396.

Kawagoe JC, Abrams AE, Walse SS (2017) Ethyl formate fumigation for the control of stinkbugs in vehicle consignments. In: Paper presented at the International Research Conference on Methyl Bromide Alternatives and Emissions Reductions (MBAO), San Diego, California, USA.

Lee BH, Kim HM, Kim BS, Yang JO, Moon YM, Ren YL (2016) Evaluation of the synergistic effect between ethyl formate and phospine for control of *Aphis gossypii* (Aphididae). J Econ Entomol 109(1): 143–147.

Lee D-H, Short BD, Joseph SV, Bergh JC, Leskey TC (2013) Review of the biology, ecology, and management of Halyomorpha halys (Hemiptera: Pentatomidae) in China, Japan, and the Republic of Korea. Environ Entomol 42(4): 627–641.

Lewis Sr. RJ (2016) Hawley's condensed chemical dictionary., John Wiley Sons, Inc., Hoboken, New Jersey. ISBN: 978-1-118-13515-0

Mahon D, Burrill PR, Ren YL (2002) Seed store disinfestation trials with VAPORMATE™ (ethyl formate + CO2). Pp. 205–209. In: Wright EJ, Webb E, Highley E, (eds), Proceedings of the Australian Postharvest Technical Conference, CSIRO Entomology, Australia.

Mahon D, Ren YL, Burrill PR (2003) Seed store disinfestations with VAPORMATE [EF + CO2]. In: Wright EJ, Webb MC, Highley E (ed), Stored grain in Australia 2003: Proceedings of the Australian Postharvest Technical Conference, CSIRO SGRL, Canberra.

Mai TP (2006). Experimental investigation of heterogeneously catalyzed hydrolysis of esters. PhD dissertation, Otto-von-Gueriske University of Magdeburg, Germany, 19th December 2006, p. 147.

Mata-Segreda JF (1999) Spontaneous hydrolysis of ethyl formate: Isobaric activation parameters. Intern J Chem Kinetics 32(1): 67–71.

Muthu M, Rajendran S, Krishnamurthy T, Narasimhan K, Rangaswamy J, Jayaram M, Majumder S (1984) Ethyl formate as a safe general fumigant. In: Developments in Agricultural Engineering (Vol. 5, pp. 369–393), Elsevier.

McKirdy H (2022) PhD thesis: Efficacy and suitability of liquid ethyl formate for insect pest management. Murdoch University, Australia.

Morrison RT, Boyd RN (1959). Functional derivatives of carboxylic acids. In: Organic Chemistry (3rd Edition, p. 675), Publishers Allyn and Bacon Incorporated, New York.

National Occupational Health and Safety Commission (NOHSC) (1995) Exposure standards for atmospheric contaminants in the occupational environment.

Navarro H, Navarro S, Inbari N (2022) Fumigation of edible cut flowers with ethyl formate mixed with CO_2. Integrated Protection of Stored Products IOBC-WPRS Bulletin 159: 243–248.

Neifert IE, Cook FC, Roark RC, Tonkin WH, Back EA, Cotton RT (1925) Fumigation against grain weevils with various volatile organic compounds. U.S Dept Agr Bull 1313: 40.

Newling WBS, Hinshelwood CN (1936) The kinetics of the acid and the alkaline hydrolysis of esters. J Chem Soc, London, Part II, 1357–1361.

Ni X, Ren YL, Forrester RI, Xue M, Mahon M (2008) Toxicity of ethyl formate to adult Sitophilus oryzae (L.), Tribolium castaneum (herbst) and Rhyzopertha dominica (F.). J Stored Prod Res 44(3): 241–246.

Nursten H (1970) Volatile compounds: The aroma of fruits. Biochemistry Fruits 1: 215–224.

Opdyke DL (1980) A collection of monographs originally appearing in food and cosmetics toxicology. Monographs on fragrance raw materials, Elsevier. https://www.sciencedirect.com/book/9780080237756/monographs-on-fragrance-raw-materials

Obretenov TSV, Lazarov K, Genov N (1972) Aromatic substances of raspberries. II. Aromatic substances of the wild raspberries *Rubus idaeus* and the Willamette and Malling Promise varieties. Nauchni Trudove - Vissh Institut po Khranitelna i Vkusova Promishlenost. Plovdiv 72: 275–279.

Peterson RJ, Chang SS (1982) Identification of volatile flavor components of fresh, frozen beef stew and a comparison of these with those of canned beef stew. J Food Sci 47: 1444–1448.

R'ady P, Arany I, Uzvolgyi E, Bojan F (1981) Activity of pyruvate kinase and lactic acid dehydrogenas e in mouse lung after transplacental exposure to carcinogenic and non-carcinogeni c chemicals. Toxicol Lett 8: 223–227.

Ren YL, Desmarchelier JM (2001). Natural occurrence of carbonyl sulphide and ethyl formate in grains. Pp. 639–649. In: Donahaye EJ, Navarro S, Leesch JG (eds), Proc 6th Intern Conf Controlled Atm Fumi Stored Prod, Fresno, USA, 2000.

Ren YL, Desmarchelier JM, Allen SE, Weller GL (2003) Commercial-scale trials on the application of ethyl formate to barley, oats and canola. CSIRO Entomology Technical Report No 93. 2003. Canberra, ACT, CSIRO Entomology.

Ren Y, Mahon D (2003) Field trials on ethyl formate for fumigation of on-farm storage. In: Wright EJ, Webb MC, Highley E (eds), Paper presented at the Proceedings of the Australian Postharvest Technical Conference, CSIRO Entomology.

Ren YL, Mahon D (2006) Fumigation trials on the application of ethyl formate to wheat, split faba beans and sorghum in small metal bins. J Stored Prod Res 42(3): 277–289.

Ren YL, Newman J, Kostas E (2019) Plant Biosecurity Cooperative Research Centre and CBH Australia Joint Fund Project (SI30068)—Final Report of Identification of Technologies for Management of Conical Snails in Harvested Bulk Grains.

Reuss R, Annis P (2002) Interaction of ethyl formate (EtF) with stored products. Pp. 533–538. In: Credland PF, Armitage DM, Bell CH,

Cogan PM Highley E (eds), Advances in Stored Product Protection, Proceedings of the 8th International Working Conference on Stored Product Protection, York, UK, CAB International, Wallingford, Oxfordshire.

Reuss R, Annis P, Khatri Y (2001) Fumigation of paddy rice and rice products with ethyl formate. Pp. 741–774. In: Donahaye EJ, Navarro S, Leesch JG (eds), (2001) Proc. Int. Conf. Controlled Atmosphere and Fumigation in Stored Products, Fresno, CA. 29 Oct.–3 Nov. 2000, Executive Printing Services, Clovis, CA, U.S.A.

Riudavets J, Damcevski KA, Annis PC (2000) Comparative responses of three psocid species Psocoptera: Liposcelididae) to five fumigants. In: Donahaye EJ, Navarro S, Leesch JG (eds), Proceedings of the International Conference of Controlled Atmosphere and Fumigation in Stored Products, Fresno, CA. 29 Oct.–3 Nov. 2000, Executive Printing Services, Clovis, CA, U.S.A.

Roark RC, Cotton RT (1929) Tests of various aliphatic compounds as fumigants. U.S. Dept Agr Tech Bull 162: 52.

Rohitha B, McDonald R, Hill R, Karl A (1993) A preliminary evaluation of some naturally occurring volatiles on codling moth eggs. In: Paper presented at the Proceedings of the New Zealand Plant Protection Conference.

Ryan R, De Lima C (2014) Ethyl formate fumigation an overview update. In: Paper presented at the Proceedings 11th international working conference on stored product protection.

Ryan RF, Dominiak BC (2021) Ethyl formate: Review of a rapid acting fumigant. Pp. 269–275. In: Jayas D, Jian F (eds), Proceedings of the 11th International Conference on Controlled Atmosphere and Fumigation in Stored Products (CAF2020), CAF Permanent Committee Secretariat.

Ryan RF, Shore WP (27 June–2 July, 2010) Pre-Mix and on-site mixing of fumigants. 10th International Working Conference on Stored Product Protection, Estoril, Portugal.

Sax NI, Lewis RJ (1989) Dangerous properties of industrial materials. Hazardous substances - Handbooks (7th Edition), Van Nostrand Reinhold, New York.

Shah NP, Amis ES (1954) The dielectric constant and salt effects upon the acid hydrolysis of ethyl formate. Analytica Chimica Acta 11: 401–411.

Simmons P, Fisher CK (1946) Ethyl formate and isopropyl formate as fumigants for packages of dry fruits. J Econ Entomol 39: 715–716.

Smyth JD (1954) Studies on tapeworm physiology. VII. Fertilization of Schistocephalus solidus in vitro. Exp Parasitol 3(1): 64–71.

U.S. Department of Commerce (1976) NTIS presents the case of the hidden research report. U.S. Dept. of Commerce, National Technical Information Service, Washington.

Vincent, L. E., and D. L. Lindgren. 1971. Fumigation of dried fruit insects with hydrogen phosphide and ethyl formate. pp. 4-5. Report of Forty-Eight Annual Date Growers Institute. https://archive.org/stream/SER71901558048/SER71901558048_djvu.txt. accessed on April 28, 2024.

Wright EJ, Ren YL, Haritos V, Damcevski K, Mahon D (2001) Update on ethyl formate: New toxicity data and application procedure. CSIRO Entomology, Stored Grain Research Laboratory, Canberra, Australia.

Chapter 13

Progress and Prospect of Nitric Oxide Fumigation for Postharvest Control of Pests and Microorganisms

Yong-Biao Liu

13.1 Introduction

Nitric oxide (NO) was discovered to be a cell messenger chemical in the 1980s, and since then it has been studied extensively and found to play various roles in a wide range of physiological processes of almost all organisms (Culotta and Koshland, 1992; Lowenstein et al., 1994; Beckman and Koppenol, 1996; Müller, 1997; Lamattina et al., 2003; Moncada and Higgs, 2006; Murad, 2011). Nitric oxide research has led to its applications in management of a wide variety of health problems (Ricciardolo et al., 2004; Coyle, 2013; Adusumilli et al., 2020). In agriculture, however, NO has been used for over 100 years in color preservation

DOI: 10.1201/9781003309888-13

of meat products (Haldane, 1901). Since 2000, NO has also been studied for postharvest quality preservation and shelf-life extension of fresh fruit and vegetables (Manjunatha et al., 2010; Madebo et al., 2022; Liu et al., 2023). In 2013, NO was reported to be a potent fumigant for postharvest pest control, extending its potential application to pest management (Liu, 2013). More recently, NO fumigation containing NO_2 was also demonstrated to be effective against microbes, further extending application of NO fumigation to microbial management (Liu et al., 2019; Oh and Liu, 2020; Oh et al., 2020). This expands potential applications of NO fumigation to control of both pests and microbes and makes NO fumigation more attractive for potential practical applications.

Methyl bromide was the dominant fumigant for postharvest pest control in the past, and global phasing-out of methyl bromide production has created an urgent need for alternative treatments for postharvest pest management. Currently phosphine and sulfuryl fluoride are the main alternative fumigants. But both have significant shortcomings and cannot meet the needs of postharvest pest control. Nitric oxide may have potential to be another alternative fumigant for postharvest control of not only pests but also microbes. Nitric oxide fumigation has advantages in high efficacy against pests, broader applications against both pests and microbes, beneficial effects on the postharvest storage life of fresh products, and lack of toxic residues. But commercialization of NO fumigation also faces challenges of the small niche market for chamber fumigation, requirement of an airtight fumigation chamber, need for a nitrogen source for establishing ultralow oxygen conditions, stringent fumigation procedures, competition from other alternative fumigants, and high cost of registration. In this chapter, the current state of NO fumigation research is reviewed and discussed, and suggestions are provided on procedures and applications for future research and development efforts on NO fumigation.

13.2 Procedures of Nitric Oxide Fumigation

Nitric oxide fumigation procedures were described previously in a video article (Liu et al., 2017) and other publications (Liu, 2013, 2015, 2017; Liu and Yang, 2016; Liu et al., 2021). Figure 13.1 presents a summary of NO fumigation procedures and its principles. Nitric oxide is a highly reactive molecule and reacts with oxygen spontaneously to form nitrogen dioxide (NO_2) (Beckman and Koppenol, 1996). Therefore, NO fumigation must be conducted under ultralow oxygen (ULO) conditions to preserve NO, as described before (Liu and Yang, 2016; Liu et al., 2017, 2021). An N_2 source is needed to flush the fumigation chamber to establish ULO conditions for NO

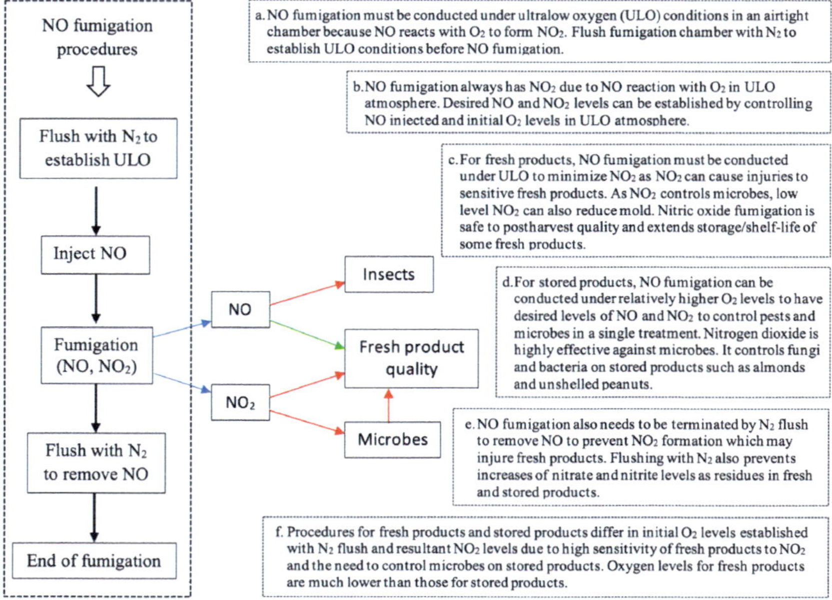

Figure 13.1 Summary of nitric oxide fumigation procedures and principles of effects. Red arrows show negative effects, and green arrow shows positive effects.

fumigation. Chambers for NO fumigation must also be airtight to prevent the leaking in of oxygen. An O_2 analyzer and an NO monitor are also necessary for NO fumigation. At the end of fumigation, the fumigation chamber should also be flushed with N_2 to reduce NO levels in the chamber and thereby minimize formation of NO_2 during ventilation. For fresh products, exposure to high levels of NO_2 can result in injuries in the form of stains or discoloration (Liu, 2016). For both fresh and stored products, exposure to NO_2 at the end of NO fumigation can lead to significant increases of nitrate (NO_3^-) and nitrite (NO_2^-) as residues (Yang and Liu, 2017, 2019b).

The initial O_2 level in the ULO atmosphere determines the NO_2 level in NO fumigation. It can be higher if products can tolerate higher levels of NO_2. Also, NO_2 has been demonstrated to be effective in controlling microbes (Liu et al., 2019; Oh and Liu, 2020; Oh et al., 2020). There is often a need to control pests and mold in stored products. Initially, O_2 can be kept to a level that will generate the desired level of NO_2 after NO injection so that NO fumigation will have the desired levels of NO and NO_2 to control pests and microbes in a single treatment. Even for fresh products such as strawberries, NO fumigation results in less mold than in controls, and NO_2 might be responsible for the mold reduction (Yang and Liu, 2018). Therefore, the initial ULO level can be flexible, to allow for the existence of a low level of NO_2 if it is safe to the postharvest quality of fresh products.

In the presence of moisture, NO_2 forms nitric acid, which is corrosive to certain metals and plastics. Therefore, chambers, tubing, and manifolds used for NO fumigation should be made of stainless steel or corrosion-resistant plastics such as polytetrafluoroethylene (PTFE). Even though all NO fumigation experiments in the past were small scale and conducted under laboratory conditions, the requirements for an airtight fumigation chamber and establishing ULO conditions are the same for large-scale commercial NO fumigations, so these experiments can be used to develop protocols for large-scale NO fumigations.

13.3 Efficacy of Nitric Oxide Fumigation Against Pests

Nitric oxide fumigation has been demonstrated to be effective against all of the 14 insects and mites tested to date (Table 13.1). There are considerable variations among different species and life stages in susceptibility to NO fumigation. In general, mobile life stages are more susceptible to NO fumigation than egg and pupa stages. Small external soft-body insects on fresh products are more susceptible than stored-product insects such as beetles and weevils and internal insects of fruits (Liu, 2013, 2015; Liu and Yang, 2016, 2018).

External pests such as the western flower thrips [*Frankliniella occidentalis* (Pergande)], lettuce aphid [*Nasonovia ribisnigri* (Mosley)], and long-tailed mealybug [*Pseudococcus longispinus* (Targioni Tozzetti)] on fresh products are highly susceptible to NO fumigation and completely controlled in a few hours at relatively low NO concentrations of ≤1% at low temperatures (Liu, 2013). Internal feeding larvae of the spotted wing drosophila [*Drosophila suzukii* (Matsumura)] in infested cherries take 8 h to control with 2.5% NO fumigation (Liu and Yang, 2016). Codling moth [*Cydia pomonella* (Linnaeus)] larvae in infested apples need 24 h of NO fumigation at 5% concentration at 2°C for complete control (Liu et al., 2016). Stored-product insects are more tolerant of NO fumigation than insects on fresh products. It takes 24 to 72 h to achieve complete control of the Indian meal moth [*Plodia interpunctella* (Hubner)], confused flour beetle (*Tribolium confusum* Jacquelin du Val), and rice weevil [*Sitophilus oryzae* (Linnaeus 1763)] with 1%–2% NO at 15°–25°C (Liu, 2013, 2015; Liu and Yang, 2016). Large lepidopteran larvae such as those of the navel orangeworm [*Amyelois transitella* (Walker 1863)] in infested pistachios were successfully controlled in 8 h with 3% NO fumigation, and eggs were controlled in 16 h of fumigation

Table 13.1 Summary of nitric oxide fumigation treatment that achieved 100% mortality of different pest species at different life stages

Species	Life Stages*	NO (%)	Time (h)	Temp (°C)	Reference
Lettuce aphid	nymph, adult	0.2	12	2	Liu, 2013
		0.5	9	2	
		1.0	3	2	
	nymph, adult (lettuce)	0.5	16	2	Yang and Liu, 2019a
Light brown apple moth	egg	3.0	12	2	Liu and Simmons, 2021
	larva, pupa	2.0	8	2	Liu, 2015
Long-tailed mealybug	nymph, adult	2.0	2	2	Liu 2013
Western flower thrips	larva, adult	0.2	8	2	Liu, 2013, 2016
		2.0	2	2	
	larva, adult (lettuce)	0.5	16	2	Yang and Liu, 2019a
Codling moth	egg, larva, pupa	2.0	48	2	Liu, 2015
	larva (apple)	5.0	24	2	Liu et al., 2016
Spotted wing drosophila	larva (cherry)	2.5	8	2	Liu, 2015
	egg/larva (strawberry)	3.0	8	2	Yang and Liu, 2018
Confused flour beetle	larva, pupa	0.5	24	20	Liu, 2013
	adult	0.5	8	20	
	egg	2.0	24	10	
Granary weevil	adult	1.0	24	25	Liu et al., 2021
Rice weevil	adult	1.0	24	25	Liu, 2013
	egg	1.0	48	25	
Indian meal moth	egg	1.0	24	20	Liu and Yang, 2016
Navel orangeworm	egg	3.0	8	25	Yang et al., 2021
		2.0	16	25	

(Continued)

Table 13.1 (Continued) Summary of nitric oxide fumigation treatment that achieved 100% mortality of different pest species at different life stages

Species	Life Stages*	NO (%)	Time (h)	Temp (°C)	Reference
	larva (walnut)	1.0	8	25	
		0.5	24	25	
	pupa	2.0	8	25	
		1.0	16	25	
Bulb mite	larva, adult (bulbs)	2.0	24	20	Liu, 2017
False spider mite	larva, adult	0.5	6	2	Liu et al., 2021
Ham mite	egg	2.0	8	25	Yang et al., 2022
	larva, adult	2.0	4	25	
	egg (ham)	0.5	48	25	
		1.0	24	25	
	larva, adult (ham)	0.5	16	25	
		1.0	8	25	

Note: *A life stage followed by a product name was either inside the product or tested together with the product.

with 2% NO (Yang et al., 2021). Light brown apple moth larvae and pupae were completely controlled in 8 h of fumigation with 2% NO at 2°C (Liu, 2015). Light brown apple moth eggs were more tolerant of NO fumigation and were completely controlled in 12 h of fumigation with 3% NO at 2°C (Liu and Simmons, 2021).

Nitric oxide fumigation is also effective against mites (Table 13.1). Bulb mites (*Rhizoglyphus spp.*) on infested peanuts were completely controlled with 2.0% NO in 24 h at 20°C (Liu, 2017). Complete control of false spider mites [*Brevipalpus phoenicis* (Geijskes)] was achieved in a 6 h fumigation with 0.5% NO at 2°C (Liu et al., 2021). Ham mite [*Tyrophagus putrescentiae* (Schrank)] eggs and larvae/adults were completely controlled in 8 and 4 h, respectively, of fumigation with 2% NO at 25°C (Yang et al., 2022). All these results show that NO fumigation has high efficacy against all pests.

Although NO fumigation has much higher efficacy against soft-bodied external pests on fresh products than against internal feeding insects, its efficacy against internal insects and eggs is a major advantage of NO over other fumigants. For example, there is no fumigant except NO that has achieved complete control of codling moth larvae in infested apples. Methyl bromide fumigation has been used in combination with post-fumigation cold storage to achieve effective control of codling moth larvae in exported apples (Hansen et al., 2000). The reason for the higher efficacy of NO fumigation compared with other fumigants for controlling internal feeding insects is that NO is a very small molecule and is a ubiquitous cell messenger chemical and, therefore, is better than other fumigants in penetrating plant tissue. Sulfuryl fluoride fumigation is not effective against insect eggs (Bell et al., 1998). Phosphine fumigation is also not very effective against eggs of some insect pests such as the light brown apple moth (Liu et al., 2013). Nitric oxide fumigation is effective against eggs of insects and mites in addition to having high efficacy against mobile life stages (Table 13.1).

13.4 Effects of NO Fumigation on Postharvest Quality of Fresh Fruit and Vegetables

Nitric oxide fumigation is safe to fresh fruit and vegetables when it is terminated properly with an N_2 flush to remove NO from the fumigation chamber. As reported by Liu (2016), 3 h fumigations with 2% NO conducted under 30–50 ppm O_2 atmospheres were safe to all fresh products tested, including lettuce, broccoli, cucumbers, peppers, tomatoes, strawberries, apples, pears, oranges, and lemons, when terminated with an N_2 flush (NO-N_2 treatment). When the NO fumigation is terminated by flushing the fumigation chamber with air to allow the NO to react with oxygen to form NO_2 (NO-Air treatment), the treatment causes stains on delicate fresh products, including leafy vegetables, broccoli, squash, and peaches, as well as some apples (Figure 13.2). In large-scale fumigations of lettuce in commercially packed cartons, 16 h of fumigation with 0.5% NO at 2°C achieved complete control of lettuce aphids and western flower thrips, and the treatment had no negative effects on the external or internal quality of lettuce 14 d after fumigation (Yang and Liu, 2019a).

It is worth noting that properly conducted NO fumigation not only is safe to product quality but also helps to extend storage/shelf-life. Nitric oxide fumigations for control of western flower thrips and spotted wing drosophila resulted in better postharvest quality of strawberries, which were significantly firmer and had a brighter, richer color than controls one week after fumigation and had reduced mold compared with controls (Liu, 2016; Yang and Liu, 2018), consistent with the extending of the postharvest life of strawberries by NO fumigation reported earlier (Wills et al., 2000). Nitric oxide fumigation for control of codling moth larvae in apples also resulted in better apple quality than controls four weeks after fumigation (Liu et al., 2016).

Figure 13.2 Effects of NO fumigation on the visual quality of fresh fruit and vegetables after 14 d cold storage. NO-N$_2$ fumigation was terminated with an N$_2$ flush; NO-air fumigation was terminated with an air flush (Liu 2016).

Many studies have shown that NO fumigation extends storage and shelf-life of fresh fruit and vegetables (Soegiarto and Wills, 2004; Manjunatha et al., 2010, 2012; Li et al., 2012; Pols et al., 2022; Liu et al., 2023). Fumigations with 5–10 ppm NO for 2 h extended postharvest life by 14% to 50% for different fresh vegetables (Soegiarto and Wills, 2004). Fumigations with 60 ppm NO for 3 h effectively suppressed ethylene formation and respiration rate and significantly helped maintain the postharvest quality of papaya fruit (Li et al., 2012). Nitric oxide fumigations at 10–300 ppm concentrations also enhanced resistance to chilling injuries in various fruits (Liu et al., 2023). In comparison with the very low concentrations of NO used for storage of fresh fruit and vegetables, NO fumigations for pest control have much higher concentrations. Published studies (Liu, 2016; Yang and Liu, 2018) showed that NO at the much higher concentrations needed for pest control also had benefits to the postharvest quality of fresh products. These benefits are worth considering when deciding what fumigation treatment should be used for postharvest pest control on fresh products.

There are many studies on the mechanisms of NO effects on fresh products (Manjunatha et al., 2010; Liu et al., 2023). Nitric oxide acts as an antagonist of ethylene and, therefore, delays ripening of fruits (Manjunatha et al., 2010). Nitric oxide also benefits the postharvest quality of fresh products through other mechanisms (Manjunatha et al., 2010, 2012; Liu et al., 2023). Beneficial effects of NO fumigation may include antimicrobial effects of NO_2, as demonstrated by reduced mold on strawberries (Yang and Liu, 2018). Therefore, NO fumigation benefits the postharvest quality of fresh products through modulating plant biochemical processes such as ethylene biosynthesis and inhibition of microbial growth by NO and NO_2.

13.5 Nitric Oxide Fumigation for Control of Microorganisms

Nitric oxide fumigation also contains NO_2 from the NO reaction with O_2 in the ULO atmosphere. The level of O_2 in the ULO determines NO_2 levels. Both NO and NO_2 are effective against microbes, but NO_2 is more effective than NO. Three-hour fumigation with 0.1% NO_2 or 1% NO completely inactivated *Aspergillus flavus* spores (Liu et al., 2019) (Table 13.2). Effective control of fungi and bacteria with NO_2 fumigation was also demonstrated on almonds and peanuts (Oh and Liu, 2020; Oh et al., 2020). Complete control of microbes on almonds was achieved in 1 d fumigation with 1% NO_2 at 25°C (Oh and Liu, 2020). There are higher loads of microbes inside peanuts than on outer surfaces (Figure 13.3). Three-day fumigation with 3% NO_2 at 25°C produced complete control of microbes on unshelled peanuts, including

Table 13.2 Effects of 3 h fumigations of *Aspergillus flavus* spores with NO and NO_2 at different concentrations on 4 d post-treatment colony formation on *Aspergillus* differentiation agar medium plates at 25°C

	Treatments						
Parameters	*Control*	*T1*	*T2*	*T3*	*T4*	*T5*	*T6*
Initial O_2 (%)	20.9	0.001	0.001	0.1	0.1	20.9	20.9
NO (%)	0	**0.1**	**1.0**	0	0.9	0	0
NO_2 (%)	0	0.001	0.001	**0.1**	0.1	0.02	**0.1**
No. of colonies	51.8 ± 1.5	41.0 ± 3.0	0	0.5 ± 0.3	0	24.6 ± 2.9	0
Reduction* (%)	0	20.8	100	99.0	100	52.5	100
Plate view	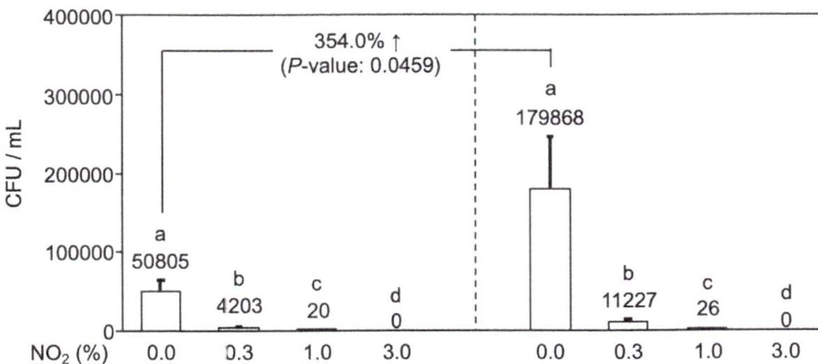						

Source: Based on data from Liu et al. (2019).
Note: *Fumigations were conducted under different oxygen levels. Reduction (%) is the reduction in the relative colony count compared with that of the control.

Figure 13.3 Colony formation units (CFU) from wash-off samples of unshelled peanuts fumigated with NO_2 for 3 d at 25°C. Samples of unshelled peanuts from fumigation treatments were washed with liquid tryptic soy broth (TSB) in bags. After a wash-off sample was taken for each treatment, peanuts were cracked open to wash off microbes inside peanuts. Wash-off samples before and after peanuts were cracked were diluted with TSB and cultured on non-selective TSB agar plates for 2 d at 32°C to count colonies and calculate the CFU. CFU data were transformed by log 10 prior to one-way ANOVA with the Tukey HSD test ($P = 0.05$). Means of CFU/mL with the same letter were not significantly different (Oh et al., 2020).

microbes inside the peanuts (Oh et al., 2020). The effects on microbes inside the unshelled peanuts indicate good penetration of NO_2 into shells,

In addition, NO fumigation has been demonstrated to be effective in controlling fungi and preserving the postharvest quality of dried apricots (Chen et al., 2019, 2021). Nitric oxide fumigation reduces mycelium biomass, damages the plasma membrane, increases cell membrane permeability, and changes the mycelium microscopic structure, thereby inhibiting fungal growth and reproduction (Chen et al., 2019). Nitric oxide fumigation reduces pest infestation as well as microbial infections on dried apricots during storage. Nitric oxide fumigation also maintains higher levels of total soluble solids, titratable acidity, and ascorbic acid as well as reducing sugar in dried apricots compared with controls. Fumigated apricots also showed reduced browning and nutrient losses compared with controls (Chen et al., 2021). All these results suggest that NO fumigation with desired levels of NO_2 can effectively control microbes on stored products.

13.6 Residues of Nitric Oxide Fumigation

In NO fumigation, NO reacts with O_2 in the ULO atmosphere to form NO_2, which can be further converted to nitrate and nitrite. Therefore, NO fumigation may leave nitrate and nitrite as residues. The levels of residues are related to the initial ULO levels in the fumigation chamber and whether the fumigation chamber is flushed properly at the end of fumigation. A total of 20 fresh and 9 stored products were tested for residues. The 20 fresh products were apples, apricots, asparagus, avocados, blueberries, broccoli, cherries, garlic, grapes, kiwis, lettuce, mangos, oranges, peaches, pears, peppers, plums, squash, strawberries, and tangerines. Most of them were fumigated for 16 h with NO at 3% at a low temperature of 2°C. For most of the 20 fresh fruits and vegetables, there were no significant increases in nitrate or nitrite levels compared with controls one day after fumigation treatment terminated with an N_2 flush (Yang and Liu, 2017). The 9 stored products tested for residues were almonds, barley, garbanzo beans, pecans, pinto beans, pistachios, rice, walnuts, and wheat. They were fumigated for 24 h with 3% NO at 25°C. For all the 9 stored products that underwent the fumigation treatment, at all posttreatment times there were no significant increases in nitrate in stored products whose fumigation treatment terminated with an N_2 flush compared with controls. And nitrite levels in all fumigated products from the treatment terminated with an N_2 flush were not significantly higher than those in controls at 14 d after fumigation (Yang and Liu, 2019b).

When NO fumigation is terminated with an air flush instead of an N_2 flush, to allow NO to fully react with O_2 to form NO_2, there are significant increases in nitrate and sometime also nitrite levels in the fumigated fresh and stored products (Yang and Liu, 2017, 2019b). The magnitudes of increases vary depending on the products. Among the fresh products, lettuce from NO fumigation without an N_2 flush had the highest increase in nitrate level, which was 2.5 times that of lettuce from NO fumigation with an N_2 flush and controls at 24 h after fumigation (Yang and Liu, 2017). For most of the 9 stored products, NO fumigation without an N_2 flush at the end of fumigation resulted in significant increases in nitrate and nitrite levels; these nitrate and nitrite levels declined over time from higher levels at 1 to 14 d after fumigation. Wheat had the highest increase in nitrate, reaching about 3 times the level of that with NO fumigation with an N_2 flush and controls (Yang and Liu, 2019b).

13.7 Discussion

Nitric oxide fumigation has the advantage over most alternative fumigants of being effective in controlling both pests and microbes in a single treatment. For pest control, NO fumigation is effective against all 14 pest species tested to date and their different life stages. The pest species tested represent different taxonomical groups, both external and internal feeders, both fresh- and stored-product pests. The effective control of all pest species suggests that NO fumigation is likely effective against all insect pests and mites, although there is a large variation in susceptibility to NO fumigation among different species and life stages. An important characteristic of the NO molecule is its small size and ability to penetrate plant tissue since it is a ubiquitous cell message molecule. Therefore, it is better than most other fumigants at controlling internal pests. One example is that 24 h of NO fumigation produces complete control of codling moth larvae in apples (Liu et al., 2016). In comparison, a combination of 55 d of cold storage and a subsequent 2 h of methyl bromide fumigation was found to yield effective control of codling moth larvae in exported apples (Hansen et al., 2000).

There are several major alternative fumigants to methyl bromide, including phosphine, sulfuryl fluoride, and ethyl formate. Phosphine is the major alternative fumigant for both fresh- and stored-product pests. However, phosphine fumigation is not effective against some pests due to tolerance or resistance. In general, phosphine fumigation also has long treatment times, which may need to be extended over 10 d to achieve effective control of some pests due to tolerance or resistance (Hole et al., 1976; Benhalima et al., 2004). Although phosphine fumigation under high oxygen conditions has higher efficacy and shorter treatment times (Liu, 2011; Liu et al., 2013), there is no indication that it will be used commercially. Sulfuryl fluoride is not effective against insect eggs (Bell et al., 1998) and is phytotoxic to fresh products (Aung et al., 2001). Therefore, it has certain limitations for postharvest pest control. Ethyl formate has high absorbing rates in fresh products and is phytotoxic to some fresh products (Stewart and Mon, 1984; Zoffoli et al., 2013; Kyung et al., 2019).

In comparison, NO fumigation is effective against all life stages and has been demonstrated to be safe to fresh products and, therefore, has the potential to be an effective alternative fumigant. However, because of stringent requirements for airtight chambers, NO fumigation is likely to be limited to chamber fumigations, which is a relatively small niche market. It is also a challenge for NO fumigation to compete with other alternative fumigants on cost, to gain market acceptance. To make NO fumigation an economically feasible alternative, it is important to expand its applications to postharvest microbial management.

Contamination by molds and mycotoxins is a major problem of stored products and may result in losses in dry matter, quality, and nutritional value. Aflatoxins of the common mold *A. flavus* are carcinogenic and pose serious health risks to consumers (Richard and Payne, 2003). Key management tools for molds include drying and modified atmosphere storage. Sulfur dioxide (SO_2) and aliphatic acids are also used as preservatives on some food products to prevent fungal growth (Magan and Aldred, 2007). For stored products, mold development is also facilitated by pest infestation. Pest infestations raise the temperature and moisture content of stored products to be more favorable for microbial infection. Pest movements also facilitate the spread of fungal spores through stored products and accelerate microbial infection; an example is the spread of *A. flavus* by navel orangeworm on almonds (Palumbo et al., 2014). Therefore, there is a need to control both pests and microbes on stored products in many cases. Nitric oxide fumigation has the potential to control insects and microbes on stored products in a single treatment.

By establishing proper initial O_2 levels, a NO fumigation treatment can have effective levels of NO for pest control and effective levels of NO_2 for microbial control. Nitrogen dioxide fumigation was demonstrated to be effective in controlling *A. flavus* spores and microbes on almonds and unshelled peanuts (Liu et al., 2019; Oh and Liu, 2020; Oh et al., 2020). A recent study showed that NO fumigation was more effective than SO_2 fumigation in controlling microbes on dried apricots (Chen et al., 2019). These results indicate good potential of NO fumigation for microbial control on stored products. The use of NO fumigation to control microbes will expand the potential market of NO fumigation and make NO fumigation more attractive for commercial development and applications.

Other benefits of NO fumigation should also be considered in determining the economic value of NO fumigation to agriculture. These benefits may include preservation of the postharvest quality of fresh fruit and vegetables and the absence of the toxic residues associated with most other fumigants. Benefits of NO fumigation to the postharvest quality of fresh products have been well demonstrated (Manjunatha et al., 2010; Liu et al., 2023) and should be considered an important factor in commercial development of NO fumigation. Currently, in addition to NO, several other chemicals are used to extend storage/shelf-life and reduce postharvest diseases in fresh products; they include 1-methylcyclopropene (1-MCP) (Satekge and Magwaza, 2022), hexanal (Song et al., 2010; Cheema et al., 2018), and diphenylamine (DPA) (EFSA, 2011; Dias et al., 2020). As NO fumigation for pest control can also benefit the postharvest quality of fresh products, it may reduce the need for postharvest application of these chemicals and its associated costs.

Nitric oxide fumigation may increase nitrate and nitrite levels in fumigated food products when fumigation is not terminated properly with N_2 flush. However, nitrate and nitrite commonly exist in food products, and their levels, which vary greatly among different food products (Weitzberg and Lundberg, 2013; Bahadoran et al., 2016), have important nutritional value (Griesenbeck et al., 2009; Hord et al., 2009; Bedale et al., 2016; Ma et al., 2018). They are also commonly added to meat products for preservation (Shakil et al., 2022). However, nitrate and nitrite at high levels can be harmful due to possible production of carcinogenic nitrosamines under certain conditions (Bedale et al., 2016). The World Health Organization (WHO) recommended daily uptake limits of 3.7 mg/kg and 0.06–0.07 mg/kg for nitrate and nitrite, respectively (Weitzberg and Lundberg, 2013). For most products, even when NO fumigation is not flushed with N_2 at the end of fumigation, NO fumigation may only result in very limited increases in nitrate and nitrite over their already existing levels. For example, although NO fumigation resulted in higher concentrations of nitrate and nitrite ions in dried apricots than in the control group, their levels were in line with the standard limits (Chen et al., 2021). Given their nutritional value, their common existence in food products, and limited increases in NO-fumigated products even when treatments were improperly terminated, there should be no concerns about residues of NO fumigation.

A major impediment to the commercial application of NO fumigation is its high cost due to complex procedures and stringent requirements for airtight fumigation chambers. The two major cost components are equipment and energy for operating the N_2 generation equipment. Nitric oxide gas is commercially available. Another factor that may need to be considered in the cost of NO fumigation is that NO fumigation will result in emission of NO_2 into the atmosphere. Nitrogen dioxide is a major air pollutant from fossil fuel–powered automobiles and power plants. Approximately 9.8 metric tons of NO_x were emitted in the United States in 2018 (US-EPA, 2022). Even though the quantity of NO_2 from NO fumigation is negligible compared to the NO_2 released from other sources, if NO fumigation is used on a large scale, a scrubbing device may still be needed to minimize NO_2 release into the atmosphere, and this may further increase the cost of NO fumigation. Even though there is a lack of suitable airtight fumigation chamber products on the market for NO fumigation, it is technically feasible to have suitable equipment, including airtight chambers, for commercial-scale NO fumigation. The cost components for NO fumigation were analyzed previously and projected to be moderate (Liu, 2015).

In summary, compared with other fumigants, NO fumigation has advantages and disadvantages as discussed above, and its prospects for commercial use are not clear at this stage. The advantages include high efficacy especially against internal pests and eggs that are hard to kill with other fumigants, preserving fresh product quality, and absence of toxic residues. Its effectiveness for microbial control on stored products is another major advantage that may help to expand applications of NO fumigation and improve its commercial prospects. However, research on microbial control with NO fumigation is still in very early stages, and more research and development efforts are needed. Given its potential in controlling postharvest pests and microorganisms and reaping other benefits, more research and development efforts are warranted to establish effective NO fumigation treatments against pests and microbes on various fresh and stored products, develop appropriate equipment and commercial-scale treatment protocols, and make NO fumigation a practical option for postharvest control of pests and microorganisms.

References

Adusumilli NC, Zhang D, Friedman JM, Friedman AJ (2020) Harnessing nitric oxide for preventing, limiting and treating the severe pulmonary consequences of COVID-19. Nitric Oxide 103:4–8.

Aung LH, Leesch JG, Jenner JF, Grafton-Cardwell EF (2001) Effects of carbonyl sulfide, methyl iodide, and sulfuryl fluoride on fruit phytotoxicity and insect mortality. Ann Appl Biol 139:93–100.

Bahadoran Z, Mirmiran P, Jeddi S, Azizi F, Ghasemi A, Hadaegh F (2016) Nitrate and nitrite content of vegetables, fruits, grains, legumes, dairy products, meats and processed meats. J Food Compo Anal 51:93–105.

Beckman JS, Koppenol WH (1996) Nitric oxide, superoxide, and peroxynitrite: The good, the bad, and the ugly. Am J Physiol 271 (Cell Physiol 40):C1424–C1437.

Bedale W, Sindelar JJ, Milkowski AL (2016) Dietary nitrate and nitrite: Benefits, risks, and evolving perceptions. Meat Sci 120:85–92.

Bell CH, Savvidou N, Wontner Smith TJ (1998) The toxicity of sulfuryl fluoride (Vikane) to eggs of insect pests of flour mills. pp. 345–350. In: Zuxun J, Quan L, Yongsheng L, Xianchang T, Lianghua G (eds.) Proc 7th Intern Working Conf Stored-Prod Prot, Beijing, China.

Benhalima H, Chaudhry MQ, Mills KA, Price NR (2004) Phosphine resistance in stored-product insects collected from various grain storage facilities in Morocco. J Stored Prod Res 40:241–249.

Cheema A, Padmanabhan P, Amer A, Parry MJ, Lim LT, Subramanian J, Paliyath G (2018) Postharvest hexanal vapor treatment delays ripening and enhances shelf life of greenhouse grown sweet bell pepper (*Capsicum annum* L.). Postharvest Biol Technol 136:80–89.

Chen Y, Deng H, Tiemur A, Wang J, Wu B (2021) Effect of nitric oxide fumigation on microorganisms and quality of dried apricots during storage. J Food Process Preserv 45:e15725.

Chen Y, Guo Q, Wei J, Zhang J, Zhang Z, Wang J-D, Wu B (2019) Inhibitory effect and mechanism of nitric oxide (NO) fumigation on fungal disease in Xinjiang Saimaiti dried apricots. LWT Food Sci Technol 116:108507.

Coyle JT (2013) Nitric oxide and symptom reduction in schizophrenia. JAMA Psych 70:664–665. (DOI:10.1001/jamapsychiatry.2013.210).

Culotta E, Koshland DE Jr (1992) NO news is good news. Sci 258: 1862–1865.

Dias C, Amaro AL, Salvador AC, Silvestre AJD, Rocha SM, Isidoro N, Pintado M (2020) Strategies to preserve postharvest quality of horticultural crops and superficial scald control: From diphenylamine antioxidant usage to more recent approaches. Antioxidants 9:356. (DOI: 10.3390/antiox9040356).

EFSA (European Food Safety Authority) (2011) Review of the existing maximum residue levels (MRLs) for diphenylamine according to Article 12 of Regulation (EC) No 396/2005. EFSA J 9:2336.

Griesenbeck JS, Steck MD, Huber JC Jr, Sharkey JR, Rene AA, Brender JD (2009) Development of estimates of dietary nitrates, nitrites, and nitrosamines for use with the short willet food frequency questionnaire. Nutrit J 8:16. (DOI: 10.1186/1475-2891-8-16).

Haldane J (1901) The red colour of salted meat. J Hyg 1:115–122.

Hansen JD, Drake SR, Moffitt HR, Robertson JL, Albano DJ, Heidt ML (2000) A two-component quarantine treatment for postharvest control of codling moth on apple cultivars intended for export to Japan and Korea. HortTechnol 10:186–194.

Hole BD, Bell CH, Mills KA, Goodship G (1976) The toxicity of phosphine to all developmental stages of thirteen species of stored product beetles. J Stored Prod Res 12:235–244.

Hord NG, Tang Y, Bryan NS (2009) Food sources of nitrates and nitrites: The physiologic context for potential health benefits. Am J Clin Nutr 90:1–10.

Kyung YJ, Kim HK, Cho SW, Kim BS, Yang JO, Lee BH, Koo HN, Kim GH (2019) Comparison of the efficacy and phytotoxicity of phosphine and ethyl formate for controlling *Pseudococcus longispinus* (Hemiptera: Pseudococcidae) and *Pseudococcus orchidicola* on imported foliage nursery plants. J Econ Entomol 112:2149–2156.

Lamattina L, Garcia-Mata C, Graziano M, Pagnussat G (2003) Nitric oxide: The versatility of an extensive signal molecule. Annu Rev Plant Biol 54:109–136.

Li XP, Wu B, Guo Q, Wang JD, Zhang P, Chen WX (2012) Effects of nitric oxide on postharvest quality and soluble sugar content in papaya fruit during ripening. J Food Process Preserv 38:591–599. (DOI: 10.1111/jfpp.12007).

Liu YB (2011) Oxygen enhances phosphine toxicity for postharvest pest control. J Econ Entomol 104:1455–1461.

Liu YB (2013) Nitric oxide as a potent fumigant for postharvest pest control. J Econ Entomol 106:2267–2274.

Liu YB (2015) Nitric oxide as a new fumigant for postharvest pest control on fresh commodities. Acta Hort 1105:321–328.

Liu YB (2016) Nitric oxide fumigation for control of western flower thrips and its safety to postharvest quality of fresh fruit and vegetables. J Asia-Pacific Entomol 19:1191–1195.

Liu YB (2017) Nitric oxide fumigation for control of bulb mites on flower bulbs. J Econ Entomol 110:2046–2051. (DOI: 10.1093/jee/tox187).

Liu Y, Chen T, Tao N, Yan T, Wang Q, Li Q (2023) Nitric oxide is essential to keep the postharvest quality of fruits and vegetables. Hort 9:135. (DOI: 10.3390/horticulturae9020135).

Liu YB, Liu SS, Simmons GS, Walse SS, Myers SW (2013) Effects of phosphine fumigation on survivorship of light brown apple moth, *Epiphyas postvittana* (Walker) (Lepidoptera: Tortricidae) eggs. J Econ Entomol 106: 1613–1618.

Liu YB, Oh S, Jurick WM II (2019) Response of *Aspergillus flavus* spores to nitric oxide fumigations in atmospheres with different oxygen concentrations. J Stored Prod Res 83:78–83.

Liu YB, Oh S, Yang X (2021) Nitric oxide fumigation for postharvest control of pests and pathogens. pp. 288–295. In: Jayas DS, Jian F (eds.) Proc 11th Intern Conf Controlled Atm Fumi Stored Prod (CAF2020), CAF Permanent Committee Secretariat, Winnipeg, Canada.

Liu YB, Simmons S (2021) Effects of nitric oxide fumigation on mortality of light brown apple moth, *Epiphyas postvittana* (Lepidoptera: Tortricidae). Agri Sci 12:1286–1294. (DOI: 10.4236/as.2021.1211082).

Liu YB, Yang X (2016) Prospect of nitric oxide as a new fumigant for postharvest pest control. pp. 161–166. In: Navarro S, Jayas DS, Alagusundaram K (eds.), Proc 10th Intern Conf Controlled Atm Fumi Stored Prod (CAF2016), CAF Permanent Committee Secretariat, Winnipeg, Canada.

Liu YB, Yang X (2018) Nitric oxide as a new fumigant for postharvest pest control. pp. 596–604. In: Adler CS, Opit G, Fürstenau B, Müller-Blenkle C, Kern P, Arthur FH, Athanassiou CG, Bartosik R, Campbell J, Carvalho MO, Chayaprasert W, Fields P, Li Z, Maier D, Nayak M, Nukenine E, Obeng-Ofori O, Phillips T, Riudavets J, Throne J, Schöller M, Stejskal V, Talwana H, Timlick B, Trematerra P (eds), Proc 12th Intern Working Conf Stored-Prod Prot, Berlin, Germany.

Liu YB, Yang X, Masuda T (2017) Procedures of laboratory fumigation for pest control with nitric oxide gas. J Vis Exp 129:e56309. (DOI:10.3791/56309).

Liu YB, Yang X, Simmons G (2016) Efficacy of nitric oxide fumigation for controlling codling moth in apples. Insects 7:71. (DOI:10.3390/insects7040071).

Lowenstein CJ, Dinerman JL, Snyder SH (1994) Nitric oxide: A physiologic messenger. Ann Intern Medicine 120:177–256.

Ma L, Hu L, Feng X, Wang S (2018) Review: Nitrate and nitrite in health and disease. Aging Dis 9:938–945.

Madebo MP, Ayalew Y, Zheng Y, Jin P (2022) Nitric oxide and its donor sodium-nitroprusside regulation of the postharvest quality and oxidative stress on fruits: A systematic review and meta-analysis, Food Rev Intern 1–29. (DOI: 10.1080/87559129.2022.2122995).

Magan N, Aldred D (2007) Post-harvest control strategies: Minimizing mycotoxins in the food chain. Intern J Food Microbiol 119:131–139.

Manjunatha G, Lokesh V, Bhagyalashmi N (2012) Nitric oxide-induced enhancement of banana fruit attributes and keeping quality. Acta Hort 934:799–806.

Manjunatha G, Lokesh V, Neelwarne B (2010) Nitric oxide in fruit ripening: Trends and opportunities. Biotechnol Adv 28:489–499.

Moncada S, Higgs EA (2006) The discovery of nitric oxide and its role in vascular biology. British J Pharmacol 147:S193–S201.

Müller U (1997) The nitric oxide system in insects. Prog Neurobiol 51:363–381.

Murad F (2011) Nitric oxide: The coming second messenger. Rambam Maimonides Med J 2:e0038. (DOI: 10.5041/RMMJ.10038).

Oh S, Liu YB (2020) Effectiveness of nitrogen dioxide fumigation for microbial control on stored almonds. J Food Prot 83:599–604.

Oh S, Singh R, Liu YB (2020) Nitrogen dioxide fumigation for microbial control on unshelled peanuts. Agri Sci 11:1159–1169.

Palumbo JD, Mahoney NE, Light DM, Siegel J, Puckett RD, Michailides TJ (2014) Spread of *Aspergillus flavus* by navel orangeworm (*Amylois transitella*) on almond. Plant Dis 98:1194–1199.

Pols S, Van de Poel B, Hertog MLATM, Nicolaï BM (2022) The regulatory role of nitric oxide and its significance for future postharvest applications. Postharvest Biol Technol 188:111869. https://doi.org/10.1016/j.postharvbio.2022.111869.

Ricciardolo FLM, Sterk PJ, Gaston B, Folkerts G (2004) Nitric oxide in health and disease of the respiratory system. Physiol Rev 84:731–765.

Richard JL, Payne GA (2003) Mycotoxins: Risks in Plant, Animal, and Human Systems. Task Force Report, No. 139, CAST (Council for Agriculture Science and Technology), Ames.

Satekge TK, Magwaza LM (2022) Postharvest application of 1-methylcyclopropene (1-MCP) on climacteric fruits: Factors affecting efficacy, Intern J Fruit Sci 22:595–607. (DOI: 10.1080/15538362.2022.2085231).

Shakil MH, Trisha AT, Rahman M, Taludar S, Kobun R, Huda N, Zaman W (2022) Nitrites in cured meats, health risk issues, alternatives to nitrites: A review. Foods 11:3355. (DOI: 10.3390/foods11213355).

Soegiarto L, Wills RBH (2004) Short term fumigation with nitric oxide gas in air to extend the postharvest life of broccoli, green bean, and Bok Choy. HortTechnol 14: 538–540.

Song J, Fan L, Forney C, Campbell-Palmer L, Fillmore S (2010) Effect of hexanal vapor to control postharvest decay and extend shelf-life of highbush blueberry fruit during controlled atmosphere storage. Can J Plant Sci 90:359–366.

Stewart JK, Mon TR (1984) Commercial-scale vacuum fumigation with ethyl formate for postharvest control of the green peach aphid (Homoptera: Aphididae) on film-wrapped lettuce. J Econ Entomol 77:569–573.

US-EPA (U.S. Environmental Protection Agency) (2022) Air Emissions Inventories, Air Pollutant Emissions Trends Data, National Tier 1 CAPS Trends. https://www.epa.gov/air-emissions-inventories/air-pollutant-emissions-trends-data.

Weitzberg E, Lundberg JO (2013) Novel aspects of dietary nitrate and human health. Annu Rev Nutr 33:129–159.

Wills RBH, Ku VVV, Leshem YY (2000) Fumigation with nitric oxide to extend the postharvest life of strawberries. Postharvest Biol Technol 18:75–79.

Yang X, Liu YB (2017) Residual analysis of nitric oxide fumigation on fresh fruit and vegetables. Postharvest Biol Technol 132:105–108.

Yang X, Liu YB (2018) Nitric oxide fumigation for control of spotted wing drosophila (Diptera: Drosophilidae) in strawberries. J Econ Entomol 111:1180–1184. (DOI: 10.1093/jee/toy074).

Yang X, Liu YB (2019a) Nitric oxide fumigation for postharvest pest control on lettuce. Pest Mgmt Sci 75:390–395. (DOI 10.1002/ps.5123).

Yang X, Liu YB (2019b) Residue analysis of nitric oxide fumigation in nine stored grain and nut products. J Stored Prod Res 84:101521.

Yang X, Liu YB, Simmons G, Light D, Haff R (2021) Nitric oxide fumigation for control of navel orangeworm, *Amyelois transitella*, on walnut. J Appl Entomol 145:270–276. (DOI: 10.1111/jen.12846).

Yang X, Liu YB, Singh R, Phillips TW (2022) Nitric oxide fumigation for control of ham mite, *Tyrophagus putrescentiae* (Sarcoptiformes: Acaridae). J Econ Entomol 115:501–507. (DOI: 10.1093/jee/toac014).

Zoffoli JP, Michelow P, Naranjo P (2013) Sensitivity of fruit species to ethyl formate fumigation under quarantine concentrations. Acta Hort 1012:763–767.

Chapter 14

Contact Insecticides

Chemicals, Diatomaceous Earth, and Amorphous Silica for Stored-Grain Protection

Manjree Agarwal

14.1 Introduction

Insects pose a significant threat to agriculture and human health, leading to substantial economic losses globally. Among insects, stored-grain insects cause significant loss worldwide, in both developed and developing countries. According to the latest database from FAO using 700 publications, the extent of these losses can vary depending on several factors, including geographical location, climate, storage conditions, and insect species. While providing precise global figures is challenging, it is estimated that stored grain losses due to insects range from 5% to 20% annually. In some regions with inadequate storage facilities or poor pest management practices, losses can be much higher, exceeding 50%. According to the database from FAO, these losses occur during storage, transportation, and processing stages, impacting food security, economic stability, and profitability of the agricultural sector (FAO, 2021). Over the years, the use of chemical pesticides has been the primary approach for controlling insect populations. Pesticide use went up by 30% between 2000 and 2020 (FAO, 2022). Under pesticides, insecticides are

DOI: 10.1201/9781003309888-14

the category used to kill insects. Insecticides are agents of chemical or biological origin that control insects. Control may result from killing the insect or otherwise preventing it from engaging in behaviors deemed destructive. Stored-grain insecticides broadly fall under two categories: fumigants and contact insecticides (also called grain protectants). Fumigants are chemical substances used to control pests, including insects, rodents, fungi, and other organisms, in enclosed spaces. As the name suggests, they work by releasing toxic fumes or gases that penetrate the target area, reaching hidden or hard-to-reach places where pests may be present. These chemicals can interfere with respiration, metabolic pathways, or nervous system functions, leading to the elimination of pests at various life stages. Fumigants are typically applied in a sealed environment, to allow the gases to disperse uniformly and maximize their efficacy. Unlike contact insecticides, a fumigant does not need to be in direct contact with insects or pests. A few examples of common fumigants are methyl bromide, phosphine, ethyl formate, and sulfuryl fluoride. Methyl bromide and phosphine are the most commonly used fumigants in agriculture and pest control, but methyl bromide has been phased out or restricted in many countries because of its ozone-depleting properties.

In contrast to fumigants, contact insecticides are a group of insecticides that are designed to kill insects by coming into direct contact with them. They work by physically damaging the insect's exoskeleton and causing the insect to dehydrate and die. They are applied as liquids or dusts to the grain for long-term protection against stored-grain pests. In the context of grain storage and silos, contact insecticides have commonly been used to protect stored grains from insect infestation since time immemorial.

This chapter provides an overview of three types of contact insecticides: chemical insecticides, diatomaceous earth, and amorphous silica.

14.2 Contact Chemical Insecticides

Contact chemical insecticides are a type of insecticide that kill pests upon direct contact. They are applied directly to the grain or to the storage facility surfaces to create a protective barrier against pests. These insecticides can be applied as a dust, a spray, or a gas. They work by disrupting the insect's nervous system or physiological functions, ultimately leading to the insect's death.

14.2.1 Types of Contact Chemical Insecticides for Grain Protection

Contact chemical insecticides can be classified into different chemical groups, such as organophosphates, pyrethroids, carbamates, and neonicotinoids. These pesticides have different modes of action and may vary in their effectiveness against different types of insects. This section provides an overview of the chemical classes commonly used as contact insecticides in stored-grain pest management.

14.2.1.1 Organophosphates

Organophosphates work by inhibiting the activity of an enzyme called acetylcholinesterase, which is essential for the proper functioning of the nervous system in insects. Malathion, chloropyrifos, and pirimiphos-methyl are organophosphates commonly used as grain protectants.

Malathion was the first protectant. According to Collins (2006), it came into widespread use in Australia in the mid-1960s and was responsible for the government's "nil tolerance" mandate. However, *Tribolium castaneum* (Herbst) and *Rhyzopertha dominica* (F.) started to develop resistance in the late 1960s (Van Graver and Winks, 1994). Resistance became so widespread that use of malathion began to decline, and alternatives were investigated (Collins, 2006). After malathion, fenitrothion, chlorpyrifos-methyl, and pirimiphos-methyl were introduced, but the insects developed resistance soon, as they belong to the same class of compounds. Arthur (1996) has provided an extensive list on resistance development to organophosphates.

14.2.1.2 Carbamates

Carbamate insecticides also target the nervous system of insects. Like organophosphates, they work by inhibiting the activity of acetylcholinesterase. Carbaryl and bendiocarb are examples of carbamate insecticides used in grain storage.

14.2.1.3 Pyrethroids

To overcome the resistance problem of organophosphates, another class of compounds was developed, called pyrethroids. Pyrethroids are synthetic chemicals modeled after the natural pyrethrins found in chrysanthemum flowers. They are widely used in grain storage due to their effectiveness against a broad range of insects. Pyrethroids act on the nervous system of insects, causing paralysis and eventually death. Pyrethroids work by binding to voltage-gated sodium channels in the nerve cells of insects, leading to hyperexcitation and paralysis. Some pyrethroids commonly used in grain protection include bioresmethrin, permethrin, cypermethrin, and deltamethrin.

Bioresmethrin was used successfully for about 12 yr in Australia until resistance was first detected in 1990 (Collins et al., 1993). Deltamethrin is still being used as broad-spectrum synthetic pyrethroid for grain protection. It exhibits excellent residual activity, meaning that it remains effective for an extended period after application. Deltamethrin can be applied directly to

461

the grain or used as a surface treatment in the storage facility. Athanassiou et al. (2004) have reported it to be effective against *Sitophilus oryzae* (L.) at the rate of 0.25 ppm and to show residual effect until 3.5 months with control of 86%. Permethrin is another widely used synthetic pyrethroid insecticide. It has a similar mode of action to deltamethrin and provides effective control against various grain pests. Permethrin can be applied as a spray or dust directly to the grain or as a surface treatment. Cyfluthrin is another synthetic pyrethroid insecticide, known for its quick knockdown and residual activity. It is effective against a wide range of stored-grain pests and can be used as a grain protectant or as a surface treatment in storage facilities. Athanassiou et al. (2004) have reported it to be effective against *S. oryzae* at the rate of 0.25 ppm and to show residual effect until 2 months with control of 83%. Previous studies have shown that deltamethrin, cypermethrin, and permethrin (Papadopoulou-Mourkidou and Tomazou, 1991) were much more effective than malathion in controlling *S. oryzae*.

14.2.1.4 Insect Growth Regulators

Insect growth regulators (IGRs) are insecticides that mimic insect molting hormones, inhibit chitin synthesis, or are ecdysteroid agonists and thereby disrupt the normal development of insects (Oberlander et al., 1997). Insect growth regulators interfere mainly with three physiological processes: growth and development of immature insects, induction of metamorphosis, or chitin synthesis in the integument. Compounds are considered low-risk chemicals, have low mammalian toxicity, and are specific to insects. There is renewed interest in using IGRs in pest management systems for stored grain (Oberlander et al., 1997). One of the most common and widely used hormone analogue under this category is methoprene.

Methoprene is a juvenile hormone analogue that has been used extensively in stored-product pest management. The formulations of methoprene originally introduced into the stored-product market in the 1980s contained the racemic mixture with both R- and S- forms, but now only the purified S-methoprene isomer is used. Methoprene has received broad attention and has been tested over decades for its direct lethal effects, but many recent studies focus more on sublethal effects (Wijayaratne et al., 2018). It has gained importance in controlling *R. dominica*, one of the most significant pests and the most difficult to control because of the resistance problem with organophosphates and pyrethroids (Oberlander et al., 1997).

It is generally effective against externally feeding stored-product insects such as *T. castaneum* (Herbst), the red flour beetle, and *Oryzaephilus surinamensis* (L.), the saw-toothed grain beetle (Mian and Mulla, 1982a, 1982b), but it does not give good control of *Sitophilus* species (Samson et al., 1990). Oberlander et al. (1997) have provided an efficacy list of methoprene against various stored-product insect species.

14.2.1.5 Neonicotinoids

Neonicotinoids are a group of neurotoxic new-generation insecticides. They have been successfully evaluated for the control of agricultural pests that are resistant to pyrethroids (Daglish and Nayak, 2012; Tsaganou et al., 2014). They are used primarily for seed coating and have been found to be effective against various stored-grain pests. One such compound used against pests is thiamethoxam. Tsaganou et al. (2021) tested thiamethoxam and found it to be effective against 7 stored-grain pests at a rate of 10 ppm and duration of 14 d without any progeny emergence.

14.2.2 Advantages of Contact Insecticides for Grain Protection

Effectiveness: Contact insecticides, when properly applied, can provide effective control of a wide range of stored-grain pests. They act quickly upon contact, killing insects directly and preventing further infestation.

Residual Activity: Many contact insecticides have residual activity, meaning that they remain active for an extended period after application. This can provide ongoing protection against pests and help maintain the quality of stored grains over time.

Broad Spectrum: Contact insecticides often have a broad spectrum of activity, meaning that they can target and control various types of pests commonly found in grain storage, such as beetles, weevils, and moths.

Versatility: Contact insecticides can be applied directly to the grain or used as surface treatments in the storage facility. This versatility allows for multiple application methods depending on the specific requirements and conditions of the storage facility.

14.2.3 Disadvantages of Contact Insecticides for Grain Protection

Resistance Development: Prolonged and repeated use of contact insecticides can lead to the development of resistance in insect populations. Insects with genetic traits that enable them to survive exposure to the insecticides can multiply, potentially rendering the insecticides less effective over time. To mitigate this risk, it is important to rotate and alternate between different insecticides with varying modes of action or to apply a combination treatment. This is discussed in more detail at the end of Section 14.5.

Environmental Impact: Contact insecticides can have environmental implications if not used properly. They may pose risks to non-target organisms, such as beneficial insects, birds, and mammals. Additionally, incorrect application or disposal of insecticides can contaminate soil, water sources, or other ecosystems. Following proper application guidelines and adhering to environmental regulations is crucial to minimize these risks.

14.2.4 Usage and Mortality Data

Specific usage and mortality data for contact insecticides used as grain protectants can vary depending on factors such as region, insect species, and individual product formulations. It is recommended to consult local agricultural authorities, extension services, or pest management professionals for the most up-to-date information relevant to your specific location.

Studies and research are conducted periodically to assess the efficacy and mortality rates of different insecticides. These studies evaluate the effectiveness of contact insecticides against specific grain pests and provide insights into their performance. However, the availability of recent data may vary based on the location and the specific insecticide in question.

To access the most current and accurate usage and mortality data for grain protectants, it is recommended to consult scientific literature, research publications, and regulatory agencies dedicated to agriculture and pest management in your region. These sources often provide comprehensive information on effectiveness, recommended dosage, application methods, and any specific guidelines for the use of contact insecticides in grain storage.

14.2.5 Requirement of Alternatives to Contact Chemical Insecticides

Indiscriminate use of contact pesticides has posed challenges in controlling pest resistance and residue problems. Despite the hazard, many countries continue to use them, but some have been proactive in taking them off the market as environmental and health regulations become stricter. Therefore, there is a need for alternative plant protection materials, of which diatomaceous earth and amorphous silica are two.

14.3 Diatomaceous Earth

For centuries, grain was protected from insect infestation by adding some form of powder or dust directly to the grain, and diatomaceous earth (DE) is one of those dusts. It is a dust composed of unicellular algae fossilized bodies called diatoms and exists in a natural state as soft chalky rock deposits formed about 30 to 80 million years ago (Round et al., 1990).

14.3.1 Physicochemical Properties of Diatomaceous Earth

Before DE is marketed and commercialized, it goes through the process of quarrying, drying, and milling. Through this process, DE is further dehydrated and its particle size reduced to between 0.5 µ and more than 100 µ (Quarles, 1992; Subramanyam and Roesli, 2000). Figure 14.1 shows the scanning electron micrograph of registered diatomaceous earth Dryacide. From the image it can be easily seen that the particle size is between 10 and 30 µ.

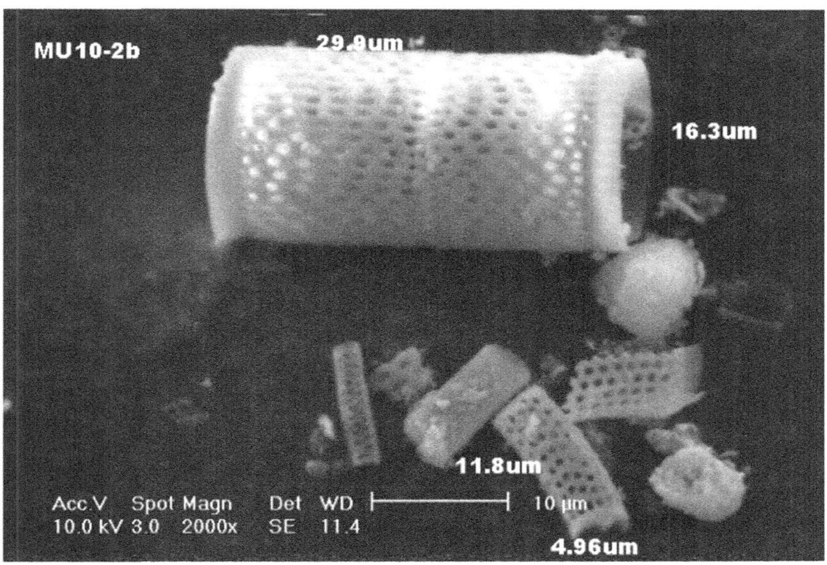

Figure 14.1 Scanning electron micrograph of registered diatomaceous earth Dryacide. (From Agarwal, unpublished micrograph.)

The composition of DE is 80%–95% amorphous silicon dioxide, and the rest is mainly crystalline silicon dioxide. The higher the percentage of amorphous silica in DE, the safer its handling.

As per the Select Committee on Generally Recognised as Safe (GRAS) Substances (2020), DE is considered safe and included under conclusion 1, which states that there is no evidence to support a hazard to the public when it is used at levels that are now current.

14.3.2 Application of Diatomaceous Earth in Grain Protection

Diatomaceous earth's natural existence and its abrasive and porous nature give it unique properties that make it a potent insecticide and desiccant. The high silica content and low toxicity to mammals make it an appealing option for grain storage. This inert material was initially used as a carrier for potent insecticides such as nicotine, rotenone, arsenic, and fluorine compounds (Chiu, 1939). The first publications about DE use in stored-grain protection came out in the early 20th century (Calvert, 1930; Chiu, 1939). Since then, hundreds of research papers and reviews have been published documenting the role of DE in grain protection against various stored-grain pests and many other pests like ants, bedbugs, textile pests, crickets, termites, earwigs, beetles, silverfish, fleas, centipedes, pillbugs, poultry mites, ticks, and snails (Korunić et al., 2016). To avoid repetition, I would like to give a few reference lists that can be referred to for additional details. The use of DE for structural treatment in stored product facilities was studied by Desmarchelier and Dines (1987), Wright (1990), Desmarchelier et al. (1993), Bridgeman (1994), McLaughlin (1994), Korunić and Fields (1995), and Korunić et al. (1996).

About twelve DE based reviews have been published so far (Quarles, 1992; Quarles and Winn, 1996; Golob, 1997; Korunić, 1998; Subramanyam and Roesli, 2000; Fields and Korunić, 2002; Nikpay, 2006; Quarles, 2007; Korunić, 2013; Shah and Khan, 2014; Korunić et al., 2016; Zeni et al., 2021), with numerous references cited within each of the reviews. Mostly, the advantages of DE are described as its being a natural inert material and a safe and effective insecticide, with a physical mode of action and long persistence, that does not leave hazardous residues; mostly the conclusion is that DE is a promising insecticide (Korunić et al., 2016).

14.3.3 Mechanism of Action

Unlike organophosphates, carbamates, and pyrethroids, DE does not affect the nervous system of the insect body. It is known to cause more physical damage to the body, which in turn causes physiological damage and eventually the death of the insect. Its main mechanisms of action are abrasion and adsorption.

An insect's body is covered with a layer of wax made up of cuticular lipids. These lipids are present on the outermost layer of an insect's exoskeleton, called its epicuticle, and serve as a barrier to water loss from the body. When the insect body comes in contact with DE, it sticks or adsorbs onto the lipids on the insect's epicuticle and prevents the lipids from restricting water loss (Golob, 1997). As a result, the insect loses body moisture through the damaged spots on its epicuticle and dies of desiccation after a period. Diatomaceous earth, being an effective desiccant, also takes up the moisture from the insect body and dehydrates it to death.

In addition to its adsorptive ability, DE also has an abrasive property. It can cause abrasion to the cuticle and digestive tract (Jackson and Webley, 1994), eventually causing loss of water. Diatomaceous earth is also known to block the breathing organs or spiracles and thus suffocate the insect to death (Jackson and Webley, 1994). Dust also gets lodged between cuticular segments, antennae, joints, and mouth parts, increasing water loss. Vrba et al. (1983) concluded that desiccation stress may lead to other physiological changes that cause death. Vincent (1971) proposed that additional water loss may occur through the tracheal system or through the cuticle due to the secretion of the hormone bursicon, which could also be contributing to insect mortality.

14.3.4 Efficacy of Diatomaceous Earth in Grain Protection

There are numerous factors that affect the efficacy of DE dust. Mentioned below are some of them:

14.3.4.1 Moisture

The presence of water adversely affects DE's performance. It is well known that diatom particles absorb both surface and air water, depending on the relative humidity present in the air. As relative humidity increases, the quantity of water absorbed by DE increases. The efficacy of DE is significantly reduced when its moisture content is increased to above 8%–10%. When particles are filled with a film of water and the holes and pores are completely saturated with water, there is no longer free space left for DE to absorb wax from the cuticle of the insect, and its insecticidal efficacy is lost entirely (Fields and Korunić, 2000). Aldryhim (1993) also reported a decrease in toxicity of Dryacide against *R. dominica* with an increase in humidity.

The presence of water also adversely affects DE's dispersion properties. As DE dries after getting wet, it tends to agglomerate and stick to the surfaces on which it dries rather than sticking to the insects that crawl over the surfaces, and as a result it becomes difficult to be picked up by passing insects. Consequently, the insect does not pick up the quantity of particles necessary to facilitate desiccation (Korunić et al., 2016).

14.3.4.2 Physical Properties of the Dust

Shape, particle size, pore diameter, surface area, purity, and size uniformity all affect the effectiveness of the dust. The efficacy of DE is a function of the physical properties attributable to the morphological characteristics of the diatoms (Korunić, 1998). Korunić (1997) took 36 different diatomaceous earths or formulations collected from the United States, Mexico, Canada, Australia, Japan, China, and Macedonia (Europe), and the results indicate that the efficacy of DE against insects depends on different properties of the diatom particles. Properties include the ability of DE particles to reduce bulk grain density (test weight) and the tendency of DE particles to adhere on the grain surface, as well as its particle size distribution, diatom shape, and pH. Ideally, active DE should have a high amorphous silicon dioxide content and a uniform particle size (less than 10 μm in diameter) containing very little clay and other impurities (Quarles, 1992). Vayias et al. (2009) have shown that DEs with particles smaller than 45 μm were more effective than DEs with larger particles against *R. dominica*, S. *oryzae*, and *Cryptolestes ferrugineus* (Stephens). Nonetheless, Baliota and Athanassiou (2020) have shown that it is the particle shape, rather than the size, that had a certain effect on the insecticidal value of DEs and that smaller particles do not necessarily mean higher efficacy.

14.3.4.3 Differences in Insect Morphology and Physiology

Insects small in size, like the rusty grain beetle, flat grain beetle, and rice saw-tooth beetle, are killed faster than bigger insects like the red flour beetle. This could be because of less moisture in them according to body size which get lost much quicker than the bulkier insect body which have much high moisture.

Insects with softer waxes were more susceptible than insects with harder waxes (Ebeling, 1971). Similarly, Agarwal (unpublished data) found that hairy insects like *Trogoderma variabile* (Ballion) and *T. granarium* (Everts) larvae are more resistant to DE. This could be because of low or no penetration of the DE dust into the insect body due to hinderance from the long surface hairs.

14.3.5 Advantages and Limitations of Diatomaceous Earth

This section highlights the advantages of DE as a contact insecticide, including its low toxicity to mammals, lack of problematic residue, and compatibility with organic farming practices. However, it also discusses the limitations and challenges associated with DE, such as variable efficacy under different environmental conditions and the need for proper application techniques.

14.3.5.1 Advantages

Nontoxic to humans and animals: Diatomaceous earth is a natural, nontoxic substance, making it safe for humans and animals. It has low mammalian toxicity [DE rat oral LD_{50} > 5,000 mg/kg of body weight (Subramanyam et al., 1994)]. It offers a chemical-free pest control option for stored grains, reducing health risks associated with chemical pesticides.

Effective against crawling insects: Diatomaceous earth is particularly effective against crawling insects commonly found in stored grains, such as grain weevils, flour beetles, and mites. When these pests encounter DE, it damages their exoskeleton, leading to dehydration and death.

Residue-free: Diatomaceous earth leaves no harmful residues on the grains, unlike contact insecticides like organophosphates and pyrethroids, ensuring that the stored food remains uncontaminated and safe for consumption.

Environmentally friendly: Unlike synthetic pesticides, DE does not harm the environment, beneficial insects, or natural predators of grain pests. It poses minimal risk to the overall ecosystem.

Long lasting: When applied properly, DE can provide extended protection against stored-grain pests. Its effectiveness does not diminish over time as long as the conditions remain dry.

14.3.5.2 Limitations

Despite witnessing a century's worth of research, the emergence of numerous commercial products, and various advantages, DE has not yet basked in the limelight of popularity. This can be attributed to the following limitations.

Reduction in grain bulk density: As DE is being applied at a high rate directly to grain, it reduces the bulk density remarkably. Korunić et al. (1998) have reported reduction in bulk density by 6 kg/hL when DE is applied at 500 ppm. As a result, the flowability of the grain decreases and the angle of repose increases while grain is being loaded for shipment. This is a major hurdle in grain trade between countries. Agarwal (unpublished data) noticed an increase in angle of repose by 25% when grain was treated with DE at 200 mg/kg. Jackson and Webley (1994) showed that at 0.1 g of Dryacide per kg of wheat, the angle of repose increased by about 8 degrees over that of untreated grain (24 degrees). To overcome this problem, slurry application has been developed, but this again reduces the efficacy of DE.

High dosage required: The high recommended dosages of DE (1 to 4 g/kg) adversely affect physical and mechanical properties of grain, including flowability and bulk density (test weight) (Quinlan and Berndt, 1966; Redlinger and Womack, 1966; Jackson and Webley, 1994), leave visible residues (Quinlan and Berndt, 1966; Jackson and Webley, 1994), affect dielectric moisture readings (Quinlan and Berndt, 1966; White et al., 1975), and produce excessive amounts of airborne dusts during handling (Quinlan and Berndt, 1966). The milling industry is reluctant to accept grain treated with DE because of possible damage to milling machinery from its abrasive nature (Subramanyam et al., 1994). As a result, in Australia, DE is used principally as a structural treatment for empty silos and not the entire grain mass because of the adverse effects DE has on the handling of the grain. And also, DE being nonfood grade, it is difficult to market the products that are treated with DE, as the food often then requires treatment to remove the insecticide before it is safe for processing and/or consumption.

Limited effect on flying insects: Diatomaceous earth's mode of action is primarily contact based, which makes it less effective against flying insects like moths that infest stored grains. Airborne insects may not come into sufficient contact with the DE for it to be fully effective.

Requirement for dry conditions: Diatomaceous earth loses its effectiveness when exposed to moisture. In high humidity or damp storage conditions, it may clump together, reducing its ability to control pests effectively.

Slow action: Diatomaceous earth does not provide an immediate knockdown effect on stored-grain pests like other chemical contact insecticides. It may take some time for the pests to come in contact with the DE and be affected by its desiccating properties, as it relies on physical action rather than chemical reactions.

Difficult application in bulk storage: Applying DE evenly in large-scale storage facilities can be challenging. Proper distribution is essential to achieve effective pest control, as uneven application may lead to pockets of untreated areas.

Potential threat to occupational health and safety: Inhalation of DE dust can cause respiratory irritation. especially if the DE is inhaled in large quantities or over an extended period. Therefore, it is essential to take precautions when applying DE in powdered form. Depending on the source and processing, DE can contain from 0.1% to 60% crystalline silica. The DEs registered as insecticides generally have less than 7% crystalline silica. Crystalline silica has been associated with silicosis and was classified by the International Agency for Research on Cancer as a probable carcinogen (International Diatom Producers Association, 1990).

Damage to machinery: The dust in the grain also causes machinery breakdowns, increased maintenance problems, and slowed grain flowability. A 33%–40% decrease in auger-loading capacity and an apparent auger power loss were reported for grain treated with the DE dusts Perma Guard and Kenite 2-1 at a rate of 3.5 g/kg of grain. Silica aerogel treatment at 0.5 g/kg also damaged machinery and reduced auger-loading capacity by 50% (Subramanyam and Roesli, 2000).

Failure to provide a long-lasting solution: The effectiveness of DE as an insecticide diminishes when it gets wet or is exposed to moisture. Hence, it may require frequent reapplication in areas prone to humidity or dampness.

Risk to beneficial insects: While DE can be selective in targeting harmful insects, it may inadvertently harm beneficial insects like bees and ladybugs if not used with caution.

14.3.6 Requirement of Alternatives to Diatomaceous Earth

Diatomaceous earth formulation can be a good alternative to pesticides, as DE is a naturally occurring material composed of fossilized remains of diatom (Korunic et al., 1998), but nevertheless, because of the above-mentioned problems large-scale commercialization has not been possible. Given the above drawbacks, there remains a requirement for a next-generation protectant, which could be synthetic amorphous silica.

14.4 Synthetic Amorphous Silica

Amorphous silica is another type of contact insecticide, made up of tiny particles of silica, a mineral that is found in rocks and sand. It consists of nano-sized primary particles, composed of nano- or micrometer-sized aggregates and of agglomerates in the micrometer-sized range. Hence these materials fall under the definition of nanostructured materials. Synthetic amorphous silica (SAS) has been widely used in topical and oral medicines, food, and cosmetics for decades without evidence of adverse human health effects. Only in the last few years has it been picked up as a game changer in controlling stored-grain pests through the patents US 2017/0071233 and EP3119202 by Ren and Agarwal (2017), who found it to be highly effective against a wide range of insects.

Because of the absence or very small amount of crystalline silica, amorphous silica is a low-risk pesticide, with a low toxicity to humans and other mammals. It is also nonflammable and nonreactive, making it safe to handle and transport. Amorphous silica can be applied as a powder or a liquid spray and can be used in a variety of settings, including silos, homes, hotels, and restaurants. Hence, it satisfies the criteria for a current-generation pesticide.

14.4.1 Manufacture and Particle Chemistry

There are three types of silica (silicon dioxide), which are all found under CAS No. 7631-86-9: crystalline silica, amorphous silica (naturally occurring or as a byproduct in the form of fused silica or silica fume), and synthetic amorphous silica (SAS).

The difference between the amorphous and crystalline silica forms arises from the connectivity of the tetrahedral units. Amorphous silica consists of a nonrepeating network of tetrahedra, where all the oxygen corners connect two neighboring tetrahedra. The amorphous structure is very "open"; that is, channels exist through which small positive ions such as Na+ and K+ can readily migrate (Fruijtier-Pölloth, 2012).

SAS can be produced either via a wet process that uses acids to produce colorless micro-porous silica known as precipitated silica and silica gel (CAS No. 112926-00-8; includes silica gel, precipitated silica, and colloidal silica) or via a thermal process (also known as pyrogenic or fumed) that uses heat to produce a very fine particulate or colloidal form (CAS No. 112945-52-5).

In general, SAS contains no detectable amounts of crystalline silica, with detection limits varying between 0.01% and 0.3% by weight, depending on the method used. Synthetic amorphous silica also contains fewer impurities than biogenic amorphous silica (e.g., DE) and can be distinguished from other forms of amorphous silica by its high chemical purity, its fine particulate nature, and characteristics of the particles observable by electron microscopy.

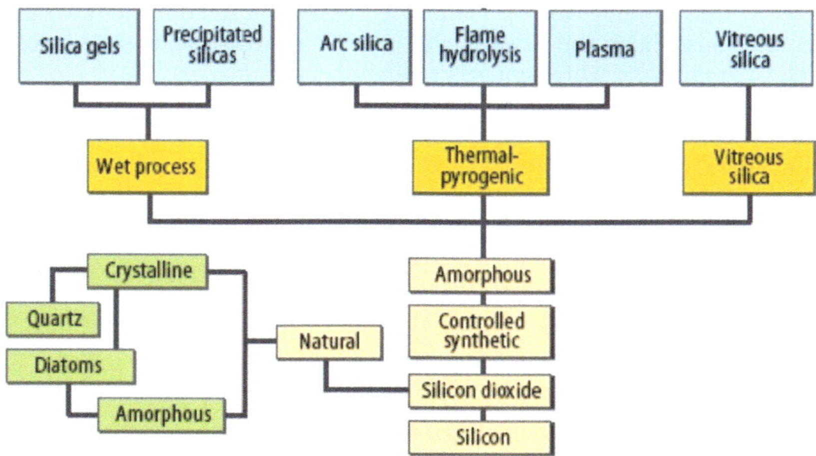

Figure 14.2 Various forms of silica. (From Reader and Nargiello 2020).

Figure 14.2 shows various forms of silica (Reader and Nargiello 2020).

14.4.2 Different Grades of Synthetic Amorphous Silica

Uses of lower grades of SAS, which may contain some impurities and in which the SAS becomes an integral part of the product matrix and no longer exists as a powder, include serving as reinforcing fillers in rubber and silicone products (e.g., tires), as functional pigments in emulsion paints and lacquers, and as an anti-blocking agent to stop plastic films from sticking to each other.

Higher-purity food grades are approved by government regulators for direct use in food products and indirect uses in food packaging like cling wraps, as well as personal health care products and pharmaceuticals (toothpaste), where they act as a thickener or maintain flow properties in powder products.

Synthetic amorphous silica may be either hydrated or nonhydrated and contains silicon and oxygen connected in a three-dimensional macro-molecular network, which imparts a general chemical inertness.

Global production of SAS products is just over 1 Mt, with about 100,000 t of pyrogenic silica, 800,000 t of precipitated silica, and about 115,000 t of silica gels. Precipitated silica forms the largest group of SAS products due to demand over the last decade for use in tires (about 30% of manufacture). Most of the output is produced in Western Europe, North America, and Japan.

14.4.3 Physicochemical Properties of Synthetic Amorphous Silica

Particle size distribution curves and the accuracy of measurements depend on the method used, sample preparation, and whether the measurement was performed in solid or liquid phase. Typical physicochemical proper-ties of the different forms have been published (Fruijtier-Pölloth, 2012), with silanol (Si OH) groups on the SAS surface making untreated SAS hydrophilic.

To make SAS hydrophobic, all forms of SAS require surface modifi-cation. A typical treating agent for surface modification is dichlorodimeth-ylsilane, which hydrolyzes to form polydimethylsiloxane. The treated SAS bears on its surface both the hydrophobic entities (polydimethlysiloxy units) and the remaining hydrophilic entities (i.e., surface silanols). The core material is still amorphous silica (Fruijtier-Pölloth, 2012).

Both manufacturing processes (precipitation and wet) result in SAS materials that are aggregates of strongly (covalently) bonded or fused pri-mary particles. Precipitated silica and silica gel contain a large amount of bound water and tend to agglomerate, causing them to have an even larger particle size (Fruijtier-Pölloth, 2012).

As part of the Plant Biosecurity Cooperative Research Centre (PBCRC) project, Ren and Agarwal (2017) compared the physicochemical characteris-tics of various SAS to DE-based registered products (Table 14.1). The SAS products selected are currently in use by the food industry in a range of applications, none of which are related to insect control. These were allo-cated an accession number from MU1 through MU9 (Table 14.1). Of the nine, four are marketed in Australia (Pirosil PS 300, Pirosil AS-70A, Pirosil PS 200, and Pirosil AS 100A) and are registered on the Australian Inventory of Chemical Substances (AICS). All SAS products were registered with a maximum residue limit (MRL) of 15 g (dust)/kg (food).

The samples were physically characterized by electron microscopy to determine particle size and particle variation. The BET surface area was

Table 14.1 Physiochemical Properties of Synthetic Amorphous Silica in Comparison to Registered Diatomaceous Earth Products

Dust Code	Silica Ingredient	Processing Method*	Charge	Average Particle Size (nm)	BET Surface Area (m^2/g)	Amorphous SiO_2 (%)	Crystalline SiO_2 (%)
MU1	SAS	Wet method	Hydrophilic	<192.9	100	>99	none
MU2	SAS	Wet method	Hydrophilic	<143.3	185	>98	none
MU3	SAS	Wet method	Hydrophilic	<131.4	80	>98	none
MU4	SAS	Wet method	Hydrophilic	<114.2	180	>98	none
MU5	SAS	Wet method	Hydrophilic	<127.0	110	>98	none
MU6	SAS	Wet method	Hydrophilic	<140.1	100	>98	none
MU7	SAS	Wet/thermal method	Hydrophilic	<116.4	265	>99	none
MU8	SAS	Wet/thermal method	Hydrophilic	<117.8	270	99.9	none
MU9	SAS	Wet/thermal method, surface modified	Hydrophobic	<116.8	280	>99	none
Dryacide	DE	N/A	Hydrophilic	>20,000	57	>90	0.1
Absorbacide	DE	N/A	Hydrophilic	>20,000	32	96	<4
Diafil 610	DE	N/A	Hydrophilic	>20,000	25	89	0.5

* Wet method includes precipitated and/or aerogel method; thermal method includes thermal and/or pyrogenic and/or fumed method.

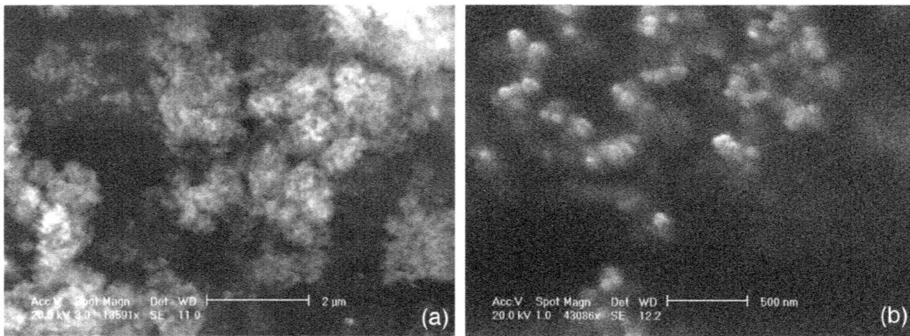

Figure 14.3 (a) Scanning electron micrograph of synthetic amorphous silica agglomerate at 13,591 magnification. (b) Scanning electron micrograph of separated synthetic amorphous silica particle at 43,086 magnifications. (Courtesy of Agarwal, unpublished micrograph).

also determined (Table 14.1). This technology, named after the authors of the theory (Brunauer, Emmett, and Teller), is a widely used technique for the calculation of effective surface area using adsorption and for explaining the physical adsorption of gas on a solid surface.

From Table 14.1 it is evident that the selected MU dusts had large specific surface area (>110 m^2/g) compared to DE products (<50 m^2/g) and a small particle size (100–200 nm) in comparison with the three commercial DE dusts (>20 µm). Figure 14.3(a) and (b) show the scanning electron micrograph of SAS dust at different magnifications.

14.4.4 Application of SAS in Grain Protection

This section covers synthetic amorphous silica, which are food grade, have less than 0.1% crystalline silica, and had their insecticidal property discovered for the very first time by Ren and Agarwal (2017). Nine SAS dusts (MU1 to MU 9) and three DE-based products were selected and compared for their efficacy at 150 and 200 mg/kg for 5 and 10 d in wheat grain at standard moisture condition against four stored-grain pests: the cigarette beetle, *Lasioderma serricorne* (F.); the rice weevil, *S. oryzae* (L.); the lesser grain borer, *R. dominica* (F.); and the rust red flour beetle, *T. castaneum* (Herbst). Researchers found that all the SAS dusts were highly effective over the DE dust type (Tables 14.2 and 14.3). Based on these bioassay results, the four best performing SAS were selected for further trials using low dosage and additional insect species and commodities.

Table 14.2 Mortality (%) of Five Tested Species of Adult Insects at 25°C at a Dose Rate of 150 mg/kg for 5 and 10 d Treatment

Dust	T. castaneum		R. dominica		S. oryzae		L. serricorne		C. pusillus	
	5 d	10 d	5 d	10 d	5 d	10 d	5 d	10 d	5 d	10 d
MU1	97.9	99.3	96.9	97	98.8	99.8	100	100	100	100
MU2	99.8	100	98.3	99.3	100	100	100	100	100	100
MU3	90.2	98	97.7	98.4	100	100	100	100	100	100
MU4	99.7	100	99.3	100	100	100	100	100	100	100
MU5	79.8	95.8	98.1	95.6	99.8	98.6	100	100	100	100
MU6	97.9	99.3	96.9	97	98.8	99.8	100	100	100	100
MU7	100	100	99.6	100	100	100	100	100	100	100
MU8	99.5	100	99.8	100	100	100	100	100	100	100
MU9	100	100	100	100	100	100	100	100	100	100
MU10 (Dryacide)	0.5	5.3	40.4	74.9	49.2	79.3	18.9	48.8	32.0	39.4
MU11 (Absorbacide)	1.3	7.2	35.6	68.3	51.1	77.9	27.5	50.0	27.4	41.2
MU12 (Diafil) 610	2.7	6.9	41.7	73.6	53	69.8	16.8	47.6	25.1	35.8

Table 14.3 Mortality (%) of Five Tested Species of Adult Insects at 25°C and a Dose Rate of 200 mg/kg for 5 and 10 d Treatment

Dust	T. castaneum		R. dominica		S. oryzae		L. serricorne		C. pusillus	
	5 d	10 d	5 d	10 d	5 d	10 d	5 d	10 d	5 d	10 d
MU1	100	100	100	100	100	100	100	100	100	100
MU2	100	100	100	100	100	100	100	100	100	100
MU3	100	100	100	100	100	100	100	100	100	100
MU4	100	100	100	100	100	100	100	100	100	100
MU5	100	100	100	100	100	100	100	100	100	100
MU6	100	100	100	100	100	100	100	100	100	100
MU7	100	100	100	100	100	100	100	100	100	100
MU8	100	100	100	100	100	100	100	100	100	100
MU9	100	100	100	100	100	100	100	100	100	100
MU10 (Dryacide)	11	15.3	39.8	67.9	42.5	68.3	28.2	48.6	31.3	43.1
MU11 (Absorbacide)	9.7	13.1	35.7	48.2	37.1	49.6	31.0	45.9	36.1	49.9
MU12 (Diafil) 610	14.0	17.2	35.5	49.1	41.5	52.8	29.3	53.9	37.5	45.2

Table 14.4 Mortality (%) of Five Tested Species of Adult Insects at 25°C and a Dose Rate of 50 mg/kg for 7 and 14 d Treatment

	T. castaneum		R. dominica		S. oryzae		L. serricorne		C. pusillus	
Dust	7 d	14 d	7d	14 d	7 d	14 d	7 d	14 d	7 d	14 d
MU4	68.2	95.3	66.9	87.5	100	100	100	100	100	100
MU7	92.1	99	97.2	98	100	100	100	100	100	100
MU8	59.1	85.6	97.2	99.5	96.5	100	100	100	100	100
MU9	70.3	88.8	97.6	99.3	93.1	99.8	100	100	100	100

Tables 14.4 and 14.5 show that even at the low dosage of 50 mg/kg of treated wheat, the SAS can kill 100% of cigarette beetles, flat grain beetles [*Cryptolestes pusillus* (Schönherr)], and rice weevils from the seventh day onward. Similar results of high efficacy were observed with other commodities like barley, canola, and oats (Agarwal, unpublished data). A long-term trial (for 6 wk) was also performed in 30 kg of wheat at a dose rate of 150 mg/kg. Table 14.6 shows 100% mortality for all five insect species used (Ren and Agarwal, 2017). After the 6 wk trial, two sets of long-term bioassays were conducted for 6 months in jars and 12 months in a mini-silo for MU 4, 7, 8, and 9, to assess its long-term efficacy and reduction in progeny production against *L. serricorne, T. castaneum, C. pusillus, S. oryzae*, and *R. dominica* at 50, 100, and 150 ppm for 6 months and at 150 mg/kg for 1 yr. For *C. pusillus* and *L. serricorne,* 100% control for 6 months can be seen even at a dosage as low as 50 mg/kg. For T. *castaneum,* 100% control for 6 months is observed from a dose rate of 100 mg/kg. And for *S.oryzae* and *R. dominica,* 100% control for 6 months is observed from a dose rate of 150 mg/kg. In the case of the 12 months mini-silo experiment, 100% control of progeny was observed for all five species, except for 87% progeny reduction for *R.*

Table 14.5 Mortality (%) of Five Tested Species of Adult Insects at 25°C and a Dose Rate of 100 mg/kg for 7 and 14 d Treatment

	T. castaneum		R. dominica		S. oryzae		L. serricorne		C. pusillus	
Dust	7 d	14 d	7 d	14 d	7 d	14 d	7 d	14 d	7 d	14 d
MU4	96.9	100	83.3	89.8	100	100	100	100	100	100
MU7	98.6	100	99	99.8	100	100	100	100	100	100
MU8	91.3	99	97	97.6	100	100	100	100	100	100
MU9	93.5	99.8	100	100	100	100	100	100	100	100

Table 14.6 Larger-Scale Treatment of Five Species of Insects at 25°C and a Dose Rate of 150 mg/kg for 6 wk Treatment

Dust	Insect Species	No. of Adults Added	Dead	Alive	Mortality (%)
MU4 dust	T. castaneum	997	1,005	0	100
	S. oryzae	980	980	0	100
	R. dominica	989	989	0	100
	L. serricorne	1,013	1,013	0	100
	C. pusillus	941	941	0	100
MU7 dust	T. castaneum	983	985	0	100
	S. oryzae	995	1,003	0	100
	R. dominica	1,001	1,001	0	100
	L. serricorn	975	975	0	100
	C. pusillus	803	803	0	100
MU8 dust	T. castaneum	996	998	0	100
	S. oryzae	957	958	0	100
	R. dominica	981	982	0	100
	L. serricorne	1,002	1,002	0	100
	C. pusillus	795	795	0	100
MU9 dust	T. castaneum	1,010	1,010	0	100
	S. oryzae	991	991	0	100
	R. dominica	996	1,001	0	100
	L. serricorne	1,000	1,000	0	100
	C. pusillus	863	863	0	100
Control	T. castaneum	990	8	1,907	-
	S. oryzae	1,521	184	2,324	-
	R. dominica	869	51	1,870	-
	L. serricorne	997	879	1,425	-
	C. pusillus	1,308	53	2,563	-

dominica treated with MU4 (Agarwal, unpublished data). Results recorded in this study show the effectiveness of SAS against stored-grain pests, which could be a game changer in protection of grains in storage.

Doaa and Nilly (2015) used fumed silica, Aerosil 200 nano particles (99.8% SiO_2), against the rice weevil, lesser grain borer, and cowpea beetle, *Callosobruchus maculatus* (F.), at $28 \pm 2°C$ and $65 \pm 5\%$ RH. Complete 100% mortality was observed with 1 g/kg for *C. maculatus*, 1.5 g//kg for

R. dominica, and 2.5 g/kg for *S. oryzae,* from the second, third, and sixth day of the treatment, respectively. This study used a dosage 10 to 30 times higher to achieve the same result reported by Ren and Agarwal (2017). This could be attributed to the lower purity and less surface area of amorphous silica in Aerosil 200 compared to MU-based SAS. Ren and Agarwal (2017) also tested SAS against the Indian meal moth, *Plodia. Interpunctella,* in almonds, peanuts, and sultanas and found that MU8 and MU9 can protect the almonds for 100 d at a dose rate of 100 g/t. When the same dust was tested by Metz (2022) for almonds and pistachios, it was found that it significantly reduced populations of both *P. interpunctella* and *T. castaneum* at all doses tested ($\geq$200 g/t) and showed higher mortality against both pests than the industry standard diatomaceous earth. The efficacy of SAS was then tested under three different temperatures (13°, 25°, 33°C) and relative humidity (32%, 55%, 75%) at a rate of 200 g/t. Both *P. interpunctella* and *T. castaneum* showed an effect on mortality dependent on climate combinations.

14.4.5 Mechanisms of Action

As SAS is a recent addition to the realm of contact pesticides, the precise mode of action remains an area of ongoing investigation. Although its mechanism shares similarities with that of DE-based dust, a comprehensive understanding is yet to emerge.

The particles constituting amorphous silica possess small size and an irregular morphology, rendering them abrasive to insects. Upon contact, these particles adhere to the insect's exoskeleton and initiate the absorption of the protective waxes and oils that shield the insect's body. This causes loss of moisture and eventual dehydration, causing the mortality of the insect.

Li et al. (2020) studied the effect of both hydrophilic and hydrophobic SAS dust on insect cuticle of *T. castaneum* and *S. oryzae* by comparing the ventral reflectance using a hyperspectral imaging technique coupled with deep learning tools. In total, 856 images were taken and subjected to back propagation neural network models. The results showed clear differentiation between the treated insects and controls.

The photos showed that treated insects often accumulated dust around their mouth parts (Figure 14.4). This accumulation potentially obstructed airways and food passages, contributing to eventual fatality. A similar observation was reported by Subramanyam and Roesli (2000).

Furthermore, Agarwal, in unpublished videography experiments, exhibited a loss of locomotive control among SAS-treated insects, ultimately leading to desiccation-induced limb and antenna fractures. An alternative hypothesis suggests that the deposition of SAS particles on insect body

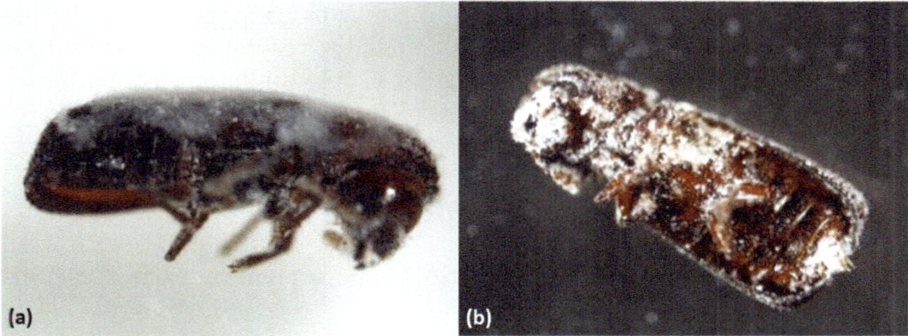

Figure 14.4 Dorsal (left image) and ventral (right image) side of *R. dominica* treated with synthetic amorphous silica, showing its deposition onto various body parts. (Photos: Manjree Agarwal.)

hairs (Figure 14.5a) may affect their sense of motion, ultimately contributing to their mortality, given the sensory role of insect hairs. Scanning electron microscopy (SEM) imagery depicted the presence of SAS particles on insect spiracles (Figure 14.5b), potentially disrupting respiratory function and causing suffocation; similar observations were made by Webb (1946).

These findings are further supported by Fruijtier-Pölloth's (2012) study, according to which SAS adsorbs to cellular surfaces and can affect membrane structures and integrity. Toxicity is linked to mechanisms of interactions with outer and inner cell membranes, signaling responses, and vesicle trafficking pathways. Interaction with membranes may induce

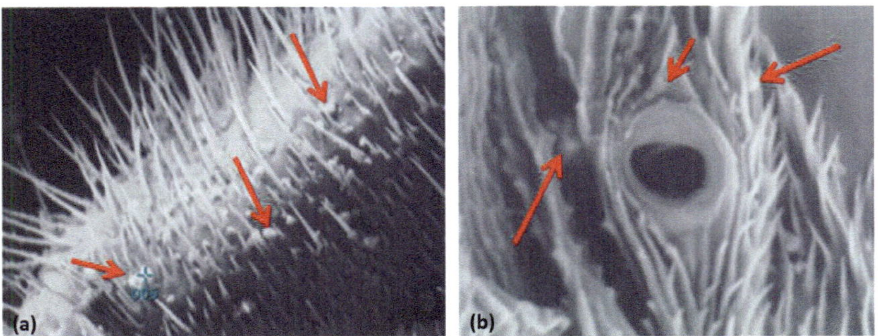

Figure 14.5 (a) Electron-imaging scan of SAS-treated insect showing deposition of SAS onto abdomen hair, indicated by red arrows. (b) Electron-imaging scan of SAS-treated insect showing deposition of SAS onto the opening of the spiracle, indicated by red arrows. (Photos: Manjree Agarwal.)

the release of endosomal substances, reactive oxygen species, cytokines, and chemokines and thus induce inflammatory responses. Despite these hypotheses, a comprehensive elucidation of SAS's mode of action necessitates further in-depth research and exploration.

14.4.6 Efficacy of SAS in Grain Protection

Just like that of DE, the efficacy of SAS is dependent on various factors, which are discussed below.

14.4.6.1 Particle Size and Charge of SAS

The particle size of SAS was observed to span from 1 to 200 nm, contrasting with the size of DE dust particles, which exceeded 10,000 nm (Table 14.1). This makes some of the SAS about 20–40 times more efficacious than the registered DE dusts, but some others, like aerosol 200 with a particle size between 5 and 50 nm, still possess efficacy close to that of DE dust (Doaa and Nilly, 2015). This shows that particle size is not the paramount determinant of SAS efficacy, but the biological activity of SAS can be related to the particle shape, charge, and surface characteristics, as mentioned by Fruijtier-Pölloth (2012). Differences in efficacy of the dust despite low particle size could be attributed to the fact that, because of its semi-nano and nano structure, there is a charge associated with SAS dust. As the insects traverse through the grain, they become magnetically receptive to the dust particles, which accumulate prominently on their body surfaces, as the insects also possess a charge. This accumulation results in a significantly elevated concentration around the insect's body, culminating in its death. Intriguingly, this implies that even when the applied dust's concentration on the grain is relatively low, the accumulated dust on the insect's body surface is substantially higher, thereby inducing mortality (Agarwal, unpublished data).

14.4.6.2 Surface Area of SAS

The smaller the particle size, the higher the surface area and the more bioefficient and bioavailable the dust is to the target insects. The BET surface area is widely used for the calculation of effective surface area and to explain the physical adsorption of gas on a solid surface. If we compare the various MU dusts, such as the top four SAS dusts (i.e., MU4, MU7, MU8, and MU9) (Table 14.1), against DE-based dusts MU10, MU11, and MU12, the BETs of DE dust are between 25 and 57 m^2/g, while they are above 200 m^2/g for the top-performing SAS dusts. The dosage to bring about 100% kill for dust with a low BET is between 1 kg and 10 kg/t, in comparison to 0.5 to 2 g/t for SAS dust with a higher BET, depending on the insect species. Thus, we can say that BET surface area is one of the key factors in determining the kill dosage, as it directly affects the bioavailability of the dust to the insect surface, this being a contact pesticide (Agarwal, unpublished data).

14.4.6.3 Moisture and Relative Humidity

Both the moisture of the grain and the relative humidity in the environment have been shown to have a negative effect on the efficacy of the SAS dust. Two level of grain moisture, 11.4% and 13.0%, and two levels of RH, 60 ± 5% and 75 ± 5%, were tested on *S. oryzae, T. castaneum, R. dominica,* and *C. pusillus* in wheat, using both MU8 and MU9, and results showed a negative effect on the efficacy of SAS with an increase in moisture content and relative humidity. Formulation MU9 functioned better than MU8 treatments at high relative humidity. The factor that probably contributed to the better efficacy of MU9 is its hydrophobic physical characteristic (Agarwal, unpublished data).

14.4.6.4 Temperature

Three levels of temperature, 15°, 25°, and 35°C, were tested against *S. oryzae, T. castaneum, R. dominica,* and *C. pusillus* in wheat with 11.4% MC, using both MU8 and MU9, and it was found that mortality of *S. oryzae* and *C. pusillus* increased with temperature but mortality of *T. castaneum* and *R. dominica* was better at 25°C than at 35°C. However, the efficacy was better at 35°C than at 15°C. This might be due to the less active behavior of insects at a temperature like 15°C compared to a warm environment. The suppression of activities reduces the opportunity for insects to pick up a lethal amount of dust to the cuticle (Agarwal, unpublished data).

14.4.6.5 Differences in Insect Morphology and Physiology

The type of insect and insect size play a significant role in the efficacy of SAS, just as they do for DE. Insects small in size are more susceptible than bigger insects. For insects small in size, like *C. pusillus, C. ferrugineus,* and *O. surinamensis,* 100% mortality was achieved at 25 mg/kg within 7 d, while *S. oryzae* and *T. castaneum* started to show 100% mortality from a 100 mg/kg dosage of SAS dust and *R. dominica* showed mortality of 83%–90%. *Rhyzopertha dominica* was found to be the hardest of the species to kill, and this can be attributed to the fact that they are slow moving or that they have a thicker cuticular wax layer (Agarwal, unpublished data).

14.4.6.6 Type of Grain Commodity

When the top SAS dusts, MU7, 8, and 9, were tested against wheat, oats, barley, and canola at 150 mg/kg, the efficacy of these dusts in wheat and barley was found to be much higher than that in oats. This can be attributed to the fact that oats have long petioles, which the insect crawls to and holds onto as a refuge from the dust. In the case of canola, the efficacy of the dust was almost negligible; this might be because the oil content of canola made the dust ineffective (Agarwal, unpublished data).

A similar result was reported by Athanassiou et al. (2003) using the DE SilicoSec against the rice weevil: the insecticidal efficacy varied significantly among barley, maize, peeled rice, and paddy rice.

14.4.7 Advantages and Limitations of Synthetic Amorphous Silica

Just like DE, SAS has its own advantages and limitations.

14.4.7.1 Advantages

Food grade option: Unlike DE, SAS comes in both food grade and nonfood grade, and the treatment plan for SAS can be established accordingly. If the SAS is just for structural treatment, the nonfood grade can be used, but if it is for treatment of food items, the food grade, which is already registered under the food grade category, can be used. Another advantage of having food grade SAS dust is that after treatment and before consumption, the commodity does not need to go through a cleaning process such as further separation, sieving, or washing to get rid of dust.

Low dosage required: Because of its high efficacy, a 40–100 times smaller dosage is required than for DE-based dust. This low dosage results in easy application with no dusty environment.

No silicosis concerns: Unlike DE-based dust, SAS does not cause any silicosis, as it does not have any crystalline silica. Also, a very low treatment dosage is required, so the application is not very dusty or harmful to the applicator.

No damage to machinery: It does not cause any damage to the application machinery, as the particle size is very fine and hence no abrasion is caused.

Long-lasting protection: When applied properly, SAS can provide protection for almost a year against stored-grain pests. Its effectiveness does not diminish over time as long as the conditions remain dry.

Fast acting effect: Unlike DE dust, SAS is very fast acting and can kill insects from 2 d onward (Ren and Agarwal, 2017).

14.4.7.2 Limitations

There is no doubt that synthetic amorphous silica can be a game changer in stored-grain protection, but it does have its own limitations, which are as follows.

Reduction in flowability and increase in angle of repose: Just like DE, SAS causes a reduction in flowability and an increase in angle of repose, neither of which is acceptable in the international grain trade. Ren and Agarwal (2017) have compared the gravity angle between untreated wheat and wheat treated with different concentrations of SAS and found an increase in the angle of repose by 9 degrees when grain was treated with SAS at 150 mg/kg. But when the concentration of SAS was reduced to 50 mg/kg, there was no effect on the angle of repose. To have no effect on flowability and the angle of repose, Agarwal (unpublished data) focused on getting maximum efficacy against the majority of the stored-grain insects with the minimum dose (<50 mg/kg) possible by deflocculating the SAS dust.

Requirement for dry conditions: In high humidity or damp storage conditions, the ability of SAS to control pests is reduced. The reduction in efficacy is much higher in the case of hydrophilic SAS dust than in the case of hydrophobic SAS dust.

Occupational health and safety concerns: Inhalation of SAS dust can cause respiratory irritation. Therefore, it is essential to take precautions by wearing the appropriate mask and overalls when applying SAS in powdered form.

Risk to beneficial insects: Synthetic amorphous silica may harm beneficial insects like bees and ladybugs if not used with caution.

14.5 Combination Treatment

As we can see from the above section, each individual class of protectants has its own advantages and disadvantages. To overcome the shortcoming of each class, combination treatments are considered. Some examples of combination treatments are the following:

Methoprene with DE: When methoprene is used in combination with other insecticides, the result is often an additive effect on insect mortality; methoprene demonstrates this effect when used with diatomaceous earth (Arthur, 2004). Different formulations of the insect growth regulator methoprene were compared to its combination with DE for control of *R. dominica*. Within each methoprene concentration, survival of the insect generally decreased with increasing concentration of commercial diatomaceous earth (DE) Protect-Its® (Arthur, 2004).

Methoprene, organophosphates, and pyrethroids: Daglish (1998) reported a combination treatment composed of an organophosphate protectant (fenitrothion, chlorpyrifos-methyl, pirimiphos-methyl, or methacrifos) and either synergized deltamethrin or the juvenile hormone analogue methoprene. Protectants were applied at registered or likely application rates and evaluated against *R. dominica*, *S. oryzae,* and *T. castaneum* in stored wheat and also against *R. dominica* in stored maize. The results reported no interactions of any significance in terms of prevention of synergism or antagonism happening between the treatments.

Pyrethroids with DE: Combinations of dust and pyrethroids have significantly enhanced the performance of each. Machekano et al. (2017) used deltamethrin at the dose rate of 0.1 and 0.05 ppm with DE and got the same efficacy as with the recommended dose of 130 ppm deltamethrin. Karimzadeh et al. (2021) have reported many other combinations of pyrethroids and DE for controlling insects.

14.6 Conclusions

Contact insecticides are one of the candidates for use as an alternative to fumigants. Each of the contact insecticides, such as chemicals, diatomaceous earth, and amorphous silica, has its own advantages and limitations. Based on their efficacy, safety, and environmental considerations, an appropriate treatment regime needs to be developed to tap their maximum potential and ensure sustainable pest control practices. Though the majority of them have developed resistance and require alternatives, synthetic amorphous silica could be a game changer in stored-grain protection. Further work is needed on its mode of action and application method.

Acknowledgments

The author wishes to acknowledge the funding body PBCRC, for funding the synthetic amorphous silica projects; Prof. Yonglin Ren and the staff of the Postharvest Biosecurity Laboratory at Murdoch University, Dr. Junxi, Ms. Suchi, Ms. Amy, Ms. Shirley, Mr. J. Newman, and Mr. C. Park, for their support in SAS-related trials; and grain bulk handlers CBH for providing the wheat.

References

Aldryhim YN (1993) Combination of classes of wheat and environmental factors affecting the efficacy of amorphous silica dust, oryacide, against *Rhyzopertha dominica* (F.). J Stored Prod Res 29: 271–275.

Arthur FH (1996) Grain protectants: Current status and prospects for the future. J Stored Prod Res 32(4): 293–302.

Arthur FH (2004) Evaluation of methoprene alone and in combination with diatomaceous earth to control *Rhyzopertha dominica* (Coleoptera: Bostrichidae) on stored wheat. J Stored Prod Res 40(5): 485–498.

Athanassiou CG, Kavallieratos NG, Tsaganou FC, Vayias BJ, Dimizas CB and Buchelos CT (2003) Effect of grain type on the insecticidal efficacy of SilicoSec against *Sitophilus oryzae* (L) (Coleoptera: Curculionidae). Crop Prot 22: 1141–1147.

Athanassiou CG, Papagregoriou AS, Buchelos CT (2004) Insecticidal and residual effect of three pyrethroids against *Sitophilus oryzae* (L.) (Coleoptera: Curculionidae) on stored wheat. J Stored Prod Res 40(3): 289–297.

Baliota GV, Athanassiou CG (2020) Evaluation of a Greek diatomaceous earth for stored product insect control and techniques that maximize its insecticidal efficacy. Appl Sci 10: 6441.

Bridgeman BW (1994) Structural treatment with amorphous silica slurry: an integral component of GRAINCO's IPM strategy. pp. 628–630. In: Highley E, Wright EJ, Banks HJ, Champ BR (eds), Pro 6th Intern Working Conf Stored-Prod Prot, Canberra, Australia.

Calvert R (1930) Diatomaceous Earth. American Chemical Society Monograph. Reprint 1976University Microfilms Ann Arbor MI pp 251.

Chiu SF (1939) Toxicity of so-called "inert" materials with the bean weevil *Acanthoscelides obtectus*. J Econ. Entomol 32(2): 240–248.

Collins PJ, Lambkin TM, Bridgeman BW, Pulvirenti C (1993) Resistance to grain protectant insecticides in coleopterous pests of stored cereals in Queensland, Australia. Eco Entomol 86: 239–245.

Collins PJ (2006) Resistance to Chemical Treatments in Insect Pests of Stored Grain and Its Management. In: Lorini L, Bacaltchuk B, Beckel H, Deckers D, Sundfeld E, Santos JP, Biagi JD, Celaro JC, D'A LR, Faroni, Bortolini LOF, Sartori MR, Elias MC, Guedes RNC, Fonseca RG, Scussel VM (eds), Proc 9th Intern Working Conf Stored-Prod Prot, Campinas, São Paulo, Brazil.

Daglish GJ (1998) Efficacy of six grain protectants applied alone or in combination against three species of Coleoptera. J Stored Prod Res 34: 263–268.

Daglish GJ, Nayak MK (2012) Potential of the neonicotinoid imidacloprid and the oxadiazine indoxacarb for controlling five coleopteran pests of stored grain. Insect Sci 19: 96–101.

Desmarchelier JM, Dines JC (1987) Dryacide treatment of stored wheat: Its efficacy against insects, and after processing. Aust J Exp Agric 27: 309–312.

Desmarchelier JM, Wright EJ, Allen SE (1993) Dryacide®: a structural treatment for stored product insects. pp. 483–485. In: Corey SA, Dall DJ, Milne WM (eds), Proc 5th Aust Appl Entomol Res Conf, Canberra, Australia.

Doaa MB, Nilly AHA (2015) Entomotoxic effect of Aerosil 200 Nano Particles against three main stored grain insects. Int J Adv Res 3(8): 1371–1376.

Ebeling W (1971) Sorptive dust for pest control. Ann Rev Entomol 16: 123–158.

Fields P, Korunić Z. (2000) The effect of grain moisture content and temperature on the efficacy of diatomaceous earths from different geographical locations against stored-product beetles. J. Stored Prod. Res., 36(1): 1–13. doi: http://dx.doi.org/10.1016/S0022-474X(99)00021-1

Fields PG, Korunić Z (2002) Post-harvest insect control with inert dusts. In Pimentel D (ed), Dekker Encyclopedia of Pest Management. Marcel Dekker, New York, pp. 650–653.

Food and Agriculture Organization (FAO) (2021). "FAO Food Loss and Food Waste Databases." [Online]. Available: https://www.fao.org/platform-food-loss-waste/flw-data/en/

Food and Agriculture Organization (FAO) (2022) "World Food and Agriculture - Statistical Yearbook 2022." pp. 1–363. [Online]. Available: https://doi.org/10.4060/cc2211en. Rome.

Fruijtier-Pölloth C (2012) The toxicological mode of action and the safety of synthetic amorphous silica – A nanostructured material. Toxicology 294: 61–79.

Golob P (1997) Current status and future perspectives for inert dust for control of stored product insects. J Stored Prod Res 33(1): 69–81.

GRAS Substances Database (2020) Generally Recognized As Safe. U.S. Department of Health and Human Services. U.S. Food and Drug Administration Silver Spring MD.

International Diatom Producers Association (IDPA) (1990) The Nature and Safe Handling of Diatomaceous Earth. Comments by W. Clark Copper 1-6.

Jackson K, Webley D (1994) Effects of Dryacide on the physical properties of grains, pulses and oilseeds. pp. 635–637. In: Highley E, Wright EJ, Banks HJ, Champ BR (eds), Pro 6th Intern Working Conf Stored-Prod Prot, Canberra, Australia.

Karimzadeh R, Salehpoor M, Saber M (2021) Initial efficacy of pyrethroids, inert dusts, their low-dose combinations and low temperature on *Oryzaephilus surinamensis* and *Sitophilus granarius*. J Stored Prod Res 91: 101780.

Korunić Z (1997) Rapid assessment of the insecticidal value of diatomaceous earths without conducting bioassays. J Stored Prod Res 33(3): 219–229.

Korunić Z (1998) Diatomaceous earth, a group of natural insecticide. J Stored Prod Res 34(2/3): 87–97.

Korunić Z (2013) Diatomaceous earths - Natural Insecticides. Pesticides & Phytomedicine (Belgrade) 28(2): 77–95.

Korunic Z, Cenkowski S, Fields P (1998) Grain bulk density as affected by diatomaceous earth and application method. Postharvest Biol Technol 13(1): 81–89.

Korunić Z, Fields P (1995) Diatomaceous Earth Insecticidal Composition. Canadian and U.S.A. Patent 5,773,017.

Korunić Z, Fields PG (2002) Post-harvest insect control with inert dusts. In Pimentel D (ed) Dekker Encyclopedia of Pest Management. Marcel Dekker, New York, pp. 650–653.

Korunić Z, Fields PG, Kovacs MIP, Noll JS, Lukow OM, Demianyk CJ, Shibley KJ (1996) The effect of diatomaceous earth on grain quality. Postharvest Biol Technol 9(3): 373–387.

Korunić Z, Rozman V, Liška A, Lucić P (2016) A review of natural insecticides based on diatomaceous earths. Poljoprivreda 22(1): 10–18. https://doi.org/10.18047/poljo.22.1.2

Li Y, Agarwal M, Cao Y, Ren YL (2020) Effect of synthetic amorphous silica (SAS) powder on the cuticle of *Tribolium castaneum* and *Sitophilus oryzae* using hyperspectral imaging technique. Pest Manag Sci 76(1): 314–323.

Machekano H, Mvumi BM, Chinwada P, Richardson-Kageler SJ, Rwafa R (2017) Efficacy of diatomaceous earths and their low-dose combinations with spinosad or deltamethrin against three beetle pests of stored-maize. J Stored Prod Res 72: 128–136.

McLaughlin A (1994) Laboratory trials on desiccant dust insecticides. pp. 638–645. In: Highley E, Wright EJ, Banks HJ, Champ BR (eds), Pro 6th Intern Working Conf Stored-Prod Prot, Canberra, Australia.

Metz RL (2022) The Effect of Synthetic Amorphous Silica for Postharvest Control of *Tribolium castaneum* and *Plodia interpunctella* under Varying Factors of Temperature and Humidity. Dissertation, California State University, Fresno, United States.

Mian LS, Mulla MS (1982a) Biological activity of IGRs against four stored-product coleopterans. J Econ Entomol 75: 80–85.

Mian LS, Mulla MS (1982b) Residual activity of insect growth regulators against stored-product beetles in grain commodities. J Econ Entomol 75: 599–603.

Nikpay A (2006) Diatomaceous earth as alternatives to chemical insecticides in stored grain. Insect Sci 13(6): 421–429.

Oberlander H, Silhacek DL, Shaaya E, Ishaaya I (1997) Current status and future perspectives of the use of insect growth regulators for the control of stored product insects. J Stored Prod Res 33: 1–6.

Papadopoulou-Mourkidou E, Tomazou T (1991) Persistence and activity of permethrin in stored wheat and its residues in wheat milling fractions. J Stored Prod Res 27(4): 249–254.

Quarles W (1992) Diatomaceous earth for pest control. IPM Practitioner 14(5/6): 1–11.

Quarles W (2007) Diatomaceous earth. PCT 32-34.

Quarles W, Winn P (1996) Diatomaceous earth and stored product pests. IPM Practitioner 18(5/6): 1–10.

Quinlan JK, Berndt WL (1966) Evaluation in Illinois of four inert dusts on stored shelled corn for protection against insects – a progress report. USDA/ARS Report No. 51-6.

Reader CJ and Nargiello M (2020): The use of engineered silica to enhance coatings. CoatingsTech 17(6): 36-45

Redlinger LM, Womack H (1966) Evaluation of four inert dusts for the protection of stored corn in Georgia from insect attack. USDA/ARS S1-7 June.

Ren YL, Agarwal M (2017) Method of controlling insects and insecticide for use therein. US 2017/0071233 and EP3119202.

Round FE, Crawford RM, Mann DG (1990) The Diatoms. Biology & Morphology of the genera. Cambridge University Press, New York, USA.

Samson PR, Parker RJ, Hall EA (1990) Efficacy of the insect growth regulators methoprene, fenoxycarb, and diflubenzuron against Rhyzopertha dominica (F.) (Coleoptera: Bostrichidae) on maize and paddy rice. J Stored Prod Res 26: 215–221.

Shah MA, Khan AA (2014) Use of diatomaceous earth for the management of stored-product pests. Int. J Pest Manage 60(2): 100–113.

Subramanyam B, Roesli R (2000) Inert dusts. In Subramanyam B, Hagstrum DW (ed), Alternatives to Pesticides in Stored-Product IPM. Kluwer Academic Publishers, Dordrecht, pp. 321–380.

Subramanyam B, Swanson CL, Madamanchi N, Norwood S (1994) Effectiveness of Insecto, a new diatomaceous earth formulation, in suppressing several stored grain insect species. pp. 650–659. In: Highley E, Wright EJ, Banks HJ, Champ BR (eds), Pro 6th Intern Working Conf Stored-Prod Prot, Canberra, Australia.

Tsaganou FC, Vassilakos TN, Athanassiou CG (2014) Knockdown and mortality of five stored product beetle species after short exposures of thiamethoxam. J Econ Entomol 107: 2222–2228.

Tsaganou FK, Vassilakos TN, Athanassiou CG (2021) Insecticidal effect of thiamethoxam against seven stored-product beetle species. J Stored Prod Res 93: 101843.

Van Graver J, Winks RG (1994) A Brief History of the Entomological Problems of Wheat Storage in Australia. pp. 1250–1258. In: Highley E, Wright EJ, Banks HJ, Champ BR (eds), Pro 6th Intern Working Conf Stored-Prod Prot, Canberra, Australia.

Vayias BJ, Athanassiou CG, Korunić Z, Rozman V (2009) Evaluation of natural diatomaceous earth deposits from south-eastern Europe for stored-grain protection: The effect of particle size. Pest Manag Sci 65: 1118–1123.

Vincent JFV (1971) Effects of burs icon on cuticular properties of Locusta migratoria migratoria. J Insect Physiol 17: 625–636.

Vrba CH, Arai HP, Nosal M (1983) The effect of silica aerogel on the mortality of *Tribolium confusum* (Duval) as a function of exposure time and food deprivation. Can J Zool 61: 1481–1486.

Webb JE (1946) The penetration of derris through the spiracles and cuticle of *Melophagus ovinus*, L. Bull Entomol Res 36: 15–22.

White GD, Bern WL, Wilson JL (1975) Evaluating diatomaceous earth, silica-Aerogel dust and malathion to protect stored wheat from insects. USDA/ARS Marketing Research Report No. 1038: 1-18.
Wijayaratne LKW, Arthur FH, Whyard S (2018) Methoprene and control of stored-product insects. J Stored Prod Res 76: 161–169.
Wright EJ (1990) A trapping method to evaluate efficacy of a structural treatment in empty silos. pp. 1455–1463. In: Fleurat-Lessard F, Ducom P (eds), Proc 5th Intern Working Conf Stored-Prod Prot, Bordeaux, France.
Zeni V, Baliota GV, Benelli G, Canale A, Athanassiou CG (2021) Diatomaceous earth for arthropod pest control: Back to the future. Molecules 26: 7487. https://doi.org/10.3390/molecules26247487.

Chapter 15

Natural Products for Fumigation and Treatment

*Samuel Adelani Babarinde and
Adeyemi Oluseye Akinyemi*

15.1 Introduction

Natural products are the outcome of nature and chemical investigations and have diverse sources including botanicals and marine and microbial organisms. They are originally produced by life but can be produced by chemical synthesis. Their uses span medicinal applications, food, industrial products, and pesticides. Eucalyptol, a monoterpenoid compound in botanical essential oils, is used in the pharmaceutical and cosmetic industries, *Mentha spicata* provides a good ingredient for flavors in toothpaste and mouthwash. Citrus peel essential oils are used in the food industry for flavor enrichment (Rajendran and Sriranjini, 2008).

The spectrum of the synthetic products registered as fumigants varies from one region of the world to another. In Australia, carbonyl sulphide and ethyl formate have been registered as fumigants (Reichmuth, 2010). In the United States, chloropicrin, propylene oxide, and methyl iodide have been registered. Methyl iodode is registered in the United States and Japan

DOI: 10.1201/9781003309888-15

to control microbes (Reichmuth, 2010). Some of these registered products have their particular advantages, including comparative economic advantage, high penetration ability, reduced residues in treated products, suitability for bulk products, and efficacy against different growth stages of arthropods. Despite the listed advantages, the search for reliable alternatives is ongoing due to safety and environmental concerns. Table 15.1 briefly summarizes the comparison of synthetic and natural products used for fumigation and produce storage.

Early natural products of botanical origin used as insecticides include nicotine obtained from *Nicotiana tabacum*, pyrethrum from *Chrysanthemum cinerariefolium,* ryanodine from *Ryania speciosa,* and azadirachtin from *Azadirachta indica* (Addor, 1995). The discovery of these early natural products of botanical origin was followed by the discovery of microbial products. The major examples in this category are abamectin, which is a fermentation product of the soil-living bacterium *Streptomyces avermitis,* and Spinosad, which is obtained from another bacterium, *Saccharopolysposa spinosa.* These products have been reported primarily as insecticides. Among the early natural products used as herbicides is bialaphos, a tripeptide produced by the bacteria *Streptomyces viridochromeogenes* and *S. hydroscopicus,* which has been reported as a nonselective, broad-spectrum herbicide (Li et al., 2003). Some species of mushrooms have also been researched as possible sources of insecticidal or microbial products. For instance, the chloroform extract of Algerian desert truffle, *Terfezia claveryi* Chatin, was assayed against the rice weevil, *Sitophilus oryzae* (L.), and the lesser grain borer, *Rhyzopertha dominica* (F.), and the rice weevil was more susceptible to it than the lesser grain borer (Neggaz et al., 2020).

To a very large extent, natural products of botanical origin have been the most researched pesticidal group; the major compounds well explored are from the families of Asteraceae, Zingiberaceae, Annonaceae, Meliaceae, Rutaceae, Apocynaceae, Solaneceae, Apiaceae, Piperaceae, Malvaceae, Canellaceae, Liliaceae, Apiaceae, Cucurbitaceae, Lauraceae, Myrtaceae, Iridaceae, Pinaceae, Anacardiaceae, and Lamiaceae. Botanicals were initially formulated as crude/aqueous extracts, powders, and ashes. These were followed by the organic extracts made of polar and nonpolar solvents and the comparatively more popular essential oils (EOs). The recent trend is the use of nanotechnology in essential oil bioassays and formulations.

This chapter focuses on the pesticidal properties of natural products, their toxicity against the target pests, the impact of the natural products on the quality of the stored produce, and challenges of natural product bioprospecting. The chapter concludes with future research needs on natural products for fumigation and treatment.

Table 15.1 Merits and Demerits of Synthetic and Natural Pesticides

Merits	Demerits
Synthetic Chemicals	
They are considered relatively cheap.	Use of pesticides increases the cost of production among farmers, especially for crops that are highly susceptible to pest attacks.
They are fast and usually have an immediate impact and can be suitable when there is a need to suppress a pest population suddenly.	Some are broad spectrum, killing beneficial and nontargeted organisms such as pollinators, biocontrol agents, and soil organisms.
They are not bulky and are suitable for bulky products.	They are poisonous to users, especially in Sub-Saharan Africa, where users do not use protective wear when spraying.
They have a high penetration ability: some act as both a contact and a systemic insecticide and can be spayed as small particles, thus penetrating even minute areas.	Some of the banned pesticides are still illegally used in some countries, thus exposing innocent people to their negative impacts.
They are accessible and most have been registered and standardized; thus users can use them based on recommendations and label.	Residues are found in crops, fruit, vegetables, and even animal products.
	They are washed into water bodies, thus contaminating water bodies.
	Some of the available pesticides are persistent and are not easily biodegradable.
	Pests can develop resistance to synthetic pesticides. Resistance has been discovered in major stored-product pests such as the red flour beetle (Attia et al., 2020).

(Continued)

Table 15.1 (Continued) Merits and Demerits of Synthetic and Natural Pesticides

Merits	Demerits
Natural Products	
They are compatible with other control methods.	Formulations are limited.
They have elective toxicity. Not being broad spectrum, they may have reduced or no mammalian toxicity.	Products suffer from instability and high volatility: often many repeated treatments are necessary for effectiveness.
They are less persistent in the environment and on treated products.	
They are acceptable in organic agriculture.	
Limited technical know-how is required for their use.	

15.2 Natural Products Used Against Storage Insect Pests

Several products of natural origin with insecticidal potential have been documented. Plant-based EOs are natural plant products that are aromatic, a mixture of fragrant substances and chemically pure volatile compounds obtained from plant materials consisting principally of lipophilic and highly volatile plant metabolites, including sesquiterpene and monoterpene (Ríos, 2016). They have been extensively researched as effective biopesticides against all kinds of storage and field insect pests including pathogens, especially in a series of laboratory studies. The EOs from *Xylopia parviflora* have been reported as insecticidal products against the cowpea seed bruchid, *Callosobruchus maculatus* (Fabricius), and the rust red flour beetle, *Tribolium castaneum* (Herbst) (Babarinde et al., 2017a, 2019). *Xylopia aethiopica* was also reported to be insecticidal against *C. maculatus*, *S. zeamais* Motschulsky, and *T. castaneum* (Babarinde et al., 2008; Akinyemi et al., 2016). Essential oils obtained from *X. parviflora* root bark and *Hoslundia opposita* leaf via hydro distillation were toxic and repellent against cowpea seed bruchid and red flour beetle (Babarinde et al.,2014; 2017b).

Magnolia coriacea and *M. macclurei* essential oils were evaluated as protectants of agricultural produce against three major stored-product insects: the red flour beetle, cigarette beetle, and booklouse. *Magnolia coriacea* and *M. macclurei* EOs showed promising contact toxicity to the cigarette beetle, with LD_{50} values of 11.7 and 12.3 µg/adult, respectively. The contact toxicity of *M. coriacea* EO to the booklouse (LC_{50} = 95.5 µg/cm^2) was much stronger than that of *M. macclurei* EO (LC_{50} = 245.4 µg/cm^2). To explore the contribution of individual compounds to the insecticidal activity of EO, chemical analysis was performed using GC-MS. Results showed that nerolidol (27.84%), agarospirol (18.34%), elixene (15.84%), and helminthogermacrene (12.69%) were major compounds of *M. coriacea* EOs; β-guaiene (60.31%) and elixene (20.42%) dominated in *M. macclurei* EO. Nerolidol and β-guaiene showed contact activity to three insect species. Nerolidol showed stronger contact toxicity to the red flour beetle and cigarette beetle than *M. coriacea* EO did; both samples were similarly toxic to the booklouse. β-guaiene was much more strongly toxic to the red flour beetle and booklouse, but weaker to the cigarette beetle than *M. macclurei* EO (Feng et al., 2022).

The insecticidal activities of the EO of *Rhynchanthus beesianus* rhizomes against adults of *Liposcelis entomophila* (Enderlein) and *T. castaneum* have also been reported (Pan et al., 2023). The EO exhibited fumigant toxicity against the adults of *L. entomophila* and *T. castaneum* with LC_{50} values of 0.57 and 4.96 mg/L air, while the two isolates methyl eugenol and α-terpineol possessed fumigant toxicity against the booklice (LC_{50} = 0.15 and 0.48 mg/L air, respectively) and the beetles (LC_{50} = 1.81 and 4.96 mg/L air, respectively). The oil also possessed contact toxicity against the booklice and the beetles, with LD_{50} values of 121.56 µg/cm^2 and 54.93 µg/adult, respectively, while the two isolates β-eudesmol and elemol showed contact toxicity against *L. entomophila* (LD_{50} = 99.21 and 35.19 µg/cm^2, respectively) and *T. castaneum* (LD_{50} = 35.26 and 8.89 µg/adult, respectively). It was therefore concluded that the oil of *R. beesianus* rhizomes and its isolates have potential as a source for natural insecticide.

The EOs extracted from the flower and root tissues of *Ferula persica* were evaluated against *T. castaneum* adults and the fourth instar larvae of *Ephestia kuehniella* Zeller. The EOs from the flower and root tissues of *F. persica* were initially prepared using the water distillation method and subsequently utilized for fumigation toxicity assays. To this end, four different series of EO concentrations (each one contained a group of six different concentrations) belonging to both tissues were applied for *T. castaneum* and *E. kuehniella*. In all cases, EO concentration substantially affected the mortality rate of both storage pests of interest. Based on the LC_{50} values of the fumigant assay, the insecticidal activity of EOs from both flower and root tissues of the plant is less toxic against *E. kuehniella* than *T. castaneum* (Chaghakaboodi et al., 2022). To a very great extent, natural products from aromatic plants are prepared as EOs and are subjected to chromatographic analysis (GC or GC-MS or GC-FID) for identification of bioactive constituents. A brief survey of common bioactive compounds found in natural products shows that the bioactive compounds belong principally to monoterpenes, sesquiterpenes, diterpenes, carbocyclic acid (saturated or unsaturated fatty acids), esters, ketones, alcohol, chromene, and furanocumarins (Table 15.2). The chemical structures of some of the compounds are presented in Figure 15.1.

Table 15.2 Major Compounds in Pesticidal Oils

Number	Compound Name	Oil Type	Classification
1	1,8-cineole	Essential oil	monoterpene
2	β-himachalene	Essential oil	sesquiterpene
3	D-limonene	Essential oil	monoterpene
4	α-pinene	Essential oil	monoterpene
5	caryophyllene	Essential oil	sesquiterpene
6	citral	Essential oil	monoterpene
7	β-pinene	Essential oil	monoterpene
8	p-cymene	Essential oil	monoterpene
9	thymol	Essential oil	monoterpene
10	carvacrol	Essential oil	monoterpene
11	β-thujaplicin	Essential oil	monoterpene
12	camphor	Essential oil	monoterpene
13	zingiberene	Essential oil	sesquiterpene
14	γ-terpine	Essential oil	monoterpene
15	α-terpineol	Essential oil	monoterpene
16	brussonol	Essential oil	diterpene
17	pentadecanoic acid	Fixed oil	carboxylic acid
18	oleic acid	Fixed oil	carboxylic acid
19	stearic acid (octadecanoic acid)	Fixed oil	carboxylic acid
20	palmitic acid	Fixed oil	carboxylic acid
21	linoleic acid	Fixed oil	carboxylic acid
22	β-phellandrene	Essential oil	monoterpene
23	nominine	Essential oil	Diterpene
24	bisabolol oxide I	Essential oil	sesquiterpene
25	nerolidol	Essential oil	sesquiterpene
26	bisabolol	Essential oil	sesquiterpene
27	Z-carveol	Essential oil	monoterpene
28	E-p-menta-1-en-9-ol	Essential oil	monoterpene
29	linalool	Essential oil	monoterpene
30	spathulenol	Essential oil	sesquiterpene
31	β-eudesmol	Essential oil	sesquiterpene
32	geranial	Essential oil	monoterpene

(Continued)

Table 15.2 (Continued) Major Compounds in Pesticidal Oils

Number	Compound Name	Oil Type	Classification
33	geranyl acetate	Essential oil	Ester
34	β-elemene	Essential oil	sesquiterpene
35	β-mycrene	Essential oil	monoterpene
36	piperitone	Essential oil	monoterpene
37	terpinen-4-ol	Essential oil	monoterpene
38	terpinolene	Essential oil	monoterpene
39	bornyl acetate	Essential oil	Ester
40	glycerol-1-palmitate	Essential oil	Ester
41	germacrene D	Essential oil	sesquiterpene
42	verbenol	Essential oil	monoterpene
43	cadinene	Essential oil	sesquiterpene
44	preconene I/precocene II	Essential oil	chromene
45	β-guainene	Essential oil	sesquiterpene
46	α-thujone	Essential oil	monoterpene
47	fenchone	Essential oil	monoterpene
48	methyl cinnamate	Essential oil	Ester
49	methyl chavicol	Essential oil	monoterpene
50	humulene	Essential oil	sesquiterpene
51	eugenol	Essential oil	monoterpene
52	α-gurjunene	Essential oil	sesquiterpene
53	isopimpinellin	Essential oil	furanocoumarin
54	geraniol	Essential oil	monoterpene
55	menthyl acetate	Essential oil	Ester
56	menthol	Essential oil	Alcohol
57	menthone	Essential oil	Ketone
58	elemol	Essential oil	sesquiterpene
59	aromadendrene	Essential oil	sesquiterpene
60	α-muurolene	Essential oil	sesquiterpene
61	menthofuran	Essential oil	monoterpene

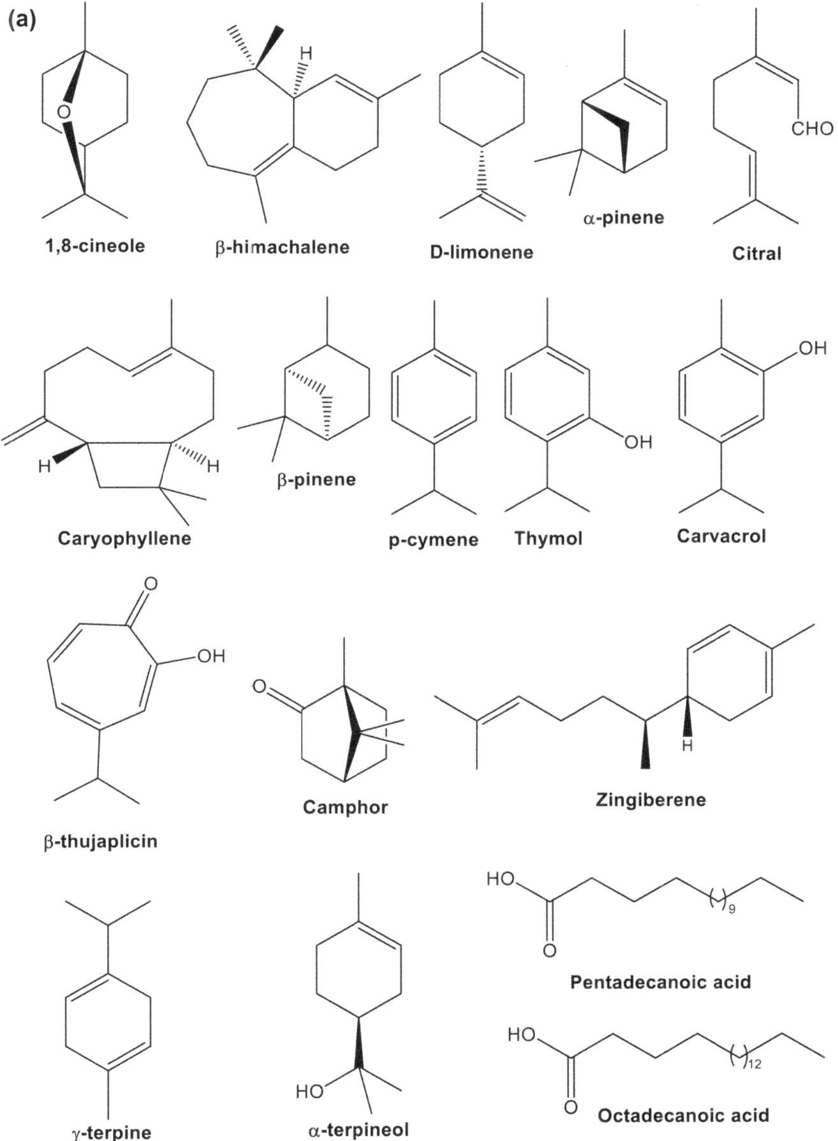

Figure 15.1 Chemical structures of selected constituents in pesticidal oils.

(Continued)

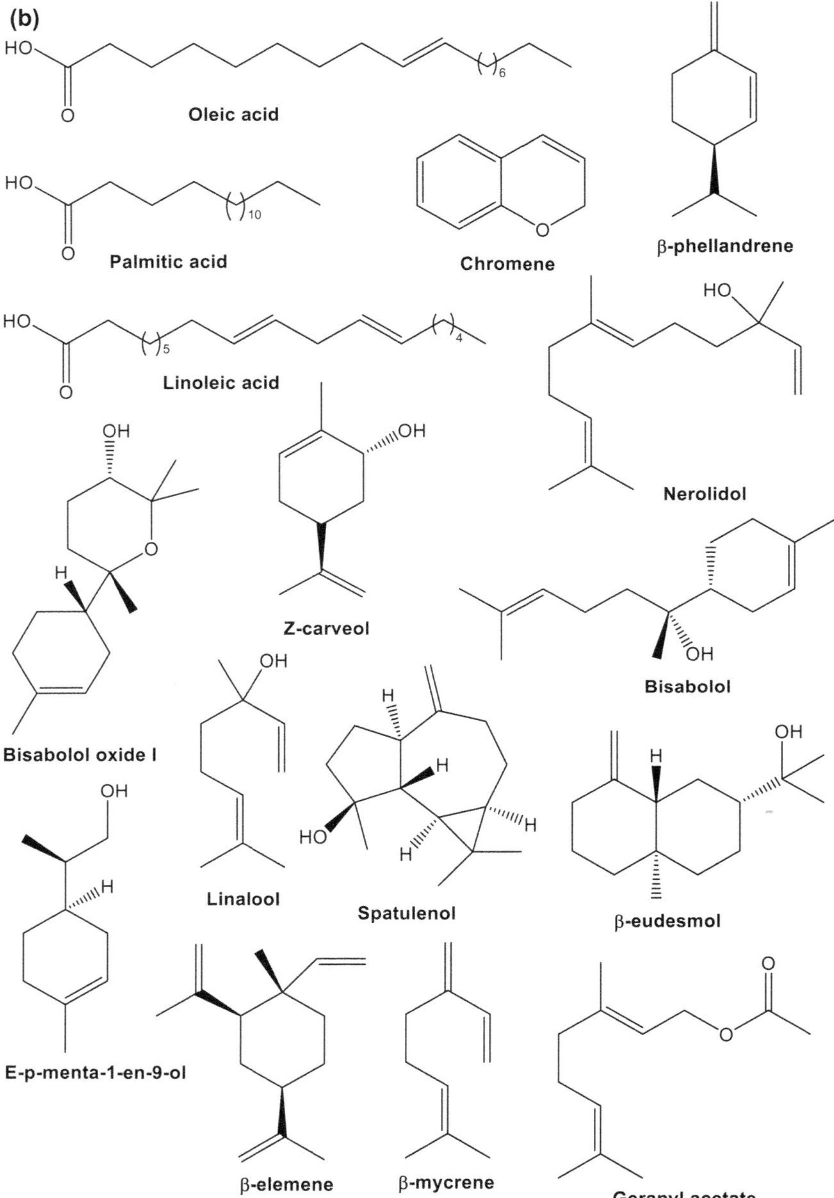

Figure 15.1 (Continued)

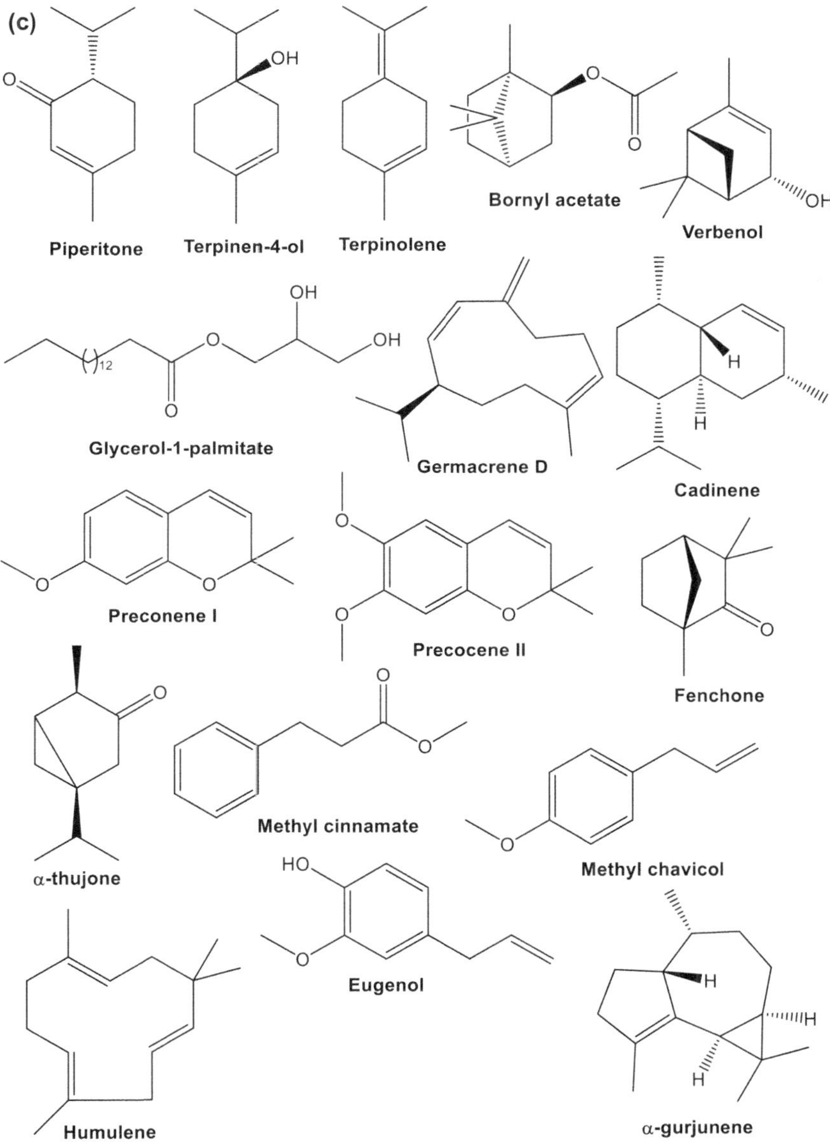

Figure 15.1 (Continued)

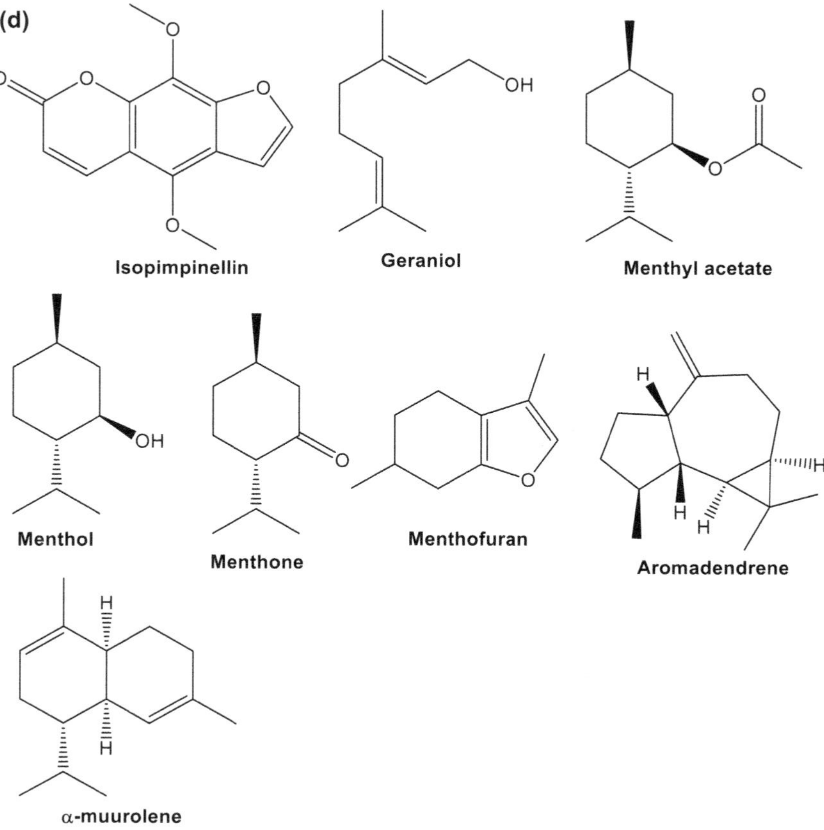

Figure 15.1 (Continued)

15.3 Use of Combinations of Natural Products

Several studies have established synergistic, additive, or even antagonistic effects of natural product mixtures. The antagonistic effects are generally not desirable to achieve the purpose of fumigation and treatment. The synergistic effect of spinosad with three botanical (*Aframomum melegueta*, *Eugenia aromatica*, and *Piper guineense*) powders was investigated against *T. castaneum* infesting melon (*Citrullus lanatus*) seeds. Treatments included sole application of each botanical powder (50 g/kg of melon seed), sole application of spinosad (SASp) (1.0 g/kg), application of a mixture of spinosad (0.5 g/kg) + a botanical powder (25 g/kg), and an untreated control. Data were collected on tenebronid mortality rate (%) (PM) and melon seed weight loss rate in % (PWL). At 3–14 days after treatment (DAT), the PM observed in melon seeds treated with SASp (90.00%–100.00%) was not significantly ($p > 0.05$) different from the PM observed in melon seeds treated with spinosad + *E. aromatica* powder (86.67%–100.00%) and spinosad + *P. guineense* powder (85.00%–100.00%). The percentage weight loss observed in melon seeds treated with spinosad + a botanical powder (1.17%–1.40%) was not significantly different from the PWL observed in seeds treated with SASp (0.42%), but it was significantly lower than the PWL (3.28%) observed in melon seeds treated with sole application of *A. melegueta* powder (Babarinde et al., 2018).

Tak and Isman (2016) investigated the insecticidal activities of lemongrass oil and citral, as well as the metabolism of citral in larvae of the cabbage looper, *Trichoplusia ni* (Hübner), in association with well-known enzyme inhibitors. Among the inhibitors tested, piperonyl butoxide showed the highest increase in toxicity, followed by triphenyl phosphate, but no synergistic interaction between the inhibitors was observed. Topical application of citral to fifth instar larvae produced mild reductions in food consumption, and frass analysis after 24 h revealed geranic acid (99.7%) and neric acid (98.8%) as major metabolites of citral. Neither citral nor any other metabolites were found following *in vivo* analysis of larvae after 24 h, and no significant effect of enzyme inhibitors was observed on diet consumption or citral metabolism.

In another experiment, evaluating the activity of of cumin seed and black pepper EO individually and in combination against the rice weevil (*S. oryzae*), cumin seed and black pepper oil individually exhibited 100% insecticidal activity (contact and fumigant) against *S. oryzae* at 24 and 168 h of exposure in the absence and presence of food, respectively. Binary mixtures of cumin seed and black pepper oil at 60:40 and 80:20 ratios showed higher toxicity than individual EOs against *S. oryzae*. It was further established that the additive interaction of binary combinations of EOs evoked complete insecticidal activity of *S. oryzae* (Remesh and Babu, 2023).

In spite of the reported cases of the additive and/or synergistic effect of natural product mixtures, antagonistic effects are not a rarity. Therefore, the joint action effect of natural product mixtures should be investigated to establish additive or synergistic effect rather than upholding the more the merrier perspective. This is a necessity because the combination of certain products could initiate antagonistic effects.

15.4 Natural Products Used Against Microbial Pests in Storage

The ability of pathogens to adapt very fast to fungicide application through mutation and to develop resistance (Kretschmer et al., 2009) makes the use of plant products important and timely. In an attempt to get safe and environmentally benign protection of agricultural products, researchers have studied some essential oils of botanical origin that have been reported effective against some microbial pests infesting stored produce. Apart from their insecticidal potential, some natural products also show effectiveness against microbial pathogens and nonpathological organisms. Natural products of plant origin have shown antifungal, antibacterial, and nematocidal effects on pests, and they have therefore been explored for the treatment of pathogenic infections on stored products. As previously mentioned, these natural products of plant origin can be applied as powders, extracts, and essential oils, and various research investigations have shown them to be effective in these various forms. Although the use of the powdered form is no longer popular due to advances in research, evidence in the literature showed that powder botanicals were researched and found to be effective in the control of stored-product pathogens. As an example, the powder form of plant-derived substances from cloves (*Eugenia caryophyllus*) inhibited the growth of mycotoxigenic molds including *Aspergillus flavus* NRRL A16900 and *A. parasiticus* NRRL 2999 (aflatoxins), *A. ochraceus* NRRL 3174 (ochratoxin), *Penicillium* species M46 (ochratoxin), *P. patulum* M59 (patulin), *P. roquefortii* M247 (patulin), and *P. citrinum* M565 (citrinin) for over 21 d, while powder of cinnamon (*Cinnamomum zeylanicum* B.) and mustard (*Sinapis alba* L.) had the same fungistatic activity for at least 6 d (Azzouz and Bullerman, 1982). In addition, the population of second-stage juveniles of the plant parasitic nematode *Melodogyne javanica* was significantly reduced by mixing the soil containing the nematode with powder of *Inula viscosa* at 0.1% (W/W) (Oka et al., 2001), and other studies have confirmed the nematocidal effects of herbs (Moosavi, 2012). The aqueous extracts of chamomile (*Anthemis nobilis* L.) and malva (*Malva sylvestris* L.) have also been reported to totally inhibit the growth of important stored-product fungi including *A. candidus*, *A. niger*, *Fusarium culmorum*, and *Penicillium* spp (Magro et al., 2006), and ethanolic extract of *Adhatoda vasica* showed a strong antimicrobial activity against *Xanthomonas oryzae*, *Pseudomonas solanacearum*, and *P. syringae*.

Essential oils of botanical origin have been investigated and found to show antibacterial and anti-fungi capacity against foodborne pathogens. Using the volatile activity method, EO on *Citrus sinensis* was found to show complete toxicity against 10 pathogenic fungi (Sharma and Tripathi, 2006), while EO formed an important barrier against food poisoning caused by contaminants such as the *Escherichia coli* associated with animal and vegetable products (Munekata et al., 2020). According to Bava et al. (2022), the efficacy of the EO of the epigeal part of fennel (*Foeniculum vulgare* sbps. *Piperitum*) and its fractions (leaves, achenes, and flowers) was evaluated through *in vitro* and semi-field experiments. Using techniques that only allowed evaporation, fennel EO and its fractions were shown to have moderate efficacy. In particular, in *in vitro* tests, at the highest concentration and exposure time, the whole plant and the leaf, achene, and flower fractions evoked acaricidal efficacies of 56% (flowers), 52% (achenes), 64% (leaves), and 68% (whole plant). In the semi-field trials, the EO of the whole plant at the highest concentration caused the death of 53% of the mites exposed to the vapors. In another study, the EOs obtained from *Eucalyptus camaldulensis* and *Citrus sinensis* were assayed as antibacterial for archeological organic materials using the fumigation method (Ali et al., 2021). The presence of microorganisms and their effect on the old manuscripts and mummy deterioration were assessed in the study. The results show that the tested EOs inhibited the mycelia growth.

While individual EOs have shown antimicrobial effects, research evidence has also demonstrated additive effects of the combination of plant-based EOs against mycotoxin-producing strains of *A. flavus* and *Fusarium graminearum* and an antagonistic effect against *A. niger* and *A. nomius* (Zimmermann et al., 2023). The increased focus and research on the use of EOs for stored-product pest control has led to research that focuses on its effectiveness considering its high level of volatility and high instability. Among the ways in which nanotechnology is being explored to encapsulate consumer-safe substances with antipathogenic properties is consideration of a novel active packaging for meat storage (Jacinto-Valderrama et al., 2023). Essential oil–loaded nano-emulsions have been used as edible coating for preservation to enhance food safety (Shen et al., 2023). More research focused on the encapsulation of EO using nanotechnology or enhancing the antimicrobial potential of these natural products without compromising safety will extend its commercialization and use beyond mere laboratory experimentation. Also, many microbial natural products with the potential to protect stored agricultural produce against insect pests have been documented (Table 15.3).

15.5 Marine Natural Products Used as Pesticidal Agents

The focus on marine natural products has been on their pharmacological properties. As pathogenic organisms evolve and develop resistance to existing pharmaceuticals, the marine environment provides novel leads against fungal, parasitic, bacterial, and viral diseases (Donia and Hamann, 2003; Van Minh et al., 2005). Above 20,000 bioactive compounds have been

Table 15.3 Performance of Selected Microbial Natural Products Against Stored-Product Insect Pests

Number	Microbial Organism	Target Pest	Crop	Rate	Efficacy	Reference
1	*Heterorhabditis indica* Poinar, Karunakar, and David (Homl strain)	*Plodia interpunctella* (Hübner)		200 and 400 IJs per *P. interpunctella* larva (4,000 and 8,000 IJs/2 mL)	60%–70% mortality, 3 d	Mbata and Shapiro-Ilan (2010)
2	*Habrobracon hebetor* (Say)	*Plodia interpunctella*		200 and 400 IJs per *P. interpunctella* larva (4,000 and 8,000 IJs/2 mL)	60%–75% mortality, 3 d	Mbata and Shapiro-Ilan (2010)
3	*Habrobracon hebetor* and *Heterorhabditis indica*	*Plodia interpunctella*		200 and 400 IJs per *P. interpunctella* larva (4,000 and 8,000 IJs/2 mL)	98.0%–99.25% mortality, 3 d	Mbata and Shapiro-Ilan (2010)
4	*Isaria fumosorosea*	*Sitophilus oryzae*	Wheat	1.81×10^7 conidia per mL on 10 g wheat and also for contact mode with and without food	85.5% mortality and 100.0% mortality for both contact modes of applications	Kavallieratos et al. (2014)
5	*Metarhizium anisopliae*	*Sitophilus oryzae*	Wheat	0.12×10^6, 0.22×10^6, 0.32×10^6, and 0.42×10^6 conidia/g wheat grains	27, 35, 47, and 54% mortality, 11 d	Abdel-Raheem et al. (2015)
6	*Metarhizium anisopliae*	*Oryzaephilus surinamensis*	Wheat	0.12×10^6, 0.22×10^6, 0.32×10^6, and 0.42×10^6 conidia/g wheat grains	40, 45, 55, and 65% mortality, 11 d	Abdel-Raheem et al. (2015)

(Continued)

Table 15.3 (Continued) Performance of Selected Microbial Natural Products Against Stored-Product Insect Pests

Number	Microbial Organism	Target Pest	Crop	Rate	Efficacy	Reference
7	Bacillus thuringensis	Rhyzopertha dominica	Wheat	Concentration = 2.5 mg/mL and application rate = 50 mL/10 g broken wheat grains	93% mortality, 7 d	Mohamed (2016)
8	Bacillus thuringensis	Callosobruchus maculatus	Cowpea	4×10^8 cells/mL	100% mortality	Malaikozhundan and Vinodhini (2018)
9	Bacillus thuringensis	Tribolium castaneum	Bean	2.0 g crystalline (Cry) proteins of Bt/10 g diet	LC_{50} value of 0.46 and LT_{50} value of 1.50	Elgizawy and Ashry (2019)
10	Heterorhabditis bacteriophora Poinar	Small and large larvae of Trogodarma granarium	Wheat	50,000 IJs/mL corresponding to application rate of 5,000 IJs/larva was applied on 10 g of wheat	76.7% and 60.0% mortality on small and large larvae, respectively, 8 d	Karanastasi et al. (2020)
11	Steinernema carpocapsae	Small and large larvae of Trogodarma granarium	Wheat	50,000 IJs/mL corresponding to application rate of 5,000 IJs/larva was applied on 10 g of wheat	91.1% and 63.3% mortality on small and large larvae, respectively, 8 d	Karanastasi et al. (2020)

(Continued)

Table 15.3 (Continued) Performance of Selected Microbial Natural Products Against Stored-Product Insect Pests

Number	Microbial Organism	Target Pest	Crop	Rate	Efficacy	Reference
12	Steinernema feltiae	Small and large larvae of Trogodarma granarium	Wheat	50,000 IJs/mL corresponding to application rate of 5,000 IJs/larva was applied on 10 g of wheat	98.9% and 87.8% mortality on small and large larvae, respectively, 8 d	Karanastasi et al. (2020)
13	Metarhizium anisopliae	Callosobruchus maculatus	Chickpeas	15 g chickpeas sprayed with 2 mL of spore suspensions with 1×10^6 or 1×10^8 conidia/mL	LT_{50} and LT_{90} values of 4.45 for 5.34 d at 26°C and 5.17 for 6.15 d at 22°C, respectively	Ozdemir et al. (2020)
14	Beauveria bassiana	Rhyzopertha dominica	Maize, rice, and wheat	1.5×10^8 and 1.5×10^{10} conidia/kg	Mortality 21 d: 44.2 and 65.3% (maize); 44.2 and 65.3% (rice); 56.4 and 76.6% (wheat)	Wakil et al. (2021)
15	Beauveria bassiana	Tribolium castaneum	Maize, rice, and wheat	1.5×10^8 and 1.5×10^{10} conidia/kg	Mortality 21 d: 56.5 and 60.2% (maize); 43.6 and 64.5% (rice); 48.1 and 68.2% (wheat)	Wakil et al. (2021)

(Continued)

Table 15.3 (Continued) Performance of Selected Microbial Natural Products Against Stored-Product Insect Pests

Number	Microbial Organism	Target Pest	Crop	Rate	Efficacy	Reference
16	Beauveria bassiana	Liposcelis paeta Pearman	Maize, rice, and wheat	1.5×10^8 and 1.5×10^{10} conidia/kg	Mortality 21 d: 62.4 and 76.7% (maize); 66.4% and 83.2% (rice); 71.2 and 88.7% (wheat)	Wakil et al. (2021)
17	Beauveria bassiana	Cryptolestes ferrugineus Stephens	Maize, rice, and wheat	1.5×10^8 and 1.5×10^{10} conidia/kg	Mortality 21 d: 57.1 and 72.7% (maize); 63.2 and 78.2% (rice); 68.2 and 81.0% (wheat)	Wakil et al. (2021)
18	Metarhizium anisopliae	Acanthoscelides obtectus	Common bean	1.7×10^5, 2.3×10^5, and 7.9×10^5 conidia/mL	59%–81% mortality	Lak et al. (2022)
19	Beauveria bassiana	Sitophilus granaries	Wheat	1×10^7 conidia/kg	41.5% mortality 14 d at 30°C	Wakil et al. (2023)
20	Beauveria bassiana	Trogodarma granarium	Wheat	1×10^7 conidia/kg	36.0% mortality 14 d at 30°C	Wakil et al. (2023)
21	Xenorhabdus nematophila	Oryzaephilus surinamensis adults	Date	1×10^8 cells/mL of water applied on 50 g of date fruits	70% mortality 21 d	Alwaneen (2023)

isolated from marine microbes (Demain, 2014). The desire to discover new drugs has led scientists from various disciplines to explore the marine resources.

Secondary metabolites from the bacterial isolates obtained from seaweed (*Enteromorpha compressa*) and jelly fish samples were extracted and tested for pesticidal activity against brine shrimp (*Artemia naupli*). The LC$_{50}$ of the extracted secondary metabolites of the isolates were found to be very high in *Micrococcus* sp. (JF2) (about 10 µg/mL), medium in *Steptomyces* sp. (Em5) (about 100 µg/mL), and low in *Vibrio* sp. (Vijayalakshmi et al., 2017). According to Joseph et al. (2010), ethanolic extracts obtained from marine sponges were insecticidal against fifth instar larvae of *Culex quinquefasciatus* (Say.) (Diptera: Culicidae) and two lepidopterous pests: *Achaea janata* (Linn.) and *Pericallia ricini* (Fab.).

Although the focus of the scientists who worked on the pesticidal potential of marine products has not been on storage pests, an earlier report by Wright (1998) indicated the biodiversity of marine habitats and the organisms found there. The chemical classes associated with the marine ecosystem include terpenes, polyketides, peptides, alkaloids, acetogenins, organosulphur compounds, peptides and proteins, shikimates, fatty acids, and xanthene. Mudianta et al. (2014) also reported the terpene chemistry of mollusks. Those reports emphasize the need to explore the various compounds found in the marine ecosystem for bioactivity against insecticidal and microbial pests of stored products. In a recent review conducted by Song et al. (2021), 72 natural compounds were reported and only three (octadecadiennoic acid and n-hexadecanoic acid produced by red alga, *Laurencia brandenii,* against *S. oryzae* and chabrolene extracted from *Coral nephthea* against *S. zeamais*) dealt with control of stored-product insects. Zodape (2014) investigated the pesticidal potential of the crude extracts of the whole body, carapace, and hepatopancrease of intertidal crab, *Atergatis integerrimus* (Lamarck), obtained in Mumbai, India, against *S. oryzae* and reported 100% mortality of the rice weevil after 24 h of treatment of the weevils.

15.6 Natural Products for Crop Quality Preservation

Table 15.4 presents basic information on some plants used for fumigant treatment of stored products. In some laboratory bioassays, however, admixing natural products with stored products such as cereals rendered the natural products less efficacious than in the absence of the stored products. The reason advanced for that observation is the high sorption of the natural product by the stored commodity. A very pertinent case is the use of EOs. In spite of that peculiarity, the relevance of using EOs in commodity storage cannot

Table 15.4 Performance of Selected Botanical Natural Products Against Stored-Product Insect Pests (Years 2001–2023)

Number	Natural Product	Application Rate	Commodity	Insects with Life Stages	Mortality	Reference
1	*Cucurbita maxima* (Duchesne)	0.42 mg/cm^2 air	Rice grains	*Rhizoperta dominica, Sitophilus oryzae* (L.), and *Tribolium castaneum* adult	100% mortality	Kim and Ahn (2001)
2	*Chenopodium ambrosioides* EO	0.2 µL/cm^2	Maize grain, wheat, green peas, mung beans, and white bean	*Callosobruchus chinensis, C. maculatus, Acanthoscelides obtectus, Sitophilus granaries* (L.), *S. zeamais,* and *Prostephanus truncatus* (Horn)	80%–100% for all except *S. zeamais* (5%) and *C. maculatus* (20%), 24 h	Tapondjou et al. (2002)
3	*Curcuma longa* L. leaves EO	Concentration of 5.2 mg/cm^2	Wheat	*Tribolium castaneum* egg, larva, and adult	Reduced oviposition and egg hatching by 72% and 80%	Tripathi et al. (2002)
4	*Artemisia absinthium* L., *Artemisia santonicum* L. EO, and *Artemisia spicigera* C. Koch	3, 6, and 9 µL oil/L air	Wheat grains	Granary weevil, *Sitophilus granarius* adult	56%–86%, 53%–83%, and 53%–90% mortality, respectively	Kordali et al. (2006)
5	*Ferula galbaniflua, Cupressus sempervirens,* and *Melaleuca viridiflora*	414.4 mg/L air	Sterilized artificial diet	*Cadra cautella* (Walker) larva	100%, 91% and 93%, mortality, respectively	Sim et al. (2006)

(Continued)

Table 15.4 (Continued) Performance of Selected Botanical Natural Products Against Stored-Product Insect Pests (Years 2001–2023)

Number	Natural Product	Application Rate	Commodity	Insects with Life Stages	Mortality	Reference
6	*Rosmarinus officinalis* EO	322.6 mg/L air	Sterilized artificial diet	*Cadra cautella* larva	100% mortality	Sim et al. (2006)
7	Cymbopogon martini EO	100 µL/L air	Stored wheat and gram (garbanzo bean, *Cicer arietinum*)	*Callosobruchus chinensis* and *Tribolium castaneum* adult	*C. martinii* oil significantly affected oviposition and adult development and caused 100% mortality of *C. chinensis*	Kumar et al. (2007)
8	*Artemisia sieberi* Besser	37 µL/L air	Beans, grains, Whole rice and wheat flour	*Callosobruchus maculatus, Sitophilus oryzae,* and *Tribolium castaneum*	100% mortality for all insects, 24 h	Negahban et al. (2007)
9	*Aegle marmelos* (L.), Correa EO	500 µg/mL (ppm)	Cowpea seeds, wheat grains, and wheat flour	*Callosobruchus chinensis, Tribolium castaneum, Rhyzopertha dominica,* and *Sitophilus oryzae*	Deterred feeding in insects and reduced grain damage as well as weight loss	Kumar et al. (2008)
10	*Citrus aurantifolia*	0.2 v/w	Dried yam	*Dinoderus porcellus* Lesne adult	100% mortality, 24 h	Oni and Ileke (2008)
11	*Coriandrum sativum* (L.)	0.08 µg/mL air	Wheat flour and yeast flour (19:1)	*Tribolium castaneum* larva, pupa, and adult	100% mortslity for all stages of development, 24 h	Islam et al. (2009)

(Continued)

Table 15.4 (Continued) Performance of Selected Botanical Natural Products Against Stored-Product Insect Pests (Years 2001–2023)

Number	Natural Product	Application Rate	Commodity	Insects with Life Stages	Mortality	Reference
12	*Hyptis spicigera* Lam. EO	10 µL oil/L air	Wheat grains	*Sitophilus zeamias* and *Tribolium castenum*	>80% mortality for *S. zeamais* <30% mortality for *T. castenum,,* 48 h	Jo et al. (2009)
13	*Eucalyptus leucoxylon* EO	370 µL/L air	Bean grains, whole rice, and wheat flour mixed with yeast (10:1, w/w), respectively	*Callosobruchus maculatus, Sitophilus oryzae,* and *Tribolium castaneum* adult	100% mortality, 24 h	Kambouzia et al. (2009)
14	*Origanum onites* L. and *Satureja thymbra* L.	9 µ L/L air	Kidney beans, mixture consisting of 1 kg wheat flour, 55 g yeast, 30 g germs of wheat, and dried apricot	*Ephestia kuehniella* (Lepidoptera: Pyralidae), *Plodia interpunctella* Hübner (Lepidoptera: Pyralidae), and *Acanthoscelides obtectus* Say (Coleoptera: Bruchidae)	100% mortality	Ayvaz et al. (2010)
15	*Ricinus communis* (L.)	2%, 5%, and 10% concentration with acetone	Not indicated	*Trogonella granarium* adult and larva	83%–100% mortality of adults; 76%–93% mortality of 1st, 2nd, 3rd, and 4th stage larva, 24 h	Jenan (2014)

(Continued)

Table 15.4 (Continued) Performance of Selected Botanical Natural Products Against Stored-Product Insect Pests (Years 2001–2023)

Number	Natural Product	Application Rate	Commodity	Insects with Life Stages	Mortality	Reference
16	Diallyl trisulfide	0.015 µL/L air	Wheat, rice, sorghum, and corn	*Sitotroga cerealella* (Olivier) female adult	Mortality of larva and pupa, 24 h	Fouad et al. (2014)
17	*Citrus reticulata* EO	71.18 µL/mL, 1.52 µL/g, and 41.92 µL/L of air	Whole corn and grain	*Sitophilus zeamais* adult	80.2% mortality	Athanassiou et al. (2014)
18	*Xylopia parviflora* (A. Rich.) Benth	0.78, 1.56 3.15, and 6.25 µL/mL air	Cowpea seed	*Callosobruchus maculatus* adult	90% mortality, 24 h	Babarinde et al. (2015)
19	*Xylopia aethiopica*	0.5 and 1 mL	Cowpea	*Callosobruchus maculatus*	Up to 62.5% mortality	Akinyemi et al. (2016)
20	*Ricinus communis* (L.)	0.5 and 0.75 mL	Bambara groundnut	*Callosobruchus maculatus*	65%–92.5% mortality	Babarinde et al. (2016a)
21	*Syzygium aromaticum* EO	15% against *T. granarium* and 10% against *S. paniceum*	Wheat seeds and wheat flour	*Trogoderma granarium* and *Stegobium paniceum* (L.) adult	100% mortality of *T. granarium* and *S. paniceum*, 24 h	Boraei (2016)
22	*Mentha spicata* EO	15% against *T. granarium* and 10% against *S. paniceum*	Wheat seeds and wheat flour	*Trogoderma granarium* and *Stegobium paniceum* adult	80% mortality of *T. granarium* and 100% mortality of *S. paniceum*, 24 h	Boraei (2016)

(Continued)

Table 15.4 (Continued) Performance of Selected Botanical Natural Products Against Stored-Product Insect Pests (Years 2001–2023)

Number	Natural Product	Application Rate	Commodity	Insects with Life Stages	Mortality	Reference
23	Mentha pulegium	45.46, 113.6, and 568.2 µL/L air	Wheat flour	Lasioderma serricorne (F.) and Tribolium castaneum	100% mortality of T. castaneum and 80%–100% mortality of L. serricorne	Salem et al. (2017)
24	Hoslundia opposita	0.67 v/w	Cowpea	Callosobruchus maculatus	76.5% progeny emergence inhibition rate	Babarinde et al. (2017b)
25	Tithonia diversifolia Hemsl.	10% w/v	Cowpea	Callosobruchus maculatus	92.6% mortality, 72 h	Green et al. (2017)
26	Vitex pseudonegundo	31.96 µL/L air	Maize grain	Sitophilous oryzae	61.7% mortality	Trivedi et al. (2018)
27	Ajania nitida		Wheat flour mixed with yeast(10:1, w/w)	T. castaneum and L. serricorne	LC_{50} = 21.07 mg/L air against T. castaneum and LC_{50} = 11.23 mg/L air against L. serricorne	Li et al. (2018)
28	Ajania nematoloba		Wheat flour mixed with yeast(10:1, w/w)	T. castaneum and L. serricorne	LC_{50} = 69.45 mg/L air against T. castaneum and no effect against L. serricorne	Li et al. (2018)

(Continued)

Table 15.4 (Continued) Performance of Selected Botanical Natural Products Against Stored-Product Insect Pests (Years 2001–2023)

Number	Natural Product	Application Rate	Commodity	Insects with Life Stages	Mortality	Reference
29	*Origanum onites, O. vulgare*	0.15 % (v/v)	Soft bread wheat and dry yeast for *Rhyzopertha dominica* and *Tribolium confusum* (Jacquelin du Val) Whole wheat for *Sitophilus granarius* and *Sitophilus oryzae*	*Rhyzopertha dominica, Tribolium confusum, Sitophilus granarius,* and *Sitophilus oryzae*	99%–100% mortality	Alkan (2020)
30	*Citrus jambhiri* Lush.	80 and 107 mL/L air	Millet seeds and maize grains	*Tribolium castaneum* and *S. zeamais*	53.75%–100.00% mortality of *Tribolium castaneum*, 24 h, and 93%–100% mortality of *S. zeamais*	Babarinde et al. (2021a)
31	*Cymbopogon citratus* and *Thuja plicata*	55.33 µL/L air	Dried cassava chips (cultivar Oko Iyawo)	*Prostephanus truncatus*	100% mortality, 6 h, and 100% mortality, 48 h	Babarinde et al. (2021b)
32	*Ferula asafoetida* and Mentha x piperit	Concentration of 2.0 mL/cm^3 air	Bull horn	*Anthrenus verbasci* (L.) larva	50% mortality, 7 d	Faheem and Abdurraheem (2022)

(Continued)

Table 15.4 (Continued) Performance of Selected Botanical Natural Products Against Stored-Product Insect Pests (Years 2001–2023)

Number	Natural Product	Application Rate	Commodity	Insects with Life Stages	Mortality	Reference
33	Citrus aurantium L.	25, 50, 100, and 200 µL/L air	Chickpea seeds	Callosobruchus maculatus male and female adult	5%–95% mortality and longevity for females is shortened by approximately 5–7 d or 3–7 d, depending on the concentration	El Kasimi et al. (2023)
34	Dillenia indica (ethyl acetate extract)	50 mg/L air	Wheat, Triticum aestivum L.; green gram, Vigna radiata (L.) R. Wilczek; and seeds	Tribolium castaneum, Sitophilus oryzae, and Rhyzopertha dominica adult	70%–85% mortality, 24 h	Singh et al. (2023)
35	Cupressus sempervirens (crude EO and nanoemulsion)	20.0 µL/L air	Wheat grains	Sitophilus oryzae adult	96% mortality with crude EO and 100% mortality with nanoemulsion	Almadiy et al. (2023)

be overemphasized. Essential oils and their constituents, due to their anti-microbial and insecticidal properties, are useful as fumigants, being highly volatile, effective in low concentrations, biodegradable, and safe. Some of them can improve the quality of the protected crops by increasing the content of antioxidants (Namiota and Bonikowski, 2021). Essential oils are capable of maintaining the level of ascorbic acid in connection with the changes in the ascorbate-glutathione pathway enzymes (Namiota and Bonikowski, 2021). Thymol has also been reported to exhibit a positive effect on retaining the ascorbic acid levels when used for the protection of *Fragaria x ananassa* Duch (Geransayeh et al., 2015). The addition of thymol, eugenol, menthol, and eucalyptol improved the firmness of *Prunus avium* L. fruits in a modified atmosphere package (Serrano et al., 2005).

When 21 EOs were evaluated as potential sprout suppressants in cv. Ranger Russet potatoes at room-temperature storage, treatment with *Artemisia herba-alba* EO was the most effective at suppressing both sprout length and sprout number over a 90 d storage period. *Cistus ladanifer, Ocimum basilicum, Ormenis mixta,* and *Salvia sclarea* EOs significantly reduced sprout length for shorter storage periods, whereas *Cinnamomum zeylanicum* (bark) and *Laurus nobilis* EOs also significantly reduced sprout number (Thoma et al., 2022). The researchers concluded that certain EOs could be used as sprout suppressants for room-temperature potato storage, providing needed alternatives for the potato industry.

As documented by Rajendran and Sriranjini (2008), the nutritive value of red gram kept for 6 months and fumigated with *M. arvensis* essential oil was not affected (Srivastava et al., 1989). Singh et al. (1995) showed that the nutritional quality of sorghum exposed to *M. arvensis* oil at the dose of 167 mL/L for 3 months was not affected. As well, the rheological properties of wheat flour that was fumigated with monoterpenes (camphor, 1,8-cineole, and carvacrol) at the dose of 200 mL/kg for 24 days at 5–10°C was not altered (Rozman et al., 2006). In terms of palatability, the bitter taste associated with some natural protectants may reduce the level of acceptance of food materials. For example, neem seed oil used to protect dried fish against *Dermestis maculatus* (De Geer) reduced the level of acceptance of the fish due to its bitter taste (Babarinde et al., 2016b).

15.7 Challenges of Natural Products Bioprospecting

The use of natural products is generally believed to be safer and more environmentally friendly than that of synthetic products. This is because they are considered easily biodegradable and less toxic to mammals and water bodies. Unlike with synthetic pesticides, limited studies have reported the potential of pests to develop resistance to the use of natural products. Also, they may be potentially compatible with other biological control methods. The use of these products is considered safe, with limited toxic residues found in human bodies when they are consumed as contaminants in food products. Different parts of the plant-based natural products, including the seed, leaves, stem, flower, and roots, have shown strong pesticidal properties. Most of them are also medicinal and have shown strong insecticidal and anti-pathological potential in laboratory studies. Although there is need to do further research on the level of toxicity, their potential safety also leverages the fact that most of these pesticidal plants also have medicinal benefits to human and some are consumed as herbs. However, there are several challenges that limit the use of natural products for pest control. Unlike the situation of the synthetic pesticides, the level of industrial development of natural products is still very low, especially in Sub-Saharan Africa. This is due to problems that limit their bioprospecting, as outlined in the following sections.

15.7.1 Instability of Essential Oils and Plant-Derived Products

Instability is probably the most important limiting factor to the use of natural products, especially plant-based EOs, as fumigants. This implies that these EOs can easily change, break down, and become less effective or even lose their toxic potential as biopesticides. Several factors, including the environmental and storage conditions, influence their stability. However, their volatility and instability are major limitations to their use for field and commercial purposes. In a classic review done by Turek and Stintzing (2013), the authors identified temperature, light, and oxygen as major factors that influenced the integrity of EOs. When stored at low temperatures rather than more elevated ones, essential oil tends to retain their primary quality with less alteration to their chemical composition (Mehdizadeh et al., 2017; Dedebas et al., 2022). This is the major reason EOs are stored in refrigeration before use. Exposure to light also limits the toxicological stability of EOs. In a study that evaluated the toxicological stability of *Ocimim basilicum* and its major components against *S. zeamais*, an important storage pest of maize, storage at a temperature less that 20°C and in the absence of light enhanced toxicological stability (Moura et al., 2021).

15.7.2 Availability of Raw Materials for Commercial Production

To use natural products as commercial fumigants will demand that their raw materials be continually available. For the natural products of botanical origin, most of the plants are usually available in the wild, especially in developing countries. Limited efforts have been made to engage in commercial production of these botanicals, but this is fundamental to achieving sustainability. Some botanical species with established pesticidal property may become endangered due to anthropogenic activities if conscious efforts are not made to domesticate them.

The fundamental approach should include sourcing the materials from their natural habitat. Botanical taxonomists should then be contacted for the authentication of the botanical materials. Consequently, empirical studies should be conducted to establish the best growing conditions, soil and nutrient requirements, and sustainable pest management strategies. Further to these should be the determination of the best postharvest handling of the botanicals, with special attention to the preservation of their seeds and propagules or both. The breeders and geneticists should also work on quality improvement based on any indices of interest highlighted by crop protectionists and agrochemical manufacturers.

15.7.3 Policies and Legislation on Natural Products

Policies that promote the use of natural products need to be developed, especially in developing countries. Though the use of natural products as fumigants is not strange to the region, decades of the use of synthetic pesticides have overshadowed the use of natural products. Policies should target ensuring safety in their formulation, storage, and usage for commercial purposes. The production of some plant EOs has been standardized in countries like the United States. However, this may not be the case in many developing countries (Stevenson et al., 2012). Policy makers should be made to realize the importance of making purposeful policies. Factually, in this context, the onus of educating the policy makers on the necessity of policies and legislation on natural products lies on crop protectionists, environmentalists, and food scientists. In many developing countries, political factors are the bane of the formulation and execution of proper policies, including those related to natural products standardization, propagation, and usage.

15.7.4 Necessity of Large Doses for Effective Stored-Product Treatment

Another major bottleneck of natural products bioprospecting is the fact that large doses are often required for efficacy. In natural products that are within the reach of resource-poor small-scale farmers and do not require industrial process, the active ingredients are often masked by their carrier materials. Consequently, application of such products requires much larger doses than are required when the industrial version of the same active product is applied. A pertinent example is the use of *Chrysanthemum ciner-ariefolium,* which is the raw material for the synthesis of pyrethrum. Using the powder, extract, ash, or EO from the plant cannot be as effective as using lambdacyhalothrin, which is the synthetic pyrethroid from the plant.

15.8 Formulation and Encapsulation of Natural Products as Fumigants

Research into the individual natural product pesticide formulations will increase storability, effectiveness, handling, and stability. This will enhance the commercialization of these products. Traditionally, formulation methods usually included the use of wettable powders, emulsifiable concentrate, encapsulation, water-dispensable granules, and capsule suspension (Seaman, 1990). Advances in research have, however, promoted encapsulation using nanotechnology, thus enhancing the stability of EOs. Polymeric nano-encapsulation preserves the EO of plants by preventing their degradation, improving water solubility, enhancing a moderated release of volatiles, and preventing unnecessary losses (Lammari et al., 2020). Using the nano-precipitation method, experimentation with encapsulated *Rosmarinus officinalis* suggests that this technique can help in development of a biopesticide with controlled release properties, reducing the concentration used and frequency of application compared to those of unencapsulated products (Khoobdel et al., 2017). Nano-encapsulation of EOs also plays an important storage role in preservation against microbes and toxins, thus enhancing their shelf-life (Gupta et al., 2023b).

15.9 Natural Products and Integrated Pest Management

The use of natural products often is less toxic and enduring than that of synthetic chemicals. It might therefore not be practical for farmers to solely depend on them for eradication of major pest problems, even though the limited level of toxicity may be beneficial in making them less broad spectrum than synthetic chemicals and thus less destructive to beneficial organisms such as other biocontrol agents. However, to achieve effective control, the volume used by farmers will also increase. Indeed, low toxicity and the increased quantity and frequency of use of these natural products due to their high biodegradability and volatility have become a major constraint to achieving their economic viability (Gupta et al., 2023a).

To promote the use of natural products for pest control among farmers, they must be incorporated as part of an integrated approach to pest management both in the field and in the store. Some stored-product pests are actually field-to-store pests, and some stored products are more vulnerable to pest attack due to primary field infestation. Therefore, controlling pests from the field is important for quality crop maintenance in the store. Natural products can function as push repellents in a push-pull pest management system that involves pushing the pests away from the farm using biopesticides and pulling them using a trap crop that surrounds the farms. This is practical in the small-scale farming system that is predominant in the developing countries. Other natural products, such as plant-based insecticides that are compatible with other biocontrol agents (not harmful or repellent to parasitoids, for example), will be effective components of an IPM management for pests when combined with other biocontrol agents such as parasitoids. In the search for alternatives to synthetic pesticides in maize (*Zea mays*, L.) storage, an integrated weevil management scheme involving varietal resistance and *X. aethiopica* seed extract was evaluated in the laboratory. Varying rates of *X. aethiopica* extract were assessed on three maize varieties (Pajo White, Tsolo Yellow, and DMR-LSR-Y) with differing susceptibility to *S. zeamais*. The different rates of *X. aethiopica* extract significantly interacted with maize varietal resistance and reduced fecundity of *S. zeamais* and maize seed weight loss due to insects' feeding. Treatment of seeds with extract at 1.0 mL/20 g seed caused significant (above 50%) mortality at 24 h posttreatment in all varieties, whereas no mortality was recorded with the control and 0.2 mL/20 g seed. The variety DMRLSR-Y was better protected than the remaining two varieties in the integrated weevil management scheme (Babarinde et al., 2008).

The assumption that biopesticides are not toxic to natural enemies is not always true. In a study that tested the toxicity of EOs against the seedpod weevil *Ceutorhynchus obstrictus* (Marsham) and a model parasitoid *Nasonia vitripennis* (Walker), even though the EO resulted in mortality of the pest, 100% mortality of the parasitoid was observed after 3 h at a low concentration of the EOs of *Thymus vulgare* and *Foeniculum vulgare* (Sulg et al., 2023). Development of IPM for insect pests must target the conservation of parasitoids and other natural enemies; therefore, there is need for more research on the compatibility of natural products with these natural enemies by determining the level of toxicity of EOs on nontarget organisms before the EOs are integrated into an IPM management program.

15.10 Biotechnology and Biopesticides

In recent years, the application of molecular biology in studying the response of insect pests to biopesticides such as essential oils, using transcriptome analysis and RNA sequencing, has become popular. This analysis reflects the gene expression of the detoxifying enzymes in insects. For example, in a study by Huang et al. (2018), transcriptome profiling was used to detect the differential gene expression of detoxifying enzymes in *S. zeamais* in response to the fumigant terpinen-4-ol. In another study on the mechanism of action of the essential oil of *Chamaecyparis obtuse* on *T. casternium*, an important stored-product pest of flour, it was found to deferentially express genes focused on chitin metabolic processes and some genes that regulate aging using transcriptome sequencing (Zhang et al., 2023). The synergistic effects of different components of an essential oil have also been predicted using RNA-Seq and qRT-PCR through the decrease in detoxification and removal of xenobiotics (You et al., 2023). Use of molecular methods in the development of varieties of plant botanicals with more stable and effective essential oils needs to be explored further.

15.11 Adaptability and Commercialization of Fumigant Findings by Local Farmers

The potential use of natural products of plant sources as fumigants against pests has been extensively researched in the last century. Various preparation forms, including powders, aqueous polar and nonpolar solvent extracts, and essential oils, have been explored and found effective. However, limited studies have evaluated the effectiveness of these botanicals in the field. Most studies have been limited to laboratory evaluations of the toxicological and repellency properties of these natural products. More quality research needs to be done on the effectiveness of these natural products for field and store applications. This will enhance commercialization of natural product–based fumigants. Apart from their use for pest management, the residual effects of the application of these natural products on health and the environment need to be empirically tested. As has been previously mentioned, the assumption that some of these natural products are not potentially toxic to mammals has been found to be wrong. For example, at some point, certain pesticidal plants that were generally considered acceptable and promoted in EU regions and even used by organic farmers were later found to be hazardous (Stevenson et al., 2012).

The challenge often is that research is not targeted toward the needs of the farmers. Also, local farmers may be unaware of the efficacy of new products. Efforts must be made by stakeholders to ensure the dissemination of findings to potential end users. There should also be room for feedback that will stimulate further research.

As previously mentioned, ensuring sustainability and a regular supply of raw materials for industrial production of bioinsecticides will demand large-scale cultivation of these botanicals. Since the level of effectiveness of these botanicals is dependent on the parts of the plant being explored, there may be a need to breed plants for such benefits.

15.12 Present Challenge and Future Needs in Natural Product Research

1. Although several thousand plants and microbial and marine products have been identified as potential sources of natural products for fumigation and stored products treatment, the need for a continuous search to discover new species cannot be overlooked. The impact of climate change has placed a premium on scientists' continuing to reevaluate some parameters to ascertain that climate change impacts have not caused a shift in some established landmarks. If climate change could influence the fecundity of pests, its potential influence on the bioactivity of natural products should also be investigated periodically.

2. Taxonomic problems are fundamental challenges that should be tackled in natural product bioprospecting. In areas where plants are identified via ethnobotanical practices, they are primarily identified via their vernacular names. Experience has shown that such an approach can be misleading, since dialect disparities, especially in the developing countries, can cause a mismatch of identities. Therefore, pesticidal plant species should be authenticated by a taxonomist and a voucher specimen should be deposited in an herbarium, where a voucher number is obtained. The use of molecular characterization could be pertinent for botanical authentication in the future, where appropriate facilities are available. Also, chromatographic procedures should be explored for the identification of marine and microbial organisms with pesticidal potential.

3. While the search for potential sources of natural products of botanical, microbial, and marine origin should continue, it is pertinent to stress that efforts should also be geared toward commercialization of the discovered products. In addition, attention should be paid to the economic implications of product development. This would enable resource-poor small-scale farmers in developing countries to adopt the new products.

4. Hitherto, there has been an assumption that natural products are selective and not broad spectrum, thereby being nontoxic to natural enemies of pests. There is the need for empirical pilot-scale studies to evaluate the susceptibility of specific biological control agents to natural products. This should be a prerequisite before any Integrated pest management scheme including natural products and biocontrol agents could be recommended for postharvest pest control.

5. Collaborative studies involving natural product chemists, crop protectionists, and food scientists should be supported by funders. Such collaborations should ensure comprehensive studies, which could offer wholesome and informed decisions on the adoption of natural products for food storage and protection.
6. Legislation and government policies should ensure preservation of the potential sources of natural products on a national and a regional basis. It is crucial to protect grossly endangered species via multifaceted anthropological activities.
7. Some suppositions on the safety of natural products used for food preservation are based on the medicinal and nutritional properties of the raw materials. Therefore, future studies should include the evaluation of the toxicity of natural products to mammals and other nontarget beneficial organisms.

Acknowledgments

The authors acknowledge Mr. Olagoke Zachaeus Olatunde, Chinese Academy of Sciences, Fuzhou, China, for his technical support in the drawing of chemical structures of the major constituents of the insecticidal oils. Mr. Ismail Ayegboyin assisted in collation of some materials used in drafting parts of the tables in the original manuscript.

References

Abdel-Raheem MA, Ismail IA, AbdelRahman RS, Farag NA, AbdelRhman IE (2015) Entomopathogenic fungi, *Beauveria bassiana* (Bals.) and *Metarhizium anisopliae* (Metsch.) as biological control agents on some stored product insects. J Entomol Zool Studies 3(6): 316–320.

Addor RW (1995) Insecticides. In: Godfrey CRA (ed), Agrochemicals from Natural Products. Marcel Dekker Inc. pp. 1–62.

Akinyemi AO, Babarinde SA, Oyerinde RM, Aderonbi AA (2016) Bioactivity of acetone and chloroform extracts of *Xylopia aethiopica* (Dunal) A. Rich seed against *Callosobruchus maculatus* Fabricius (Coleoptera: Bruchidae. JBAPN 6(5–6): 412–423.

Ali MO, Taha AS, AboElgat WAA, Fatma EL (2021) Applications of natural essential oils as antibacterial for archaeological organic materials. Intern J Adv Sci Res Innov 4(2): 8–25.

Alkan M (2020) Chemical composition and insecticidal potential of different *Origanum* spp. (Lamiaceae) essential oils against four stored product pests. Turk Entomol Derg 44(2): 149–163.

Almadiy AA, Nenaah GE, Albogami BZ, Shawer DM, Alasmari S (2023) *Cupressus sempervirens* essential oil, nanoemulsion, and major terpenes as sustainable green pesticides against the rice weevil. Sustainability 15(10): 8021.

Alwaneen WS (2023) The aptness of entompathogenic bacteria against saw toothed grain beetle (*Oryzaephilus surinamensis* [L.]) (Coleoptera: Silvanidae) in dates under laboratory conditions. J King Saud Univ Sci 35(4): 102641.

Athanassiou CG, Kavallieratos NG, Lazzari FA (2014) Insecticidal effect of Keepdry® for the control of *Sitophilus oryzae* (L.) (Coleoptera: Curculionidae) and *Rhyzopertha dominica* (F.) (Coleoptera: Bostrychidae) on wheat under laboratory conditions. J Stored Prod Res 59: 133–139.

Attia MA, Wahba TF, Shaarawy N, Moustafa FI, Guedes RN, Dewer Y (2020) Stored grain pest prevalence and insecticide resistance in Egyptian populations of the red flour beetle *Tribolium castaneum* (Herbst) and the rice weevil *Sitophilus oryzae* (L.). J Stored Prod Res 87: 101611.

Ayvaz A, Sagdic O, Karaborklu S, Ozturk I (2010) Insecticidal activity of the essential oils from different plants against three stored-product insects. J Insect Sci 10(1): 21.

Azzouz MA, Bullerman LB (1982) Comparative antimycotic effects of selected herbs, spices, plant components and commercial antifungal agents. J Food Prot 45(14): 1298–1301.

Babarinde SA, Adebayo MA, Oduyemi K (2008) Integrating varietal resistance with *Xylopia aethiopica* (Dunal) A. Richard seed extract for the management of *Sitophilus zeamais* Motschulsky in stored maize. Afr J Biotechnol 7(8): 1187–1191.

Babarinde SA, Adebayo TA, Usman LA, Ameen OM, Akinyemi AO, Onajole OT, Adekale O (2016b) Preservation of smoked African catfish, *Clarias gariepinus* Burchell against *Dermestes maculatus* De Geer (Coleoptera: Dermestidae) using neem seed oil-iodized salt mixtures. Agric Conspec Sci 81(4): 235–240.

Babarinde SA, Akinyemi AO, Odewole AF, Yisa S, Oke OO, Adeyemo AO (2016a) Phytochemical composition and insecticidal properties of mechanically extracted castor, seed oil against cowpea seed bruchid (*Callosobruchus maculatus* Fabricius) infesting bambara groundnut. Elixir Agric 92: 39086–39092.

Babarinde SA, Akinyemi AO, Usman LA, Odewole AF, Sangodele AO, Iyiola OO, Olalere OD (2014) Toxicity and repellency of *Hoslundia opposita* Vahl (Lamiaceae) leaves' essential oil against rust-red flour beetle, *Tribolium castaneum* Herbst (Coleoptera: Tenebrionidae). Nat Prod Res 28(6): 365–371.

Babarinde SA, Kemabonta KA, Aderanti IA, Kolawole FC, Adeleye AD (2018) Synergistic effect of spinosad with selected botanical powders as biorational insecticides against adults of *Tribolium castaneum* Herbst, 1797 (Coleoptera: Tenebrionidae). J Agric Sci 63(1): 39–51.

Babarinde SA, Kemabonta KA, Olatunde OZ, Ojutiku EO, Adeniyi AK (2021a) Composition and toxicity of rough lemon (*Citrus jambhiri* Lush.) rind essential oil against red flour beetle. *Acta Ecologica Sinica* 41(4): 325–331.

Babarinde SA, Ottun AT, Olaniran OA, Adebayo TA, Oderinde AE, Odewole AF, Alao FO (2021b) Acute toxicity of two bio-fumigants against larger grain borer, *Prostephanus truncatus* (Horn) (Coleoptera: Bostrichidae). Pp. 246–252 In: Jayas DS, Jian F (eds) Proceedings of the 11th International Conference on Controlled Atmosphere and Fumigation in Stored Products (CAF2020), CAF Permanent Committee Secretariat, Winnipeg, Canada.

Babarinde SA, Pitan OO (2019) Preliminary screening of selected tropical botanicals as cowpea protectants against cowpea seed bruchid, *Callosobruchus maculatus* Fabricius (Coleoptrea: Chrysomelidae: Bruchineae). Muniz Entomol. Zool 14(2): 389–94.

Babarinde SA, Pitan OOR, Ajala MO, Olatunde GO (2017a) Insectifugal and insecticidal potentials of two tropical botanical essential oils against cowpea seed bruchid. Environ Sci Pollut Res 24(24): 19785–19794.

Babarinde SA, Pitan OOR, Olatunde GO, Ajala MO (2015) First report of toxicity of *Xylopia parviflora* (A. Rich.) Benth (Annonaceae) root bark's essential oil against cowpea seed bruchid, *Callosobruchus maculatus* Fabricius (Coleoptera: Chrysomelidae: Bruchinae). Nat Prod Res 29(4): 349–352.

Babarinde SA, Pitan OOR, Olatunde GO, Ajala MO (2017b) Chemical composition of the essential oil of Nigeria grown *Hoslundia opposita* Vahl (Lamiaceae) dried leaves and its bioactivity against cowpea seed bruchid. Chem Biodivers 14(6) DOI: 10.1002/cbdv.201600418.

Bava R, Castagna F, Palma E, Musolino V, Carresi C, Cardamone A, Lupia C, Marrelli M, Conforti F, Roncada P, Musella V (2022) Phytochemical profile of *Foeniculum vulgare* Subsp. piperitum essential oils and evaluation of acaricidal efficacy against *Varroa destructor* in *Apis mellifera* by *in vitro* and semi-field fumigation tests. Vet Sci 9(12): 684.

Boraei DM (2016) Toxicity of two fixed plant oils by using a new fumigant method against *Trogoderma granarium* Everts and *Stegopium paniceum* (L.). J Plant Prot Pathol 7(12): 791–796.

Chaghakaboodi Z, Nasiri J, Farahani S (2022) Fumigation toxicity of the essential oils of *Ferula persica* against *Tribolium castaneum* and *Ephestia kuehniella*. Agrotecch Ind Crop 2(3): 123–130.

Dedebas T, Ekici L, Sagdic O (2022) Influence of heating on thermal stability of essential oils during storage. J Essent Oil Bear Pl 25(3): 611–625.

Demain AL (2014) Importance of microbial natural products and the need to revitalize their discovery. J Ind Microbiol 41(2): 185–201.

Donia M, Hamann MT (2003) Marine natural products and their potential applications as anti-infective agents. Lancet Infect 3(6): 338–348.

Ebadi MT, Azizi M, Sefidkon F, Ahmadi N (2015) Influence of different drying methods on drying period, essential oil content and composition of *Lippia citriodora* Kunth. J Appl Res Med Aromat Plants 2(4): 182–187.

Elgizawy KK, Ashry NM (2019) Efficiency of *Bacillus thuringiensis* strains and their cry proteins against the red flour beetle, *Tribolium castaneum* (Herbst) (Coleoptera: Tenebrionidae). Egypt J Biol Pest Control 29(1): 1–9.

El Kasimi R, Douiri F, Haddi K, Boughdad A (2023) Bioactivity of essential oil from *Citrus aurantium* peel against the pulse beetle *Callosbruchus maculatus* F. on chickpea. Agriculture 13(2): 232.

Faheem F, Abdurraheem K (2022) Bioactivity of three indigenous medicinal plants (*Ferula asafoetida*, *Syzygium aromaticum*, and *Mentha x piperita*) as fumigants for the control of the skin and skin-product pest (*Anthrenus verbasci*) in museums, libraries, and archives. Int J Conserv Sci 13(2): 565–578.

Feng YX, Lu XX, Du YS, Zheng Y, Zeng D, Du SS (2022) Sesquiterpenoid-rich essential oils from two Magnolia plants: Contact and repellent activity to three stored-product insects. J Oleo Sci 71(3): 435–443.

Fouad HA, Faroni LRDA, de Souza Tavares W, Ribeiro RC, de Sousa Freitas S, Zanuncio JC (2014) Botanical extracts of plants from the Brazilian Cerrado for the integrated management of *Sitotroga cerealella* (Lepidoptera: Gelechiidae) in stored grain. J Stored Prod Res 57: 6–11.

Geransayeh M, Sepahvand S, Abdossi V, Nezhad RA (2015) Effect of thymol treatment on decay, postharvest life and quality of strawberry (*Fragaria ananassa*) Fruit cv.'Gaviota. Int J Agron Agric Res 6(4): 151–162.

Gerwick BC, Sparks TC (2014) Natural products for pest control: An analysis of their role, value and future. Pest Manag Sci 70(8): 1169–1185.

Green PW, Belmain SR, Ndakidemi PA, Farrell IW, Stevenson PC (2017) Insecticidal activity of *Tithonia diversifolia* and *Vernonia amygdalina*. Ind Crop Prod 110: 15–21.

Gupta V, Singh PP, Kumar A, Kumar M, Raghuvanshi TS, Prakash B (2023b) Nanoencapsulated plant essential oils as a shelf-life enhancer for herbal raw materials. In: Thomas S, Omowunmi Oyedeji A, Samuel Oluwafemi O, Rose Jaquilin PJ (eds), Nanotechnology in Herbal Medicine. Woodhead Publishing. pp. 491–513.

Gupta I, Singh R, Muthusamy S, Sharma M, Grewal K, Singh HP, Batish DR (2023a) Plant essential oils as biopesticides: Applications, mechanisms, innovations, and constraints. Plants 12(16): 2916.

Huang Y, Liao M, Yang Q, Xiao J, Hu Z, Zhou L, Cao H (2018) Transcriptome profiling reveals differential gene expression of detoxification enzymes in *Sitophilus zeamais* responding to terpinen-4-ol fumigation. Pesticide Biochem Physiol 149: 44–53.

Islam MS, Hasan MM, Xiong W, Zhang SC, Lei CL (2009) Fumigant and repellent activities of essential oil from *Coriandrum sativum* (L.) (Apiaceae) against red flour beetle *Tribolium castaneum* (Herbst) (Coleoptera: Tenebrionidae). J Pest Sci 82: 171–177.

Jacinto-Valderrama RA, Andrade CT, Pateiro M, Lorenzo JM, Conte-Junior CA (2023) Recent trends in active packaging using nanotechnology to inhibit oxidation and microbiological growth in muscle foods. Foods 12(19): 3662.

Jenan U (2014) Fumigant toxicity of *Ricinus communis* L. oil on adults and larva of some stored product insects. J Nat Sci Res 4(4): 26–29.

Jo O, Onek LA, Deng LA, Omolo EO (2009) Insecticidal potency of *Hyptis spicigera* preparations against *Sitophilus zeamais* (l.) and *Tribolium castaneum* (Herbst) on stored maize grains. Afr J Agric Res (3): 187–192.

Joseph B, Sujatha S, Jeevitha MV (2010) Screening of pesticidal activities of some marine sponge extracts against chosen pests. J Biopestic 3(2): 495–498.

Kambouzia J, Negahban M, Moharramipour S (2009) Fumigant toxicity of *Eucalyptus leucoxylon* against stored product insects. Am Eurasian J Sustain 3(2): 229–233.

Karanastasi E, Kavallieratos NG, Boukouvala MC, Christodoulopoulou AD, Papadopoulou AA (2020) Effect of three entomopathogenic nematode species to *Trogoderma granarium* Everts (Coleoptera: Dermestidae) larvae on stored-wheat. J Stored Prod Res 88: 101641.

Kavallieratos NG, Athanassiou CG, Aountala MM, Kontodimas DC (2014) Evaluation of the entomopathogenic fungi *Beauveria bassiana*, *Metarhizium anisopliae*, and *Isaria fumosorosea* for control of *Sitophilus oryzae*. J Food Protect 77(1): 87–93.

Khoobdel M, Ahsaei SM, Farzaneh M (2017) Insecticidal activity of polycaprolactone nanocapsules loaded with *Rosmarinus officinalis* essential oil in *Tribolium castaneum* (Herbst). Entomol Res 47(3): 175–184.

Kim DH, Ahn YJ (2001) Contact and fumigant activities of constituents of *Foeniculum vulgare* fruit against three coleopteran stored-product insects. Pest Manag Sci 57(3): 301–306.

Kordali S, Aslan I, Calmasur O, Cakir A (2006) Toxicity of essential oils isolated from three *Artemisia* species and some of their major components to granary weevil, *Sitophilus granarius* (L.) (Coleoptera: Curculionidae). Ind Crop Prod 23(2): 162–170.

Kretschmer M, Leroch M, Mosbach A, Walker AS, Fillinger S, Mernke D, Hahn M (2009) Fungicide-driven evolution and molecular basis of multidrug resistance in field populations of the grey mould fungus Botrytis cinerea. PLoS Pathol 5(12): e1000696.

Kumar R, Kumar A, Prasal CS, Dubey NK, Samant R (2008) Insecticidal activity *Aegle marmelos* (L.) Correa essential oil against four stored grain insect pests. Int J Food Safety 10: 39–49.

Kumar R, Srivastava M, Dubey NK (2007) Evaluation of *Cymbopogon martinii* oil extract for control of postharvest insect deterioration in cereals and legumes. J Food Prot 70(1): 172–8.

Lak F, Zandi-Sohani N, Ghodoum Parizipour MH, Ebadollahi A (2022) Synergic effects of some plant-derived essential oils and Iranian isolates of entomopathogenic fungus *Metarhizium anisopliae* Sorokin to control *Acanthoscelides obtectus* (Say) (Coleoptera: Chrysomelidae). Front Plant Sci 13: 1075761.

Lammari N, Louaer O, Meniai AH, Elaissari A (2020) Encapsulation of essential oils via nanoprecipitation process: Overview, progress, challenges, and prospects. Pharmaceutics 12(5): 431.

Li Y, Sun Z, Zhuang X, Xu L, Chen S, Li M (2003). Research progress on microbial herbicides. Crop Prot 22(2): 247–52.

Li Y, Yan SS, Wang JJ, Li LY, Zhang J, Wang K, Liang JY (2018) Insecticidal activities and chemical composition of the essential oils of *Ajania nitida* and *Ajania nematoloba* from China. J Oleo Sci 67(12): 1571–7.

Magro A, Carolino M, Bastos M, Mexia A (2006) Efficacy of plant extracts against stored products fungi. Rev Iberoam Micol 23(3): 176–178.

Malaikozhundan B, Vinodhini J (2018) Biological control of the pulse beetle, *Callosobruchus maculatus* in stored grains using the entomopathogenic bacteria, *Bacillus thuringiensis*. Microb Pathogenesis 114: 139–46.

Mbata GN, Shapiro-Ilan DI (2010) Compatibility of *Heterorhabditis indica* (Rhabditida: Heterorhabditidae) and *Habrobracon hebetor* (Hymenoptera: Braconidae) for biological control of *Plodia interpunctella* (Lepidoptera: Pyralidae). Biol Control 54(2): 75–82.

Mehdizadeh L, Ghasemi Pirbalouti A, Moghaddam M (2017) Storage stability of essential oil of cumin (*Cuminum cyminum* L.) as a function of temperature. Int J Food Prop 20(sup2): 1742–50.

Mohamed GS (2016) Pathogenicity of entomopathogenic fungus *Beauveria bassia*na and bacterium *Bacillus thuringiensis* var. kurstaki against the lesser grain borer *Rhyzopertha dominica* F. (Coleoptera: Bostrichidae) under laboratory conditions. JBAM (Egypt) 7: 39–44.

Moosavi MR (2012) Nematicidal effect of some herbal powders and their aqueous extracts against *Meloidogyne javanica* [efecto nematicida de algunos polvos herbales y sus extractos acuosos sobre *Meloidogyne javanica*.]. Nematropica 1: 48–56.

Moura ED, Faroni LR, Heleno FF, Rodrigues AA (2021) Toxicological stability of *Ocimum basilicum* essential oil and its major components in the control of *Sitophilus zeamais*. Molecules 26(21): 6483.

Mudianta IW, White AM, Suciati, Katavic PL, Krishnaraj RR, Winters AE, Mollo E, Cheney KL, Garson MJ (2014) Chemoecological studies on marine natural products: Terpene chemistry from marine mollusks. Pure Appl Chem 86(6): 995–1002.

Munekata PE, Pateiro M, Rodríguez-Lázaro D, Domínguez R, Zhong J, Lorenzo JM (2020) The role of essential oils against pathogenic *Escherichia coli* in food products. Microorganisms 8(6): 924.

Namiota M, Bonikowski R (2021) The current state of knowledge about essential oil fumigation for quality of crops during postharvest. Int J Mol Sci 22(24): 13351.

Negahban M, Moharramipour S, Sefidkon F (2007) Fumigant toxicity of essential oil from *Artemisia sieberi* Besser against three stored-product insects. J Stored Prod Res 43(2): 123–8.

Neggaz S, Chenni M, Zitouni-Haouar FE, Fernandez X (2020) Mycochemical composition and insecticidal bioactivity of Algerian desert truffles extract against two stored-product insects: *Sitophilus oryzae* (L.) (Coleoptera: Curculionidae) and *Rhyzopertha dominica* (F.) (Coleoptera: Bostrychidae). Biotech 10(11): 481.

Oka Y, Ben-Daniel BH, Cohen Y (2001) Nematicidal activity of powder and extracts of *Inula viscosa*. Nematol 3(8): 735–42.

Oni MO, Ileke KD (2008) Fumigant toxicity of four botanical plant oils on survival, egg laying and progeny development of the dried yam beetle, *Dinoderus porcellus* (Coleoptera: Bostrichidae). J Agric Res 4(2): 31–6.

Ozdemir IO, Tuncer C, Erper I, Kushiyev R (2020) Efficacy of the entomopathogenic fungi; *Beauveria bassiana* and *Metarhizium anisopliae* against the cowpea weevil, *Callosobruchus maculatus* F. (Coleoptera: Chrysomelidae: Bruchinae). Egypt J Biol Pest Control 30(1): 1–5.

Pan X, Xiao H, Hu X, Liu ZL (2023) Insecticidal activities of the essential oil of *Rhynchanthus beesianus* rhizomes and its constituents against two species of grain storage insects. Z Naturforsch C 78(1–2): 83–89.

Rajendran S, Sriranjini V (2008) Plant products as fumigants for stored-product insect control. J Stored Prod Res 44(2): 126–135.

Reichmuth C (2010) Fumigants in stored product protection. Julius-Kühn-Archiv 429: 56.

Remesh AV, Babu CV (2023) Fumigant and contact toxicities of individual and additive combinations of biorational-essential oils for control of rice weevil (*Sitophilus oryzae*). Nat Prod Res 37(16): 2748–2752.

Ríos JL (2016) Essential oils: What they are and how the terms are used and defined. In: Preedy VR (ed), Essential Oils in Food Preservation, Flavor and Safety. Academic Press. pp. 3–10.

Rozman V, Kalinovic I, Liska A, (2006) Bioactivity of 1,8-cineole, camphor and carvacrol against rusty grain beetle (*Cryptolestes ferrugineus* Steph.) on stored wheat. In: Lorini I, Bacaltchuk B, Beckel H, Deckers D, Sundfeld E, dos Santos JP, Biagi JD, Celaro JC, Faroni LRD'A, Bartolini L. de OF, Sartori MR, Elias MC, Guedes RNC, De-Fonseca RG, Scussel VM (eds), Proceedings of the Ninth International Working Conference on Stored Product Protection, 15–18 October 2006, Sao Paulo, Brazil, Brazilian Post-harvest Association, Campinas, Brazil, pp. 687–694.

Salem N, Bachrouch O, Sriti J, Msaada K, Khammassi S, Hammami M, Selmi S, Boushih E, Koorani S, Abderreba M, Limam F, Mediouni Ben Jemaa J (2017) Fumigant and repellent potentials of *Ricinus communis* and *Mentha pulegium* essential oils against *Tribolium castaneum* and *Lasioderma serricorne*. Int J Food Prop 20(sup3): S2899–S2913.

Seaman D (1990) Trends in the formulation of pesticides—an overview. Pestic Sci 29(4): 437–449.

Serrano, M., Martinez-Romero, D., Castillo, S., Guillén, F., & Valero, D. (2005). The use of natural antifungal compounds improves the beneficial effect of MAP in sweet cherry storage. Innov Food Sci Emerg Technol 6(1): 115–123.

Sharma N, Tripathi A (2006) Fungitoxicity of the essential oil of *Citrus sinensis* on post-harvest pathogens. World J Microbiol Biotechnol 22: 587–593.

Shen C, Chen W, Li C, Aziz T, Cui H, Lin L (2023) Topical advances of edible coating based on the nanoemulsions encapsulated with plant essential oils for foodborne pathogen control. Food Control 145: 109419.

Sim MJ, Choi DR, Ahn YJ (2006) Vapor phase toxicity of plant essential oils to *Cadra cautella* (Lepidoptera: Pyralidae). J Econ Entomol 99(2): 593–598.

Singh KD, Koijam AS, Bharali R, Rajashekar Y (2023) Insecticidal and biochemical effects of *Dillenia indica* L. leaves against three major stored grain insect pests. Front Plant Sci 14: 1135946.

Singh M, Srivastava S, Srivastava RP, Chauhan SS (1995) Effect of Japanese mint (*Mentha arvensis*) oil as fumigant on nutritional quality of stored sorghum. Plant Foods Hum Nutr 47: 109–114.

Song C, Yang J, Zhang M, Ding G, Jia C, Qin J, Guo L (2021) Marine natural products: The important resource of biological insecticide. Chem Biodivers 18(5): e2001020.

Srivastava S, Gupta KC, Agrawal A (1989) Japanese mint oil as fumigant and its effect on insect infestation, nutritive value and germinability of pigeonpea seeds during storage. Seed Sci Res 17: 96–98.

Stevenson PC, Nyirenda SP, Mvumi BM, Sola P, Kamanula JF, Sileshi G, Belmain SR (2012) Pesticidal plants: A viable alternative insect pest management approach for resource-poor farming in Africa. In: Koul O, Khokhar S, Dhaliwal DS, Singh R (eds), Biopesticides in Environment and Food Security. Scientific Publishers, Jodhpur. pp. 212–238.

Sulg S, Kaasik R, Kallavus T, Veromann E (2023) Toxicity of essential oils on cabbage seedpod weevil (*Ceutorhynchus obstrictus*) and a model parasitoid (*Nasonia vitripennis*). Front Agron 5: 1107201.

Tak JH, Isman MB (2016) Metabolism of citral, the major constituent of lemongrass oil, in the cabbage looper, *Trichoplusia ni*, and effects of enzyme inhibitors on toxicity and metabolism. Pestic Biochem Phys 133: 20–25.

Tapondjou LA, Adler CLAC, Bouda H, Fontem DA (2002) Efficacy of powder and essential oil from *Chenopodium ambrosioides* leaves as post-harvest grain protectants against six-stored product beetles. J Stored Prod Res 38(4): 395–402.

Thoma JL, Cantrell CL, Zheljazkov VD (2022) Effects of essential oil fumigation on potato sprouting at room-temperature storage. Plants 11(22): 3109.

Tripathi AK, Prajapati V, Verma N, Bahl JR, Bansal RP, Khanuja SPS, Kumar S (2002) Bioactivities of the leaf essential oil of *Curcuma longa* (var. ch-66) on three species of stored-product beetles (Coleoptera). J Econ Entomol 95(1): 183–189.

Trivedi A, Nayak N, Kumar J (2018) Recent advances and review on use of botanicals from medicinal and aromatic plants in stored grain pest management. J Entomol Zool Stud 6(3): 295–300.

Turek C, Stintzing FC (2013) Stability of essential oils: A review. Comp Reviews Food Sci Food Safety 12(1): 40–53.

Van Minh C, Van Kiem P, Dang NH (2005) Marine natural products and their potential application in the future. ASEAN J Sci Technol Dev 22(4): 297–311.

Vijayalakshmi S, Theenadhayalan G, Murugesh S (2017) Assessment of pesticidal activity of marine bacterial extracts using brine shrimp (*Artemia salina*) lethality bioassay. Proceedings of the International Conference on Creating and Enabling Future through Science, Technology and Innovation: Dynamics and Challenges for Development Endeavors 18-19 May 2017, Wollega University, Nekemte, Ethiopia.

Wakil W, Kavallieratos NG, Nika EP, Qayyum MA, Yaseen T, Ghazanfar MU, Yasin M (2023) Combinations of *Beauveria bassiana* and spinetoram for the management of four important stored-product pests: Laboratory and field trials. Environ Sci Pollut Res 30(10): 27698–27715.

Wakil W, Schmitt T, Kavallieratos NG (2021) Mortality and progeny production of four stored-product insect species on three grain commodities treated with *Beauveria bassiana* and diatomaceous earths. J Stored Prod Res 93: 101738.

Wright AE (1988) Isolation of marine natural products. In: Cannell JP (ed) Natural Products Isolation. Humana Press, Totowa, New Jersey, USA. 365–408.

You CX, Liu J, Li X, Zhang WJ, Yu XX, He Q, Liu N, Pan YY, Dai KD, Jiang C (2023) Cocktail effect and synergistic mechanism of two components of *Perilla frutescens* essential oil, perillaldehyde and carvone, against *Tribolium castaneum*. Ind Crops Prod 195: 116433.

Zhang L, Zhang Y, He Y, Dai H, Shu Z, Zhang W, Bi J (2023) The component of the *Chamaecyparis obtusa* essential oil and insecticidal activity against *Tribolium castaneum* (Herbst). Pestic Biochem Phys 195: 105546.

Zimmermann RC, Poitevin CG, da Luz TS, Mazarotto EJ, Furuie JL, Martins CE, do Amaral W, Cipriano RR, da Rosa JM, Pimentel IC, Zawadneak MA (2023) Antifungal activity of essential oils and their combinations against storage fungi. Environ Sci Pollut Res 30(16): 48559–48570.

Zodape, G. V. (2014). Studies on the pesticidal activities of bioactive compounds of intertidal crab Atergatis integerrimus (Lamark) of west coast of Mumbai. Bionano Front 2(2): 248–289.

Chapter 16

Engineering Aspects of Grain Fumigation

Ronald T. Noyes, Chris Newman, Thierry Ducom, Valerie Ducom, and Patrick Ducom

16.1 Introduction

Fumigation of grain and seed storages and transports has increased rapidly in the past 60–70 years in most countries due to increased production of cereal grains, oilseeds, and pulses (collectively referred to as grains) to feed growing populations. Following World War II, agricultural production grew in North America, Europe, and developing countries from improved soil fertility with use of commercial agricultural chemicals, expanded field planting including crop density, improved moisture management and seed varieties, and improved harvesting mechanization.

The shift to narrow row spacing of soybeans and corn and adoption of east-west row orientation reduced sunlight to soil between rows, sharply diminishing weed emergence. Minimum tillage also diminished weed emergence, as well as equipment and fuel cost and soil compaction. Improved herbicides reduced cultivation, further advancing crop production.

However, drying and storage technology lagged behind grain production, especially in developing countries with grain stored in 50-kg bags in warehouses and open-roofed shelters, resulting in 20%–35% annual losses of food grains from mold spoilage and pest invasion in storages.

DOI: 10.1201/9781003309888-16

The shift to field-shelling of corn, using 2-, 4-, and 5-row picker-sheller was quickly surpassed by combine manufacturers integrating picker-sheller technology into small-grain combines with interchangeable 2 row corn heads introduced in 1954. U. S. combines manufacturers rapidly expanded production with 6-8- and 10-row corn-heads, mounted on wheat combines, which dramatically increased corn harvest capacity."Corn directly harvested by combines exceeded that harvested by pickers in 1965 - -". Combine corn heads reached a sales plateau of 26,000 units in 1966. (Quick and Buchele, 1978) The rapid adoption of corn combines was a major initiative driving grain drying and storage research and development. (McKenzie, 1987).

Shelled corn requires much more drying energy than other field grains. In the mid-1960s through the 1980s, research and extension leaders at U.S. land grant colleges focused on farm and elevator grain drying and storage technologies. From 1960 to 1990, Purdue University's Bruce A. McKenzie, Extension Agricultural Engineer, was the national leader in developing and promoting highly efficient grain drying, handling, and storage center designs for farms and elevators, including future expansion options (McKenzie, 1987).

Concrete silo annexes of 10–30 silos were built at country elevators across the corn, wheat, and soybean regions of the United States from the 1960s through the 1980s. Major U.S. grain companies built terminal storage facilities with 100 to 750 silos per annex, ranging from 0.054 to 0.41 Mt (2 to 15 million bu) per annex. Farmland Industries built four terminal annexes, two 0.41 Mt and two 0.273 Mt each (two 15 Mbu and two 10 Mbu), totaling 1.36 Mt (50 Mbu) of storage in Oklahoma's central wheat region at Enid, Oklahoma.

Rapid growth in storage systems in North and South America, Europe, and other grain production regions of the world brought rapid grain insect population pressures. Stored-product insect control and management were major research and regulatory expansion factors in many countries, requiring improved storage sanitation and pesticide application practices.

From the 1950s through the 1980s, malathion was a primary insect pesticide used in farm and elevator storage. Poor application and sanitation techniques resulted in insect resistance, reducing malathion's effectiveness. Commercial elevators used methyl bromide (MB) for decades, but because of health risks to workers and its ozone-depleting potential, by the 1990s there was strong pressure on the U.S. Environmental Protection Agency (EPA) and governing agencies of most other countries to stop its use. In 2003, methyl bromide was banned in the United States and 182 other countries by the Montreal Protocol, except for a few special outdoor field practices such as horticultural crop and golf course greens treatments.

Recently, the Office of Food for Peace, Agricultural Commodity Protection by Phosphine Fumigation of USAID (USAID, 2013) and the USDA Agricultural Marketing Service have made concerted efforts to increase protection of phosphine fumigants by providing better application guidelines through extension programs and training emphasis at land grant university programs. The USDA Fumigation Handbook provides the current status of basic phosphine fumigation guidelines for university researchers and state department of agriculture training and safety applications. (USDA, 2023

This chapter is organized around five distinct topics:

1. manifolding multiple grain storages for CLF (Section 16.2),
2. ground-level fumigant dosage cabinets (Section 16.3),
3. practical storage sealing methods for CLF (Section 16.4),
4. alternative fumigation monitoring and methods (Section 16.5), and
5. CLF adoption and innovations in countries outside the United States (Section 16.6).

16.2 Phosphine Fumigation Advancements

16.2.1 1980 J-System Patent

When James S. Cook, phosphine research director at Degesch America, received his 1980 U.S. patent *Low Airflow Fumigation Method*, he patented a phosphine (PH_3) gas recirculation fumigation method which he labeled "the J-System" (Cook, 1980). Cook's patent also documented phosphine's lower spontaneous explosive limit (LEL) at 17,900 ppm (1.79%) concentration. Occupational Safety and Health Administration (OSHA) regulations now confirm this LEL valu. (Krischik et al,1995)

Cook's J-System, an engineering approach to fumigation, defined lower and upper flow rates from 0.062 to 0.62 $m^3/h^{-1}t^{-1}$ (0.001–0.01 ft^3/min/ bu) as producing satisfactory results for recirculation of a phosphine-air mixture, to yield lethal concentrations throughout the fumigated structures. However, from a practical standpoint, through CLF systems operating at Oklahoma grain elevators, including the first Oklahoma CLF system (Figure 16.1), Noyes demonstrated that airflow rates above and below Cook's specified upper and lower flow rate guidelines also produced satisfactory results (Noyes, 1993, Noyes et al, 1995)

When these phosphine safety guidelines were made public in the early 1980s, phosphine became the preferred fumigant for grain insect control and management in the United States and other countries. Phosphine research became the focus of USDA and university grain insect scientists for the next two decades. United States land grant university extension and

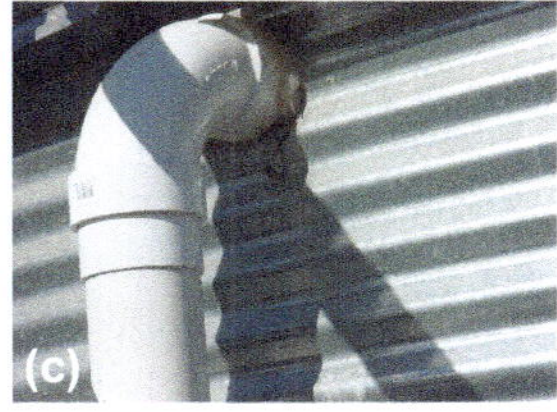

Figure 16.1 (a) First CLF model. (b) In first CLF model, CLF upper suction (15 cm) and lower pressure (10 cm) pipes. (c) Caulked U-shaped suction pipe elbows.

research faculty worked on grain insect management, emphasizing sanitation of facilities to limit insect habitats and food supplies.

Major grain insect research sites from 1980 to 2010 were the U.S. Grain Market Production and Research Center (GMPRC), Manhattan, Kansas; the National Grain Insect Research Center (NGIRC), Savannah, Georgia; Purdue University's Postharvest Education and Research Center (PHERC); Oklahoma State University's Stored Product Research and Education Center (SPREC); Kansas State University; Iowa State University; North Dakota State University; and the University of Minnesota. Canadian University of Manitoba's research group also concentrated on grain insect management. Australia's CSIRO, Canberra developed valuable phosphine research on unique storage conditions. The Western Australian government required grain companies to install sealed storage bins to meet Australian requirements for exporting insect-free wheat.

16.2.2 Development of Closed Loop Fumigation (CLF)

In late 1987, Noyes evaluated two J-Systems installed on two diverse grain storage structures at the Fairfax Elevator facility in Kansas City. One was a 10,000-t (350,000-bu) welded-steel tank, while the second was a 565-t (20,000-bu) concrete silo. The steel tank used two 0.062 kW (1/12 hp) J-System blowers, each providing 296 m³/h (175 ft³/min) gas flow; 592 m³/h (350 ft³/min) manifolded to six aeration fans. The steel tank gas flow was 0.059 (m³/h)/t [0.001 (ft³/min)/bu]. The silo gas flow was 0.52 (m³/h)/t [0.0087 (ft³/min)/bu]. Both J-Systems worked well.

In 1988, after observing the J-System at Fairfax Elevator, Noyes developed a multi-bin gas recirculation system for an Oklahoma elevator, combining two 5,000-t steel bins as Oklahoma's first manifolded PH₃ gas recirculation system (Figure 16.1a), creating two 10,000-t storage volumes, each with a Cincinnati Fan PB-9A 0.75 kW (1.0 hp) cast aluminum centrifugal blower (Cincinnati Fan, 2023) (Figure 16.1b), which provided 0.085

(m³/h)/t [0.0014 (ft³/min)/bu], 30% above Cook's minimum recommended gas flow. This new multi-bin manifolded fumigation system was named closed loop fumigation (CLF) (Noyes, 1995).

This Oklahoma elevator CLF model could have manifolded all four 5,000-t bins as one fumigation volume using a PB-10A 1.0 kW blower to circulate PH_3 at about 0.062 (m³/h)/t [0.001 (ft³/min)/bu] through 20,000 t of wheat, just like the four steel bins shown in the model CLF elevator (Figure 16.2 in Section 16.2.6) at an Oklahoma elevator 10 years later. Higher gas flow rates using larger blowers were available design options at that time (Cincinnati Fan, 2023).

From 1990 through 2003, Noyes designed several CLF elevator systems each year as Oklahoma elevator managers saw the benefits of faster, more precise fumigations at the elevators using CLF systems. By 1995, Noyes also was receiving requests for CLF systems to fumigate concrete silo annexes. By the late 1990s, requests for flat storage CLF systems were being received. He used recirculation blowers with airflow rates ranging from 0.208 to 0.104 (m³/h)/t [0.0035 to 0.001 (ft³/min)/bu] through CLF storages.

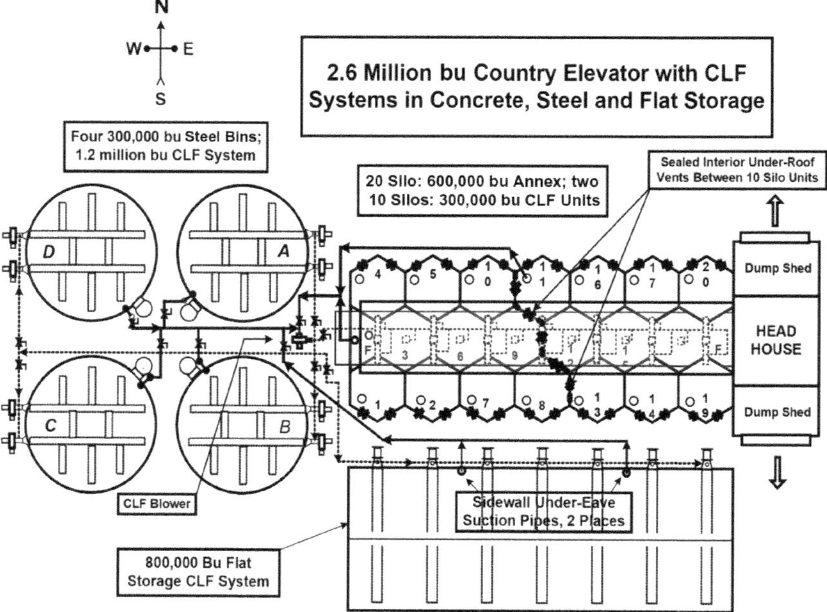

Figure 16.2 Oklahoma elevator using a CLF blower for four 8,160-t (300,000 bu) steel bins, twenty 816-t (30,000 bu) concrete silos, and one 21,760-t (800,000 bu) flat storage.

These Oklahoma CLF grain elevator systems have confirmed that storage units of variable sizes can be satisfactorily fumigated as one storage volume, allowing more precise fumigation of an entire grain storage system at one time, based on more uniform gas concentrations that were monitored during and at the end of fumigations. The CLF model demonstrated that manifolded storage systems containing a variety of connected bin sizes could be fumigated much more efficiently and with better efficacy than individual grain storage bins.

16.2.3 Early Improvements of CLF

An important steel bin suction piping change from the first CLF model system (Figure 16.1a) involved connecting headspace suction pipes through the bin wall just below the roof using two 90° Schedule 40 polyvinyl chloride (PVC) elbows (Figure 16.1c), instead of laying pipes up to mid-roof and penetrating bin roofs. Electric drill hole saws cut smooth precise holes through bin walls near roof access doors. Pipes were mounted on wall brackets or clamped to outside wall ladder siderails.

In the 1990s, many Oklahoma elevators implemented CLF principles of economic bin sealing and manifolding all steel bins and all concrete silos in a concrete annex to form 10,000 t to 50,000 t (350,000 bu to 1.75 Mbu) fumigation units. Design and economic data developed from these CLF systems were used in annual Oklahoma Elevator Management Workshops (Noyes et al., 1995).

Closed loop fumigation systems typically use one recirculation blower for 2–10 steel bins and 5–40 concrete silos as a CLF unit. Steel bin and concrete silo storage facilities of 5,000 t to 100,000 t (185,000 bu to 3.7 Mbu) are fumigated as one unit with one or two CLF blowers. Larger steel bin and silo storage complexes may need multiple CLF blowers and complex manifolds. The model Oklahoma elevator in Figure 16.2 could have used two or three blowers, sized for the three different grain volumes, but was designed to operate with only one CLF blower. Conversely, the first Oklahoma model CLF system, shown in Figure 16.1a and b, could have also operated very well with one CLF blower, instead of two.

16.2.4 Developing CLF for Small Flat Storages

From 1998 to 2000, several grain companies requested CLF systems for flat storage warehouses. Mid-Oklahoma (Mid-OK) Coop in Kingfisher had the best sealed warehouse in Oklahoma. Two men worked 7 d to seal their 20,000-t (750,000-bu) warehouse and install two CLF blowers; each blower supplied gas to half of their aeration ducts. In 2000, their peak gas concentration was 2,050 ppm 24 h after dosage. Seven days later, with brief (15 min) daily gas recirculation to increase headspace gas concentrations, their headspace gas level was 850 ppm. In 2001, Mid-OK Coop reduced the dosage 50%. Their 24 h gas reading was 1,050 ppm; after 7 d, headspace gas was 450 ppm. Their sealing was holding up well.

16.2.5 Designing CLF Manifolds for Large Flat Storages

Multiple CLF blowers were used on large warehouses. The largest CLF system in Oklahoma was a 90,000-t (3.5-Mbu) 45 m × 150 m (150 ft × 500 ft) warehouse operated by the Peavy Grain Company, Tulsa Port of Catoosa. This structure, with 18 aeration fans and ducts equally spaced per side, used three 15 cm (6 in) ID × 50 m (165 ft) pipe manifolds per side, with one 2.25 kW (3.0 hp) blower connected to six aeration fans/ducts centered on each 50 m section. All three 50 m manifolds were connected by PVC ball valves so that the company could operate the manifolds on both sides using only two CLF blowers per side, if needed.

One year, Peavey Company's middle CLF blower failed; they completed the fumigation using two blowers on one side and three blowers on the other. Another year, their warehouse was only two-thirds filled from one end. They fumigated that year using the end and middle 50 m manifolds on both sides with four CLF blowers, with excellent efficacy. This demonstrates the flexibility and management capability of CLF systems. The Peavey Elevator Superintendent said that using the CLF system in 2000 was the first time he was able to control grain insects since they started using the warehouse more than a decade earlier.

16.2.6 Model Oklahoma CLF Elevator

In the late 1990s, a CLF system was designed for a 70,700-t (2.6-Mbu) Oklahoma elevator that stored grain in concrete silos, steel bins, and a flat storage warehouse (Figure 16.2). The elevator economically sealed all three types of storage and had installed a CLF manifold system in it operated with one 3.7 kW (5.0 hp) Cincinnati Fan Company PB-14A centrifugal blower (Cincinnati Fan, 2023), rated at 2,900 m³/h (1,700 ft³/min).

The blower was connected by manifold pressure and suction piping valves to the warehouse, four steel bins, and 20 concrete silos as three separate fumigated volumes. This blower delivered gas flow to their 32,640-t steel bin storage at 0.089 (m³/h)/t [0.0014 (ft³/min)/bu], to their 21,760-t flat storage at 0.133 (m³/h)/t [0.0021 (ft³/min)/bu], and to their 20 concrete silos (16,320 t) at 0.178 (m³/h)/t [0.0028 (ft³/min)/bu].

After each storage system reaches peak uniform gas concentration (1.5–2.5 d after dosage), headspace gas concentration (HGC) is monitored in each bin or silo for the initial 2–3 d to determine which bin or silo has the lowest gas reading. When surface gas reaches the peak HGC in the target bin, the CLF blower is shut off. Throughout the elevator fumigation, the CLF blower can be switched to alternative manifolds to check and maintain desired HGCs in each of the three storage systems (Noyes et al., 1998). Workers could use their hand-held phosphine monitor to check for gas leaks in all storage units during each fumigation, mark leak points, and reseal those leaks during or after the fumigation, wearing personal protective equipment (PPE) as needed when sealing is done during the fumigation.

16.3 Ground-Level Dosage Cabinets for CLF

A grain storage colleague in Western Australia has worked since the early 2000s with many grain companies and farmers who are using ground-level AlP reaction chambers (C. Newman, personal communication, January 26, 2023), which Noyes calls gas release cabinets (GRCs). Several companies in Australia produce and supply these chambers to Australian grain companies and farmers. These chambers have support brackets for perforated bottom trays sized to release gas dosages for a range of grain volumes.

16.3.1 Pellet, Tablet, or Blanket Dosages in GRCs

The AlP GRCs are designed for airflow through dosage trays for unhampered release of phosphine gas into solar thermosiphon or fan-powered recirculation fumigation piping systems. All chambers are designed with open pipes welded on ends or sides.

If a blower motor stops on powered systems, gas is pushed through the both open pipes into the base and top of storages by its vapor pressure while gas generation continues in the chamber. Some chambers (Figure 16.3a–e) have blowers mounted directly on them, while others (Figure 16.4a and b) are connected to thermosiphon systems or external blowers. These systems provide continuous gas flow into storage.

Fumigators use ground-level AlP GRCs for the improvement in safety and efficiency that comes from not having to climb bins to place dosages in the top of grain bins. Solar thermosiphon systems work well with AlP GRCs, since the reversing flow provides excellent mixing of phosphine gas as it generates at remote sites without electric power.

Even when thermosiphon gas movement stops after sunset, open piping at both ends of AlP GRCs allows generating PH_3 gas to flow from the

Figure 16.3 (a) Blower mounted chamber. (b) Screen bottom dosage trays. (c) Blanket dosage application. (d) Sealed chamber with flash-proof blower and flex hoses at both ends. (e) Chamber with blower.

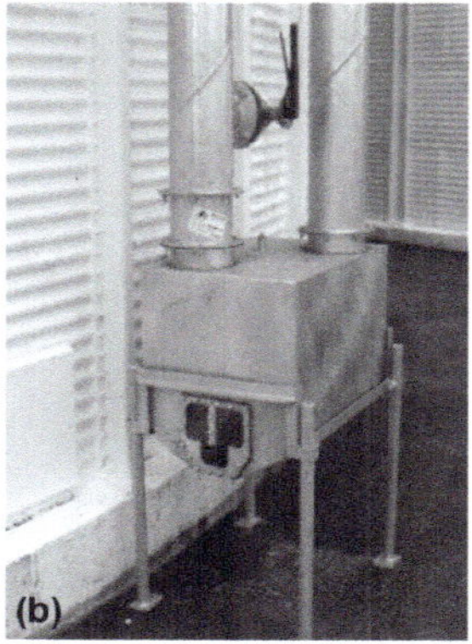

Figure 16.4 (a) Thermosiphon chamber. (b) Chamber with cleanout.

chamber in both directions from the increased vapor pressure. In properly designed PH_3 fumigation recirculation systems using government regulated dosages, there is no risk of reaching the phosphine gas LEL (17,900 ppm) concentration, as the gas will continuously move into the base and headspace of silos.

Even when silos are filled level with the top of sidewalls, headspace volumes are typically 5% or more of total empty silo volumes. Field grain contains 37%–42% interstitial air void space between seeds or kernels, so gas entering headspace or aeration ducts at the base of grain storages will flow easily into the grain bulk.

16.3.2 Ground-Level Dosage Cabinet Functions

Ground-level GRCs should be standard equipment in current and future CLF systems. Using GRCs improves fumigation safety, efficiency, and efficacy. The GRC in Figure 16.5 was a prototype cabinet developed by Noyes at Oklahoma State University (OSU) in 2000–2001.

Figure 16.5a shows an OSU fumigation workshop for Oklahoma grain elevator workers. Dummy pellets were used to demonstrate loading the dosage onto 8 shelves. The clear window (Figure 16.5a) was designed to allow fumigators to observe the progress of dosage generation during the fumigation.

The cabinet was designed to top load the dosage through a clear tube with two valves (Figure 16.5a) onto each of the 8 trays, then spread the dosage into shallow layers (3–4 pellets or 2 tablets deep) using a built-in "rake" system (Figure 16.5b) on each tray. The two upper valves formed an airlock during redosing to selected shelves when needed during the fumigation.

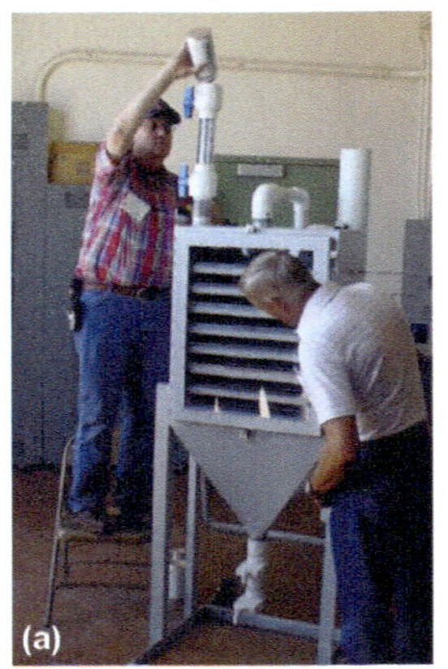

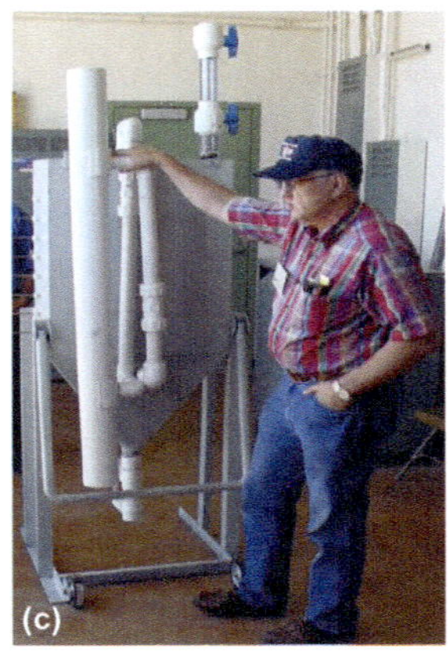

Figure 16.5 (a) Filling gas release cabinets trays. (b) Demonstrating dosage tray leveling rake. (c) Describing closed loop fumigation manifold and pipes diverting suction airflow through GRC trays. (d) CLF dosage cabinet piping layout demonstration. (e and f) CLF dosage cabinet and blower showing flexible tubing manifold connections.

Once the dosage was placed, the front door latched securely, and fumigation placards posted, the CLF blower would be started immediately. The fumigator could visually watch the progress of pellet/tablet reduction and ash development as gas was generated.

The CLF blower (Figure 16.5d–f) should be started as soon as the front door of the GRC is closed and secured after dosage. All GRCs should have a locking latch so that the door cannot be opened during the fumigation. Only the fumigator or licensed assistants should be authorized to open GRCs after the fumigation is completed.

In Figure 16.5d and e, the CLF blower suction port is connected by a black flexible tube to the top of the vertical PVC pipe on the GRC back. The bottom of the vertical PVC pipe is attached to a black flexible tube which is designed to pull gas from bin headspaces. Although this was a demonstration, it illustrates the economic material and assembly-time value of using black flexible tubing as part of the piping in primary pressure and suction manifolds around an irregular layout configuration of steel grain bins. Black flexible tubing is very economical compared to assembling straight PVC pipes and fittings. Management of CLF systems is discussed in more detail in Section 16.3.3, Section 16.3.4 and Section 16.3.5.

Part of the return suction air flows continuously through the GRC to recover gas as it generates. During fumigations using a cabinet like the one in Figure 16.5d–f, the return air is pulled through a small PVC pipe from the base of the large vertical PVC manifold shown in Figure 16.5c and d into the top center of the cabinet (Figure 16.5a–c). This suction air flows down through all dosage trays, then out through the back of the cabinet through a second small PVC pipe (Figure 16.5c) from below the dosage trays. The workshop speaker is holding this pipe where it reconnects near the top of the large PVC manifold.

Although this cabinet was not humidity system equipped, it had the potential to precisely add humidity through a pressure-regulated micro-mist nozzle with a manual shutoff valve, controlled by a humidistat-activated solenoid valve to cycle moisture into the airstream, increasing gas generation as needed. The mist nozzle would be installed in the inlet airstream to increase air humidity through dosage trays.

The PH_3 concentration increase is monitored periodically during the gas generation period. The continuous controlled humidity level in the GRC would be designed to accelerate gas generation by an estimated 300%–500% compared to ambient air moisture absorption by phosphine pellets or tablets. Peak gas concentrations could be achieved in 6–12 h using pellets with continuous controlled humidity in the GRC, compared to 24–36 h with ambient air dosage. Although tablet gas generation typically requires about 50% more time as tablets contain five times the mass (3 g vs 0.6 g) and release five times the gas of pellets, with controlled humidity, tablet gas generation with added humidity would be similar to pellet gas generation.

16.3.3 CLF Design and Operating Principles

Although individual storage volumes in grain elevators or farm storage systems may vary in size, when all storage structures are sustainably sealed, 5–10 steel bins or 10–30 (or more) silos can be satisfactorily manifolded together as one CLF volume. Closed loop fumigations are typically completed much more efficiently, with better efficacy and safety than individual storage unit fumigation.

Phosphine dosages sized for the desired peak gas concentration of combined CLF storage volumes can be conveniently placed in ground-level GRCs which are incorporated directly into the CLF suction manifold airstream.

Gas release cabinets must always have unrestricted gas flow through the cabinet—no valves or blockages can be installed in inlet or outlet pipes. Ground-level GRCs improve fumigation time, save labor, and improve worker safety by minimizing climbing inside the tops of storages during fumigations.

Gas release cabinets with open inlet and outlet piping have been used to fumigate thousands of grain storage silos in Western Australia for the past 15–20 years. Grain companies in the United States and other grain countries should be using GRCs with CLF recirculation systems and monitoring for improved safety, efficiency, and efficacy.

When GRCs are not used for dosages, either phosphine from canisters of pellets or tablets or phosphine blankets can be placed in the tops of one or more bins to complete the dosage of a CLF unit, but GRCs pay for themselves in a short time through better grain management.

Once the dosage is placed in a GRC or in bin, silo, or warehouse headspaces and doors are sealed, the CLF blower must be started immediately and should operate continuously until peak PH_3 gas concentration is achieved. If peak concentrations are not monitored, the blower should be operated without stopping for at least 36 h for pellet dosages or 48 h for tablet dosages.

Once the blower is stopped, grain bulk in each CLF storage unit holds the PH_3 gas reserve for that storage unit during the remainder of the fumigations.

Operate the CLF blower(s) to circulate PH_3 gas until all unit headspaces reach uniform peak concentration (UPC)—typically 1.5–2.5 d. Stop the CLF blower and record the HGC at that point, keeping in mind that continuous CLF blower operation significantly increases gas leakage.

If a blower malfunctions and stops before UPC is reached, gas from a GRC will flow slowly under low pressure through the suction and pressure manifolds into storage headspaces and bases due to phosphine's differential vapor pressure. Gas concentrations will not rise to the LEL of 17,900 ppm in a properly designed and operated CLF system.

Monitoring all unit HGCs. Decide what target low gas concentration (target LGC) is acceptable.

Considering that stored grain entomologists recommend 200 ppm for at least 100 h to kill all life stages of all stored grain insects, 200 ppm is used in this example. Assume that the dosage results in an initial UPC of 500 ppm. Allowing a margin of 100 ppm above the kill level of 200 ppm, set the target LGC at 200 + 100 = 300 ppm.

Once the target LGC of 300 ppm is reached and the bin number recorded, start the blower.

Monitor the HGC in the target bin. When the target bin HGC stops climbing, *stop* the blower. Record the new peak HGC. Note and record the blower running time required to reach the new (lower) peak HGC. Record the difference between the new HGC and the target LGC.

Once a new peak HGC is reached, monitor the target bin gas concentration; do not restart the blower until headspace gas drops to the selected target LGC.

Continue monitoring and operating the blower as needed to maintain the HGC. If the HGC drops to the LGC level or lower, consider adding supplemental dosage. Always keep HGCs above 200 ppm.

If a thunderstorm or strong windstorm occurs during a CLF, monitor the HGC of all units to determine if the HGC in any bin has dropped below the LGC or below the kill level (200 ppm), and try to determine if storage leaks have occurred. A new low target bin may have developed during or immediately after a storm. Strong barometric pressure drops will cause major gas leakage and corresponding ambient air entrance, with large reductions of headspace gas concentrations.

In warm sunny climates, strong sunlight can heat one side of steel storage structures while shaded sides may be much cooler, creating strong internal air currents circulating upward near walls in the grain bulk and downward through the inner grain bulk, causing HGCs to drop significantly. Headspace gas concentrations of the low target bin should be checked every 3–4 h unless all the CLF bins are well sealed. To counter these gas losses, operate the CLF blower only long enough to generate new peak HGCs. If HGCs have dropped below or near the kill level, redose and continue CLF.

16.3.4 Functioning of CLF Systems

Closed loop fumigation systems operate best when gas is "pushed" upward from base aeration ducts of each storage structure because storage bases are typically tighter than storage headspaces. From the CLF blower outlet a 10–15 cm ID pipe, (typically Schedule 40 exterior grade PVC pipe) is used in most primary manifolds. Smaller lateral PVC pipes transfer gas to the base of all storage bins included in each CLF system unit. Schedule 40 PVC pipe is ideal—it is easy to work with and has gastight elbows, tees, ball valves, and other fittings for main manifolds and lateral pipes connected to each storage unit base and top. However, flexible corrugated nonperforated black drainage hose is often installed in steel bin and concrete silo systems where using curved manifold or lateral piping through irregular-sized bin layouts is simpler than plumbing several PVC pipe sections together. However, flex tubing systems are typically faster and cost less to assemble.

Ball valves, installed in all lateral pressure and suction pipes of each storage unit, are closed to isolate empty bins during fumigations to prevent gas from circulating through and leaking from empty bins. Ball valve handles are in line when open and turned 90 degrees to the pipe when closed, so there's no confusion when closing pipes to isolate empty bins. However, filling empty bins with phosphine gas periodically may be important to control insects that may be in residual grain if elevator housekeeping is not thorough.

Gas is drawn from the storage headspace of steel bins and warehouses through two 90-degree PVC elbows which are connected through steel bin and warehouse sidewalls to form an inverted U pipe assembly attaching the suction pipe from storage headspace to the CLF blower suction manifold. The two elbows should "sandwich/clamp" bin walls and be caulked (Figure 16.1c) as they are glued together for a tight seal during assembly through hole-saw openings in steel bin and warehouse walls. Suction pipe holes should be cut just below the roof eave close to bin-roof access doors for ease in cutting holes and installing suction pipe assemblies. The suction pipe inside elbow inlet should be 10–15 cm above the grain surface to avoid grain being drawn into the suction manifold. Vertical suction pipes that hang close to the service ladder can be attached to ladder side rails or be supported by side-wall brackets.

16.3.5 Gas Movement Through Bulk Grain

Static pressure drops rapidly as gas flows from manifold and transition pipes into aeration ducts, then pressure drops to even lower as gas spreads across storage bases and flows upward through kernel interstitial openings (typically about 40% open space in grain bulks). As low pressure slowly pushes gas upward, pressure in the upper grain bulk gradually transitions from positive to negative.

To understand gas movement through typical CLF systems, let's analyze gas velocity through the four steel bins, 20 silos, and flat storage in Figure 16.2, with their central CLF blower and interconnected gas manifold configuration.

16.3.5.1 Gas Movement Through Steel Bins

Assume the four 8,160-t (300,000-bu) bins are 27.5 m (90 ft) in diameter and 14.6 m (48 ft) sidewall grain depth. The cross-section area of a 27.5-m (90-ft) diameter bin is 594 m² (6362 ft²). The 40% grain void space provides a 238 m² (2,545 ft²) airflow pathway up through the grain bulk. When a steel bin is filled level with the top of the sidewall, grain fills about 95% of the bin volume, leaving 5% of headspace air volume (Navarro and Noyes, 2002, p. 444).

At 0.498–0.747 kPa (2–3 in) static pressure, the 3.7 kW (5 hp) Cincinnati PB14A blower (Cincinnati Fan, 2023) discharges about 48.1 m³/min (1,700 ft³/min) air/gas into a 25.4 cm (10 in) ID PVC pipe which provides a 507 cm² manifold with a gas velocity of 15.86 m/s (52.0 f/s). The gas flows into four 15.2 cm (6 in) ID laterals, delivering 11.9 m³/min (425 ft³/min) to each bin. As the gas flows from aeration fans into aeration floor ducts, it slows, and pressure drops dramatically as it spreads across the bin floor, then flows upward through the 236 m² of grain bulk void space. The average gas velocity up through the grain bulk is 0.0504 m/min (0.165 ft/min), 19.8 min/m, or about 5 cm/min. Travel time for one gas change through 14.8 m of grain depth is 294 min, or 4.9 h, making for about 4.9 gas changes per day.

As gas emerges from the grain surface, it flows toward the suction pipe inlet in a shallow layer across surface grain below stagnant headspace gas and is pulled by vacuum into the suction pipe inlet. If gas leaks from the storage base, an equal volume of air must be drawn into the headspace as the blower volume remains constant, thus diluting gas concentration.

16.3.5.2 Gas Movement Through Concrete Silos

When the CLF manifold is switched from the steel bins to the concrete silos, the steel bins manifold valve is closed and the concrete silo manifold valve is opened before the CLF blower is started. A 20.3 cm (8 in) ID PVC pipe manifold is routed into the silo annex basement and is placed below the drag conveyor (like the plumbing shown in Figure 16.2a–d) that receives grain from the first 10 silos. Then the pressure manifold is reduced to a 15.2 cm (6 in) ID manifold beneath the drag conveyor under the remaining 10 silos.

A 5.0 cm (2 in) ID PVC pipe lateral with ball valve connects to each silo hopper (Figure 16.2b–d) through a perforated adapter manifold mounted over an opening cut in the grain discharge spout directly above the slide gate. With 2,900 m³/h (1,700 ft³/min) flowing through 20 silos, each silo has 145 m³/h (85 ft³/min) flowing up through 29.2 m² (314 ft²) × 0.40 (40% void space) = 11.7 m² (126 ft²), so gas flow is 145 (m³/h)/11.7 m² [85 (ft³/min)/126 ft²] = 12.4 m/h (0.675 ft/min). With grain depth of 38.1 m (125 ft), gas travel time is 38.1/12.4 (125/0.675) = 3.07 h/change (185 min/change), or 7.7 gas changes per day.

Fumigating the 20 silos as one 16,300-t (600,00-bu) CLF grain bulk should operate like the four steel grain bins holding 33,000 t (1.2 Mbu).

Once peak uniform concentration has been reached, stop the blower, and monitor the headspace gas of each of the two groups of silos during the next 24 h. Select the lower HGC level of the two groups of silos, then operate the blower and monitor the HGC for several minutes until it reaches the highest gas reading—the new HGC. Since the silo grain volume is half the steel bin volume, gas flow is twice as fast through silos, so CLF blowers can be operated about half as long on silos as on steel bins to reset the new HGC.

16.3.5.3 Gas Movement Through Warehouses

This warehouse holds 22,000 t (800,000 bu) of wheat, but grain bulk is much shallower and with its irregular (shallow sidewall to deep center) grain bulk profile, it is more difficult to manage than steel bins and concrete silos. But, CLF will provide excellent efficacy with proper management. Since the exact volume of this flat storage is known, for this example let us assume sidewall height, warehouse width, and grain surface slope, then calculate aeration duct lengths and spacings, and gas flow parameters.

Assume 5.5 m (18 ft) sidewall grain height, 36.6 m (120 ft) width, 26-degree grain surface slope resulting in 304,880 m^3 (800,000 bu = 1,000,000 ft^3) grain volume.

Volume per unit of length: 36.6 m/2 (120 ft/2) × sine 26° = 18.3 m × 0.4384 = 8.02 m (60 ft × 0.4384 = 26.1 ft). Round off to 8.0 m (26 ft) center depth above sidewall grain depth. Grain volume per foot of warehouse length: 36.6 m × 5.5 m + 18.3 m × 8 m = 201.3 m^2 + 146.4 m^2 = 347.7 m^3/m (120 ft ×18 ft + 60 ft × 26 ft = 2,160 ft^2 + 1,560 ft^2 = 3,720 ft^2/ft). Warehouse length: 28,317 m^3/347.7 m^3/m = 81.44 m (1,000,000 ft^3/3,720 ft^3/ft = 268.8 ft). Round up 81.44 m to 85 m (268.8 ft to 280 ft) length to account for end slopes to grain pile for warehouse doors. Assume 8 aeration ducts @ 9.15 m (30 ft) spacings with 42.7 m (110 ft) perforated ducts across the building. From the 20.3 cm (8 in) ID PVC pressure manifold pipe, 7.6 cm (3 in) PVC laterals connected to the 8 aeration ducts.

The warehouse grain bed "gas profile" is complex and variable: 5.5 m (18 ft) wall depth sloping up to 13.4 m (44 ft) depth at 18.3 m (60 ft) from the wall at the warehouse centerline, with gas spreading 4.6 m (15 ft) laterally, then flowing upward. Gas spreads and begins rising fast near the walls, but steadily slows through the increased resistance of deeper grain toward the warehouse center. With 2,900 m^3/h (1,700 ft^3/min) gas flow distributed into 8 laterals, each lateral receives about 362.5 m^3/h (212 ft^3/min). The warehouse floor area is 36.6 m × 85 m (120 ft × 280 ft) = 3,111 m^2 (33,600 ft^2). The airflow void space directly above the floor is 3,111 m^2 × 0.40 (33,600 × 0.40) = 1,244 m^2 (13,440 ft^2). The average upward gas flow through the 1,244 m^2 (13,440 ft^2) void space would be 2,900 (m^3/h)/1,244 m^2 [1,700 (ft^3/min)/13,440 ft^2] = 2.33 m/h (0.127 ft/min).

This flat storage warehouse has variable depths with very low gas flow rates at low static pressure. Actual gas flow rates will not vary significantly. Gas flowing through 5.5 m (18 ft) of grain near the wall will emerge in about 2.4 h, while the center gas will surface in about 5.8 h; the average "dwell time" for gas will be the average of these two times, or $(5.8 + 2.4)/2 = 8.2/2 = 4.1$ h per cycle.

Management of the CLF system for the warehouse will be similar to that for CLF in bins or silos. Once the peak gas concentration is reached, stop the blower. Monitor headspace gas concentrations at several locations along the length of the building. Based on the peak gas concentration, determine an amount of gas concentration drop between high and low gas levels to maintain. For example, if the initial peak concentration was 540 ppm, a 100 ppm drop to 440 ppm might be a suitable point to operate the blower 10–20 min to raise the gas level back to a new HGC peak, which might be 520 ppm in a tightly sealed structure.

16.3.6 Managing CLF Gas Concentrations

16.3.6.1 Uniform Peak Gas Concentrations

Once gas reaches initial peak concentration after dosage (typically in 1.5–2.5 d), stop the CLF blower and record the peak headspace gas concentration (HGC) by sampling gas immediately from suction pipes of each bin. Stopping gas recirculation once concentration is uniform eliminates continuous gas leakage. This allows each bin to "store" gas in its grain bulk with minimal loss during each CLF. Based on initial peak gas HGCs, calculate a low gas level (LGL) when the CLF blower is to be restarted to push gas up from the grain bulk to establish new HGCs because as gas leaks during fumigations, HGCs gradually declines. However, gas concentrations in each bin bulk will not change much even when the blower is only operated 15–30 min every day or two to recharge HGCs in each bin, because the blower is only pushing gas upward in the top meter or two of the grain bulk while slowly diluting the bottom few meters of the grain bulk as it recirculates headspace gas into the storage base.

Shortly before stopping the blower the first time, record the peak HGC for all bins. To check HGCs of each bin, operate the blower 2–3 min and sample each bin's suction pipe at the bin base before it connects to the primary suction manifold. Recheck/record HGCs in all bins daily. The bin with the lowest gas concentration should be monitored daily to determine when to restart the blower to recharge HGCs.

16.3.6.2 Monitoring Gas Levels During Stormy Weather

During thunderstorms, high winds with large barometric pressure drops will cause gas to be sucked out of bin headspaces on one side of storage, an effect that is offset by ambient air flow into the headspace from the opposite side. Gas losses are very large during these stormy periods. Headspace gas concentrations must be checked during or immediately after stormy weather as gas concentrations may drop well below target LGLs, requiring the CLF blower to be operated longer than normal to reset HGCs. If storms persist or storages are not well sealed, substantial redosing may be required. This is why an excellent monitoring system is needed for precise fumigation operation. Monitoring systems pay for themselves with properly conducted fumigation practices.

16.3.7 Gas Monitoring Equipment Use

Monitoring gas levels is essential in managing CLF systems. A high-quality portable phosphine monitor is needed to document phosphine gas peak uniformity, daily HGC variations, and major gas losses during thunderstorms to determine blower restart conditions. Hand-held phosphine gas monitors are needed to locate gas leaks during fumigations to improve sealing.

Gas sampling ports for HGCs and LGLs should be conveniently located at ground level on CLF suction manifolds for warehouses, steel bin groups, and concrete silo annexes. Flat storage buildings will lose more gas during low barometric events (high wind and thunderstorms) than steel bins or concrete silos, as they are harder to seal and have far more exposed sidewall and roof surface area. If one area of a warehouse has a lower gas concentration than other sections, use that reading as your LGL sampling location for blower restarts, but try to improve sealing to reduce gas losses at these locations.

All grain storages have large cross-sections with grain kernel interstitial openings. Most commercial grains contain about 40% air void space, so the gas-air mixture velocity from base to headspace is extremely slow. For example, commercial aeration typically provides about 0.62 $(m^3/h)/t$ [0.1 $(ft^3/min)/bu$] flow rate. Aeration air travel time (dwell time) through full grain bins at this airflow rate is about 5 min from base to grain surface. Using James Cook's (1980) patent guidelines, CLF gas flows range from 0.062 to 0.0062 $(m^3/h)/t$ (0.01 to 0.001) [1/100th–1/1000th] cfm/bu, or from 1/10 to 1/100 as fast as commercial aeration.

Closed loop fumigation gas dwell times range from about 50 to 500 min per gas exchange. Thus, CLF gas pressures during recirculation are extremely low; there is little pressure against bin sidewalls as gas flows upward to the headspace and across surface grain to the outlet suction pipe during recirculation. With typical CLF gas flow rates of 0.21–0.124 $(m^3/h)/t$ (1/300th–1/500th cfm/bu), dwell times range from about 2.5 to 4 h.

Phosphine pellets typically release about 90% of gas in 24–36 h, based on air humidity and temperature. Phosphine tablets require about 50%–70% longer. Closed loop fumigation blowers are started as soon as dosage is distributed in storages or is placed in a GRC, which is connected with open inlet and outlet pipes to the CLF blower. Thus, CLF gas circuits are open through all fumigated storages, eliminating any blockage, so when commercially recommended dosages are used, gas concentrations are far below LEL in CLF systems. Once the gas dosage has been released and concentrations are uniformly stable through all storage units, the CLF blower is stopped. Each grain unit bulk retains its own gas storage.

The concentration remains stable except for minor gas leakage from storage extremities or during major atmospheric weather extremes. Most of the leakage will normally occur from the top of storages, thus the need to monitor HGC levels to develop a history of gas leakage rates for each CLF storage unit during fumigations. The target minimum HGC is determined for the CLF unit with the lowest HGC level (most leakage). When HGC reaches the low target level (LTL), the CLF blower is operated for a few minutes until the HGC stabilizes at the desired high target level (HTL).

16.4 Practical Storage Sealing Methods for CLF

16.4.1 Sealing Storages for CLF

Sealing leaky bins, silos, and warehouses is vital to successful CLF operation. Figure 16.6a illustrates available off-the-shelf materials used for affordable grain storage sealing in Oklahoma CLF storages. These materials include expanding impervious adhesive foams, medium to strong tackiness adhesive sprays (red-topped cans), elastomeric concrete foundation paint and nylon mesh (Kool Seal, Figure 16.6a), heavy plastic contractor bags, high-quality duct tape, exterior grade silicone caulk, and 6 mil polyethylene

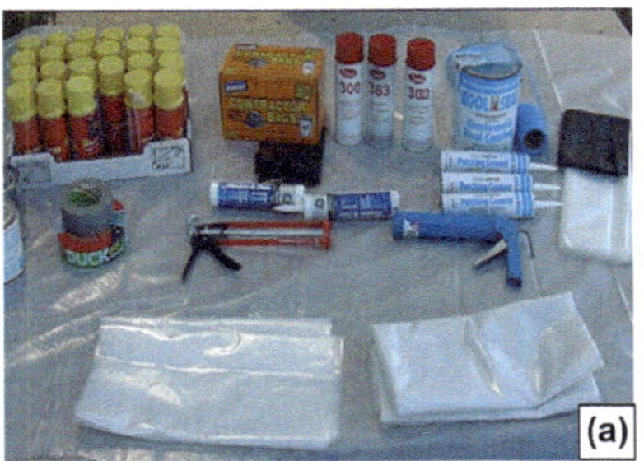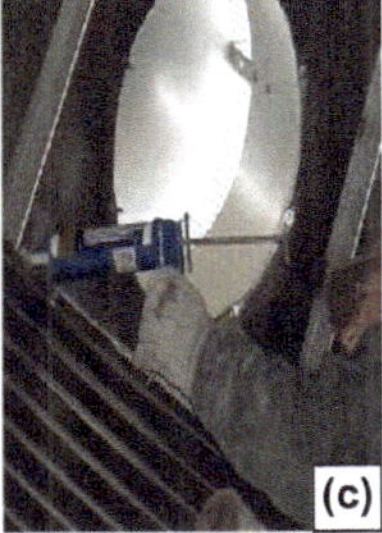

Figure 16.6 (a) Practical, economical sealing materials: elastomeric paint, expanding impervious foam, exterior grade silicone caulk, high-quality plastic sheet, and duct tape. (b) Foam for wide roof gaps. (c) Silicone caulk for narrow roof gaps.

plastic sheeting. Note that adhesive foam products like Great Stuff have an expansion ratio of three to five times their original volume. Practice with small test batches, releasing it slowly until you determine how fast and how much it expands before it becomes firm. Most expansion occurs in the first 10 to 15 min. Apply it in shallow (1–2 cm) layers; wait 10 to 15 min between layers to allow gradual filling of large spaces.

Given below are primary practical sealing areas in priority sequence of major gas loss sealing importance:

Steel Bins: Roof-wall gaps, roof vents, fill conveyors and down spouts, discharge conveyors, roof and side doors, concrete base–sidewall seams.
Concrete Silos: Under-roof wall vents, fill points, unload conveyors, roof deck–wall gap, manhole doors.
Warehouses: Drive- and walk-in doors, windows, roof vents, fill points, discharge conveyors, roof leaks, concrete base–sidewall seams.

16.4.2 Sealing Steel Bin Roof to Sidewall Air Gaps

The air gaps between the roof and the sidewall are the largest gas leak areas that must be sealed in bolted steel bins (Figure 16.6b and c). Roof panel ridge stiffeners at sidewalls (Figure 16.7a) and fill cap flashing rings require critical sealing. Roof panel ridge ends under the fill ring flashing collect dust, fines, and other airborne debris, which must be cleaned out and sealed with expanding foam sealers like Great Stuff (Figure 16.6a and b). Roof panel ridge stiffeners at the sidewalls need to be filled with adhesive spray foam (Figure 16.6b). Steel (or wooden) plates that profile roof rib openings (Figure 16.7a) can be caulked around the edges for a tight seal, or openings can be foam sealed using a temporary blocking material until the foam hardens.

Fill cap openings and roof doors (Figures 16.7b and 16.8a) are sealed using 6 mil plastic sheeting attached to raised flange openings and sides

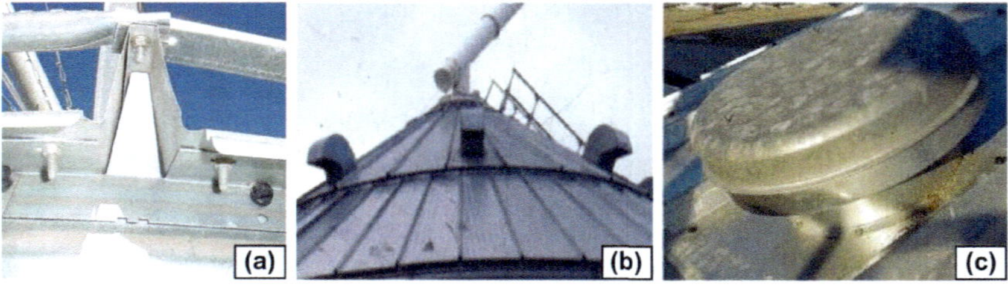

Figure 16.7 (a) Caulk edge of steel roof ridge plate, (b) Fill cap and down-spout sealing. (c) Double bag roof vents.

Figure 16.8 Seal door flanges with (a) adhesive spray and plastic sheeting or (b) elastomeric paint base seal.

with adhesive spray. Plastic 6 mil sheets should be oversized by 20–30 cm so that sheet edges can overlap below the door and fill ring lips. Wrap excess plastic down the sides and bind with duct tape, sealing plastic to steel surfaces. Seal roof vents (Figure 16.7b and c) using doubled industrial plastic bags, duct tape around vent bases, and cross-tape tops of bags to provide wind resistance. Seal the concrete foundation to the steel wall base junction (Figure 16.8b) with an exterior elastomeric paint for a durable flexible seal. Reseal concrete bin bases annually.

Many steel bins have fill spouts or auger discharge spouts installed through bin fill caps (Figure 16.7b). Some downspouts have an extended "Y" with removable cleanout plates, so the spout can be "stuffed" or "foamed" to block gas flow up spouts but the blockage can be removed after fumigation. Spouts with flange ring connectors above fill caps (which can be loosened) allow thin plates to be inserted between flange rings to block spout gas loss. If spouts cannot be sealed above the bin, spouts must be sealed inside the bin using adhesive spray, 6 mil poly sheeting, or industrial plastic bags and duct tape.

16.4.3 Sealing Roof and Sidewall Doors and Fan Inlets

Bin roof doors and bin side doors are large openings where large volumes of gas can escape if they are not well sealed. These openings have flanges that can be sealed using adhesive spray, 6 mil plastic sheeting, and duct tape. Cut the plastic sheeting at least 20–30 cm wider than the outer dimensions of the roof and sidewall door flanges (Figure 16.8a).

Spray the flange surfaces with a medium to strong tackiness industrial adhesive (the cans labeled 300 and 303 in Figure 16.6a), then lay the plastic sheet over the opening with the oversized sheet centered on the door flanges. Lightly press the plastic sheet against the flanges so that the adhesive bonds. Read the label on how quickly tackiness sets up. As soon as the plastic sheet bonds, tuck the excess plastic edges under the flange, and wrap duct tape tightly around the plastic so that the plastic sheeting is held tightly across the flanges. Close and latch the doors to protect the plastic sheeting from windblown debris that may penetrate the plastic, causing gas leaks.

The primary sealing area of aeration fans is the fan inlet (Figure 16.9a). Measure the diameter of the fan inlet orifice's outer edges, then increase the diameter of the 6 mil polyethylene sheet material by 20–30 cm to allow material to wrap behind or under the orifice ring. Aeration fan guards are bolted near the outer edges of the fan inlet orifice. Silicon-caulk the bolts and holes before applying plastic sheeting over the inlet orifice. Apply duct tape in a cross (X) pattern across the fan guard plastic, covering to keep it from flapping and being perforated during strong winds or fumigations.

Foam the motor shaft holes through fan housings to prevent gas leaks (Figure 16.9b). Inspect aeration fan housings for holes, wiring conduit entry points, cracks, gaps at the mounting face to the bin or aeration ducts, or flange connections from the fan outlet to aeration ducts, and then seal these leak points with caulk, adhesive foam, plastic sheeting and adhesive spray, or other suitable methods. Centrifugal fans mounted on concrete pads collect moisture and trash, which can rust through. Apply expanding foam under fan housings or caulk around the housing base on concrete to seal fan housings.

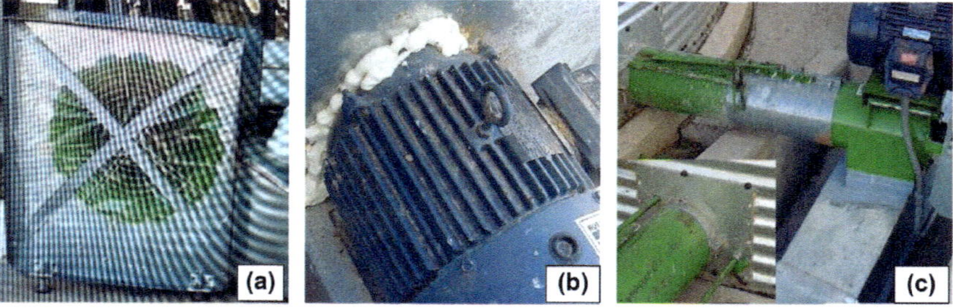

Figure 16.9 (a) Sealed fan inlet with plastic and X-braced. (b) Fan motor shaft foam sealed. (c) Caulk auger push rods and foam inside tube.

16.4.4 Sealing Discharge Conveyors

Steel bin, concrete silo annex, or warehouse discharge conveyors (Figure 16.9c) are hard to seal. Remove the U-trough conveyor cover close to the wall, then add about 20–30 cm of expanding foam to the U-trough under the bin or warehouse wall. While the foam is still expanding, replace the cover panel so that the foam pushes up and seals against the cover. After fumigation, cut out most of the foam; the remaining foam will be removed as grain discharges.

Most round tube auger conveyors have flange rings close to the foundation of storage structures. Disconnect flange rings and loosen bearings on the shaft to allow the auger tube to slide forward 10–15 cm. Fill the auger tube cavity toward the bin or warehouse foundation with a few centimeters of expanding foam, then reconnect the flange ring and tighten the shaft bearing. The 10–15 cm of foam thickness will block gas flow, and the foam will shear away when the auger is started after the fumigation. It may also be possible to fill the end of the auger housing with foam just ahead of the auger discharge to form an airtight plug if foam can be sprayed into the 20–30 cm of the auger ahead of the discharge outlet.

The discharge auger in Figure 16.9c has a silver connecting band that can be loosened and slid forward enough to allow a short section of auger tube to be filled with expanding foam. If foam plugging of discharge conveyors is not feasible, the outlet of the conveyor including motor and drive must be "bagged" with two to three industrial trash bags layered and sealed tightly with duct tape so that there is a plastic to steel tubing seal under the duct tape.

16.4.5 Sealing Flat Storage Warehouses

Flat storage warehouses vary more in gas leakage than steel bins or silos. The best warehouses for sealing are those with long vertical wall sheets attached to horizontal steel stringers between the sidewall and roof frames. Walking inside warehouses on sunny days will indicate leakiness with the number and size of visible light spots. If there are a lot of light spots, the best option may be to coat and seal the inside walls with a thin tough exterior elastomeric spray or roller paint. Large roof eave, roof deck, and conveyor inlet and discharge openings should be sealed as they are in steel bins.

16.4.6 Sealing Concrete Silos

Most concrete silo leaks are at the top and bottom. Solid reinforced concrete sidewalls generally have few leaks, except for openings built for specific purposes, such as access doors, grain discharge chutes, under roof deck vent openings, or spout outlets. Seal those with adhesive sprays, plastic sheeting, or bags and duct tape.

Roof to sidewall air gaps are primary problems on concrete silos where roof decks were precast, then set on top of concrete silo walls without a filler material between roof and walls. Usually, these junctions are not caulked or sealed. Sealing the cracks between the silo roof and wall and external vent openings is best done when the silo is filled to within 1.5–2 m of the roof so that boards or tarps on the grain surface can be used as work platforms. If roof deck wall gaps are extremely close (0.25–0.5 cm), silicone caulking will work. For larger roof-wall gaps (0.5–1 cm), use foam spray to reduce material and labor.

Precast silo roofs on 1950–1980 U.S. concrete silos were set on concrete sidewalls, leaving narrow (0.5–1.0 cm) irregular gaps between the roof and sidewall which are hard to seal. These gaps fill with grain dust and particles, becoming plugged after many years. Elevator staff should inspect roof gaps when the silos are full (so that they can stand on tarps or boards on the surface grain) and fill the gaps with silicone caulk to improve roof headspace seals.

Use expanding foam spray on exterior vents (Figure 16.11a), building up shallow layers. An alternative is to cut plywood, cardboard (Figure 16.10b), or steel plates anchored with J bolt hooks attached to vertical bars, and then spray silicone or adhesive foam around the edges, building up a thick layer of foam or caulk between the concrete and plate materials. Another option is to cut a close-fitting plate and use a silicone bead or adhesive spray around the edges to push and secure it in place when the adhesive dries or the caulk sets. Discharge conveyors will be sealed like conveyors from steel bins or warehouses. Figures 16.10c and 16.11a illustrate methods of filling external vent openings. Figure 16.11b shows silo annex basement grain chute sealing points.

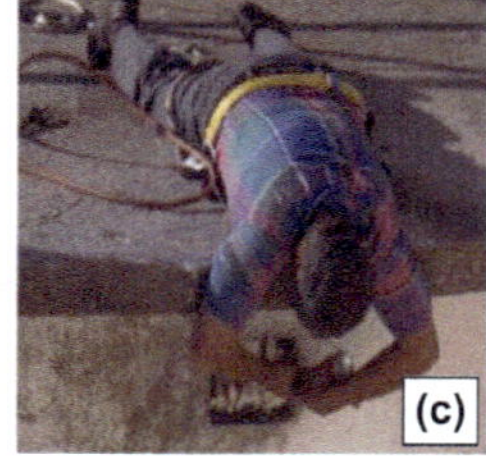

Figure 16.10 (a) Foam filling roof deck vents. (b) Roof vent sealing tools. (c) Roof vent foam sealing.

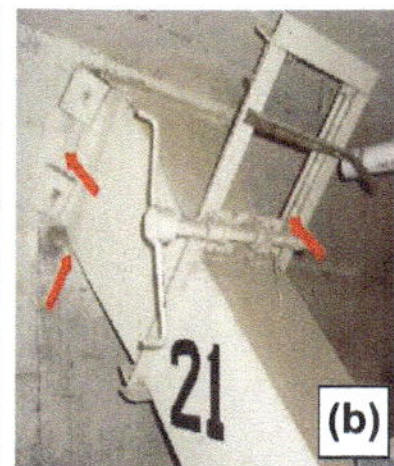

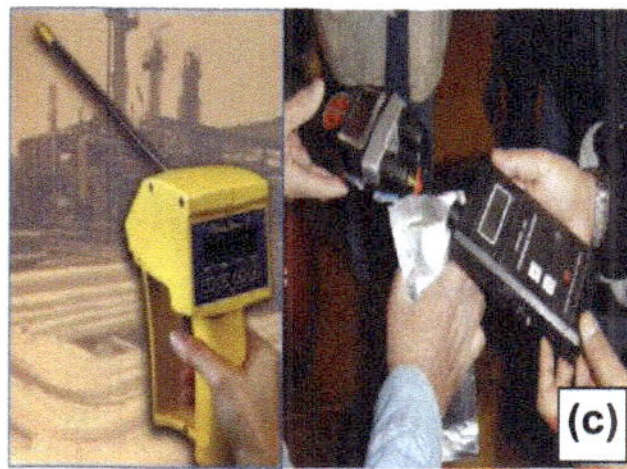

Figure 16.11 (a) Foam against cardboard. (b) Caulking silo spouts. (c) PH$_3$ gas meters needed for assessing sealing efficacy.

16.4.7 Storage Leakage Monitoring and Inspections

Regardless of the type of storage, continued surveillance for leaks to plug to improve storage seals improves fumigation. Finding gas leaks pays dividends in gas savings and higher fumigation efficacy. The best investment for sealing is a high-quality PH$_3$ electronic tester such as the PortaSens D-16 Phosphine Gas Detector (Gas-Sensing, 2023) (Figure 16.11c), which allows for hundreds of quick samples. During fumigations, "sniffing" for gas leaks around discharge conveyors, aeration fans and ducts, the base of storages, sidewall and roof openings that have been sealed, fill and discharge conveyors, bin sidewall corrugated sheet junctions, fill rings and roof overhangs, and bin base to concrete junctions is vital to improving storage sealing. Enter empty storage units on sunny days to look for light points in the roofs and walls, then seal them.

16.4.8 CLF System Benefits

Although the initial installation costs of CLF systems may be substantial, especially if outside contractors are used, the financial payback period can be relatively short. Worker satisfaction is also improved because CLF creates (1) much faster response and purge times than probe-tarp fumigation; (2) reduced fumigation labor; (3) improved worker safety; (4) reduced housekeeping; (5) reduced grain shrinkage and operating expense from no "turning" in concrete facilities; (6) lower fumigant dosages; (7) less gas release to environment; and (8) high efficacy, which minimizes insect PH$_3$ resistance.

16.4.9 CLF Plumbing Systems

In CLF systems, several combined storages are connected by pressure and suction manifolds, like the upper and lower pipes connected to the CLF

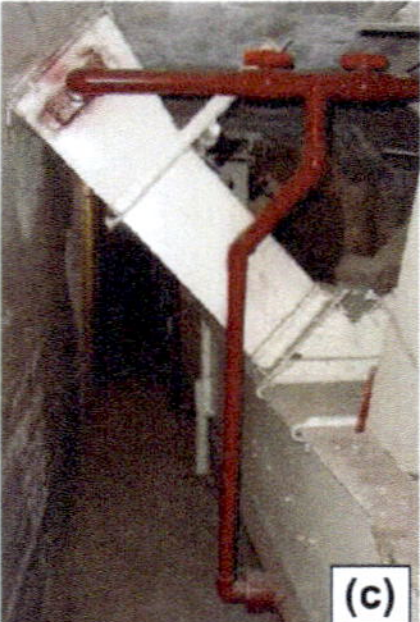

Figure 16.12 (a) CLF blower in silo annex. (b) Pressure manifold. (c) Lateral pipe to side silos. (d) Lateral pipe to center silos.

blower on steel bins (Figure 16.1a and b). Pressure manifold pipes distribute gas from the blower (Figure 16.12a and b) into the base of each storage through small lateral pipes (Figure 16.12c–d), which tee off from the pressure manifold connecting to the bottoms of silos. Lateral pipe shut-off valves (Figure 16.12c–d) allow empty storage bins to be bypassed. In concrete silos, pressure pipes connect to grain discharge spouts through perforated boxes (Figure 16.12c), so gas flows from the silo base up to the headspace. Blower suction pulls gas from all silo headspaces back to the CLF blower inlet (white pipe, Figure 16.12a).

16.4.10 CLF Plumbing Materials

Schedule 40 UV-resistant PVC pipe is the typical piping material used for CLF plumbing where straight piping is logical. Black flexible corrugated drainage tubing is often very practical and economical, as it comes in long sections and follows the contour of storage bins and tanks, minimizing PVC pipe fitting costs and labor. Gas velocity in CLF systems is quite low, so friction losses from flex tubing corrugations are negligible. Adapting connectors between PVC pipes and flexible drainage hoses is relatively simple.

16.4.11 CLF Blowers

Cast aluminum blowers work best for CLF systems. Cincinnati Fan Co. PB series 3450 RPM cast aluminum blowers (Figure 16.13a) were used for CLF systems in all Oklahoma elevators. Figure 16.13b illustrates an aluminum

Figure 16.13 (a) Cast aluminum CLF blowers. (b) CLF blower recirculating to 6 silos.

blower circulating gas through six concrete silos. These blowers maintain excellent flows through a range of gas pressures, providing stable gas flows through CLF systems. They are weather and spark resistant and not subject to chemical reaction with phosphine gas.

Closed loop fumigation blower performance data for the range of recirculation blowers are listed by Navarro and Noyes (2002). Manifold pressures for a wide range of airflow rates per 3 m (10 ft) lengths of 07.9, 11.8, 15.6, 23.6, 31.5 mm (2, 3, 4, 6, 8-in) ID PVC pipe is listed in Table 1 of Jones et al. (2017). These data are useful in selecting CLF pipe and blower sizes. Degesch America and Chicago Blower Company also market cast aluminum blowers comparable to Cincinnati Fan's PB Serie blowers.

16.4.12 Venting Fumigants with CLF Blowers

Once fumigation is complete, bins or silos must be adequately vented before workers can reenter storage units or grain can be shipped. For storages with aeration systems, unseal roof doors and vents and operate the aeration fans for 3–6 h. Aeration fans should be resealed immediately after the storage is vented, to block insect infestation in the storage base.

When using CLF blowers to purge silos or bins, disconnect the blower suction piping and then operate the blower to circulate ambient air to purge the storages. Open all roof vents or hatches so that exhaust gas is immediately diluted with air moving across the tops of the storage units. Storages without aeration fans can be purged much faster and to much lower gas levels by using CLF blowers than by conventional gravity draft venting.

When CLF blowers are used for venting, 3–5 d of continuous blower operation should be used with open roof doors and vents. Storage bin, warehouse, or silo base openings should remain sealed after the fumigant gas is purged, to minimize insect reinfestation (Noyes et al., 1989).

16.5 CLF Adoption and Innovations in Countries Outside the United States

16.5.1 CLF Adoption Throughout China

Within a few days after my CLF PowerPoint and poster presentations at the 1998 International Working Conference for Stored Product Protection (IWCSPP) in Beijing, China (Noyes et al., 1998), the China Oils and Seeds Committee adopted my CLF system into their construction designs for all concrete silos (squats) and warehouses built in 19 new state grain depots across China, which were completed between 1998 and 2000. In November 1999, Dr. Shlomo Navarro and I were invited to lecture at two-day seminars in Nanjing and Beijing on the topic of aeration and CLF.

We included current photos of China's new CLF and aeration systems (Noyes and Navarro, 1999) in our seminars. We also presented a five-hour summary of aeration and CLF to more than 600 delegates of the National Conference of Grain Depot Directors, Managers and Research Scientists in Beijing immediately after our second two-day seminar in Beijing. After inspecting two of their state grain depot storages at Nanjing and Beijing, I believe China has some of the tightest, lowest leakage, and spoilage-free storage of grains of any country, except Western Australia with its half-life pressure standards for all steel silos.

16.5.2 CLF and Associated Innovations in Australia

16.5.2.1 Sealing Grain Storage Commences in Western Australia

In Western Australia, retro sealing of central grain storages operated by Cooperative Bulk Handling (CBH) commenced in the early 1980s under direction from CSIRO Stored Grain Laboratory (Banks et al., 1979). Since that initial period of sealing existing storages, all new fixed storages have been sealed during construction to a specified standard and now comprise approximately 9 Mt of sealed steel silos (Feineler, 2022).

When the sealing of the central storage system was in process, CBH requested that the Department of Agriculture in Western Australia contact grain silo manufacturers to encourage them to commence sealing their silos, with the aim of reducing the background pool of insects that could infest grain entering the central handling system (Feineler, 2022).

This started the production of sealed transportable farm silos, and within a few years all silo manufacturers offered a sealed silo option. Since the early 1990s, competition between manufacturers to achieve a sealed silo and promote their product as gastight has driven the change to the point where now 100% of transportable silos produced in Western Australia are sealed to the gastight standard recommended by CSIRO (Newman, 1997).

Most of the grain storage in Western Australia is in the form of factory-assembled silos ranging in size from 15 to 120 t and transported to farms on trucks (Figure 16.14). The smaller sizes of grain silos on Western Australian farms are a direct consequence of a very efficient and fast CBH grain receival system, which obviates the need for farmers to store grain on farms for later delivery.

16.5.2.2 Australian Sealing Standards

The half-life pressure test $(P_{1/2})$, devised by CSIRO, allows for some small leaks in the silo but adequately describes the gastightness needed to enable a full lethal fumigation using phosphine gas. The $P_{1/2}$ for farm silos is stated in millimeters of water gauge (w.g.), which allows a simple test in the absence of expensive pressure gauges. For small farm silos, a $P_{1/2}$ from 25 mm to 12 mm w.g. (250 to 125 Pa) in 3 min or longer in a full silo is recommended (Graver, 2004).

In the eastern states of Australia (Queensland, New South Wales, Victoria), there is a greater number of large flat floor silos (>1,000 t)

Figure 16.14 Loading 90-t silos onto a truck for delivery to farms.

563

on farms to allow for more flexibility in delivery at harvest, due to the slower intake at the central grain receiving points. Many of these flat floor silos are imported and were designed for aeration and not fumigation. There was no identifiable standard for a gastight silo to enable the purchaser to have some surety that they could be used for an effective fumigation.

An industry benchmark for pressure testing sealable, gastight silos was created by Standards Australia, to provide the purchaser with confidence that the silo could be effectively fumigated. The Australian Standard AS 2628, created in 2010, requires that a newly constructed silo retain a half-life pressure for 5 min from a 25 mm head of water gauge (250 Pa). A silo that has been established and used on a farm should be able to retain a half-life pressure for 3 min from a 25 mm head of water gauge. The lower standard for silos established on the farm allows for some deterioration of the gastightness from movement of the fabric during loading and out-loading and some deterioration of the seals during handling of the sealing plates. Research conducted in the 1990s demonstrated that an effective fumigation could be conducted in a farm silo tested at a $P_{1/2}$ of 3 min (Newman et al., 2004). The higher standard at manufacturing or construction of the silo ensures a high-quality product with a long-lasting seal.

For large horizontal stores, a different pressure standard is applied. Most have a flexible steel sheet roof, and the pressure test for small silos could cause damage to the roof. The pressure test may be 150 Pa (15 mm) for 15 min but will vary according to company specifications.

16.5.2.3 Australian Thermosiphon Recirculation

Thermosiphon is a method of passive heat exchange, by which liquid or air is circulated without a mechanical pump. In the grain storage context, it consists of a pipe connected to the peak and base of the structure. Air in the pipe, expanding under passive heat exchange from the sun or high ambient temperature, becomes less dense and buoyant and moves upward in the pipe, causing air to be drawn into the pipe at the base of the silo. The process is continual, provided there is a temperature differential between the ambient air and the internal commodity, and will circulate the internal air, mixed with any gas that may be introduced into the system.

A reversal of airflow will occur when ambient temperatures fall below commodity temperatures, and air moves down the external pipe and upward through the commodity profile. A study by one of the authors (Newman), conducted in 2005, showed the air movement and the cessation of air movement in the pipe (numbers and lines in italics in Table 16.1) once the ambient and commodity temperature reached uniformity.

Table 16.1 Air Movement Speed and Cessation of Air Movement with a Thermosiphon System

Weather	Time	Ambient °C	Ambient Wind Speed (kph)	Airspeed Tube (m/s) (32 mm orifice)	Airspeed in 90 mm Pipe (m/s)	Volume Flow in 90 mm Pipe (L/min)	m^3/h	m^3/d
Fine	12:30	31	7.2	0.55	0.068	25.95	1.55	37.2
Fine	13:30	34	8	0.6	0.074	28.24	1.69	40.6
Fine	14:30	36	11	0.45	0.056	21.37	1.28	30.7
Fine	15:30	35.5	22	0.48	0.059	22.521	1.35	32.4
Fine	16:30	36	13	0.45	0.056	21.37	1.28	30.7
Cloud	17:30	35	16	0.29	0.035	13.6	0.8	19.2
Cloud	18:00	34	18	0.29	0.035	13.6	0.8	19.2
Cloud	18:30	33.5	15	0.21	0.026	9.9	0.59	16
Cloud	19:00	33	7	0.19	0.023	8.7	0.53	12.7
Cloud	19:30	32	6	0.11	0.014	5.3	0.32	7.7
Cloud	20:00	31	5.5	0.09	0.011	4.2	0.25	6
*Cloud	21:21	30	0	0		0		
*Cloud	22:00	29	0	0		0		
*Cloud	23:00	28	0	0		0		
*Part cloud	0:00	26	3	0		0		
Part cloud	2:00	24	6	0.04	0.005	1.9	0.11	2.6
Part cloud	4:00	24	5	0.06	0.007	2.67	0.16	3.8
Part cloud	6:00	23	0	0.1	0.012	4.6	0.27	6.5
Cloud	7:00	26	2	0.05	0.006	2.3	0.14	3.4
Cloud	8:00	28	8	0.08	0.01	3.8	0.23	5.5
Cloud	9:00	30	15	0.19	0.023	8.7	0.53	12.7
Cloud	10:00	31	20	0.15	0.018	6.8	0.41	9.8
Rain	11:00	27	9	0.04	0.005	1.9	0.11	2.6
Rain	11:30	27	7	0		0		

*29°C

Note: Thermosiphon air speed test, February 2, 2005, E Popanyinning, Western Australia. Silo pressure test >180 s, silo 90 m^3 and barley at 9.6% MC.

Sealed silos must be fitted with a pressure relief system to avoid structural damage from thermal expansion and contraction caused by daily temperature fluctuations. The most efficient pressure relief system should be attached to the headspace, where air expansion and contraction are the greatest. An oil-bath valve is connected into the headspace pipe at ground level for ease of observation and servicing. It allows air into and out of the headspace and prevents insects from entering.

Phosphine is a penetrating gas that diffuses through airspace, so a pipe attached to the wall of the silo and connected into the headspace and base provides a continuous conduit for phosphine to circulate. Eventually the concentration of phosphine in the silo air will reach its maximum combination in the entire airspace, but the recirculation in the thermosiphon pipe will continue day and night until the silo seal is broken.

Airflow in the thermosiphon pipe is very low and can be difficult to measure, but phosphine gas can be easily measured and will demonstrate airflow however small. An electronic phosphine monitor plumbed into the thermosiphon pipe and left switched on will show escalating values as the gas moves up the pipe as the day warms.

Fumigation of the commodity should be conducted soon after harvest to eliminate any stored-grain insects that may have flown in during the loading process. This gives sufficient time for the fumigation to commence and the $C \times T$ (CTP values) to be achieved before a later outturn. If a defined outturn date requires an exact start and finish time, a powered AlP application device would provide more precise timing. The powered equipment can be connected into the thermosiphon pipe to ensure rapid injection of the gas into the silo to achieve the threshold concentration. When all the gas has completed release from the AlP or from cylinders, the powered system can be disconnected, and the thermosiphon will circulate the silo atmosphere until outturn.

Thermosiphon systems are dependent on differential temperatures between ambient air and commodity the wider the difference between the two, the faster the air will move through the external pipe. The time to threshold concentration in several silos of different sizes is listed in Table 16.2. The time to threshold varies between 24 and 99 h.

16.5.2.4 Recirculation of Fumigant Gases in Gastight Grain Storage Structures

The larger flat storages (> 500 t) require some form of recirculation of the internal atmosphere to achieve a uniform lethal gas in order to ensure elimination of all life stages of the inhabiting insects. This can be done by using

Table 16.2 Time to >200 ppm under Thermosiphon Distribution

Location	Silo (tonnes)	Time to 200 ppm (h)	Grain Temperature (°C)
Wyatt 1/05	75	48	
House 2/06	75	29	28
Patten 3/05	75	60	
Newman A 2/14	75	72	35
Newman B 2/14	75	24	29
Newman C 2/14	75	33	
Moylan 1/10	90	39	
Moylan 7/10	90	72	
Roberts 32 6/10	120	48	
Roberts 39 6/10	120	51	
Roberts 8/08	1,200	34	30
Arthur River 2 2/12	1,200	48	
Arthur River 3 2/12	1,200	50	
Roberts 1 3/11	1,200	83	22
Roberts 2 3/11	1,200	56	25
Roberts 3 3/11	1,200	56	28
Roberts 4 3/11	1,200	82	24
Roberts 2 6/10	1,200	**24**	23
Roberts 4 6/10	1,200	43	24
Arthur River 1 2/10	1,200	36	Top load
Roberts 8/08	1,200	42	
Roberts 1 2/12	1,200	**99**	22
Roberts 2 2/12	1,200	76	22
Roberts 4 2/12	1,200	91	22
Average		**54**	

a powered blower and piping system that recirculates the gas continuously from either the top or the bottom of the storage (Figure 16.15).

Smaller storages (<500 t) also require recirculation, and this can be achieved either by power or by solar air movement created in external piping that is connected from the headspace to the base of the storage (Figure 16.16).

Figure 16.15 Blower powers flat storage gas recirculation; vertical gas vent pipe.

The original concept and design of a ground-level AlP reaction chamber were created by Don Bird of Bird's Silos of Popanyinning, Western Australia. The concept was subsequently adopted by other silo manufacturers into 75- to 120-t silos around Australia, with individual modifications to the reaction chamber (Figure 16.17).

The design of the pipework on a 75-t silo is shown in Figure 16.17. In this design, combining the AlP reaction chamber at the bottom of the cone with the lower sealing plate, the fumigator can sit under the cone base and

Figure 16.16 Black solar thermosiphon pipe recirculates gas to headspace or base.

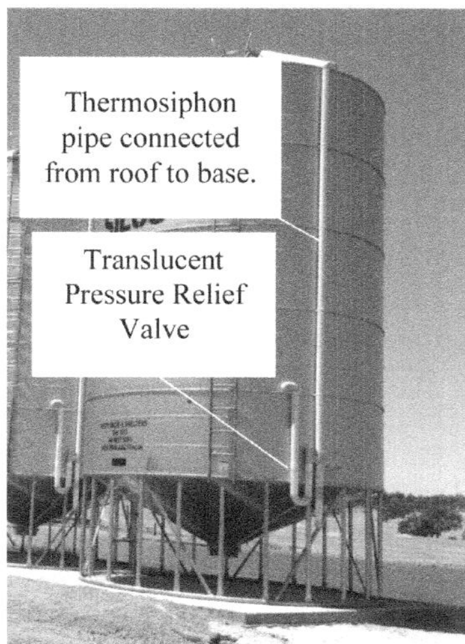

Figure 16.17 Original design by Bird's Silos, with translucent pressure relief valve.

load the AlP reaction chamber with either "bag chain" formulation or tablets, as shown in Figure 16.18a and b.

16.5.2.5 Ground-Level Phosphine Reaction Chambers

The label application rate for aluminum phosphide in Australia is 1.5 g/m^3 or approximately 2 tablets/t. Aluminum phosphide containers for farm use contain 100 tablets, and bag chain containers are the equivalent of 100 tablets.

Figure 16.18 Loading the base reaction chamber with (a) AlP bag chains or (b) tablets.

Each 3 g tablet liberates 1 g of phosphine gas (PH_3). The reaction chambers are of sufficient size to accommodate 240 tablets if the silo is 120-t capacity.

Depending on the background level of phosphine resistance, the concentration needed will vary between 200 and 300 ppm/m^3 for 7–10 d. An alternative standard with very low levels of insect resistance, adopted by CBH of Western Australia, is 100 ppm/m^3 for 14 d.

An alternative design for a ground-level AlP reaction chamber is attached under the cone of the silo (Figure 16.19) and a sealing plate is lifted to place the AlP and retrieve the aluminum hydroxide powder after phosphine release. The released gas enters the cone base via a square rolled hollow section tube, which emerges inside the silo under a shedder plate to prevent grain from entering the reaction chamber.

In another design (Figure 16.20), the AlP reaction chamber is attached to the cone base and equipped with a hinged sealable door enabling placement and retrieval of the AlP. The internal exit for the phosphine gas is the right-angle 40 mm pipe section, which emerges beneath a perforated aeration duct. The translucent pressure relief valve allows instant checking of oil levels and can be used as a manometer. This silo is also fitted with a translucent pressure relief valve for ease of pressure testing and checking oil level.

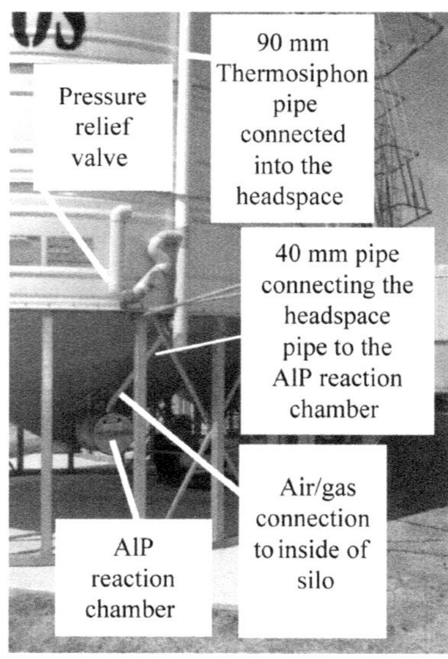

Figure 16.19 Alternative ground-level AlP reaction chamber attached to the silo cone.

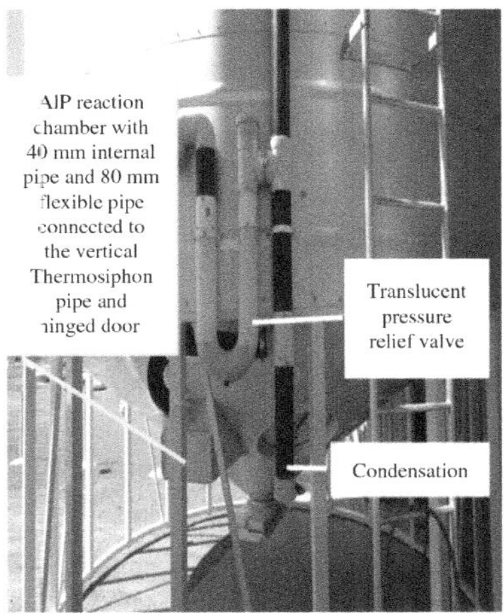

Figure 16.20 Alternative ground-level hinged door AIP reaction chamber.

16.5.2.6 Ground-Level Phosphine Application in 1200-t Flat Floor Silos

An experimental design was created as part of thermosiphon experiments on 1,200-t silos (Newman et al., 2012). This enabled 3 m AIP "blankets" to be probed into the base of a 1,200-t silo. The silos used contained V-configuration aeration ducts (Figure 16.21).

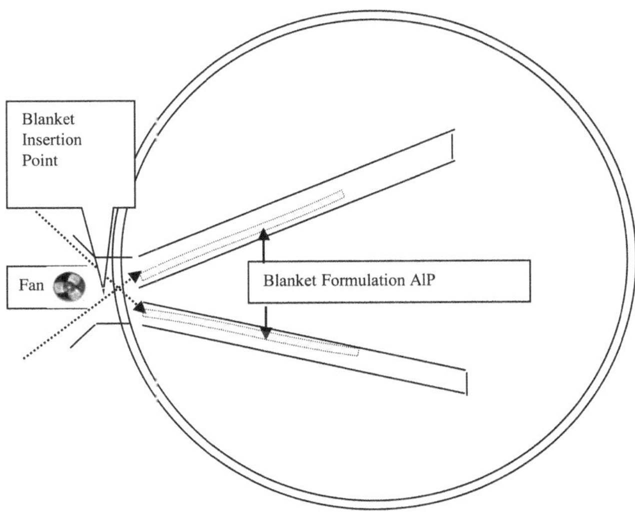

Figure 16.21 Plan of aeration ducts used as phosphine reaction chambers.

Doors were created on each side of the aeration transition section and the blankets slid into the aeration using a long probe. A section of rope was tied to the end of each blanket before insertion to allow retrieval when the fumigation was complete.

16.5.2.7 Energy-Powered AlP Ground-Level Application Equipment

When there is permanent electricity at the silo site, a blower-powered ground-level application dosage chamber can be used, like those in Figure 16.3a, d, or e. This enables grain operators to plan the start and finish to the fumigation period and provides a date for outload of the commodity.

16.6 Dynamic Fumigation: The Era of Continuous Phosphine Monitoring

16.6.1 Historical Background

Phosphine fumigations in large or nongastight structures pose the challenge of obtaining and maintaining gas homogenization at an efficient concentration level for a minimum amount of time defined by grain temperature. We cannot be sure whether the fumigation was successful without measuring the gas in key locations (Figure 16.22). The important advantage of continuous gas concentration measurements is the opportunity to adjust the technique in real time to guarantee a successful fumigation

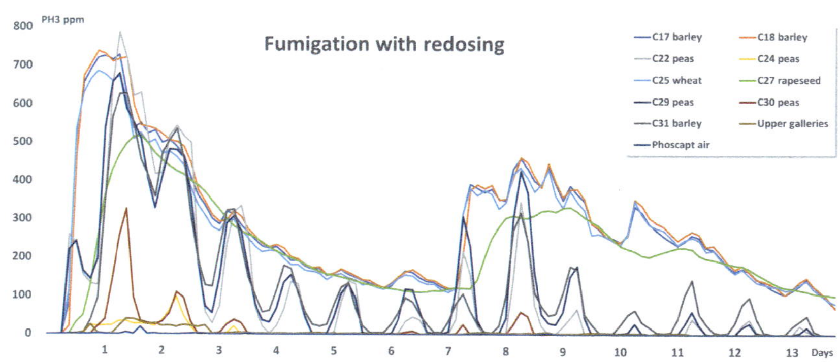

Figure 16.22 Gas monitoring on multiple lines, showing redosing.

with just the right amount of gas, even in the most complicated situations. The PhosCapt® phosphine monitors, the result of more than 30 years of research, have enabled the development of innovative dynamic fumigation techniques.

For a long time, measurement methods were either too cumbersome, expensive, or difficult to implement in the field. Until the 1980s, the only means of measurement in the field relied on Draeger® or Auer® detector tubes. In 1986, for the first time, a new principle appeared, based on the fact that phosphine is a reducing gas and, therefore, an electrochemical cell could react to it (Ducom and Bourges, 1987). There were electronic devices for measuring the reducing gas CO, and the manufacturer, Herrmann-Moritz, suggested that Ducom try a CO device, which, with a few modifications, worked very well on phosphine. This discovery opened the way to precise, easy-to-use, inexpensive devices. It was finally possible to take a lot of measurements at a low cost.

Thanks to the first microcomputers in the 1980s, it became possible to consider multi-point measurements, automation, data recording, and real-time data transmission. In 1987, the first automaton for taking multi-point phosphine measurements was created at the LNDS laboratory in Bordeaux, France (Figure 16.23). Its data could be accessed remotely using a Minitel connection (ancestor of the Internet in France). The PhosCapt®-MP phosphine monitor, commercialized in 2015, resulted from this research (Figure 16.24).

Using these real-time data (sent by email using a cellular network), one can accurately adjust the concentration in all locations using different techniques by adapting the gas flow to the different parts of the system. In a CLF system, one can decide exactly when and how long the blower needs to run. After initial peak uniform concentration is reached in CLF, the blower is off most of the time.

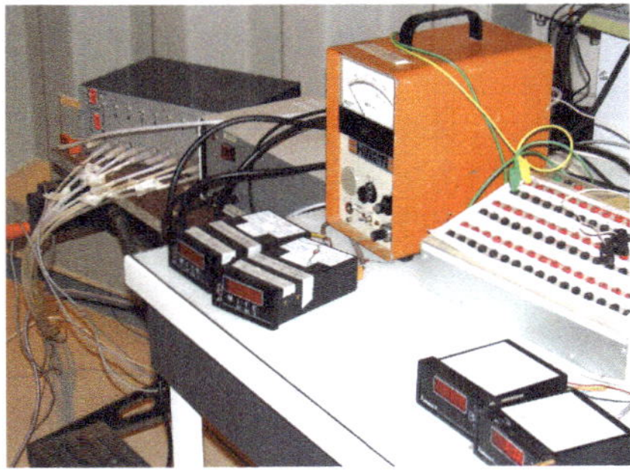

Figure 16.23 First phosphine system in 1987, using modified Herrmann-Moritz CO meter.

Figure 16.24 Four 9 Line PhosCapt-MP in 2015.

In all cases, the possibility to quickly react to the evolution in concentration creates a dynamic gas diffusion system. The main goal is to maintain minimum required concentrations in every location throughout the fumigation.

16.6.2 Monitoring: The Key to Effective Fumigations

In France, very few silos are sufficiently gastight to ensure good fumigation (i.e., to homogenize the gas to a uniform concentration and keep it in place). One exception is concrete silos, which are generally sufficiently gastight if they are closed at the top. Workers can just check for and seal leak sources such as manholes, extractors, and grain inlets. Traditionally, phosphine generators are placed at the top of the cell or sometimes at the bottom, because even though homogenization is slower, it is more uniform (Ducom et al., 2021). Active recirculation can help but might lead to more leakage. Metal silos are generally not made to be gastight. They are often open between the walls and the roof to allow dust to escape. There are also silos or flat storages, where cells are often impossible to keep gastight, even with tarping, which may also be too complex or expensive. In these conditions, it is necessary to implement a fumigation system, which does not require sealing.

A solution known as SIROFLO® was proposed at CSIRO by Winks (1993). It is a continuous pressurized phosphine distribution system that distributes gas upward throughout the grain. Winks showed that exposure time is far more critical than gas concentration. To control major grain-infesting species, it is essential to significantly prolong the usual exposure time of 5 d. In these conditions, high phosphine concentrations are not needed; in fact, the gas is more effective at much lower than normal concentrations. To overcome egg and pupal survival, gas release must be continued long enough that sufficient gas is still available when the insects hatch.

In situations where sealing is possible but not totally gastight, users often prefer recirculation using a ventilation system, like CLF (Noyes et al., 1998), which is one of the most advanced recirculation systems as it combines multiple storages in one system (described in Section 16.2.6). To take advantage of these gas diffusion techniques, it is highly recommended to have precise knowledge of the gas concentration in key locations so as to monitor possible leaks: inside the grain at different levels, in the headspaces, near the openings (extractors, ventilation doors), but also outside the fumigated enclosure. In a 2020 study (Ducom et al., 2021), the continuous measurements in the middle of a large silo revealed remarkable daily oscillations with a high amplitude: the concentration was lowest in the morning and highest in the evening. This showed the importance of taking multiple measurements every day to take into consideration these variations, which are often due to atmospheric events (wind, temperature).

16.6.3 Dynamic Ground-Level Gas Generation

A new approach to phosphine fumigation technique is a dynamic gas diffusion process. A dynamic and efficient approach to phosphine fumigation of grain storages should have four basic components.

1. Storage structures are sealed using practical, affordable sealing methods on all major and most minor leak sources so that fumigation gas leakage is minimal.
2. A ground-level dosage generation delivery system safely delivers an efficient quantity of phosphine gas within a relatively short time from the start of the generation process, with the ability to redose easily.
3. The CLF phosphine gas recirculation system provides fumigant gas efficiently to all manifold-connected storage bins simultaneously (in parallel) through pressure and suction manifolds with plumbing shut-off valves on entry and exit plumbing of each bin, which allows empty bins to be isolated from continuous gas movement, blocking empty bin volume and continuous gas leakage. Once uniform peak concentration is monitored in headspaces, CLF blowers are stopped, thus maintaining high gas concentrations (reserve gas storage) in each grain bulk. When headspace gas drops to the low target level, blowers are operated for a few minutes to recharge headspace gas concentrations, keeping more than 200 ppm in all grain.

4. An economical, full coverage phosphine monitoring system allows fumigators to react in real time because they receive continuous gas concentration measurements at key locations in all fumigated storages and work areas where leaks can occur. Monitoring helps determine when grain bins and work areas are clear at the end of fumigations.

In France, phosphine can be generated only using metal phosphides; liquid phosphine in cylinders has not yet been approved. A technique has been developed that consists in passing a controlled flow of air into hydrolysis chambers containing a selected number of Phostoxin[R] BagBlanket bag rolls (Figure 16.25).

This enables one to inject gas with or without recirculation. The phosphine concentration leaving the chambers varies depending on the air humidity, the fan flow rate, the quantity of AlP product loaded in the chamber, and whether or not recirculation is used. Continuous gas monitoring is essential to dynamically adjust the different parameters: the fan flow rate, the quantity of gas generated, the adjustment of valves in the ventilation ducts, and the sealing, if needed and possible. This technique also requires continuous concentration monitoring in the hydrolysis chamber to avoid excessively high concentrations, which could lead to accidents. Thus, there is a security system to prevent overly high concentrations.

Figure 16.25 Two hydrolysis chambers in Buchères.

This technique makes it possible to fumigate silos or warehouses that cannot be sealed.

To detect gas leaks with low concentration measurements (< 20 ppm), the PhosCapt®-MP automatically switches between high and low concentration sensors so that it can measure very low concentrations with precise accuracy (0.01 ppm) when required. This is also very important during the degassing phase, when low-level monitoring can ensure that the gas concentrations are low enough for the release of the grain to be safe and fumigation safety placards can be removed.

16.6.4 Four Examples of Fumigations in France Using Continuous Monitoring

1. **Baziège (2019):** Fumigation of four 10,200-m³ concrete bins with ten PhosCapt® monitors to study phosphine distribution using two application methods.
2. **Roncenay (2023):** Fumigation of eight 2,300-m³ concrete bins with CLF-type recirculation using a manifold.
3. **Pomacle (2023):** Fumigation of three 13,000-m³ nongastight metal bins without tarping, using Siroflo-like gas diffusion without recirculation.
4. **Buchères (2022):** Siroflo-like fumigation without recirculation or tarping of a total of 150,000 m³ in two ground sheds containing 75,000 t of grain.

The three monitored fumigations at Roncenay, Pomacle, and Buchères were in different storage types, using different phosphine gas delivery methods, based on the types of containment of the bulk grain at each site. *Roncenay* storage was in concrete silos, with poured reinforced concrete sidewalls. *Pomacle* stored grain in large steel silos with a significant open space between the steel sidewall and overhanging roof. The *Buchères* storage was in two large ground sheds, each containing several separate flat-floored storage bins, with a large open headspace between all bin grain surfaces and the roof, as shown later in Figures 16.32 and 16.33. Storage sealing conditions varied widely. Even though storage bins at Roncenay and Pomacle looked similar at each site, gas leakage between storage bins resulted in significant variations in the gas levels in grain near the top of each cell. Data below the grain surface for the cells at each of the sites are shown in Figure 16.26.

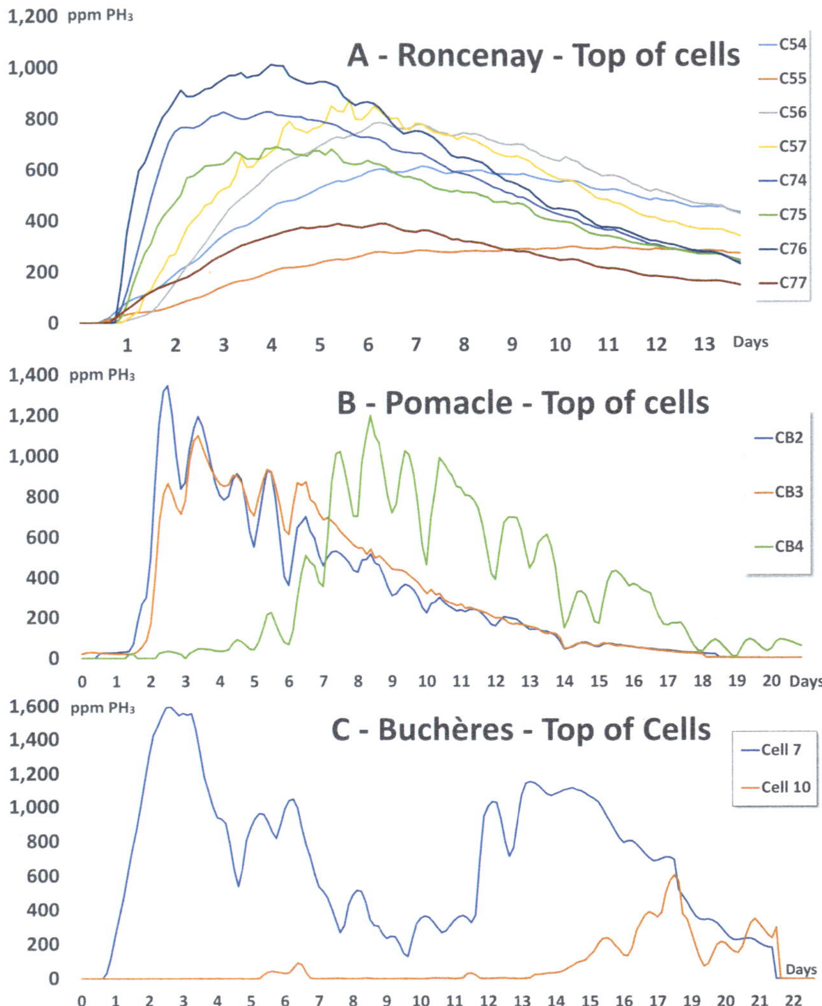

Figure 16.26   Phosphine levels in grain at top of cells at Roncenay, Pomacle, and Buchères.

16.6.4.1 Baziège, August 2019

The four silos were fumigated in passive mode, meaning that no active diffusion or recirculation was used (Figure 16.27). This fumigation was the subject of a study (Ducom et al., 2021) aimed at characterizing the differences in phosphine penetration and distribution into a grain mass using two

Figure 16.27 Four concrete 10,200-m³ bins with 30,000 t of durum wheat (*T* = 27°–30°C).

types of applications, one from the top of the silo (Silos A and B) and the other from the bottom (Silos C and D). The silos were partially sealed. Silos B and D had a double dose.

Ten PhosCapt®-MP monitors took a total of 30,000 measurements from 104 locations over 37 days, sending emails every 3 h.

Data showed that applying gas from the bottom gave total efficacy at all levels, estimated by the threshold of 200 ppm for 6–7 d (144 h). On the other hand, gassing from the top gave no efficacy throughout the silo, even at double the dose. For the first time, thanks to the 30,000+ measurements, it was possible to visualize PH$_3$ distribution in all its complexity (Figure 16.28).

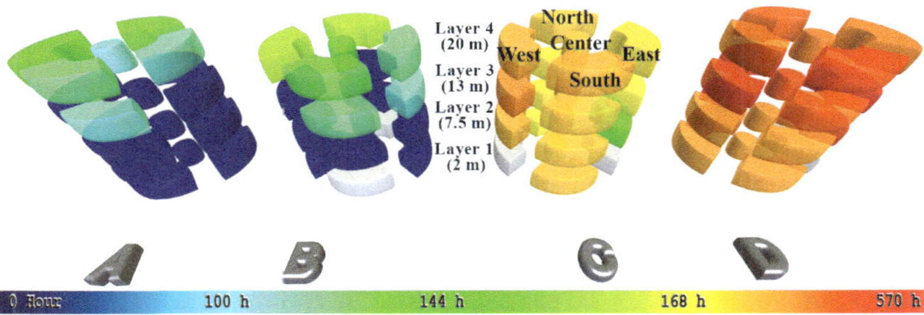

Figure 16.28 200 ppm exposure time 3D cartography (grey = data unavailable).

Figure 16.29 Six of eight concrete 2,300-m³ conical-bottom bins containing 500–1,900 t of peas.

16.6.4.2 Roncenay, August 2023

Active gas diffusion was used with four hydrolysis chambers and recirculation in 8 silos with conical bottoms infested by the pea weevil (Figure 16.29). Valves on the ventilation ducts were used to manually adjust the gas flow. Dosage was 1.85 g/m³ and temperature during fumigation varied between 8° and 21°C. A recharge was ready in case the concentration showed a significant drop, falling below 200 ppm.

One PhosCapt®-MP with one line in the grain at the top of each cell and one line in the upper gallery to detect leaks took 1,000 measurements at 9 locations, 8 times a day, during 14 d.

The concentrations were relatively uniform after the first five days, except for cells C55 and C77, but all cells remained well above 200 ppm during the long gas exposure time, which exceeded 10 days. No recharge was necessary. Leakage in the gallery was very low, which is consistent with good results in the cells. No adult bruchid appeared, but it was too early to confirm total effectiveness (at the time of writing, two months after fumigation).

16.6.4.3 Pomacle, August 2023

Three out of eight bins were fumigated without roof sealing (each bin had a roof-sidewall air gap) using active gas diffusion without recirculation, with

580

Figure 16.30 Three 13,000-m³ nongastight metal bins fumigated at 25°C.

one hydrolysis chamber per bin (Figure 16.30). Each of the cell's 16 ventilation ducts was connected to a manifold using flexible tubing from the hydrolysis chamber. Valves on the ventilation ducts were used to manually adjust gas flow with a dosage of 3 g/m³ and temperature during fumigation around 25°C.

One PhosCapt®-MP monitor (Figure 16.31) took a total of 1,000 measurements at 6 locations in the three bins, 8 times a day, during the 21 d fumigation. One line was placed below the grain surface at the top of each cell and one line was in each of the bottom galleries to detect leaks.

The objective was to obtain the longest continuous gas exposure time (over 15 d in this case). The chart shows gas concentrations are relatively uniform, remaining above 200 ppm in all three cells for more than 10 d (Figure 16.28b).

In the galleries, the PH_3 concentrations were quite high, especially at the start of the hydrolysis: 150 to 200 ppm. In fact, the sealing around Cell

Figure 16.31 PhosCapt®-MP mounted in weatherproof cabinet beside bin.

Figure 16.32 150,000 m³ in two ground sheds containing 75,000 t of wheat (temperature 25°C).

4 air entry doors was not tight enough. When the gallery doors were tightened, it took 2 d for the gas concentration in Cell 4 to reach the target level. After 5–6 d, concentrations in the galleries stabilized at about 40–60 ppm. After the fumigation, no live insects were found.

16.6.4.4 Buchères, October 2022

Eight hydrolysis chambers connected to the ventilation system slowly injected gas inside 100,000 m³ of grain without tarping or recirculation (Figure 16.32). The dosage was 3 g/m³ and temperature during fumigation was 25°C.

The ventilation flow was modulated based on air humidity and the PH_3 gas concentration measurements in each cell. A recharge was needed after 10 d of fumigation with a slightly smaller dosage: 2 to 3 g/m³.

In Cell 7, the concentration was maintained above 200 ppm for 10 d. Redosing on the 10th day prolonged this gas exposure to 20 d. In Cell 10, even after redosing, the gas concentration was still close to zero. It was necessary to adjust the ventilation duct valves to route enough gas to Cell 10 to maintain a good concentration for 8 d (Figure 16.26c). Before the fumigation, there was a very significant rice weevil infestation in all the cells. After

Figure 16.33 Three PhosCapt®-MPs monitoring 29 lines.

fumigation, sieving revealed no living insects. In the following 7 mo, no insects were found during quality checks.

Three PhosCapt®-MPs (Figure 16.33) took more than 5,000 gas concentration measurements at 29 locations over 25 d. The fumigation manager received the results by email every 3 h. At the end of the fumigation, an XLS file was downloaded from each device.

Because of the configuration of the open top cells in the two flat storages, the sample probes were 'thrown' into the open cell grain surfaces (Figure 16.34) from the walkway beside the cells.

Figure 16.34 Monitoring probes thrown into the grain like javelins.

16.7 Conclusions

Based on the material described in this chapter, the following conclusions can be drawn:

Practical sealing of leaks in all storage structures is critical for effective fumigation.

Closed loop fumigation manifolding multiple storage units of different sizes and types can be implemented in the field.

Ground-based phosphine gas release chambers enhance safety and efficacy of fumigation.

Thermosyphon is a useful passive system for enhancing distribution of fumigants, particularly in remote areas where mechanical blowers could not be installed due to lack of electric power.

Small solar-powered blowers can be productive for positive recirculation of gas at much lower levels than aeration gas flow rates.

Dynamic fumigation with continuous monitoring even in leaky bins can ensure efficacy; however, it increases chemical use.

Continuous monitoring of phosphine in surrounding areas enhances worker safety.

References

Banks HJ, Annis PC, Wiseman JR (1979). Permanent Sealing of 16,000 Tonne Capacity Shed for Fumigation of Modified Atmosphere Storage of Grain. CSIRO Australia. Division of Entomology. Report No.0. 12. p. 16.

Cincinnati Fan (2023). PB Series Cast Aluminum Pressure Blowers, Mason, OH 45040-9135, Cat. No. PB-0811, Supersedes PB-1206, p. 8, http://www.cincinnatifan.com, accessed September 17, 2023.

Cook JS (1980). Low Airflow Fumigation Method. U.S. Patent No. 4,200,657. P. O. Box 5421, Houston, Texas 77021. Issued April 29.

Ducom P, Bourges C (1987). Determination of phosphine concentrations with an electrochemical cell. In: Donahaye E and Navarro S (Ed.) Proc. 4th Int. Work. Conf. Stored-Product Protection, Tel Aviv, Israel, pp. 630–635.

Ducom V, Simioni F, Mercadal D, Busson D (2021). A comparative study of phosphine distribution using two application methods with 30,000 gas concentration measurements. In: Jayas DS and Jian F (Eds.) Proceedings of the 11th International Conference on Controlled Atmosphere and Fumigation in Stored Products (CAF2020), CAF Permanent Committee Secretariat, Winnipeg, Canada, pp. 63–70.

Feineler G (2022). Personal Communication. Manager, Project Delivery, Cooperative Bulk Handling (CBH), Western Australia.

Gas-Sensing (2023). PortaSens D-16, 0-1000 ppm, Email: support@gas-sensing.com, accessed October 29, 2023.

Graver JEVS (2004). Guide to Fumigation Under Gas-Proof Sheets, FAO Website, p. 17; 9.19.2023.

Jones C, Hardin J, Bonjour E (2017). Design of Closed-Loop Fumigation Systems for Grain Storage Structures, Extension Circular BAE-1111, Oklahoma State University, Stillwater Oklahoma.

Krischik V, Cuperus G, Galliart D (1995). Stored Grain Management, Circular E-912, Oklahoma State University, p. 140.

McKenzie BA (1987). Grain Drying, Handling and Storage Handbook, MWPS-13, Midwest Plan Service, Ames, IA, p. 88.

Navarro S, Noyes RT (2002). The Mechanics and Physics of Modern Aeration Management, CRC Press, Boca Raton, Florida, p. 444.

Newman CR (1997). The response of the silo manufacturing industry in Western Australia. In: Donahaye EJ, Navarro S, and Varnava A (Eds.) Proceedings of an International Conference on Controlled Atmosphere and Fumigation in Stored Products, Nicosia, Cyprus, 21–26 April 1996, Printco Ltd., Nicosia, Cyprus, pp. 393–406.

Newman CR et al. (2012). Investigation into the use of thermosiphon pipes to distribute phosphine gas through grain silos from a ground level introduction point. In: Navarro S, Banks HJ, Jayas DS, Bell CH, Noyes RT, Ferizli AG, Emekci M, Isikber AA, and Alagusundaram K (Eds.) 9th International Conference on Controlled Atmosphere and Fumigation in Stored Products. Antalya, Turkey, p. 557.

Newman CR, Daglish GJ, Wallbank BE, Burrill PR (2004). Investigation into control of phosphine resistant *Rhyzopertha dominica* in sealed farm silos fumigated with phosphine. In: Donahaye EJ, Navarro S, Bell C, Jayas D, Noyes R, Phillips TW (Eds.) (2007) Proc. Int. Conf. Controlled Atmosphere and Fumigation in Stored Products, Gold-Coast Australia 8-13th August 2004. FTIC Ltd. Publishing, Israel, pp. 275–282.

Noyes RT, Clary BL, Stringer ME (1989). Closed loop fumigation. In: Proc Fumigation Workshop, Circular E-888, Coop Extension Service, Oklahoma State Univ., Stillwater, Oklahoma, pp. 103–112.

Noyes RT, Kenkel P, Tate G (1995). Closed loop fumigation. Stored Grain Management, OSU Circular E-912, First Revision, Oklahoma State University, Stillwater, Oklahoma, pp. 153–161.

Noyes RT, Navarro S (1999). Development of Cooperative Programs with Postharvest Authorities in China. Report submitted to USDA, Foreign Agricultural Services (FAS), Research and Scientific Exchange Division (RSED), p. 18.

Noyes RT, Phillips TW, Cuperus GW, Bonjour E (1998). Advances in recirculation fumigation in the U.S.A. In: Zuxun J, Quan L, Yongsheng L, Xianchang T, and Lianghua G (Eds.) Proc 7th International Working Conference for Stored Product Protection, Beijing, China, pp. 454–462.

Noyes, RT (1993). Closed Loop Fumigation Design and Management, Workshop Paper, Indiana Stored Grain Pest Management Workshop, Purdue University, Lafayette, IN, September 9, p. 18.

Quick GR, Buchele WF (1978). The Grain Harvesters, American Society of Agricultural Engineers, St. Joeph Michigan, pp. 211–224.

USAID (2013). Office of Food for Peace. Agricultural Commodity Protection by Phosphine Fumigation: Programmatic Environmental Assessment, (November 2013), p. 128.

USDA (2023). Fumigation Handbook (2023), Agricultural Marketing Service, USDA, Washington, DC, October 1, 2023, p. 71.

Winks RG (1993). The development of siroflo® in Australia. In: Navarro S and Donahaye E (Eds.) Proc. Int. Conf. Controlled Atmosphere and Fumigation in Grain Storages, Winnipeg, Canada, June 1992, Caspit Press Ltd., Jerusalem, pp. 399–410.

Chapter 17

Modeling the Concentration of Phosphine and Insect Mortality with Computational Fluid Dynamics

Paraskevi Agrafioti, Efstathios Kaloudis, Dimitrios Kateris, and Christos G. Athanassiou

17.1 Introduction

Phosphine is widely recognized as the most commonly used fumigant for controlling stored-product insects worldwide (Collins, 2010; Warrick, 2011). Its popularity stems from its versatility, cost-effectiveness, and efficacy in treating a wide range of commodities, including bulked grains, dried fruits, tobacco, and nuts (Bell, 2000; Emery et al., 2003; Kaur and Nayak, 2015). Due to the withdrawal of methyl bromide (MB) (UNEP, 1995), the use of phosphine has increased gradually since 2005 (Bell, 2000; Chaudhry, 2000). Moreover, phosphine has gained international acceptance as a low-residue

DOI: 10.1201/9781003309888-17

treatment, making it a preferred choice for grain preservation and storage (Nayak et al., 2015).

However, despite its widespread application, the continuous and extensive use of phosphine has resulted in the emergence of resistance among many stored-product insect species, posing a significant challenge to effective control of pests (refer to Chapter 2). Resistant populations of several stored-product insects have been reported in many parts of the world, including the United States (Saglam et al., 2015; Cato et al., 2017; Afful et al., 2018), Australia (Jagadeesan et al., 2012; Nayak et al., 2013; Holloway et al., 2016), India (Kaur et al., 2015), Pakistan (Ahmad et al., 2013), Morocco (Benhalima et al., 2004), China (Song et al., 2011), Brazil (Pimentel and Guedes, 2010), and Europe (Sakka et al., 2018; Agrafioti et al., 2019). This widespread resistance development underscores the need for alternative strategies to combat insect infestations and preserve the effectiveness of phosphine-based fumigations.

In addition to resistance, another factor that can impede the success of phosphine fumigations is the inefficient monitoring of gas concentrations (Flinn et al., 2003; Phillips et al., 2012; Brabec et al., 2019). Traditional monitoring methods, such as digital monitors and glass tubes, have limitations in terms of accuracy, ease of use, and the frequency of measurements. These methods often require trained personnel and multiple measurements, which may not be feasible in real-world fumigation scenarios. Consequently, concentration measurements are typically taken at daily intervals or less frequently, hindering the ability to obtain real-time monitoring data. The quantification of gas concentrations is further complicated by the spatial and temporal variations in phosphine distribution within fumigated structures.

Recent advancements in monitoring techniques have enabled continuous and accurate measurement of phosphine concentrations during fumigations (Brabec et al., 2019). One promising technology that has emerged is wireless phosphine sensors. These sensors, equipped with electrochemical technology, provide high accuracy in detecting gas concentrations and offer wireless connectivity for real-time data transmission (Brabec et al., 2019). The use of wireless sensors allows for more frequent data collection, providing insights into the dynamic changes in phosphine distribution and facilitating the development of precision fumigation strategies.

Athanassiou et al. (2016) evaluated, for the first time, phosphine presence and distribution by using phosphine wireless sensors, which had been placed within the fumigated area in commercial containers with dried figs. That work clearly illustrated that gas concentration was uneven and fluctuated dramatically within the same day. Recently, Agrafioti et al. (2018a, b, 2020b) utilized wireless phosphine sensors to evaluate the distribution and efficacy of phosphine in different commercial facilities such as horizontal warehouses, ship holds, silos, storage under tarpaulin, and containers, and they reported that fumigation processes in containers achieved better results than those in other facilities. Brabec et al. (2019) evaluated the accuracy and responsiveness of wireless sensor devices in grain silos and confirmed that the sensors provided a more detailed picture of the fumigation process than did conventional monitoring devices (glass tubes or digital machines). In another study, Aulicky et al. (2015) showed that during a commercial fumigation in a flour mill, phosphine concentration varied remarkably among the different floors of the facility. Furthermore, Brabec et al. (2019), by using phosphine sensors, found that there were considerable variations in phosphine concentrations among locations within the same silo. Additional studies have documented that underdosing and uneven distribution may lead to resistance development (Phillips et al., 2012; Nayak et al., 2013; Boac et al., 2014).

Despite the facilitation of continuous monitoring and enhanced usability through the utilization of wireless sensors, it remains imperative to acquire comprehensive knowledge regarding the dispersion of phosphine, along with the resulting insect mortality, across the entirety of the fumigated area. Mathematical modeling, particularly computational fluid dynamics (CFD), offers a powerful tool to simulate and predict phosphine distribution in different storage structures (Mills et al., 2001; Chayaprasert et al., 2006; Boac et al., 2014). Computational fluid dynamics models can simulate complex thermal and fluid flow phenomena, enabling the analysis of gas dispersion, airflow patterns, and concentration gradients within fumigated environments. Indicatively, Mills et al. (2001) validated this model against average phosphine concentration measurements and showed a positive pressure system for combatting dilution during phosphine fumigations in bulk grains. Furthermore, Boac et al. (2014) studied phosphine distribution in bunker storage structures in conjunction with the effect of wind. In that study, the authors found that phosphine distribution was characterized by increasing phosphine concentration dispersal over the grain stack surface enclosed underneath a tarp (Boac et al., 2014). Isa et al. (2016) worked with mathematical models that could predict fumigant distribution in leaky silos using fan-forced fumigation. In that work, the authors studied the fan-forced fumigation via an inlet at the base of the silo, showing that leak locations can significantly impact the fumigant distribution.

A modeling approach able to combine the distribution of phosphine and the prediction of insect mortality is CFD, which has been successfully evaluated in the case of fumigations in silos by Agrafioti et al. (2020a). In that work, it was found that phosphine distribution in cylindrical grain silos can be notably improved through the operation of a recirculation system, an approach that was validated with field trials (Agrafioti et al., 2020a). Apart from silos, there are several reports on other types of storage structures, such as food processing facilities (Chayaprasert et al., 2006), bulk grain structures (Plumier et al., 2018), and bunkers (Boac et al., 2014). Phosphine distribution in different structures has been evaluated by many research groups through different mathematical models (Daglish and Pavic, 2008; Darby, 2008; Darby et al., 2009).

The transportation and storage of grain and other agricultural commodities involve various methods and structures, each accounting for a significant percentage or volume of the total storage and transportation capacity. Silos are commonly used for storing grain, with a substantial portion of the total grain storage capacity allocated to these structures. These tall, vertical structures can hold thousands of tonnes of grain, providing efficient storage with minimal footprint and reasonable cost per tonne of storage. Warehouses also play a crucial role in storing agricultural commodities, including grain. They accommodate a substantial portion of the overall storage capacity, with their large storage areas. Shipping containers are primarily used for the international transportation of grain and agricultural commodities. While they may not individually account for a significant volume of storage, they enable the efficient and secure movement of commodities across vast distances. Each shipping container has a standardized capacity, typically ranging from 6.1 m to 12.2 m (20 to 40 feet) in length, and can hold several tonnes of grain.

Railcars, particularly hopper railcars, are instrumental in the long-distance transportation of grain and agricultural commodities. These specialized railcars are designed to carry bulk commodities and can transport large volumes of grain in a single trip. Transportation by rail accounted for ~25% of total U.S. grain transport (Association of American Railroads, 2021). Rail is the predominant method for long-distance transport of grain, with the average hauling distance for corn and wheat railcars being 1,817 and 1,667 km (1,129 and 1,037 miles), respectively, in 2009 (Prater and Sparger, 2013). In the United States, class 1 railroads moved ~ 1,500,000 carloads of grains and oilseeds and moved ~750,000 carloads of processed-grain products in 2021. Corn had the highest rail volume moved with ~691,000 carloads, while wheat had ~305,000 carloads (Association of American Railroads, 2021). Bulk grain and flour are typically shipped by rail in hopper-bottom cars. These railcars have openings in the top for loading commodities and sloped floors so that commodities can be unloaded through doors in the bottom of the car. Boxcars have flat floors and are used for shipping pallets of packaged grain and grain products; they typically have sliding doors on each side of the car to facilitate loading and unloading.

This chapter aims to address the aforementioned gaps in knowledge and provide insights into the distribution of phosphine in different structures, such as cylindrical silos, containers, and railcars. The research involves the integration of CFD simulations with the use of wireless phosphine sensors to validate and refine the models. By correlating phosphine distribution patterns with insect mortality, the chapter aims to enhance our understanding of the factors influencing the effectiveness of phosphine fumigations. Ultimately, the findings of this study can contribute to the development of precision fumigation strategies that optimize phosphine distribution, minimize resistance development, and improve pest control outcomes in various storage and transportation contexts (Brabec et al., 2019; Agrafioti et al., 2020a, b).

17.2 Methodology

17.2.1 CFD-Simulation Technique

In the context of phosphine fumigations, it is crucial to guarantee that the concentration of phosphine surpasses the predetermined parts per million (ppm) thresholds throughout the entire storage area, aiming to achieve complete eradication of insects with a mortality rate of 99.9%. In this particular investigation, CFD was employed to characterize the dispersion of gas in correlation with the mortality of insects.

Computational fluid dynamics constitutes the computational examination of systems featuring fluid dynamics, heat transfer, and associated phenomena, such as chemical reactions, through computer-based simulation. This methodology wields considerable efficacy, spanning a broad spectrum of industrial and nonindustrial applications (Versteeg and Malalasekera, 2007). Noteworthy examples are the following:

- Aerodynamics of aircraft and vehicles, including the study of lift and drag
- Hydrodynamics of ships
- Power plants, with a focus on combustion in internal combustion engines and gas turbines
- Turbomachinery, examining fluid flows within rotating passages, diffusers, and related components
- Electrical and electronic engineering, addressing the cooling of equipment, including microcircuits
- Chemical process engineering, investigating mixing and separation processes, as well as polymer molding
- Environmental considerations in both external and internal building environments, involving wind loading and heating/ventilation
- Marine engineering, pertaining to the assessment of loads on offshore structures
- Environmental engineering, encompassing the distribution of pollutants and effluents
- Hydrology and oceanography, with a focus on flows in rivers, estuaries, and oceans
- Meteorology, particularly in the context of weather prediction
- Biomedical engineering, specifically the study of blood flows within arteries and veins

Phenomena involving fluid flows and their associated behaviors are typically described using partial differential equations or integro-differential equations. In general, these equations defy analytical solutions, except in specific circumstances. To derive an approximate solution, numerical methods are employed, necessitating the utilization of discretization techniques to approximate the differential equations as a system of algebraic equations. This system can then be computationally solved. The approximations are typically applied to finite regions in space and/or time, resulting in a numerical solution that provides outcomes at discrete spatial and temporal points (Ferziger et al., 2002). A typical CFD code consist of three principal components: a preprocessor, a solver, and a postprocessor (Versteeg and Malalasekera, 2007).

17.2.1.1 Preprocessor

Preprocessing entails the input of a flow problem into a CFD program through a user-friendly interface and the subsequent transformation of this input into a format suitable for utilization by the solver. User actions during preprocessing encompass the following:

- Defining the geometry of the region of interest, referred to as the computational domain
- Grid generation, which subdivides the domain into nonoverlapping subdomains using a grid or mesh composed of cells or control volumes
- Selecting the physical and chemical phenomena to be modeled
- Defining fluid properties
- Specifying appropriate boundary conditions for cells coinciding with or touching the domain boundary

The solution to a flow problem, involving parameters such as velocity, pressure, and temperature, is defined at nodes within each cell. The accuracy of a CFD solution is primarily determined by the number of cells in the grid. Generally, a larger number of cells leads to improved solution accuracy. Both the accuracy of the solution and the computational resources, such as hardware and calculation time, are contingent on the grid's resolution.

17.2.1.2 Solver

The numerical algorithm in the solver encompasses the following steps:

- Integrating the governing equations of fluid flow across all finite control volumes within the domain
- Discretization, which involves converting the resultant integral equations into a system of algebraic equations
- Solving the algebraic equations through an iterative method

17.2.1.3 Postprocessor

Leading CFD packages now feature versatile data visualization tools, including

- Display of domain geometry and grid
- Generation of vector plots
- Creation of line and shaded contour plots
- Production of 2D and 3D surface plots
- Particle tracking for visualizing fluid flows
- Dynamic result animation

17.2.2 Governing Equations

The CFD solver used in this study was implemented through the utilization of OpenFoam, freely available open-source software developed by the OpenFOAM Foundation, Ltd., London. The software, originally introduced by Weller et al. (1998), possesses the capability to effectively resolve the transport equations associated with incompressible fluid flow, heat transfer, and mass transfer, while also taking into consideration the effects of porous media. Additional insights into the creation of CFD models and the execution of CFD simulations in OpenFOAM can be found in Holzmann (2016). In this publication, the author offers a comprehensive overview of the fundamental mathematical principles underpinning computational fluid dynamics, elucidating the derivation of conservation equations through the finite volume approach. Specifically, the author expounds on the development of mass and momentum equations, as well as various energy equations. Given the prevalence of turbulence in engineering applications, the work further explores Reynolds-averaging methods, delving into the explanation of eddy-viscosity theory and the deduction of equations governing turbulent kinetic energy and dissipation.

The following transport equations are fundamental in describing the physical phenomena relevant to the present investigation:

$$\nabla u = 0 \tag{17.1a}$$

$$\frac{\partial u}{\partial t} + \frac{1}{\phi} u \nabla u = -\phi \nabla p + \nu \nabla^2 u - \phi \frac{\nu}{K} u - \phi \frac{F_e}{\sqrt{K}} \vee u \vee u + \phi g \beta \left(T - T_{\text{ref}} \right)$$
$$+ \ \phi g \beta_c \left(C - C_{\text{ref}} \right) \tag{17.1b}$$

$$\frac{\partial T}{\partial t} + \phi \frac{\left(\rho C_p \right)_f}{\left(\rho C_p \right)_{\text{eff}}} u \nabla T = \frac{k_{\text{eff}}}{\left(\rho C_p \right)_{\text{eff}}} \nabla^2 T \tag{17.1c}$$

$$\phi \frac{\partial C}{\partial t} + \phi u \nabla C = \phi \nabla^2 \left(\frac{D_{\mathrm{m}}}{\tau} C \right) - \phi B_1 C + B_2 q \tag{17.1d}$$

$$\frac{\partial q}{\partial t} = -B_3 q + \phi B_4 C \tag{17.1e}$$

In the set of Equations 17.1a–e, the velocity vector u represents the flow velocity, while p, T, and C denote the pressure, temperature, and concentration of phosphine in the air, respectively. The binary diffusion coefficient (D_{m}) is utilized to account for the diffusion of phosphine in the mixture ($\mathrm{m^2\ s^{-1}}$). The momentum equations incorporate the buoyancy forces generated by both temperature and concentration gradients, employing the Boussinesq approximation. This approximation assumes that the variation of density (ρ) with temperature (T) is linear, represented by the equation $\rho = \rho_{\mathrm{eff}} - \rho_{\mathrm{eff}} \beta (T - T_{\mathrm{ref}})$. The volumetric coefficient of thermal expansion (β) and the species expansion coefficient (β_c) for ideal gases can be determined using Equations 17.2a and b, respectively.

$$\beta = \frac{-1}{\rho} \left(\frac{\partial \rho}{\partial T} \right)_{\mathrm{p}} = \frac{1}{T} \tag{17.2a}$$

$$\beta_c = \frac{-1}{\rho} \left(\frac{\partial \rho}{\partial C} \right)_{\mathrm{p}} = \frac{1}{\rho_{\mathrm{air}}} \left(\frac{MW_{\mathrm{air}}}{MW_{\mathrm{gas}}} - 1 \right) \tag{17.2b}$$

where MW_{air} and MW_{gas} are the molecular weights of air and phosphine gas, respectively.

To account for the influence of porosity on regular molecular diffusion, Shen and Chen (2007) suggest scaling the diffusion coefficient with tortuosity. Consequently, an effective diffusivity coefficient can be defined as follows:

$$D_{\mathrm{eff}} = \frac{D_{\mathrm{m}}}{\tau} \tag{17.3}$$

Neethirajan et al. (2008) calculated $\tau = 2.4$ for wheat.

There is a strong correlation of the degassing rate to air temperature (T_{air}) and relative humidity (RH), and improper calculation of the degassing rates could lead a fumigation to fail. According to Xianchang (1994), the decomposition times of aluminum phosphide tablets ranged from 36 to 204 h. Regarding the CFD model, the generation of phosphine gas was simulated as a time-dependent function, incorporating the degassing rates obtained from the manufacturer of the metal phosphides (Figure 17.1).

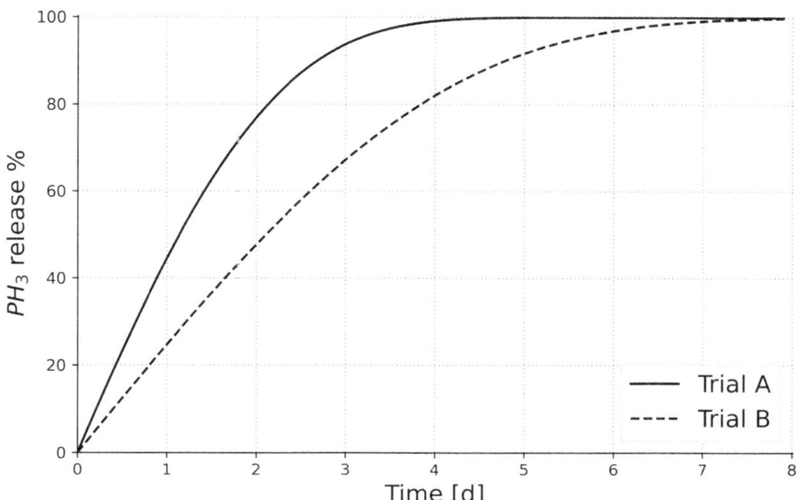

Figure 17.1 Time evolution of magnesium phosphide plates releasing phosphine gas (as a percentage of the total quantity available) for two trials. The gas release in trial A was considerably faster due to the higher ambient temperatures (Brabec et al., 2022).

17.2.3 Porous Media Approach

To consider the impact of grains on gas flow, it was assumed that the grains possess characteristics of a porous medium. The flow within these porous layers is mathematically described by the Darcy-Brinkman formulation. The relationship between the geometric function F_e, the permeability K of the porous medium, and the porosity φ is established based on Ergun's experimental investigations.

$$F_e = \frac{1.75}{\sqrt{150\phi^3}} \tag{17.4a}$$

$$K = \frac{\phi^3 d_p^2}{150\left(1-\phi\right)^2} \tag{17.4b}$$

The effective properties $\left(\rho C_p\right)_{eff}$ and k_{eff} are calculated as a function of the fluid and porous material:

$$\left(\rho C_p\right)_{eff} = \left(1-\varphi\right)\left(\rho C_p\right)_{solid} + \varphi\left(\rho C_p\right)_f \tag{17.5a}$$

$$k_{eff} = \left(1-\varphi\right)k_{solid} + \varphi k_f \tag{17.5b}$$

17.2.4 Sorption Effects

The adsorption of phosphine by grains occurs at varying rates depending on the type of grain. This sorption process has the potential to decrease the concentration of fumigation doses to sublethal levels before disinfestation of the grain is achieved. Therefore, there is a need for a predictive model to estimate the losses of fumigant due to sorption. Darby (2008) proposed a model that establishes a relationship between the fumigant concentration in the spaces between grains, denoted as C, and the average concentration of fumigant within the grain kernels, represented as q. This relationship is described by Equations 17.1d and e, which indicate that phosphine is absorbed into the grain and simultaneously degrades in the surrounding air. The coefficients B_1, B_2, B_3, and B_4 in these equations are independent of the concentrations C and q.

$$B_1 = \frac{S_{sorp}k_f}{B_{fill}} \tag{17.6a}$$

$$B_2 = \frac{S_{sorp}k_f}{B_{fill}F} \tag{17.6b}$$

$$B_3 = \frac{S_{sorp}k_f}{(1-\phi)F} + k_{bind} \tag{17.6c}$$

$$B_4 = \frac{S_{sorp}k_f}{(1-\phi)} \tag{17.6d}$$

$$B_{fill} = \varphi + \frac{1-R_{fill}}{R_{fill}} \tag{17.6e}$$

where S_{sorp} is the specific adsorption surface area, k_f is a linear mass transfer coefficient, F is the partition relation coefficient, and k_{bind} is the coefficient for irreversible reaction/binding of the adsorbed fumigant in the grain kernel. For wheat, the above parameters have the following values: $S_{sorp}k_f$ = 0.0125, F = 0.3, and k_{bind} = 0.0569.

17.2.5 Boundary Conditions

To ensure an accurate assessment of the interaction between the storage (computational domain) and its surroundings, the convective boundary conditions described by Barreto et al. (2013) were employed for the evaluation of phosphine concentration and heat transfer, respectively.

$$-D_m \frac{\partial C}{\partial x}\bigg|_{x=0} = h_m(C - C_{amb}) \tag{17.7a}$$

$$-k\frac{\partial T}{\partial x}\bigg|_{x=0} = h_c\left(T - T_{amb}\right) - \alpha_h G + \epsilon\sigma\left(T^4 - T_{sky}^4\right) \tag{17.7b}$$

Coefficients h_m, h are a function of silo geometry (cylinder, orthogonal), fluid medium (air, water), and fluid velocity (e.g., wind velocity). In Equation 17.7b, the second term on the right-hand side is the heat gain due to solar radiation and the third term is the net radiation heat loss rate for a hot object that is radiating energy to its cooler surroundings (Adelard et al., 1998):

$$T_{sky} = 0.0552 T_{amb} \sqrt{T_{amb}} \tag{17.8a}$$

$$h_c = 10.45 - U_{wind} + 10\sqrt{U_{wind}} \tag{17.8b}$$

Figure 17.2 presents typical time series of ambient temperature, wind velocity, and solar radiation that are used as inputs for the simulations.

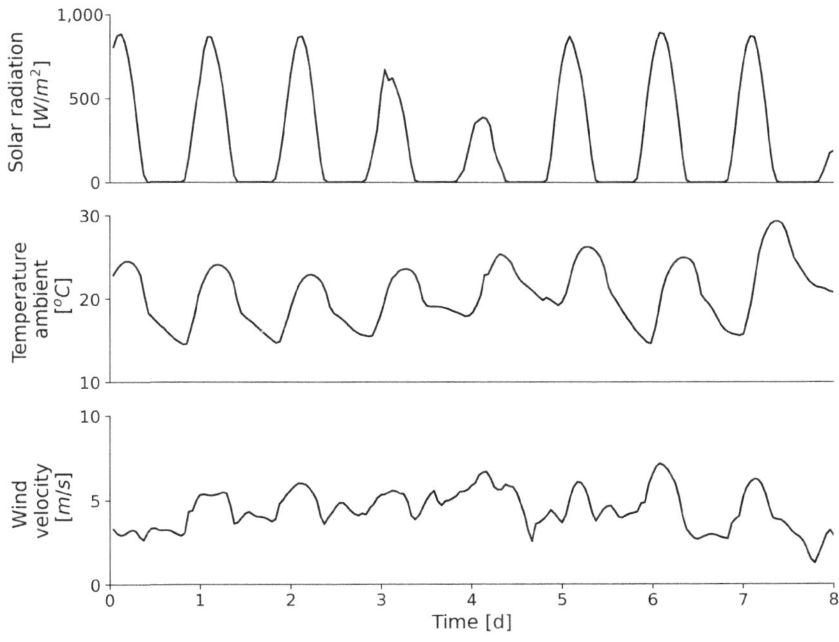

Figure 17.2 Time variation of ambient conditions (solar radiation, temperature, and wind velocity). Similar data are provided to the model as boundary conditions, depending on the weather conditions.

17.2.6 Meshing/Discretization

Meshing, in the context of CFD, refers to the process of dividing the computational domain into smaller subregions called elements or cells. These elements collectively form a grid or mesh, which discretizes the domain and enables the numerical analysis of fluid flow and other physical phenomena. The purpose of meshing is to approximate the continuous fluid domain with a finite number of discrete elements. This allows for the application of numerical methods and algorithms to solve the governing equations of fluid flow within each element. Meshing plays a crucial role in CFD simulations, as it affects the accuracy, stability, and computational efficiency of the solution. The mesh must adequately represent the geometry of the domain, capturing all relevant details and features, such as solid boundaries, obstacles, and complex flow regions. It should also appropriately resolve the flow physics of interest, ensuring that the discretization is fine enough to capture important flow characteristics while avoiding excessive computational costs. Different meshing techniques exist, such as structured, unstructured, and hybrid meshes, each with its own advantages and limitations (Tucker, 2016). Mesh quality, including element shape, size, and distribution, is an essential consideration to minimize errors and improve solution accuracy. In the subsequent examples, structured computational mesh was utilized to ensure enhanced accuracy. All cells within the mesh were hexahedra, which facilitated computational efficiency and reliable numerical analysis. Additionally, grid-clustering techniques were applied near the sides of the domain to effectively capture substantial gradients present in the flow. Examples for a silo, a container, and a railcar are given in Figures 17.3, 17.4, and 17.5, respectively.

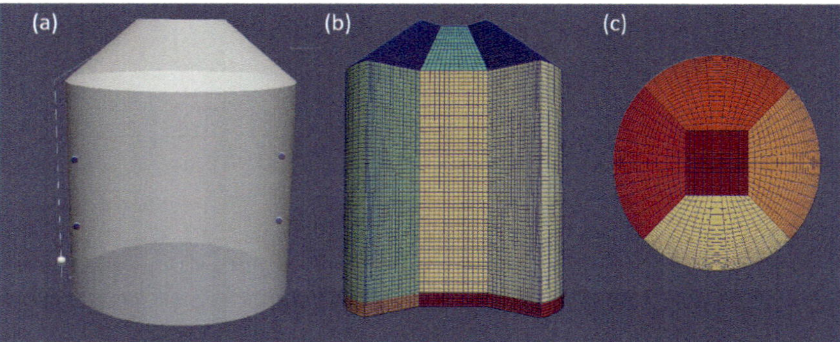

Figure 17.3 (a) The three-dimensional model of the cylindrical silo. Dots show the locations of the four phosphine sensors. (b) The side view of computational zones and mesh used for the silo simulation. (c) The bottom view of computational zones and mesh used for the silo simulation.

(a)

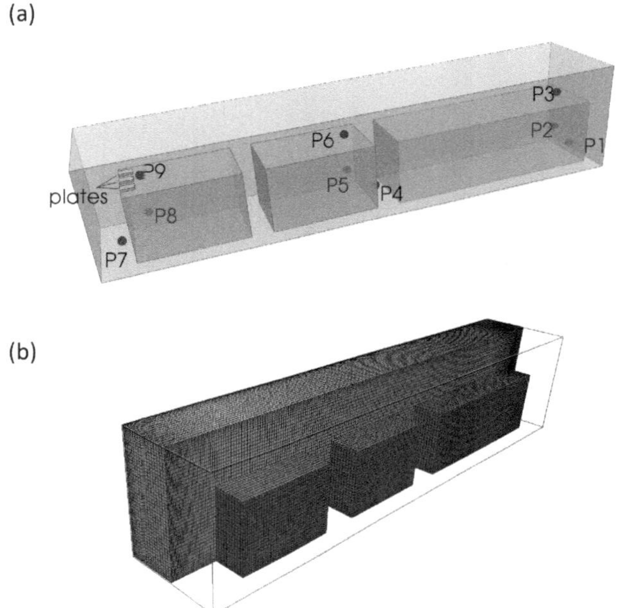

(b)

Figure 17.4 (a) The three-dimensional model of a container with currants. Dots (P1–P9) depict the locations of the wireless phosphine sensors. Furthermore, the position of the magnesium phosphide plates is shown. (b) The mesh used for the simulation of the shipping container. The mesh resolution is increased near the walls of the container. The slice view reveals the pallets with currants (porous material) and their respective mesh.

17.2.7 Insect Mortality

The impact of phosphine on the mortality of grain insects is well established and attributed to two factors: the concentration level of phosphine and the duration of exposure (Winks, 1984, 1985; Collins et al., 2005; Phillips et al., 2012; Athanassiou et al., 2019). To quantify insect mortality, an indicator function denoted as $IM(x, t)$ can be defined, where x represents the spatial coordinates and t represents the time variable.

$$I(x,t) = \frac{1}{K_1} \int_0^t C(x,t)^{K_2} \, dt \qquad (17.9)$$

The empirical constants K_1 and K_2 are species- and strain-specific and vary depending on the insect under consideration. The indicator function $IM(x, t)$ incorporates the duration of exposure to phosphine that an

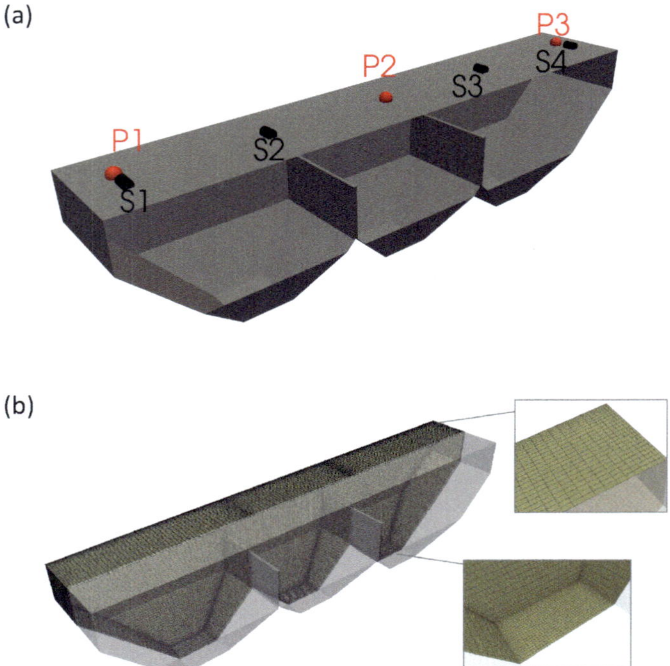

Figure 17.5 (a) The three-dimensional model of the hopper car considered in this work. The railcar had three hoppers divided by two partitions (bulkheads). Black cylinders (S1–S4) depict the locations of the wireless phosphine sensors, whereas the positions of the phosphine tablets are shown by red spheres (P1–P3). (b) The mesh used for the simulation of the railcar. The mesh resolution is increased near the walls. The slice view reveals the two partitions and their respective mesh.

insect has undergone. At a particular location within the grain, when $IM(x, t)$ is less than 1, it indicates the presence of surviving insects. Conversely, when $IM(x, t)$ is greater than 1, it suggests that at least 99.9% of the insect population has been eradicated. In the current simulations, we have utilized $K_1 = 4.04$ and $K_2 = 0.6105$, which correspond to *Rhyzopertha dominica* (F.) (Coleoptera: Bostrychidae). These values have been adopted from Collins et al. (2005) and are incorporated into the model.

17.2.8 Measurement of Phosphine Concentration

In order to validate the CFD model, it is essential to compare its results with experimental measurements. Monitoring of phosphine concentration can be achieved through the utilization of wireless sensors, such as the ones marketed by Centaur Analytics Inc. (California). These sensors are based on electrochemical technology, ensuring high accuracy in phosphine concentration detection. Equipped with wireless connectivity, these sensors have the capability to transmit data at frequent intervals, such as every 2 h, from within stored grain. The transmitted data are received in real time by a cloud platform, where they are downloaded and subjected to further analysis. The accuracy of these sensors has been successfully assessed by Brabec et al. (2019) (also refer to Chapter 16).

17.3 Phosphine Distribution and Insect Mortality

The most commonly used protocol to determine the phosphine resistance is the FAO (Food and Agriculture Organization) method, which is based on 30–50 ppm (depending on the species) and has the individuals exposed for 20 h (FAO, 1975). This method has been used with slight modifications, especially in terms of the concentrations (50–200 ppm) and the exposure interval (mainly after 3 d) at which mortality is recorded, in order to diagnose resistance in field populations by many research groups (Opit et al., 2012; Agrafioti et al., 2018; Afful et al., 2018). Another protocol, developed by the Detia Degesch GmbH company (Laudenbach, Germany) and called the phosphine tolerance test, is based on exposure intervals, which are usually in minutes (for less than 15 min), to concentrations that are as high as 3,000 ppm; it has been applied to 13 different species (Steuerwald et al., 2006; Agrafioti et al., 2019; Athanassiou et al., 2019; Cato et al., 2019). This quick diagnostic can be operated on site at the industrial level, without the need for highly specialized personnel. Another quick diagnostic method, which can be utilized by the industry, has been developed by Nayak et al. (2013, 2019) for different stored-product insects. The tobacco industry has established a protocol for the susceptibility to phosphine of the cigarette or tobacco beetle, *Lasioderma serricorne* (F.) (Coleoptera: Anobiidae), and the tobacco moth, *Ephestia elutella* (Hübner) (Lepidoptera: Pyralidae). This protocol was developed by researchers of the Cooperation Center for Scientific Research Relative to Tobacco (CORESTA) (CORESTA, 2019). For instance, in the case of *L. serricorne*, the CORESTA protocol recommends using 200 ppm for 4 d for the susceptible populations at >20°C and 700 ppm for 10 d at 20°–25°C for the resistant populations. More information about the phosphine resistance and the protocols that are used are presented in the Chapter 2.

Phosphine distribution in different structures has been evaluated by many research groups through different mathematical models (Daglish and Pavic, 2008; Darby, 2008; Darby et al., 2009). For instance, Mills et al. (2001) validated the CFD model against average phosphine concentration measurements and showed a positive pressure system for combatting dilution during phosphine fumigation in bulk grains. Another study evaluated the phosphine distribution in bunker structures in conjunction with the effect of wind (Boac et al., 2014). The CFD model, developed by Isa et al. (2016), was used in a simulation of phosphine distribution in a leaky cylindrical silo. In that study, they found that leaks near the base of the silo serve to inhibit the even distribution of fumigant within the silo. Furthermore, Agrafioti et al. (2020a) studied the distribution of phosphine in cylindrical silos when CFD was used for precision fumigation and for the first time correlated insect mortality with real field fumigation trials. Specifically, the scenario that provided the best results was to use the recirculation system from the beginning of the fumigation for approximately 50 h, until the concentration reached over 300 ppm and all phosphine sensors placed in the silo had gas equilibrium. In this case, as shown in Figure 17.6, small discrepancies were observed between the fourth and fifth day as the sensor recorded some local minimum and maximum values. In that study, the overall performance of the CFD model was considered satisfactory, ensuring the validity of the phosphine concentration predictions

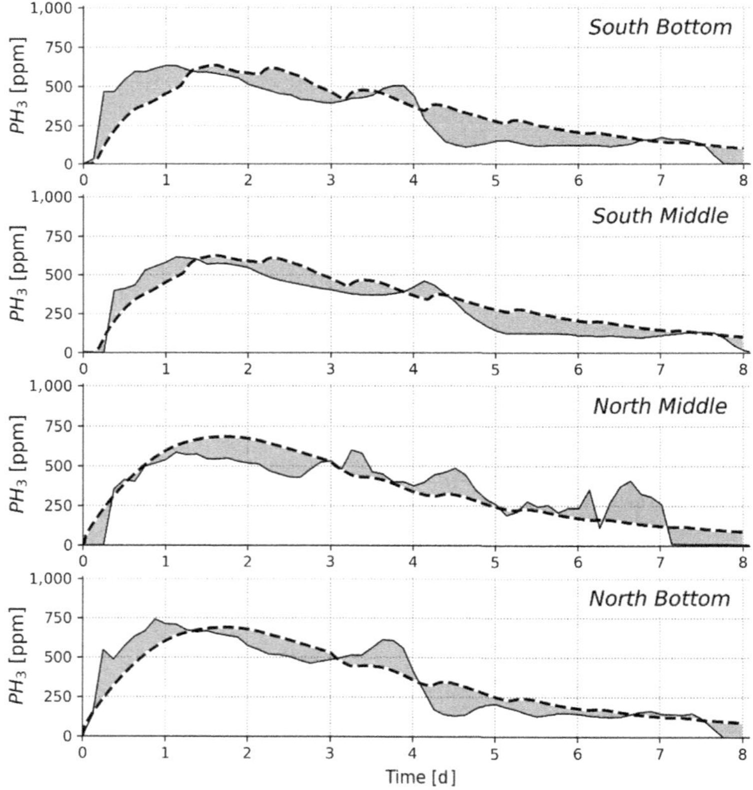

Figure 17.6 Phosphine concentration (ppm) comparison of sensor data (solid lines) with simulation predictions (dashed lines) at four locations inside a silo, from up to down: south bottom, south middle, north middle, and north bottom.

for the entire silo space as presented in Figure 17.7. The field tests confirm the simulation results, as all insects were found completely controlled. Moreover, for both species (*R. dominica* and *O. surinamensis*) complete mortality was obtained, regardless of the resistance levels of the populations tested.

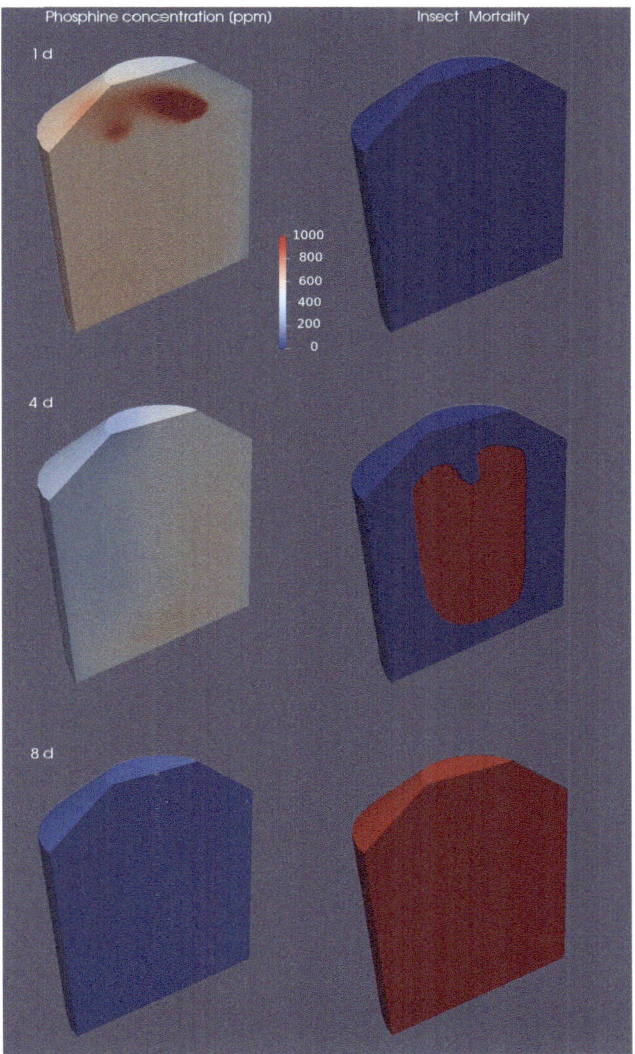

Figure 17.7 Phosphine concentration and insect mortality profiles at three time instances in a silo. On the right, red color indicates zones with 99.9% insect mortality.

Another study focused on the prediction of phosphine distribution in containers using the CFD method, based on the correlation of insect mortality with field fumigation trials (Agrafioti et al., 2021). Specifically, the dimensional model of the container is presented in Figure 17.4. In Figure 17.8, the black lines indicate sensors that were placed inside the currant

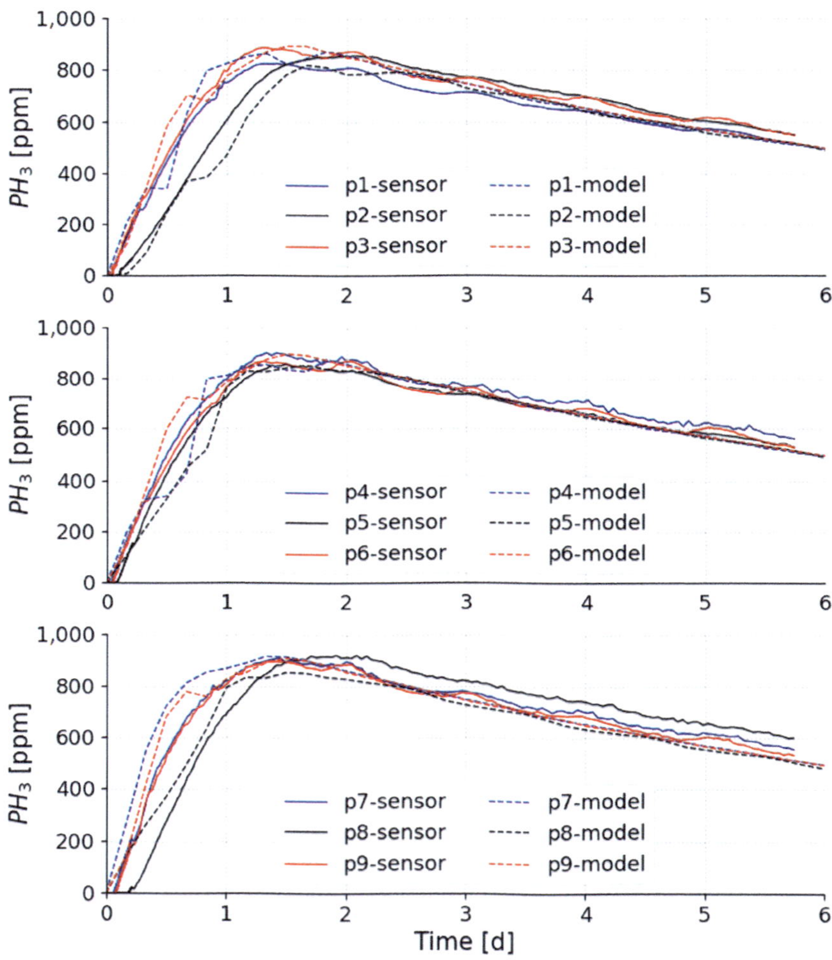

Figure 17.8 Phosphine concentration (ppm) comparison of sensor data (solid lines) with simulation predictions (dashed lines) at nine positions inside the shipping container filled with currant pallets. Locations P1–P9 are shown in Figure 17.4.

boxes (sensors P2, P5, and P8). As expected, there was a time delay for phosphine to reach these places due to the limited gas permeability of the cartons. This phenomenon was visible during the first 24 h of the treatment, and as time progressed, an equilibrium was reached between the concentrations on the inside and the outside of the boxes filled with currants (Figure 17.9).

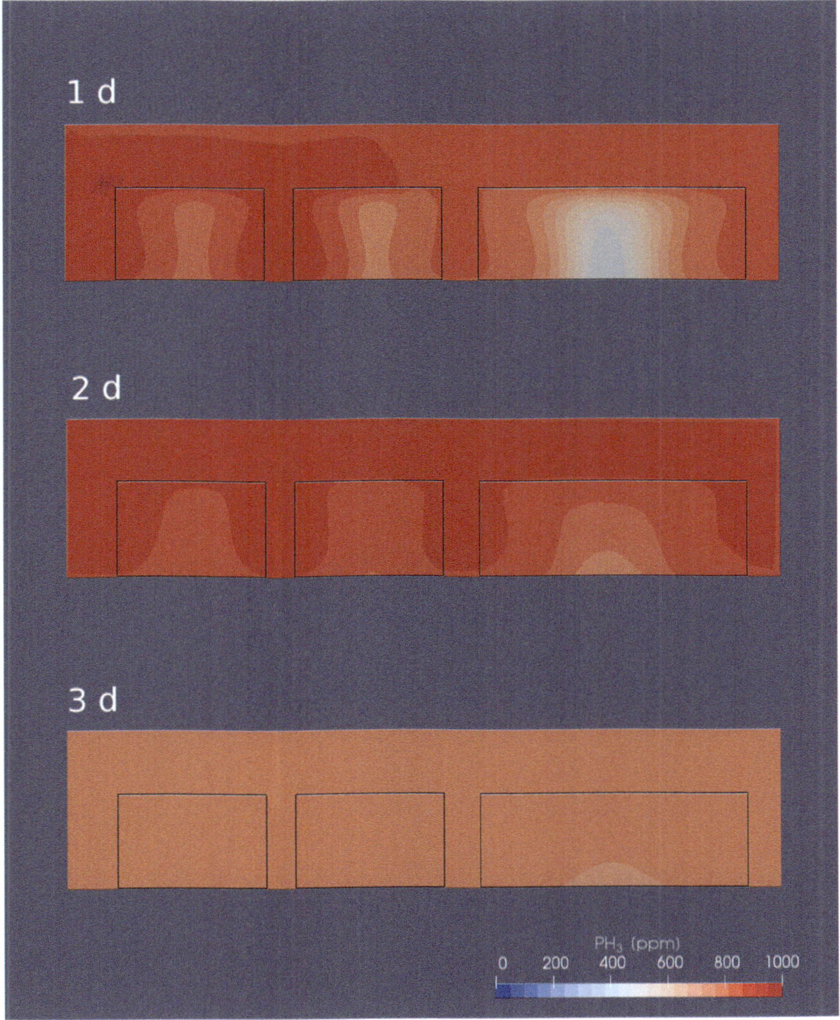

Figure 17.9 Estimated phosphine concentration profiles (two-dimensional, midsection view) at three time instances for a shipping container.

A useful augmentation of the phosphine concentration profiles was the prediction of insect control using the insect mortality model. Figure 17.10 presents two-dimensional (midsection view) insect mortality profiles, at three time instances, for a shipping container, depicting the areas in which *R. dominica* could not survive the fumigation. After 4.5 d, there

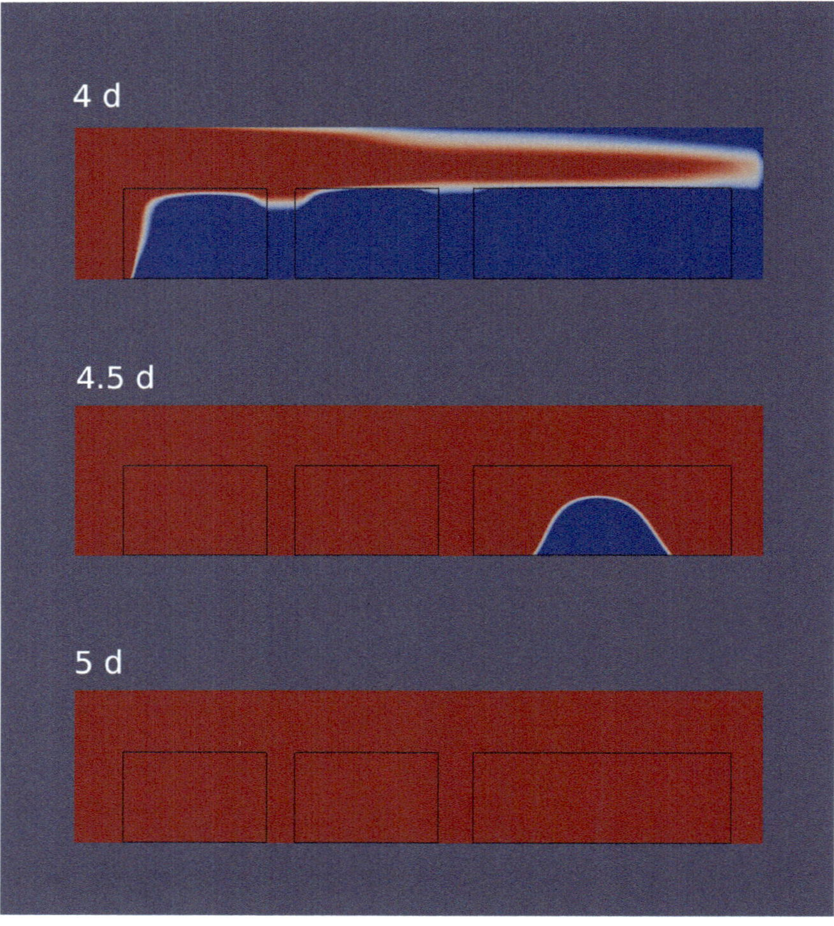

Figure 17.10 Insect mortality profiles (two-dimensional, midsection view) at three time instances for a shipping container. Red color indicates zones with 99.9% insect mortality. Black boxes show the pallets filled with currants.

is an area in the middle of the largest boxes at the back of the container which complete insect control has not yet reached. Finally, at the end of the fifth day, all areas in the container were treated successfully, leaving no alive insects (Figure 17.10). The field tests confirmed the simulation results, as all insects (*R. dominica* and *O. surinamensis*) were found completely controlled.

Another structure that was evaluated using CFD method was railcars (Brabec et al., 2022). In that study, the first railcar filled with corn grits was studied with the aim of validating the accuracy of the model with railcar sensor data. The hopper-bottom railcars held phosphine gas at the top for the entire 8 d (Figure 17.5). The results of that study are presented in Figures 17.11 and 17.12. The top layers of the corn grits were monitored for over 8 d. In Figure 17.11, sensors showed that phosphine concentration had similar results for all positions in the headspace. The four sensors across the top and in the headspace had good gas retention and distribution. The data indicated that the concentration was rapidly increasing during the first 24 h. The CFD model was shown to provide estimates of the phosphine concentration and distribution, which matched well with the data, validating the CFD model as a useful tool.

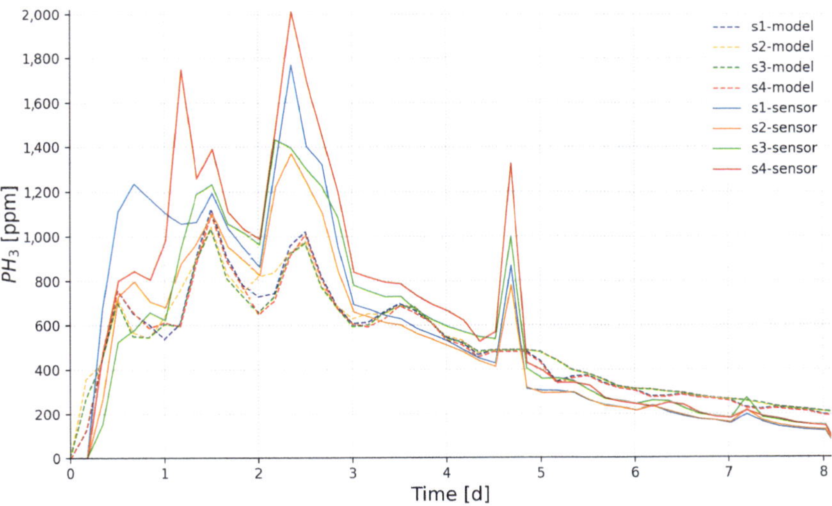

Figure 17.11 Phosphine concentration (ppm) comparison of railcar sensor data (solid lines) and CFD simulation predictions (dashed lines) at four positions inside the hopper car. Sensor locations are shown in Figure 17.5.

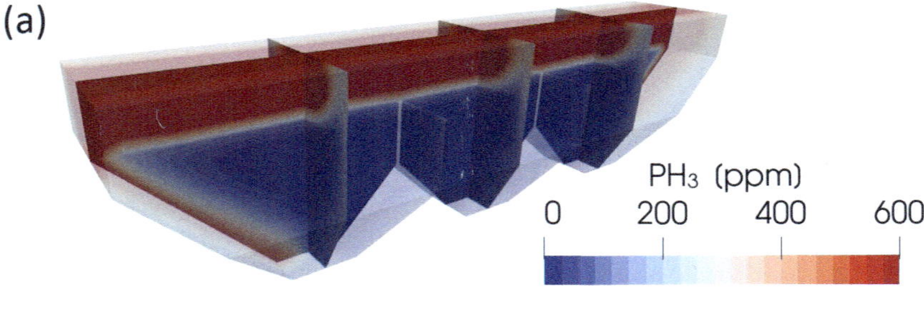

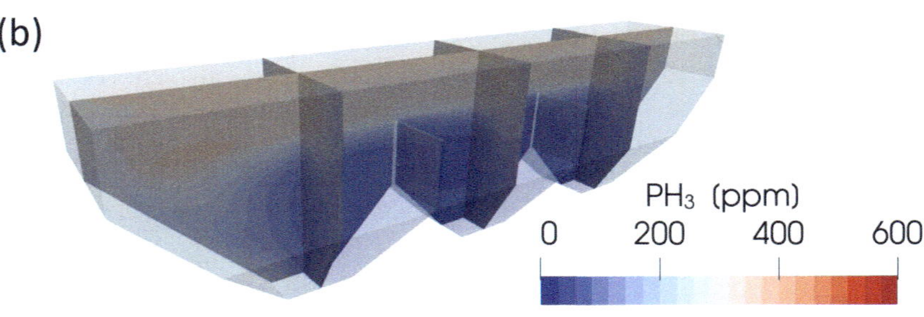

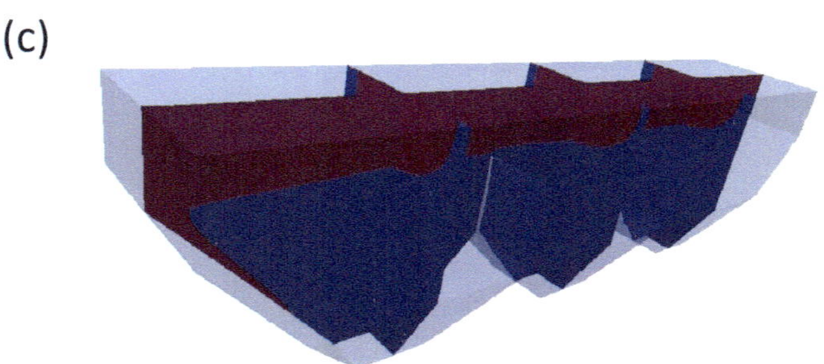

Figure 17.12 Estimated phosphine concentration profiles after (a) 2 d and (b) 6 d. Three-dimensional models of midsections of the railcar. The concentration levels can be interpreted with the color legend video link: https://youtu.be/b0p20S9zbxg. (c) 99% mortality profiles for both adult insects and eggs at the end of the fumigation process. Three-dimensional model of railcar midsections for insect mortality. Red color indicates zones with 99.9% insect mortality for both adults and eggs. Eggs are more difficult to kill. The blue zone represents the potential survival of eggs.

17.4 Conclusions/Future

Phosphine demonstrates exceptional efficacy in the control of stored-product insects across a broad spectrum of storage facilities and commodities (Collins, 2010; Athanassiou et al., 2016; Agrafioti et al., 2020a, b). This chapter presented studies that assessed the spatio-temporal dispersion of phosphine within commercial containers, grain silos, and rail hopper cars through the utilization of modeling techniques. The accuracy of these models has been evaluated by comparing their predictions with sensor data. The results affirm the utility of CFD models in predicting gas distribution and, consequently, insect mortality during fumigation processes.

Among the principal causes of failure of commercial fumigations are the application of phosphine in structures with leaks and poor seals (Agrafioti et al., 2020b) and the insufficient consideration of phosphine absorption by the stored commodities (Reddy et al., 2007). Beyond the forecasted concentrations, it is essential to recognize that the success of a fumigation operation hinges on the effective control of target insect species throughout periods exceeding the duration of the fumigation process, such as by addressing offspring survival. Based on the aforementioned factors and prior research findings, the implementation of best management practices in phosphine usage, particularly in terms of achieving appropriate concentrations and exposure durations, must be a pivotal component of resistance management protocols. In other words, the survival of insects in commercial fumigations may primarily result from inadequate Ct (concentration × time) treatments applied to the fumigated space, rather than from the presence of resistant insect species, as previously observed in various facilities (Agrafioti et al., 2020b).

Fumigant concentrations can also be influenced by several other factors, including environmental conditions, such as temperature, humidity (Reed and Pan, 2000; Darby, 2011), and wind speed (Cryer, 2008), and sorption by the commodity (Daglish and Pavic, 2008). Natural convection currents significantly impact gas distribution in all structures, potentially reducing gas concentrations in leaky structures (Kaloudis et al., 2018). Despite the apparent challenges posed by sorption during fumigation, it appears that this phenomenon prolongs the proximity of gas to the fumigated commodity, facilitating gradual desorption activity (Plumier and Maier, 2018b).

In summary, field fumigation trials underscore the effectiveness of phosphine sensors in conjunction with the associated CFD-based model for measuring phosphine concentrations, positioning computational modeling as a valuable tool in future integrated pest management–based programs during the postharvest phases of durable agricultural commodities (Jayas, 2023). Future research should concentrate on expanding the model's applicability to various storage structures, particularly in vessels or other in-transit fumigations. Furthermore, future research endeavors could enhance the modeling approach, considering advancements such as the utilization of the lattice Boltzmann method to enhance computational efficiency and facilitating rapid computations on parallel architectures, spanning multicore CPUs, GPUs, heterogeneous clusters, and supercomputers.

Acknowledgments

It was carried out as part of the project "PrecisionFEEDProtect: Precision protection of stored feed from entomological infestations using innovative technologies" (Project code: KMP6-0077613) under the framework of the Action "Investment Plans of Innovation" of the Operational Program "Central Macedonia 2021-2027," that is co-funded by the European Regional Development Fund and Greece.

References

Adelard L, Pignolet-Tardan F, Mara T, Lauret P, Garde F, Boyer H (1998) Sky temperature modelization and applications in building simulation. Renew Energ 15: 418–430.

Afful E, Elliot B, Nayak MK, Phillips TW (2018) Phosphine resistance in North American of the lesser grain borer, *Rhyzopertha dominica* (F.) (Coleoptera: Bostrychidae). J Econ Entomol 111: 463–469.

Agrafioti P, Athanassiou CG, Nayak MK (2019) Detection of phosphine resistance in major stored-product insects in Greece and evaluation of field resistance test kit. J Stored Prod Res 82: 40–47.

Agrafioti P, Athanassiou C, Sotiroudas V (2018a) Lessons learned for phosphine distribution and efficacy by using wireless phosphine sensors. Engineering for Stored Product protection and pest Prevention. 12th international Working Conference on Stored Product Protection in Berlin, Germany, October 7–11, 2018.

Agrafioti P, Athanassiou C, Sotiroudas V (2018b) Lessons learned for phosphine distribution and efficacy by using wireless phosphine sensors. In C. S. Adler, G. Opit, B. Fürstenau, C. Müller-Blenkle, P. Kern, F. H. Arthur, C. G. Athanassiou, R. Bartosik, J. Campbell, M. O. Carvalho,W. Chayaprasert, P. Fields, Z. Li, D. Maier, M. Nayak, E. Nukenine, D. Obeng-Ofori, T. Phillips, J. Riudavets, J. Throne, M. Schöller, V. Stejskal, H. Talwana, B. Timlick and P. Trematerra (eds.), Proceedings in: 12th International Working Conference on Stored Product Protection (IWCSPP) in Berlin, Germany, October 7–11, p. 79. https://ojs.openagrar.de/index.php/JKA/article/download/10691/9756

Agrafioti P, Kaloudis E, Bantas S, Sotiroudas V, Athanassiou CG (2020a) Modeling the distribution of phosphine and insect mortality in cylindrical grain silos with computational fluid dynamics: Validation with field trials. Comp Electron Agric 173: 105383.

Agrafioti P, Kaloudis E, Bantas S, Sotiroudas V, Athanassiou CG (2021) Phosphine distribution and insect mortality in commercial metal shipping containers using wireless sensors and CFD modeling. Comp Electron Agric 184: 106087.

Agrafioti P, Sotiroudas V, Kaloudis E, Bantas S, Athanassiou CG (2020b) Real time monitoring of phosphine and insect mortality in different storage facilities. J Stored Prod Res 89: 101726.

Ahmad A, Ahmed M, Ali QM, Abbas M, Arif S (2013) Monitoring of resistance against phosphine in stored grain insect pests in Sindh. Middle - East J Sci Res 16: 1501–1507.

Association of American Railroads (2021) Grain: How freight rail moves wheat, soy & corn. https://www.aar.org/article/freight-rail-grain/

Athanassiou CG, Kavallieratos NG, Brabec DL, Oppert B, Guedes RNC, Campbell JF (2019) From immobilization to recovery: Towards the development of a rapid diagnostic indicator for phosphine resistance. J Stored Prod Res 80: 28–33.

Athanassiou CG, Rumbos CI, Sakka M, Sotiroudas V (2016) Insecticidal efficacy of phosphine fumigation at low pressure against major stored-product insect species in a commercial dried fig facility. Crop Prot 90: 177–185.

Aulicky R, Stejskal V, Frydova B, Athanassiou CG (2015) Susceptibility of two strains of the confused flour beetle (Coleoptera: Tenebrionidae) following phosphine structural mill fumigation: Effects of concentration, temperature, and flour deposits. J Econ Entomol 108: 2823–2830.

Barreto AA, Abalone R, Gastón A, Bartosik R (2013) Analysis of storage conditions of a wheat silo-bag for different weather conditions by computer simulation. Biosyst Eng 116: 497–508.

Bell CH (2000) Fumigation in the 21st century. Crop Prot 19: 563–569.

Benhalima H, Chaudhry MQ, Mills KA, Price NR (2004) Phosphine resistance in stored-product insects collected from various grain storage facilities in Morocco. J Stored Prod Res 40: 241–249.

Boac JM, Casada ME, Lawrence J, Plumier B, Maier DE, Ambrose RPK (2014) Modeling phosphine distribution in grain storage bunker. In 11th International Working Conference on Stored Product Protection, pp. 256–263. http://spiru.cgahr.ksu.edu/proj/iwcspp/pdf2/11/046.pdf

Brabec D, Campbell J, Arthur F, Casada M, Tilley D, Bantas S (2019) Evaluation of wireless phosphine sensors for monitoring fumigation gas in wheat stored in farms bins. Insects 10: 121.

Brabec D, Kaloudis E, Athanassiou CG, Campbell J, Agrafioti P, Scheff DS, Bantas S, Sotiroudas V (2022) Fumigation monitoring and modeling of hopper-bottom railcars loaded with corn grits. J Biosyst Eng 47: 358–369.

Cato AJ, Elliot B, Nayak MK, Phillips TW (2017) Geographic variation in phosphine resistance among North American populations of the red flour beetle (Coleoptera: Tenebrionidae). J Econ Entomol 110: 1359–1365.

Cato A, Afful E, Nayak MK, Phillips TW (2019) Evaluation of knockdown bioassay methods to assess phosphine resistance in the red flour beetle, *Tribolium castaneum* (Herbst) (Coleoptera: Tenebrionidae). Insects 10: 140.

Chaudhry MQ (2000) Phosphine resistance: A growing threat to an ideal fumigant. Pesticide Outlook, 11 (3), pp. 88–91.

Chayaprasert W, Maier DE, Ileleji KE, Murthy JY (2006) Modeling the structural fumigation of flour mills and food processing facilities. In Proceedings of the 9th International Working Conference on Stored-Product Protection, Fumigation and Controlled Atmosphere, p. 6156. http://spiru.cgahr.ksu.edu/proj/iwcspp/pdf2/9/6156.pdf

Collins PJ (2010). Research on stored product protection in Australia: A review of pas, present and future directions. In M. O. Carvalho, P. G. Fields, C. S. Adler, F. H. Arthur, C. G. Athanassiou, J. F. Campbell, F. Fleurat-Lessard, P. W. Flinn, R. J. Hodges, A. A. Isikber, S. Navarro, R.T. Noyes, J. Riudavets, K. K. Sinha, G. R. Throne B. H. Timlick, P. Trematerra and N. D. G. White (eds.), Proceeding of the 10th International Working Conference on Stored product Protection, 27 June- 2July 2010, Estoril, Portugal. Juluis Kuhn-Institute, Berlin, Germany, pp. 3–13.

Collins PJ, Daglish GJ, Pavic H, Kopittke RA (2005) Response of mixed-age cultures of phosphine-resistant and susceptible strains of lesser grain borer, *Rhyzopertha dominica*, to phosphine at a range of concentrations and exposure periods. J Stored Prod Res 41: 373–385.

Cryer SA (2008) Predicted gas loss of sulfuryl fluoride and methyl bromide during structural fumigation. J Stored Prod Res 44: 1–10.

Daglish GJ, Pavic H (2008) Effect of phosphine dose on sorption in wheat. Pest Manag Sci 64: 513–518.

Darby JA (2008) A kinetic model of fumigant sorption by grain using batch experimental data. Pest Manag Sci 64: 519–526.

Darby JA (2011) Ensuring effective phosphine application in grain stores. Final Report CRC50091. Cooperative Research Centre for National Plant Biosecurity, Canberra, Australia.

Darby JA, Wilis T, Damcevski K (2009) Modelling the kinetics of ethyl formate sorption by wheat using batch experiment. Pest Manag Sci 65: 982–990.

Emery NR, Collins JP, Wallbank EB (2003) Monitoring and managing phosphine resistance in Australia. In Proceedings of the Australian Postharvest Technical Conference, Canberra.

Food and Agriculture Organization (FAO) (1975) Tentative method for adults of some major pest species of stored cereals with methyl bromide and phosphine, FAO method no. 16 FAO Plant Prot Bull 23: 12–25.

Ferziger JH, Perić M, Street RL (2019) Computational Methods for Fluid Dynamics. Springer.

Flinn PW, Hagstrum DW, Reed C, Phillips TW (2003) United Stated department of agriculture-agricultural research service stored-grain areawide integrated pest management program. Pest Manag Sci 59: 614–618.

Holloway JC, Falk MG, Emery RN, Collins PJ, Nayak MK (2016) Resistance to phosphine in *Sitophilus oryzae* in Australia: A national analysis of trends and frequencies over time and geographical spread. J Stored Prod Res 69: 129–137.

Holzmann T (2016) Mathematics, Numerics, Derivations and OpenFOAM®. Holzmann CFD, Loeben, Germany.

Isa ZM, Farrell TW, Fulford GR, Kelson NA (2016) Mathematical modelling and numerical simulation of phosphine flow during grain fumigation in leaky cylindrical silos. J Stored Prod Res 67: 28–40.

Jagadeesan R, Collins PJ, Daglish GJ, Ebert PR, Schlipalius DI (2012) Phosphine resistance in the rust red flour beetle, *Tribolium castaneum* (Herbst) (Coleoptera: Tenebrionidae): Inheritance, gene interactions and fitenss costs. PLoS One 7: e31582.

Jayas DS (2023) Mathematical models for managing stored grains globally for enhanced food security. J Appl Zool Res 34(1): 1–10.

Kaloudis E, Bantas S, Athanassiou C, Agrafioti P, Sotiroudas V (2018) Modeling distribution of phosphine in cylindrical grain silos with CFD methods for precision fumigation. In C. S. Adler, G. Opit, B. Fürstenau, C. Müller-Blenkle, P. Kern, F. H. Arthur, C. G. Athanassiou, R. Bartosik, J. Campbell, M. O. Carvalho, W. Chayaprasert, P. Fields, Z. Li, D. Maier, M. Nayak, E. Nukenine, D. Obeng-Ofori, T. Phillips, J. Riudavets, J. Throne, M. Schöller, V. Stejskal, H. Talwana, B. Timlick and P. Trematerra (eds.), proceedings in: 12th International Working Conference on Stored Product Protection (IWCSPP) in Berlin, Germany, October 7–11, p. 243. https://ojs.openagrar.de/index.php/JKA/article/download/10771/9840

Kaur R, Nayak MK (2015) Developing effective fumigation protocols to manage strongly phosphine-resistant *Cryptolestes ferrugineus* (Stephens) (Coleoptera: Laemophloeidae). Pest Manag Sci 71: 1297–1302.

Kaur R, Subbarayalu M, Jagadeesan R, Daglish GJ, Nayak MK, Naik HR, Ramasamy S, Subramanian C, Ebert PR, Schlipalius DI (2015) Phosphine resistance in India is characterized by a dihydrolipoamide dehydrogenase variant that is otherwise unobserved in eukaryotes. Heredity 115: 188–194.

Mills KA, Wontner-Smith TJ, Bartlett DI, Harral BB (2001) A new positive pressure system for combating dilution during phosphine fumigations of bulk grain. In 2000: International Conference Controlled Atmosphere and Fumigation in Stored Products, pp. 405–420.

Nayak MK, Collins PJ, Holloway JK, Emery RN, Pavic H, Bartlet J (2013) Strong resistance to phosphine in the rusty grain beetle, *Cryptolestes ferrugineus* (Stephens) (Coleoptera: Laemophloeidae): Its characterization, a rapid assay for diagnosis and its distribution in Australia. Pest Manag Sci 69: 48–53.

Nayak MK, Daglish GJ, Phillips TW (2015) Managing resistance to chemical treatments in stored products pests. Stewart Postharvest Rev 11(1): 2015

Nayak MK, Kaur R, Jagadeesan R, Pavic H, Phillips TW, Daglish GJ (2019) Development of a quick knockdown test for diagnosing resistance to phosphine in Sitophilus oryzae (Coleoptera: Curculionidae), a major pest of stored products. J Econ Entomol 112: 1975–1982.

Neethirajan S, Jayas DS, White NDG, Zhang H (2008) Investigation of 3D geometry of bulk wheat and pea pores using X-ray computed tomography images. Comp Electron Agric 63: 104–111.

Opit GP, Phillips TW, Aikins MJ, Hasan MM (2012) Phosphine resistance in *Tribolium castaneum* and *Rhyzopertha dominica* from stored wheat in Oklahoma. J Econ Entomol 105: 1107–1114.

Phillips TW, Thoms EM, Demark J, Walse DS (2012) Fumigation. Ch. 14 In D. W., Hahstrum, T. W. Phillips and G. Cuperus (eds.), Stored Product Protection. Kansas State University, Manhattan, Kansas.

Pimentel MAG, Guedes RNC (2010) Spread of phosphine among Brazilian populations of three species of stored products insects. Neotrop Entomol 39: 101–107.

Plumier B, Maier D (2018a) Use of a 3D finite element model to predict post fumigation phosphine desorption. In C. S. Adler, G. Opit, B. Fürstenau, C. Müller-Blenkle, P. Kern, F. H. Arthur, C. G. Athanassiou, R. Bartosik, J. Campbell, M. O. Carvalho, W. Chayaprasert, P. Fields, Z. Li, D. Maier, M. Nayak, E. Nukenine, D. Obeng-Ofori, T. Phillips, J. Riudavets, J. Throne, M. Schöller, V. Stejskal, H. Talwana, B. Timlick and P. Trematerra (eds.), Proceedings in: 12th International Working Conference on Stored Product Protection (IWCSPP) in Berlin, Germany, October 7–11, p. 80.

Plumier BM, Maier DE (2018b) Sensitivity analysis of a fumigant movement and loss model for bulk stored grain to predict effects of environmental conditions and operational variables on fumigation efficacy. J Stored Prod Res 78: 18–26.

Plumier BM, Schramm M, Maier DE (2018) Developing and verifying a fumigant loss model for bulk stored grain to predict phosphine concentrations by taking into account fumigant leakage and sorption. J Stored Prod Res 77: 197–204.

Prater ME, Sparger A (2013) Grain and oilseed shipment sizes and distance hauled by rail. U.S. Department of Agriculture, Agricultural Marketing Service. Agricultural Marketing Service, Washington, DC. https://www.ams.usda.gov/sites/default/files/media/Grain%20 and%20Oilseed%20Shipment%20Sizes%20and%20Distance%20 Hauled%20by%20Rail.pdf

Reddy PV, Rajashekar Y, Begum K, Leelaja BC, Rajendran S (2007) The relation between phosphine sorption and terminal gas concentrations in successful fumigation of food commodities. Pest Manag Sci 63: 96–103.

Reed C, Pan H (2000) Loss of phosphine from unsealed bins of wheat at six combinations of grain temperature and grain moisture content. J Stored Prod Res 36: 263–279.

Saglam O, Edde PA, Phillips TW (2015) Resistance of *Lasioderma serricorne* (Coleoptera: Anobidae) to fumigation with phosphine. J Econ Entomol 108: 2489–2495.

Sakka M, Riga M, Vontas J, Gotze C, Allegra J, Jakob G, Athanassiou C (2018) Evaluation of tolerance/resistance to phosphine of stored product beetle populations from Europe, by using different diagnostic methods. In C. S. Adler, G. Opit, B. Fürstenau, C. Müller-Blenkle, P. Kern, F. H. Arthur, C. G. Athanassiou, R. Bartosik, J. Campbell, M. O. Carvalho, W. Chayaprasert, P. Fields, Z. Li, D. Maier, M. Nayak, E. Nukenine, D. Obeng-Ofori, T. Phillips, J. Riudavets, J. Throne, M. Schöller, V. Stejskal, H. Talwana, B. Timlick and P. Trematerra (eds.), Proceedings in: 12th International Working Conference on Stored Product Protection (IWCSPP) in Berlin, Germany, October 7–11, p. 155.

Shen L, Chen Z (2007) Critical review of the impact of tortuosity on diffusion. Chem Eng Sci 62: 3748–3755.

Song X, Wang P, Zhang H (2011) Phosphine resistance in *Rhyzopertha dominica* (Fabricius) (Coleoptera: Bostrichidae) from different geographical populations in China. African J Biotech 10: 16367–16373.

Steuerwald R, Dierks-Lange H, Schmitt S (2006) Rapid bioassay for determining the phosphine tolerance Proceedings of the 9th International Working Conference on Stored Product Protection. Vol 4: 306–311.

Tucker PG (2016) Advanced Computational Fluid and Aerodynamics (Vol. 54). Cambridge University Press.

UNEP (1995) Montreal Protocol on Substances that Deplete the Ozone Layer. Report of the Methyl Bromide Technical Options Committee. Assessment. UNEP, Nairobi, Kenya, p. 304.

Versteeg HK, Malalasekera W (2007) An Introduction to Computational Fluid Dynamics: the Finite Volume Method. Pearson Education.

Warrick C (2011) Fumigating with Phosphine, Other Fumigants and Controlled Atmosphere. Technical report. Grains Research and Development Cooperation, pp. 1–16.

Weller HG, Tabor G, Jasak H, Fureby C (1998) A tensorial approach to computational continuum mechanics using object-oriented techniques. Comput Phys 12: 620–631.

Winks RG (1984) The toxicity of phosphine to adults of *Tribolium castaneum* (Herbst): Time as a dosage factor. J Stored Prod Res 20: 45–56.

Winks RG (1985) The toxocity of phosphine to adults of *Tribolium castaneum* (Herbst): Phosphine-induced narcosis. J Stored Prod Res 21: 25–29.

Xianchang, T. (1994). Evolution of phosphine from aluminium phosphide formulations at various temperatures and humidities. In *Proceeding of the 6th international working conference on stored product protection*. 17-23 April 1994, Canberra, Australia.

Chapter 18

Safety During Fumigation

Edmond L. Bonjour and Carol L. Jones

18.1 Introduction

Integrated pest management (IPM) is an important strategy for maintaining the quality of stored products. Fumigation is a key component of stored-product IPM. Proper care and procedures are needed when using potentially dangerous insecticides and working in dangerous environments to keep workers in the stored-product industry safe (Bonjour and Jones, 2021). This chapter will cover some of the more important safety aspects when using solid aluminum and magnesium phosphide formulations, the cylinderized phosphine products ECO2FUME® Fumigant Gas and VAPORPH3OS® Phosphine Fumigant, and the sulfuryl fluoride product ProFume® Gas Fumigant in the stored grain industry in the United States. (Rules and regulations covering other countries are too varied to discuss in this chapter.) Several product labels were consulted for this safety information.

18.2 Restricted-Use Insecticides

Fumigants, both phosphine products and sulfuryl fluoride, for stored grain protection are restricted-use pesticides due to the high acute inhalation toxicity of the gas. These insecticides are available for retail sale only to

DOI: 10.1201/9781003309888-18

licensed dealers and certified applicators. Application is permissible only by certified applicators or persons under their direct supervision and only for uses covered by the certified applicator's certification and permitted in the applicator's manual.

18.3 Safety Data Sheet and Product Labeling

All Safety Data Sheets (SDS) and product labeling must be on site and easily accessible by all employees and visitors at the facility. Product labeling includes what is found on the product container along with the applicator's manual. The SDS provides detailed safety and health information not found on container labels. The label is a legal document. Using the product in any manner inconsistent with the label and labeling is illegal.

Before fumigation begins, carefully read and review both the label and the applicator's manual. Also, become familiar with and comply with all applicable local, state, and federal laws.

18.4 Lockout Tagout Procedures

The OSHA lockout tagout (LOTO) standard for the control of hazardous energy, Title 29 Code of Federal Regulations (CFR) Part 1910.147, was developed in 1982 by the U.S. Occupational Safety and Health Administration (OSHA) to help protect workers who service and maintain equipment from unexpected startup that could cause injury. It went into effect in 1989 (Occupational Safety & Health Administration [OSHA], 1989).

Prior to performing a fumigation, steps need to be taken to ensure that no one will potentially be harmed by energized equipment, especially if the fumigator and team must enter a structure. Therefore, all equipment should be locked out and de-energized so that it cannot be unexpectedly started while someone is inside a structure. Use a designated lock, hasp, or circuit breaker lockout for this purpose. The lockout device should be strong enough to prevent removal with the use of excessive force (Figure 18.1). Tagout is a warning system that is used when there is no way to secure the startup equipment; the tag has words such as "Danger, Do Not Operate." The tag must be constructed and printed so that exposure to weather conditions or damp locations will not cause it to deteriorate or the words to become illegible. Locked-out equipment should also have a tag indicating who performed the LOTO, date and time, and any additional details. An annual review of LOTO procedures is highly recommended.

Figure 18.1   Lockout tagout.

18.5 First Aid Measures

Knowing the appropriate first aid is important when exposure to a fumigant occurs. In all cases, contact a doctor or poison control center for treatment advice, and have the product container, label, and applicator's manual available when calling or going for treatment.

If a fumigant gas is inhaled, move the person to fresh air. If the person is not breathing, call an emergency responder or an ambulance, then give the person artificial respiration.

If the fumigant product is swallowed, have the person sip a glass of water if the person can swallow. Do not induce vomiting unless told to do so by a doctor or poison control center. In all cases, do not give anything by mouth to a person who is unconscious.

If the phosphine product gets on clothing, take off the contaminated clothing immediately and allow the clothes to aerate in a ventilated area prior to laundering.

If the product gets on skin, rinse the skin immediately with plenty of water for 15–20 min.

If the product enters eyes, hold the eye open and rinse slowly and gently with cool water for 15–20 min. Remove contact lenses, if present, after the first 5 min, then continue rinsing the eye.

18.6 Grain Condition and Risk

There is a direct correlation between grain condition and safety hazards for workers in and around storage areas. Even tiny amounts of moldy grain, old grain, insect-infested grain, and trash can cause clumping to occur in the grain. Grain can stick to the walls of the structure. Bridges can form in the top layer of grain. During unloading, bridges and lumps will not go through the reclaim system. When employees enter the structure, entrapment and engulfment can occur when the bridge breaks through or the wall of grain on the bin wall breaks loose. OSHA has strict laws about entering bins (OSHA, 1996). Full-body harnesses and lifelines are required anytime the grain is over the entrant's head, is more than waist deep, or provides the potential for the entrant to become entrapped or engulfed. For fumigators, this risk exists when they enter the structure to check for structure integrity from the inside, seal the structure, or treat it with fumigant. These same precautions apply to flat storage structures. Grain bins should be considered permit-required confined spaces and should be identified for workers and others who might enter bins. Other structures, such as boot pits, should be inspected and appropriately labeled if they meet the OSHA requirements for permit-required confined spaces (OSHA, 2011b).

Air quality can be compromised by out-of-condition grain. Carbon dioxide (CO_2) and methane (CH_4) gases are products of grain spoilage. This is motivation for monitoring air quality before entering the structure. Even when oxygen (O_2) appears to be sufficient, other harmful gases may be present. Running aeration fans prior to entry can help remove these gases. However, testing the air using a monitor is the best method of ensuring a safe environment. Continue to monitor air quality while working in a grain bin because pockets of high CO_2 may exist. Solid fumigant products may not have activated in dry environments and may begin to produce toxic gas when disturbed.

Bin entry permits are required when entering a bin, silo, or tank unless the employer or employer's representative is present during the entire operation (OSHA, 2011a). If the plane between the inside of a bin and the outside environment is crossed, a permit is required. Bin entry permits are a safety checklist for the grain structure. Air monitoring, available safety equipment, grain levels and conditions, and entrant and attendant names with signatures are all part of the bin entry permit. Grain moves and conditions can change between entries, making it essential that bin entry permit steps be performed and documented.

While fumigation is never performed alone, bin entry should never be performed alone either. An attendant trained in rescue and emergency procedures must remain at the entry point and in contact with the person entering and working in the bin. That attendant never leaves this position until the entry is complete and the entrant is outside of the bin again.

18.7 Spoiled Grain Detection

Methods to detect grain spoilage and mold include temperature cables, CO_2 sensors, visual inspection, and detection of odors from the bin access points. Temperature will rise around areas in the grain that have biological activity such as insects or mold present. Each node on the temperature cable will sense about a meter around that node. Carbon dioxide sensors can detect biological activity when the fan starts. Any rise over the normal CO_2 range (about 400–600 ppm) of the bin indicates that someplace in the bin there are insects or mold is forming, and an increase in CO_2 levels can cause an issue for fumigators. During visual inspection, look for a whitish dull appearance in the grain. Also look for clumping and grain stuck to walls or in mounds. Out-of-condition grain may also emit an odor. Take care when looking into a bin in case the odors are dangerous. Smell only from outside the bin at the bin entrance. It is important to train employees on these methods of spoilage detection. They may be the first ones to detect a change in the grain during their regular work activities. The earlier a problem is detected, the easier it is to remedy.

18.8 Sealing

It is critical to thoroughly seal the structure to be fumigated so as to contain the fumigant gas at the appropriate level for the designated time (Bonjour et al., 2014). Several products may be used to seal a structure: closed-cell foam sealant, duct tape or pressure-sensitive paper tape, silicone sealant, heavy-weight (at least 6 mil) plastic, and elastomeric sealer (Figure 18.2). Permanent areas that can be sealed include eaves and bases of structures, aeration fan motors and ducts, seams and missing hardware, and areas around unloading equipment. Areas that will be sealed temporarily include bin door entries, roof ventilation, manhole covers, unloading spouts, and aeration fan intakes.

18.9 Closed Loop Fumigation

Controlling where fumigant travels in the bin and evacuation of the gas is a challenge and concern for fumigators. Closed loop systems take gas out of the headspace of the bin and return it to the bottom of the bin (Jones et al.,

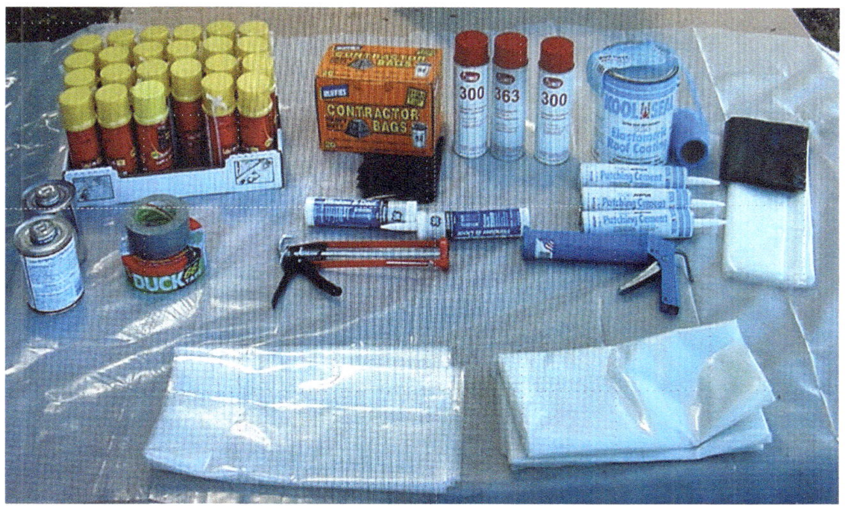

Figure 18.2 Sealing materials.

2011) (Figure 18.3). For systems to be successful, extensive sealing is criti-cal. A small fan removes the gas from the headspace and places the gas either directly into the bottom of the bin below the perforated flooring or into the aeration ductwork at the bottom of the bin. It is especially important to seal the aeration fan and all vents to prevent the closed loop fan from forcing gas out of these areas. Closed loop systems have been shown to reduce the amount of fumigant required to maintain effective gas concentrations by as much as 75% while providing better distribution of gas than conventional methods of fumigation (Jones et al., 2008).

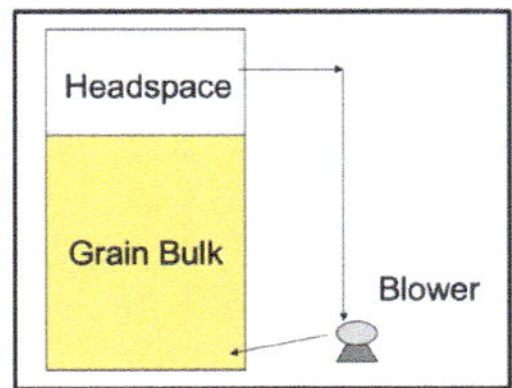

Figure 18.3 Closed loop system.

Figure 18.4 Phosphine pellets and tablets.

18.10 Protective Clothing

Protective clothing is necessary when using solid phosphine formulations (Figure 18.4). Wear dry gloves of cotton, preferably, or other material if contact with pellets, tablets, or dust is likely to occur (Figure 18.5). Gloves should remain dry during use. Do not use gloves to wipe sweat off your brow or other body parts. Always wash your hands thoroughly after handling phosphine products. Used gloves and other clothing that may be contaminated should be aerated in a well-ventilated area prior to laundering.

Figure 18.5 Wearing cotton gloves to apply pellets.

For the cylinderized phosphine products ECO2FUME® Fumigant Gas (Figure 18.6) and VAPORPH3OS® Phosphine Fumigant (Figure 18.7), wear leather work gloves or leather-faced cotton gloves when connecting cylinders to or disconnecting cylinders from the dispensing equipment to prevent the potential for frostbite. Always wear safety glasses when working with pressurized equipment to prevent freezing or cryogenic burns to the eyes by rapidly evaporating liquid.

When using sulfuryl fluoride, do not wear gloves or rubber boots, which could trap the liquid fumigant against your skin. Clothes or shoes that have become contaminated with liquid should not be reused until thoroughly ventilated. During the application, wear a loose-fitting or well-ventilated long-sleeve shirt, long pants, shoes, and socks. Also, wear splash-resistant goggles or a full-face shield during the introduction of the fumigant or when working around any lines containing the product under pressure (Figure 18.8).

When handling compressed gas cylinders, it is recommended to wear steel-toed safety shoes because of the weight of the cylinders.

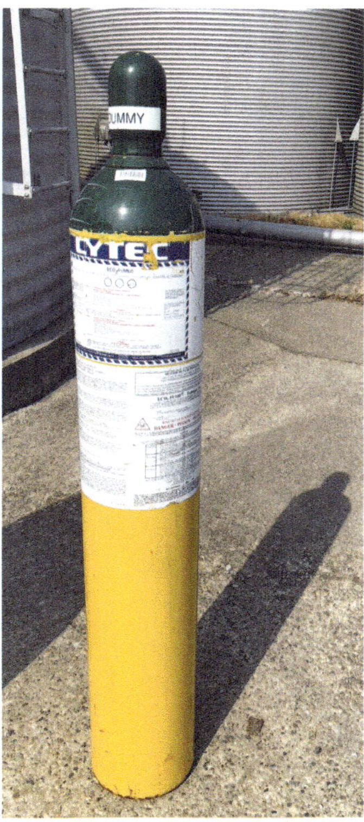

Figure 18.6 ECO2FUME® Fumigant Gas.

Figure 18.7 VAPORPH3OS® Phosphine Fumigant (red-topped cylinder).

Figure 18.8 Wearing a full-face shield to release ProFume® Gas Fumigant.

18.11 Respiratory Protection

Respiratory protection is required when concentration levels of phosphine are unknown or when concentrations exceed permissible exposure limits of 0.3 ppm. A full-face gas mask/phosphine canister combination approved by the National Institute for Occupational Safety and Health/Mine Safety and Health Administration (NIOSH/MSHA) may be used at phosphine levels from above 0.3 ppm up to 15 ppm (Figure 18.9). If the phosphine level is above 15 ppm or is unknown, a NIOSH/MSHA-approved self-contained breathing apparatus (SCBA) must be worn (Figure 18.10). If phosphine is to be applied from within the structure to be fumigated, an approved full-face gas mask – phosphine canister combination or SCBA must be available at the site of application in case it is needed. Respiratory protection must also be available for applications from outside the area to be fumigated, such as the addition of tablets or pellets to automatic dispensing devices and outdoor applications.

For sulfuryl fluoride, if the concentration in the fumigated area does not exceed 1 ppm, no respiratory protection is required. When the concentration is above 1 ppm or when the concentration is unknown, wear a NIOSH-approved positive-pressure SCBA, approval number prefix TC-13F, or a combination air-supplied/SCBA respirator. The SCBA must be at the fumigation location and operations before the fumigation begins.

Figure 18.9 Full-face gas mask with phosphine canister.

Figure 18.10   Self-contained breathing apparatus.

18.12 Certified Applicator

A certified applicator is a member of the fumigation crew who has successfully completed proper training and has been approved by the state to purchase and apply fumigants. Whenever a fumigation is to be performed, a certified applicator must be physically present and be responsible for and maintain visual and/or voice contact with all fumigation workers during the opening of the product containers, during the application of the fumigant, and during the initial opening of the fumigation structure for ventilation. During the application of the fumigant, at least two persons, a certified applicator and a trained person or two trained persons under the direct supervision of the certified applicator, must be present. Monitoring must be conducted to characterize the application and determine the fumigator's exposure. Document the fumigant levels and retain documents for two years. Before reentry into a structure, the facility must be ventilated until the phosphine level is 0.3 ppm or below or the sulfuryl fluoride level is 0.1 ppm or below.

After the fumigation is complete and the structure is ready to be ventilated, perform the ventilation at the highest point on the structure. Utilizing this technique keeps the highest concentrations of fumigant directed away from workers and bystanders, and the concentration is quickly diluted in the atmosphere. Running fans or an aeration system is an excellent method to attain quick ventilation and is essential where cross ventilation is poor.

Persons with documented training in the handling of phosphine products must be responsible for receiving, ventilating, and removing placards from vehicles that have been fumigated in transit by rail. Transporting containers or vehicles under fumigation over public roads is prohibited. The trained person must be trained by a certified applicator following the Environmental Protection Agency's (EPA) accepted product applicator's manual, which must precede or be attached to the outside of the transport vehicle. When training has been completed and the employee demonstrates safety knowledge proficiency, the training date must be recorded and maintained in the employee's safety training record for a minimum of three years. Refresher training must be done annually and documented.

In-transit fumigation with sulfuryl fluoride of vehicles such as containers, trucks, railcars, surface ships, and other transport vehicles or vessels is prohibited. Vehicles and vessels may be fumigated with sulfuryl fluoride if they are stationary for the entire fumigation process, including ventilation.

18.13 Gas Detection Equipment

Structures storing grains may develop hazardous atmospheres because of gases given off by decaying grain or from a previous fumigation that was not ventilated properly after the fumigation was complete. Before entering any storage structure, it is important to determine that there is sufficient O_2 present, and that the phosphine level is 0.3 ppm or lower and the sulfuryl fluoride level is 1 ppm or lower. If a fumigant is detected or there is a lack of O_2, the facility must be ventilated to ensure that the toxic gases are reduced to nonhazardous levels and that adequate O_2 levels are maintained (Bonjour, 2022).

18.13.1 Oxygen Monitoring

Normal O_2 levels in the atmosphere range between 19% and 23%; when levels are outside that range, workers can experience problems. Electronic O_2 sensors help prevent these dangers by accurately monitoring O_2 levels and triggering an alarm when the level drops below 19.5%, the current OSHA-mandated level (OSHA, 1996). Using an O_2 monitor helps determine that the working environment is safe to enter (Figure 18.11).

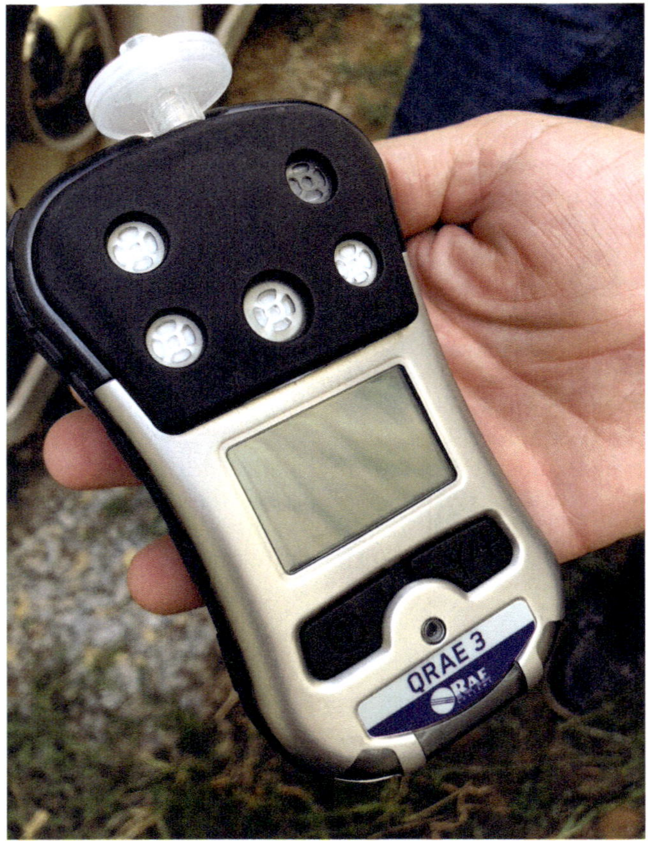

Figure 18.11 QRAE 3 gas detector with oxygen sensor.

18.13.2 *Phosphine Monitoring*

For phosphine gas detection, there are glass detection tubes or electronic gas monitors to determine the amount of phosphine gas present in the structure. Ideally, sampling hoses should be installed prior to the introduction of fumigant. Glass detection tubes (Figure 18.12) with a sampling pump are portable, are simple to use and require little training, only detect one gas, are relatively inexpensive, and are accurate. Glass tubes have an expiration date and must not be used if expired. Expired tubes may be retained, labeled "for training purposes only," and used to train new employees, but they cannot be used to determine the amount of phosphine present during

Figure 18.12 Monitoring phosphine with a glass detection tube.

a fumigation. They should be stored in an area away from in-date tubes. Electronic gas monitors (Figure 18.13) are portable, may be complex to use and require extensive training, may be able to read multiple gases, are relatively rapid, are expensive, have a digital readout, must be calibrated according to the manufacturer's instructions, and must be bump tested before each use or, at a minimum, every 30 d.

Electronic gas detection monitors for phosphine must be calibrated against a known standard. Calibration can be done in house with the required gas and testing equipment (regulators and adaptors) or the monitor can be sent back to the distributor or manufacturer for calibration. The calibration gas must be introduced at the proper pressure and flow rate. Overpressurization can damage the sensor. Monitors should be calibrated

Figure 18.13 Monitoring phosphine with an electronic gas monitor.

at the altitude at which they will be used, as changes in atmospheric pressure will influence the instrument's response. For some instruments with an active sampling pump, the pump must be disconnected from the sensor and the gas flow rate set to match the sampling rate of the pump.

A bump test must be performed before every use of an electronic gas detection monitor for phosphine. A bump test kit (Figure 18.14) verifies calibration by exposing the instrument to a known concentration of test gas. Instruments should be zeroed before the bump test is performed to give a more accurate picture of the bump test results. When a bump test is performed, the test gas concentration should be high enough to trigger the instrument's alarm. If the bump test does not trigger the alarm, a full calibration must be performed.

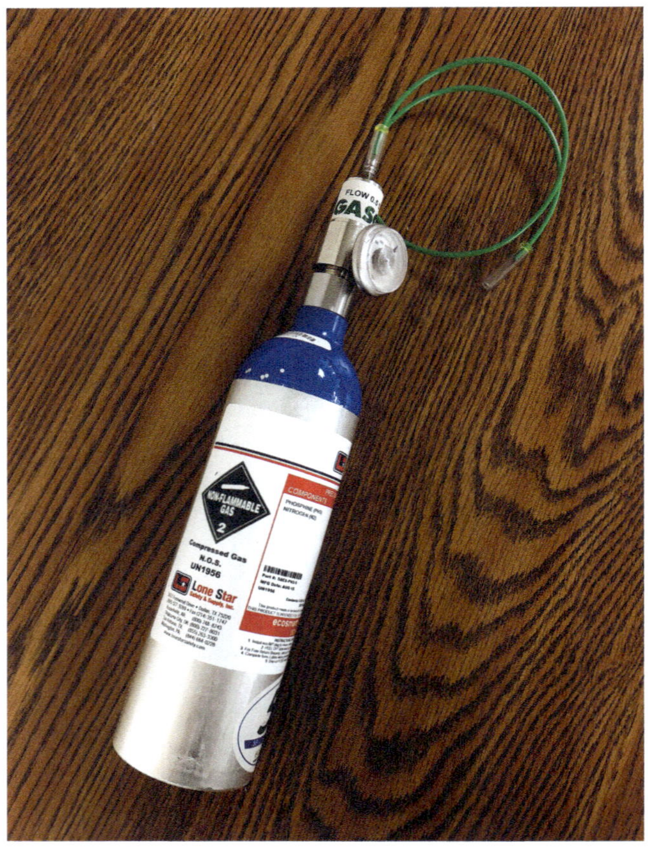

Figure 18.14   Bump test kit.

Some confusion exists regarding proper calibration procedures and frequency for phosphine monitors. To clarify the issue, the International Safety Equipment Association (ISEA) updated its position statement on instrument calibration in 2010, stating "a bump test or full calibration check of direct-reading portable gas monitors should be conducted before each day's use in accordance with the manufacturer's instructions, using an appropriate test gas." If the instrument fails a bump test, it must be adjusted through a full calibration before it is used. The ISEA recommends more frequent testing if environmental conditions that could affect instrument performance are suspected, such as sensor poisons or excessively dusty conditions. Always refer to the monitor manufacturer's information for the specific requirements for that monitor.

18.13.3 Sulfuryl Fluoride Monitoring

For sulfuryl fluoride, use only a detection device of sufficient sensitivity, such as the INTERSCAN gas analyzer (Model GF 1900) (Interscan Corporation, Simi Valley, California) or MIRAN vapor analyzer (SapphIRe) (Thermo Fisher Scientific, Franklin, Massachusetts), to confirm a concentration of sulfuryl fluoride less than 1 ppm (Figure 18.15). If sulfuryl fluoride concentrations exceed 1 ppm, the sealing of the isolated space from the fumigated area is not working and the immediate space must be evacuated unless the sulfuryl fluoride concentration is continuously monitored to prevent exposure greater than 1 ppm. An SCBA must be worn by the applicator if the concentration exceeds 1 ppm.

The INTERSCAN must be calibrated within one month prior to use as a clearance device. All other detection devices must be calibrated according to manufacturer recommendations. Small sample cylinders containing known concentrations of sulfuryl fluoride are available for calibrating

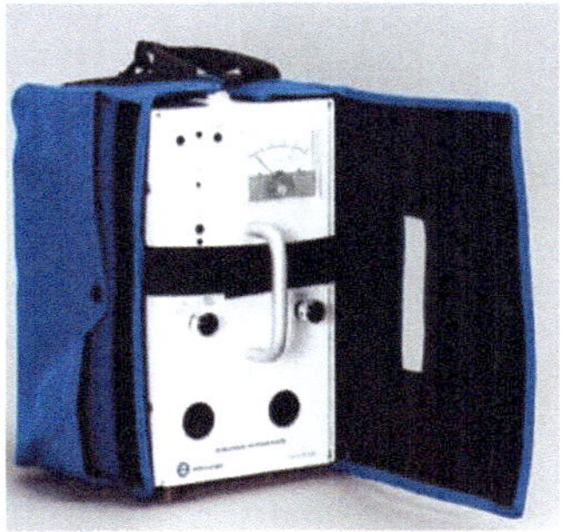

Figure 18.15 INTERSCAN gas analyzer.

equipment. Specially designed plastic sample bags are used to introduce the gas/air mixture into the INTERSCAN. Adjustments to the instrument can then be made to accurately measure the known concentration.

MIRAN gas analyzers are factory calibrated and do not need frequent recalibration. They should be checked by the electronic method or by the closed loop method and sent in when recalibration is indicated.

There may be other detection devices developed as technology advances. Contact your sulfuryl fluoride distributor for the latest information on detection devices.

18.14 Placarding of Fumigated Areas

Placarding of structures and areas to be fumigated is essential to provide a safe environment for all personnel and bystanders. All entrances to the fumigated area must be placarded. The placard (Figure 18.16) must be made of substantial material that can be expected to withstand adverse weather conditions, must be securely affixed to the structure,

Figure 18.16 Phosphine placard.

must remain legible during the entire fumigation period, and must bear the following wording:

- The signal words DANGER/PELIGRO and the SKULL AND CROSSBONES symbol in red
- The statement "Structure and/or commodity under fumigation, DO NOT ENTER/NO ENTRE"
- The statement "This sign may only be removed by a certified applicator or a person with documented training after the structure and/or commodity is completely aerated (contains 0.3 ppm or less of phosphine gas, or less than or equal to 1 ppm of sulfuryl fluoride gas). If incompletely aerated commodity is transferred to a new storage structure, the new structure must also be placarded if it contains more than 0.3 ppm phosphine or more than 1 ppm sulfuryl fluoride. Worker exposure during this transfer must not exceed allowable limits."
- Date and time of fumigation
- Name and EPA registration number of fumigant
- Name, address, and telephone number of the fumigation company and/or applicator
- A 24-h emergency response telephone number

Placards should be placed in advance of the fumigation to keep unauthorized personnel away from the structure. For railroad hopper cars, placards must be placed on both sides of the car near the ladders and next to the top hatches into which the fumigant is introduced. Do not remove placards until the commodity or area has been ventilated.

18.15 Storage

No fumigants should be stored in areas that could be occupied by people or where water, food, or feed could become contaminated. Store products in a dry, well-ventilated area away from heat, under lock and key, and at a distance from buildings where people or domestic animals may be located (Bonjour, 2023). As always, keep them out of the reach of children.

Make sure the storage facility is securely locked and posted as a pesticide storage area. The storage must be marked with the following signs: Danger/Poison with skull and cross bones (Figure 18.17), phosphine or sulfuryl fluoride NFPA chemical label (Figures 18.18 and 18.19), and Authorized Personnel Only.

Phosphine pellets and tablets supplied in gastight aluminum sealed flasks should be used completely after the flask is opened. The storage shelf-life of a flask is nearly unlimited if the aluminum seal is not removed

Figure 18.17   Danger/Poison skull and crossbones.

Figure 18.18   Phosphine NFPA chemical label.

Figure 18.19   Sulfuryl fluoride NFPA chemical label.

or damaged. Regularly check the containers for leaks and for cans that are damaged or corroded.

Solid phosphine formulations should not be stored where temperatures exceed 54°C and cylinderized formulations should not exceed 51°C.

Cylinders of sulfuryl fluoride and both cylinderized phosphine products must always be tightly secured in an upright or vertical position to prevent them from being knocked over. The value cap and safety cylinder cap must be securely installed when the cylinder is not connected to dispensing equipment. It is advisable to check the valves regularly for leaks. Leaks can cause dangerous concentrations of fumigant to accumulate in a closed storage facility. Gas detection devices are valuable tools to use for checking your storage areas for any leaks of fumigants. If a leak is detected, be sure to use the appropriate respiratory protection.

It is highly recommended that both full, partially used, and empty cylinders be stored in an outdoor storage structure in a dedicated and properly designed storage area. It should have a firm and level surface, preferably of reinforced concrete, which is well drained. Make sure that the storage area is away from ventilation intakes of buildings. Ideally, once the cylinder is empty, contact your distributor for proper return instructions.

18.16 Educating Stored Product Workers

Grain elevator and fumigation workshops are critical in helping keep stored product managers and workers educated and up to date on safety requirements and the availability and proper usage of insecticides for controlling insects in stored products (Bonjour, 2020). Certified applicators have a requirement to obtain CEUs or to retest every few years to maintain their certification. Certified applicators and others working in this industry benefit from hearing regularly from university personnel and industry specialists working in the stored product area during these workshops. Fact sheets, news releases, and industry magazine articles are important for the stored product industry to inform their personnel of safety regulations, new insecticides, and good practices to utilize in this important agricultural endeavor.

Following all the necessary safety precautions when using fumigants should ensure that everyone is able to go home safely at the end of the day.

References

Bonjour EL (2020) Safety education is critical for grain fumigators. Oklahoma Cooperative Extension Service, Fact Sheet, EPP-7104.

Bonjour EL (2022) Gas monitoring is a must – Keep safe by watching oxygen and phosphine fumigant levels. Grain Journal, Sep/Oct issue. pp. 84–86.

Bonjour EL (2023) Storing phosphine fumigants – Designate an outdoor storage structure in a low-traffic area. Grain Journal, May/June issue. pp. 54–55.

Bonjour EL, Jones CL (2021) Safety education – Critical for grain fumigators, pp. 91–97. In: Jayas DS, Jian F (eds.), Proceedings of the 11th International Conference on Controlled Atmosphere and Fumigation in Stored Products (CAF2020), CAF Permanent Committee Secretariat, Winnipeg, Canada.

Bonjour EL, Jones CL, Beeby RL (2014) Improving phosphine fumigation by sealing and using a closed-loop system, pp. 337–341. In: Athanassiou CG, Kavallieratos NG, Weintraub PG (eds.), Proc Conf of the Intern Org for Biol and Integrated Control, Western Palaearctic Region Section, Working Group on Integrated Protection of Stored Products, Talence, France.

Jones CL, Bonjour EL, Beeby RL, Noyes RT, Phillips TW (2008) Closed loop fumigation of a small rural concrete elevator in a growing urban setting, pp. 365–368. In: Daolin G, Navarro S, Jian Y, Cheng T, Zuxun J, Yue L, Yang L, Haipeng W (eds.), Proc 8th Intern Conf Controlled Atm Fumi Stored Prod, Chengdu, China.

Jones J, Hardin J, Bonjour E (2011) Design of closed-loop fumigation systems for grain storage structures. Oklahoma Cooperative Extension Service, Fact Sheet, BA-1111.

Occupational Safety & Health Administration (2011a) Grain handling facilities, Standard Number 1910.272. https://www.osha.gov/laws-regs/regulations/standardnumber/1910/1910.272. Accessed 10 May 2023

Occupational Safety & Health Administration (2011b) Permit-required confined spaces, Standard Number 1910.146. https://www.osha.gov/laws-regs/regulations/standardnumber/1910/1910.146. Accessed 10 May 2023

Occupational Safety & Health Administration (1989) The control of hazardous energy (lockout/tagout), Standard Number 1910.147. https://www.osha.gov/laws-regs/regulations/standardnumber/1910/1910.147. Accessed 10 May 2023

Occupational Safety & Health Administration (1996) Regulations (Grain Handling, OSHA 3103). https://www.osha.gov/sites/default/files/publications/osha3103.pdf. Accessed 10 May 2023

Index

Note: Locators in *italics* represent figures and **bold** indicate tables in the text.

Dihydronicotinamide adenine
dinucleotide (NADH), 293
Diphenylamine (DPA), 453
Direct placement, 212
Direct release method, 216–217
Discharge conveyors, sealing, 557
Discriminating dose (DD), 47, 49
DLD, *see* Dihydrolipoamide
dehydrogenase
DNA, *see* Deoxyribonucleic acid
Douglas fir, *see Pseudotsuga menziesii*
DPA, *see* Diphenylamine
Dried-fruit beetle, *see Carpophilus
hemipterus*
Drosophila suzukii, 443
Dryacide, 425, 465, *465*, 467
Dry grains, storage of, 100, 101
Dryocoetes hectographus, 320, **325**
Dry paddy, storage of, 107
Durable stored products, 8, 30
Dynamic fumigation; *see also* Fumigation
continuous monitoring, fumigations in
France using, 577
Baziège (2019), 578–579
Buchères (2022), 582–583
Pomacle (2023), 580–582
Roncenay (2023), 580
dynamic ground-level gas generation,
575–577
historical background, 572–574
monitoring, 574–575

E

EC, *see* Electrochemistry
Ecdysone, 17
ECHA, *see* European Chemical Agency
ECO$_2$FUME Fumigant Gas, 616, 623, *623*
Ectomyelois ceratoniae, **342**
EDN, *see* Ethanedinitrile
EF, *see* Ethyl formate
Effective stored-product treatment,
necessity of large doses for,
520–521
Electrochemistry (EC), 219
Elytra, 9
EMATE, 203, **204**
Embryo development, 11
EMC control, *see* Equilibrium moisture
content control
Endrosis sarcitrella, **15, 342**

Energy-powered AlP ground-level
application equipment, 572
Enteromorpha compressa, 510
Environmental Protection Agency (EPA),
535, 627
EOs, *see* Essential oils
EPA, *see* Environmental Protection
Agency
Ephestia sp.
E. cautella, 80, **80**, 81, *81*, 82, *82*, 86,
90, **257, 345**, 382
E. elutella, **15, 244, 250, 342**, 382, 601
E. figulilella, 411
E. kuehniella, **15, 22, 257, 260, 342,
345, 384**, 388, 389, 496
Epicocum nigrum, **349**
Equilibrium moisture content (EMC)
control, 361
Escherichia coli, 292, 293, 298, 504
Essential oils (EOs), 493, 495, 496, 505
and plant-derived products, 519
Ethanedinitrile (EDN), 56, 205, 312, 313,
314
applications of, 314–316
bark beetles and weevils, 319–320
phytosanitary approvals, 324
properties of, 313–314
regulatory status, 323
EDN phytosanitary approvals, 324
ISPM-28 and -15 status, 324
termites
West Indian drywood termite
(Cryptotermes brevis), 321
terrestrial invertebrates
pinewood nematode
(*Bursaphelenchus xylophilus*),
321–322
sirex woodwasp (*Sirex noctilio*),
322–323
wood and timber pests, efficacy with,
316–323
wood borers
Asian longhorn beetle
(*Anoplophora glabripennis*),
318–319
burnt-pine longhorn beetle
(*Arhopalus ferus*), 316–317
European house borer (*Hylotrupes
bajulus* L.), 317–318
Japanese pine sawyer
(*Monochamus alternatus*), 317